Uitgeschieden
in juni 2025

CONTINUUM MECHANICS

D. S. Chandrasekharaiah
Department of Mathematics
Bangalore University
Central College Campus
Bangalore, 560 001
India

Lokenath Debnath
Department of Mathematics and
Department of Mechanical and Aerospace Engineering
University of Central Florida
Orlando, Florida 32816-1364
U.S.A.

CONTINUUM MECHANICS

D. S. Chandrasekharaiah
Bangalore University
Bangalore, India

Lokenath Debnath
University of Central Florida
Orlando, Florida

ACADEMIC PRESS

Boston   San Diego   New York
London   Sydney   Tokyo   Toronto

This book is printed on acid-free paper. ∞

Copyright © 1994 by Academic Press, Inc.

All rights reserved.
No part of this publication may be reproduced or
transmitted in any form or by any means, electronic
or mechanical, including photocopy, recording, or
any information storage and retrieval system, without
permission in writing from the publisher.

ACADEMIC PRESS, INC.
525 B Street, Suite 1900, San Diego, CA 92101-4495

United Kingdom Edition published by
ACADEMIC PRESS LIMITED
24–28 Oval Road, London NW1 7DX

Library of Congress Cataloging-in-Publication Data

Chandrasekharaiah, D. S., 1943–
  Continuum mechanics / D. S. Chandrasekharaiah, Lokenath Debnath.
    p. cm.
  Includes bibliographical references and index.
  ISBN 0-12-167880-6 (acid-free)
  1. Continuum mechanics. I. Debnath, Lokenath. II. Title.
QA808.2.C482 1994
531—dc20   93-41278
                                                               CIP

PRINTED IN THE UNITED STATES OF AMERICA

94 95 96 97 BB 9 8 7 6 5 4 3 2 1

To our children
PRASHANTH, PRATHIBHA, POORNIMA
and
JAYANTA

TABLE OF CONTENTS

Preface   xi

1
SUFFIX NOTATION   1

1.1 Introduction   1
1.2 Range and summation conventions   2
1.3 Free and dummy suffixes   3
1.4 Summary of results in vector algebra   6
1.5 Summary of results in matrix algebra   12
1.6 The symbol δ_{ij} 16
1.7 The symbol ε_{ijk} 19
1.8 Exercises   29

2
ALGEBRA OF TENSORS   33

2.1 Introduction   33
2.2 Coordinate transformations   33
2.3 Cartesian tensors   39
2.4 Properties of tensors   46
2.5 Isotropic tensors   58
2.6 Isotropic tensor of order four   61
2.7 Tensors as linear operators   66
2.8 Transpose of a tensor   69

viii   Contents

2.9   Symmetric and skew tensors   72
2.10 Dual vector of a skew tensor   74
2.11 Invariants of a tensor   77
2.12 Deviatoric tensors   81
2.13 Eigenvalues and eigenvectors   83
2.14 Polar decomposition   94
2.15 Exercises   101

3
CALCULUS OF TENSORS   109

3.1   Introduction   109
3.2   Scalar, vector and tensor functions   109
3.3   Comma notation   113
3.4   Gradient of a scalar, divergence and curl of a vector   117
3.5   Gradient of a vector, divergence and curl of a tensor   125
3.6   Integral theorems for vectors   136
3.7   Integral theorems for tensors   143
3.8   Exercises   151

4
CONTINUUM HYPOTHESIS   155

4.1   Introduction   155
4.2   Notion of a continuum   157
4.3   Configuration of a continuum   157
4.4   Mass and density   159
4.5   Descriptions of motion   160
4.6   Material and spatial coordinates   163

5
DEFORMATION   167

5.1   Introduction   167
5.2   Deformation gradient tensor   167
5.3   Stretch and rotation   180
5.4   Strain tensors   188
5.5   Strain-displacement relations   200
5.6   Infinitesimal strain tensor   205
5.7   Infinitesimal stretch and rotation   211
5.8   Compatibility conditions   217
5.9   Principal strains   227

5.10 Strain-deviator   230
5.11 Exercises   232

6
MOTION   241

6.1 Introduction   241
6.2 Material and local time-derivatives   242
6.3 Stretching and vorticity   256
6.4 Path lines, stream lines, and vortex lines   266
6.5 Transport formulas   273
6.6 Circulation and vorticity   278
6.7 Exercises   284

7
STRESS   293

7.1 Introduction   293
7.2 Body forces and surface forces   293
7.3 Stress components   296
7.4 Stress tensor   298
7.5 Normal and shear stresses   304
7.6 Principal stresses   306
7.7 Stress-deviator   312
7.8 Boundary condition for the stress tensor   314
7.9 Piola–Kirchhoff stress tensors   316
7.10 Exercises   318

8
FUNDAMENTAL LAWS OF CONTINUUM MECHANICS   325

8.1 Introduction   325
8.2 Conservation of mass   326
8.3 Balance of linear momentum   331
8.4 Balance of angular momentum   336
8.5 General solutions of the equation of equilibrium   338
8.6 Balance of energy   344
8.7 Entropy inequality   351
8.8 Constitutive equations   355
8.9 Exercises   357

9
EQUATIONS OF LINEAR ELASTICITY   363

9.1 Introduction   363
9.2 Generalized Hooke's law   364
9.3 Physical meanings of elastic moduli   368
9.4 Governing equations   376
9.5 Boundary value problems   383
9.6 Uniqueness of solution (static case)   385
9.7 Uniqueness of solution (dynamic case)   387
9.8 Navier's equation   390
9.9 Displacement formulation   398
9.10 Stress formulation   400
9.11 Beltrami–Michell equation   402
9.12 Some static problems   407
9.13 Elastic waves   431
9.14 Exercises   450

10
EQUATIONS OF FLUID MECHANICS   461

10.1 Introduction   461
10.2 Viscous and nonviscous fluids   461
10.3 Stress tensor for a nonviscous fluid   463
10.4 Governing equations for a nonviscous fluid flow   464
10.5 Initial and boundary conditions   473
10.6 Euler's equation of motion   477
10.7 Equation of motion of an elastic fluid   485
10.8 Bernoulli's equations   492
10.9 Water waves   500
10.10 Stress tensor for a viscous fluid   508
10.11 Shear viscosity and bulk viscosity   514
10.12 Governing equations for a viscous fluid flow   518
10.13 Initial and boundary conditions   523
10.14 Navier–Stokes equation   526
10.15 Some viscous flow problems   538
10.16 Exercises   562

Answers/Hints to Selected Exercises   569

Bibliography   581

Index   585

5.10 Strain-deviator 230
5.11 Exercises 232

6
MOTION 241

6.1 Introduction 241
6.2 Material and local time-derivatives 242
6.3 Stretching and vorticity 256
6.4 Path lines, stream lines, and vortex lines 266
6.5 Transport formulas 273
6.6 Circulation and vorticity 278
6.7 Exercises 284

7
STRESS 293

7.1 Introduction 293
7.2 Body forces and surface forces 293
7.3 Stress components 296
7.4 Stress tensor 298
7.5 Normal and shear stresses 304
7.6 Principal stresses 306
7.7 Stress-deviator 312
7.8 Boundary condition for the stress tensor 314
7.9 Piola–Kirchhoff stress tensors 316
7.10 Exercises 318

8
FUNDAMENTAL LAWS OF CONTINUUM MECHANICS 325

8.1 Introduction 325
8.2 Conservation of mass 326
8.3 Balance of linear momentum 331
8.4 Balance of angular momentum 336
8.5 General solutions of the equation of equilibrium 338
8.6 Balance of energy 344
8.7 Entropy inequality 351
8.8 Constitutive equations 355
8.9 Exercises 357

9
EQUATIONS OF LINEAR ELASTICITY   363

9.1 Introduction   363
9.2 Generalized Hooke's law   364
9.3 Physical meanings of elastic moduli   368
9.4 Governing equations   376
9.5 Boundary value problems   383
9.6 Uniqueness of solution (static case)   385
9.7 Uniqueness of solution (dynamic case)   387
9.8 Navier's equation   390
9.9 Displacement formulation   398
9.10 Stress formulation   400
9.11 Beltrami–Michell equation   402
9.12 Some static problems   407
9.13 Elastic waves   431
9.14 Exercises   450

10
EQUATIONS OF FLUID MECHANICS   461

10.1 Introduction   461
10.2 Viscous and nonviscous fluids   461
10.3 Stress tensor for a nonviscous fluid   463
10.4 Governing equations for a nonviscous fluid flow   464
10.5 Initial and boundary conditions   473
10.6 Euler's equation of motion   477
10.7 Equation of motion of an elastic fluid   485
10.8 Bernoulli's equations   492
10.9 Water waves   500
10.10 Stress tensor for a viscous fluid   508
10.11 Shear viscosity and bulk viscosity   514
10.12 Governing equations for a viscous fluid flow   518
10.13 Initial and boundary conditions   523
10.14 Navier–Stokes equation   526
10.15 Some viscous flow problems   538
10.16 Exercises   562

Answers/Hints to Selected Exercises   569

Bibliography   581

Index   585

CHAPTER 1
SUFFIX NOTATION

1.1
INTRODUCTION

The language of tensors is best suited for the development of the subject of continuum mechanics. The compactness as well as the efficiency of the tensor notation is very useful for the study of this subject and gives the subject a great beauty. The use of Cartesian tensors is sufficient for the development of the theory of continuum mechanics; for the solution of specific problems, orthogonal curvilinear coordinates suitable to the geometry of the problem may lead to simplification of the analysis. We therefore use Cartesian tensors in the main body of the text, and in the first three chapters we present a detailed and self-contained account of Cartesian tensors primarily tailored to the needs of continuum mechanics. A shorthand notation, known as the suffix notation (or subscript notation or index notation), employed in the treatment of Cartesian tensors is introduced in this chapter. It is assumed that the reader has a basic knowledge of vector algebra, matrix theory and three-dimensional analytic geometry.

1.2
RANGE AND SUMMATION CONVENTIONS

Consider the following system of algebraic equations

$$a_{11}x_1 + a_{12}x_2 + a_{13}x_3 = b_1$$
$$a_{21}x_1 + a_{22}x_2 + a_{23}x_3 = b_2 \qquad (1.2.1)$$
$$a_{31}x_1 + a_{32}x_2 + a_{33}x_3 = b_3$$

We can write these equations as

$$a_{i1}x_1 + a_{i2}x_2 + a_{i3}x_3 = b_i, \qquad i = 1, 2, 3 \qquad (1.2.2)$$

By using summation sign, this can be rewritten as

$$\sum_{k=1}^{3} a_{ik}x_k = b_i, \qquad i = 1, 2, 3 \qquad (1.2.3)$$

We say that the values 1, 2, 3 form the *range of the suffixes i and k*.

Let us adopt the following convention.

CONVENTION 1 All the suffixes we employ have the range 1, 2, 3. (This is known as the *range convention*.)

Then (1.2.3) may be shortened to

$$\sum_{k} a_{ik}x_k = b_i \qquad (1.2.4)$$

We note that the suffixes i and k play different roles in (1.2.4); although summation is indicated over the suffix k, the suffix i is left "free." We observe that whereas the suffix k (over which summation is taken) is *repeated* in the term included under the summation sign, the suffix i appears *exactly once* in every term. This observation prompts us to adopt another convention stated as follows.

CONVENTION 2 Whenever a suffix is repeated (once) in a term, summation is required to be taken over that suffix. (This is known as the *summation convention*.)

Then, we can write $\sum_k a_{ik}x_k$ as just $a_{ik}x_k$. Consequently, (1.2.4) can be rewritten as

$$a_{ik}x_k = b_i \qquad (1.2.5)$$

This is a concise form of the system of equations (1.2.1).

An equation such as (1.2.5) containing suffixes for which the range and summation conventions are applicable is referred to as an equation written in the *suffix notation, subscript notation* or *index notation*. Depending on the context, such an equation may be viewed either as a representation (in a concise form) of a system of equations or as a representative (typical equation) of the system.

It should be emphasized that, according to the summation convention, a summation over a suffix is implied only if the suffix is repeated (only once) in the *same term*. Thus, the symbol $a_k + b_k$ *does not* represent the sum $\sum_{k=1}^{3}(a_k + b_k)$ and the symbol $a_{ii}x_i$ makes no sense. Repetition of a suffix more than once (in a term) is prohibited under the summation convention.

The summation convention is useful in writing a double sum, a triple sum, etc., also in a short form. For example, we can write

$$\sum_{i=1}^{3}\sum_{j=1}^{3} a_{ij}b_{ij} \quad \text{simply as} \quad a_{ij}b_{ij}$$

which has nine terms. Similarly, the symbol $a_{ij}b_{jk}c_{ki}$ represents the triple sum

$$\sum_{i=1}^{3}\sum_{j=1}^{3}\sum_{k=1}^{3} a_{ij}b_{jk}c_{ki}$$

which contains 27 terms.

1.3
FREE AND DUMMY SUFFIXES

It has been noted that (1.2.5) is a concise form of the system of equations (1.2.1). A suffix such as k in (1.2.5) that is summed over is called a *summation suffix* or a *dummy suffix*. A suffix such as i in (1.2.5) that is free of summation is called a *free suffix* or a *live suffix*.

Since a dummy suffix just indicates summation, the letter used to denote it is of no consequence, because expressions such as $a_k b_k$ and $a_m b_m$ represent the same sum as $a_1 b_1 + a_2 b_2 + a_3 b_3$. As such, a dummy suffix may be replaced by any other suffix within the provisions of the summation convention. For example, in the expression $a_{ik}x_k$ the suffix k may be changed to any suffix, say m, other than i. (Note that changing k to i leads to $a_{ii}x_i$, which makes no sense under the summation convention!) The concise form (1.2.5) of equations (1.2.1) may therefore be written also as

$$a_{im}x_m = b_i \tag{1.3.1}$$

1 SUFFIX NOTATION

On the other hand, (1.2.5) has the same meaning as

$$a_{jk}x_k = b_j \tag{1.3.2}$$

because, by the range and summation conventions, (1.3.2) actually stands for

$$\sum_{k=1}^{3} a_{jk}x_k = b_j, \quad j = 1, 2, 3$$

which are precisely equations (1.2.1). In (1.3.2), j is a free suffix and this equation may be obtained from (1.2.5) by changing the free suffix i to j in *every* term of (1.2.5). This illustrates the fact that a free suffix may also be changed to any other suffix (of course within the provisions of the summation convention) provided the same change is made in every term.

Often we deal with systems of equations whose concise form requires the use of more than one free suffix. For example, consider the following system of equations:

$$\begin{aligned}
a_{11}x_1^2 + a_{12}x_2x_1 + a_{13}x_3x_1 &= b_{11} \\
a_{11}x_1x_2 + a_{12}x_2^2 + a_{13}x_3x_2 &= b_{12} \\
a_{11}x_1x_3 + a_{12}x_2x_3 + a_{13}x_3^2 &= b_{13} \\
a_{21}x_1^2 + a_{22}x_2x_1 + a_{23}x_3x_1 &= b_{21} \\
a_{21}x_1x_2 + a_{22}x_2^2 + a_{23}x_3x_2 &= b_{22} \\
a_{21}x_1x_3 + a_{22}x_2x_3 + a_{23}x_3^2 &= b_{23} \\
a_{31}x_1^2 + a_{32}x_2x_1 + a_{33}x_3x_1 &= b_{31} \\
a_{31}x_1x_2 + a_{32}x_2^2 + a_{33}x_3x_2 &= b_{32} \\
a_{31}x_1x_3 + a_{32}x_2x_3 + a_{33}x_3^2 &= b_{33}
\end{aligned} \tag{1.3.3}$$

By using the range and summation conventions, the first three equations in (1.3.3) may be represented as

$$a_{1k}x_k x_j = b_{1j} \tag{1.3.4}$$

Similarly, the next three and the last three equations in (1.3.3) can be represented, respectively, as

$$a_{2k}x_k x_j = b_{2j} \tag{1.3.5}$$

$$a_{3k}x_k x_j = b_{3j} \tag{1.3.6}$$

Equations (1.3.4) to (1.3.6) may be put together as

$$a_{ik}x_k x_j = b_{ij} \tag{1.3.7}$$

PREFACE

Solid and fluid mechanics are two major subjects studied by all students of applied mathematics, physics and engineering. Traditionally, these two subjects are taught separately by two different specialists whose approach, orientation and notation are in general different. In such separate treatments, it has not always been clear to students that the fundamental ideas and general principles are indeed common to these subjects. The modern trend is therefore to make a unified presentation of the ideas and general principles common to all branches of solid and fluid mechanics under the general heading of Continuum Mechanics. This unified course develops the fundamentals and foundations more carefully than the traditional separate courses where normal tendency is to put emphasis on applications. Once familiar with the basic concepts and general principles of continuum mechanics, the student will find little difficulty in specializing in various branches of solid and fluid mechanics at a later stage.

There appear to be many books available for use by students studying continuum mechanics. Some are excellent but too sophisticated and terse for the beginner. Some are too elementary or have only limited scope in their contents. While teaching continuum mechanics, the authors have found difficulty over the choice of textbooks to accompany the lectures. They have felt the need of a detailed and self-contained textbook primarily intended for the beginners. This book is an attempt to meet this need. It is based upon courses of lectures given by the authors over a number of years to the first year graduate students in Bangalore University, Calcutta University, East Carolina University and the University of Central Florida.

The book assumes only a limited knowledge of mechanics, and the material in it has been selected to introduce the reader to the fundamental ideas, general principles and applications of continuum mechanics. Despite its bulk the book is genuinely an introduction to continuum mechanics; hence no attempt is made to present a detailed account of solid and fluid mechanics except for the formulation of their governing equations and immediate simple applications. It is hoped the book will prepare the reader for further study of various branches of solid and fluid mechanics including nonlinear elasticity, plasticity, thermoelasticity, viscoelasticity and non-Newtonian fluid dynamics.

A good knowledge of vectors and tensors is essential for a full appreciation of continuum mechanics. A simple and self-contained presentation of these topics primarily tailored to the needs of continuum mechanics is therefore included in the first three chapters of the book. Since the Cartesian tensor formulation is sufficient for the development of continuum mechanics at an elementary level, we have limited our discussion of tensors to Cartesian tensors only. Bearing in mind the mathematical background and skill of the students for whom the book is primarily intended, we have made only minimal use of abstract mathematics. The reader is assumed to be familiar with traditional mathematics including matrices, geometry, differential and integral calculus, and ordinary and partial differential equations, in three-dimensional space.

Chapters 4–8 discuss the fundamental concepts, general principles and major results of nonlinear continuum mechanics in a detailed and systematic way. Chapter 4 introduces the continuum hypothesis, basic definitions and the meanings of the Lagrangian and Eulerian formulations of continuum mechanics.

The study of deformation of a continuum is the major topic of Chapter 5. Stretch and strain tensors are introduced and their respective geometrical significances explained. The strain-displacement relations in the general (nonlinear) and linearized forms are obtained. The compatibility condition for the linearized case is derived. Principal strains and principal directions of strain are discussed in some detail.

Chapter 6 deals with the instantaneous motion of a continuum. The concept of material derivative is defined and the velocity and acceleration vectors are introduced. The stretching tensor and the vorticity tensor/vector are discussed along with their physical significance. The transport formulas are then proved. The concepts of path lines, stream lines, vortex lines and circulation are introduced for subsequent references in fluid mechanics.

The seventh chapter is concerned with the concept of stress in a continuum. Based upon the Cauchy's stress principle, the stress vector and the stress tensor are defined and their relationship discussed. In addition, the normal stress, the shear stress, the principal stresses and the principal directions of stress are defined and their basic properties examined. The Piola-Kirchhoff stress tensors are also introduced.

The field equations of continuum mechanics are presented in Chapter 8. The equation of continuity and the equations of motion and equilibrium are obtained by using the laws of balance of mass and momentum. Some general solutions of the equilibrium equation in terms of various stress functions are presented. The first law of thermodynamics is used to establish the energy equation. The Clausius–Duhem inequality is obtained by the use of the second law of thermodynamics. It is pointed out that all the field equations are applicable to all continua representing solids, liquids and gases regardless of their internal physical structure. The crucial need for the so-called constitutive equations which distinguish one class of materials from the other while studying the individual branches of continuum mechanics is indicated. This is the key chapter in the sense that every specialized branch of solid or fluid mechanics is just an offshoot of this chapter. Discussion of the constitutive theory falls beyond the scope of the book.

The last two chapters are devoted to the development of the governing equations of two basic areas of continuum mechanics: linear elasticity and mechanics of nonviscous and Newtonian viscous fluids. Chapter nine deals with the fundamental equations of the linear theory of elastic solids. The constitutive equation for a linear elastic solid (generalized Hooke's law) is postulated and then specialized to homogeneous and isotropic solids. The governing equations of elastostatics and elastodynamics are derived and the uniqueness of solutions established. Some standard elastostatic problems including extension, bending and torsion of beams and the pressure-vessel problems are discussed. Finally, wave propagation problems including plane waves, Rayleigh waves and Love waves are studied in some detail.

The final chapter deals with the fundamental equations of fluid mechanics. Based upon the appropriate constitutive relations, the Euler's equation for a non-viscous fluid, and the Navier–Stokes equation for a viscous fluid are derived and their consequences studied. Some standard viscous flow problems are considered. Further, a brief introduction to water waves is given.

Throughout the book, major emphasis is given to the logical development of the fundamental principles and unified treatment of solid and fluid mechanics. All the mathematical preliminaries are presented in Chapters 1 through 3 in order to develop a systematic theory of continuum mechanics. However, it is not necessary for the reader to know everything contained in these Chapters before taking up the study of continuum mechanics which begins with Chapter 4. One can start with Chapter 4 after having just a broad review of Chapters 1 to 3 and return to appropriate sections of these chapters for a detailed study as and when the need arises. The theory of continuum mechanics dealt with in Chapters 4 to 8 is essentially a systematic mathematical theory. Special effort is made to present this theory in concise and clear terms from a mathematical point of view; physical considerations and motivating arguments are not emphasized beyond a certain point. However, the last two chapters, which provide links with the traditional developments of the theory of elasticity and fluid mechanics, are applied-oriented and contain sufficient physical explanations.

The book contains over 250 worked examples and over 500 exercises. Some of these are elementary and some are challenging. These should help the student in the process of understanding and mastering so analytical a subject as continuum mechanics. Answers and hints to some selected exercises are provided at the end of the book.

This is a text book designed for use by the beginners in continuum mechanics. We have therefore made no attempt to present any new material as such. Also, we have refrained from burdening the reader with historical notes and references to original sources. Those interested in an advanced treatment of the topics covered in the book along with full references may consult the encyclopedia articles of Gurtin, Serrin, Sneddon and Berry, Truesdell and Noll, and Truesdell and Toupin listed in the Bibliography given at the end of the book. The Bibliography also includes some other works recommended for further study and reference. Many of the results and problems presented in the book are either motivated by or borrowed from the works cited in the Bibliography. We wish to acknowledge our indebtness to these works.

In preparing the book, the authors have been encouraged by and have benefited from the helpful comments/criticisms of a number of students and faculty members of several universities in the United States and India. Comments and suggestions made by anonymous reviewers have helped to improve the quality of the book. The authors are thankful to all these individuals for their interest in the book.

A major portion of the present version of the book was written when the first author was a visiting Fulbright scholar at the University of Central Florida, Orlando, under the Indo-American Fellowship Program. He records his grateful thanks to the agencies sponsoring the program.

Our special thanks to Jackie Callahan and June Wingler who typed the manuscript and cheerfully put up with constant revisions and changes. In spite of the best efforts of everyone involved, doubtless some typographical errors remain. We do hope that these are both few and obvious, and will cause minimal confusion. Finally, we thank Brian Miller, Editor, and the staff of Academic Press for their assistance and cooperation.

D. S. Chandrasekharaiah
Lokenath Debnath

This is a concise form of equations (1.3.3) written in the suffix notation. Note that this concise form contains two free suffixes, i and j, and one dummy suffix, k.

Since every suffix has the range 1, 2, 3 according to the range convention, an equation that contains one free suffix represents three equations, as in the case of (1.2.5); an equation that contains two free suffixes represents nine equations as in the case of (1.3.7), and so forth. An equation containing N free suffixes represents 3^N equations.

The concepts of free and dummy suffixes as well as the range and summation conventions adopted to systems of equations, as just described, may be employed to other mathematical systems also. Thus, the set of three numbers (a_1, a_2, a_3) can be represented as just (a_i). The 3×3 matrix

$$\begin{bmatrix} a_{11} & a_{12} & a_{13} \\ a_{21} & a_{22} & a_{23} \\ a_{31} & a_{32} & a_{33} \end{bmatrix}$$

may be represented as just (a_{ij}) or $[a_{ij}]$.

In general, the symbol $a_{ijk...}$ involving N free suffixes $i, j, k \ldots$ represents 3^N numbers.

EXAMPLE 1.3.1 If $a_i = \alpha_{ij} b_j$ and $b_i = \beta_{ij} c_j$, write down a_i in terms of c_i.

Solution It is given that

$$a_i = \alpha_{ij} b_j \tag{1.3.8}$$

and

$$b_i = \beta_{ij} c_j \tag{1.3.9}$$

If we change the free suffix from i to j and the dummy suffix from j to k in (1.3.9), we get

$$b_j = \beta_{jk} c_k \tag{1.3.10}$$

This has the same meaning as that of (1.3.9). Substituting for b_j from (1.3.10) in (1.3.8) gives

$$a_i = \alpha_{ij} \beta_{jk} c_k \tag{1.3.11}$$

This is the expression for a_i in terms of c_i. ■

EXAMPLE 1.3.2 Show that

(i) $$a_{ij} b_{ij} = a_{ji} b_{ji} \tag{1.3.12}$$

(ii) $$(a_{ijk} + a_{jki} + a_{kij}) x_i x_j x_k = 3 a_{ijk} x_i x_j x_k \tag{1.3.13}$$

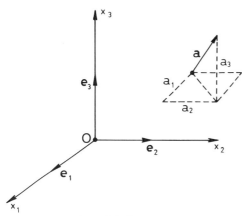

Figure 1.2. Base vectors.

combination of $\mathbf{e}_1, \mathbf{e}_2, \mathbf{e}_3$ as follows:

$$\mathbf{a} = a_1 \mathbf{e}_1 + a_2 \mathbf{e}_2 + a_3 \mathbf{e}_3 \tag{1.4.1}$$

Here a_1, a_2, a_3 are real numbers representing the projections of $\mathbf{a}$ on the axes; these are called the *components* of $\mathbf{a}$ along $\mathbf{e}_1$, $\mathbf{e}_2$ and $\mathbf{e}_3$, respectively.

Since

$$\mathbf{e}_1 = 1\mathbf{e}_1 + 0\mathbf{e}_2 + 0\mathbf{e}_3$$

the components of $\mathbf{e}_1$ are $(1, 0, 0)$. Similarly, the components of $\mathbf{e}_2$ are $(0, 1, 0)$ and those of $\mathbf{e}_3$ are $(0, 0, 1)$.

By employing the summation convention, expression (1.4.1) can be rewritten as

$$\mathbf{a} = a_k \mathbf{e}_k \tag{1.4.2}$$

The three components a_1, a_2, a_3 of $\mathbf{a}$ are referred to briefly as a_i. However, the symbol a_i is also frequently used to denote a typical component of $\mathbf{a}$; so a_i is referred to as the ith component of $\mathbf{a}$.

We denote the ith component of $\mathbf{a}$ also by $[\mathbf{a}]_i$; thus

$$[\mathbf{a}]_i = a_i \tag{1.4.3}$$

If α is a real number, then the vector $\alpha \mathbf{a}$ is given by

$$\alpha \mathbf{a} = (\alpha a_1) \mathbf{e}_1 + (\alpha a_2) \mathbf{e}_2 + (\alpha a_3) \mathbf{e}_3 = \alpha a_k \mathbf{e}_k \tag{1.4.4}$$

Thus, the components of $\alpha \mathbf{a}$ are αa_i; that is,

$$[\alpha \mathbf{a}]_i = \alpha a_i = \alpha [\mathbf{a}]_i \tag{1.4.5}$$

As a direct consequence of this, it follows that $-a_i$ are components of $-\mathbf{a}$.

Solution (i) In $a_{ij}b_{ij}$ both i and j are dummy suffixes. Hence i may be replaced by any suffix and j may be replaced by any suffix. In particular, we may replace i by j, and j by i. Thus

$$a_{ij}b_{ij} = a_{ji}b_{ji}$$

(ii) We first note that

$$(a_{ijk} + a_{jki} + a_{kij})x_i x_j x_k = a_{ijk}x_i x_j x_k + a_{jki}x_i x_j x_k + a_{kij}x_i x_j x_k \quad (1.3.14)$$

In $a_{jki}x_i x_j x_k$, all of i, j, k are dummy suffixes. As such i may be replaced by k, j may be replaced by i and k may be replaced by j. Thus,

$$a_{jki}x_i x_j x_k = a_{ijk}x_k x_i x_j = a_{ijk}x_i x_j x_k$$

Similarly,

$$a_{kij}x_i x_j x_k = a_{ijk}x_j x_k x_i = a_{ijk}x_i x_j x_k$$

Thus, all the three terms in the righthand side of (1.3.14) are equal; their sum is therefore $3a_{ijk}x_i x_j x_k$. This proves (1.3.13). ■

EXAMPLE 1.3.3 Suppose that the equations

$$a_{ip}a_{jp}c_j = b_{ij}c_j \quad (1.3.15)$$

hold for arbitrary c_i. Show that

$$a_{ip}a_{jp} = b_{ij} \quad (1.3.16)$$

Solution Since equations (1.3.15) hold for arbitrary c_i, these hold when $c_1 = 1$, $c_2 = 0$, $c_3 = 0$, and we obtain

$$a_{ip}a_{1p} = b_{i1} \quad (1.3.17)$$

Similarly, taking $c_1 = c_3 = 0$, $c_2 = 1$; and $c_1 = c_2 = 0$, $c_3 = 1$ in (1.3.15), we get, respectively,

$$a_{ip}a_{2p} = b_{i2}; \qquad a_{ip}a_{3p} = b_{i3} \quad (1.3.18)$$

Clearly, (1.3.17) and (1.3.18) constitute the required result (1.3.16). ■

1.4 SUMMARY OF RESULTS IN VECTOR ALGEBRA

A *vector* is an entity that has two characteristics: magnitude and direction. Force and velocity are two typical examples of a vector. Geometrically, a vector is represented by a directed line segment, with the length of the segment representing the magnitude of the vector and the direction of the segment indicating the direction of the vector. Evidently, the magnitude of a vector is a nonnegative real number.

Vectors are denoted by boldface letters in print (like **a**) or letters with a superposed arrow in manuscripts (like $\vec{a}$). In this text, only lowercase letters are used to denote vectors, reserving capital letters for tensors (to be introduced later). The magnitude of a vector **a** is denoted by $|\mathbf{a}|$ or a. A vector **a** is called a *unit vector* if $|\mathbf{a}| = 1$. A unit vector directed along a vector **a** is denoted by $\hat{\mathbf{a}}$. A vector whose magnitude is 0 is called the *zero vector*; it is denoted by **0** (or $\vec{0}$).

Two vectors are said to be *equal* if they have the same magnitude and the same direction. If **a** is a vector and α is a real number, then $\alpha \mathbf{a}$, called the *scalar multiple* of α and **a**, is a vector whose magnitude is $|\alpha| |\mathbf{a}|$ and whose direction is that of **a** or opposite to **a** accordingly as $\alpha > 0$ or $\alpha < 0$. If $\alpha = 0$, then $\alpha \mathbf{a} = \mathbf{0}$. The vector $(-1)\mathbf{a}$ is denoted by $-\mathbf{a}$. Every vector **a** can be written as $\mathbf{a} = |\mathbf{a}|\hat{\mathbf{a}}$.

Two vectors are said to be *collinear* if their directions are either the same or opposite. If **a** and **b** are collinear vectors, then $\mathbf{a} = \alpha \mathbf{b}$ for some real number α.

The *sum* $\mathbf{a} + \mathbf{b}$ and the *difference* $\mathbf{a} - \mathbf{b}$ of vectors **a** and **b** are defined by the parallelogram law (see Figure 1.1). It is easy to verify that $\mathbf{b} + \mathbf{a} = \mathbf{a} + \mathbf{b}$, $\mathbf{b} - \mathbf{a} = -(\mathbf{a} - \mathbf{b})$, $\mathbf{a} + \mathbf{0} = \mathbf{a}$, and $\mathbf{a} - \mathbf{a} = \mathbf{0}$.

Suppose that there is a righthanded rectangular system of Cartesian axes with a fixed origin O. We refer to these axes as the x_1, x_2, x_3-axes, or briefly the x_i-axes. Also, we denote the unit vectors directed along the positive x_1, x_2, x_3-axes by $\mathbf{e}_1, \mathbf{e}_2, \mathbf{e}_3$, respectively (see Figure 1.2). Then $\mathbf{e}_i$ are called the *base vectors* of the x_i-system.

By virtue of the definition of equality of vectors and the parallelogram law of addition of vectors, every vector **a** can be expressed as a linear

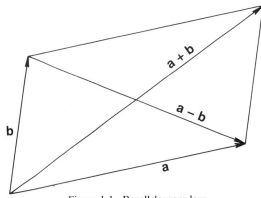

Figure 1.1. Parallelogram law.

1.4 SUMMARY OF RESULTS IN VECTOR ALGEBRA 9

Let **b** be a vector with components b_i; that is,

$$\mathbf{b} = b_1 \mathbf{e}_1 + b_2 \mathbf{e}_2 + b_3 \mathbf{e}_3 = b_k \mathbf{e}_k \qquad (1.4.6)$$

Then, by the definition of equality of vectors, it follows that $\mathbf{a} = \mathbf{b}$ if and only if $a_1 = b_1$, $a_2 = b_2$, $a_3 = b_3$, or briefly

$$a_i = b_i \qquad (1.4.7)$$

Thus, the *vector equation* $\mathbf{a} = \mathbf{b}$ is equivalent to the equations $a_i = b_i$ written in the suffix notation.

By the parallelogram law of addition of vectors, it follows that the sum $\mathbf{a} + \mathbf{b}$ and the difference $\mathbf{a} - \mathbf{b}$ of $\mathbf{a}$ and $\mathbf{b}$ are given by

$$\mathbf{a} \pm \mathbf{b} = (a_1 \pm b_1)\mathbf{e}_1 + (a_2 \pm b_2)\mathbf{e}_2 + (a_3 \pm b_3)\mathbf{e}_3 = (a_k \pm b_k)\mathbf{e}_k \qquad (1.4.8)$$

The components of $\mathbf{a} \pm \mathbf{b}$ are thus $a_i \pm b_i$; that is,

$$[\mathbf{a} \pm \mathbf{b}]_i = a_i \pm b_i = [\mathbf{a}]_i \pm [\mathbf{b}]_i \qquad (1.4.9)$$

A direct consequence of (1.4.9) is that the components of the zero vector **0** are all 0.

The *scalar product* or the *dot product* $\mathbf{a} \cdot \mathbf{b}$ of $\mathbf{a}$ and $\mathbf{b}$ is a real number defined by

$$\mathbf{a} \cdot \mathbf{b} = |\mathbf{a}||\mathbf{b}| \cos \theta \qquad (1.4.10)$$

where θ is the angle between the directions of $\mathbf{a}$ and $\mathbf{b}$. Evidently, $\mathbf{b} \cdot \mathbf{a} = \mathbf{a} \cdot \mathbf{b}$.

Two vectors are said to be *orthogonal* if the angle between their directions is $\pi/2$. It follows that $\mathbf{a}$ and $\mathbf{b}$ are orthogonal if and only if $\mathbf{a} \cdot \mathbf{b} = 0$.

Obviously, $\mathbf{a} \cdot \mathbf{0} = 0$ for every vector $\mathbf{a}$. Also, if $\mathbf{a} \cdot \mathbf{b} = 0$ for every vector $\mathbf{a}$, then $\mathbf{b}$ must be the zero vector. This means that the zero vector is the only vector that is orthogonal to every vector.

In terms of components, $\mathbf{a} \cdot \mathbf{b}$ is given as follows:

$$\mathbf{a} \cdot \mathbf{b} = a_1 b_1 + a_2 b_2 + a_3 b_3 = a_k b_k \qquad (1.4.11)$$

In particular, $\mathbf{a} \cdot \mathbf{e}_1 = a_1$, $\mathbf{a} \cdot \mathbf{e}_2 = a_2$, $\mathbf{a} \cdot \mathbf{e}_3 = a_3$, or briefly,

$$\mathbf{a} \cdot \mathbf{e}_i = a_i = [\mathbf{a}]_i \qquad (1.4.12)$$

Note that for specified $\mathbf{e}_i$ equation (1.4.1) determines $\mathbf{a}$ when a_i are known and equations (1.4.12) determine a_i when $\mathbf{a}$ is known. Thus, for specified $\mathbf{e}_i$ a vector $\mathbf{a}$ is completely determined by the ordered triplet $(a_i) = (a_1, a_2, a_3)$. This triplet serves as a representation of $\mathbf{a}$ in the x_i-system.

From expressions (1.4.10) and (1.4.11), it follows that

$$\mathbf{a} \cdot \mathbf{a} = |\mathbf{a}|^2 = a_1^2 + a_2^2 + a_3^2 = a_k a_k \qquad (1.4.13)$$

From expressions (1.4.10) and (1.4.12), we find that

$$a_i = |\mathbf{a}| \cos \theta_i \tag{1.4.14}$$

where θ_i is the angle that (the direction of) $\mathbf{a}$ makes with the x_i-axis. The numbers $\cos \theta_i$ are therefore the *direction cosines* of $\mathbf{a}$. From (1.4.14) we note that the numbers a_i are proportional to these direction cosines; a_i are therefore the *direction ratios* of $\mathbf{a}$. Thus, for any vector $\mathbf{a}$ the components a_i are its direction ratios.

In particular, if $\mathbf{a}$ is a unit vector, we have $|\mathbf{a}| = 1$, and expression (1.4.14) shows that a_i are the direction cosines of $\mathbf{a}$. Thus, for any unit vector $\mathbf{a}$, the components a_i are its direction cosines. It follows that the direction cosines of a vector are the components of the unit vector directed along the vector.

The *vector product* or the *cross product* $\mathbf{a} \times \mathbf{b}$ of $\mathbf{a}$ and $\mathbf{b}$ (Figure 1.3) is the vector defined by

$$\mathbf{a} \times \mathbf{b} = |\mathbf{a}| |\mathbf{b}| |\sin \theta| \mathbf{n} \tag{1.4.15}$$

where θ is the angle between the directions of $\mathbf{a}$ and $\mathbf{b}$ and $\mathbf{n}$ is the *unit vector* that is perpendicular to both $\mathbf{a}$ and $\mathbf{b}$ and that is in the direction of the advancement of a righthanded screw (corkscrew) rotated from $\mathbf{a}$ to $\mathbf{b}$. It follows that (i) $\mathbf{a} \times \mathbf{a} = \mathbf{0}$, (ii) $\mathbf{b} \times \mathbf{a} = -(\mathbf{a} \times \mathbf{b})$, and (iii) $\mathbf{a}$ and $\mathbf{b}$ are collinear if and only if $\mathbf{a} \times \mathbf{b} = \mathbf{0}$.

Obviously, $\mathbf{a} \times \mathbf{0} = \mathbf{0}$ for every vector $\mathbf{a}$. Also, if $\mathbf{a} \times \mathbf{b} = \mathbf{0}$ for every vector $\mathbf{a}$, then $\mathbf{b}$ must be the zero vector. Hence, the zero vector is the only vector that is collinear with every vector.

In terms of components, the cross product $\mathbf{a} \times \mathbf{b}$ is given by

$$\mathbf{a} \times \mathbf{b} = (a_2 b_3 - a_3 b_2)\mathbf{e}_1 + (a_3 b_1 - a_1 b_3)\mathbf{e}_2 + (a_1 b_2 - a_2 b_1)\mathbf{e}_3 \tag{1.4.16}$$

From (1.4.15), we find that $|\mathbf{a} \times \mathbf{b}|$ represents the area of the parallelogram whose sides are represented by $\mathbf{a}$ and $\mathbf{b}$.

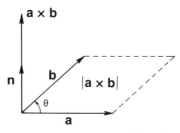

Figure 1.3. Vector product of $\mathbf{a}$ and $\mathbf{b}$.

1.4 SUMMARY OF RESULTS IN VECTOR ALGEBRA 11

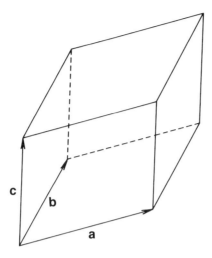

Figure 1.4. Parallelopiped determined by vectors **a**, **b**, **c**.

If **c** is a vector with components c_i, then the *scalar triple product* **a** · (**b** × **c**) is the scalar product of **a** and **b** × **c**; it is given by

$$\mathbf{a} \cdot (\mathbf{b} \times \mathbf{c}) = a_1(b_2 c_3 - b_3 c_2) + a_2(b_3 c_1 - b_1 c_3) + a_3(b_1 c_2 - b_2 c_1)$$

$$= \begin{vmatrix} a_1 & a_2 & a_3 \\ b_1 & b_2 & b_3 \\ c_1 & c_2 & c_3 \end{vmatrix} \quad (1.4.17)$$

It can be easily verified that

$$\mathbf{a} \cdot (\mathbf{b} \times \mathbf{c}) = (\mathbf{a} \times \mathbf{b}) \cdot \mathbf{c} \quad (1.4.18)$$

The absolute value of **a** · (**b** × **c**) represents the volume of the parallelopiped whose adjacent edges with a common corner are represented by **a**, **b**, **c** (see Figure 1.4).

It follows that **a**, **b**, **c** are *coplanar* (that is, **a**, **b**, **c** are all parallel to the same plane) if and only if **a** · (**b** × **c**) = 0. In such a situation, each of **a**, **b**, **c** can be expressed as a linear combination of the other two; for example, **a** = α**b** + β**c** for some real numbers α and β. The products **a** · (**b** × **c**) and (**a** × **b**) · **c** are also denoted by [**a**, **b**, **c**].

The *vector triple products* **a** × (**b** × **c**) and (**a** × **b**) × **c** are the vector products of the vectors **a** and **b** × **c** and **a** × **b** and **c**, respectively, in the order indicated. These two products are different in general; in fact, the products have the following explicit forms:

$$\mathbf{a} \times (\mathbf{b} \times \mathbf{c}) = (\mathbf{a} \cdot \mathbf{c})\mathbf{b} - (\mathbf{a} \cdot \mathbf{b})\mathbf{c} \tag{1.4.19}$$

$$(\mathbf{a} \times \mathbf{b}) \times \mathbf{c} = (\mathbf{a} \cdot \mathbf{c})\mathbf{b} - (\mathbf{b} \cdot \mathbf{c})\mathbf{a} \tag{1.4.20}$$

It can be proven that $\mathbf{a} \times (\mathbf{b} \times \mathbf{c}) = (\mathbf{a} \times \mathbf{b}) \times \mathbf{c}$ if and only if $\mathbf{b} \times (\mathbf{c} \times \mathbf{a}) = \mathbf{0}$.

1.5
SUMMARY OF RESULTS IN MATRIX ALGEBRA

By a *matrix* we mean a rectangular array of elements. The elements may be real or complex numbers or any other mathematical objects. A matrix having m rows and n columns (where m and n are positive integers) is referred to as an $m \times n$ matrix; if $m = n$, the matrix is called a *square matrix of order n*. In this text, we will be mainly dealing with square matrices of order 3 whose elements are real numbers. We denote such a matrix by $[a_{ij}]$ whose explicit form is as follows:

$$[a_{ij}] = \begin{bmatrix} a_{11} & a_{12} & a_{13} \\ a_{21} & a_{22} & a_{23} \\ a_{31} & a_{32} & a_{33} \end{bmatrix} \tag{1.5.1}$$

Note that for a given i and given j, a_{ij} denotes the element in the ith row and the jth column in the matrix $[a_{ij}]$. This element is referred to as the (ij)th or *typical element* of the matrix. In particular, the elements a_{11}, a_{22} and a_{33} are called the *diagonal elements* of $[a_{ij}]$. If in a square matrix all elements except the diagonal elements are 0, the matrix is called a *diagonal matrix*.

Two matrices $[a_{ij}]$ and $[b_{ij}]$ are said to be *equal* if their corresponding elements are equal; that is, $a_{11} = b_{11}, a_{12} = b_{12}, \ldots, a_{33} = b_{33}$, or, briefly, $a_{ij} = b_{ij}$. Thus, the *matrix equation* $[a_{ij}] = [b_{ij}]$ is *equivalent* to the equations $a_{ij} = b_{ij}$ in the suffix notation.

The matrix obtained by interchanging the rows and the columns of a matrix $[a_{ij}]$ is called the *transpose* of $[a_{ij}]$ denoted by $[a_{ij}]^T$. Thus,

$$[a_{ij}]^T = \begin{bmatrix} a_{11} & a_{21} & a_{31} \\ a_{12} & a_{22} & a_{32} \\ a_{13} & a_{23} & a_{33} \end{bmatrix} \tag{1.5.2}$$

1.5 SUMMARY OF RESULTS IN MATRIX ALGEBRA 13

It is evident that the typical element of $[a_{ij}]^T$ is a_{ji}; therefore,

$$[a_{ij}]^T = [a_{ji}] \tag{1.5.3}$$

It is trivial to verify that

$$([a_{ij}]^T)^T = [a_{ij}] \tag{1.5.4}$$

The determinant formed by the elements a_{ij} of a matrix $[a_{ij}]$ is called the *determinant of the matrix* $[a_{ij}]$; it is denoted by $\det[a_{ij}]$. Thus,

$$\det[a_{ij}] = \begin{vmatrix} a_{11} & a_{12} & a_{13} \\ a_{21} & a_{22} & a_{23} \\ a_{31} & a_{32} & a_{33} \end{vmatrix} \tag{1.5.5}$$

Since the value of a determinant remains unchanged if its rows and columns are interchanged, it follows that

$$\det[a_{ij}] = \det[a_{ji}] = \det[a_{ij}]^T \tag{1.5.6}$$

The matrix obtained by multiplying every element of a matrix $[a_{ij}]$ by a number α is called the *scalar multiple* of α and $[a_{ij}]$; denoted by $\alpha[a_{ij}]$. Thus,

$$\alpha[a_{ij}] = [\alpha a_{ij}] = \begin{bmatrix} \alpha a_{11} & \alpha a_{12} & \alpha a_{13} \\ \alpha a_{21} & \alpha a_{22} & \alpha a_{23} \\ \alpha a_{31} & \alpha a_{32} & \alpha a_{33} \end{bmatrix} \tag{1.5.7}$$

Evidently,

$$[\alpha a_{ij}]^T = [\alpha a_{ji}] = [\alpha a_{ij}]^T \tag{1.5.8}$$

Also,

$$\det[\alpha a_{ij}] = \alpha^3 \det[a_{ij}] \tag{1.5.9}$$

In particular, the scalar multiple of -1 and $[a_{ij}]$ is called the *negative* of $[a_{ij}]$, denoted by $-[a_{ij}]$. Thus,

$$-[a_{ij}] = [-a_{ij}] \tag{1.5.10}$$

A matrix $[a_{ij}]$ is said to be *symmetric* or *skew-symmetric*, accordingly, as $[a_{ij}]^T = \pm[a_{ij}]$, or, equivalently, $a_{ji} = \pm a_{ij}$. Thus $[a_{ij}]$ is symmetric if and only if $a_{21} = a_{12}$, $a_{31} = a_{13}$ and $a_{32} = a_{23}$, and skew-symmetric if and only if $a_{11} = a_{22} = a_{33} = 0$, $a_{21} = -a_{12}$, $a_{31} = -a_{13}$ and $a_{32} = -a_{23}$.

Given the matrices $[a_{ij}]$ and $[b_{ij}]$, the matrix obtained by adding the corresponding elements of $[a_{ij}]$ and $[b_{ij}]$ is called the *sum* of $[a_{ij}]$ and $[b_{ij}]$, denoted by $[a_{ij}] + [b_{ij}]$. Thus, by definition,

$$[a_{ij}] + [b_{ij}] = [a_{ij} + b_{ij}] = [b_{ij}] + [a_{ij}] \qquad (1.5.11)$$

Similarly, the matrix obtained by subtracting the elements of $[b_{ij}]$ from the corresponding elements of $[a_{ij}]$ is called the *difference* of $[a_{ij}]$ and $[b_{ij}]$ in that order; it is denoted by $[a_{ij}] - [b_{ij}]$. Thus, by definition,

$$[a_{ij}] - [b_{ij}] = [a_{ij} - b_{ij}] \qquad (1.5.12)$$

It is trivial that

$$[b_{ij}] - [a_{ij}] = -\{[a_{ij}] - [b_{ij}]\} \qquad (1.5.13)$$

For any matrix $[a_{ij}]$, the matrix $[a_{ij}] - [a_{ij}]$ is obviously a matrix all of whose elements are equal to 0. Such a matrix is called the *zero matrix*; we denote it by $[0]$. Thus,

$$[a_{ij}] - [a_{ij}] = [0] \qquad (1.5.14)$$

where

$$[0] = \begin{bmatrix} 0 & 0 & 0 \\ 0 & 0 & 0 \\ 0 & 0 & 0 \end{bmatrix} \qquad (1.5.15)$$

For any two matrices $[a_{ij}]$ and $[b_{ij}]$, we have

$$([a_{ij}] \pm [b_{ij}])^T = [a_{ij}]^T \pm [b_{ij}]^T \qquad (1.5.16)$$

For any matrix $[a_{ij}]$, it can be verified that $[a_{ij}] + [a_{ij}]^T$ is a symmetric matrix, and $[a_{ij}] - [a_{ij}]^T$ is a skew-symmetric matrix. Also,

$$[a_{ij}] = \tfrac{1}{2}([a_{ij}] + [a_{ij}]^T) + \tfrac{1}{2}([a_{ij}] - [a_{ij}]^T) \qquad (1.5.17)$$

The first matrix in the righthand side of (1.5.17), namely $\tfrac{1}{2}([a_{ij}] + [a_{ij}]^T)$, is symmetric and the second matrix, namely $\tfrac{1}{2}([a_{ij}] - [a_{ij}]^T)$, is skew-symmetric. As such, every matrix $[a_{ij}]$ can be represented uniquely as a sum of a symmetric matrix and a skew-symmetric matrix. The matrix $\tfrac{1}{2}([a_{ij}] + [a_{ij}]^T)$ is called the *symmetric part* of $[a_{ij}]$ and the matrix $\tfrac{1}{2}([a_{ij}] - [a_{ij}]^T)$ the *skew-symmetric part*.

Given the matrices $[a_{ij}]$ and $[b_{ij}]$, the *product* of $[a_{ij}]$ and $[b_{ij}]$ in that order is defined as a matrix $[c_{ij}]$ where

$$c_{ij} = \sum_{k=1}^{3} a_{ik} b_{kj} = a_{ik} b_{kj} \qquad (1.5.18)$$

1.3 FREE AND DUMMY SUFFIXES

This is a concise form of equations (1.3.3) written in the suffix notation. Note that this concise form contains two free suffixes, i and j, and one dummy suffix, k.

Since every suffix has the range 1, 2, 3 according to the range convention, an equation that contains one free suffix represents three equations, as in the case of (1.2.5); an equation that contains two free suffixes represents nine equations as in the case of (1.3.7), and so forth. An equation containing N free suffixes represents 3^N equations.

The concepts of free and dummy suffixes as well as the range and summation conventions adopted to systems of equations, as just described, may be employed to other mathematical systems also. Thus, the set of three numbers (a_1, a_2, a_3) can be represented as just (a_i). The 3×3 matrix

$$\begin{bmatrix} a_{11} & a_{12} & a_{13} \\ a_{21} & a_{22} & a_{23} \\ a_{31} & a_{32} & a_{33} \end{bmatrix}$$

may be represented as just (a_{ij}) or $[a_{ij}]$.

In general, the symbol $a_{ijk...}$ involving N free suffixes $i, j, k \ldots$ represents 3^N numbers.

EXAMPLE 1.3.1 If $a_i = \alpha_{ij} b_j$ and $b_i = \beta_{ij} c_j$, write down a_i in terms of c_i.

Solution It is given that

$$a_i = \alpha_{ij} b_j \tag{1.3.8}$$

and

$$b_i = \beta_{ij} c_j \tag{1.3.9}$$

If we change the free suffix from i to j and the dummy suffix from j to k in (1.3.9), we get

$$b_j = \beta_{jk} c_k \tag{1.3.10}$$

This has the same meaning as that of (1.3.9). Substituting for b_j from (1.3.10) in (1.3.8) gives

$$a_i = \alpha_{ij} \beta_{jk} c_k \tag{1.3.11}$$

This is the expression for a_i in terms of c_i. ∎

EXAMPLE 1.3.2 Show that

(i) $$a_{ij} b_{ij} = a_{ji} b_{ji} \tag{1.3.12}$$

(ii) $$(a_{ijk} + a_{jki} + a_{kij}) x_i x_j x_k = 3 a_{ijk} x_i x_j x_k \tag{1.3.13}$$

Solution (i) In $a_{ij}b_{ij}$ both i and j are dummy suffixes. Hence i may be replaced by any suffix and j may be replaced by any suffix. In particular, we may replace i by j, and j by i. Thus

$$a_{ij}b_{ij} = a_{ji}b_{ji}$$

(ii) We first note that

$$(a_{ijk} + a_{jki} + a_{kij})x_i x_j x_k = a_{ijk}x_i x_j x_k + a_{jki}x_i x_j x_k + a_{kij}x_i x_j x_k \quad (1.3.14)$$

In $a_{jki}x_i x_j x_k$, all of i, j, k are dummy suffixes. As such i may be replaced by k, j may be replaced by i and k may be replaced by j. Thus,

$$a_{jki}x_i x_j x_k = a_{ijk}x_k x_i x_j = a_{ijk}x_i x_j x_k$$

Similarly,

$$a_{kij}x_i x_j x_k = a_{ijk}x_j x_k x_i = a_{ijk}x_i x_j x_k$$

Thus, all the three terms in the righthand side of (1.3.14) are equal; their sum is therefore $3a_{ijk}x_i x_j x_k$. This proves (1.3.13). ∎

EXAMPLE 1.3.3 Suppose that the equations

$$a_{ip}a_{jp}c_j = b_{ij}c_j \quad (1.3.15)$$

hold for arbitrary c_i. Show that

$$a_{ip}a_{jp} = b_{ij} \quad (1.3.16)$$

Solution Since equations (1.3.15) hold for arbitrary c_i, these hold when $c_1 = 1$, $c_2 = 0$, $c_3 = 0$, and we obtain

$$a_{ip}a_{1p} = b_{i1} \quad (1.3.17)$$

Similarly, taking $c_1 = c_3 = 0$, $c_2 = 1$; and $c_1 = c_2 = 0$, $c_3 = 1$ in (1.3.15), we get, respectively,

$$a_{ip}a_{2p} = b_{i2}; \qquad a_{ip}a_{3p} = b_{i3} \quad (1.3.18)$$

Clearly, (1.3.17) and (1.3.18) constitute the required result (1.3.16). ∎

1.4
SUMMARY OF RESULTS IN VECTOR ALGEBRA

A *vector* is an entity that has two characteristics: magnitude and direction. Force and velocity are two typical examples of a vector. Geometrically, a vector is represented by a directed line segment, with the length of the segment representing the magnitude of the vector and the direction of the

1.4 SUMMARY OF RESULTS IN VECTOR ALGEBRA

segment indicating the direction of the vector. Evidently, the magnitude of a vector is a nonnegative real number.

Vectors are denoted by boldface letters in print (like **a**) or letters with a superposed arrow in manuscripts (like $\vec{a}$). In this text, only lowercase letters are used to denote vectors, reserving capital letters for tensors (to be introduced later). The magnitude of a vector **a** is denoted by $|\mathbf{a}|$ or a. A vector **a** is called a *unit vector* if $|\mathbf{a}| = 1$. A unit vector directed along a vector **a** is denoted by **â**. A vector whose magnitude is 0 is called the *zero vector*; it is denoted by **0** (or $\vec{0}$).

Two vectors are said to be *equal* if they have the same magnitude and the same direction. If **a** is a vector and α is a real number, then $\alpha\mathbf{a}$, called the **scalar multiple** of α and **a**, is a vector whose magnitude is $|\alpha||\mathbf{a}|$ and whose direction is that of **a** or opposite to **a** accordingly as $\alpha > 0$ or $\alpha < 0$. If $\alpha = 0$, then $\alpha\mathbf{a} = \mathbf{0}$. The vector $(-1)\mathbf{a}$ is denoted by $-\mathbf{a}$. Every vector **a** can be written as $\mathbf{a} = |\mathbf{a}|\mathbf{\hat{a}}$.

Two vectors are said to be *collinear* if their directions are either the same or opposite. If **a** and **b** are collinear vectors, then $\mathbf{a} = \alpha\mathbf{b}$ for some real number α.

The *sum* $\mathbf{a} + \mathbf{b}$ and the *difference* $\mathbf{a} - \mathbf{b}$ of vectors **a** and **b** are defined by the parallelogram law (see Figure 1.1). It is easy to verify that $\mathbf{b} + \mathbf{a} = \mathbf{a} + \mathbf{b}$, $\mathbf{b} - \mathbf{a} = -(\mathbf{a} - \mathbf{b})$, $\mathbf{a} + \mathbf{0} = \mathbf{a}$, and $\mathbf{a} - \mathbf{a} = \mathbf{0}$.

Suppose that there is a righthanded rectangular system of Cartesian axes with a fixed origin O. We refer to these axes as the x_1, x_2, x_3-axes, or briefly the x_i-axes. Also, we denote the unit vectors directed along the positive x_1, x_2, x_3-axes by $\mathbf{e}_1$, $\mathbf{e}_2$, $\mathbf{e}_3$, respectively (see Figure 1.2). Then $\mathbf{e}_i$ are called the *base vectors* of the x_i-system.

By virtue of the definition of equality of vectors and the parallelogram law of addition of vectors, every vector **a** can be expressed as a linear

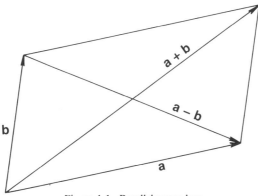

Figure 1.1. Parallelogram law.

8   1 SUFFIX NOTATION

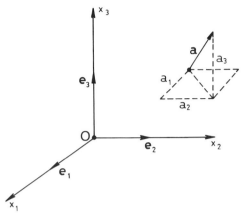

Figure 1.2. Base vectors.

combination of e_1, e_2, e_3 as follows:

$$\mathbf{a} = a_1\mathbf{e}_1 + a_2\mathbf{e}_2 + a_3\mathbf{e}_3 \tag{1.4.1}$$

Here a_1, a_2, a_3 are real numbers representing the projections of **a** on the axes; these are called the *components* of **a** along $\mathbf{e}_1$, $\mathbf{e}_2$ and $\mathbf{e}_3$, respectively.

Since

$$\mathbf{e}_1 = 1\mathbf{e}_1 + 0\mathbf{e}_2 + 0\mathbf{e}_3$$

the components of $\mathbf{e}_1$ are $(1, 0, 0)$. Similarly, the components of $\mathbf{e}_2$ are $(0, 1, 0)$ and those of $\mathbf{e}_3$ are $(0, 0, 1)$.

By employing the summation convention, expression (1.4.1) can be rewritten as

$$\mathbf{a} = a_k \mathbf{e}_k \tag{1.4.2}$$

The three components a_1, a_2, a_3 of **a** are referred to briefly as a_i. However, the symbol a_i is also frequently used to denote a typical component of **a**; so a_i is referred to as the ith component of **a**.

We denote the ith component of **a** also by $[\mathbf{a}]_i$; thus

$$[\mathbf{a}]_i = a_i \tag{1.4.3}$$

If α is a real number, then the vector $\alpha\mathbf{a}$ is given by

$$\alpha\mathbf{a} = (\alpha a_1)\mathbf{e}_1 + (\alpha a_2)\mathbf{e}_2 + (\alpha a_3)\mathbf{e}_3 = \alpha a_k \mathbf{e}_k \tag{1.4.4}$$

Thus, the components of $\alpha\mathbf{a}$ are αa_i; that is,

$$[\alpha\mathbf{a}]_i = \alpha a_i = \alpha[\mathbf{a}]_i \tag{1.4.5}$$

As a direct consequence of this, it follows that $-a_i$ are components of $-\mathbf{a}$.

1.4 SUMMARY OF RESULTS IN VECTOR ALGEBRA

Let **b** be a vector with components b_i; that is,

$$\mathbf{b} = b_1 \mathbf{e}_1 + b_2 \mathbf{e}_2 + b_3 \mathbf{e}_3 = b_k \mathbf{e}_k \tag{1.4.6}$$

Then, by the definition of equality of vectors, it follows that $\mathbf{a} = \mathbf{b}$ if and only if $a_1 = b_1$, $a_2 = b_2$, $a_3 = b_3$, or briefly

$$a_i = b_i \tag{1.4.7}$$

Thus, the *vector equation* $\mathbf{a} = \mathbf{b}$ is equivalent to the equations $a_i = b_i$ written in the suffix notation.

By the parallelogram law of addition of vectors, it follows that the sum $\mathbf{a} + \mathbf{b}$ and the difference $\mathbf{a} - \mathbf{b}$ of **a** and **b** are given by

$$\mathbf{a} \pm \mathbf{b} = (a_1 \pm b_1)\mathbf{e}_1 + (a_2 \pm b_2)\mathbf{e}_2 + (a_3 \pm b_3)\mathbf{e}_3 = (a_k \pm b_k)\mathbf{e}_k \tag{1.4.8}$$

The components of $\mathbf{a} \pm \mathbf{b}$ are thus $a_i \pm b_i$; that is,

$$[\mathbf{a} \pm \mathbf{b}]_i = a_i \pm b_i = [\mathbf{a}]_i \pm [\mathbf{b}]_i \tag{1.4.9}$$

A direct consequence of (1.4.9) is that the components of the zero vector **0** are all 0.

The *scalar product* or the *dot product* $\mathbf{a} \cdot \mathbf{b}$ of **a** and **b** is a real number defined by

$$\mathbf{a} \cdot \mathbf{b} = |\mathbf{a}||\mathbf{b}| \cos \theta \tag{1.4.10}$$

where θ is the angle between the directions of **a** and **b**. Evidently, $\mathbf{b} \cdot \mathbf{a} = \mathbf{a} \cdot \mathbf{b}$.

Two vectors are said to be *orthogonal* if the angle between their directions is $\pi/2$. It follows that **a** and **b** are orthogonal if and only if $\mathbf{a} \cdot \mathbf{b} = 0$.

Obviously, $\mathbf{a} \cdot \mathbf{0} = 0$ for every vector **a**. Also, if $\mathbf{a} \cdot \mathbf{b} = 0$ for every vector **a**, then **b** must be the zero vector. This means that the zero vector is the only vector that is orthogonal to every vector.

In terms of components, $\mathbf{a} \cdot \mathbf{b}$ is given as follows:

$$\mathbf{a} \cdot \mathbf{b} = a_1 b_1 + a_2 b_2 + a_3 b_3 = a_k b_k \tag{1.4.11}$$

In particular, $\mathbf{a} \cdot \mathbf{e}_1 = a_1$, $\mathbf{a} \cdot \mathbf{e}_2 = a_2$, $\mathbf{a} \cdot \mathbf{e}_3 = a_3$, or briefly,

$$\mathbf{a} \cdot \mathbf{e}_i = a_i = [\mathbf{a}]_i \tag{1.4.12}$$

Note that for specified $\mathbf{e}_i$ equation (1.4.1) determines **a** when a_i are known and equations (1.4.12) determine a_i when **a** is known. Thus, for specified $\mathbf{e}_i$ a vector **a** is completely determined by the ordered triplet $(a_i) = (a_1, a_2, a_3)$. This triplet serves as a representation of **a** in the x_i-system.

From expressions (1.4.10) and (1.4.11), it follows that

$$\mathbf{a} \cdot \mathbf{a} = |\mathbf{a}|^2 = a_1^2 + a_2^2 + a_3^2 = a_k a_k \tag{1.4.13}$$

From expressions (1.4.10) and (1.4.12), we find that

$$a_i = |\mathbf{a}| \cos \theta_i \qquad (1.4.14)$$

where θ_i is the angle that (the direction of) $\mathbf{a}$ makes with the x_i-axis. The numbers $\cos \theta_i$ are therefore the *direction cosines* of $\mathbf{a}$. From (1.4.14) we note that the numbers a_i are proportional to these direction cosines; a_i are therefore the *direction ratios* of $\mathbf{a}$. Thus, for any vector $\mathbf{a}$ the components a_i are its direction ratios.

In particular, if $\mathbf{a}$ is a unit vector, we have $|\mathbf{a}| = 1$, and expression (1.4.14) shows that a_i are the direction cosines of $\mathbf{a}$. Thus, for any unit vector $\mathbf{a}$, the components a_i are its direction cosines. It follows that the direction cosines of a vector are the components of the unit vector directed along the vector.

The *vector product* or the *cross product* $\mathbf{a} \times \mathbf{b}$ of $\mathbf{a}$ and $\mathbf{b}$ (Figure 1.3) is the vector defined by

$$\mathbf{a} \times \mathbf{b} = |\mathbf{a}| |\mathbf{b}| |\sin \theta| \mathbf{n} \qquad (1.4.15)$$

where θ is the angle between the directions of $\mathbf{a}$ and $\mathbf{b}$ and $\mathbf{n}$ is the *unit vector* that is perpendicular to both $\mathbf{a}$ and $\mathbf{b}$ and that is in the direction of the advancement of a righthanded screw (corkscrew) rotated from $\mathbf{a}$ to $\mathbf{b}$. It follows that (i) $\mathbf{a} \times \mathbf{a} = \mathbf{0}$, (ii) $\mathbf{b} \times \mathbf{a} = -(\mathbf{a} \times \mathbf{b})$, and (iii) $\mathbf{a}$ and $\mathbf{b}$ are collinear if and only if $\mathbf{a} \times \mathbf{b} = \mathbf{0}$.

Obviously, $\mathbf{a} \times \mathbf{0} = \mathbf{0}$ for every vector $\mathbf{a}$. Also, if $\mathbf{a} \times \mathbf{b} = \mathbf{0}$ for every vector $\mathbf{a}$, then $\mathbf{b}$ must be the zero vector. Hence, the zero vector is the only vector that is collinear with every vector.

In terms of components, the cross product $\mathbf{a} \times \mathbf{b}$ is given by

$$\mathbf{a} \times \mathbf{b} = (a_2 b_3 - a_3 b_2)\mathbf{e}_1 + (a_3 b_1 - a_1 b_3)\mathbf{e}_2 + (a_1 b_2 - a_2 b_1)\mathbf{e}_3 \qquad (1.4.16)$$

From (1.4.15), we find that $|\mathbf{a} \times \mathbf{b}|$ represents the area of the parallelogram whose sides are represented by $\mathbf{a}$ and $\mathbf{b}$.

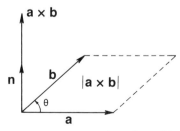

Figure 1.3. Vector product of $\mathbf{a}$ and $\mathbf{b}$.

1.4 SUMMARY OF RESULTS IN VECTOR ALGEBRA 11

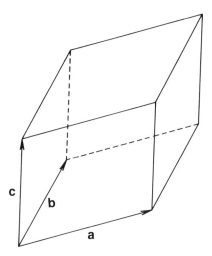

Figure 1.4. Parallelopiped determined by vectors **a**, **b**, **c**.

If **c** is a vector with components c_i, then the *scalar triple product* **a** · (**b** × **c**) is the scalar product of **a** and **b** × **c**; it is given by

$$\mathbf{a} \cdot (\mathbf{b} \times \mathbf{c}) = a_1(b_2 c_3 - b_3 c_2) + a_2(b_3 c_1 - b_1 c_3) + a_3(b_1 c_2 - b_2 c_1)$$

$$= \begin{vmatrix} a_1 & a_2 & a_3 \\ b_1 & b_2 & b_3 \\ c_1 & c_2 & c_3 \end{vmatrix} \quad (1.4.17)$$

It can be easily verified that

$$\mathbf{a} \cdot (\mathbf{b} \times \mathbf{c}) = (\mathbf{a} \times \mathbf{b}) \cdot \mathbf{c} \quad (1.4.18)$$

The absolute value of **a** · (**b** × **c**) represents the volume of the parallelopiped whose adjacent edges with a common corner are represented by **a**, **b**, **c** (see Figure 1.4).

It follows that **a**, **b**, **c** are *coplanar* (that is, **a**, **b**, **c** are all parallel to the same plane) if and only if **a** · (**b** × **c**) = 0. In such a situation, each of **a**, **b**, **c** can be expressed as a linear combination of the other two; for example, **a** = α**b** + β**c** for some real numbers α and β. The products **a** · (**b** × **c**) and (**a** × **b**) · **c** are also denoted by [**a**, **b**, **c**].

The *vector triple products* **a** × (**b** × **c**) and (**a** × **b**) × **c** are the vector products of the vectors **a** and **b** × **c** and **a** × **b** and **c**, respectively, in the order indicated. These two products are different in general; in fact, the

12 1 SUFFIX NOTATION

products have the following explicit forms:

$$\mathbf{a} \times (\mathbf{b} \times \mathbf{c}) = (\mathbf{a} \cdot \mathbf{c})\mathbf{b} - (\mathbf{a} \cdot \mathbf{b})\mathbf{c} \qquad (1.4.19)$$

$$(\mathbf{a} \times \mathbf{b}) \times \mathbf{c} = (\mathbf{a} \cdot \mathbf{c})\mathbf{b} - (\mathbf{b} \cdot \mathbf{c})\mathbf{a} \qquad (1.4.20)$$

It can be proven that $\mathbf{a} \times (\mathbf{b} \times \mathbf{c}) = (\mathbf{a} \times \mathbf{b}) \times \mathbf{c}$ if and only if $\mathbf{b} \times (\mathbf{c} \times \mathbf{a}) = \mathbf{0}$.

1.5
SUMMARY OF RESULTS IN MATRIX ALGEBRA

By a *matrix* we mean a rectangular array of elements. The elements may be real or complex numbers or any other mathematical objects. A matrix having m rows and n columns (where m and n are positive integers) is referred to as an $m \times n$ matrix; if $m = n$, the matrix is called a *square matrix of order n*. In this text, we will be mainly dealing with square matrices of order 3 whose elements are real numbers. We denote such a matrix by $[a_{ij}]$ whose explicit form is as follows:

$$[a_{ij}] = \begin{bmatrix} a_{11} & a_{12} & a_{13} \\ a_{21} & a_{22} & a_{23} \\ a_{31} & a_{32} & a_{33} \end{bmatrix} \qquad (1.5.1)$$

Note that for a given i and given j, a_{ij} denotes the element in the ith row and the jth column in the matrix $[a_{ij}]$. This element is referred to as the (ij)th or *typical element* of the matrix. In particular, the elements a_{11}, a_{22} and a_{33} are called the *diagonal elements* of $[a_{ij}]$. If in a square matrix all elements except the diagonal elements are 0, the matrix is called a *diagonal matrix*.

Two matrices $[a_{ij}]$ and $[b_{ij}]$ are said to be *equal* if their corresponding elements are equal; that is, $a_{11} = b_{11}$, $a_{12} = b_{12}, \ldots, a_{33} = b_{33}$, or, briefly, $a_{ij} = b_{ij}$. Thus, the *matrix equation* $[a_{ij}] = [b_{ij}]$ is *equivalent* to the equations $a_{ij} = b_{ij}$ in the suffix notation.

The matrix obtained by interchanging the rows and the columns of a matrix $[a_{ij}]$ is called the *transpose* of $[a_{ij}]$ denoted by $[a_{ij}]^T$. Thus,

$$[a_{ij}]^T = \begin{bmatrix} a_{11} & a_{21} & a_{31} \\ a_{12} & a_{22} & a_{32} \\ a_{13} & a_{23} & a_{33} \end{bmatrix} \qquad (1.5.2)$$

1.5 SUMMARY OF RESULTS IN MATRIX ALGEBRA 13

It is evident that the typical element of $[a_{ij}]^T$ is a_{ji}; therefore,

$$[a_{ij}]^T = [a_{ji}] \tag{1.5.3}$$

It is trivial to verify that

$$([a_{ij}]^T)^T = [a_{ij}] \tag{1.5.4}$$

The determinant formed by the elements a_{ij} of a matrix $[a_{ij}]$ is called the *determinant of the matrix* $[a_{ij}]$; it is denoted by $\det[a_{ij}]$. Thus,

$$\det[a_{ij}] = \begin{vmatrix} a_{11} & a_{12} & a_{13} \\ a_{21} & a_{22} & a_{23} \\ a_{31} & a_{32} & a_{33} \end{vmatrix} \tag{1.5.5}$$

Since the value of a determinant remains unchanged if its rows and columns are interchanged, it follows that

$$\det[a_{ij}] = \det[a_{ji}] = \det[a_{ij}]^T \tag{1.5.6}$$

The matrix obtained by multiplying every element of a matrix $[a_{ij}]$ by a number α is called the *scalar multiple* of α and $[a_{ij}]$; denoted by $\alpha[a_{ij}]$. Thus,

$$\alpha[a_{ij}] = [\alpha a_{ij}] = \begin{bmatrix} \alpha a_{11} & \alpha a_{12} & \alpha a_{13} \\ \alpha a_{21} & \alpha a_{22} & \alpha a_{23} \\ \alpha a_{31} & \alpha a_{32} & \alpha a_{33} \end{bmatrix} \tag{1.5.7}$$

Evidently,

$$[\alpha a_{ij}]^T = [\alpha a_{ji}] = [\alpha a_{ij}]^T \tag{1.5.8}$$

Also,

$$\det[\alpha a_{ij}] = \alpha^3 \det[a_{ij}] \tag{1.5.9}$$

In particular, the scalar multiple of -1 and $[a_{ij}]$ is called the *negative* of $[a_{ij}]$, denoted by $-[a_{ij}]$. Thus,

$$-[a_{ij}] = [-a_{ij}] \tag{1.5.10}$$

A matrix $[a_{ij}]$ is said to be *symmetric* or *skew-symmetric*, accordingly, as $[a_{ij}]^T = \pm[a_{ij}]$, or, equivalently, $a_{ji} = \pm a_{ij}$. Thus $[a_{ij}]$ is symmetric if and only if $a_{21} = a_{12}$, $a_{31} = a_{13}$ and $a_{32} = a_{23}$, and skew-symmetric if and only if $a_{11} = a_{22} = a_{33} = 0$, $a_{21} = -a_{12}$, $a_{31} = -a_{13}$ and $a_{32} = -a_{23}$.

Given the matrices $[a_{ij}]$ and $[b_{ij}]$, the matrix obtained by adding the corresponding elements of $[a_{ij}]$ and $[b_{ij}]$ is called the *sum* of $[a_{ij}]$ and $[b_{ij}]$, denoted by $[a_{ij}] + [b_{ij}]$. Thus, by definition,

$$[a_{ij}] + [b_{ij}] = [a_{ij} + b_{ij}] = [b_{ij}] + [a_{ij}] \qquad (1.5.11)$$

Similarly, the matrix obtained by subtracting the elements of $[b_{ij}]$ from the corresponding elements of $[a_{ij}]$ is called the *difference* of $[a_{ij}]$ and $[b_{ij}]$ in that order; it is denoted by $[a_{ij}] - [b_{ij}]$. Thus, by definition,

$$[a_{ij}] - [b_{ij}] = [a_{ij} - b_{ij}] \qquad (1.5.12)$$

It is trivial that

$$[b_{ij}] - [a_{ij}] = -\{[a_{ij}] - [b_{ij}]\} \qquad (1.5.13)$$

For any matrix $[a_{ij}]$, the matrix $[a_{ij}] - [a_{ij}]$ is obviously a matrix all of whose elements are equal to 0. Such a matrix is called the *zero matrix*; we denote it by $[0]$. Thus,

$$[a_{ij}] - [a_{ij}] = [0] \qquad (1.5.14)$$

where

$$[0] = \begin{bmatrix} 0 & 0 & 0 \\ 0 & 0 & 0 \\ 0 & 0 & 0 \end{bmatrix} \qquad (1.5.15)$$

For any two matrices $[a_{ij}]$ and $[b_{ij}]$, we have

$$([a_{ij}] \pm [b_{ij}])^T = [a_{ij}]^T \pm [b_{ij}]^T \qquad (1.5.16)$$

For any matrix $[a_{ij}]$, it can be verified that $[a_{ij}] + [a_{ij}]^T$ is a symmetric matrix, and $[a_{ij}] - [a_{ij}]^T$ is a skew-symmetric matrix. Also,

$$[a_{ij}] = \tfrac{1}{2}([a_{ij}] + [a_{ij}]^T) + \tfrac{1}{2}([a_{ij}] - [a_{ij}]^T) \qquad (1.5.17)$$

The first matrix in the righthand side of (1.5.17), namely $\tfrac{1}{2}([a_{ij}] + [a_{ij}]^T)$, is symmetric and the second matrix, namely $\tfrac{1}{2}([a_{ij}] - [a_{ij}]^T)$, is skew-symmetric. As such, every matrix $[a_{ij}]$ can be represented uniquely as a sum of a symmetric matrix and a skew-symmetric matrix. The matrix $\tfrac{1}{2}([a_{ij}] + [a_{ij}]^T)$ is called the *symmetric part* of $[a_{ij}]$ and the matrix $\tfrac{1}{2}([a_{ij}] - [a_{ij}]^T)$ the *skew-symmetric part*.

Given the matrices $[a_{ij}]$ and $[b_{ij}]$, the *product* of $[a_{ij}]$ and $[b_{ij}]$ in that order is defined as a matrix $[c_{ij}]$ where

$$c_{ij} = \sum_{k=1}^{3} a_{ik} b_{kj} = a_{ik} b_{kj} \qquad (1.5.18)$$

1.5 SUMMARY OF RESULTS IN MATRIX ALGEBRA

This product is denoted by $[a_{ij}][b_{ij}]$. Thus,

$$[a_{ij}][b_{ij}] = [a_{ik}b_{kj}] \tag{1.5.19}$$

In the same way

$$[b_{ij}][a_{ij}] = [b_{ik}a_{kj}] \tag{1.5.20}$$

From (1.5.19) and (1.5.20), it is evident that $[a_{ij}][b_{ij}] \neq [b_{ij}][a_{ij}]$ in general.

The product $[a_{ij}][a_{ij}]$ is called the *square* of the matrix $[a_{ij}]$, denoted by $[a_{ij}]^2$. Thus

$$[a_{ij}]^2 = [a_{ij}][a_{ij}] = [a_{ik}a_{kj}] \tag{1.5.21}$$

Higher powers of $[a_{ij}]$ are defined in an analogous way.

By virtue of expression (1.5.19), it follows that, for any two matrices $[a_{ij}]$ and $[b_{ij}]$, we have

$$[a_{ij}][b_{ij}]^T = [a_{ik}b_{jk}] \tag{1.5.22}$$

$$[a_{ij}]^T[b_{ij}] = [a_{ki}b_{kj}] \tag{1.5.23}$$

$$([a_{ij}][b_{ij}])^T = [a_{jk}b_{ki}] = ([b_{ij}]^T[a_{ij}]^T) \tag{1.5.24}$$

$$\det([a_{ij}][b_{ij}]) = (\det[a_{ij}])(\det[b_{ij}]) \tag{1.5.25}$$

Also, for any three matrices $[a_{ij}]$, $[b_{ij}]$ and $[c_{ij}]$, we have

$$[a_{ij}]([b_{ij}][c_{ij}]) = ([a_{ij}][b_{ij}])[c_{ij}] \tag{1.5.26}$$

$$[a_{ij}]([b_{ij}] \pm [c_{ij}]) = [a_{ij}][b_{ij}] \pm [a_{ij}][c_{ij}] \tag{1.5.27}$$

$$([a_{ij}] \pm [b_{ij}])[c_{ij}] = [a_{ij}][c_{ij}] \pm [b_{ij}][c_{ij}] \tag{1.5.28}$$

Given a matrix $[a_{ij}]$, if there exists a matrix, denoted by $[a_{ij}]^{-1}$, such that

$$[a_{ij}][a_{ij}]^{-1} = [a_{ij}]^{-1}[a_{ij}] = [I] \tag{1.5.29}$$

where $[I]$ is the matrix given by

$$[I] = \begin{bmatrix} 1 & 0 & 0 \\ 0 & 1 & 0 \\ 0 & 0 & 1 \end{bmatrix} \tag{1.5.30}$$

then we say that $[a_{ij}]$ is an *invertible matrix* or a *nonsingular matrix*; the matrix $[a_{ij}]^{-1}$ is then called the (multiplicative) *inverse* of $[a_{ij}]$. Otherwise, $[a_{ij}]$ is said to be a *singular matrix*. It can be proved that $[a_{ij}]$ is nonsingular if and only if $\det[a_{ij}] \neq 0$.

The matrix $[I]$ given by (1.5.30) is called the *unit matrix* or the *identity matrix* (of order 3). It is trivial that $\det[I] = 1$; as such $[I]$ is nonsingular.

16 1 SUFFIX NOTATION

The inverse of $[I]$ is itself. Also, for any matrix $[a_{ij}]$, we have

$$[I][a_{ij}] = [a_{ij}][I] = [a_{ij}] \tag{1.5.31}$$

From results (1.5.25) and (1.5.29), it follows that for any non-singular matrix $[a_{ij}]$,

$$\det[a_{ij}]^{-1} = \frac{1}{\det[a_{ij}]} \tag{1.5.32}$$

It can be proven that if $[a_{ij}]$ and $[b_{ij}]$ are nonsingular matrices, then $[a_{ij}][b_{ij}]$ is also nonsingular and

$$([a_{ij}][b_{ij}])^{-1} = [b_{ij}]^{-1}[a_{ij}]^{-1} \tag{1.5.33}$$

A matrix $[a_{ij}]$ is said to be *orthogonal* if it is nonsingular and its inverse is equal to its transpose; that is, if

$$[a_{ij}]^T[a_{ij}] = [a_{ij}][a_{ij}]^T = [I] \tag{1.5.34}$$

It follows that for an orthogonal matrix $[a_{ij}]$,

$$\det[a_{ij}] = \pm 1 \tag{1.5.35}$$

An orthogonal matrix $[a_{ij}]$ for which $\det[a_{ij}] = 1$ is called a *proper orthogonal matrix*.

1.6
THE SYMBOL δ_{ij}

Recall that the base vectors $\mathbf{e}_1$, $\mathbf{e}_2$ and $\mathbf{e}_3$ are the unit vectors directed along the x_1, x_2 and x_3 axes, respectively. Since the axes are assumed to be rectangular, expression (1.4.10) yields the following relations:

$$\begin{aligned}\mathbf{e}_1 \cdot \mathbf{e}_1 = \mathbf{e}_2 \cdot \mathbf{e}_2 = \mathbf{e}_3 \cdot \mathbf{e}_3 = 1 \\ \mathbf{e}_1 \cdot \mathbf{e}_2 = \mathbf{e}_2 \cdot \mathbf{e}_1 = \mathbf{e}_1 \cdot \mathbf{e}_3 = \mathbf{e}_3 \cdot \mathbf{e}_1 = \mathbf{e}_2 \cdot \mathbf{e}_3 = \mathbf{e}_3 \cdot \mathbf{e}_2 = 0\end{aligned} \tag{1.6.1}$$

These nine relations may be represented in a condensed form by defining a symbol δ_{ij} as follows:

$$\delta_{ij} = \begin{cases} 1 & \text{if } i \text{ and } j \text{ take the } \textit{same} \text{ value} \\ 0 & \text{if } i \text{ and } j \text{ take } \textit{different} \text{ values} \end{cases} \tag{1.6.2}$$

That is,

$$\delta_{11} = \delta_{22} = \delta_{33} = 1, \quad \delta_{12} = \delta_{21} = \delta_{13} = \delta_{31} = \delta_{23} = \delta_{32} = 0$$

Clearly, a condensed form of relations (1.6.1) is

$$\mathbf{e}_i \cdot \mathbf{e}_j = \delta_{ij} \qquad (1.6.3)$$

The symbol δ_{ij} defined by (1.6.2) is known as the *Kronecker δ symbol*, named after the German mathematician Leopold Kronecker (1823–1891). The following property is inherent in the definition of δ_{ij}:

$$\delta_{ij} = \delta_{ji} \qquad (1.6.4)$$

It is important to note that, by summation convention,

$$\delta_{ii} = \delta_{11} + \delta_{22} + \delta_{33} = 3 \qquad (1.6.5)$$

From (1.6.2), it is evident that δ_{ij} are the elements of the unit matrix $[I]$ defined by (1.5.30). Thus

$$[I] = [\delta_{ij}] = \begin{bmatrix} 1 & 0 & 0 \\ 0 & 1 & 0 \\ 0 & 0 & 1 \end{bmatrix} \qquad (1.6.6)$$

Consequently,

$$\det(\delta_{ij}) = 1 \qquad (1.6.7)$$

EXAMPLE 1.6.1 Write down the following equations in a condensed form:

$$a_{11} = \alpha(b_{11} + b_{22} + b_{33}) + \beta b_{11}$$
$$a_{22} = \alpha(b_{11} + b_{22} + b_{33}) + \beta b_{22}$$
$$a_{33} = \alpha(b_{11} + b_{22} + b_{33}) + \beta b_{33}$$
$$a_{12} = \beta b_{12}, \quad a_{23} = \beta b_{23}, \quad a_{31} = \beta b_{31}$$
$$a_{21} = \beta b_{21}, \quad a_{32} = \beta b_{32}, \quad a_{13} = \beta b_{13}$$

Solution We have

$$a_{11} = \alpha(b_{11} + b_{22} + b_{33}) + \beta b_{11} = \alpha b_{kk} + \beta b_{11}$$
$$= \alpha \delta_{11} b_{kk} + \beta b_{11}, \quad \text{because} \quad \delta_{11} = 1$$

Expressions for a_{22} and a_{33} may be written similarly. Also,

$$a_{12} = \beta b_{12} = \alpha \delta_{12} b_{kk} + \beta b_{12},$$

because $\delta_{12} = 0$. Expressions for a_{23}, a_{31}, a_{21}, a_{32}, a_{13} may be written similarly.

18 1 **SUFFIX NOTATION**

It is now obvious that the given equations may be expressed in the following condensed form:
$$a_{ij} = \alpha \delta_{ij} b_{kk} + \beta b_{ij}$$ ■

EXAMPLE 1.6.2 Given that
$$a_{ij} = \alpha \delta_{ij} b_{kk} + \beta b_{ij} \quad (1.6.8)$$
where $\beta \neq 0$, $3\alpha + \beta \neq 0$, find b_{ij} in terms of a_{ij}.

Solution From (1.6.8), we get
$$a_{pp} = \alpha \delta_{pp} b_{kk} + \beta b_{pp} = (3\alpha + \beta) b_{pp},$$
because $\delta_{pp} = 3$ and $b_{kk} = b_{pp}$
so that
$$b_{pp} = \frac{1}{3\alpha + \beta} a_{pp} \quad (1.6.9)$$

It also follows from (1.6.8) that
$$b_{ij} = \frac{1}{\beta} [a_{ij} - \alpha \delta_{ij} b_{pp}] \quad (1.6.10)$$

Substituting for b_{pp} from (1.6.9) in (1.6.10), we obtain
$$b_{ij} = \frac{1}{\beta} \left[a_{ij} - \frac{\alpha}{3\alpha + \beta} \delta_{ij} a_{pp} \right] \quad ■ \quad (1.6.11)$$

EXAMPLE 1.6.3 Prove the following:

(i) $\delta_{ij} a_j = a_i$,   (ii) $\delta_{ij} a_{jk} = a_{ik}$,   (iii) $\delta_{ij} a_{kj} = a_{ki}$

(iv) $\delta_{ij} a_{ij} = a_{ii}$,   (v) $\delta_{ij} \delta_{jk} a_{km} = a_{im}$.

(Results of this type are called *substitution properties* of δ_{ij}.)

Solution (i) Put $b_i = \delta_{ij} a_j$. Then
$$b_1 = \delta_{11} a_1 + \delta_{12} a_2 + \delta_{13} a_3 = a_1$$
Similarly, $b_2 = a_2$, $b_3 = a_3$. Thus, $b_i = a_i$. That is, $\delta_{ij} a_j = a_i$.

(ii) Put $b_{ik} = \delta_{ij} a_{jk}$. Then
$$b_{11} = \delta_{1j} a_{j1} = \delta_{11} a_{11} + \delta_{12} a_{21} + \delta_{13} a_{31} = a_{11}$$
Similarly, $b_{22} = a_{22}$, $b_{33} = a_{33}$. Also,
$$b_{12} = \delta_{1j} a_{j2} = \delta_{11} a_{12} + \delta_{12} a_{22} + \delta_{13} a_{32} = a_{12}$$

Similarly, $b_{21} = a_{21}$, $b_{13} = a_{13}$, $b_{31} = a_{31}$, $b_{23} = a_{23}$, $b_{32} = a_{32}$. Thus, $b_{ik} = a_{ik}$. That is, $\delta_{ij} a_{jk} = a_{ik}$.

(iii) Put $c_{ik} = a_{ki}$. Then result (ii) yields $\delta_{ij} c_{jk} = c_{ik}$. That is, $\delta_{ij} a_{kj} = a_{ki}$.

(iv) $\delta_{ij} a_{ij} = a_{ii}$, by result (iii).

(v) $\delta_{ij} \delta_{jk} a_{km} = \delta_{ik} a_{km} = a_{im}$, by result (ii). ■

EXAMPLE 1.6.4 Prove the matrix identity (1.5.31).

Solution Since $[I] = [\delta_{ij}] = [\delta_{ji}]$, expression (1.5.19) and results (ii) and (iii) of Example 1.6.3 yield

$$[I][a_{ij}] = [\delta_{ik} a_{kj}] = [a_{ij}] = [\delta_{jk} a_{ik}] = [a_{ik} \delta_{kj}] = [a_{ij}][I] \quad ■$$

1.7
THE SYMBOL ε_{ijk}

It has been assumed that the coordinate system for which $\mathbf{e}_i$ are base vectors is a righthanded system; see Section 1.4. Consequently, by virtue of expression (1.4.15), the vectors $\mathbf{e}_i$ obey the following relations, in addition to (1.6.1):

$$\begin{aligned} \mathbf{e}_1 \times \mathbf{e}_2 &= -(\mathbf{e}_2 \times \mathbf{e}_1) = \mathbf{e}_3 \\ \mathbf{e}_2 \times \mathbf{e}_3 &= -(\mathbf{e}_3 \times \mathbf{e}_2) = \mathbf{e}_1 \\ \mathbf{e}_3 \times \mathbf{e}_1 &= -(\mathbf{e}_1 \times \mathbf{e}_3) = \mathbf{e}_2 \end{aligned} \quad (1.7.1)$$

From relations (1.6.1) and (1.7.1), we obtain the following 27 relations:

$$\begin{aligned} \mathbf{e}_1 \cdot (\mathbf{e}_2 \times \mathbf{e}_3) &= \mathbf{e}_2 \cdot (\mathbf{e}_3 \times \mathbf{e}_1) = \mathbf{e}_3 \cdot (\mathbf{e}_1 \times \mathbf{e}_2) = 1 \\ \mathbf{e}_1 \cdot (\mathbf{e}_3 \times \mathbf{e}_2) &= \mathbf{e}_2 \cdot (\mathbf{e}_1 \times \mathbf{e}_3) = \mathbf{e}_3 \cdot (\mathbf{e}_2 \times \mathbf{e}_1) = -1 \\ \mathbf{e}_1 \cdot (\mathbf{e}_1 \times \mathbf{e}_2) &= \mathbf{e}_2 \cdot (\mathbf{e}_2 \times \mathbf{e}_3) = \mathbf{e}_3 \cdot (\mathbf{e}_3 \times \mathbf{e}_1) = 0 \\ \mathbf{e}_1 \cdot (\mathbf{e}_2 \times \mathbf{e}_1) &= \mathbf{e}_2 \cdot (\mathbf{e}_3 \times \mathbf{e}_2) = \mathbf{e}_3 \cdot (\mathbf{e}_1 \times \mathbf{e}_3) = 0 \\ \mathbf{e}_1 \cdot (\mathbf{e}_1 \times \mathbf{e}_3) &= \mathbf{e}_2 \cdot (\mathbf{e}_2 \times \mathbf{e}_1) = \mathbf{e}_3 \cdot (\mathbf{e}_3 \times \mathbf{e}_2) = 0 \\ \mathbf{e}_1 \cdot (\mathbf{e}_3 \times \mathbf{e}_1) &= \mathbf{e}_2 \cdot (\mathbf{e}_1 \times \mathbf{e}_2) = \mathbf{e}_3 \cdot (\mathbf{e}_2 \times \mathbf{e}_3) = 0 \\ \mathbf{e}_1 \cdot (\mathbf{e}_2 \times \mathbf{e}_2) &= \mathbf{e}_2 \cdot (\mathbf{e}_3 \times \mathbf{e}_3) = \mathbf{e}_3 \cdot (\mathbf{e}_1 \times \mathbf{e}_1) = 0 \\ \mathbf{e}_1 \cdot (\mathbf{e}_3 \times \mathbf{e}_3) &= \mathbf{e}_2 \cdot (\mathbf{e}_1 \times \mathbf{e}_1) = \mathbf{e}_3 \cdot (\mathbf{e}_2 \times \mathbf{e}_2) = 0 \\ \mathbf{e}_1 \cdot (\mathbf{e}_1 \times \mathbf{e}_1) &= \mathbf{e}_2 \cdot (\mathbf{e}_2 \times \mathbf{e}_2) = \mathbf{e}_3 \cdot (\mathbf{e}_3 \times \mathbf{e}_3) = 0 \end{aligned} \quad (1.7.2)$$

1 SUFFIX NOTATION

These 27 relations may be represented in a condensed form by defining a symbol ε_{ijk} as follows:

$$\varepsilon_{ijk} = \begin{cases} 1, & \text{if } i, j, k \text{ take values in the cyclic order,} \\ -1, & \text{if } i, j, k \text{ take values in the acylic order,} \\ 0, & \text{if two or all of } i, j, k \text{ take the same value.} \end{cases} \quad (1.7.3)$$

That is,

$$\varepsilon_{123} = 1, \quad \varepsilon_{132} = -1, \quad \varepsilon_{111} = \varepsilon_{112} = 0, \quad \text{etc.}$$

It is obvious that a condensed form of relations (1.7.2) is

$$\mathbf{e}_i \cdot (\mathbf{e}_j \times \mathbf{e}_k) = \varepsilon_{ijk} \quad (1.7.4)$$

The symbol ε_{ijk} defined by (1.7.3) is known as the *Levi-Civita ε-symbol*, named after the Italian mathematician Tullio Levi-Civita (1873–1941). The ε-symbol is also referred to as the *permutation symbol, the alternating symbol* or *the alternator*.

The following facts can be easily verified:

$$\varepsilon_{ijk} = \tfrac{1}{2}(i - j)(j - k)(k - i) \quad (1.7.5)$$

$$\varepsilon_{ijk} = \varepsilon_{jki} = \varepsilon_{kij} = -\varepsilon_{ikj} = -\varepsilon_{jik} = -\varepsilon_{kji} \quad (1.7.6)$$

$$\mathbf{e}_i \times \mathbf{e}_j = \varepsilon_{ijk} \mathbf{e}_k \quad (1.7.7)$$

The ε-symbol is useful in expressing the vector product $\mathbf{a} \times \mathbf{b}$ and the scalar triple product $[\mathbf{a}, \mathbf{b}, \mathbf{c}]$ in the suffix notation as described in the following.

For any vectors $\mathbf{a}$ and $\mathbf{b}$ with components a_i and b_i, respectively, we have

$$\mathbf{a} \times \mathbf{b} = (a_i \mathbf{e}_i) \times (b_j \mathbf{e}_j) = a_i b_j (\varepsilon_{ijk} \mathbf{e}_k)$$

using (1.7.7)

$$= \varepsilon_{jki} a_j b_k \mathbf{e}_i = \varepsilon_{ijk} a_j b_k \mathbf{e}_i, \quad (1.7.8)$$

on changing dummy suffixes and using (1.7.6). Thus, the symbol $\varepsilon_{ijk} a_j b_k$ denotes the ith component of the vector $\mathbf{a} \times \mathbf{b}$; that is,

$$[\mathbf{a} \times \mathbf{b}]_i = \varepsilon_{ijk} a_j b_k = \varepsilon_{ijk} [\mathbf{a}]_j [\mathbf{b}]_k \quad (1.7.9)$$

Since $\mathbf{a} \times \mathbf{a} = \mathbf{0}$, it follows that

$$\varepsilon_{ijk} a_j a_k = 0 \quad (1.7.10)$$

This expression can also be verified directly.

For any vector **c** with components c_i we have

$$[\mathbf{a}, \mathbf{b}, \mathbf{c}] = (\mathbf{a} \times \mathbf{b}) \cdot \mathbf{c} = [\mathbf{a} \times \mathbf{b}]_i [\mathbf{c}]_i = \varepsilon_{ijk} a_j b_k c_i \quad (1.7.11)$$

on using (1.7.9). Interchanging the dummy suffixes and using (1.7.6), (1.7.11) can be rewritten as

$$[\mathbf{a}, \mathbf{b}, \mathbf{c}] = \varepsilon_{ijk} a_i b_j c_k \quad (1.7.12)$$

Recalling that $[\mathbf{a}, \mathbf{b}, \mathbf{c}] = \mathbf{a} \cdot (\mathbf{b} \times \mathbf{c})$ also, we find from (1.4.17) and (1.7.12) that

$$\varepsilon_{ijk} a_i b_j c_k = \begin{vmatrix} a_1 & a_2 & a_3 \\ b_1 & b_2 & b_3 \\ c_1 & c_2 & c_3 \end{vmatrix} \quad (1.7.13)$$

This expression can also be verified directly.

EXAMPLE 1.7.1 Show that

$$\varepsilon_{ijk} = \begin{vmatrix} \mathbf{e}_i \cdot \mathbf{e}_1 & \mathbf{e}_i \cdot \mathbf{e}_2 & \mathbf{e}_i \cdot \mathbf{e}_3 \\ \mathbf{e}_j \cdot \mathbf{e}_1 & \mathbf{e}_j \cdot \mathbf{e}_2 & \mathbf{e}_j \cdot \mathbf{e}_3 \\ \mathbf{e}_k \cdot \mathbf{e}_1 & \mathbf{e}_k \cdot \mathbf{e}_2 & \mathbf{e}_k \cdot \mathbf{e}_3 \end{vmatrix} = \begin{vmatrix} \delta_{i1} & \delta_{i2} & \delta_{i3} \\ \delta_{j1} & \delta_{j2} & \delta_{j3} \\ \delta_{k1} & \delta_{k2} & \delta_{k3} \end{vmatrix} \quad (1.7.14)$$

Solution We begin with expression (1.4.17). Noting that for any vector **a** with components a_i, we have $a_i = \mathbf{a} \cdot \mathbf{e}_i$, this expression can be rewritten as

$$\mathbf{a} \cdot (\mathbf{b} \times \mathbf{c}) = \begin{vmatrix} \mathbf{a} \cdot \mathbf{e}_1 & \mathbf{a} \cdot \mathbf{e}_2 & \mathbf{a} \cdot \mathbf{e}_3 \\ \mathbf{b} \cdot \mathbf{e}_1 & \mathbf{b} \cdot \mathbf{e}_2 & \mathbf{b} \cdot \mathbf{e}_3 \\ \mathbf{c} \cdot \mathbf{e}_1 & \mathbf{c} \cdot \mathbf{e}_2 & \mathbf{c} \cdot \mathbf{e}_3 \end{vmatrix} \quad (1.7.15)$$

Setting $\mathbf{a} = \mathbf{e}_i$, $\mathbf{b} = \mathbf{e}_j$, $\mathbf{c} = \mathbf{e}_k$ in (1.7.15) and using (1.7.4), it turns out that

$$\varepsilon_{ijk} = \begin{vmatrix} \mathbf{e}_i \cdot \mathbf{e}_1 & \mathbf{e}_i \cdot \mathbf{e}_2 & \mathbf{e}_i \cdot \mathbf{e}_3 \\ \mathbf{e}_j \cdot \mathbf{e}_1 & \mathbf{e}_j \cdot \mathbf{e}_2 & \mathbf{e}_j \cdot \mathbf{e}_3 \\ \mathbf{e}_k \cdot \mathbf{e}_1 & \mathbf{e}_k \cdot \mathbf{e}_2 & \mathbf{e}_k \cdot \mathbf{e}_3 \end{vmatrix}$$

which, on using (1.6.3), gives

$$\varepsilon_{ijk} = \begin{vmatrix} \delta_{i1} & \delta_{i2} & \delta_{i3} \\ \delta_{j1} & \delta_{j2} & \delta_{j3} \\ \delta_{k1} & \delta_{k2} & \delta_{k3} \end{vmatrix} \quad \blacksquare$$

22   1 SUFFIX NOTATION

EXAMPLE 1.7.2 Show that $a_{ij} = a_{ji}$ if and only if $\varepsilon_{ijk} a_{jk} = 0$.

Solution For $i = 1$, we have

$$\varepsilon_{ijk} a_{jk} = \varepsilon_{1jk} a_{jk} = \sum_{j=1}^{3} \sum_{k=1}^{3} \varepsilon_{1jk} a_{jk}$$

$$= \sum_{j=1}^{3} (\varepsilon_{1j1} a_{j1} + \varepsilon_{1j2} a_{j2} + \varepsilon_{1j3} a_{j3})$$

$$= \sum_{j=1}^{3} (\varepsilon_{1j2} a_{j2} + \varepsilon_{1j3} a_{j3})$$

$$= \varepsilon_{132} a_{32} + \varepsilon_{123} a_{23} = a_{23} - a_{32}$$

by using the definition of the ε symbol.

Similarly, we get

$$\varepsilon_{ijk} a_{jk} = a_{31} - a_{13} \quad \text{for } i = 2$$

$$\varepsilon_{ijk} a_{jk} = a_{12} - a_{21} \quad \text{for } i = 3$$

Thus

$$\varepsilon_{ijk} a_{jk} = \begin{cases} a_{23} - a_{32} & \text{for } i = 1 \\ a_{31} - a_{13} & \text{for } i = 2 \\ a_{12} - a_{21} & \text{for } i = 3 \end{cases} \quad (1.7.16)$$

If $\varepsilon_{ijk} a_{jk} = 0$, then from (1.7.16) it follows that $a_{23} = a_{32}$, $a_{31} = a_{13}$ and $a_{12} = a_{21}$; that is $a_{ij} = a_{ji}$.

Conversely, if $a_{ij} = a_{ji}$, then $a_{12} = a_{21}$, $a_{13} = a_{31}$ and $a_{23} = a_{32}$. Consequently, the lefthand side of (1.7.16) is 0 for all i; that is, $\varepsilon_{ijk} a_{jk} = 0$. ∎

EXAMPLE 1.7.3 (i) Show that

$$\varepsilon_{pqr} a_{ip} a_{jq} a_{kr} = \begin{vmatrix} a_{i1} & a_{i2} & a_{i3} \\ a_{j1} & a_{j2} & a_{j3} \\ a_{k1} & a_{k2} & a_{k3} \end{vmatrix} \quad (1.7.17)$$

(ii) Deduce that

$$\det(a_{ij}) = \varepsilon_{pqr} a_{1p} a_{2q} a_{3r} = \varepsilon_{pqr} a_{p1} a_{q2} a_{r3} \quad (1.7.18)$$

Solution (i) We have

$$\varepsilon_{pqr}a_{ip}a_{jq}a_{kr} = \sum_{p=1}^{3}\sum_{q=1}^{3}\sum_{r=1}^{3}\varepsilon_{pqr}a_{ip}a_{jq}a_{kr}$$

$$= \sum_{p=1}^{3}\sum_{q=1}^{3}a_{ip}a_{jq}(\varepsilon_{pq1}a_{k1} + \varepsilon_{pq2}a_{k2} + \varepsilon_{pq3}a_{k3})$$

$$= \sum_{p=1}^{3}a_{ip}\{a_{k1}(\varepsilon_{p21}a_{j2} + \varepsilon_{p31}a_{j3}) + a_{k2}(\varepsilon_{p12}a_{j1} + \varepsilon_{p32}a_{j3})$$

$$+ a_{k3}(\varepsilon_{p13}a_{j1} + \varepsilon_{p23}a_{j2})\}$$

(other terms being 0 by the definition of the ε-symbol)

$$= a_{k1}(\varepsilon_{321}a_{i3}a_{j2} + \varepsilon_{231}a_{i2}a_{j3}) + a_{k2}(\varepsilon_{312}a_{i3}a_{j1} + \varepsilon_{132}a_{i1}a_{j3})$$

$$+ a_{k3}(\varepsilon_{213}a_{i2}a_{j1} + \varepsilon_{123}a_{i1}a_{j2})$$

(other terms being 0)

$$= a_{i1}(a_{j2}a_{k3} - a_{j3}a_{k2}) + a_{i2}(a_{k1}a_{j3} - a_{k3}a_{j1})$$

$$+ a_{i3}(a_{k2}a_{j1} - a_{k1}a_{j2})$$

(on using that $\varepsilon_{123} = 1$, $\varepsilon_{132} = -1$, etc.)

$$= \begin{vmatrix} a_{i1} & a_{i2} & a_{i3} \\ a_{j1} & a_{j2} & a_{j3} \\ a_{k1} & a_{k2} & a_{k3} \end{vmatrix}$$

(ii) In the above result if we set $i = 1$, $j = 2$, $k = 3$, we obtain

$$\varepsilon_{pqr}a_{1p}a_{2q}a_{3r} = \det(a_{ij}) \tag{1.7.19}$$

Suppose we set $b_{ij} = a_{ji}$. Then

$$\varepsilon_{pqr}a_{p1}a_{q2}a_{r3} = \varepsilon_{pqr}b_{1p}b_{2q}b_{3r}$$

$$= \det(b_{ij}), \quad \text{by (1.7.19)}$$

$$= \det(a_{ji}) = \det(a_{ij}) \tag{1.7.20}$$

because $\det(a_{ij}) = \det(a_{ji})$. Expressions (1.7.19) and (1.7.20) are the required results in (1.7.18). ∎

EXAMPLE 1.7.4 (i) Let $D = \det(a_{ij})$. Verify that

$$\varepsilon_{ijk}\varepsilon_{pqr}D = \begin{vmatrix} a_{ip} & a_{iq} & a_{ir} \\ a_{jp} & a_{jq} & a_{jr} \\ a_{kp} & a_{kq} & a_{kr} \end{vmatrix} \tag{1.7.21}$$

(ii) Hence deduce the following results:

(a) $$\varepsilon_{ijk}\varepsilon_{pqr} = \begin{vmatrix} \delta_{ip} & \delta_{iq} & \delta_{ir} \\ \delta_{jp} & \delta_{jq} & \delta_{jr} \\ \delta_{kp} & \delta_{kq} & \delta_{kr} \end{vmatrix} \tag{1.7.22}$$

(b) $\quad\varepsilon_{ijk}\varepsilon_{pqk} = \delta_{ip}\delta_{jq} - \delta_{iq}\delta_{jp}$ (1.7.23)

(c) $\quad\varepsilon_{ijk}\varepsilon_{pjk} = 2\delta_{ip}$ (1.7.24)

(d) $\quad\varepsilon_{ijk}\varepsilon_{ijk} = 6$ (1.7.25)

Solution (i) If (at least) two of i, j, k or two of p, q, r are equal, then both sides of (1.7.21) are 0. If i, j, k and p, q, r are both cyclic or acyclic, then each side of (1.7.21) is equal to D. If i, j, k are cyclic but p, q, r are acyclic or vice versa, then each side of (1.7.21) is equal to $-D$. Thus, in all possible cases, result (1.7.21) is verified.

(ii) (a) If we set $a_{ij} = \delta_{ij}$ in (1.7.21) and recall that $\det(\delta_{ij}) = 1$, we readily get (1.7.22).

(b) It follows from (1.7.22) that

$$\varepsilon_{ijk}\varepsilon_{pqk} = \begin{vmatrix} \delta_{ip} & \delta_{iq} & \delta_{ik} \\ \delta_{jp} & \delta_{jq} & \delta_{jk} \\ \delta_{kp} & \delta_{kq} & \delta_{kk} \end{vmatrix}$$

$$= \delta_{ip}(\delta_{jq}\delta_{kk} - \delta_{jk}\delta_{kq}) + \delta_{iq}(\delta_{jk}\delta_{kp} - \delta_{jp}\delta_{kk})$$
$$+ \delta_{ik}(\delta_{jp}\delta_{kq} - \delta_{jq}\delta_{kp})$$

$$= \delta_{ip}(3\delta_{jq} - \delta_{jq}) + \delta_{iq}(\delta_{jp} - 3\delta_{jp}) + (\delta_{jp}\delta_{iq} - \delta_{jq}\delta_{ip})$$

$$= \delta_{ip}\delta_{jq} - \delta_{iq}\delta_{jp}$$

(c) Hence

$$\varepsilon_{ijk}\varepsilon_{pjk} = \delta_{ip}\delta_{jj} - \delta_{ij}\delta_{jp} = 3\delta_{ip} - \delta_{ip} = 2\delta_{ip}$$

(d) Consequently

$$\varepsilon_{ijk}\varepsilon_{ijk} = 2\delta_{ii} = 6 \quad\blacksquare$$

Note: The identity (1.7.21) and its particular cases (1.7.22) to (1.7.25) are of great utility in subsequent discussions. In particular, the identity (1.7.23) is known as the *ε-δ identity* or the *permutation identity*. Since a cyclic permutation of suffixes does not change the sign of the permutation symbol, this identity may also be expressed in the following alternative forms:

$$\varepsilon_{ijk}\varepsilon_{pqk} = \varepsilon_{ijk}\varepsilon_{kpq} = \varepsilon_{kij}\varepsilon_{kpq} = \varepsilon_{jki}\varepsilon_{qkp} = \delta_{ip}\delta_{jq} - \delta_{iq}\delta_{jp} \quad (1.7.23)'$$

EXAMPLE 1.7.5 Show that

$$\det(a_{ij}) = \tfrac{1}{6}\varepsilon_{ijk}\varepsilon_{pqr}a_{ip}a_{jq}a_{kr} = \tfrac{1}{6}\varepsilon_{ijk}\varepsilon_{pqr}a_{pi}a_{qj}a_{rk} \quad (1.7.26)$$

$$= \tfrac{1}{6}(a_{ii}a_{jj}a_{kk} + 2a_{ij}a_{jk}a_{ki} - 3a_{ij}a_{ji}a_{kk}) \quad (1.7.27)$$

If $D = \det(a_{ij})$, deduce that

$$\varepsilon_{ijk}D = \varepsilon_{pqr}a_{ip}a_{jq}a_{kr} = \varepsilon_{pqr}a_{pi}a_{qj}a_{rk} \quad (1.7.28)$$

Solution Expanding the determinant on the righthand side of (1.7.17), we obtain

$$\varepsilon_{pqr}a_{ip}a_{jq}a_{kr} = a_{i1}(a_{j2}a_{k3} - a_{j3}a_{k2}) + a_{i2}(a_{j3}a_{k1} - a_{j1}a_{k3})$$
$$+ a_{i3}(a_{j1}a_{k2} - a_{k1}a_{j2}) \quad (1.7.29)$$

Now

$$\varepsilon_{ijk}a_{i1}(a_{j2}a_{k3} - a_{j3}a_{k2}) = \varepsilon_{ijk}a_{i1}a_{j2}a_{k3} - \varepsilon_{ijk}a_{i1}a_{j3}a_{k2}$$
$$= \varepsilon_{ijk}a_{i1}a_{j2}a_{k3} - \varepsilon_{ikj}a_{i1}a_{j2}a_{k3}$$

(on interchanging the dummy suffixes j and k in the second term)

$$= 2\varepsilon_{ijk}a_{i1}a_{j2}a_{k3} \quad (1.7.30)$$

since $\varepsilon_{ijk} = -\varepsilon_{ikj}$. Similarly,

$$\varepsilon_{ijk}a_{i2}(a_{k1}a_{j3} - a_{k3}a_{j1}) = 2\varepsilon_{ijk}a_{i2}a_{k1}a_{j3} \quad (1.7.31)$$

Interchanging the dummy suffixes i and j, j and k, and k and i in $\varepsilon_{ijk}a_{i2}a_{k1}a_{j3}$ and noting that $\varepsilon_{ijk} = \varepsilon_{kij}$, we find that

$$\varepsilon_{ijk}a_{i2}a_{k1}a_{j3} = \varepsilon_{kij}a_{i1}a_{j2}a_{k3} = \varepsilon_{ijk}a_{i1}a_{j2}a_{k3}$$

Hence, (1.7.31) becomes

$$\varepsilon_{ijk}a_{i2}(a_{k1}a_{j3} - a_{k3}a_{j1}) = 2\varepsilon_{ijk}a_{i1}a_{j2}a_{k3} \quad (1.7.32)$$

In the same way we obtain

$$\varepsilon_{ijk}a_{i3}(a_{k2}a_{j1} - a_{k1}a_{j2}) = 2\varepsilon_{ijk}a_{i1}a_{j2}a_{k3} \quad (1.7.33)$$

From (1.7.29), (1.7.30), (1.7.32) and (1.7.33), we get

$$\varepsilon_{ijk}\varepsilon_{pqr}a_{ip}a_{jq}a_{kr} = 6\varepsilon_{ijk}a_{i1}a_{j2}a_{k3} \quad (1.7.34)$$

Noting that $\varepsilon_{ijk}\varepsilon_{pqr}a_{ip}a_{jq}a_{kr} = \varepsilon_{ijk}\varepsilon_{pqr}a_{pi}a_{qj}a_{kr}$ (interchanging the dummies i and p, j and q, and k and r) and noting from (1.7.18) that $\varepsilon_{ijk}a_{i1}a_{j2}a_{k3} = \det(a_{ij})$, we obtain (1.7.26) from (1.7.34).

Next, it follows from (1.7.22) that

$$\varepsilon_{ijk}\varepsilon_{pqr}a_{ip}a_{jq}a_{kr} = (a_{ip}a_{jq}a_{kr}) \begin{vmatrix} \delta_{ip} & \delta_{iq} & \delta_{ir} \\ \delta_{jp} & \delta_{jq} & \delta_{jr} \\ \delta_{kp} & \delta_{kq} & \delta_{kr} \end{vmatrix}$$

$$= \{\delta_{ip}(\delta_{jq}\delta_{kr} - \delta_{kq}\delta_{jr}) + \delta_{iq}(\delta_{jr}\delta_{kp} - \delta_{jp}\delta_{kr})$$
$$+ \delta_{ir}(\delta_{jp}\delta_{kq} - \delta_{kp}\delta_{jq})\}a_{ip}a_{jq}a_{kr} \quad (1.7.35)$$

Using the substitution property of δ_{ij} and changing the dummy suffixes appropriately, we find that

$$\delta_{ip}(\delta_{jq}\delta_{kr} - \delta_{jr}\delta_{kq})a_{ip}a_{jq}a_{kr} = a_{ii}a_{jj}a_{kk} - a_{ij}a_{ji}a_{kk} \quad (1.7.36)$$

$$\delta_{iq}(\delta_{jr}\delta_{kp} - \delta_{jp}\delta_{kr})a_{ip}a_{jq}a_{kr} = a_{ij}a_{jk}a_{ki} - a_{ij}a_{ji}a_{kk} \quad (1.7.37)$$

$$\delta_{ir}(\delta_{jp}\delta_{kq} - \delta_{kq}\delta_{jp})a_{ip}a_{jq}a_{kr} = a_{ij}a_{jk}a_{ki} - a_{ij}a_{ji}a_{kk} \quad (1.7.38)$$

Substituting (1.7.36)–(1.7.38) in the righthand side of (1.7.35) and using the resulting expression in (1.7.26) we arrive at result (1.7.27).

Finally, multiplying (1.7.26) throughout by ε_{ijk} and using (1.7.25), we obtain (1.7.28). ∎

EXAMPLE 1.7.6 Given the matrix $[a_{ij}]$, consider the matrix $[a_{ij}^*]$ where

$$a_{ij}^* = \tfrac{1}{2}\varepsilon_{ipq}\varepsilon_{jrs}a_{pr}a_{qs} \quad (1.7.39)$$

Show that

$$[a_{ij}][a_{ij}^*]^T = [a_{ij}^*]^T[a_{ij}] = D[I] \quad (1.7.40)$$

where $D = \det(a_{ij})$. Deduce that, if $D \neq 0$,

$$[a_{ij}]^{-1} = \frac{1}{D}[a_{ji}^*] \quad (1.7.41)$$

Solution By virtue of (1.5.22), we have

$$[a_{ij}][a_{ij}^*]^T = [a_{ik}a_{jk}^*], \quad [a_{ij}^*]^T[a_{ij}] = [a_{ki}^*a_{kj}] \quad (1.7.42)$$

Using (1.7.39), (1.7.28) and (1.7.24) we find that

$$a_{ik}a_{jk}^* = \tfrac{1}{2}\varepsilon_{jpq}\varepsilon_{krs}a_{ik}a_{pr}a_{qs} = \tfrac{1}{2}\varepsilon_{jpq}(D\varepsilon_{ipq}) = D\delta_{ij} \qquad (1.7.43)$$

$$a_{ki}^*a_{kj} = \tfrac{1}{2}\varepsilon_{kpq}\varepsilon_{irs}a_{pr}a_{qs}a_{kj} = \tfrac{1}{2}\varepsilon_{jrs}\varepsilon_{irs}D = D\delta_{ij} \qquad (1.7.44)$$

Using (1.7.43) and (1.7.44) in (1.7.42) and noting that $[\delta_{ij}] = [I]$, we get (1.7.40).

If $D \neq 0$, then the matrix $[a_{ij}]$ is invertible, and it follows from (1.7.40) that

$$[a_{ij}]^{-1}[a_{ij}][a_{ij}^*]^T = D[a_{ij}]^{-1}[I]$$

which on using (1.5.29) and (1.5.31) yields

$$[a_{ij}^*]^T = D[a_{ij}]^{-1}$$

This is the expression (1.7.41). ∎

Note: The matrix $[a_{ij}^*]$ is called the *adjugate* or *cofactor* and its transpose $[a_{ij}^*]^T = [a_{ji}^*]$ is called the *adjoint* of the matrix $[a_{ij}]$. Expression (1.7.41) is useful to compute the inverse of a nonsingular matrix.

EXAMPLE 1.7.7 Given the vectors **a**, **b**, **c**:

(i) Write down the ith component of $\mathbf{a} \times (\mathbf{b} \times \mathbf{c})$.
(ii) Use the vector identity (1.4.19) to prove the ε-δ identity (1.7.23).
(iii) Use the ε-δ identity to prove the identity (1.4.19).

Solution (i) If we put $\mathbf{d} = \mathbf{b} \times \mathbf{c}$, then by virtue of expression (1.7.9), we obtain

$$d_k = [\mathbf{b} \times \mathbf{c}]_k = \varepsilon_{kpq}b_p c_q \qquad (1.7.45)$$

Also,

$$[\mathbf{a} \times (\mathbf{b} \times \mathbf{c})]_i = [\mathbf{a} \times \mathbf{d}]_i = \varepsilon_{ijk}a_j d_k$$

which by use of (1.7.45), becomes

$$[\mathbf{a} \times (\mathbf{b} \times \mathbf{c})]_i = \varepsilon_{ijk}\varepsilon_{kpq}a_j b_p c_q \qquad (1.7.46)$$

Thus, $\varepsilon_{ijk}\varepsilon_{kpq}a_j b_p c_q$ is the ith component of $\mathbf{a} \times (\mathbf{b} \times \mathbf{c})$.

(ii) It follows from the vector identity (1.4.19) that

$$[\mathbf{a} \times (\mathbf{b} \times \mathbf{c})]_i = (a_j c_j)b_i - (a_j b_j)c_i \qquad (1.7.47)$$

Using (1.7.46), (1.7.47) becomes

$$\varepsilon_{ijk}\varepsilon_{kpq}a_j b_p c_q = a_j c_j b_i - a_j b_j c_i \qquad (1.7.48)$$

28 1 SUFFIX NOTATION

By using the substitution properties of δ_{ij}, we find that

$$a_j c_j b_i = a_j(\delta_{jq} c_q)(\delta_{ip} b_p) = \delta_{ip}\delta_{jq} a_j b_p c_q$$
$$a_j b_j c_i = a_j(\delta_{jp} b_p)(\delta_{iq} c_q) = \delta_{iq}\delta_{jp} a_j b_p c_q$$

By use of these expressions, (1.7.48) becomes

$$(\varepsilon_{ijk}\varepsilon_{kpq} - \delta_{ip}\delta_{jq} + \delta_{iq}\delta_{jp}) a_j b_p c_q = 0 \tag{1.7.49}$$

This should hold for arbitrary a_i, b_i and c_i. Hence

$$\varepsilon_{ijk}\varepsilon_{kpq} - \delta_{ip}\delta_{jq} + \delta_{iq}\delta_{jp} = 0$$

This is the ε-δ identity (1.7.23). (This serves as an alternative proof for the ε-δ identity.)

(iii) If the ε-δ identity (1.7.23) holds, then (1.7.49) is valid for arbitrary a_i, b_i, c_i. But, (1.7.49) is equivalent to (1.7.47). Thus, (1.7.23) yields (1.7.47), which is nothing but (1.4.19). ∎

EXAMPLE 1.7.8 Prove the vector identity

$$(\mathbf{a} \times \mathbf{b}) \cdot (\mathbf{c} \times \mathbf{d}) = (\mathbf{a} \cdot \mathbf{c})(\mathbf{b} \cdot \mathbf{d}) - (\mathbf{a} \cdot \mathbf{d})(\mathbf{b} \cdot \mathbf{c}) \tag{1.7.50}$$

Deduce that

$$\tfrac{1}{2}[(\mathbf{a} \times \mathbf{b}) \cdot (\mathbf{b} \times \mathbf{c}) + (\mathbf{a} \cdot \mathbf{b})(\mathbf{b} \cdot \mathbf{c})] = (\mathbf{a} \cdot \mathbf{b})(\mathbf{b} \cdot \mathbf{c}) - \tfrac{1}{2}b^2(\mathbf{a} \cdot \mathbf{c}). \tag{1.7.51}$$

Solution By use of (1.4.11), (1.7.9), (1.7.23) and the substitution property of δ_{ij}, we find that

$$\begin{aligned}(\mathbf{a} \times \mathbf{b}) \cdot (\mathbf{c} \times \mathbf{d}) &= [\mathbf{a} \times \mathbf{b}]_k [\mathbf{c} \times \mathbf{d}]_k \\ &= (\varepsilon_{kpq} a_p b_q)(\varepsilon_{kmn} c_m d_n) \\ &= (\delta_{pm}\delta_{qn} - \delta_{pn}\delta_{qm})(a_p b_q c_m d_n) \\ &= (a_m b_n - a_n b_m)(c_m d_n) \\ &= (a_m c_m)(b_n d_n) - (a_n d_n)(b_m c_m) \\ &= (\mathbf{a} \cdot \mathbf{c})(\mathbf{b} \cdot \mathbf{d}) - (\mathbf{a} \cdot \mathbf{d})(\mathbf{b} \cdot \mathbf{c}),\end{aligned}$$

which is (1.7.50). As a particular case, we get

$$(\mathbf{a} \times \mathbf{b}) \cdot (\mathbf{b} \times \mathbf{c}) = (\mathbf{a} \cdot \mathbf{b})(\mathbf{b} \cdot \mathbf{c}) - (\mathbf{a} \cdot \mathbf{c})(\mathbf{b} \cdot \mathbf{b}) \tag{1.7.52}$$

Adding $(\mathbf{a} \cdot \mathbf{b})(\mathbf{b} \cdot \mathbf{c})$ to both sides of (1.7.52) gives

$$(\mathbf{a} \times \mathbf{b}) \cdot (\mathbf{b} \times \mathbf{c}) + (\mathbf{a} \cdot \mathbf{b})(\mathbf{b} \cdot \mathbf{c}) = 2(\mathbf{a} \cdot \mathbf{b})(\mathbf{b} \cdot \mathbf{c}) - (\mathbf{a} \cdot \mathbf{c})b^2$$

which is (1.7.51). ∎

1.8 EXERCISES

1. State which of the following expressions are meaningful in the suffix notation. Write out the unabridged versions of the meaningful expressions:

(i) a_{ii} (ii) $a_{ij}b_j$ (iii) $a_{ij}b_i$
(iv) $a_{ii}b_i$ (v) $a_{ii}b_{jj}$ (vi) $a_{ii}b_{ii}$
(vii) $a_{rs}b_{sr}$ (viii) $a_{rs}b_{ss}$ (ix) $a_{ijk}b_{ik}$

2. Which of the following expressions have the same meaning?

$$a_{ij}b_j, \quad a_{rs}b_s, \quad a_{pq}b_p, \quad a_{ij}b_ib_j, \quad a_{pq}b_pb_q, \quad a_{sr}b_sb_r$$

3. State which of the following equations are meaningful in the suffix notation.

(i) $x_i = \alpha_{ij}y_j$ (ii) $y_j = \alpha_{ik}x_k$
(iii) $a_{ij} = \alpha_{ii}\alpha_{jj}$ (iv) $a_{ij} = \alpha_{ip}\alpha_{jp}$
(v) $a_{ij} = \alpha_{im}\alpha_{jn}b_{mn}$ (vi) $a_{rs} = \alpha_{ir}\alpha_{js}b_{ij}$
(vii) $a_{ij} = \alpha_{im}\alpha_{jn}b_{rs}$ (viii) $a_{pq} = \alpha_{pr}\alpha_{qs}b_{ss}$

4. Which of the following expressions have the same meaning as $a_i = \alpha_{pi}b_p$?

(i) $a_p = \alpha_{pi}b_i$ (ii) $a_p = \alpha_{ip}b_i$
(iii) $a_m = \alpha_{jm}b_j$ (iv) $a_r = \alpha_{sr}b_s$
(v) $a_k = \alpha_{km}b_m$ (vi) $a_k = \alpha_{rk}b_r$

5. Write down the following system of equations in the suffix notation:

$$a_{11} = b_{11}, \quad a_{22} = b_{22}, \quad a_{33} = b_{33}, \quad a_{12} = \tfrac{1}{2}(b_{12} + b_{21}) = a_{21},$$
$$a_{13} = \tfrac{1}{2}(b_{13} + b_{31}) = a_{31}, \quad a_{23} = \tfrac{1}{2}(b_{23} + b_{32}) = a_{32}$$

6. If $a_{ij} = -a_{ji}$, show that $a_{ij}x_ix_j = 0$ for all x_i.

7. If $a_{ij} = a_{ji}$ and $b_{ij} = -b_{ji}$, show that $a_{ij}b_{ij} = 0$.

8. If $a_{ij} = a_{ji}$ and $b_{ij} = \tfrac{1}{2}(c_{ij} + c_{ji})$, show that $a_{ij}b_{ij} = a_{ij}c_{ij}$.

9. If $a_{ij} = \tfrac{1}{2}(b_{ij} + b_{ji})$ and $c_{ij} = \tfrac{1}{2}(b_{ij} - b_{ji})$, show that $a_{ij}c_{ij} = 0$.

10. In the system of axes with base vectors $\mathbf{e}_i$, show that $\mathbf{a} = (\mathbf{a} \cdot \mathbf{e}_i)\mathbf{e}_i$ for every vector $\mathbf{a}$.

11. Prove the expressions (1.4.16) through (1.4.18).

12. The absolute value of the scalar triple product $[\mathbf{a}, \mathbf{b}, \mathbf{c}]$ represents the volume of the parallelopiped whose adjacent edges with a common corner are represented by $\mathbf{a}, \mathbf{b}, \mathbf{c}$. Justify this statement.

13. Prove the identities (1.4.19) and (1.4.20). Deduce that $\mathbf{a} \times (\mathbf{b} \times \mathbf{c}) = (\mathbf{a} \times \mathbf{b}) \times \mathbf{c}$ if and only if $\mathbf{b} \times (\mathbf{c} \times \mathbf{a}) = \mathbf{0}$. Interpret the condition geometrically.

14. Given the matrix

$$[a_{ij}] = \begin{bmatrix} 1 & 0 & -1 \\ -1 & 2 & 0 \\ 3 & 0 & -1 \end{bmatrix}$$

find a_{ii}, $a_{ij}a_{ji}$ and $a_{ij}a_{ij}$.

15. Represent the following matrix as a sum of a symmetric matrix and a skew-symmetric matrix:

$$[a_{ij}] = \begin{bmatrix} 2 & 0 & 4 \\ -6 & 8 & 0 \\ 8 & 10 & -8 \end{bmatrix}$$

16. For the matrix given in the previous exercise, compute $[a_{ij}][a_{ij}]^T$, $[a_{ij}]^T[a_{ij}]$ and $[a_{ij}]^2$.

17. Prove the expressions (1.5.4), (1.5.6), (1.5.16), (1.5.24) to (1.5.28), (1.5.33) and (1.5.35).

18. Prove that a matrix $[a_{ij}]$ has a multiplicative inverse if and only if $\det[a_{ij}] \neq 0$.

19. Verify that the following matrices are nonsingular. Find their inverses.

(i) $\begin{bmatrix} 1 & 0 & 0 \\ 0 & 0 & 2 \\ 0 & 2 & 3 \end{bmatrix}$ (ii) $\begin{bmatrix} 2 & 2 & 0 \\ -1 & 1 & 0 \\ 0 & 0 & -1 \end{bmatrix}$

20. Verify that the following matrices are orthogonal.

(i) $\begin{bmatrix} 1 & 0 & 0 \\ 0 & 0 & 1 \\ 0 & 1 & 0 \end{bmatrix}$ (ii) $\begin{bmatrix} \cos\theta & -\sin\theta & 0 \\ \sin\theta & \cos\theta & 0 \\ 0 & 0 & 1 \end{bmatrix}$

(iii) $\begin{bmatrix} \frac{1}{\sqrt{2}} & \frac{1}{\sqrt{2}} & 0 \\ -\frac{1}{\sqrt{2}} & \frac{1}{\sqrt{2}} & 0 \\ 0 & 0 & -1 \end{bmatrix}$ (iv) $\begin{bmatrix} 0 & \frac{1}{\sqrt{2}} & -\frac{1}{\sqrt{2}} \\ \frac{1}{\sqrt{2}} & \frac{1}{2} & \frac{1}{2} \\ \frac{1}{\sqrt{2}} & -\frac{1}{2} & -\frac{1}{2} \end{bmatrix}$

21. Show that
  (i) $\delta_{ij}\delta_{ij} = 3$
  (ii) $\delta_{ij}\delta_{jk}\delta_{ik} = 3$
  (iii) $\delta_{ik}\delta_{jm}\delta_{ij} = \delta_{km}$

22. Simplify the following
  (i) $\delta_{ij}(a_{ij} - a_{ji})$
  (ii) $\delta_{ip}\delta_{jq}a_p b_j c_q$
  (iii) $(\delta_{ij} + a_{ij})(\delta_{ij} - a_{ij})$

23. Write down the following equations in matrix forms:
  (i) $\alpha_{ip}\alpha_{jp} = \delta_{ij}$
  (ii) $\alpha_{pi}\alpha_{pj} = \delta_{ij}$
  (iii) $a_{ij} = \alpha\delta_{ij}b_{kk} + \beta b_{ij}$

24. Write down the following sets of equations in condensed forms:
  (i) $a_{11} = a_{22} = a_{33} = -p$
   $a_{12} = a_{21} = a_{23} = a_{32} = a_{13} = a_{31} = 0$
  (ii) $a_{111} + a_{122} + a_{133} + b_1 = 0$,
   $a_{211} + a_{222} + a_{233} + b_2 = 0$,
   $a_{311} + a_{322} + a_{333} + b_3 = 0$.

25. If $a_i = \varepsilon_{ijk} b_{jk}$, evaluate a_1, a_2, a_3.

26. If $a_{ik} = \varepsilon_{ijk} b_j$, show that $a_{ik} = -a_{ki}$.

27. Prove the following:
  (i) $\varepsilon_{kki} = 0$
  (ii) $\delta_{ij}\varepsilon_{ijk} = 0$
  (iii) $\varepsilon_{pqr}\varepsilon_{rqs} = -2\delta_{ps}$
  (iv) $\varepsilon_{pqr}\varepsilon_{rqp} = -6$

28. Verify the relation (1.7.7), and using this relation deduce the relation (1.7.4).

29. Show that
$$\varepsilon_{ijk} a_i b_j = \begin{bmatrix} a_1 & a_2 & a_3 \\ b_1 & b_2 & b_3 \\ \delta_{k1} & \delta_{k2} & \delta_{k3} \end{bmatrix}$$

30. Show that vectors with components a_i, b_i, c_i are coplanar if and only if $\varepsilon_{ijk} a_i b_j c_k = 0$.

31. If $a_{ij} = -a_{ji}$ and $b_i = \frac{1}{2}\varepsilon_{ijk} a_{jk}$, show that $a_{pq} = \varepsilon_{pqr} b_r$.

32. If **n** is a unit vector with components n_i and **v** is a vector with components v_i, show that
$$v_i = v_k n_k n_i + \varepsilon_{ijk}\varepsilon_{krs} n_j v_r n_s$$

33. Write down the following equations in the suffix notation:
   (i) $\mathbf{x} = \alpha(\mathbf{a} \times \mathbf{b}) + \beta\mathbf{a}$
   (ii) $(\mathbf{a} - \mathbf{b}) \cdot (\mathbf{c} \times \mathbf{d}) = 0$
   (iii) $\mathbf{x} = \mathbf{b} + \alpha\mathbf{c} \times \{(\mathbf{a} - \mathbf{b}) \times \mathbf{c}\}$
   (iv) $(\mathbf{a} \times \mathbf{b})^2 = a^2 b^2 - (\mathbf{a} \cdot \mathbf{b})^2$

34. Prove the following vector identities by using the suffix notation:
   (i) $(\mathbf{a} \times \mathbf{b}) \times \mathbf{c} = (\mathbf{a} \cdot \mathbf{c})\mathbf{b} - (\mathbf{b} \cdot \mathbf{c})\mathbf{a}$
   (ii) $(\mathbf{a} \times \mathbf{b}) \times (\mathbf{c} \times \mathbf{d}) = [\mathbf{a}, \mathbf{b}, \mathbf{d}]\mathbf{c} - [\mathbf{a}, \mathbf{b}, \mathbf{c}]\mathbf{d} = [\mathbf{c}, \mathbf{d}, \mathbf{a}]\mathbf{b} - [\mathbf{c}, \mathbf{d}, \mathbf{b}]\mathbf{a}$
   (iii) $[\mathbf{b} \times \mathbf{c}, \ \mathbf{c} \times \mathbf{a}, \ \mathbf{a} \times \mathbf{b}] = [\mathbf{a}, \mathbf{b}, \mathbf{c}]^2$

35. Deduce the ε-δ identity by using the vector identity (1.4.20).

CHAPTER 2
ALGEBRA OF TENSORS

2.1
INTRODUCTION

This chapter is devoted to the study of some algebraic aspects of Cartesian tensors. The concept of a Cartesian tensor is introduced through certain rules of coordinate transformations, and these rules are employed to establish some basic algebraic properties of tensors. Second-order tensors are interpreted as linear operators on vectors; these are studied in some detail with the aid of matrices.

2.2
COORDINATE TRANSFORMATIONS

Let us consider a righthanded system of rectangular Cartesian axes with a fixed origin O and denote the coordinates of a general point P with respect to (w.r.t.) this system by (x_1, x_2, x_3), or briefly, (x_i). Also, let the position vector of P w.r.t. O be denoted by $\mathbf{x}$. Then, x_i are the components of $\mathbf{x}$ along the axes, and by virtue of (1.4.1) we have

$$\mathbf{x} = x_1 \mathbf{e}_1 + x_2 \mathbf{e}_2 + x_3 \mathbf{e}_3 = x_p \mathbf{e}_p \qquad (2.2.1)$$

34 2 ALGEBRA OF TENSORS

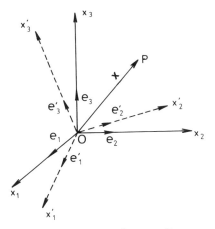

Figure 2.1. Two sets of axes at O.

Now, let us suppose that there is another righthanded system of rectangular Cartesian axes with the *same* (fixed) origin O and denote the coordinates of P w.r.t. this "new" coordinate system by (x_i'). See Figure 2.1. The new system of axes may be regarded as having been obtained by rotating the "old" system of axes (namely, the x_i-axes) about some line in space through O. The position vector $\mathbf{x}$ of P has the following representation in the new system

$$\mathbf{x} = x_1'\mathbf{e}_1' + x_2'\mathbf{e}_2' + x_3'\mathbf{e}_3' = x_p'\mathbf{e}_p' \qquad (2.2.2)$$

where $\mathbf{e}_1'$ is the unit vector directed along the positive x_1'-axis and so forth. Since the x_i' system is also rectangular and righthanded, $\mathbf{e}_i'$ satisfy equations that are analogous to those satisfied by $\mathbf{e}_i$, namely (1.6.3) and (1.7.4). Thus,

$$\mathbf{e}_i' \cdot \mathbf{e}_j' = \delta_{ij} \qquad (2.2.3)$$

$$[\mathbf{e}_i', \mathbf{e}_j', \mathbf{e}_k'] = \varepsilon_{ijk} \qquad (2.2.4)$$

From (2.2.1) and (2.2.2), we note that the vector $\mathbf{x}$ is represented as the ordered triplet (x_i) in the old system of axes and the ordered triplet (x_i') in the new system. These two triplets are different from one another as long as the two coordinate systems are noncoincident and $\mathbf{x} \neq \mathbf{0}$. We now proceed to obtain relations connecting x_i and x_i'.

We first introduce a matrix $[\alpha_{ij}]$ whose elements α_{ij} are defined as

$$\alpha_{ij} = \mathbf{e}_i' \cdot \mathbf{e}_j = \cos(x_i', x_j) \qquad (2.2.5)$$

where $\cos(x_i', x_j)$ denotes the cosine of the angle between the positive x_i' axis and the positive x_j axis. Thus, with respect to the old system of axes, $\alpha_{11}, \alpha_{12}, \alpha_{13}$ are the direction cosines of the x_1' axis; $\alpha_{21}, \alpha_{22}, \alpha_{23}$ are the

2.2 COORDINATE TRANSFORMATIONS

Table 2.1. Table of Direction Cosines

	$\mathbf{e}_1$	$\mathbf{e}_2$	$\mathbf{e}_3$
$\mathbf{e}'_1$	α_{11}	α_{12}	α_{13}
$\mathbf{e}'_2$	α_{21}	α_{22}	α_{23}
$\mathbf{e}'_3$	α_{31}	α_{32}	α_{33}

direction cosines of the x'_2 axis; and $\alpha_{31}, \alpha_{32}, \alpha_{33}$ are the direction cosines of the x'_3 axis. The relations (2.2.5) may be displayed in a tabular form as shown in Table 2.1, which is referred to as the *table of direction cosines*.

Taking the scalar product with $\mathbf{e}'_i$ on both sides of (2.2.1) and using (2.2.5) as well as the fact that $\mathbf{x} \cdot \mathbf{e}'_i = x'_i$ (which follows from (2.2.2)) we get

$$x'_i = \alpha_{i1} x_1 + \alpha_{i2} x_2 + \alpha_{i3} x_3 = \alpha_{ip} x_p \qquad (2.2.6)$$

Similarly, we obtain, from (2.2.1) to (2.2.5),

$$x_i = \mathbf{x} \cdot \mathbf{e}_i = (x'_p \mathbf{e}'_p) \cdot \mathbf{e}_i = x'_p (\mathbf{e}'_p \cdot \mathbf{e}_i) = \alpha_{pi} x'_p \qquad (2.2.7)$$

When the orientations of the new axes w.r.t. the old axes are known, the coefficients α_{ij} are known. Expressions (2.2.6) then determine x'_i in terms of x_i, and (2.2.7) determine x_i in terms of x'_i. In other words, (2.2.6) represent the law that transforms the triplet (x_i) to the triplet (x'_i) and (2.2.7) represent the inverse law.

Substituting for x_p from (2.2.7) in (2.2.6) gives $\alpha_{ip} \alpha_{jp} x'_j = x'_i$. Since $x'_i = \delta_{ij} x'_j$, we get $\alpha_{ip} \alpha_{jp} x'_j = \delta_{ij} x'_j$. This relation holds for arbitrary x'_j. It therefore follows that (see Example 1.3.3)

$$\alpha_{ip} \alpha_{jp} = \delta_{ij} \qquad (2.2.8)$$

Similarly, substitution of x'_p in (2.2.7) from (2.2.6) leads to

$$\alpha_{pi} \alpha_{pj} = \delta_{ij} \qquad (2.2.9)$$

When written in the unabridged form, the nine relations in (2.2.8) read as follows

$$\begin{aligned}
\alpha_{11}^2 + \alpha_{12}^2 + \alpha_{13}^2 &= 1 \\
\alpha_{21}^2 + \alpha_{22}^2 + \alpha_{23}^2 &= 1 \\
\alpha_{31}^2 + \alpha_{32}^2 + \alpha_{33}^2 &= 1 \\
\alpha_{11} \alpha_{21} + \alpha_{12} \alpha_{22} + \alpha_{13} \alpha_{23} &= 0 \\
\alpha_{21} \alpha_{31} + \alpha_{22} \alpha_{32} + \alpha_{23} \alpha_{33} &= 0 \\
\alpha_{31} \alpha_{11} + \alpha_{32} \alpha_{12} + \alpha_{33} \alpha_{13} &= 0
\end{aligned} \qquad (2.2.10)$$

2 ALGEBRA OF TENSORS

The first three relations in (2.2.10) show that the sums of the squares of the direction cosines of the x_1', x_2' and x_3' axes (w.r.t. x_i axes) are each equal to 1, as expected. The last three equations in (2.2.10) verify the fact that the x_1', x_2' and x_3' axes are mutually orthogonal. The corresponding relations for the direction cosines of the x_1, x_2 and x_3 axes w.r.t. the x_i' axes follow from (2.2.9). The relations (2.2.8) and (2.2.9) are referred to as the *orthonormal relations* for α_{ij}.

In matrix notation, relations (2.2.8) and (2.2.9) may be represented, respectively, as follows:

$$[\alpha_{ij}][\alpha_{ij}]^T = [I] \qquad (2.2.11)$$

$$[\alpha_{ij}]^T[\alpha_{ij}] = [I] \qquad (2.2.12)$$

These expressions show that the matrix $[\alpha_{ij}]$ is nonsingular and that $[\alpha_{ij}]^{-1} = [\alpha_{ij}]^T$; in other words, the matrix $[\alpha_{ij}]$ is orthogonal. Therefore, the transformation laws (2.2.6) and (2.2.7), determined by $[\alpha_{ij}]$, are called *orthogonal transformations*. The matrix $[\alpha_{ij}]$ is referred to as the *matrix of the transformation* from the x_i system to the x_i' system.

EXAMPLE 2.2.1 The x_i' system is obtained by rotating the x_i system about the x_3 axis through an angle θ in the sense of the righthanded screw. Find the transformation matrix. If a point P has coordinates (1, 1, 1) in the x_i system, find its coordinates in the x_i' system. If a point Q has coordinates (1, 1, 1) in the x_i' system, find its coordinates in the x_i system.

Solution Figure 2.2 illustrates how the x_i' system is related to the x_i system. From this figure we readily get Table 2.2 of direction cosines for the given transformation (on using Table 2.1). Hence the matrix of the given

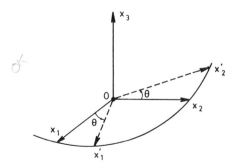

Figure 2.2. Rotation about the x_3 axis.

2.2 COORDINATE TRANSFORMATIONS

Table 2.2. Direction Cosines for Example 2.2.1

.	e_1	e_2	e_3
e'_1	$\cos\theta$	$\sin\theta$	0
e'_2	$-\sin\theta$	$\cos\theta$	0
e'_3	0	0	1

transformation is

$$[\alpha_{ij}] = \begin{bmatrix} \cos\theta & \sin\theta & 0 \\ -\sin\theta & \cos\theta & 0 \\ 0 & 0 & 1 \end{bmatrix} \quad (2.2.13)$$

The coordinates of the point P in the x_i system are given as $x_1 = 1$, $x_2 = 1$, $x_3 = 1$. Its coordinates x'_i in the x'_i system may be obtained by employing the transformation law (2.2.6). Using the values of α_{ij} given by (2.2.13), the law (2.2.6) yields

$$\begin{aligned} x'_1 &= \alpha_{1p}x_p = \alpha_{11}x_1 + \alpha_{12}x_2 + \alpha_{13}x_3 = \cos\theta + \sin\theta \\ x'_2 &= \alpha_{2p}x_p = \alpha_{21}x_1 + \alpha_{22}x_2 + \alpha_{23}x_3 = \cos\theta - \sin\theta \\ x'_3 &= \alpha_{3p}x_p = \alpha_{31}x_1 + \alpha_{32}x_2 + \alpha_{33}x_3 = 1 \end{aligned} \quad (2.2.14)$$

Thus, the coordinates of P in the x'_i system are $(\cos\theta + \sin\theta, \cos\theta - \sin\theta, 1)$.

The coordinates of the point Q in the x'_i system are given as $x'_1 = 1$, $x'_2 = 1$, $x'_3 = 1$. Its coordinates x_i in the x_i system follow from the law (2.2.7), which, on using (2.2.13), yields

$$\begin{aligned} x_1 &= \alpha_{p1}x'_p = \alpha_{11}x'_1 + \alpha_{21}x'_2 + \alpha_{31}x'_3 = \cos\theta - \sin\theta \\ x_2 &= \alpha_{p2}x'_p = \alpha_{12}x'_1 + \alpha_{22}x'_2 + \alpha_{32}x'_3 = \cos\theta + \sin\theta \\ x_3 &= \alpha_{p3}x'_p = \alpha_{13}x'_1 + \alpha_{23}x'_2 + \alpha_{33}x'_3 = 1 \end{aligned} \quad (2.2.15)$$

Hence, the coordinates of Q in the x_i system are $(\cos\theta - \sin\theta, \cos\theta + \sin\theta, 1)$. ∎

EXAMPLE 2.2.2 Complete Table 2.3 of direction cosines.

Solution By data, $\alpha_{11} = \alpha_{21} = \frac{1}{2}\sqrt{\frac{3}{2}}$, $\alpha_{12} = \alpha_{22} = \alpha_{13} = -\alpha_{23} = 1/2\sqrt{2}$. The orthonormal relations (2.2.10) now yield the following conditions on $\alpha_{31}, \alpha_{32}, \alpha_{33}$.

$$\alpha_{31}^2 + \alpha_{32}^2 + \alpha_{33}^2 = 1$$
$$\sqrt{3}\,\alpha_{31} + \alpha_{32} + \alpha_{33} = 0$$
$$\sqrt{3}\,\alpha_{31} + \alpha_{32} - \alpha_{33} = 0$$

2 ALGEBRA OF TENSORS

Table 2.3. Direction Cosines for Example 2.2.2

	e_1	e_2	e_3
e_1'	$\dfrac{1}{2}\sqrt{\dfrac{3}{2}}$	$\dfrac{1}{2\sqrt{2}}$	$\dfrac{1}{2\sqrt{2}}$
e_2'	$\dfrac{1}{2}\sqrt{\dfrac{3}{2}}$	$\dfrac{1}{2\sqrt{2}}$	$-\dfrac{1}{2\sqrt{2}}$
e_3'	?	?	?

Solving these equations, we obtain

$$\alpha_{31} = \pm\frac{1}{2}, \quad \alpha_{32} = \mp\frac{\sqrt{3}}{2}, \quad \alpha_{33} = 0$$

and the Table 2.3 is completed. ∎

EXAMPLE 2.2.3 Show that

$$[\mathbf{e}_i', \mathbf{e}_j', \mathbf{e}_k'] = \begin{vmatrix} \alpha_{i1} & \alpha_{i2} & \alpha_{i3} \\ \alpha_{j1} & \alpha_{j2} & \alpha_{j3} \\ \alpha_{k1} & \alpha_{k2} & \alpha_{k3} \end{vmatrix} \qquad (2.2.16)$$

Hence, deduce that

$$\det[\alpha_{ij}] = 1 \qquad (2.2.17)$$

Solution For any vector $\mathbf{a}$, we have

$$\mathbf{a} = (\mathbf{a} \cdot \mathbf{e}_1)\mathbf{e}_1 + (\mathbf{a} \cdot \mathbf{e}_2)\mathbf{e}_2 + (\mathbf{a} \cdot \mathbf{e}_3)\mathbf{e}_3$$

Accordingly,

$$\mathbf{e}_i' = (\mathbf{e}_i' \cdot \mathbf{e}_1)\mathbf{e}_1 + (\mathbf{e}_i' \cdot \mathbf{e}_2)\mathbf{e}_2 + (\mathbf{e}_i' \cdot \mathbf{e}_3)\mathbf{e}_3$$
$$= \alpha_{i1}\mathbf{e}_1 + \alpha_{i2}\mathbf{e}_2 + \alpha_{i3}\mathbf{e}_3$$

on using the definition of α_{ij}. Similarly,

$$\mathbf{e}_j' = \alpha_{j1}\mathbf{e}_1 + \alpha_{j2}\mathbf{e}_2 + \alpha_{j3}\mathbf{e}_3$$
$$\mathbf{e}_k' = \alpha_{k1}\mathbf{e}_1 + \alpha_{k2}\mathbf{e}_2 + \alpha_{k3}\mathbf{e}_3$$

Hence, by virtue of expression (1.4.17), we obtain

$$[\mathbf{e}_i', \mathbf{e}_j', \mathbf{e}_k'] = \begin{vmatrix} \alpha_{i1} & \alpha_{i2} & \alpha_{i3} \\ \alpha_{j1} & \alpha_{j2} & \alpha_{j3} \\ \alpha_{k1} & \alpha_{k2} & \alpha_{k3} \end{vmatrix}$$

In particular,

$$[\mathbf{e}'_1, \mathbf{e}'_2, \mathbf{e}'_3] = \begin{vmatrix} \alpha_{11} & \alpha_{12} & \alpha_{13} \\ \alpha_{21} & \alpha_{22} & \alpha_{23} \\ \alpha_{31} & \alpha_{32} & \alpha_{33} \end{vmatrix} = \det[\alpha_{ij}] \quad (2.2.18)$$

On the other hand,

$$[\mathbf{e}'_1, \mathbf{e}'_2, \mathbf{e}'_3] = (\mathbf{e}'_1 \times \mathbf{e}'_2) \cdot \mathbf{e}'_3 = \mathbf{e}'_3 \cdot \mathbf{e}'_3 = 1. \quad (2.2.19)$$

From (2.2.18) and (2.2.19), we get

$$\det[\alpha_{ij}] = [\mathbf{e}'_1, \mathbf{e}'_2, \mathbf{e}'_3] = 1 \quad \blacksquare$$

Note: Expressions (2.2.11), (2.2.12) and (2.2.17) show that $[\alpha_{ij}]$ is a proper orthogonal matrix. The coordinate transformations determined by α_{ij} are therefore called *proper orthogonal transformations*. Only such coordinate transformations are dealt with in our further discussions; we will refer to them simply as *coordinate transformations* (with no adjective).

2.3 CARTESIAN TENSORS

In the previous section, we saw that if (x_i) are the coordinates of a point in one coordinate system and (x'_i) are the coordinates of the same point in any other coordinate system with the same fixed origin, then x_i and x'_i are related by equations (2.2.6) and (2.2.7). This is equivalent to saying that if the position vector $\mathbf{x}$ of a point is represented as the ordered triplet (x_i) in one coordinate system and as the ordered triplet (x'_i) in another coordinate system, then x_i and x'_i transform to each other according to the rules given by (2.2.6) and (2.2.7). We now show that precisely the same transformation rules hold for the components of an *arbitrary* vector when it is referred to two systems of coordinates (with the same fixed origin).

Let $\mathbf{a}$ be a vector having components a_i along the x_i axes and components a'_i along the x'_i axes, so that $\mathbf{a}$ is represented as the ordered triplet (a_i) in the coordinate system (x_i) and the ordered triplet (a'_i) in the coordinate system (x'_i). Then we have

$$\mathbf{a} = a_1\mathbf{e}_1 + a_2\mathbf{e}_2 + a_3\mathbf{e}_3 = a_p\mathbf{e}_p \quad (2.3.1)$$

$$\mathbf{a} = a'_1\mathbf{e}'_1 + a'_2\mathbf{e}'_2 + a'_3\mathbf{e}'_3 = a'_p\mathbf{e}'_p \quad (2.3.2)$$

Taking the scalar product with $\mathbf{e}'_i$ on both sides of (2.3.1), and noting that $\mathbf{a} \cdot \mathbf{e}'_i = a'_i$, we get

$$a'_i = \alpha_{ip} a_p \quad (2.3.3)$$

where α_{ij} are as defined by (2.2.5). Similarly, taking the scalar product with $\mathbf{e}_i$ on both sides of (2.3.2) and noting that $\mathbf{a} \cdot \mathbf{e}_i = a_i$, we obtain

$$a_i = \alpha_{pi} a'_p \qquad (2.3.4)$$

Relations (2.3.3) and (2.3.4) are the rules that determine a'_i in terms of a_i and vice-versa. Evidently, these relations are analogous to (2.2.6) and (2.2.7). (Indeed, if we set $\mathbf{a} = \mathbf{x}$, relations (2.3.3) and (2.3.4) reduce to (2.2.6) and (2.2.7), respectively.)

The foregoing analysis may be summarized thusly. With respect to a (every) system of axes, a vector $\mathbf{a}$ may be represented as an ordered triplet of real numbers; if (a_i) and (a'_i) are the representations of a vector in the x_i and x'_i systems, respectively, then a_i and a'_i obey the transformation rules (2.3.3) and (2.3.4). This result describes a characteristic property of vectors, and the property may be used to define a vector in an alternative way. When defined in this "new" way, a vector is referred to as a *Cartesian tensor of order 1*.

Definition A Cartesian tensor of order 1 (or a vector) is an entity that may be represented as an ordered triplet of (real) numbers in every Cartesian coordinate system with the property that if the ordered triplet (a_i) is the representation of the entity in the x_i system and the ordered triplet (a'_i) is the representation of the entity in the x'_i system, then a_i and a'_i obey the following transformation rules:

$$a'_i = \alpha_{ip} a_p; \qquad a_i = \alpha_{pi} a'_p \qquad (2.3.5\text{a, b})$$

Then a_i and a'_i are called the *components of a first-order Cartesian tensor* in the x_i and x'_i systems, respectively.

2.3.1 TENSOR PRODUCT OF VECTORS

Consider now a vector represented as the triplet (b_i) in the x_i system and the triplet (b'_i) in the x'_i system. Then

$$b'_i = \alpha_{ip} b_p; \qquad b_i = \alpha_{pi} b'_p \qquad (2.3.6)$$

As usual, let $\mathbf{a}$ denote the vector for which the transformation rules (2.3.5) hold and $\mathbf{b}$ denote the vector for which (2.3.6) hold.

From (2.3.5) and (2.3.6) we obtain

$$\begin{aligned}a'_i b'_j &= (\alpha_{ip} a_p)(\alpha_{jq} b_q) = \alpha_{ip} \alpha_{jq} a_p b_q \\ a_i b_j &= (\alpha_{pi} a'_p)(\alpha_{qj} b'_q) = \alpha_{pi} \alpha_{qj} a'_p b'_q\end{aligned} \qquad (2.3.7)$$

Note that $[a_i b_j]$ is a 3×3 matrix in the x_i system and $[a'_i b'_j]$ is a 3×3 matrix in the x'_i system; the elements of these two matrices transform to each other according to the rules given by (2.3.7). If we treat $[a_i b_j]$ and $[a'_i b'_j]$ as representations of a certain entity in the x_i and x'_i systems, respectively, then that entity is called the *outer product* or *tensor product* of the vectors **a** and **b** in that order, denoted by $\mathbf{a} \otimes \mathbf{b}$. This product serves as an example of what is called a *Cartesian tensor of order 2*.

2.3.2 CARTESIAN TENSOR OF SECOND ORDER

Definition A Cartesian tensor of order 2 is an entity that may be represented as a 3×3 (real) matrix in every Cartesian coordinate system with the property that, if the matrix $[a_{ij}]$ is the representation of the entity in the x_i system and $[a'_{ij}]$ is the representation of the entity in the x'_i system, then a_{ij} and a'_{ij} obey the following transformation rules:

$$a'_{ij} = \alpha_{ip}\alpha_{jq}a_{pq}; \qquad a_{ij} = \alpha_{pi}\alpha_{qj}a'_{pq} \qquad (2.3.8\text{a, b})$$

Then a_{ij} and a'_{ij} are called the *components of a second-order Cartesian tensor* in the x_i and x'_i systems, respectively. Note that the transformation rules (2.3.8) are motivated by and generalizations of the rules (2.3.7).

2.3.3 CARTESIAN TENSORS OF HIGHER ORDER

Cartesian tensors of higher orders are defined by generalizing the transformation rules (2.3.8) in a natural way. The definition of a *Cartesian tensor of order N*, where N is a positive integer, is given as follows.

Definition A Cartesian tensor of order N, where N is a positive integer, is an entity that may be represented as a set of 3^N real numbers in every Cartesian coordinate system with the property that if $(a_{ijk...})$ is the representation of the entity in the x_i-system and $(a'_{ijk...})$ is the representation of the entity in the x'_i system, then $a_{ijk...}$ and $a'_{ijk...}$ obey the following transformation rules:

$$\underbrace{a'_{ijk...}}_{N \text{ suffixes}} = \underbrace{\alpha_{ip}\alpha_{jq}\alpha_{kr}\cdots}_{N \text{ factors}} \underbrace{a_{pqr...}}_{N \text{ suffixes}} \qquad (2.3.9\text{a})$$

$$\underbrace{a_{ijk...}}_{N \text{ suffixes}} = \underbrace{\alpha_{pi}\alpha_{qj}\alpha_{rk}\cdots}_{N \text{ factors}} \underbrace{a'_{pqr...}}_{N \text{ suffixes}} \qquad (2.3.9\text{b})$$

Then $a_{ijk...}$ and $a'_{ijk...}$ are called the *components of a Cartesian tensor of order N* in the x_i and x'_i systems, respectively.

We have seen that if **a** and **b** are two vectors, then the tensor product $\mathbf{a} \otimes \mathbf{b}$, represented as the matrix $[a_i b_j]$ in the x_i system and the matrix $[a'_i b'_j]$ in the x'_i system, is an example of a Cartesian tensor of order 2. One may show that if **c** is a vector with components c_i and c'_i in the x_i and x'_i systems, respectively, then the entity $\mathbf{a} \otimes \mathbf{b} \otimes \mathbf{c}$ represented as $(a_i b_j c_k)$ in the x_i system and $(a'_i b'_j c'_k)$ in the x'_i system is a Cartesian tensor of order 3. Examples of Cartesian tensors of higher orders may be constructed in an analogous way. ∎

2.3.4 SCALAR INVARIANTS

Quite often we deal with quantities that may be represented by single numbers not dependent on any coordinate system. Temperature, density, mass, etc., are examples of such quantities; these are referred to as *scalar invariants*. Hence, if a scalar invariant is represented as a number a in the x_i system and a number a' in the x'_i system, then $a' = a$. An entity obeying a transformation law of this type is referred to as a *Cartesian tensor of order 0*. Thus, a Cartesian tensor of order 0 is precisely a scalar invariant.

In the preceding paragraphs we defined Cartesian tensors of order $0, 1, 2, \ldots$ through appropriate transformation rules. The adjective *Cartesian* was used to emphasize the fact that the transformation rules pertained to Cartesian coodinate systems. It should be borne in mind that for an entity to be a Cartesian tensor, first of all it should have a representation as a set of real numbers in every Cartesian coordinate system and next, the individual numbers constituting the set should obey the appropriate transformation rules. Then, these individual numbers are called the *Cartesian tensor components* in the coordinate system in which they are defined. For a Cartesian tensor of order N (≥ 0), there exist 3^N components in every Cartesian coordinate system. While the tensor components are always defined with respect to one or another coordinate system, the tensor as such is an entity by itself; it depends on no coordinate system whatsoever.

2.3.5 TERMINOLOGY AND NOTATION

Henceforth, a Cartesian tensor will be referred to simply as a *tensor*. Further, the term *tensor* will be used only when the order is higher than or equal to 2; a tensor of order 1 will be referred to as a *vector* and a tensor of order 0 as a *scalar invariant* or just a *scalar*.

Like vectors, tensors are also denoted by boldface letters (in print) or letters with an arrow above them (in manuscripts). If not used with proper care, this notation often leads to confusion, particularly when vectors and tensors are treated together. In order to avoid such possible confusion, we shall consistently denote a tensor by an uppercase letter with an overscript indicating its order, in constrast to a vector always denoted by a lowercase letter. Thus, a symbol such as $\mathbf{a}$ (or $\vec{a}$) will denote a *vector* as usual, while a symbol such as $\overset{N}{\mathbf{A}}$ $\left(\text{or } \overset{N}{\vec{A}}\right)$ will denote a *tensor* of order N (≥ 2). In

2.3 CARTESIAN TENSORS 43

continuum mechanics, we deal mainly with second-order tensors, in addition to scalars and vectors. Therefore, we relax the notation for second-order tensors a little. We will denote a second-order tensor by an uppercase letter *without* the superscript 2 (which will be automatically understood); in other words, we will use just $\mathbf{A}$ (or $\vec{A}$) for $\overset{2}{\mathbf{A}}$ (or $\overset{2}{\vec{A}}$). Also, when we refer to a tensor without explicitly indicating its order, it will be understood that the order is 2.

If the vector defined by the transformation rules (2.3.5) is denoted as usual by $\mathbf{a}$, then we write

$$[\mathbf{a}]_i = a_i \qquad (2.3.10)$$

to mean that the ith (typical) component of $\mathbf{a}$ in the x_i system is a_i; see expression (1.4.3).

If the tensor defined by the transformation rules (2.3.8) is denoted by $\mathbf{A}$, then the matrix $[a_{ij}]$ that represents the tensor in the x_i system is called the *matrix of the tensor* $\mathbf{A}$ in the x_i system, denoted by $[\mathbf{A}]$. Thus, by definition,

$$[\mathbf{A}] = [a_{ij}] \qquad (2.3.11)$$

so that the elements of the matrix $[\mathbf{A}]$ are precisely the components of the tensor $\mathbf{A}$ in the x_i system. We write

$$[\mathbf{A}]_{ij} = a_{ij} \qquad (2.3.12)$$

to mean that the ijth (typical) component of $\mathbf{A}$ in the x_i system is a_{ij}.

Thus, for the tensor product $\mathbf{a} \otimes \mathbf{b}$ of vectors $\mathbf{a}$ and $\mathbf{b}$,

$$[\mathbf{a} \otimes \mathbf{b}]_{ij} = a_i b_j = [\mathbf{a}]_i [\mathbf{b}]_j \qquad (2.3.13)$$

Similarly, we write

$$\left[\overset{N}{\mathbf{A}}\right]_{ijk\ldots} = a_{ijk\ldots} \qquad (2.3.14)$$

to mean that the $(ijk\ldots)$th (typical) component of $\overset{N}{\mathbf{A}}$ in the x_i system is $a_{ijk\ldots}$.

It has been pointed out that the components of a vector or a tensor are always associated with a system of axes. However, for brevity, we will make statements like "$a_{ijk\ldots}$ are the components of a tensor" to actually mean that "$a_{ijk\ldots}$ are the components of a tensor in the x_i system." (Similar terminology will be used for vectors as well.)

Although tensors are denoted by capital letters, no confusion should arise if the components of tensors are denoted by lowercase letters, like those of vectors. For, the components of vectors appear with a single suffix, such as a_i and those of tensors appear with multiple suffixes, such as a_{ij} in the case of a second-order tensor and a_{ijk} in the case of a third-order tensor. The number of free suffixes appended to a symbol indicates the actual order of the tensor being dealt with. This is indeed one of the greatest advantages of the suffix notation in the treatment of tensors. In what follows, unless stated to the contrary, the components of a tensor, denoted by an uppercase letter, will be denoted by the corresponding small letter appended with an appropriate number of suffixes, and vice-versa.

EXAMPLE 2.3.1 In the x_i system, a vector **a** has components

$$a_1 = -1, \quad a_2 = 0, \quad a_3 = 1 \tag{2.3.15}$$

and a tensor **A** has its matrix

$$[a_{ij}] = \begin{bmatrix} 0 & 1 & 0 \\ -1 & 0 & 2 \\ 0 & -2 & 0 \end{bmatrix} \tag{2.3.16}$$

The x_i' system is obtained by rotating the x_i system about the x_3 axis through an angle of 45° in the sense of the righthanded screw. Find the components of **a** and the matrix of **A** in the x_i' system.

Solution For the given transformation, Table 2.4 is the table of direction cosines (see Fig. 2.2 and Table 2.2 with $\theta = 45°$). If a_i' are the components of the given vector **a** in the x_i' system, the transformation rule (2.3.5a) for a vector yields

$$a_i' = \alpha_{ip} a_p = \alpha_{i1} a_1 + \alpha_{i2} a_2 + \alpha_{i3} a_3 = -\alpha_{i1} + \alpha_{i3}$$

using (2.3.15). Therefore, on using Table 2.4,

$$a_1' = -\alpha_{11} + \alpha_{13} = -\frac{1}{\sqrt{2}}$$

$$a_2' = -\alpha_{21} + \alpha_{23} = \frac{1}{\sqrt{2}}$$

$$a_3' = -\alpha_{31} + \alpha_{33} = 1$$

Thus, the components of **a** in the x_i' system are

$$a_1' = -\frac{1}{\sqrt{2}}, \quad a_2' = \frac{1}{\sqrt{2}}, \quad a_3' = 1 \tag{2.3.17}$$

Table 2.4. Direction Cosines of Example 2.3.1

	e_1	e_2	e_3
e_1'	$1/\sqrt{2}$	$1/\sqrt{2}$	0
e_2'	$-1/\sqrt{2}$	$1/\sqrt{2}$	0
e_3'	0	0	1

Next, if a'_{ij} are the components of the given tensor **A** in the x'_i system, the transformation law (2.3.8a) for second-order tensor yields

$$a'_{ij} = \alpha_{ip}\alpha_{jq}a_{pq}$$
$$= \alpha_{ip}(\alpha_{j1}a_{p1} + \alpha_{j2}a_{p2} + \alpha_{j3}a_{p3})$$
$$= \alpha_{j1}(\alpha_{i1}a_{11} + \alpha_{i2}a_{21} + \alpha_{i3}a_{31}) + \alpha_{j2}(\alpha_{i1}a_{12} + \alpha_{i2}a_{22} + \alpha_{i3}a_{32})$$
$$+ \alpha_{j3}(\alpha_{i1}a_{13} + \alpha_{i2}a_{23} + \alpha_{i3}a_{33})$$
$$= -\alpha_{j1}\alpha_{i2} + \alpha_{j2}\alpha_{i1} - 2\alpha_{j2}\alpha_{i3} + 2\alpha_{j3}\alpha_{i2}$$

on using (2.3.16). Therefore, on using Table 2.4, we get

$$a'_{11} = -\alpha_{11}\alpha_{12} + \alpha_{12}\alpha_{11} - 2\alpha_{12}\alpha_{13} + 2\alpha_{13}\alpha_{12} = 0$$
$$a'_{12} = 1, \quad a'_{13} = \sqrt{2}, \quad a'_{21} = -1, \quad a'_{22} = 0, \quad a'_{23} = \sqrt{2}$$
$$a'_{31} = -\sqrt{2}, \quad a'_{32} = -\sqrt{2}, \quad a'_{33} = 0$$

Thus, in the x'_i system, the matrix of **A** is

$$[a'_{ij}] = \begin{bmatrix} 0 & 1 & \sqrt{2} \\ -1 & 0 & \sqrt{2} \\ -\sqrt{2} & -\sqrt{2} & 0 \end{bmatrix} \quad \blacksquare \quad (2.3.18)$$

EXAMPLE 2.3.2 Show that each of the two rules of transformation given in (2.3.8) follows from the other.

Solution First, consider the rule (2.3.8a); namely, $a'_{ij} = \alpha_{ip}\alpha_{jq}a_{pq}$. This yields

$$\alpha_{im}\alpha_{jn}a'_{ij} = \alpha_{im}\alpha_{jn}\alpha_{ip}\alpha_{jq}a_{pq}$$
$$= \delta_{mp}\delta_{nq}a_{pq}$$

using (2.2.9);

$$= a_{mn}$$

Thus, $a_{mn} = \alpha_{im}\alpha_{jn}a'_{ij}$. This is precisely the rule (2.3.8b).

Next, let us start with the rule (2.3.8b); namely, $a_{ij} = \alpha_{pi}\alpha_{qj}a'_{pq}$. This yields

$$\alpha_{mi}\alpha_{nj}a_{ij} = \alpha_{mi}\alpha_{nj}\alpha_{pi}\alpha_{qj}a'_{pq}$$
$$= \delta_{mp}\delta_{nq}a'_{pq}$$

using (2.2.8);

$$= a'_{mn}$$

Thus $a'_{mn} = \alpha_{mi}\alpha_{nj}a_{ij}$. This is precisely the rule (2.3.8a). $\blacksquare$

46 2 ALGEBRA OF TENSORS

Note: In the same way, one may show that each of the two transformation rules governing a vector and a tensor of (any) order N (>2) may be obtained from the other. Consequently, while treating a tensor, it is sufficient to consider *only one* of the two transformation rules governing the tensor.

EXAMPLE 2.3.3 Let **a** and **b** be vectors with components a_i and b_i and **A** be a tensor with components a_{ij}. Show that $a_i b_i$ and a_{ii} are scalar invariants.

Solution If a_i' and b_i' are components of **a** and **b** in the x_i' system, we have, by the transformation rule of vector components,

$$a_i' b_i' = (\alpha_{ip} a_p)(\alpha_{iq} b_q) = (\alpha_{ip} \alpha_{iq}) a_p b_q = \delta_{pq} a_p b_q = a_p b_p = a_i b_i \quad (2.3.19)$$

Thus, $a_i b_i$ has the same value in all coordinate systems; it is therefore a scalar invariant. Note that this scalar invariant is actually the *scalar product* of **a** and **b**, namely $\mathbf{a} \cdot \mathbf{b}$.

If a_{ij}' are the components of **A** in the x_i' system, we have, by the transformation rule of tensor components of order 2, $a_{ij}' = \alpha_{ip} \alpha_{jq} a_{pq}$. Hence

$$a_{ii}' = \alpha_{ip} \alpha_{iq} a_{pq} = \delta_{pq} a_{pq} = a_{pp} = a_{ii} \quad (2.3.20)$$

Thus, a_{ii} has the same value in all coordinate systems; it is therefore a scalar invariant. This invariant is called the *trace* of **A**, denoted by tr **A**.

2.4
PROPERTIES OF TENSORS

We now proceed to obtain some basic algebraic properties of tensors.

2.4.1 ZERO TENSOR

It is obvious that if all the components of a vector are (equal to) 0 in one coordinate system, then they are 0 in all coordinate systems. (Because of this property, the **0** vector is often defined as the vector all of whose components are 0 in a coordinate system.) This property is true for tensors of all orders, and this is an important property. Proof of the property follows here; the definition of zero tensor will follow immediately thereafter.

PROPERTY 1 If all components of a tensor are 0 in one coordinate system, then they are 0 in all coordinate systems.

2.4 PROPERTIES OF TENSORS

Proof Let a_{ij} and a'_{ij} be components of a second-order tensor **A** in two coordinate systems x_i and x'_i, respectively. Suppose that $a_{ij} = 0$. Then the transformation rule yields

$$a'_{ij} = \alpha_{ip}\alpha_{jq}a_{pq} = 0 \tag{2.4.1}$$

This proves the property for second-order tensors. The proof for higher order tensors is similar. ∎

Definition A tensor of order N all of whose components are 0 in a coordinate system is called the *zero tensor* of order N, denoted by **0**.

2.4.2 EQUALITY OF TENSORS

If the corresponding components of two vectors **a** and **b** are equal in a coordinate system, then the components of their difference **a** − **b** are all **0** in that coordinate system. Consequently, the components of **a** − **b** are 0 in all coordinate systems. (Because of this property, we say that **a** and **b** are equal, that is **a** = **b**, if the corresponding components of **a** and **b** are equal in a coordinate system.) This property is true for tensors of all orders, and this is another important property. Proof of the property follows here; the definition of equality of tensors will follow immediately thereafter.

PROPERTY 2 If the corresponding components of two tensors of the same order are equal in one coordinate system, then they are equal in all coordinate systems.

Proof Let **A** and **B** be two second-order tensors and a_{ij} and b_{ij} be their components in the x_i system. Suppose $a_{ij} = b_{ij}$. If a'_{ij} and b'_{ij} are the components of **A** and **B** in the x'_i system, the transformation rule gives

$$a'_{ij} = \alpha_{ip}\alpha_{jq}a_{pq} = \alpha_{ip}\alpha_{jq}b_{pq}$$

(since $a_{pq} = b_{pq}$)

$$= b'_{ij} \tag{2.4.2}$$

This proves the property for second-order tensors. The proof for higher order tensors is analogous. ∎

Definition Two tensors of the same order whose corresponding components are equal in a coordinate system (and hence in all coordinate systems, by the property just proven) are called *equal tensors*.

Note: If $\overset{N}{\mathbf{A}}$ and $\overset{N}{\mathbf{B}}$ are equal tensors, we write $\overset{N}{\mathbf{A}} = \overset{N}{\mathbf{B}}$. It is obvious that $\overset{N}{\mathbf{A}} = \overset{N}{\mathbf{B}}$ if and only if $\overset{N}{\mathbf{A}} - \overset{N}{\mathbf{B}}$ is the zero tensor (of order N).

2.4.3 TENSOR EQUATIONS

From the definition of equality of tensors it follows that if $a_{ijk...}$ and $b_{ijk...}$ are components of two tensors $\overset{N}{\mathbf{A}}$ and $\overset{N}{\mathbf{B}}$, then the 3^N equations

$$a_{ijk...} = b_{ijk...} \tag{2.4.3}$$

are equivalent to the equation

$$\overset{N}{\mathbf{A}} = \overset{N}{\mathbf{B}} \tag{2.4.4}$$

Thus, in order to prove that two tensors are equal, it is sufficient to show that their corresponding components are equal in any one of the coordinate systems. Equations of the type (2.4.3) and (2.4.4) are called *tensor equations*. It is to be emphasized that equations of the type (2.4.3) actually imply the equality of two tensors. Equations like (2.4.4) are referred to as tensor equations expressed in the *direct notation*, while those like (2.4.3) are referred to as tensor equations in the *suffix notation*.

2.4.4 SCALAR MULTIPLE OF A TENSOR

If α is a scalar and a_i are components of a vector **a**, we have seen that αa_i are components of the vector $\alpha\mathbf{a}$, called the *scalar multiple* of α and **a**. This property and notation may be extended to tensors as well, as follows.

PROPERTY 3 If α is a scalar and $a_{ijk...}$ are components of a tensor $\overset{N}{\mathbf{A}}$, then $\alpha a_{ijk...}$ are components of a tensor of order N. (This tensor is denoted by $\alpha\overset{N}{\mathbf{A}}$ and referred to as the *scalar multiple* of α and $\overset{N}{\mathbf{A}}$.)

Proof Consider a second-order tensor **A** with components a_{ij} and put $c_{ij} = \alpha a_{ij}$ in all coordinate systems. This means that $c'_{ij} = \alpha' a'_{ij}$ in the x'_i system.

Since α is a scalar, we have $\alpha' = \alpha$, so that

$$\begin{aligned} c'_{ij} &= \alpha a'_{ij} \\ &= \alpha \alpha_{ip}\alpha_{jq}a_{pq} = \alpha_{ip}\alpha_{jq}(\alpha a_{pq}) \\ &= \alpha_{ip}\alpha_{jq}c_{pq} \end{aligned} \tag{2.4.5}$$

Thus, $c_{ij} = \alpha a_{ij}$ obey the transformation rule of a second-order tensor; therefore αa_{ij} are components of a second-order tensor.

This proves the property for second-order tensors. The proof for higher-order tensors is analogous. ∎

2.4 PROPERTIES OF TENSORS 49

Note: From the property 3, it follows that

$$[\alpha \overset{N}{\mathbf{A}}]_{ijk...} = \alpha[\overset{N}{\mathbf{A}}]_{ijk...} \qquad (2.4.6)$$

The scalar multiple of -1 and $\overset{N}{\mathbf{A}}$, namely, $(-1)\overset{N}{\mathbf{A}}$, is denoted by $-\overset{N}{\mathbf{A}}$, referred to as the *negative* of $\overset{N}{\mathbf{A}}$.

2.4.5 SUM AND DIFFERENCE OF TENSORS

If a_i are components of a vector $\mathbf{a}$ and b_i are components of a vector $\mathbf{b}$, we have seen that $a_i + b_i$ are components of the vector sum $\mathbf{a} + \mathbf{b}$ and $a_i - b_i$ are components of the vector difference $\mathbf{a} - \mathbf{b}$. This property and notation may be extended to tensors as well, as follows.

PROPERTY 4 Let $a_{ijk...}$ be components of a tensor $\overset{N}{\mathbf{A}}$ and $b_{ijk...}$ be components of a tensor $\overset{N}{\mathbf{B}}$. Then $(a_{ijk...} + b_{ijk...})$ are components of a tensor of the same order N (called the *tensor sum* $\overset{N}{\mathbf{A}} + \overset{N}{\mathbf{B}}$). Also, $(a_{ijk...} - b_{ijk...})$ are components of a tensor of the same order (called the *tensor difference* $\overset{N}{\mathbf{A}} - \overset{N}{\mathbf{B}}$).

Proof Consider two second-order tensors $\mathbf{A}$ and $\mathbf{B}$ with components a_{ij} and b_{ij}, respectively. Put $c_{ij} = a_{ij} + b_{ij}$ in all coordinate systems. Then,

$$\begin{aligned} c'_{ij} &= a'_{ij} + b'_{ij} \\ &= \alpha_{ip}\alpha_{jq}a_{pq} + \alpha_{ip}\alpha_{jq}b_{pq} = \alpha_{ip}\alpha_{jq}(a_{pq} + b_{pq}) \\ &= \alpha_{ip}\alpha_{jq}c_{pq} \end{aligned} \qquad (2.4.7)$$

Thus, $c_{ij} = a_{ij} + b_{ij}$ obey the transformation rule of a second-order tensor; therefore $a_{ij} + b_{ij}$ are components of a second-order tensor. Similarly, $a_{ij} - b_{ij}$ are components of a second-order tensor.

This proves the property for second-order tensors. The proof for higher order tensors is analogous. ∎

Note: From the property 4, it follows that

$$[\overset{N}{\mathbf{A}} \pm \overset{N}{\mathbf{B}}]_{ijk...} = [\overset{N}{\mathbf{A}}]_{ijk...} \pm [\overset{N}{\mathbf{B}}]_{ijk...} \qquad (2.4.8)$$

It is easy to see that

$$\overset{N}{\mathbf{A}} + \overset{N}{\mathbf{B}} = \overset{N}{\mathbf{B}} + \overset{N}{\mathbf{A}} \qquad (2.4.9)$$

and

$$\overset{N}{\mathbf{A}} - \overset{N}{\mathbf{B}} = -(\overset{N}{\mathbf{B}} - \overset{N}{\mathbf{A}}) \qquad (2.4.10)$$

Remark: The tensor sum and tensor difference are defined for tensors of the same order.

2.4.6 TENSOR MULTIPLICATION

If **a** and **b** are vectors with components a_i and b_i (respectively), we have seen that $a_i b_j$ are components of a second-order tensor, called the *tensor product* $\mathbf{a} \otimes \mathbf{b}$. Such products may also be obtained by combining a vector and a tensor or two tensors; the outcome is a tensor whose order is the sum of the original orders. This process of obtaining new tensors from given vectors or tensors is called *tensor multiplication*; the process is illustrated in the following.

PROPERTY 5 (i) If a_{ij} are components of a second-order tensor **A** and b_i are components of a vector **b**, then $a_{ij} b_k$ are components of a third-order tensor (known as the *tensor product* of **A** and **b** in that order, denoted $\mathbf{A} \otimes \mathbf{b}$).

(ii) If a_{ij} and b_{ij} are components of two second-order tensors **A** and **B**, respectively, then $a_{ik} b_{mj}$ are components of a fourth-order tensor (called the *tensor product* of **A** and **B** in that order, denoted $\mathbf{A} \otimes \mathbf{B}$).

Proof (i) Note that $(a_{ij} b_k)$ is a system of $3^3 = 27$ numbers. Put $c_{ijk} = a_{ij} b_k$ in all coordinate systems. Then

$$c'_{ijk} = a'_{ij} b'_k = (\alpha_{ip} \alpha_{jq} a_{pq})(\alpha_{kr} b_r)$$
$$= \alpha_{ip} \alpha_{jq} \alpha_{kr}(a_{pq} b_r) = \alpha_{ip} \alpha_{jq} \alpha_{kr} c_{pqr} \quad (2.4.11)$$

Thus, the numbers c_{ijk} obey the transformation rule of a third-order tensor. Hence $c_{ijk} = a_{ij} b_k$ are components of a third-order tensor.

(ii) Note that $(a_{ik} b_{mj})$ is a system of $3^4 = 81$ numbers. Put $c_{ikmj} = a_{ik} b_{mj}$ in all coordinate systems. Then

$$c'_{ikmj} = a'_{ik} b'_{mj} = (\alpha_{ip} \alpha_{kq} a_{pq})(\alpha_{mr} \alpha_{js} b_{rs})$$
$$= \alpha_{ip} \alpha_{kq} \alpha_{mr} \alpha_{js} a_{pq} b_{rs} = \alpha_{ip} \alpha_{kq} \alpha_{mr} \alpha_{js} c_{pqrs} \quad (2.4.12)$$

Thus, the numbers c_{ikmj} obey the transformation rule of a fourth-order tensor. Hence $c_{ikmj} = a_{ik} b_{mj}$ are components of a fourth-order tensor. ∎

Note: From the property 5, it follows that

$$[\mathbf{A} \otimes \mathbf{b}]_{ijk} = a_{ij} b_k = [\mathbf{A}]_{ij} [\mathbf{b}]_k \quad (2.4.13)$$

$$[\mathbf{A} \otimes \mathbf{B}]_{ikmj} = a_{ik} b_{mj} = [\mathbf{A}]_{ik} [\mathbf{B}]_{mj} \quad (2.4.14)$$

2.4.7 CONTRACTION

Let us again look at the tensor product $\mathbf{a} \otimes \mathbf{b}$ of two vectors $\mathbf{a}$ and $\mathbf{b}$. Suppose that we change the suffix j to the suffix i in the symbol $a_i b_j$ representing the components of $\mathbf{a} \otimes \mathbf{b}$. Then $a_i b_j$ becomes $a_i b_i$, which represents a scalar, namely $\mathbf{a} \cdot \mathbf{b}$. Thus, the replacement of j by i in the symbol $a_i b_j$ reduces the second-order tensor $\mathbf{a} \otimes \mathbf{b}$ to a scalar. Such a process is called a *contraction operation*. Also, the tensor resulting from a contraction operation is called a *contraction* of the original tensor. Contraction operations are applicable to tensors of all orders (higher than 1); each such operation reduces the order of a tensor by 2. Some illustrations follow.

PROPERTY 6 If a_{ij} and b_{ij} are components of second-order tensors $\mathbf{A}$ and $\mathbf{B}$ and c_i are components of a vector $\mathbf{c}$, then (i) $a_{ij} c_j$ are components of a vector, called the *product of* $\mathbf{A}$ *and* $\mathbf{c}$ in that order and denoted $\mathbf{Ac}$; (ii) $a_{ik} b_{kj}$ are components of a second-order tensor, called the *product of* $\mathbf{A}$ *and* $\mathbf{B}$ in that order and denoted $\mathbf{AB}$; (iii) $a_{ij} b_{ij}$ is a scalar, called the *scalar product of* $\mathbf{A}$ *and* $\mathbf{B}$ and denoted $\mathbf{A} \cdot \mathbf{B}$.

Proof (i) We note that, by property 5(i), $a_{ij} c_k$ are components of a third-order tensor and $a_{ij} c_j$ can be obtained from $a_{ij} c_k$ by a contraction operation (namely, changing k to j). We have to show that $a_{ij} c_j$ are components of a vector.

Since $a_{ij} c_k$ are components of a tensor, we have

$$a'_{ij} c'_k = \alpha_{ip} \alpha_{jq} \alpha_{kr} a_{pq} c_r$$

Hence

$$a'_{ij} c'_j = \alpha_{ip} \alpha_{jq} \alpha_{jr} a_{pq} c_r$$

$$= \alpha_{ip} \delta_{qr} a_{pq} c_r = \alpha_{ip} a_{pq} c_q \quad (2.4.15)$$

This transformation rule shows that $a_{ij} c_j$ are indeed components of a vector. The result illustrates that a third-order tensor reduces to a vector as a result of a contraction operation.

(ii) We note that, by property 5(ii), $a_{ik} b_{mj}$ are components of a fourth-order tensor, and that $a_{ik} b_{kj}$ can be obtained from $a_{ik} b_{mj}$ by a contraction operation (namely, changing m to k). We have to show that $a_{ik} b_{kj}$ are components of a second-order tensor.

Since $a_{ik} b_{mj}$ are components of a fourth-order tensor, we have

$$a'_{ik} b'_{mj} = \alpha_{ip} \alpha_{kq} \alpha_{mr} \alpha_{js} a_{pq} b_{rs}$$

52   2 ALGEBRA OF TENSORS

Hence
$$a'_{ik}b'_{kj} = \alpha_{ip}\alpha_{kq}\alpha_{kr}\alpha_{js}a_{pq}b_{rs}$$
$$= \alpha_{ip}\delta_{qr}\alpha_{js}a_{pq}b_{rs}$$
$$= \alpha_{ip}\alpha_{js}a_{pq}b_{qs} \qquad (2.4.16)$$

This transformation rule shows that $a_{ik}b_{kj}$ are indeed components of a second-order tensor. The result illustrates that a fourth-order tensor reduces to a second-order tensor as a result of a contraction operation.

(iii) We note that $a_{ij}b_{ij}$ can be obtained from $a_{ik}b_{mj}$ by two contraction operations (namely, replacing k by j and m by i). Whereas $a_{ik}b_{mj}$ are components of a fourth-order tensor, we have to show that $a_{ij}b_{ij}$ is a scalar.

Since $a_{ik}b_{mj}$ are components of a fourth-order tensor, we have
$$a'_{ik}b'_{mj} = \alpha_{ip}\alpha_{kq}\alpha_{mr}\alpha_{js}a_{pq}b_{rs}$$
Hence
$$a'_{ij}b'_{ij} = \alpha_{ip}\alpha_{jq}\alpha_{ir}\alpha_{js}a_{pq}b_{rs}$$
$$= \delta_{pr}\delta_{qs}a_{pq}b_{rs}$$
$$= a_{pq}b_{pq} = a_{ij}b_{ij} \qquad (2.4.17)$$

Thus $a_{ij}b_{ij}$ is a scalar. This result illustrates that a fourth-order tensor reduces to a scalar (zeroth order tensor) as a result of two contraction operations. ∎

Note: From the property 6, it follows that
$$[\mathbf{Ac}]_i = a_{ij}c_j = [\mathbf{A}]_{ij}[\mathbf{c}]_j \qquad (2.4.18)$$
$$[\mathbf{AB}]_{ij} = a_{ik}b_{kj} = [\mathbf{A}]_{ik}[\mathbf{B}]_{kj} \qquad (2.4.19)$$
$$\mathbf{A} \cdot \mathbf{B} = a_{ij}b_{ij} = [\mathbf{A}]_{ij}[\mathbf{B}]_{ij} \qquad (2.4.20)$$

It is not hard to verify that
$$[\mathbf{AB}] = [\mathbf{A}][\mathbf{B}], \quad \mathbf{AB} \neq \mathbf{BA}, \quad \mathbf{A}\cdot\mathbf{B} = \mathbf{B}\cdot\mathbf{A} \qquad (2.4.21)$$

The product $\mathbf{AA}$ is called the *square* of the tensor $\mathbf{A}$ and is denoted by $\mathbf{A}^2$. The *cube* $\mathbf{A}^3$ of $\mathbf{A}$ is defined by $\mathbf{A}^3 = (\mathbf{A}^2)\mathbf{A}$. Higher powers of $\mathbf{A}$ are defined in an analogous way. Thus,
$$[\mathbf{A}^2] = [a_{ik}a_{kj}], \qquad [\mathbf{A}^3] = [a_{ik}a_{kp}a_{pj}] \qquad (2.4.22)$$

and so forth.

2.4 PROPERTIES OF TENSORS

The following identities can be easily verified.

(i) $\quad\quad\quad\quad\quad\quad (\alpha \mathbf{A})\mathbf{a} = \alpha(\mathbf{A}\mathbf{a}) \quad\quad\quad\quad\quad\quad$ (2.4.23)

(ii) $\quad\quad\quad\quad\quad\quad (\mathbf{A} \pm \mathbf{B})\mathbf{a} = \mathbf{A}\mathbf{a} \pm \mathbf{B}\mathbf{a} \quad\quad\quad\quad\quad\quad$ (2.4.24)

(iii) $\quad\quad\quad\quad\quad\quad (\mathbf{A}\mathbf{B})\mathbf{a} = \mathbf{A}(\mathbf{B}\mathbf{a}) \quad\quad\quad\quad\quad\quad$ (2.4.25)

(iv) $\quad\quad\quad\quad\quad\quad (\mathbf{A}\mathbf{B})\mathbf{C}\mathbf{a} = \mathbf{A}[\mathbf{B}(\mathbf{C}\mathbf{a})] \quad\quad\quad\quad\quad\quad$ (2.4.26)

(v) $\quad\quad\quad\quad [\mathbf{A}(\mathbf{B}\mathbf{C})]\mathbf{a} = \mathbf{A}[(\mathbf{B}\mathbf{C})\mathbf{a}] = \mathbf{A}[\mathbf{B}(\mathbf{C}\mathbf{a})] \quad\quad\quad$ (2.4.27)

2.4.8 QUOTIENT LAWS

If a_i and b_i are components of vectors **a** and **b** we have noted that $a_i b_i$ is a scalar, namely, $\mathbf{a} \cdot \mathbf{b}$, and that if a_{ij} are components of a tensor **A**, then $a_{ij} b_j$ are components of a vector, namely, **Ab**. Partial converses of these results are obtained in the following.

PROPERTY 7 (i) Let (a_i) be an ordered triplet related to the x_i system. For an arbitrary vector with components b_i, if $a_i b_i$ is a scalar, then a_i are components of a vector.

(ii) Let $[a_{ij}]$ be a 3×3 matrix related to the x_i system. For an arbitrary vector with components b_i, if $a_{ij} b_j$ are components of a vector, then a_{ij} are components of a tensor.

Proof (i) Since $a_i b_i$ is a scalar, we have $a_i b_i = a_i' b_i'$. Since b_i are components of a vector, we get $a_i b_i = a_i'(\alpha_{ip} b_p)$; that is,

$$(a_p - \alpha_{ip} a_i') b_p = 0 \quad\quad (2.4.28)$$

Since b_i are arbitrary, it follows from (2.4.28) that $a_p = \alpha_{ip} a_i'$. Thus a_i obey the transformation rule of a vector. Hence a_i are components of a vector.

(ii) Put $c_i = a_{ij} b_j$ in all coordinate systems. Then $c_i' = a_{ij}' b_j'$, or

$$\alpha_{ip} c_p = a_{ij}'(\alpha_{jq} b_q) \quad\quad (2.4.29)$$

because b_i and c_i are components of vectors, by data. Multiplying (2.4.29) throughout by α_{ir} and noting that $\alpha_{ip}\alpha_{ir} = \delta_{pr}$ and that $\delta_{pr} c_p = c_r = a_{rq} b_q$, we obtain

$$(a_{rq} - \alpha_{ir}\alpha_{jq} a_{ij}') b_q = 0 \quad\quad (2.4.30)$$

Since b_i are arbitrary, it follows from (2.4.30) that $a_{rq} = \alpha_{ir}\alpha_{jq} a_{ij}'$. Thus, a_{ij} obey the transformation rule of a second-order tensor. Hence a_{ij} are components of a second-order tensor. ∎

54 2 ALGEBRA OF TENSORS

Note: The two results proved above may be employed as tests to decide whether a given ordered triplet/(3×3) matrix represents a vector/second-order tensor. Analogous results may be established for systems of the form (a_{ijk}), (a_{ijkm}) and so on. Such results are all referred to as *quotient laws*. Some other useful quotient laws are contained in Examples 2.4.4, 2.4.5 and 2.4.7.

EXAMPLE 2.4.1 (i) Show that δ_{ij} are components of a second-order tensor.
(ii) Deduce that $\delta'_{ij} = \delta_{ij}$.

Solution (i) Since $\delta_{ij} = \mathbf{e}_i \cdot \mathbf{e}_j$, $[\delta_{ij}]$ may be treated as a 3×3 matrix related to the x_i system. Also, for any vector **b** with components b_i, we have $\delta_{ij} b_j = b_i$, which are components of a vector, namely, **b** itself. Hence, by a quotient law (Property 7(ii)) it follows that δ_{ij} are components of a tensor.
(ii) Consequently,

$$\delta'_{ij} = \alpha_{ip}\alpha_{jq}\delta_{pq} = \alpha_{ip}\alpha_{jp} = \delta_{ij} \quad \blacksquare \tag{2.4.31}$$

Note: The tensor whose components are δ_{ij} is referred to as the *unit tensor* or the *identity tensor*, denoted by **I**. Expression (2.4.31) shows that the components of **I** remain unaltered under all coordinate transformations. In other words, the unit matrix of order 3, namely, $[I] = [\delta_{ij}]$, is the matrix of **I** in all Cartesian coordinate systems.

EXAMPLE 2.4.2 Prove the following identities:

(i) $\quad\quad\quad\quad\quad\quad\quad\quad \mathbf{Ia} = \mathbf{a}$ (2.4.32)

(ii) $\quad\quad\quad\quad\quad\quad\quad \mathbf{IA} = \mathbf{AI} = \mathbf{A}$ (2.4.33)

(iii) $\quad\quad\quad\quad\quad\quad \mathbf{I} \cdot \mathbf{A} = \mathbf{A} \cdot \mathbf{I} = \mathrm{tr}\,\mathbf{A}$ (2.4.34)

(iv) $\quad\quad \mathbf{I} \cdot \mathbf{a} \otimes \mathbf{b} = \mathbf{a} \otimes \mathbf{b} \cdot \mathbf{I} = \mathrm{tr}(\mathbf{a} \otimes \mathbf{b}) = \mathbf{a} \cdot \mathbf{b}$ (2.4.35)

(v) $\quad\quad\quad\quad\quad\quad\quad \mathbf{e}_k \otimes \mathbf{e}_k = \mathbf{I}$ (2.4.36)

Solution (i) Since **I** is a tensor and **a** is a vector, **Ia** is a vector, by property 6. Since the components of **I** are δ_{ij}, expression (2.4.18) yields

$$[\mathbf{Ia}]_i = \delta_{ij} a_j = a_i = [\mathbf{a}]_i$$

This proves (2.4.32).

(ii) By property 6, we note that **IA** and **AI** are tensors. Also, expression (2.4.19) yields

$$[\mathbf{IA}]_{ij} = \delta_{ik} a_{kj} = a_{ij} = [\mathbf{A}]_{ij}$$
$$[\mathbf{AI}]_{ij} = a_{ik} \delta_{kj} = a_{ij} = [\mathbf{A}]_{ij}$$

This proves (2.4.33).

2.4 PROPERTIES OF TENSORS 55

(iii) By property 6, we note that $\mathbf{I} \cdot \mathbf{A}$ and $\mathbf{A} \cdot \mathbf{I}$ are scalars. Also, expression (2.4.20) yields

$$\mathbf{I} \cdot \mathbf{A} = \delta_{ij} a_{ij} = a_{ii} = a_{ij}\delta_{ij} = \mathbf{A} \cdot \mathbf{I}$$

Since $a_{ii} = \text{tr } \mathbf{A}$ (see Example 2.3.3), identity (2.4.34) is proved.

(iv) Taking $\mathbf{A} = \mathbf{a} \otimes \mathbf{b}$ in (2.4.34), we get

$$\mathbf{I} \cdot (\mathbf{a} \otimes \mathbf{b}) = (\mathbf{a} \otimes \mathbf{b}) \cdot \mathbf{I} = \text{tr}(\mathbf{a} \otimes \mathbf{b})$$

On the other hand, by using (2.4.20), we get

$$\mathbf{I} \cdot (\mathbf{a} \otimes \mathbf{b}) = \delta_{ij}(a_i b_j) = a_i b_i = \mathbf{a} \cdot \mathbf{b}$$

Thus, (2.4.35) is proved.

(v) We first note that

$$\mathbf{e}_k \otimes \mathbf{e}_k = \mathbf{e}_1 \otimes \mathbf{e}_1 + \mathbf{e}_2 \otimes \mathbf{e}_2 + \mathbf{e}_3 \otimes \mathbf{e}_3$$

Thus, $\mathbf{e}_k \otimes \mathbf{e}_k$ is a sum of tensor products of vectors. Therefore, it is a tensor. Since $[\mathbf{a}]_i = \mathbf{a} \cdot \mathbf{e}_i$ for any vector $\mathbf{a}$, we have

$$[\mathbf{e}_k]_i = \mathbf{e}_k \cdot \mathbf{e}_i = \delta_{ki}$$

Hence, by using expression (2.3.13), we obtain

$$[\mathbf{e}_k \otimes \mathbf{e}_k]_{ij} = [\mathbf{e}_k]_i [\mathbf{e}_k]_j = \delta_{ki}\delta_{kj} = \delta_{ij} = [\mathbf{I}]_{ij}$$

This proves (2.4.36). ∎

EXAMPLE 2.4.3 Prove the following identities:

(i) $\qquad (\mathbf{a} \otimes \mathbf{b})\mathbf{c} = (\mathbf{b} \cdot \mathbf{c})\mathbf{a}$ (2.4.37)

(ii) $\qquad (\mathbf{a} \otimes \mathbf{b} - \mathbf{b} \otimes \mathbf{a})\mathbf{c} = \mathbf{c} \times (\mathbf{a} \times \mathbf{b})$ (2.4.38)

(iii) $\qquad \mathbf{A}(\mathbf{a} \otimes \mathbf{b}) = (\mathbf{A}\mathbf{a}) \otimes \mathbf{b}$ (2.4.39)

(iv) $\qquad (\mathbf{a} \otimes \mathbf{b})(\mathbf{c} \otimes \mathbf{d}) = (\mathbf{b} \cdot \mathbf{c})(\mathbf{a} \otimes \mathbf{d})$ (2.4.40)

Solution (i) Taking $\mathbf{A} = \mathbf{a} \otimes \mathbf{b}$ in expression (2.4.18) and using (2.3.13), we get

$$[(\mathbf{a} \otimes \mathbf{b})\mathbf{c}]_i = [\mathbf{a} \otimes \mathbf{b}]_{ij}[\mathbf{c}]_j = a_i b_j c_j = (\mathbf{b} \cdot \mathbf{c})a_i = [(\mathbf{b} \cdot \mathbf{c})\mathbf{a}]_i$$

This proves (2.4.37).

(ii) Consequently,

$$(\mathbf{a} \otimes \mathbf{b} - \mathbf{b} \otimes \mathbf{a})\mathbf{c} = (\mathbf{a} \otimes \mathbf{b})\mathbf{c} - (\mathbf{b} \otimes \mathbf{a})\mathbf{c}$$
$$= (\mathbf{b} \cdot \mathbf{c})\mathbf{a} - (\mathbf{a} \cdot \mathbf{c})\mathbf{b} = \mathbf{c} \times (\mathbf{a} \times \mathbf{b})$$

which is (2.4.38).

(iii) Taking $\mathbf{B} = \mathbf{a} \otimes \mathbf{b}$ in (2.4.19) and using (2.3.13) and (2.4.18), we get

$$[\mathbf{A}(\mathbf{a} \otimes \mathbf{b})]_{ij} = [\mathbf{A}]_{ik}[\mathbf{a} \otimes \mathbf{b}]_{kj} = a_{ik}a_k b_j$$
$$= [\mathbf{Aa}]_i [\mathbf{b}]_j = [(\mathbf{Aa}) \otimes \mathbf{b}]_{ij}$$

This proves (2.4.39).

(iv) Taking $\mathbf{A} = \mathbf{a} \otimes \mathbf{b}$ and $\mathbf{B} = \mathbf{c} \otimes \mathbf{d}$ in (2.4.19) and using (2.3.13), we get

$$[(\mathbf{a} \otimes \mathbf{b})(\mathbf{c} \otimes \mathbf{d})]_{ij} = [\mathbf{a} \otimes \mathbf{b}]_{ik}[\mathbf{c} \otimes \mathbf{d}]_{kj} = a_i b_k c_k d_j$$
$$= (\mathbf{b} \cdot \mathbf{c})a_i d_j = (\mathbf{b} \cdot \mathbf{c})[\mathbf{a} \otimes \mathbf{d}]_{ij}$$

This proves (2.4.40). ∎

EXAMPLE 2.4.4 Let $[a_{ij}]$ be a 3×3 matrix related to the x_i system. For two arbitrary vectors with components b_i and c_i, if $a_{ij}b_i c_j$ is a scalar, show that a_{ij} are components of a second-order tensor.

Solution Since $a_{ij}b_i c_j$ is a scalar, we have

$$a_{ij}b_i c_j = a'_{ij}b'_i c'_j \qquad (2.4.41)$$

Since b_i and c_i are vector components, (2.4.41) yields, on noting that $a_{ij}b_i c_j = a_{pq}b_p c_q$,

$$(a_{pq} - \alpha_{ip}\alpha_{jq}a'_{ij})b_p c_q = 0$$

Consequently, since b_i and c_i are arbitrary, we should have

$$a_{pq} = \alpha_{ip}\alpha_{jq}a'_{ij}$$

Hence a_{ij} are components of a second-order tensor. ∎

EXAMPLE 2.4.5 (i) Let (a_{ijk}) be a system of $3^3 = 27$ numbers related to the x_i axes. For arbitrary vectors with components b_i, c_i and d_i, if $a_{ijk}b_i c_j d_k$ is a scalar, show that a_{ijk} are components of a third-order tensor.

(ii) Show that ε_{ijk} are components of a third-order tensor.

(iii) Deduce that $\varepsilon'_{ijk} = \varepsilon_{ijk}$.

Solution (i) Since $a_{ijk}b_i c_j d_k$ is a scalar, we have

$$a_{ijk}b_i c_j d_k = a'_{ijk}b'_i c'_j d'_k \qquad (2.4.42)$$

Since b_i, c_i and d_i are components of vectors, (2.4.42) yields, on noting that

$$a_{ijk}b_i c_j d_k = a_{pqr}b_p c_q d_r$$

the following expression:

$$(a_{pqr} - \alpha_{ip}\alpha_{jq}\alpha_{kr}a'_{ijk})b_p c_q d_r = 0$$

Consequently, since b_i, c_i and d_i are arbitrary, we should have

$$a_{pqr} = \alpha_{ip}\alpha_{jq}\alpha_{kr}a'_{ijk}$$

2.4 PROPERTIES OF TENSORS

This transformation rule shows that a_{ijk} are components of a third-order tensor.

(ii) Note that (ε_{ijk}) is a system of 27 numbers. Since $\varepsilon_{ijk} = [\mathbf{e}_i, \mathbf{e}_j, \mathbf{e}_k]$, the system (ε_{ijk}) may be treated as one related to the x_i axes. Also, for any three vectors, $\mathbf{b}, \mathbf{c}, \mathbf{d}$, $[\mathbf{b}, \mathbf{c}, \mathbf{d}] = \varepsilon_{ijk} b_i c_j d_k$ is a scalar representing the volume of the parallelopiped determined by the vectors $\mathbf{b}, \mathbf{c}, \mathbf{d}$. Hence, from the result proven in (i), it follows that ε_{ijk} are components of a third-order tensor.

(iii) Consequently,

$$\varepsilon'_{ijk} = \alpha_{ip}\alpha_{jq}\alpha_{kr}\varepsilon_{pqr}$$

$$= \begin{vmatrix} \alpha_{i1} & \alpha_{i2} & \alpha_{i3} \\ \alpha_{j1} & \alpha_{j2} & \alpha_{j3} \\ \alpha_{k1} & \alpha_{k2} & \alpha_{k3} \end{vmatrix}$$

see (1.7.17);

$$= [\mathbf{e}'_i, \mathbf{e}'_j, \mathbf{e}'_k]$$

see (2.2.16). By use of expression (2.2.4), we now get

$$\varepsilon'_{ijk} = \varepsilon_{ijk} \quad \blacksquare \qquad (2.4.43)$$

Note: The tensor whose components are ε_{ijk} is referred to as the *permutation tensor* or the *alternating tensor*. The result proven in (iii) means that, like the components of the identity tensor, the components of the permutation tensor also remain unaltered under all coordinate transformations.

EXAMPLE 2.4.6 By using a transformation law for a vector, show that the vector product $\mathbf{a} \times \mathbf{b}$ of two vectors $\mathbf{a}$ and $\mathbf{b}$ is a vector.

Solution We recall that

$$[\mathbf{a} \times \mathbf{b}]_i = \varepsilon_{ijk} a_j b_k$$

and set $c_i = \varepsilon_{ijk} a_j b_k$ in all coordinate systems. Then

$$c'_i = \varepsilon'_{ijk} a'_j b'_k$$

By using the fact that ε_{ijk} are components of a tensor and a_i, b_i are components of vectors, this becomes

$$c'_i = (\alpha_{ip}\alpha_{jq}\alpha_{kr}\varepsilon_{pqr})(\alpha_{jm}a_m)(\alpha_{kn}b_n)$$

$$= \alpha_{ip}\delta_{qm}\delta_{rn}\varepsilon_{pqr}a_m b_n = \alpha_{ip}\varepsilon_{pmn}a_m b_n$$

$$= \alpha_{ip} c_p$$

This shows that $c_i = \varepsilon_{ijk} a_j b_k$ are components of a vector; in other words $\mathbf{a} \times \mathbf{b}$ is a vector. $\blacksquare$

58 2 ALGEBRA OF TENSORS

EXAMPLE 2.4.7 (i) If c_{ijkm} and a_{ij} are components of tensors of orders 4 and 2, respectively, show that $c_{ijkm}a_{km}$ are components of a second-order tensor.

(ii) Let (c_{ijkm}) be a system of 81 numbers related to the x_i axes. For an arbitrary second-order tensor with components a_{ij} if $c_{ijkm}a_{km}$ are components of a second-order tensor, show that c_{ijkm} are components of a fourth-order tensor.

Solution (i) Put $d_{ij} = c_{ijkm}a_{km}$ in all coordinate systems. Then $d'_{ij} = c'_{ijkm}a'_{km}$. Since c_{ijkm} and a_{ij} are tensor components, this relation yields

$$d'_{ij} = (\alpha_{ip}\alpha_{jq}\alpha_{kr}\alpha_{ms}c_{pqrs})(\alpha_{kh}\alpha_{mn}a_{hn})$$
$$= \alpha_{ip}\alpha_{jq}\delta_{rh}\delta_{sn}c_{pqrs}a_{hn}$$
$$= \alpha_{ip}\alpha_{jq}c_{pqrs}a_{rs}$$
$$= \alpha_{ip}\alpha_{jq}d_{pq}$$

This transformation rule shows that $d_{ij} = c_{ijkm}a_{km}$ are components of a second-order tensor.

(ii) Put $d_{ij} = c_{ijkm}a_{km}$ in all coordinate systems so that $d'_{ij} = c'_{ijkm}a'_{km}$ as previously. But, now, a_{ij} and d_{ij} are tensor components. As such,

$$\alpha_{ip}\alpha_{jq}d_{pq} = c'_{ijkm}\alpha_{kr}\alpha_{ms}a_{rs} \tag{2.4.44}$$

Multiplying (2.4.44) throughout by $\alpha_{ih}\alpha_{jn}$ and noting that $\alpha_{ih}\alpha_{ip} = \delta_{hp}$, $\alpha_{jn}\alpha_{jq} = \delta_{nq}$ and $\delta_{hp}\delta_{nq}d_{pq} = d_{hn}$, we get

$$d_{hn} = \alpha_{ih}\alpha_{jn}\alpha_{kr}\alpha_{ms}c'_{ijkm}a_{rs} \tag{2.4.45}$$

Since $d_{hn} = c_{hnrs}a_{rs}$, we get from (2.4.45),

$$(c_{hnrs} - \alpha_{ih}\alpha_{jn}\alpha_{kr}\alpha_{ms}c'_{ijkm})a_{rs} = 0 \tag{2.4.46}$$

Since a_{rs} are arbitrary, it follows from (2.4.46) that

$$c_{hnrs} = \alpha_{ih}\alpha_{jn}\alpha_{kr}\alpha_{ms}c'_{ijkm}$$

This transformation rule shows that c_{ijkm} are indeed components of a fourth-order tensor. ∎

Note: The result proved in (ii) is a quotient law for a fourth-order tensor.

2.5
ISOTROPIC TENSORS

A tensor whose components remain unchanged under all coordinate transformations is called an *isotropic tensor*. In other words, a tensor

2.5 ISOTROPIC TENSORS

$\overset{N}{\mathbf{A}}$ having components $a_{ijk...}$ in the x_i system and $a'_{ijk...}$ in x'_i system is isotropic if and only if

$$a'_{ijk...} = a_{ijk...} \tag{2.5.1}$$

for all choices of the x_i and x'_i systems.

We have seen that the identity tensor and the permutation tensor possess this property; these are therefore examples of isotropic tensors. The definition of a scalar itself reveals that a scalar is an isotropic tensor. The geometrical meaning of components of a nonzero vector suggests that the components *do* change under coordinate transformations; as such, a nonzero vector cannot be an isotropic tensor. On the other hand, zero tensors of all orders are isotropic.

It is easy to verify that a scalar multiple of an isotropic tensor is an isotropic tensor and that the sum and the difference of two isotropic tensors is an isotropic tensor. Since the identity tensor **I** is isotropic, it follows that every scalar multiple of **I** is an isotropic tensor of order 2. The converse of this result, namely, that every second-order isotropic tensor is a scalar multiple of **I**, is also true. The following theorem establishes this fact.

THEOREM 2.5.1 If a_{ij} are components of an isotropic tensor (of second order), then $a_{ij} = \alpha \delta_{ij}$ for some scalar α.

Proof Since the given tensor is isotropic, we have $a'_{ij} = a_{ij}$ for *all* choices of the x'_i system.

We first consider the coordinate transformation defined by

$$\mathbf{e}'_1 = \mathbf{e}_2, \quad \mathbf{e}'_2 = \mathbf{e}_3, \quad \mathbf{e}'_3 = \mathbf{e}_1 \tag{2.5.2}$$

For this transformation we have

$$\alpha_{12} = \alpha_{23} = \alpha_{31} = 1; \quad \text{all other } \alpha_{ij} = 0 \tag{2.5.3}$$

see Figure 2.3 and Table 2.5.

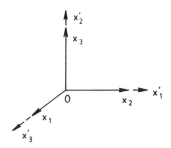

Figure 2.3. Theorem 2.5.1.

2 ALGEBRA OF TENSORS

Table 2.5. Theorem 2.5.1

·	e_1	e_2	e_3
e'_1	0	1	0
e'_2	0	0	1
e'_3	1	0	0

Using (2.5.3), the transformation rule of a second-order tensor together with the fact that $a'_{ij} = a_{ij}$ gives

$$a_{11} = a'_{11} = \alpha_{1p}\alpha_{1q}a_{pq} = a_{22} \qquad (2.5.4)$$

Similarly, we find that

$$a_{22} = a_{33} = a_{11} \qquad (2.5.5)$$

$$a_{12} = a_{23} = a_{31}; \qquad a_{21} = a_{32} = a_{13} \qquad (2.5.6)$$

Next, consider the coordinate transformation defined by

$$e'_1 = e_2, \qquad e'_2 = -e_1, \qquad e'_3 = e_3 \qquad (2.5.7)$$

This transformation corresponds to the rotation of the x_i axes about the x_3 axis through an angle of 90°. Hence

$$\alpha_{12} = \alpha_{33} = 1, \qquad \alpha_{21} = -1; \qquad \text{all other } \alpha_{ij} = 0 \qquad (2.5.8)$$

see Figure 2.2 and Table 2.2 with $\theta = 90°$.

Using (2.5.8), the transformation rule of a second-order tensor together with the fact that $a'_{ij} = a_{ij}$ gives

$$a_{13} = a'_{13} = \alpha_{1p}\alpha_{3q}a_{pq} = a_{23} \qquad (2.5.9)$$

$$a_{23} = a'_{23} = \alpha_{2p}\alpha_{3q}a_{pq} = -a_{13} \qquad (2.5.10)$$

These equations show that

$$a_{13} = a_{23} = 0 \qquad (2.5.11)$$

Equations (2.5.6) now yield

$$a_{12} = a_{21} = a_{31} = a_{32} = 0 \qquad (2.5.12)$$

Expressions (2.5.11) and (2.5.12) show that all the nondiagonal terms of the matrix $[a_{ij}]$ are 0. Also, the diagonal terms are equal, by (2.5.5). If we set these diagonal terms equal to α, say, then we have $a_{ij} = \alpha\delta_{ij}$. This completes the proof. ∎

Note: Since the permutation tensor is isotropic, every scalar multiple of this tensor is an isotropic tensor of order 3. The converse of this result, namely, that if a_{ijk} are components of an isotropic tensor of third order, then $a_{ijk} = \alpha \varepsilon_{ijk}$ for some scalar α is also true. The proof is analogous to that of Theorem 2.5.1.

2.6
ISOTROPIC TENSOR OF ORDER 4

Since δ_{ij} are components of an isotropic tensor, it is easy to verify that the fourth-order tensors whose components are $\delta_{ij}\delta_{km}$, $\delta_{ik}\delta_{jm}$ and $\delta_{im}\delta_{jk}$ are also isotropic. Consequently, it follows that if

$$a_{ijkm} = \alpha\delta_{ij}\delta_{km} + \beta\delta_{ik}\delta_{jm} + \gamma\delta_{im}\delta_{jk}$$

where α, β, γ are scalars, then a_{ijkm} are components of an isotropic tensor. The converse of this result is also true; it is proved in the following theorem.

THEOREM 2.6.1 If a_{ijkm} are components of a fourth-order isotropic tensor, then

$$a_{ijkm} = \alpha\delta_{ij}\delta_{km} + \beta\delta_{ik}\delta_{jm} + \gamma\delta_{im}\delta_{jk} \tag{2.6.1}$$

for some scalars α, β, γ.

Proof If the given tensor is the zero tensor, the proof is trivial. We therefore suppose that at least 1 of the 81 components a_{ijkm} is *not* 0.

Since the proof is lengthy, we will give it in four stages. In the first stage, we will classify the 81 components of the tensor into five different classes (classes I to V) depending upon the nature of the suffixes present in the components. In the second stage of the proof, we will consider some particular coordinate transformations and show that the components belonging to each of the classes I–IV are equal among themselves. In the third stage, we will show, again by considering some particular transformations, that all the components belonging to class V are identically 0. In the last stage of the proof, we deduce the representation (2.6.1). The arguments throughout essentially depend on the fact that $a'_{ijkm} = a_{ijkm}$ for all coordinate transformations, whatsoever.

First Stage of the Proof. The 81 components a_{ijkm} may be classified into the following five classes:

Class I. Components in which all suffixes are the same: $a_{1111}, a_{2222}, a_{3333}$.

62   2 ALGEBRA OF TENSORS

Class II. Components in which the first two suffixes are the same and the last two suffixes are the same, second and third suffixes being different: $a_{1122}, a_{2233}, a_{3311}, a_{2211}, a_{3322}, a_{1133}$.

Class III. Components in which the first and the third suffixes are the same, and the second and fourth suffixes are the same, first and second suffixes being different: $a_{1212}, a_{2323}, a_{3131}, a_{2121}, a_{3232}, a_{1313}$.

Class IV. Components in which the first and the fourth suffixes are the same, and the second and third suffixes are same, first and the second suffixes being different: $a_{1221}, a_{2332}, a_{3113}, a_{2112}, a_{3223}, a_{1331}$.

Class V. The remaining 60 components (in which one of the suffixes is different from all the other three).

Second Stage of the Proof. After this classification, we now proceed to consider some particular transformations. The first of these transformations is

$$\mathbf{e}'_1 = \mathbf{e}_2, \qquad \mathbf{e}'_2 = \mathbf{e}_3, \qquad \mathbf{e}'_3 = \mathbf{e}_1 \qquad (2.6.2)$$

For this transformation we have, from (2.5.3),

$$\alpha_{12} = \alpha_{23} = \alpha_{31} = 1; \quad \text{other } \alpha_{ij} = 0 \qquad (2.6.3)$$

The transformation rule for a fourth-order tensor together with the fact that $a'_{ijkm} = a_{ijkm}$ yields, on using (2.6.3), the following

(i) $\qquad a_{1111} = a'_{1111} = \alpha_{1p}\alpha_{1q}\alpha_{1r}\alpha_{1s}a_{pqrs} = a_{2222} \qquad (2.6.4)$

Similarly,

$$a_{2222} = a_{3333} \qquad (2.6.5)$$

(ii) $\qquad a_{1122} = a'_{1122} = \alpha_{1p}\alpha_{1q}\alpha_{2r}\alpha_{2s}a_{pqrs} = a_{2233} \qquad (2.6.6)$

Similarly,

$$a_{2233} = a_{3311} \qquad (2.6.7)$$

(iii) $\qquad a_{2211} = a'_{2211} = \alpha_{2p}\alpha_{2q}\alpha_{1r}\alpha_{1s}a_{pqrs} = a_{3322} \qquad (2.6.8)$

Similarly,

$$a_{3322} = a_{1133} \qquad (2.6.9)$$

(iv) $\qquad a_{1212} = a'_{1212} = \alpha_{1p}\alpha_{2q}\alpha_{1r}\alpha_{2s}a_{pqrs} = a_{2323} \qquad (2.6.10)$

Similarly,

$$a_{2323} = a_{3131} \qquad (2.6.11)$$

(v) $\qquad a_{2121} = a'_{2121} = \alpha_{2p}\alpha_{1q}\alpha_{2r}\alpha_{1s}a_{pqrs} = a_{3232} \qquad (2.6.12)$

Similarly,

$$a_{3232} = a_{1313} \qquad (2.6.13)$$

2.6 ISOTROPIC TENSOR OF ORDER 4

(vi)
$$a_{1221} = a'_{1221} = \alpha_{1p}\alpha_{2q}\alpha_{2r}\alpha_{1s}a_{pqrs} = a_{2332} \tag{2.6.14}$$

Similarly,
$$a_{2332} = a_{3113} \tag{2.6.15}$$

(vii)
$$a_{2112} = a'_{2112} = \alpha_{2p}\alpha_{1q}\alpha_{1r}\alpha_{2s}a_{pqrs} = a_{3223} \tag{2.6.16}$$

Similarly,
$$a_{3223} = a_{1331} \tag{2.6.17}$$

The next coordinate transformation which we consider is

$$\mathbf{e}'_1 = -\mathbf{e}_3, \quad \mathbf{e}'_2 = -\mathbf{e}_2, \quad \mathbf{e}'_3 = -\mathbf{e}_1 \tag{2.6.18}$$

For this transformation, we have (see Figure 2.4 and Table 2.6)

$$\alpha_{13} = \alpha_{22} = \alpha_{31} = -1, \quad \text{other } \alpha_{ij} = 0 \tag{2.6.19}$$

Using (2.6.19), the transformation rule for a fourth-order tensor together with the fact that $a_{ijkm} = a'_{ijkm}$ yields

$$a_{1122} = a'_{1122} = \alpha_{1p}\alpha_{1q}\alpha_{2r}\alpha_{2s}a_{pqrs} = a_{3322} \tag{2.6.20}$$

Similarly,
$$a_{1212} = a_{3232} \tag{2.6.21}$$

$$a_{1221} = a_{3223} \tag{2.6.22}$$

Relations (2.6.4) and (2.6.5) show that all the components belonging to class I are equal; that is,

$$a_{1111} = a_{2222} = a_{3333} = \eta, \text{ say} \tag{2.6.23}$$

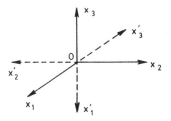

Figure 2.4. Theorem 2.6.1, Stage 2.

Table 2.6. Theorem 2.6.1, Stage 2

.	$\mathbf{e}_1$	$\mathbf{e}_2$	$\mathbf{e}_3$
$\mathbf{e}'_1$	0	0	−1
$\mathbf{e}'_2$	0	−1	0
$\mathbf{e}'_3$	−1	0	0

64 2 ALGEBRA OF TENSORS

Relations (2.6.6) to (2.6.9) and (2.6.20) show that all the components belonging to class II are equal; that is,

$$a_{1122} = a_{2233} = a_{3311} = a_{2211} = a_{3322} = a_{1133} = \alpha, \text{ say} \quad (2.6.24)$$

Relations (2.6.11) to (2.6.13) and (2.6.21) show that all the components belonging to class III are equal; that is,

$$a_{1212} = a_{2323} = a_{3131} = a_{3232} = a_{1313} = a_{2121} = \beta, \text{ say} \quad (2.6.25)$$

Relations (2.6.14) to (2.6.17) and (2.6.22) show that all the components belonging to class IV are equal; that is,

$$a_{1221} = a_{2332} = a_{3113} = a_{2112} = a_{3223} = a_{1331} = \gamma, \text{ say} \quad (2.6.26)$$

From their definitions it is evident that η, α, β and γ have the same values in all coordinate systems.

Third Stage of the Proof. We now attend to the components belonging to class V, in which one of the suffixes is different from the other three. We first take up the components of the form a_{1ijk} or a_{i1jk} or a_{ij1k} or a_{ijk1}, where i, j, k are different from 1. Let us consider the transformation defined by

$$\mathbf{e}'_1 = \mathbf{e}_1, \quad \mathbf{e}'_2 = -\mathbf{e}_2, \quad \mathbf{e}'_3 = -\mathbf{e}_3 \quad (2.6.27)$$

For this transformation, we have (see Figure 2.5 and Table 2.7)

$$\alpha_{11} = \alpha_{33} = 1, \quad \alpha_{22} = -1; \quad \text{other } \alpha_{ij} = 0 \quad (2.6.28)$$

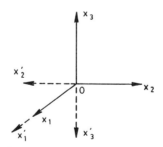

Figure 2.5. Theorem 2.6.1, Stage 3.

Table 2.7. Theorem 2.6.1, Stage 3

.	$\mathbf{e}_1$	$\mathbf{e}_2$	$\mathbf{e}_3$
$\mathbf{e}'_1$	1	0	0
$\mathbf{e}'_2$	0	−1	0
$\mathbf{e}'_3$	0	0	−1

2.6 ISOTROPIC TENSOR OF ORDER 4

On using (2.6.28), the transformation rule for a fourth-order tensor and the fact that $a'_{ijkm} = a_{ijkm}$ yield

$$a_{1ijk} = a'_{1ijk} = \alpha_{1p}\alpha_{iq}\alpha_{jr}\alpha_{ks}a_{pqrs}$$
$$= \alpha_{iq}\alpha_{jr}\alpha_{ks}a_{1qrs} = -a_{1ijk}, \quad \text{for } i,j,k \neq 1$$

Thus
$$a_{1ijk} = 0, \quad \text{for } i,j,k \neq 1 \quad (2.6.29)$$

Similarly,
$$a_{i1jk} = a_{ij1k} = a_{ijk1} = 0, \quad \text{for } i,j,k \neq 1 \quad (2.6.30)$$

By considering the transformations

$$\mathbf{e}'_1 = -\mathbf{e}_1, \quad \mathbf{e}'_2 = \mathbf{e}_2, \quad \mathbf{e}'_3 = -\mathbf{e}_3 \quad (2.6.31)$$

$$\mathbf{e}'_1 = -\mathbf{e}_1, \quad \mathbf{e}'_2 = -\mathbf{e}_2, \quad \mathbf{e}'_3 = \mathbf{e}_3 \quad (2.6.32)$$

we may show that

$$a_{2ijk} = a_{i2jk} = a_{ij2k} = a_{ijk2} = 0, \quad \text{for } i,j,k \neq 2 \quad (2.6.33)$$

$$a_{3ijk} = a_{i3jk} = a_{ij3k} = a_{ijk3} = 0, \quad \text{for } i,j,k \neq 3 \quad (2.6.34)$$

Thus, all components belonging to class V are identically 0. Consequently, at least one of the components belonging to classes I to IV is nonzero; that is, at least one of $\eta, \alpha, \beta, \gamma$ defined by (2.6.23) through (2.6.26) should be nonzero.

Fourth Stage of the Proof. We now introduce the numbers c_{ijkm} defined as

$$c_{ijkm} = a_{ijkm} - (\alpha\delta_{ij}\delta_{km} + \beta\delta_{ik}\delta_{jm} + \gamma\delta_{im}\delta_{jk}) \quad (2.6.35)$$

Obviously, c_{ijkm} form the components of a fourth-order isotropic tensor. Using relations (2.6.23) through (2.6.26) and the fact that all a_{ijkm} belonging to class V are identically 0, it is easy to verify from (2.6.35) that

$$c_{1111} = c_{2222} = c_{3333} = \eta - (\alpha + \beta + \gamma); \quad \text{other } c_{ijkm} = 0 \quad (2.6.36)$$

Finally, we consider the transformation given by

$$\mathbf{e}'_1 = \frac{1}{\sqrt{2}}(\mathbf{e}_1 + \mathbf{e}_2), \quad \mathbf{e}'_2 = \frac{1}{\sqrt{2}}(\mathbf{e}_2 - \mathbf{e}_1), \quad \mathbf{e}'_3 = \mathbf{e}_3 \quad (2.6.37)$$

This transformation corresponds to the rotation of the x_i axes about the x_3 axis through an angle of 45°, and we have (see Figure 2.2 and Table 2.2 with $\theta = 45°$)

$$\alpha_{11} = \alpha_{12} = -\alpha_{21} = \alpha_{22} = \frac{1}{\sqrt{2}}, \quad \text{other } \alpha_{ij} = 0 \quad (2.6.38)$$

66 2 ALGEBRA OF TENSORS

The transformation rule for a fourth-order tensor and the fact that $c_{ijkm} = c'_{ijkm}$ yield

$$c_{1111} = c'_{1111} = \alpha_{1p}\alpha_{1q}\alpha_{1r}\alpha_{1s}c_{pqrs}$$

$$= (\alpha_{11}^4 + \alpha_{12}^4)c_{1111}$$

by using (2.6.36);

$$= \tfrac{1}{2}c_{1111}$$

by using (2.6.38). Therefore $c_{1111} = 0$. Consequently, (2.6.36) yields $c_{2222} = c_{3333} = 0$. Thus *all* the 81 numbers c_{ijkm} defined by (2.6.35) vanish identically, and we get the representation (2.6.1). The proof of the theorem is complete. ∎

It may be mentioned that the representation (2.6.1) is useful in constitutive modeling; see, for example, sections 9.2 and 10.10.

EXAMPLE 2.6.1 The components a_{ij} and b_{ij} of two tensors **A** and **B** are related through the expressions $a_{ij} = c_{ijkm}b_{km}$. If c_{ijkm} are components of an isotropic tensor of order 4 and $b_{ij} = b_{ji}$, show that

$$a_{ij} = \lambda b_{kk}\delta_{ij} + 2\mu b_{ij} \qquad (2.6.39)$$

for some scalars λ and μ.

Solution If c_{ijkm} are components of an isotropic tensor of order 4, $a_{ij} = c_{ijkm}b_{km}$ becomes, on using (2.6.1)

$$a_{ij} = (\alpha\delta_{ij}\delta_{km} + \beta\delta_{ik}\delta_{jm} + \gamma\delta_{im}\delta_{jk})b_{km}$$

$$= \alpha\delta_{ij}\delta_{kk} + \beta b_{ij} + \gamma b_{ji} \qquad (2.6.40)$$

If $b_{ij} = b_{ji}$, (2.6.40) reduces to (2.6.39) on setting $\alpha = \lambda$ and $\tfrac{1}{2}(\beta + \gamma) = \mu$. ∎

2.7
TENSORS AS LINEAR OPERATORS

If a_{ij} are components of a second-order tensor **A** and c_i are components of a vector **c**, it has been shown that $a_{ij}c_j$ are components of a vector, denoted by **Ac** (see Section 2.4, property 6). Suppose we redesignate this vector as **d**; that is we set

$$\mathbf{Ac} = \mathbf{d} \qquad (2.7.1)$$

For a given **A**, the equation (2.7.1) defines a transformation that transforms the vector **c** to the vector **d**, with **A** effecting the transformation operation. The tensor **A** may therefore be thought of as an operator that acts on a vector **c** and generates a vector **d** = **Ac**.

For example, the identity (2.4.32), namely **Ia** = **a**, shows that the tensor **I** operating on **a** generates **a** itself. This explains why **I** is called the *identity tensor*.

We now prove a theorem that demonstrates a characteristic property of second order tensors.

THEOREM 2.7.1 If **A** is a second-order tensor, then it is a linear operator on vectors and its components are given by

$$a_{ij} = \mathbf{e}_i \cdot \mathbf{A}\mathbf{e}_j \qquad (2.7.2)$$

Conversely, if **A** is a linear operator on vectors and a_{ij} are defined by (2.7.2), then a_{ij} are components of a second-order tensor.

Proof First suppose that **A** is a second-order tensor with components a_{ij}. Consider any two scalars α, β and any two vectors **b** and **c**. Then **Ab**, **Ac** and **A**(α**b** + β**c**) are vectors. Also,

$$[\mathbf{A}(\alpha\mathbf{b} + \beta\mathbf{c})]_j = a_{ij}[\alpha\mathbf{b} + \beta\mathbf{c}]_j = a_{ij}(\alpha b_j + \beta c_j)$$

$$= \alpha(a_{ij}b_j) + \beta(a_{ij}c_j) = \alpha[\mathbf{A}\mathbf{b}]_i + \beta[\mathbf{A}\mathbf{c}]_i$$

$$= [\alpha(\mathbf{A}\mathbf{b}) + \beta(\mathbf{A}\mathbf{c})]_i$$

so that

$$\mathbf{A}(\alpha\mathbf{b} + \beta\mathbf{c}) = \alpha(\mathbf{A}\mathbf{b}) + \beta(\mathbf{A}\mathbf{c}) \qquad (2.7.3)$$

This expression shows that **A** is a *linear operator* on vectors.

Further,

$$a_{ij}b_j = [\mathbf{A}\mathbf{b}]_i = \mathbf{e}_i \cdot \mathbf{A}\mathbf{b}$$

$$= \mathbf{e}_i \cdot \mathbf{A}(b_j\mathbf{e}_j) = \mathbf{e}_i \cdot b_j(\mathbf{A}\mathbf{e}_j)$$

by property (2.7.3);

$$= (\mathbf{e}_i \cdot \mathbf{A}\mathbf{e}_j)b_j \qquad (2.7.4)$$

Since this is true for any vector **b**, expression (2.7.2) is proven.

Conversely, suppose **A** is a linear operator on vectors and a_{ij} are defined by (2.7.2). From (2.7.2) we note that a_{ij} are defined with respect to the x_i system and are uniquely determined by **A** with respect to this system. Let a'_{ij} be the corresponding numbers determined by **A** and defined with respect to the x'_i system; that is,

$$a'_{ij} = \mathbf{e}'_i \cdot \mathbf{A}\mathbf{e}'_j \qquad (2.7.5)$$

68 2 ALGEBRA OF TENSORS

Since $\mathbf{e}'_i = (\mathbf{e}'_i \cdot \mathbf{e}_p)\mathbf{e}_p = \alpha_{ip}\mathbf{e}_p$, (2.7.5) yields

$$a'_{ij} = (\alpha_{ip}\mathbf{e}_p) \cdot \mathbf{A}(\alpha_{jq}\mathbf{e}_q)$$
$$= (\alpha_{ip}\mathbf{e}_p) \cdot (\alpha_{jq}\mathbf{A}\mathbf{e}_q)$$

because $\mathbf{A}$ is a linear operator;

$$= \alpha_{ip}\alpha_{jq}(\mathbf{e}_p \cdot \mathbf{A}\mathbf{e}_q) = \alpha_{ip}\alpha_{jq}a_{pq}$$

by (2.7.2). This transformation rule shows that a_{ij} are components of a second-order tensor. The proof of the theorem is complete. ∎

By virtue of this theorem, the terms *linear operator on a vector* and *second-order tensor* are treated as synonyms.

EXAMPLE 2.7.1 A transformation A transforms every vector into a fixed vector. Show that A is not a tensor.

Solution Let $\mathbf{a}$ be the vector into which all vectors are transformed under A. Then, for any scalar α and any vector $\mathbf{c}$, we have $A\mathbf{c} = \mathbf{a}$ and $A(\alpha\mathbf{c}) = \mathbf{a}$, so that $A(\alpha\mathbf{c}) = A\mathbf{c}$. Since α is arbitrary, $A\mathbf{c} \neq \alpha A\mathbf{c}$, in general. Hence $A(\alpha\mathbf{c}) \neq \alpha A\mathbf{c}$, in general. As such, A is not a linear operator and hence not a tensor. ∎

EXAMPLE 2.7.2 Show that two tensors $\mathbf{A}$ and $\mathbf{B}$ are equal if and only if (i) $\mathbf{A}\mathbf{e}_k = \mathbf{B}\mathbf{e}_k$, or (ii) $\mathbf{A}\mathbf{a} = \mathbf{B}\mathbf{a}$ for any vector $\mathbf{a}$.

Solution (i) Let a_{ij} and b_{ij} be components of $\mathbf{A}$ and $\mathbf{B}$, respectively. We note that

$$[\mathbf{A}\mathbf{e}_k]_i = a_{ij}[\mathbf{e}_k]_j = a_{ij}(\mathbf{e}_k \cdot \mathbf{e}_j) = a_{ij}\delta_{kj} = a_{ik}$$

Similarly, $[\mathbf{B}\mathbf{e}_k]_i = b_{ik}$. Hence the condition $a_{ik} = b_{ik}$ holds if and only if $[\mathbf{A}\mathbf{e}_k]_i = [\mathbf{B}\mathbf{e}_k]_i$. That is, $\mathbf{A} = \mathbf{B}$ if and only if $\mathbf{A}\mathbf{e}_k = \mathbf{B}\mathbf{e}_k$.

(ii) Let a_i be components of $\mathbf{a}$. Then, since $\mathbf{A}$ and $\mathbf{B}$ are linear operators, we have

$$\mathbf{A}\mathbf{a} = \mathbf{A}(a_k\mathbf{e}_k) = a_k(\mathbf{A}\mathbf{e}_k)$$
$$\mathbf{B}\mathbf{a} = \mathbf{B}(a_k\mathbf{e}_k) = a_k(\mathbf{B}\mathbf{e}_k)$$

Suppose that $\mathbf{A}\mathbf{a} = \mathbf{B}\mathbf{a}$. Then, $a_k(\mathbf{A}\mathbf{e}_k) = a_k(\mathbf{B}\mathbf{e}_k)$. Since $\mathbf{a}$ is arbitrary, it follows that $\mathbf{A}\mathbf{e}_k = \mathbf{B}\mathbf{e}_k$. Hence $\mathbf{A} = \mathbf{B}$ by what has been proven already. The converse is trivially true. ∎

EXAMPLE 2.7.3 Show that every tensor **A** with components a_{ij} can be represented in the form

$$\mathbf{A} = a_{ij}\mathbf{e}_i \otimes \mathbf{e}_j \tag{2.7.6}$$

Solution Let **c** be any vector with components c_i, Then

$$\mathbf{Ac} = \mathbf{A}(c_j\mathbf{e}_j) = c_j(\mathbf{Ae}_j)$$

because **A** is a linear operator;

$$= c_j(\mathbf{e}_i \cdot \mathbf{Ae}_j)\mathbf{e}_i$$

because any vector $\mathbf{a} = (\mathbf{e}_i \cdot \mathbf{a})\mathbf{e}_i$;

$$= a_{ij}c_j\mathbf{e}_i$$

using (2.7.2);

$$= a_{ij}c_k\delta_{kj}\mathbf{e}_i = a_{ij}c_k(\mathbf{e}_k \cdot \mathbf{e}_j)\mathbf{e}_i$$

$$= a_{ij}c_k(\mathbf{e}_i \otimes \mathbf{e}_j)\mathbf{e}_k$$

by (2.4.37);

$$= a_{ij}(\mathbf{e}_i \otimes \mathbf{e}_j)\mathbf{c} \tag{2.7.7}$$

Since **c** is an arbitrary vector, (2.7.7) yields (2.7.6). ∎

2.8
TRANSPOSE OF A TENSOR

In the remaining part of this chapter we consider second-order tensors only. Given a 3×3 matrix $[a_{ij}]$, the transpose of $[a_{ij}]$ has been defined by $[a_{ij}]^T = [a_{ji}]$. If $[a_{ij}]$ is the matrix of a tensor **A**, a natural question that arises is, is $[a_{ij}]^T$ also a matrix of a tensor? The answer is in the affirmative; the justification is as follows.

Suppose that a_{ij} are tensor components. Put $[a_{ij}]^T = [b_{ij}]$ in all coordinate systems so that $b_{ij} = a_{ji}$ and $b'_{ij} = a'_{ji}$. Then

$$b'_{ij} = a'_{ji} = \alpha_{jm}\alpha_{in}a_{mn} = \alpha_{in}\alpha_{jm}b_{nm}$$

This transformation rule shows that b_{ij} are components of a tensor. Equivalently, $[b_{ij}] = [a_{ji}] = [a_{ij}]^T$ is the matrix of a tensor. This new tensor is called the *transpose* of the tensor **A**, denoted $\mathbf{A}^T$. Thus, if **A** is a tensor with matrix $[a_{ij}]$, then $\mathbf{A}^T$, the transpose of **A**, is the tensor with matrix $[a_{ij}]^T$; that is,

$$[\mathbf{A}^T] = [\mathbf{A}]^T \tag{2.8.1}$$

2 ALGEBRA OF TENSORS

The following relations, which hold for any number α and any tensors $\mathbf{A}$ and $\mathbf{B}$, may be verified:

(i) $\qquad (\mathbf{A}^T)^T = \mathbf{A}$ \hfill (2.8.2)

(ii) $\qquad (\alpha \mathbf{A})^T = \alpha \mathbf{A}^T$ \hfill (2.8.3)

(iii) $\qquad (\mathbf{A} \pm \mathbf{B})^T = \mathbf{A}^T \pm \mathbf{B}^T$ \hfill (2.8.4)

(iv) $\qquad (\mathbf{AB})^T = \mathbf{B}^T \mathbf{A}^T$ \hfill (2.8.5)

If v_i are components of a vector $\mathbf{v}$, it is easy to verify that

$$[\mathbf{A}^T \mathbf{v}]_i = a_{ji} v_j = [\mathbf{A}]_{ji} [\mathbf{v}]_j \qquad (2.8.6)$$

2.8.1 INVERSE OF A TENSOR AND THE ORTHOGONAL TENSOR

Recall that a matrix $[a_{ij}]$ is said to be *invertible* (or *nonsingular*) if there is a matrix $[a_{ij}]^{-1}$ such that

$$[a_{ij}][a_{ij}]^{-1} = [a_{ij}]^{-1}[a_{ij}] = [I] \qquad (2.8.7)$$

Then $[a_{ij}]^{-1}$ is called the *inverse of* $[a_{ij}]$. Further, $[a_{ij}]$ is said to be *orthogonal* if $[a_{ij}]$ is invertible and $[a_{ij}]^{-1} = [a_{ij}]^T$. These definitions can be extended to tensors as follows.

A tensor $\mathbf{A}$ is said to be an *invertible tensor* if there exists a tensor, which we denote by $\mathbf{A}^{-1}$, such that

$$\mathbf{A}\mathbf{A}^{-1} = \mathbf{A}^{-1}\mathbf{A} = \mathbf{I} \qquad (2.8.8)$$

Then $\mathbf{A}^{-1}$ is called the *inverse* of $\mathbf{A}$. Further, $\mathbf{A}$ is said be an *orthogonal tensor* if $\mathbf{A}$ is invertible and $\mathbf{A}^{-1} = \mathbf{A}^T$. Thus $\mathbf{A}$ is an orthogonal tensor if and only if

$$\mathbf{A}\mathbf{A}^T = \mathbf{A}^T\mathbf{A} = \mathbf{I} \qquad (2.8.9)$$

Conditions (2.8.8) and (2.8.9) show that a tensor $\mathbf{A}$ is invertible if and only if its matrix $[\mathbf{A}]$ is invertible with $[\mathbf{A}]^{-1} = [\mathbf{A}^{-1}]$ and that $\mathbf{A}$ is orthogonal if an only if $[\mathbf{A}]$ is orthogonal.

It is easy to verify that the following relations are true for any invertible tensors $\mathbf{A}$ and $\mathbf{B}$:

(i) $\qquad (\mathbf{A}^{-1})^{-1} = \mathbf{A}$ \hfill (2.8.10)

(ii) $\qquad (\mathbf{A}^{-1})^T = (\mathbf{A}^T)^{-1}$ \hfill (2.8.11)

(iii) $\qquad \mathbf{A}^{-1}(\mathbf{A}\mathbf{a}) = \mathbf{a}; \quad \mathbf{A}(\mathbf{A}^{-1}\mathbf{a}) = \mathbf{a}$ \hfill (2.8.12)

$\qquad$ for any vector $\mathbf{a}$;

(iv) $\qquad (\mathbf{AB})^{-1} = \mathbf{B}^{-1}\mathbf{A}^{-1}$ \hfill (2.8.13)

EXAMPLE 2.8.1 Let **A** be a second-order tensor. For any two vectors **a** and **b**, show that

$$\mathbf{a} \cdot \mathbf{A}\mathbf{b} = \mathbf{b} \cdot \mathbf{A}^T\mathbf{a} \qquad (2.8.14)$$

Solution Let a_i, b_i and a_{ij} be components of **a**, **b** and **A** respectively. Then $\mathbf{c} = \mathbf{A}\mathbf{b}$ is a vector whose components are $c_i = a_{ij}b_j$, and $\mathbf{d} = \mathbf{A}^T\mathbf{a}$ is a vector whose components are $d_i = a_{ji}a_j$. Therefore

$$\mathbf{a} \cdot \mathbf{A}\mathbf{b} = \mathbf{a} \cdot \mathbf{c} = a_i c_i$$
$$= a_i a_{ij} b_j = a_j a_{ji} b_i$$

(interchanging the dummy suffixes i and j);

$$= b_i d_i = \mathbf{b} \cdot \mathbf{d} = \mathbf{b} \cdot \mathbf{A}^T \mathbf{a} \qquad \blacksquare$$

EXAMPLE 2.8.2 Prove that for any two vectors **a** and **b**,

$$(\mathbf{a} \otimes \mathbf{b})^T = \mathbf{b} \otimes \mathbf{a} \qquad (2.8.15)$$

Solution We have

$$[(\mathbf{a} \otimes \mathbf{b})^T]_{ij} = [\mathbf{a} \otimes \mathbf{b}]_{ji} = a_j b_i = b_i a_j = [\mathbf{b} \otimes \mathbf{a}]_{ij}$$

which yields (2.8.15). ∎

EXAMPLE 2.8.3 Show that a tensor **Q** is orthogonal if and only if $\mathbf{Q}\mathbf{a} \cdot \mathbf{Q}\mathbf{b} = \mathbf{a} \cdot \mathbf{b}$ for all vectors **a** and **b**. Deduce that if **Q** is an orthogonal tensor, then $|\mathbf{Q}\mathbf{a}| = |\mathbf{a}|$ for all vectors **a**.

Solution With the aid of (2.8.14), we note that

$$(\mathbf{Q}\mathbf{a} \cdot \mathbf{Q}\mathbf{b}) - (\mathbf{a} \cdot \mathbf{b}) = (\mathbf{a} \cdot \mathbf{Q}^T\mathbf{Q}\mathbf{b}) - (\mathbf{a} \cdot \mathbf{b}) = \mathbf{a} \cdot (\mathbf{Q}^T\mathbf{Q} - \mathbf{I})\mathbf{b}$$

It now follows that $\mathbf{Q}\mathbf{a} \cdot \mathbf{Q}\mathbf{b} = \mathbf{a} \cdot \mathbf{b}$ for arbitrary **a** and **b** if and only if $\mathbf{Q}^T\mathbf{Q} = \mathbf{I}$ or, equivalently, **Q** is orthogonal.

If **Q** is an orthogonal tensor, then, by the result just proven, we have $\mathbf{Q}\mathbf{a} \cdot \mathbf{Q}\mathbf{a} = \mathbf{a} \cdot \mathbf{a}$; or equivalently, $|\mathbf{Q}\mathbf{a}| = |\mathbf{a}|$ for every vector **a**. ∎

EXAMPLE 2.8.4 Let **A** be a tensor. For any vectors, **u**, **a**, **b** and **n**, show that

$$[(\mathbf{u} \cdot \mathbf{a})(\mathbf{A}^T\mathbf{b})] \cdot \mathbf{n} = \mathbf{a} \cdot [(\mathbf{u} \otimes \mathbf{A}\mathbf{n})\mathbf{b}] \qquad (2.8.16)$$

Solution Let u_i, a_i, b_i and n_i be components of **u**, **a**, **b** and **n**, respectively, and a_{ij} be components of **A**. Then we get

$$[(\mathbf{u} \cdot \mathbf{a})(\mathbf{A}^T\mathbf{b})] \cdot \mathbf{n} = (u_i a_i)(a_{jk} b_j) n_k = a_i(u_i a_{jk} n_k) b_j$$
$$= [\mathbf{a}]_i [\mathbf{u} \otimes \mathbf{A}\mathbf{n}]_{ij} [\mathbf{b}]_j = \mathbf{a} \cdot [(\mathbf{u} \otimes \mathbf{A}\mathbf{n})\mathbf{b}] \qquad \blacksquare$$

72 2 ALGEBRA OF TENSORS

EXAMPLE 2.8.5 If a_{ij} are components of a tensor $\mathbf{A}$ and b_i are components of a vector $\mathbf{b}$, show that $\varepsilon_{irs}b_r a_{js}$ are components of a tensor. If this tensor is denoted by $\mathbf{b} \wedge \mathbf{A}$, show that, for any vector $\mathbf{a}$,

(i) $\qquad (\mathbf{b} \wedge \mathbf{A})\mathbf{a} = \mathbf{b} \times (\mathbf{A}^T \mathbf{a}) \qquad$ (2.8.17)

(ii) $\qquad (\mathbf{b} \wedge \mathbf{A})^T \mathbf{a} = \mathbf{A}(\mathbf{a} \times \mathbf{b}) \qquad$ (2.8.18)

Solution (i) Let $c_{ij} = \varepsilon_{irs} b_r a_{js}$ in all coordinate systems. Then

$$c'_{ij} = \varepsilon'_{irs} b'_r a'_{js}$$
$$= \alpha_{im}\alpha_{rn}\alpha_{sp}\alpha_{rq}\alpha_{jk}\alpha_{st}\varepsilon_{mnp}b_q a_{kt}$$
$$= \alpha_{im}\alpha_{jk}\delta_{nq}\delta_{pt}\varepsilon_{mnp}b_q a_{kt}$$
$$= \alpha_{im}\alpha_{jk}\varepsilon_{mnp}b_n a_{kp} = \alpha_{im}\alpha_{jk} c_{mk}$$

This transformation rule shows that $c_{ij} = \varepsilon_{irs} b_r a_{js}$ are components of a tensor. If this tensor is denoted by $\mathbf{b} \wedge \mathbf{A}$, we have

$$[\mathbf{b} \wedge \mathbf{A}]_{ij} = \varepsilon_{irs} b_r a_{js} \qquad (2.8.19)$$

Consequently,

$$[(\mathbf{b} \wedge \mathbf{A})\mathbf{a}]_i = \varepsilon_{irs} b_r a_{js} a_j = \varepsilon_{irs}[\mathbf{b}]_r [\mathbf{A}^T \mathbf{a}]_s = [\mathbf{b} \times \mathbf{A}^T \mathbf{a}]_i$$

from which (2.8.17) follows.

(ii) Also,

$$[(\mathbf{b} \wedge \mathbf{A})^T \mathbf{a}]_i = \varepsilon_{jrs} b_r a_{is} a_j = a_{is}(\varepsilon_{sjr} a_j b_r)$$
$$= [\mathbf{A}]_{is}[\mathbf{a} \times \mathbf{b}]_s = [\mathbf{A}(\mathbf{a} \times \mathbf{b})]_i$$

from which (2.8.18) follows. ∎

EXAMPLE 2.8.6 Let $[a_{ij}]$ be a matrix related to x_i system. For an arbitrary vector with components b_i, if $a_{ji} b_j$ are components of a vector, show that a_{ij} are components of a tensor.

Solution Let $c_{ij} = a_{ji}$. Then $[c_{ij}] = [a_{ij}]^T$ is a matrix related to the x_i system. By data, $c_{ij} b_j = a_{ji} b_j$ are components of a vector. Hence, by a quotient law (see Section 2.4, property 7(ii)), c_{ij} are components of a tensor. Consequently, $c_{ji} = a_{ij}$ are also components of a tensor. ∎

2.9
SYMMETRIC AND SKEW TENSORS

If $\mathbf{A}$ is a (second-order) tensor, we have seen that the transpose of $\mathbf{A}$, namely, $\mathbf{A}^T$, is a tensor whose matrix is $[\mathbf{A}]^T$. If $\mathbf{A}$ is such that $\mathbf{A}^T = \mathbf{A}$, we say that $\mathbf{A}$ is a *symmetric tensor*; if $\mathbf{A}^T = -\mathbf{A}$, we say that $\mathbf{A}$ is a

2.9 SYMMETRIC AND SKEW TENSORS

skew-symmetric tensor or just *skew tensor*. It follows that **A** is symmetric or skew accordingly as the matrix [**A**] is symmetric or skew-symmetric. Equivalently, if a_{ij} are the components of **A**, then **A** is symmetric or skew accordingly as $a_{ji} = a_{ij}$ or $a_{ji} = -a_{ij}$. It is easily seen that the identity tensor is a symmetric tensor and the tensor considered in Example 2.3.1 is a skew tensor. It is obvious that only six of the nine components of a symmetric tensor are independent. In the case of a skew tensor, three components represented as diagonal elements of the matrix of the tensor are 0 and only three of the remaining six components are independent.

For any tensor **A**, we have from (2.8.2) and (2.8.4)

$$(\mathbf{A} + \mathbf{A}^T)^T = \mathbf{A}^T + \mathbf{A} = \mathbf{A} + \mathbf{A}^T \qquad (2.9.1)$$

$$(\mathbf{A} - \mathbf{A}^T)^T = \mathbf{A}^T - \mathbf{A} = -(\mathbf{A} - \mathbf{A}^T) \qquad (2.9.2)$$

Accordingly, the tensor $\mathbf{A} + \mathbf{A}^T$ is always symmetric and the tensor $\mathbf{A} - \mathbf{A}^T$ is always skew.

Also, we note that

$$\mathbf{A} = \tfrac{1}{2}(\mathbf{A} + \mathbf{A}^T) + \tfrac{1}{2}(\mathbf{A} - \mathbf{A}^T) \qquad (2.9.3)$$

for any tensor **A**. Evidently, the first term in the righthand side of (2.9.3) is a symmetric tensor and the second term is a skew tensor. Thus, every tensor **A** may be represented as a sum of a symmetric tensor and a skew tensor. It can be proven that such a representation is unique. The tensor $\tfrac{1}{2}(\mathbf{A} + \mathbf{A}^T)$ is referred to as the *symmetric part* of **A** and denoted sym **A**. The tensor $\tfrac{1}{2}(\mathbf{A} - \mathbf{A}^T)$ is referred to as the *skew part* of **A** and denoted skw **A**. Thus, for every tensor **A**,

$$\text{sym } \mathbf{A} = \tfrac{1}{2}(\mathbf{A} + \mathbf{A}^T) \qquad (2.9.4)$$

$$\text{skw } \mathbf{A} = \tfrac{1}{2}(\mathbf{A} - \mathbf{A}^T) \qquad (2.9.5)$$

$$\mathbf{A} = \text{sym } \mathbf{A} + \text{skw } \mathbf{A} \qquad (2.9.6)$$

If a_{ij} are components of **A**, it is obvious that $\tfrac{1}{2}(a_{ij} + a_{ji})$ are components of sym **A** and $\tfrac{1}{2}(a_{ij} - a_{ji})$ are components of skw **A**.

EXAMPLE 2.9.1 If **A** is a symmetric tensor with components a_{ij} and **B** is a skew tensor with components b_{ij}, then show that

$$\mathbf{A} \cdot \mathbf{B} = a_{ij} b_{ij} = 0 \qquad (2.9.7)$$

Solution In the expression $a_{ij} b_{ij}$, both the suffixes i and j are dummies. As such, $a_{ij} b_{ij} = a_{ji} b_{ji}$. Since **A** is symmetric and **B** is skew, we have $a_{ji} = a_{ij}$ and $b_{ji} = -b_{ij}$. Hence

$$a_{ij} b_{ij} = a_{ji} b_{ji} = a_{ij}(-b_{ij}) = -a_{ij} b_{ij}$$

so that $a_{ij}b_{ij} = 0$. Since $\mathbf{A} \cdot \mathbf{B} = a_{ij}b_{ij}$, see (2.4.20), the result (2.9.7) is proven. ∎

EXAMPLE 2.9.2 If $\mathbf{A}$ is a symmetric tensor and $\mathbf{B}$ is any tensor, show that

$$\mathbf{A} \cdot \mathbf{B} = \mathbf{A} \cdot (\text{sym } \mathbf{B}) \quad (2.9.8)$$

Solution We have

$$\mathbf{A} \cdot \mathbf{B} = \mathbf{A} \cdot (\text{sym } \mathbf{B} + \text{skw } \mathbf{B}) = \mathbf{A} \cdot \text{sym } \mathbf{B} + \mathbf{A} \cdot \text{skw } \mathbf{B} \quad (2.9.9)$$

Since the tensor skw $\mathbf{B}$ is skew, and the tensor $\mathbf{A}$ is symmetric by data, we have $\mathbf{A} \cdot \text{skw } \mathbf{B} = 0$ by Example 2.9.1, and we get (2.9.8) from (2.9.9). ∎

2.10
DUAL VECTOR OF A SKEW TENSOR

One of the most important properties of a skew tensor is the existence of what is called the *dual vector* of the tensor. The property may be stated in the form of the following theorem.

THEOREM 2.10.1 Given a skew tensor $\mathbf{A}$, there exists a unique vector $\boldsymbol{\omega}$ such that

$$\mathbf{A}\mathbf{u} = \boldsymbol{\omega} \times \mathbf{u} \quad (2.10.1)$$

for every vector $\mathbf{u}$. Conversely, given a vector $\boldsymbol{\omega}$, there exists a unique skew tensor $\mathbf{A}$ such that (2.10.1) holds for every vector $\mathbf{u}$. (The vector $\boldsymbol{\omega}$ is called the *dual vector* or the *axial vector* of the tensor $\mathbf{A}$.)

Proof Let $[\mathbf{A}]_{ij} = a_{ij}$. Since $\mathbf{A}$ is skew, we have $a_{ij} = -a_{ji}$, and we may write

$$a_{ij} = \tfrac{1}{2}(a_{ij} - a_{ji}) = \tfrac{1}{2}(\delta_{ip}\delta_{jq} - \delta_{jp}\delta_{iq})a_{pq}$$
$$= \tfrac{1}{2}\varepsilon_{ijk}\varepsilon_{kpq}a_{pq} \quad (2.10.2)$$

on using the substituting property of δ_{ij} and the ε-δ identity.

Let

$$\omega_k = -\tfrac{1}{2}\varepsilon_{kpq}a_{pq} \quad (2.10.3)$$

Then ω_k are components of a vector, say, $\boldsymbol{\omega}$, and this vector is determined uniquely by $\mathbf{A}$. Substituting for ω_k from (2.10.3) in (2.10.2), we get

$$a_{ij} = -\varepsilon_{ijk}\omega_k \quad (2.10.4)$$

2.10 DUAL VECTOR OF A SKEW TENSOR

For any vector **u** with components u_i, (2.10.4) yields

$$a_{ij}u_j = -\varepsilon_{ijk}\omega_k u_j = \varepsilon_{ijk}\omega_j u_k \qquad (2.10.5)$$

which is precisely (2.10.1) in the direct notation.

Thus, given a skew tensor **A**, there exists a unique vector ω, whose components are defined by (2.10.3), such that (2.10.1) holds for an arbitrary vector **u**. The existence as well as the uniqueness of the dual vector ω of **A** has thus been established.

Conversely, given a vector ω, suppose we define a_{ij} by (2.10.4). Then, a_{ij} are components of a skew tensor, say, **A**, and this tensor is determined uniquely by ω. Since (2.10.4) yields (2.10.1), it follows that **A** is the tensor of which ω is the dual vector. The proof of the theorem is complete. ∎

Note: Expanding the righthand side of (2.10.3) with the aid of the definition of the ε-symbol, we get

$$\omega = -(a_{23}\mathbf{e}_1 + a_{31}\mathbf{e}_2 + a_{12}\mathbf{e}_3) \qquad (2.10.6)$$

Similarly, we find from (2.10.4),

$$[\mathbf{A}] = [a_{ij}] = \begin{bmatrix} 0 & -\omega_3 & \omega_2 \\ \omega_3 & 0 & -\omega_1 \\ -\omega_2 & \omega_1 & 0 \end{bmatrix} \qquad (2.10.7)$$

Expression (2.10.6) is used to find ω when **A** is known, and (2.10.7) is used to find the matrix of **A** when ω is known.

EXAMPLE 2.10.1 Find the dual vector of the skew part of the tensor **A** whose matrix is

$$[\mathbf{A}] = \begin{bmatrix} 1 & -2 & 0 \\ 0 & 1 & 2 \\ 2 & 0 & -1 \end{bmatrix}$$

Solution The transpose of the given matrix is

$$[\mathbf{A}]^T = \begin{bmatrix} 1 & 0 & 2 \\ -2 & 1 & 0 \\ 0 & 2 & -1 \end{bmatrix}$$

76   2 ALGEBRA OF TENSORS

Hence

$$[\text{skw } \mathbf{A}] = \tfrac{1}{2}([\mathbf{A}] - [\mathbf{A}]^T) = \begin{bmatrix} 0 & -1 & -1 \\ 1 & 0 & 1 \\ 1 & -1 & 0 \end{bmatrix}$$

By using (2.10.6), we get the dual vector $\boldsymbol{\omega}$ of skw $\mathbf{A}$ as

$$\boldsymbol{\omega} = -\mathbf{e}_1 - \mathbf{e}_2 + \mathbf{e}_3 \quad \blacksquare$$

EXAMPLE 2.10.2   If $\boldsymbol{\omega}$ is the dual vector of a skew tensor $\mathbf{A}$, show that

$$|\boldsymbol{\omega}| = \frac{1}{\sqrt{2}}|\mathbf{A}| \qquad (2.10.8)$$

where $|\boldsymbol{\omega}|$ is, as usual, the magnitude of $\boldsymbol{\omega}$, and $|\mathbf{A}| = \sqrt{\mathbf{A} \cdot \mathbf{A}}$.

Solution   If ω_i are components of $\boldsymbol{\omega}$ and a_{ij} are components of $\mathbf{A}$, we get from expression (2.10.3),

$$\omega_k \omega_k = \tfrac{1}{4}\varepsilon_{kpq}\varepsilon_{krs}a_{pq}a_{rs}$$

which, on using the ε-δ identity, becomes

$$\omega_k \omega_k = \tfrac{1}{4}(\delta_{pr}\delta_{qs} - \delta_{ps}\delta_{qr})a_{pq}a_{rs}$$
$$= \tfrac{1}{4}(a_{pq}a_{pq} - a_{pq}a_{qp}) = \tfrac{1}{2}a_{pq}a_{pq} \qquad (2.10.9)$$

When written in the direct notation, (2.10.9) reads

$$|\boldsymbol{\omega}|^2 = \tfrac{1}{2}\mathbf{A} \cdot \mathbf{A} \qquad (2.10.10)$$

from which (2.10.8) is immediate. $\blacksquare$

EXAMPLE 2.10.3   Let $\mathbf{Q}$ be an orthogonal tensor and $\mathbf{a}$ be a vector such that $\mathbf{Qa} = \mathbf{a}$. Show that (i) $\mathbf{Q}^T\mathbf{a} = \mathbf{a}$ and (ii) the dual vector of skw $\mathbf{Q}$ is collinear with $\mathbf{a}$.

Solution   (i) By data, $\mathbf{a} = \mathbf{Qa}$. Hence

$$\mathbf{Q}^T\mathbf{a} = \mathbf{Q}^T(\mathbf{Qa}) = (\mathbf{Q}^T\mathbf{Q})\mathbf{a}$$

But, since $\mathbf{Q}$ is an orthogonal tensor, $\mathbf{Q}^T\mathbf{Q} = \mathbf{I}$. Also, $\mathbf{Ia} = \mathbf{a}$. Therefore,

$$\mathbf{Q}^T\mathbf{a} = \mathbf{a} \qquad (2.10.11)$$

(ii) Since $\mathbf{Qa} = \mathbf{Q}^T\mathbf{a} = \mathbf{a}$, we have

$$(\text{skw } \mathbf{Q})\mathbf{a} = \tfrac{1}{2}(\mathbf{Q} - \mathbf{Q}^T)\mathbf{a} = \tfrac{1}{2}(\mathbf{Qa} - \mathbf{Q}^T\mathbf{a}) = 0 \qquad (2.10.12)$$

On the other hand, if $\boldsymbol{\omega}$ is the dual vector of skw $\mathbf{Q}$, we have

$$(\text{skw } \mathbf{Q})\mathbf{a} = \boldsymbol{\omega} \times \mathbf{a} \qquad (2.10.13)$$

From (2.10.12) and (2.10.13), we get $\boldsymbol{\omega} \times \mathbf{a} = \mathbf{0}$; that is, $\boldsymbol{\omega}$ is collinear with $\mathbf{a}$. ∎

2.11
INVARIANTS OF A TENSOR

Given a tensor $\mathbf{A}$ with components a_{ij}, we have noted in Example 2.3.3 that a_{ii} (which is the contraction of a_{ij}) is a scalar invariant called the *trace* of $\mathbf{A}$ and denoted tr $\mathbf{A}$. Besides this invariant, two other invariants associated with $\mathbf{A}$ are encountered. Keeping this in view, tr $\mathbf{A}$ is often referred to as the *first invariant* of $\mathbf{A}$ and denoted also by $I_\mathbf{A}$. Thus,

$$I_\mathbf{A} = \text{tr } \mathbf{A} = a_{ii} \qquad (2.11.1)$$

The second and the third invariants of $\mathbf{A}$ are introduced in the following.

We have noted that the square of a tensor $\mathbf{A}$, defined by $\mathbf{A}^2 = \mathbf{A}\mathbf{A}$, is also a tensor and that the components of $\mathbf{A}^2$ are $a_{ik}a_{kj}$; see (2.4.22). Hence tr $\mathbf{A}^2$, given by

$$\text{tr } \mathbf{A}^2 = a_{ik}a_{ki} \qquad (2.11.2)$$

is a scalar. Consequently, $\frac{1}{2}[(\text{tr } \mathbf{A})^2 - \text{tr } \mathbf{A}^2]$ is also a scalar. This scalar is called the *second invariant* of $\mathbf{A}$ and denoted $II_\mathbf{A}$. Thus,

$$II_\mathbf{A} = \tfrac{1}{2}[(\text{tr } \mathbf{A})^2 - \text{tr } \mathbf{A}^2] \qquad (2.11.3)$$

When written in the suffix notation, this expression reads

$$II_\mathbf{A} = \tfrac{1}{2}[a_{ii}a_{kk} - a_{ik}a_{ki}] \qquad (2.11.4)$$

Writing the righthand side of (2.11.4) in the unabridged form we get

$$II_\mathbf{A} = \begin{vmatrix} a_{11} & a_{12} \\ a_{21} & a_{22} \end{vmatrix} + \begin{vmatrix} a_{11} & a_{13} \\ a_{31} & a_{33} \end{vmatrix} + \begin{vmatrix} a_{22} & a_{23} \\ a_{32} & a_{33} \end{vmatrix} \qquad (2.11.5)$$

Some authors take $\frac{1}{2}[\mathbf{A} \cdot \mathbf{A} - (\text{tr } \mathbf{A})^2]$ as the second invariant of $\mathbf{A}$, denoted $\bar{II}_\mathbf{A}$. Thus,

$$\bar{II}_\mathbf{A} = \tfrac{1}{2}[\mathbf{A} \cdot \mathbf{A} - (\text{tr } \mathbf{A})^2] \qquad (2.11.6)$$

or, in the suffix notation,

$$\bar{II}_\mathbf{A} = \tfrac{1}{2}[a_{ij}a_{ij} - a_{ii}a_{jj}] \qquad (2.11.7)$$

For a symmetric tensor $\mathbf{A}$, $\bar{II}_\mathbf{A} = -II_\mathbf{A}$.

2 ALGEBRA OF TENSORS

It has been noted that the cube of a tensor $\mathbf{A}$, defined by $\mathbf{A}^3 = \mathbf{A}^2\mathbf{A}$, is also a tensor and its components are $a_{ik}a_{km}a_{mj}$; see (2.4.22). Hence tr $\mathbf{A}^3$, given by

$$\operatorname{tr} \mathbf{A}^3 = a_{ik}a_{km}a_{mi} \qquad (2.11.8)$$

is a scalar. Consequently, it follows that

$$\tfrac{1}{6}[(\operatorname{tr} \mathbf{A})^3 + 2\operatorname{tr} \mathbf{A}^3 - 3(\operatorname{tr} \mathbf{A}^2)(\operatorname{tr} \mathbf{A})]$$

is also a scalar. This scalar is called the *third invariant* of $\mathbf{A}$ and denoted $III_\mathbf{A}$. Thus,

$$III_\mathbf{A} = \tfrac{1}{6}[(\operatorname{tr} \mathbf{A})^3 + 2\operatorname{tr} \mathbf{A}^3 - 3(\operatorname{tr} \mathbf{A}^2)(\operatorname{tr} \mathbf{A})] \qquad (2.11.9)$$

When written in the suffix notation, this expression reads

$$III_\mathbf{A} = \tfrac{1}{6}[a_{ii}a_{jj}a_{kk} + 2a_{ik}a_{km}a_{mi} - 3a_{ik}a_{ki}a_{jj}] \qquad (2.11.10)$$

The three invariants $I_\mathbf{A}$, $II_\mathbf{A}$ and $III_\mathbf{A}$ are usually referred to as the *fundamental* or *principal invariants* of $\mathbf{A}$. It is easy to verify that $\mathbf{A}$ and $\mathbf{A}^T$ have the same principal invariants. The significance of these invariants will become clear in Section 2.13; see equation (2.13.2) in particular.

2.11.1 DETERMINANT OF A TENSOR

Using expression (1.7.27), we note that the righthand side of (2.11.10) represents $\det(a_{ij})$. Consequently, by use of (1.7.18), we obtain the following simpler representation for $III_\mathbf{A}$:

$$III_\mathbf{A} = \det[a_{ij}] = \varepsilon_{ijk}a_{i1}a_{j2}a_{k3} = \varepsilon_{ijk}a_{1i}a_{2j}a_{3k} \qquad (2.11.11)$$

This expression shows that, although the individual tensor components a_{ij} change from one set of axes to the other, the determinant of $[a_{ij}]$ as a whole retains the same value $III_\mathbf{A}$ in all sets of axes. The invariant $III_\mathbf{A}$ is (therefore) referred to as the *determinant of the tensor* $\mathbf{A}$ and denoted $\det \mathbf{A}$. Thus, the determinant of $\mathbf{A}$ is nothing but the determinant of the matrix $[\mathbf{A}]$ in any of the coordinate systems.

It has been noted that $\mathbf{A}$ is invertible if and only if $[\mathbf{A}] = [a_{ij}]$ is invertible and that $[a_{ij}]$ is invertible if and only if $\det[a_{ij}] \neq 0$. Hence, $\mathbf{A}$ is an invertible tensor if and only if $\det \mathbf{A} \neq 0$.

From the properties of matrices and determinants, it readily follows that

(i) $\qquad \det \mathbf{I} = 1$
(ii) $\qquad \det \mathbf{A} = \det \mathbf{A}^T$
(iii) $\qquad \det(\alpha\mathbf{A}) = \alpha^3 \det \mathbf{A}$ $\qquad$ (2.11.12)
(iv) $\qquad \det(\mathbf{AB}) = (\det \mathbf{A})(\det \mathbf{B})$

for any tensors $\mathbf{A}$ and $\mathbf{B}$ and any number α.

2.11 INVARIANTS OF A TENSOR

If $\mathbf{A}$ is invertible we find from (i) and (iv) of (2.11.12), on setting $\mathbf{B} = \mathbf{A}^{-1}$, that

$$\det \mathbf{A}^{-1} = \frac{1}{\det \mathbf{A}} \qquad (2.11.13)$$

From (ii) of (2.11.12), and (2.11.13), we find that $\det \mathbf{A} = \pm 1$ for an orthogonal tensor $\mathbf{A}$. An orthogonal tensor $\mathbf{A}$ for which $\det \mathbf{A} = 1$ is called a *proper orthogonal tensor*.

EXAMPLE 2.11.1 Let $\mathbf{a}$ and $\mathbf{b}$ be arbitrary vectors. Show that the second and third principal invariants of $\mathbf{a} \otimes \mathbf{b}$ are 0. If $\mathbf{a}$ and $\mathbf{b}$ are orthogonal, show that the first principal invariant of $\mathbf{a} \otimes \mathbf{b}$ is also 0.

Solution If a_i and b_i are components of $\mathbf{a}$ and $\mathbf{b}$, then $a_i b_j$ are components of $\mathbf{a} \otimes \mathbf{b}$ so that

$$I_{\mathbf{a} \otimes \mathbf{b}} = \operatorname{tr} \mathbf{a} \otimes \mathbf{b} = a_i b_i = \mathbf{a} \cdot \mathbf{b} \qquad (2.11.14)$$

If $\mathbf{a}$ and $\mathbf{b}$ are orthogonal, then $\mathbf{a} \cdot \mathbf{b} = 0$ and we get $I_{\mathbf{a} \otimes \mathbf{b}} = 0$.
Using (2.11.4), we get

$$II_{\mathbf{a} \otimes \mathbf{b}} = \tfrac{1}{2}[(a_i b_i)(a_k b_k) - (a_i b_k)(a_k b_i)] = 0 \qquad (2.11.15)$$

Also, using (2.11.10) we get

$$III_{\mathbf{a} \otimes \mathbf{b}} = \tfrac{1}{6}[(a_i b_i)(a_j b_j)(a_k b_k) + 2(a_i b_k)(a_k b_m)(a_m b_i) - 3(a_i b_k)(a_k b_i)(a_j b_j)]$$
$$= 0 \qquad \blacksquare \qquad (2.11.16)$$

Note: This example illustrates an interesting fact that the principal invariants of a nonzero tensor can all be 0.

EXAMPLE 2.11.2 For a skew tensor $\mathbf{A}$, show that $I_\mathbf{A} = III_\mathbf{A} = 0$ and $II_\mathbf{A} = |\boldsymbol{\omega}|^2$, where $\boldsymbol{\omega}$ is the dual vector of $\mathbf{A}$.

Solution Since $\mathbf{A}$ is skew, all the diagonal elements of its matrix $[a_{ij}]$ are equal to 0. Hence $\operatorname{tr} \mathbf{A} = a_{ii} = 0$; that is,

$$I_\mathbf{A} = 0 \qquad (2.11.17)$$

Expressions (2.11.9) and (2.11.10) then give

$$III_\mathbf{A} = \tfrac{1}{3} \operatorname{tr} \mathbf{A}^3 = \tfrac{1}{3} a_{ik} a_{km} a_{mi} \qquad (2.11.18)$$

Since $a_{ij} = -a_{ji}$, this yields

$$III_\mathbf{A} = -\tfrac{1}{3} a_{ki} a_{mk} a_{im} = -\tfrac{1}{3} a_{ik} a_{km} a_{mi}$$

interchanging dummy suffixes k and m;

$$= -III_\mathbf{A}$$

so that
$$III_A = 0 \qquad (2.11.19)$$

Also, using (2.11.17), expression (2.11.4) gives
$$II_A = -\tfrac{1}{2}a_{ik}a_{ki} = \tfrac{1}{2}a_{ik}a_{ik} = \tfrac{1}{2}(A \cdot A) \qquad (2.11.20)$$

Expression (2.10.10) now yields
$$II_A = |\omega|^2 \qquad \blacksquare \qquad (2.11.21)$$

EXAMPLE 2.11.3 Let a_{ij} be components of a tensor A. If
$$a^*_{ij} = \tfrac{1}{2}\varepsilon_{ipq}\varepsilon_{jrs}a_{pr}a_{qs} \qquad (2.11.22)$$
show that a^*_{ij} are components of a tensor. If this tensor is denoted by A^*, prove the following:

(i)
$$A(A^*)^T = (A^*)^T A = (\det A)I \qquad (2.11.23)$$

(ii) If A is invertible, then
$$A^{-1} = \frac{1}{\det A}(A^*)^T \qquad (2.11.24)$$

(iii) For all vectors a, b
$$A^*(a \times b) = Aa \times Ab \qquad (2.11.25)$$

(iv) For all vectors a, b, c,
$$(A^*)^T a \cdot (b \times c) = a \cdot Ab \times Ac \qquad (2.11.26)$$

Here A^* is called the *adjugate* or *cofactor* of A, and $(A^*)^T$ is called the *adjoint* of A.

Solution Since a_{ij} and ε_{ijk} are components of tensors, the righthand side of (2.11.22) represents a contraction of a product of tensors. As such a^*_{ij} are components of a tensor, A^*. Consequently, expressions (2.11.23) and (2.11.24) follow from expressions (1.7.40) and (1.7.41).

Further, for any vectors a and b with components a_i and b_i, we have
$$[A^*(a \times b)]_i = a^*_{ij}[a \times b]_j = a^*_{ij}(\varepsilon_{jmn}a_m b_n)$$
$$= \tfrac{1}{2}\varepsilon_{ipq}\varepsilon_{jrs}a_{pr}a_{qs}\varepsilon_{jmn}a_m b_n$$
using (2.11.22);
$$= \tfrac{1}{2}\varepsilon_{ipq}(\delta_{rm}\delta_{sn} - \delta_{rn}\delta_{sm})a_{pr}a_{qs}a_m b_n$$

using the ε-δ identity;

$$= \tfrac{1}{2}\varepsilon_{ipq}(a_{pr}a_r)(a_{qs}b_s) - \tfrac{1}{2}\varepsilon_{ipq}(a_{pr}b_r)(a_{qs}a_s)$$
$$= \tfrac{1}{2}[\mathbf{Aa} \times \mathbf{Ab}]_i - \tfrac{1}{2}[[\mathbf{Ab} \times \mathbf{Aa}]_i = [\mathbf{Aa} \times \mathbf{Ab}]_i$$

which proves (2.11.25).

By use of (2.8.14), we have

$$(\mathbf{A}^*)^T\mathbf{a} \cdot (\mathbf{b} \times \mathbf{c}) = \mathbf{a} \cdot \mathbf{A}^*(\mathbf{b} \times \mathbf{c})$$

which by use of (2.11.25), yields (2.11.26). ∎

Note: From (2.11.22) and (2.11.24), it follows that the components a_{ij}^{-1} of $\mathbf{A}^{-1}$ are given by

$$a_{ij}^{-1} = \frac{1}{\det \mathbf{A}} a_{ji}^* = \frac{1}{2(\det \mathbf{A})} \varepsilon_{jpq}\varepsilon_{irs}a_{pr}a_{qs} \qquad (2.11.27)$$

2.12
DEVIATORIC TENSORS

Given a tensor $\mathbf{A}$, suppose we define a new tensor $\mathbf{A}^{(d)}$ as follows:

$$\mathbf{A}^{(d)} = \mathbf{A} - \tfrac{1}{3}(\operatorname{tr} \mathbf{A})\mathbf{I} \qquad (2.12.1)$$

If a_{ij} are components of $\mathbf{A}$, then from (2.12.1) it follows that the components $a_{ij}^{(d)}$ of $\mathbf{A}^{(d)}$ are given by

$$a_{ij}^{(d)} = a_{ij} - \tfrac{1}{3}a_{kk}\delta_{ij} \qquad (2.12.2)$$

Hence $a_{ii}^{(d)} = a_{ii} - \tfrac{1}{3}(3a_{kk}) = 0$; that is,

$$\operatorname{tr} \mathbf{A}^{(d)} = 0 \qquad (2.12.3)$$

If we set

$$\alpha = \tfrac{1}{3} \operatorname{tr} \mathbf{A} = \tfrac{1}{3}a_{kk} \qquad (2.12.4)$$

then expression (2.12.1) can be rewritten as

$$\mathbf{A} = \alpha\mathbf{I} + \mathbf{A}^{(d)} \qquad (2.12.5)$$

Thus, every tensor $\mathbf{A}$ can be represented as a sum of two tensors; one of these, $\alpha\mathbf{I}$, is an isotropic tensor, and the other, $\mathbf{A}^{(d)}$, is a tensor whose trace is 0. Expressions (2.12.1) and (2.12.4) show that $\mathbf{A}^{(d)}$ and α are uniquely determined by $\mathbf{A}$. As such, the representation (2.12.5) is unique. The part $\alpha\mathbf{I}$ of this representation is called the *spherical part* of $\mathbf{A}$ and the part $\mathbf{A}^{(d)}$ is called the *deviatoric part* or the *deviator part* of $\mathbf{A}$. It is obvious that the spherical part of a tensor is always symmetric and that the deviator part is symmetric (skew) if an only if the tensor is symmetric (skew). The deviator part of a given tensor is sometimes referred to as the *deviatoric tensor* associated with the given tensor.

From (2.12.3), we note that the first invariant of $\mathbf{A}^{(d)}$ is always 0; that is,

$$I_{\mathbf{A}^{(d)}} = 0 \tag{2.12.6}$$

By use of expressions (2.11.3) and (2.11.9), we find that the second and the third invariants of $\mathbf{A}^{(d)}$ are given by

$$II_{\mathbf{A}^{(d)}} = -\tfrac{1}{2}\operatorname{tr}(\mathbf{A}^{(d)})^2 \tag{2.12.7}$$

$$III_{\mathbf{A}^{(d)}} = \tfrac{1}{3}\operatorname{tr}(\mathbf{A}^{(d)})^3 \tag{2.12.8}$$

To suppress the negative sign appearing in the righthand side of (2.12.7), some authors take $-II_{\mathbf{A}^{(d)}}$ as the second invariant of $\mathbf{A}^{(d)}$.

EXAMPLE 2.12.1 Find the spherical and deviatoric parts of the tensor $\mathbf{A}$ whose matrix is

$$[\mathbf{A}] = [a_{ij}] = \begin{bmatrix} 1 & 1 & 1 \\ 1 & 1 & 1 \\ 1 & 1 & 1 \end{bmatrix}$$

Solution We readily see that $\alpha = \tfrac{1}{3}\operatorname{tr}\mathbf{A} = 1$, and

$$[a_{ij}^{(d)}] = [a_{ij}] - \alpha[\delta_{ij}] = [a_{ij} - \alpha\delta_{ij}]$$

$$= \begin{bmatrix} 0 & 1 & 1 \\ 1 & 0 & 1 \\ 1 & 1 & 0 \end{bmatrix} \tag{2.12.9}$$

Thus, the spherical part of the given tensor is $\mathbf{I}$, and the deviatoric part is the tensor whose matrix is $[a_{ij}^{(d)}]$ given by (2.12.9). ∎

EXAMPLE 2.12.2 Show that the deviator part of a tensor $\mathbf{A}$ is $\mathbf{0}$ if and only if $\mathbf{A}$ is isotropic.

Solution If $\mathbf{A}$ is istropic, then by Theorem 2.5.1, $\mathbf{A} = \beta\mathbf{I}$ for some scalar β, so that $\operatorname{tr}\mathbf{A} = 3\beta$. Expression (2.12.1) then gives $\mathbf{A}^{(d)} = \mathbf{0}$. Conversely, if $\mathbf{A}^{(d)} = \mathbf{0}$, then (2.12.5) shows that $\mathbf{A}$ is a scalar multiple of $\mathbf{I}$. Hence $\mathbf{A}$ is isotropic. ∎

EXAMPLE 2.12.3 Show that

$$II_{\mathbf{A}^{(d)}} = \tfrac{1}{2}\{\tfrac{1}{3}(\operatorname{tr}\mathbf{A})^2 - \operatorname{tr}\mathbf{A}^2\} = II_{\mathbf{A}} - \tfrac{1}{3}I_{\mathbf{A}}^2 \qquad (2.12.10)$$

Solution From (2.12.7) and (2.12.2), we get

$$II_{\mathbf{A}^{(d)}} = -\tfrac{1}{2}a_{ik}^{(d)}a_{ki}^{(d)}$$
$$= -\tfrac{1}{2}(a_{ik} - \tfrac{1}{3}a_{mn}\delta_{ik})(a_{ki} - \tfrac{1}{3}a_{mn}\delta_{ki})$$
$$= \tfrac{1}{2}\{\tfrac{1}{3}a_{kk}^2 - a_{ik}a_{ki}\} = \tfrac{1}{2}\{\tfrac{1}{3}(\operatorname{tr}\mathbf{A})^2 - \operatorname{tr}\mathbf{A}^2\} \qquad (2.12.11)$$

With the aid of (2.11.1) and (2.11.4), we find that

$$(\tfrac{1}{3}a_{kk}^2 - a_{ik}a_{ki}) = \{\tfrac{1}{3}I_{\mathbf{A}}^2 + 2II_{\mathbf{A}} - I_{\mathbf{A}}^2\} = 2(II_{\mathbf{A}} - \tfrac{1}{3}I_{\mathbf{A}}^2) \qquad (2.12.12)$$

Expressions (2.12.11) and (2.12.12) together yield (2.12.10). ∎

2.13
EIGENVALUES AND EIGENVECTORS

It has been noted that a (second-order) tensor **A** operating on a vector **v** gives rise to a vector **Av**. A case of great interest is the one in which **Av** is collinear with **v** so that **Av** can be written as

$$\mathbf{A}\mathbf{v} = \Lambda\mathbf{v} \qquad (2.13.1)$$

for some real number Λ. A unit vector **v** possessing this property, for a given tensor **A**, is called an *eigenvector* or a *principal vector* of **A**, and the real number Λ is called the corresponding *eigenvalue* or the *principal value*. Also, the direction of **v** is called a *principal direction* of **A**.

It is evident that, if **v** is an eigenvector of a tensor **A** and Λ is the corresponding eigenvalue, then $-\mathbf{v}$ is also an eigenvector of **A** corresponding to the same eigenvalue Λ. Thus, more than one eigenvector can correspond to the same eigenvalue. But, for a given eigenvector there corresponds only one eigenvalue.

As an example, we may note that (since $\mathbf{I}\mathbf{v} = \mathbf{v}$ for every vector **v**) every unit vector is an eigenvector of the identity tensor **I**, and the number 1 is the eigenvalue corresponding to all these eigenvectors. In other words, for the tensor **I**, every direction is a principal direction and all these directions correspond to the principal value 1.

It may be noted that according to the definition of an *eigenvalue* and *eigenvector* just given, an eigenvalue is a *real* number and an eigenvector is a *unit* vector. The terms eigenvalue and eigenvector are used in this sense throughout our discussion.

2.13.1 EXISTENCE OF EIGENVALUES

The following theorem ensures that every tensor has an eigenvalue and hence an eigenvector.

THEOREM 2.13.1 A number Λ is an eigenvalue of a tensor $\mathbf{A}$ if and only if it is a real root of the cubic equation

$$\Lambda^3 - I_\mathbf{A}\Lambda^2 + II_\mathbf{A}\Lambda - III_\mathbf{A} = 0 \tag{2.13.2}$$

where $I_\mathbf{A}$, $II_\mathbf{A}$ and $III_\mathbf{A}$ are the principal invariants of $\mathbf{A}$.

Proof Let $[\mathbf{v}]_i = v_i$ and $[\mathbf{A}]_{ij} = a_{ij}$. Then the vector equation (2.13.1) reads as follows in the suffix notation:

$$a_{ij}v_j = \Lambda v_i, \quad \text{or} \quad (a_{ij} - \Lambda\delta_{ij})v_j = 0 \tag{2.13.3}$$

When written in the expanded form, this equation yields the following system of three equations:

$$\begin{aligned}(a_{11} - \Lambda)v_1 + a_{12}v_2 + a_{13}v_3 &= 0 \\ a_{21}v_1 + (a_{22} - \Lambda)v_2 + a_{23}v_3 &= 0 \\ a_{31}v_1 + a_{32}v_2 + (a_{33} - \Lambda)v_3 &= 0\end{aligned} \tag{2.13.4}$$

The determinant of the coefficients of these equations is

$$\det(\mathbf{A} - \Lambda\mathbf{I}) \equiv \begin{vmatrix} a_{11} - \Lambda & a_{12} & a_{13} \\ a_{21} & a_{22} - \Lambda & a_{23} \\ a_{31} & a_{32} & a_{33} - \Lambda \end{vmatrix}$$

$$\equiv -\Lambda^3 + I_\mathbf{A}\Lambda^2 - II_\mathbf{A}\Lambda + III_\mathbf{A} \tag{2.13.5}$$

With this information, we now suppose that Λ is an eigenvalue of $\mathbf{A}$. Then there exists a unit vector $\mathbf{v}$ such that (2.13.1) holds. Consequently, the system of equations (2.13.4) has a nontrivial solution. Therefore, the determinant of the coefficients of the system is equal to 0. Expression (2.13.5) now shows that Λ is a root of equation (2.13.2). Since Λ is a real number, being an eigenvalue of $\mathbf{A}$, it is a real root of equation (2.13.2).

Conversely, suppose that Λ is a real root of the cubic equation (2.13.2). Then the determinant of the coefficients of equations (2.13.4) is 0 for this Λ, and consequently the system (2.13.4) has a nontrivial solution, say u_i. This implies that

$$\mathbf{A}\mathbf{u} = \Lambda\mathbf{u} \tag{2.13.6}$$

where $\mathbf{u}$ is the vector whose components are u_i. If we set $\mathbf{v} = \mathbf{u}/|\mathbf{u}|$, equation (2.13.6) becomes (2.13.1) and Λ is therefore an eigenvalue of $\mathbf{A}$.

Thus, Λ is an eigenvalue of $\mathbf{A}$ if and only if it is a real root of equation (2.13.2). The proof of the theorem is complete. ∎

It is important to note that, since (2.13.2) is a cubic equation with real coefficients, it has at least one real root for Λ; consequently, $\mathbf{A}$ *has at least one eigenvalue* (and therefore at least one eigenvector). Further, since the coefficients $I_\mathbf{A}$, $II_\mathbf{A}$ and $III_\mathbf{A}$ in equation (2.13.2) are invariant under coordinate transformations, the roots of this equation are also invariant under coordinate transformations. In other words, the cubic equation (2.13.2) yields the same three roots in every choice of the coordinate axes. This means that the eigenvalues of a tensor are independent of the coordinate system chosen for the purpose of their computation.

The polynomial on the righthand side of (2.13.5) is called the *characteristic polynomial* of the tensor $\mathbf{A}$. Also, the cubic equation (2.13.2) (or its negative version) is called the *characteristic equation* of $\mathbf{A}$. Since $\mathbf{A}$ and $\mathbf{A}^T$ have the same principal invariants, it follows that the characteristic equation of $\mathbf{A}^T$ is identical with that of $\mathbf{A}$. Hence $\mathbf{A}$ and $\mathbf{A}^T$ have the same eigenvalues. From equation (2.13.2) it also follows that 0 is an eigenvalue of $\mathbf{A}$ if and only if $III_\mathbf{A} \equiv \det \mathbf{A} = 0$.

It is obvious that $\mathbf{A}$ cannot have more than one eigenvalue if the cubic (2.13.2) has a complex root. If $\mathbf{A}$ is symmetric, we will prove below that all the three roots of (2.13.2) are real and each of these roots is an eigenvalue of $\mathbf{A}$.

2.13.2 EIGENVALUES OF A SYMMETRIC TENSOR

THEOREM 2.13.2 If $\mathbf{A}$ is a symmetric tensor, then all the three roots of the characteristic equation of $\mathbf{A}$ are real, and therefore $\mathbf{A}$ has exactly three (not necessarily distinct) eigenvalues.

Proof We have noted that at least one of the roots of the characteristic equation (2.13.2) is real and this real root is an eigenvalue of $\mathbf{A}$. Denote this eigenvalue by Λ_1 and let $\mathbf{v}_1$ be a corresponding eigenvector. Then, if v_{1i} are components of $\mathbf{v}_1$ and $a_{ij} = a_{ji}$ are components of $\mathbf{A}$, we have

$$a_{ij}v_{1j} = \Lambda_1 v_{1i} \quad (2.13.7)$$

Introduce a coordinate system x_i' such that the x_1' axis is along the vector $\mathbf{v}_1$; that is, $\mathbf{e}_1' = \mathbf{v}_1$. For this system, we have

$$\alpha_{1i} = \mathbf{e}_1' \cdot \mathbf{e}_i = \mathbf{v}_1 \cdot \mathbf{e}_i = v_{1i} \quad (2.13.8)$$

If a_{ij}' are components of $\mathbf{A}$ in the x_i' system, we have

$$a_{11}' = \alpha_{1p}\alpha_{1q}a_{pq} = a_{pq}v_{1p}v_{1q} = \Lambda_1 v_{1p}v_{1p} = \Lambda_1 \quad (2.13.9)$$

86   2 ALGEBRA OF TENSORS

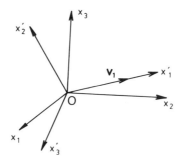

Figure 2.6. Coordinate system x'_i.

In obtaining (2.13.9), we have used (2.13.7), (2.13.8) and the fact the $\mathbf{v}_1$ is a unit vector. Also,

$$a'_{12} = \alpha_{1p}\alpha_{2q}a_{pq} = \alpha_{1q}\alpha_{2p}a_{qp}$$
$$= \alpha_{2p}a_{pq}v_{1q}$$

using (2.13.8) and that $a_{pq} = a_{qp}$;

$$= \Lambda_1 \alpha_{2p} v_{1p}$$

using (2.13.7);

$$= \Lambda_1 \alpha_{2p}\alpha_{1p} = \Lambda_1 \delta_{21}$$

using (2.13.8) and the orthogonality of α_{ij};

$$= 0 \tag{2.13.10}$$

Similarly,

$$a'_{13} = 0 \tag{2.13.11}$$

In view of (2.13.9), (2.13.10) and (2.13.11), the characteristic equation of $\mathbf{A}$ in the x'_i system is given by

$$\begin{vmatrix} \Lambda_1 - \Lambda & 0 & 0 \\ 0 & a'_{22} - \Lambda & a'_{23} \\ 0 & a'_{23} & a'_{33} - \Lambda \end{vmatrix} = 0$$

which simplifies to

$$(\Lambda_1 - \Lambda)[\Lambda^2 - (a'_{22} + a'_{33})\Lambda + a'_{22}a'_{33} - a'^2_{23}] = 0 \tag{2.13.12}$$

Evidently, one of the roots of this equation is Λ_1 and the other two roots, say, Λ_2 and Λ_3, are determined by the quadratic equation

$$\Lambda^2 - (a'_{22} + a'_{33})\Lambda + (a'_{22}a'_{33} - a'^2_{23}) = 0 \tag{2.13.13}$$

The discriminant D of this equation is

$$D = (a'_{22} + a'_{33})^2 - 4(a'_{22}a'_{33} - a'^2_{23})$$
$$= (a'_{22} - a'_{33})^2 + 4a'^2_{23} \geq 0$$

Therefore Λ_2 and Λ_3 are also real.

Thus, all the three roots of the characteristic equation of $\mathbf{A}$ are real. Since each real root of the characteristic equation is an eigenvalue of $\mathbf{A}$ by Theorem 2.13.1, it follows that $\mathbf{A}$ has exactly three eigenvalues. Further, since the three roots Λ_1, Λ_2 and Λ_3 are not necessarily distinct, the three eigenvalues of $\mathbf{A}$ need not be different from one another. From equation (2.13.2) it follows that

$$I_\mathbf{A} = \Lambda_1 + \Lambda_2 + \Lambda_3 \qquad (2.13.14\text{a})$$

$$II_\mathbf{A} = \Lambda_1\Lambda_2 + \Lambda_2\Lambda_3 + \Lambda_3\Lambda_1, \qquad (2.13.14\text{b})$$

$$III_\mathbf{A} = \Lambda_1\Lambda_2\Lambda_3 \qquad \blacksquare \qquad (2.13.14\text{c})$$

2.13.3 PROPERTIES OF EIGENVECTORS

The next three theorems provide some important properties of eigenvectors of a symmetric tensor.

THEOREM 2.13.3 Eigenvectors corresponding to two distinct eigenvalues of a symmetric tensor $\mathbf{A}$ are orthogonal. (Or, the principal directions corresponding to distinct principal values of $\mathbf{A}$ are orthogonal.)

Proof Let Λ_1 and Λ_2 be two distinct eigenvalues of $\mathbf{A}$, and $\mathbf{v}_1$ and $\mathbf{v}_2$ be the corresponding eigenvectors. Then we have

$$\mathbf{A}\mathbf{v}_1 = \Lambda_1\mathbf{v}_1; \qquad \mathbf{A}\mathbf{v}_2 = \Lambda_2\mathbf{v}_2 \qquad (2.13.15)$$

Using (2.13.15), (2.8.14) and the fact that $\mathbf{A}$ is symmetric, we find that

$$\Lambda_1\mathbf{v}_1 \cdot \mathbf{v}_2 = (\mathbf{A}\mathbf{v}_1) \cdot \mathbf{v}_2 = (\mathbf{A}\mathbf{v}_2) \cdot \mathbf{v}_1 = \Lambda_2\mathbf{v}_2 \cdot \mathbf{v}_1$$

Since $\Lambda_1 \neq \Lambda_2$, it follows that $\mathbf{v}_1 \cdot \mathbf{v}_2 = 0$. The vectors $\mathbf{v}_1$ and $\mathbf{v}_2$ are thus orthogonal. $\blacksquare$

THEOREM 2.13.4 For a symmetric tensor $\mathbf{A}$, there exist (at least) three eigenvectors that are mutually orthogonal. (Or, a symmetric tensor has at least three mutually perpendicular principal directions.)

Proof Since $\mathbf{A}$ is a symmetric tensor, it has got exactly three eigenvalues, Λ_1, Λ_2, Λ_3, by Theorem 2.13.2. Let $\mathbf{v}_1$, $\mathbf{v}_2$ and $\mathbf{v}_3$ be the corresponding eigenvectors. Since Λ_1, Λ_2 and Λ_3 are not necessarily distinct, the following three cases arise.

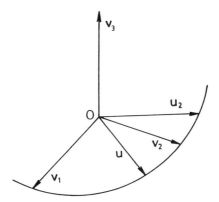

Figure 2.7. Theorem 2.13.4, case (ii).

Case (i). Suppose Λ_1, Λ_2 and Λ_3 are all distinct. Then, by Theorem 2.13.3, we have $\mathbf{v}_1 \cdot \mathbf{v}_2 = \mathbf{v}_2 \cdot \mathbf{v}_3 = \mathbf{v}_3 \cdot \mathbf{v}_1 = 0$. Thus, $\mathbf{v}_1$, $\mathbf{v}_2$ and $\mathbf{v}_3$ are mutually orthogonal.

Case (ii). Suppose two of the three eigenvalues are equal; say, $\Lambda_1 = \Lambda_2 \neq \Lambda_3$. Then $\mathbf{v}_1 \cdot \mathbf{v}_3 = \mathbf{v}_2 \cdot \mathbf{v}_3 = 0$, by Theorem 2.13.3; thus $\mathbf{v}_3$ is orthogonal to both $\mathbf{v}_1$ and $\mathbf{v}_2$ (see Figure 2.7).

Take any unit vector $\mathbf{u}$ coplanar with $\mathbf{v}_1$ and $\mathbf{v}_2$. Then, $\mathbf{u} = \alpha \mathbf{v}_1 + \beta \mathbf{v}_2$ for some numbers α and β, and we have

$$\mathbf{A}\mathbf{u} = \mathbf{A}[\alpha \mathbf{v}_1 + \beta \mathbf{v}_2]$$
$$= \alpha \mathbf{A}\mathbf{v}_1 + \beta \mathbf{A}\mathbf{v}_2$$

by the linear property of $\mathbf{A}$;

$$= \alpha(\Lambda_1 \mathbf{v}_1) + \beta(\Lambda_2 \mathbf{v}_2) = \Lambda_1(\alpha \mathbf{v}_1 + \beta \mathbf{v}_2)$$

since $\Lambda_1 = \Lambda_2$;

$$= \Lambda_1 \mathbf{u}$$

Thus $\mathbf{u}$ is an eigenvector corresponding to Λ_1. Since $\mathbf{u}$ is arbitrary, it follows that every unit vector coplanar with $\mathbf{v}_1$ and $\mathbf{v}_2$ is an eigenvector corresponding to Λ_1, and therefore there are infinitely many coplanar eigenvectors. Of these vectors choose a vector $\mathbf{u}_2$ that is orthogonal to $\mathbf{v}_1$. Then

$$\mathbf{v}_1 \cdot \mathbf{u}_2 = \mathbf{u}_2 \cdot \mathbf{v}_3 = \mathbf{v}_3 \cdot \mathbf{v}_1 = 0$$

showing that there exist (at least) three mutually orthogonal eigenvectors.

Case (iii). Suppose all the three eigenvalues are equal; that is, $\Lambda_1 = \Lambda_2 = \Lambda_3$. Then the characteristic equation should be of the following form:

$$(\Lambda_1 - \Lambda)^3 = 0$$

or

$$\begin{vmatrix} \Lambda_1 - \Lambda & 0 & 0 \\ 0 & \Lambda_1 - \Lambda & 0 \\ 0 & 0 & \Lambda_1 - \Lambda \end{vmatrix} = 0 \qquad (2.13.16)$$

Comparing the determinant in the lefthand side of (2.13.16) with that in (2.13.5), we find that $a_{11} = a_{22} = a_{33} = \Lambda_1$, other $a_{ij} = 0$. Thus $\mathbf{A} = \Lambda_1 \mathbf{I}$ so that for any vector $\mathbf{u}$, we have $\mathbf{Au} = \Lambda_1 \mathbf{u}$. Hence, $\mathbf{A}$ has infinitely many eigenvectors, all of which correspond to the same eigenvalue Λ_1. Of these vectors we can certainly choose (at least) three vectors $\mathbf{v}_1$, $\mathbf{v}_2$ and $\mathbf{v}_3$ that are mutually orthogonal. Thus, in this case also, there exist three mutually orthogonal eigenvectors for $\mathbf{A}$. This completes the proof. ∎

THEOREM 2.13.5 Given a symmetric tensor $\mathbf{A}$, there exists a (at least one) coordinate system with respect to which the matrix of $\mathbf{A}$ is diagonal. (Or, given a symmetric tensor $\mathbf{A}$, if the coordinate axes are chosen along the principal directions of $\mathbf{A}$, then with respect to these axes the matrix of $\mathbf{A}$ is diagonal.) (Such axes are called *principal axes* of $\mathbf{A}$.)

Proof By Theorem 2.13.4, there exist at least three eigenvectors $\mathbf{v}_1, \mathbf{v}_2, \mathbf{v}_3$ of $\mathbf{A}$ that are mutually orthogonal. Since $-\mathbf{v}$ is also an eigenvector when $\mathbf{v}$ is an eigenvector, we may choose these three mutually orthogonal eigenvectors in such a way that they form base vectors of a righthanded rectangular coordinate system.

We recall from (2.7.2) that, with respect to the base vectors $\mathbf{e}_i$, the components a_{ij} of $\mathbf{A}$ are given by $a_{ij} = \mathbf{e}_i \cdot \mathbf{Ae}_j$. Hence, with respect to the coordinate system with $\mathbf{v}_i$ as base vectors, we have

$$\begin{aligned} a_{11} &= \mathbf{v}_1 \cdot \mathbf{Av}_1 = \mathbf{v}_1 \cdot \Lambda_1 \mathbf{v}_1 = \Lambda_1 \\ a_{12} &= \mathbf{v}_1 \cdot \mathbf{Av}_2 = \mathbf{v}_1 \cdot \Lambda_2 \mathbf{v}_2 = 0 \end{aligned} \qquad (2.13.17)$$

Similarly,

$$a_{22} = \Lambda_2, \qquad a_{33} = \Lambda_3, \qquad a_{13} = a_{23} = 0 \qquad (2.13.18)$$

In view of (2.13.17) and (2.3.18), it follows that, with respect to the coordinate system having $\mathbf{v}_i$ as base vectors, the matrix of $\mathbf{A}$ is

$$[a_{ij}] = \begin{bmatrix} \Lambda_1 & 0 & 0 \\ 0 & \Lambda_2 & 0 \\ 0 & 0 & \Lambda_3 \end{bmatrix} \qquad (2.13.19)$$

which is a diagonal matrix and the proof is complete. ∎

90 2 ALGEBRA OF TENSORS

Note that the diagonal elements of the matrix (2.3.19) are the eigenvalues of **A**.

Summary: The results proved in Theorems 2.13.2 to 2.13.5 may be summarized as follows. A symmetric tensor possesses exactly three eigenvalues that are not necessarily distinct. The principal directions corresponding to distinct eigenvalues are orthogonal. There exist at least three mutually orthogonal principal directions. When the axes are chosen along these directions, the matrix of the tensor is purely diagonal with the eigenvalues as the diagonal elements.

EXAMPLE 2.13.1 Find the eigenvalues and eigenvectors of the tensor **A** whose matrix is

$$[\mathbf{A}] = \begin{bmatrix} 2 & 0 & 0 \\ 0 & 3 & 4 \\ 0 & 4 & -3 \end{bmatrix}$$

Solution The characteristic equation of the given tensor **A** is

$$\begin{vmatrix} 2 - \Lambda & 0 & 0 \\ 0 & 3 - \Lambda & 4 \\ 0 & 4 & -3 - \Lambda \end{vmatrix} = 0$$

Expanding the determinant we get the cubic equation

$$(2 - \Lambda)(\Lambda^2 - 25) = 0$$

whose roots are

$$\Lambda_1 = 2, \quad \Lambda_2 = 5, \quad \Lambda_3 = -5$$

These are the eigenvalues of the given tensor **A**.

To find an eigenvector **v** of **A**, associated with the eigenvalue $\Lambda_1 = 2$, we substitute for a_{ij} from the given matrix and take $\Lambda = 2$ in equations (2.13.4). Thus, we obtain the following three equations for the three components v_i of **v**:

$$2v_1 = 2v_1$$
$$3v_2 + 4v_3 = 2v_2$$
$$4v_2 - 3v_3 = 2v_3$$

The first of these equation is identically satisfied for *any* v_1, and the last two equations give $v_2 = v_3 = 0$. Since **v** has to be a unit vector, we must have $v_i v_i = 1$ from which it follows that $v_1 = \pm 1$. Thus, an eigenvector associated with the eigenvalue $\Lambda = 2$ is $\mathbf{v}_1 = \pm \mathbf{e}_1$.

For $\Lambda = \Lambda_2 = 5$, equations (2.13.4) yield $3v_1 = 0$, $v_2 - 2v_3 = 0$. These equations together with the condition $v_i v_i = 1$ give $v_1 = 0$, $v_2 = 2v_3 = \pm 2/\sqrt{5}$. Thus, an eigenvector associated with the eigenvalue $\Lambda_2 = 5$ is $\mathbf{v}_2 = \pm(1/\sqrt{5})(2\mathbf{e}_2 + \mathbf{e}_3)$.

Similarly, an eigenvector associated with the eigenvalue $\Lambda_3 = -5$ can be found to be $\mathbf{v}_3 = \pm(1/\sqrt{5})(\mathbf{e}_2 - 2\mathbf{e}_3)$. ■

EXAMPLE 2.13.2 Show that for an orthogonal tensor $\mathbf{Q}$, the only eigenvalues are ± 1. Show further that if $\mathbf{Q}$ is proper, then $+1$ is an eigenvalue, and that if $\mathbf{Q}$ is not proper, then -1 is an eigenvalue.

Solution Let Λ be an eigenvalue of $\mathbf{Q}$ with $\mathbf{v}$ as a corresponding eigenvector. Then

$$\Lambda^2 = \Lambda^2(\mathbf{v} \cdot \mathbf{v}) = \Lambda\mathbf{v} \cdot \Lambda\mathbf{v} = \mathbf{Q}\mathbf{v} \cdot \mathbf{Q}\mathbf{v}$$

Since $\mathbf{Q}\mathbf{v} \cdot \mathbf{Q}\mathbf{v} = \mathbf{v} \cdot \mathbf{v}$ (see Example 2.8.3), it follows that $\Lambda^2 = \mathbf{v} \cdot \mathbf{v} = 1$. Thus, an eigenvalue of $\mathbf{Q}$ should be ± 1.

Now, suppose that $\mathbf{Q}$ is proper; that is, $\det \mathbf{Q} = 1$. Then

$$\det(\mathbf{Q} - \mathbf{I}) = \det(\mathbf{Q} - \mathbf{I})^T = \det(\mathbf{Q}^T - \mathbf{I})$$
$$= \det(\mathbf{Q}^T - \mathbf{Q}^T\mathbf{Q}) = \det \mathbf{Q}^T(\mathbf{I} - \mathbf{Q})$$
$$= (\det \mathbf{Q}^T) \det(\mathbf{I} - \mathbf{Q}) = \det(\mathbf{I} - \mathbf{Q})$$
$$= -\det(\mathbf{Q} - \mathbf{I})$$

so that $\det(\mathbf{Q} - \mathbf{I}) = 0$, showing that 1 is an eigenvalue of $\mathbf{Q}$.

Next, suppose that $\mathbf{Q}$ is not proper; that is, $\det \mathbf{Q} = -1$. Then we obtain, following the steps of the previous case,

$$\det(\mathbf{Q} + \mathbf{I}) = \det(\mathbf{Q}^T) \det(\mathbf{Q} + \mathbf{I}) = -\det(\mathbf{Q} + \mathbf{I})$$

so that $\det(\mathbf{Q} + \mathbf{I}) = 0$, showing that -1 is an eigenvalue of $\mathbf{Q}$. ■

EXAMPLE 2.13.3 If $\mathbf{A}$ is a skew tensor, show that (i) $\mathbf{A}$ has only one eigenvalue, and (ii) there exists a unit vector $\mathbf{v}$ such that $\mathbf{A}\mathbf{v} = \mathbf{0}$.

Solution In Example 2.11.2, it has been shown that for a skew tensor $\mathbf{A}$, we have $I_\mathbf{A} = III_\mathbf{A} = 0$ and $II_\mathbf{A} = |\boldsymbol{\omega}|^2$, where $\boldsymbol{\omega}$ is the dual vector of $\mathbf{A}$. Hence in this case the characteristic equation (2.13.2) reduces to

$$\Lambda^3 + |\boldsymbol{\omega}|^2\Lambda = 0$$

This equation has only one real root, equal to 0. Thus $\mathbf{A}$ has only one eigenvalue, namely, 0. A corresponding eigenvector $\mathbf{v}$ is such that $\mathbf{A}\mathbf{v} = 0\mathbf{v} = \mathbf{0}$. ■

92 2 ALGEBRA OF TENSORS

EXAMPLE 2.13.4 Let $\mathbf{A}$ be tensor and Λ be an eigenvalue of $\mathbf{A}$ with $\mathbf{v}$ as a corresponding eigenvector. For any positive integer n, show that Λ^n is an eigenvalue of $\mathbf{A}^n$ with $\mathbf{v}$ continuing as a corresponding eigenvector.

Deduce that if $\mathbf{A}$ is invertible, then Λ^{-n}, $\Lambda \neq 0$, is an eigenvalue of $\mathbf{A}^{-n} = (\mathbf{A}^{-1})^n$ with $\mathbf{v}$ continuing as a corresponding eigenvector.

Solution Given that $\mathbf{Av} = \Lambda \mathbf{v}$, we have to show that

$$\mathbf{A}^n \mathbf{v} = \Lambda^n \mathbf{v} \tag{2.13.20}$$

for all positive integers n. Let us employ the method of induction.

For $n = 1$, (2.13.20) is true by data. We assume that (2.13.20) is true for $n = r$; that is,

$$\mathbf{A}^r \mathbf{v} = \Lambda^r \mathbf{v}$$

Then $\mathbf{A}^{r+1}\mathbf{v} = \mathbf{A}(\mathbf{A}^r \mathbf{v}) = \mathbf{A}(\Lambda^r \mathbf{v}) = \Lambda^r(\mathbf{Av}) = \Lambda^r(\Lambda \mathbf{v}) = \Lambda^{r+1}\mathbf{v}$.

This shows that (2.13.20) is valid for $n = r + 1$ as well. Hence, by induction, (2.13.20) is valid for all positive integers n.

If $\mathbf{A}$ is invertible, then $\mathbf{A}$ has a nonzero eigenvalue Λ and equation $\mathbf{Av} = \Lambda \mathbf{v}$ gives

$$\mathbf{A}^{-1}\mathbf{v} = \frac{1}{\Lambda}\mathbf{v} \tag{2.13.21}$$

showing that $1/\Lambda$ is an eigenvalue of $\mathbf{A}^{-1}$ with $\mathbf{v}$ as a corresponding eigenvector. Applying the result (2.13.20) to $\mathbf{A}^{-1}$ and denoting $(\mathbf{A}^{-1})^n$ by $\mathbf{A}^{-n}$, we obtain

$$\mathbf{A}^{-n}\mathbf{v} = \Lambda^{-n}\mathbf{v} \tag{2.13.22}$$

This shows that Λ^{-n} is an eigenvalue of $\mathbf{A}^{-n}$ with $\mathbf{v}$ continuing as a corresponding eigenvector. ∎

EXAMPLE 2.13.5 Let $\mathbf{A}$ be a symmetric tensor with Λ_i as eigenvalues and $\mathbf{v}_i$ as corresponding eigenvectors. Show that $\mathbf{A}$ can be represented as

$$\mathbf{A} = \sum_{k=1}^{3} \Lambda_k (\mathbf{v}_k \otimes \mathbf{v}_k) \tag{2.13.23}$$

This is known as the *spectral representation* of $\mathbf{A}$.

Solution In Example 2.7.3, it has been shown that every tensor $\mathbf{A}$ has the following representation in a coordinate system having $\mathbf{e}_i$ as base vectors: $\mathbf{A} = a_{ij}\mathbf{e}_i \otimes \mathbf{e}_j$. Hence, with respect to a coordinate system with the eigenvectors $\mathbf{v}_i$ as base vectors, a symmetric tensor $\mathbf{A}$ has the representation

$$\mathbf{A} = a_{ij}\mathbf{v}_i \otimes \mathbf{v}_j \tag{2.13.24}$$

2.13 EIGENVALUES AND EIGENVECTORS

Also, with respect to this coordinate system, the matrix of $\mathbf{A}$ is given by (2.3.19). Using (2.3.19), the representation (2.13.24) becomes

$$\mathbf{A} = \Lambda_1(\mathbf{v}_1 \otimes \mathbf{v}_1) + \Lambda_2(\mathbf{v}_2 \otimes \mathbf{v}_2) + \Lambda_3(\mathbf{v}_3 \otimes \mathbf{v}_3),$$

which is (2.13.23). ■

EXAMPLE 2.13.6 Show that for every tensor $\mathbf{A}$,

$$\mathbf{A}^3 - I_\mathbf{A}\mathbf{A}^2 + II_\mathbf{A}\mathbf{A} - III_\mathbf{A}\mathbf{I} = \mathbf{0} \qquad (2.13.25)$$

This result is known as the *Cayley-Hamilton theorem*.

Solution Consider the tensor $\mathbf{A} - \alpha\mathbf{I}$, where α is an arbitrary scalar, and denote its adjoint by $\mathbf{H}$. Then (2.11.23) and (2.13.5) yield

$$\mathbf{H}(\mathbf{A} - \alpha\mathbf{I}) = \{\det(\mathbf{A} - \alpha\mathbf{I})\}\mathbf{I}$$
$$= (-\alpha^3 + I_\mathbf{A}\alpha^2 - II_\mathbf{A}\alpha + III_\mathbf{A})\mathbf{I} \qquad (2.13.26)$$

The form of the righthand side of (2.13.26) suggests that $\mathbf{H}$ must be of the form

$$\mathbf{H} = \alpha^2 \mathbf{E} + \alpha \mathbf{F} + \mathbf{G} \qquad (2.13.27)$$

for some tensor $\mathbf{E}, \mathbf{F}, \mathbf{G}$. Substituting for $\mathbf{H}$ from (2.13.27) in the lefthand side of (2.13.26) and equating the corresponding terms (tensors), we get

$$\mathbf{E} = \mathbf{I}, \quad \mathbf{EA} - \mathbf{F} = I_\mathbf{A}\mathbf{I}, \quad \mathbf{G} - \mathbf{FA} = II_\mathbf{A}\mathbf{I}, \quad \mathbf{GA} = III_\mathbf{A}\mathbf{I}$$
$$(2.13.28)$$

Using these relations, we find that

$$\mathbf{A}^3 - I_\mathbf{A}\mathbf{A}^2 + II_\mathbf{A}\mathbf{A} - III_\mathbf{A}\mathbf{I} = \mathbf{A}^3 - (\mathbf{A} - \mathbf{F})\mathbf{A}^2 + (\mathbf{G} - \mathbf{FA})\mathbf{A} - \mathbf{GA}$$
$$(2.13.29)$$

Evidently, the righthand side of (2.13.29) is the zero tensor, and (2.13.25) is proven. ■

Note: When expressed in words, the identity (2.13.25) reads: Every tensor satisfies its own characteristic equation.

EXAMPLE 2.13.7 (i) If Λ is an eigenvalue of a tensor $\mathbf{A}$ with $\mathbf{v}$ as a corresponding eigenvector, show that $(\Lambda - \frac{1}{3}\operatorname{tr}\mathbf{A})$ is an eigenvalue of $\mathbf{A}^{(d)}$, with $\mathbf{v}$ as a corresponding eigenvector.

(ii) Deduce that, if $\mathbf{A}$ is symmetric and Λ_i are the eigenvalues of $\mathbf{A}$, then the eigenvalues of $\mathbf{A}^{(d)}$ are

$$\left.\begin{array}{l} \Lambda_1^{(d)} = \frac{1}{3}(2\Lambda_1 - \Lambda_2 - \Lambda_3) \\ \Lambda_2^{(d)} = \frac{1}{3}(2\Lambda_2 - \Lambda_3 - \Lambda_1) \\ \Lambda_3^{(d)} = \frac{1}{3}(2\Lambda_3 - \Lambda_1 - \Lambda_2) \end{array}\right\} \qquad (2.13.30)$$

(iii) Further, show that

$$II_{\mathbf{A}^{(d)}} = -\tfrac{1}{6}\{(\Lambda_1 - \Lambda_2)^2 + (\Lambda_2 - \Lambda_3)^2 + (\Lambda_3 - \Lambda_1)^2\}$$
$$III_{\mathbf{A}^{(d)}} = \tfrac{1}{27}(2\Lambda_1 - \Lambda_2 - \Lambda_3)(2\Lambda_2 - \Lambda_3 - \Lambda_1)(2\Lambda_3 - \Lambda_1 - \Lambda_2)$$
(2.13.31)

Solution (i) If Λ is an eigenvalue of $\mathbf{A}$ with $\mathbf{v}$ as a corresponding eigenvector, (2.12.1) yields

$$\mathbf{A}^{(d)}\mathbf{v} = \{\Lambda - \tfrac{1}{3}(\text{tr } \mathbf{A})\}\mathbf{v}$$

showing that $\{\Lambda - \tfrac{1}{3}(\text{tr } \mathbf{A})\}$ is an eigenvalue of $\mathbf{A}^{(d)}$ with $\mathbf{v}$ as a corresponding eigenvector.

(ii) If $\mathbf{A}$ is symmetric, then $\mathbf{A}^{(d)}$ is also symmetric, and from the result just proven it follows that the principal directions of $\mathbf{A}^{(d)}$ are coincident with those of $\mathbf{A}$ and that if Λ_i are the eigenvalues of $\mathbf{A}$, then $\Lambda_i^{(d)} = \{\Lambda_i - \tfrac{1}{3}(\text{tr } \mathbf{A})\}$ are the eigenvalues of $\mathbf{A}^{(d)}$. Since tr $\mathbf{A} = \Lambda_1 + \Lambda_2 + \Lambda_3$ by (2.13.14a), we get

$$\Lambda_i^{(d)} = \{\Lambda_i - \tfrac{1}{3}(\Lambda_1 + \Lambda_2 + \Lambda_3)\} \qquad (2.13.32)$$

from which expressions in (2.13.30) follow.

(iii) By virtue of the expressions (2.13.14b, c), we have

$$II_{\mathbf{A}^{(d)}} = \Lambda_1^{(d)}\Lambda_2^{(d)} + \Lambda_2^{(d)}\Lambda_3^{(d)} + \Lambda_3^{(d)}\Lambda_1^{(d)}$$
$$III_{\mathbf{A}^{(d)}} = \Lambda_1^{(d)}\Lambda_2^{(d)}\Lambda_3^{(d)}$$
(2.13.33)

Substituting for $\Lambda_i^{(d)}$ from expressions (2.13.30) into these, we readily obtain the expressions (2.13.31). ∎

2.14
POLAR DECOMPOSITION

In Section 2.9, it has been shown that every tensor $\mathbf{A}$ can be represented as a sum of two tensors, one symmetric and the other skew. Another useful representation of a tensor involving products of tensors is obtained in this section. We need some definitions and preliminary results before taking up the derivation of the representation.

2.14.1 POSITIVE DEFINITE TENSOR

A tensor $\mathbf{A}$ is said to be *positive definite* if

$$\mathbf{a} \cdot \mathbf{A}\mathbf{a} > 0 \qquad (2.14.1)$$

for every nonzero vector $\mathbf{a}$.

2.14 POLAR DECOMPOSITION

A trivial example of a positive definite tensor is the identity tensor **I**. In fact, every invertible tensor **A** yields two positive definite symmetric tensors: $\mathbf{A}^T\mathbf{A}$ and $\mathbf{A}\mathbf{A}^T$. This is proved in the following theorem.

THEOREM 2.14.1 For any tensor **A**, the tensors $\mathbf{A}\mathbf{A}^T$ and $\mathbf{A}^T\mathbf{A}$ are symmetric and if **A** is invertible, $\mathbf{A}\mathbf{A}^T$ and $\mathbf{A}^T\mathbf{A}$ are positive definite.

Proof For any tensor **A**, we have

$$(\mathbf{A}^T\mathbf{A})^T = \mathbf{A}^T(\mathbf{A}^T)^T = \mathbf{A}^T\mathbf{A}$$
$$(\mathbf{A}\mathbf{A}^T)^T = (\mathbf{A}^T)^T\mathbf{A}^T = \mathbf{A}\mathbf{A}^T$$
(2.14.2)

Hence, $\mathbf{A}^T\mathbf{A}$ and $\mathbf{A}\mathbf{A}^T$ are symmetric.

If **A** is invertible, then for any nonzero vector **a**, we have $\mathbf{A}\mathbf{a} \neq \mathbf{0}$. Because, if $\mathbf{A}\mathbf{a} = \mathbf{0}$, the $\mathbf{A}^{-1}(\mathbf{A}\mathbf{a}) = \mathbf{0}$ or $\mathbf{a} = \mathbf{0}$. Similarly, $\mathbf{A}^T\mathbf{a} \neq \mathbf{0}$. Consequently, we obtain, on using (2.8.14),

$$\mathbf{a} \cdot (\mathbf{A}^T\mathbf{A})\mathbf{a} = \mathbf{a} \cdot \mathbf{A}^T(\mathbf{A}\mathbf{a}) = \mathbf{A}\mathbf{a} \cdot \mathbf{A}\mathbf{a} > 0$$
$$\mathbf{a} \cdot (\mathbf{A}\mathbf{A}^T)\mathbf{a} = \mathbf{a} \cdot \mathbf{A}(\mathbf{A}^T\mathbf{a}) = \mathbf{A}^T\mathbf{a} \cdot \mathbf{A}^T\mathbf{a} > 0$$
(2.14.3)

Hence, $\mathbf{A}^T\mathbf{A}$ and $\mathbf{A}\mathbf{A}^T$ are positive definite. This completes the proof of the theorem. ■

The next theorem contains a characteristic property of positive definite tensors.

THEOREM 2.14.2 A positive definite tensor has only positive eigenvalues.

Proof Consider an eigenvalue Λ and an associated eigenvector **v** of a positive definite tensor **A**. Then,

$$\mathbf{v} \cdot \mathbf{A}\mathbf{v} = \mathbf{v} \cdot \Lambda\mathbf{v} = \Lambda(\mathbf{v} \cdot \mathbf{v}) = \Lambda$$

Since $\mathbf{a} \cdot \mathbf{A}\mathbf{a} > 0$ for any **a**, it follows that $\Lambda > 0$, and the proof is complete. ■

COROLLARY A positive definite symmetric tensor is invertible.

Proof Since a symmetric tensor has three eigenvalues, from Theorem 2.14.2 it follows that all the three eigenvalues of a positive definite symmetric tensor are positive. Consequently, the determinant of such a tensor is positive, and therefore, the tensor is invertible. ■

2.14.2 SQUARE ROOT OF A TENSOR

Given a tensor $\mathbf{A}$, if there exists a tensor $\mathbf{B}$ such that $\mathbf{B}^2 = \mathbf{A}$, then $\mathbf{B}$ is called a *Square root* of $\mathbf{A}$, denoted $\sqrt{\mathbf{A}}$.

For example a square root of the unit tensor $\mathbf{I}$ is the tensor whose matrix is

$$[\sqrt{\mathbf{I}}] = \begin{bmatrix} 1 & 0 & 0 \\ 0 & -1 & 0 \\ 0 & 0 & 1 \end{bmatrix}$$

THEOREM 2.14.3 Every positive definite symmetric tensor has a unique square root that itself is symmetric and positive definite.

Proof Let $\mathbf{A}$ be a positive definite symmetric tensor. Then the eigenvalues Λ_i of $\mathbf{A}$ are all positive and the corresponding eigenvectors $\mathbf{v}_i$ may be chosen to be mutually orthogonal. Also, $\mathbf{A}$ has the following spectral representation; see (2.13.23).

$$\mathbf{A} = \sum_{k=1}^{3} \Lambda_k (\mathbf{v}_k \otimes \mathbf{v}_k) \tag{2.14.4}$$

Let us now introduce a tensor $\mathbf{B}$ by

$$\mathbf{B} = \sum_{k=1}^{3} \sqrt{\Lambda_k} (\mathbf{v}_k \otimes \mathbf{v}_k) \tag{2.14.5}$$

The fact that $\mathbf{B}$ is a tensor is obvious from the nature of the righthand side of (2.14.5). By use of (2.8.15) we find from (2.14.5) that $\mathbf{B}^T = \mathbf{B}$; hence $\mathbf{B}$ is symmetric. Also, for any nonzero vector $\mathbf{a}$, we get

$$\mathbf{B}\mathbf{a} = \sum_{k=1}^{3} \sqrt{\Lambda_k} (\mathbf{v}_k \otimes \mathbf{v}_k)\mathbf{a} = \sum_{k=1}^{3} \sqrt{\Lambda_k} (\mathbf{v}_k \cdot \mathbf{a})\mathbf{v}_k$$

by using (2.4.37). Hence

$$\mathbf{a} \cdot \mathbf{B}\mathbf{a} = \sum_{k=1}^{3} \sqrt{\Lambda_k} (\mathbf{v}_k \cdot \mathbf{a})^2 > 0$$

Thus, $\mathbf{B}$ is positive definite.

Further,

$$\mathbf{B}^2 = \left[\sum_{k=1}^{3} \sqrt{\Lambda_k}(\mathbf{v}_k \otimes \mathbf{v}_k)\right]\left[\sum_{m=1}^{3} \sqrt{\Lambda_m}(\mathbf{v}_m \otimes \mathbf{v}_m)\right]$$

$$= \sum_{k=1}^{3} \sum_{m=1}^{3} \sqrt{\Lambda_k}\sqrt{\Lambda_m}(\mathbf{v}_k \otimes \mathbf{v}_k)(\mathbf{v}_m \otimes \mathbf{v}_m)$$

$$= \sum_{k=1}^{3} \sum_{m=1}^{3} \sqrt{\Lambda_k}\sqrt{\Lambda_m}(\mathbf{v}_k \cdot \mathbf{v}_m)(\mathbf{v}_k \otimes \mathbf{v}_m) \tag{2.14.6}$$

see (2.4.40). Noting that $\mathbf{v}_k \cdot \mathbf{v}_m = 0$ for $k \neq m$ and $\mathbf{v}_k \cdot \mathbf{v}_k = 1$, and using (2.14.4), we find from (2.14.6) that

$$\mathbf{B}^2 = \sum_{k=1}^{3} \Lambda_k (\mathbf{v}_k \otimes \mathbf{v}_k) = \mathbf{A} \qquad (2.14.7)$$

Thus, $\mathbf{B}$ is a positive definite symmetric tensor whose square is $\mathbf{A}$. That is, $\mathbf{B}$ is a square root of $\mathbf{A}$.

Suppose that there is another positive definite symmetric tensor $\mathbf{B}'$ whose square is also $\mathbf{A}$. If μ_i are the eigenvalues of $\mathbf{B}'$ and $\mathbf{w}_i$ are corresponding eigenvectors, then, from Example 2.13.4, it follows that μ_i^2 are eigenvalues of $\mathbf{A}$ and $\mathbf{w}_i$ are corresponding eigenvectors. Thus, $\mu_i^2 = \Lambda_i$ and $\mathbf{w}_i = \mathbf{v}_i$. As such, the spectral representation for $\mathbf{B}'$ is

$$\mathbf{B}' = \sum_{k=1}^{3} \sqrt{\Lambda_k} (\mathbf{v}_k \otimes \mathbf{v}_k)$$

Comparing this with (2.14.5), we find that $\mathbf{B} = \mathbf{B}'$. Thus $\mathbf{B}$ is unique, and the proof of the theorem is complete. ∎

COROLLARY If $\mathbf{A}$ is a positive definite symmetric tensor, then $\sqrt{\mathbf{A}}$ is invertible.

Proof The result follows from Theorem 2.14.3 and corollary to Theorem 2.14.2. ∎

We are now ready to establish the representation of a tensor as indicated in the beginning of the section.

2.14.3 POLAR DECOMPOSITION THEOREM

THEOREM 2.14.4 Every invertible tensor $\mathbf{A}$ can be represented in the form

$$\mathbf{A} = \mathbf{QU} = \mathbf{VQ} \qquad (2.14.8)$$

where $\mathbf{Q}$ is an orthogonal tensor and $\mathbf{U}$ and $\mathbf{V}$ are positive definite symmetric tensors such that $\mathbf{U}^2 = \mathbf{A}^T\mathbf{A}$ and $\mathbf{V}^2 = \mathbf{A}\mathbf{A}^T$. Furthermore, the representations are unique.

Proof Since $\mathbf{A}$ is invertible, the tensors $\mathbf{A}^T\mathbf{A}$ and $\mathbf{A}\mathbf{A}^T$ are symmetric and positive definite (by Theorem 2.14.1). Hence, these have unique positive definite symmetric square roots which are invertible (by Theorem 2.14.3 and its corollary). Let us denote the square root of $\mathbf{A}^T\mathbf{A}$ by $\mathbf{U}$ and define the tensor $\mathbf{Q}$ by

$$\mathbf{Q} = \mathbf{A}\mathbf{U}^{-1} \qquad (2.14.9)$$

2 ALGEBRA OF TENSORS

Then
$$Q^T Q = (AU^{-1})^T(AU^{-1}) = (U^T)^{-1}A^T A U^{-1}$$
$$= U^{-1} U^2 U^{-1} = I \qquad (2.14.10)$$

Hence Q is an orthogonal tensor. From expression (2.14.9), we also get
$$A = QU \qquad (2.14.11)$$

This is the first of the two representations specified in (2.14.8). To prove the uniqueness of this representation, we assume that A has another representation
$$A = \bar{Q}\bar{U} \qquad (2.14.12)$$
where $\bar{Q}$ is orthogonal and $\bar{U}$ is symmetric and positive definite. Then
$$A^T A = (\bar{Q}\bar{U})^T(\bar{Q}\bar{U}) = \bar{U}(\bar{Q}^T \bar{Q})\bar{U} = \bar{U}^2 \qquad (2.14.13)$$
showing that $\bar{U}$ is a square root of $A^T A$. But $A^T A$ has a unique square root U. As such $\bar{U} = U$. From (2.14.12) and (2.14.9) we then get
$$\bar{Q} = A\bar{U}^{-1} = AU^{-1} = Q \qquad (2.14.14)$$

Thus, the representation (2.14.11) is unique.

Next, let us define V by
$$V = QUQ^T \qquad (2.14.15)$$
Then
$$V^2 = (QUQ^T)(QUQ^T) = QU^2 Q^T = (QU)(QU)^T = AA^T \qquad (2.14.16)$$

Thus, V is the square root of AA^T; as such it is symmetric and positive definite. Also,
$$A = QU = (QUQ^T)Q = VQ \qquad (2.14.17)$$

This is the second of the representations specified in (2.14.8). Expression (2.14.15) shows that V is uniquely determined by Q and U. Hence this representation is also unique. The proof of the theorem is complete. ∎

Note: The representation $A = QU$ is called the *right polar decomposition* of A, and the representation $A = VQ$ the *left polar decomposition*. The name *polar decomposition* stems from the analogy between these representations and the polar representation $z = re^{i\theta}$ of a complex number z having r as its modulus and θ as the argument.

2.14 POLAR DECOMPOSITION

EXAMPLE 2.14.1 Find the square root of the tensor **B** whose matrix is

$$[\mathbf{B}] = \begin{bmatrix} 1 & 0 & 0 \\ 0 & 4 & 6 \\ 0 & 6 & 13 \end{bmatrix} \quad (2.14.18)$$

Solution We note that **B** is a symmetric tensor. It is easy to verify that it is positive definite also. If **U** is the square root of **B**, we have $\mathbf{U}^2 = \mathbf{B}$. Hence the components u_{ij} of **U** are given by $u_{ik}u_{kj} = b_{ij}$, where b_{ij} are components of **B**. Using (2.14.18), we find that

$$u_{1k}u_{1k} = 1, \quad u_{1k}u_{2k} = 0, \quad u_{1k}u_{3k} = 0$$
$$u_{2k}u_{2k} = 4, \quad u_{2k}u_{3k} = 6, \quad u_{3k}u_{3k} = 13$$

Solving these equations, we obtain

$$u_{11} = 1, \quad u_{23} = u_{32} = 2, \quad u_{33} = 3; \quad \text{and other } u_{ij} = 0$$

Thus,

$$[\mathbf{U}] = \begin{bmatrix} 1 & 0 & 0 \\ 0 & 0 & 2 \\ 0 & 2 & 3 \end{bmatrix} \quad (2.14.19)$$

This is the matrix of $\sqrt{\mathbf{B}}$. ∎

EXAMPLE 2.14.2 Find the polar decompositions of the tensor **A** whose matrix is

$$[\mathbf{A}] = \begin{bmatrix} 0 & 0 & -2 \\ 1 & 0 & 0 \\ 0 & 2 & 3 \end{bmatrix} \quad (2.14.20)$$

Solution For the given **A**, we readily note that $\det \mathbf{A} \neq 0$. Hence **A** is invertible.

From (2.14.20), we find that

$$[\mathbf{A}^T\mathbf{A}] = [\mathbf{A}^T][\mathbf{A}] = \begin{bmatrix} 1 & 0 & 0 \\ 0 & 4 & 6 \\ 0 & 6 & 13 \end{bmatrix} \quad (2.14.21)$$

If **U** is the square root of $\mathbf{A}^T\mathbf{A}$, the use of the result proved in Example 2.14.1 gives

$$[\mathbf{U}] = \begin{bmatrix} 1 & 0 & 0 \\ 0 & 0 & 2 \\ 0 & 2 & 3 \end{bmatrix} \qquad (2.14.22)$$

The inverse of the matrix [**U**] can be found as

$$[\mathbf{U}]^{-1} = [\mathbf{U}^{-1}] = \begin{bmatrix} 1 & 0 & 0 \\ 0 & -\tfrac{3}{4} & \tfrac{1}{2} \\ 0 & \tfrac{1}{2} & 0 \end{bmatrix} \qquad (2.14.23)$$

If we set $\mathbf{Q} = \mathbf{A}\mathbf{U}^{-1}$, then we get, from (2.14.20) and (2.14.23),

$$[\mathbf{Q}] = [\mathbf{A}][\mathbf{U}^{-1}] = \begin{bmatrix} 0 & -1 & 0 \\ 1 & 0 & 0 \\ 0 & 0 & 1 \end{bmatrix} \qquad (2.14.24)$$

With [**U**] and [**Q**] determined by (2.14.22) and (2.14.24), the polar decomposition $\mathbf{A} = \mathbf{Q}\mathbf{U}$ is obtained.

Since $\mathbf{A} = \mathbf{V}\mathbf{Q}$ is the other polar decomposition, we have

$$[\mathbf{V}] = [\mathbf{A}][\mathbf{Q}]^{-1} = [\mathbf{A}][\mathbf{Q}]^T$$

Using (2.14.20) and (2.4.24), we obtain

$$[\mathbf{V}] = \begin{bmatrix} 0 & 0 & -2 \\ 0 & 1 & 0 \\ -2 & 0 & 3 \end{bmatrix} \qquad (2.14.25)$$

With [**Q**] and [**V**] known from (2.14.24) and (2.14.25), the decomposition $\mathbf{A} = \mathbf{V}\mathbf{Q}$ is also obtained. ∎

EXAMPLE 2.14.3 Let **A** be an invertible tensor and $\mathbf{A} = \mathbf{Q}\mathbf{U} = \mathbf{V}\mathbf{Q}$ be the polar decompositions of **A**. Show that **U** and **V** have identical eigenvalues. Also, find the eigenvectors of **V** in terms of those of **U**. Deduce the representations of **Q** and **A** in terms of the eigenvalues and eigenvectors of **U** and **V**.

Solution We first note that **U** and **V** are symmetric tensors, and each of these has three eigenvalues and mutually orthogonal eigenvectors. Let Λ be an eigenvalue of **U** and **v** be a corresponding eigenvector. Then

$$(\mathbf{QU})\mathbf{v} = \mathbf{Q}(\mathbf{Uv}) = \mathbf{Q}(\Lambda\mathbf{v}) = \Lambda(\mathbf{Qv}) \qquad (2.14.26)$$

Since $\mathbf{QU} = \mathbf{VQ}$, (2.14.26) yields

$$\mathbf{V}(\mathbf{Qv}) = (\mathbf{VQ})\mathbf{v} = (\mathbf{QU})\mathbf{v} = \Lambda(\mathbf{Qv}) \qquad (2.14.27)$$

This shows that Λ is also an eigenvalue of **V** with **Qv** is a corresponding eigenvector.

Hence, if Λ_i are the eigenvalues of **U** and $\mathbf{v}_i$ are corresponding eigenvectors, then Λ_i are also eigenvalues of **V** with $\mathbf{Qv}_i$ as corresponding eigenvectors.

If $\mathbf{v}_i$ are taken as base vectors, we have from (2.4.36) and (2.13.23),

$$\mathbf{I} = \mathbf{v}_k \otimes \mathbf{v}_k, \qquad \mathbf{U} = \sum_{k=1}^{3} \Lambda_k(\mathbf{v}_k \otimes \mathbf{v}_k) \qquad (2.14.28)$$

Hence

$$\mathbf{Q} = \mathbf{QI} = \mathbf{Q}(\mathbf{v}_k \otimes \mathbf{v}_k) = (\mathbf{Qv}_k) \otimes \mathbf{v}_k \qquad (2.14.29)$$

$$\mathbf{A} = \mathbf{QU} = \mathbf{Q} \sum_{k=1}^{3} \Lambda_k(\mathbf{v}_k \otimes \mathbf{v}_k)$$

$$= \sum_{k=1}^{3} \Lambda_k \mathbf{Q}(\mathbf{v}_k \otimes \mathbf{v}_k) = \sum_{k=1}^{3} \Lambda_k(\mathbf{Qv}_k \otimes \mathbf{v}_k) \qquad (2.14.30)$$

These are the representations for **Q** and **A** in terms of the eigenvalues and eigenvectors of **U** and **V**. ∎

2.15

EXERCISES

1. Find the matrix $[\alpha_{ij}]$ if the x_i' system is obtained by rotating the x_i system in the sense of a righthanded screw through an angle
   (a) 90° about the x_1 axis.
   (b) 60° about the x_2 axis.

2. Compute α_{ij} for a pair of coordinate systems related by $\mathbf{e}_1' = -\mathbf{e}_3$, $\mathbf{e}_2' = \mathbf{e}_2$, $\mathbf{e}_3' = \mathbf{e}_1$. By what kind of rotation of the x_i system the x_i' system can be obtained? If a point P has coordinates $(1, 1, 0)$ in the x_i system, find its coordinates in the x_i' system. Also, if a point Q has coordinates $(1, 0, 1)$ in the x_i' system, find its coordinates in the x_i system.

2 ALGEBRA OF TENSORS

3. Complete the following table of direction cosines:

·	$\mathbf{e}_1$	$\mathbf{e}_2$	$\mathbf{e}_3$
$\mathbf{e}'_1$	$1/\sqrt{3}$	$1/\sqrt{3}$	?
$\mathbf{e}'_2$	0	$1/\sqrt{2}$	?
$\mathbf{e}'_3$	$-2/\sqrt{6}$	$1/\sqrt{6}$	?

4. The x'_i system is obtained by rotating the x_i system about the x_1 axis through an angle of 60° in the sense of a righthanded screw. If a plane has the equation $x_1 + 2x_2 - x_3 = 1$ in the x_i system, find its equation in the x'_i system.

5. Show that

$$\varepsilon_{ijk}\varepsilon_{pqr} = \begin{vmatrix} \alpha_{ip} & \alpha_{iq} & \alpha_{ir} \\ \alpha_{jp} & \alpha_{jq} & \alpha_{jr} \\ \alpha_{kp} & \alpha_{kq} & \alpha_{kr} \end{vmatrix}$$

Deduce that

$$\alpha_{ij} = \tfrac{1}{2}\varepsilon_{ipq}\varepsilon_{jrs}\alpha_{pr}\alpha_{qs}$$

6. Show that

$$\alpha_{ip}\alpha_{jq}\alpha_{kp}\alpha_{rs}\delta_{ij}\delta_{kr} = \delta_{qs}$$

7. The x'_i system is obtained by rotating the x_i system through an angle of 90° about the x_3 axis in the sense of a righthanded screw. A certain entity, which can be represented as an ordered triplet in every coordinate system, is represented by the triplet $(1, 2, 3)$ in the x_i system and by the triplet $(2, -1, 3)$ in the x'_i system. Verify whether this entity is a vector.

8. Show that $\mathbf{e}'_i = \alpha_{ij}\mathbf{e}_j$; $\mathbf{e}_i = \alpha_{ji}\mathbf{e}'_j$. Hence deduce the relations (2.2.8) and (2.2.9).

9. Using (2.2.8) and (2.2.9), deduce the relations (2.3.3) from (2.3.4) and vice-versa.

10. In the x_i system, a vector $\mathbf{a}$ has components

$$a_1 = 0, \quad a_2 = \frac{1}{\sqrt{2}}, \quad a_3 = -\frac{1}{\sqrt{2}}$$

and a tensor $\mathbf{A}$ has its matrix

$$[a_{ij}] = \begin{bmatrix} 1 & 0 & -1 \\ 0 & 1 & 0 \\ -1 & 0 & -1 \end{bmatrix}$$

If the x'_i system is as defined in Exercise 4, find the components of $\mathbf{a}$ and $\mathbf{A}$ in the x'_i system.

11. If α and β are scalars and a_{ij} and b_{ij} are components of tensors, show that $\alpha a_{ij} \pm \beta b_{ij}$ are components of a tensor.

2.15 EXERCISES

12. If a_i are components of a vector and b_{ij} are components of a tensor, show that $a_i b_{ij}$ are components of a vector.

13. Let a_i be components of a vector and b_{ij} be elements of a matrix, referred to the x_i system. If $a_i b_{ij}$ are components of a vector, show that b_{ij} are components of a tensor.

14. If a_{ij} and b_{ijk} are tensor components, show that so are $a_{ij} b_{krs}$.

15. Assuming that $\delta_{ij} = \delta'_{ij}$, show that δ_{ij} are components of a tensor.

16. Assuming that $\varepsilon_{ijk} = \varepsilon'_{ijk}$, show that ε_{ijk} are components of a tensor.

17. If a_i are components of a vector, show that $\varepsilon_{ijk} a_k$ are components of a skew tensor.

18. If a_{ij} are components of a tensor, show that $\varepsilon_{ijk} a_{jk}$ are components of a vector.

19. If a_{ijk} are components of a third-order tensor, show that $\varepsilon_{ijk} a_{ijk}$ is a scalar invariant.

20. If a_{ijkm} are components of a fourth-order tensor, show that a_{ijkk} are components of a second-order tensor.

21. If $\mathbf{a}$ is a unit vector, show that the tensors $\mathbf{a} \otimes \mathbf{a}$ and $\mathbf{I} - \mathbf{a} \otimes \mathbf{a}$ are equal to their own squares.

22. Prove the following identities:

(i) $(\alpha \mathbf{A})\mathbf{a} = \alpha(\mathbf{A}\mathbf{a})$

(ii) $(\mathbf{A} \pm \mathbf{B})\mathbf{a} = \mathbf{A}\mathbf{a} \pm \mathbf{B}\mathbf{a}$

(iii) $(\mathbf{A}\mathbf{B})\mathbf{a} = \mathbf{A}(\mathbf{B}\mathbf{a})$

(iv) $(\mathbf{A}\mathbf{B})(\mathbf{C}\mathbf{a}) = \mathbf{A}[\mathbf{B}(\mathbf{C}\mathbf{a})]$

(v) $[\mathbf{A}(\mathbf{B}\mathbf{C})]\mathbf{a} = \mathbf{A}[(\mathbf{B}\mathbf{C})\mathbf{a}] = \mathbf{A}[\mathbf{B}(\mathbf{C}\mathbf{a})]$

23. If $\mathbf{A}$ is a tensor such that $\mathbf{A} \cdot \mathbf{B} = 0$ for every tensor $\mathbf{B}$, show that $\mathbf{A}$ is the zero tensor.

24. Show that there is no pair of vectors $\mathbf{a}$ and $\mathbf{b}$ such that $\mathbf{a} \otimes \mathbf{b} = \mathbf{I}$.

25. Show that

(i) $(\mathbf{a} \otimes \mathbf{b}) \cdot (\mathbf{c} \otimes \mathbf{d}) = (\mathbf{a} \cdot \mathbf{c})(\mathbf{b} \cdot \mathbf{d})$ (ii) $(\mathbf{e}_i \otimes \mathbf{e}_j) \cdot (\mathbf{e}_k \otimes \mathbf{e}_m) = \delta_{ik} \delta_{jm}$

26. Prove the following identities:

(i) $\mathbf{a} \cdot \mathbf{A}\mathbf{b} = \mathbf{A} \cdot (\mathbf{a} \otimes \mathbf{b})$

(ii) $(\mathbf{a} \otimes \mathbf{b})\mathbf{A} = \mathbf{a} \otimes (\mathbf{A}^T \mathbf{b})$

(iii) $(\mathbf{A}\mathbf{e}_p) \otimes \mathbf{e}_p = \mathbf{A}$

27. If $\mathbf{b}$ is a vector such that $\mathbf{a} \otimes \mathbf{b} = 0$ for every vector $\mathbf{a}$, show that $\mathbf{b} = \mathbf{0}$.

28. If $\mathbf{a}$ is a unit vector, show that $\mathbf{I} - 2\mathbf{a} \otimes \mathbf{a}$ is an orthogonal tensor. Is this a proper orthogonal tensor?

2 ALGEBRA OF TENSORS

29. A linear transformation $\mathbf{A}$ transforms the base vectors $\mathbf{e}_i$ as follows: $\mathbf{Ae}_1 = 2\mathbf{e}_2$, $\mathbf{Ae}_2 = \mathbf{e}_1 + \mathbf{e}_3$, $\mathbf{Ae}_3 = \mathbf{e}_1$. Find the components of $\mathbf{A}$. Also find the vector to which $\mathbf{v} = \mathbf{e}_1 - \mathbf{e}_2 + \mathbf{e}_3$ is transformed under $\mathbf{A}$.

30. A transformation $\mathbf{A}$ transforms every vector $\mathbf{a}$ to the vector $\alpha\mathbf{a}$ where α is a scalar. Show that $\mathbf{A}$ is an isotropic tensor.

31. Suppose $\mathbf{A}$ is an entity defined by $\mathbf{A} = a_{ij}\mathbf{e}_i \otimes \mathbf{e}_j$, where a_{ij} are real numbers. Show that $\mathbf{A}$ is a tensor with a_{ij} as components.

32. Show that a vector is isotropic if and only if it is the zero vector.

33. Let a_{ijk} be components of a tensor of order 3. Show that this tensor is isotropic if and only if $a_{ijk} = \alpha\varepsilon_{ijk}$ for some scalar α.

34. Verify the identities (2.8.2) to (2.8.5) and (2.8.10) to (2.8.13).

35. Let $\mathbf{A}, \mathbf{B}, \mathbf{C}$ be tensors. Show that

$$\mathbf{A} \cdot (\mathbf{BC}) = (\mathbf{B}^T\mathbf{A}) \cdot \mathbf{C} = (\mathbf{AC}^T) \cdot \mathbf{B}$$

36. If the matrix of a tensor is symmetric in one coordinate system, show that it is symmetric in all coordinate systems.

37. If the matrix of a tensor is skew-symmetric in one coordinate system, show that it is skew-symmetric in all coordinate systems.

38. The product $\mathbf{b} \wedge \mathbf{A}$ of a vector $\mathbf{b}$ and a tensor $\mathbf{A}$ is defined in Example 2.8.5. Show that $(\mathbf{b} \wedge \mathbf{I})\mathbf{c} = \mathbf{b} \times \mathbf{c}$ for any vector $\mathbf{c}$.

39. If $\mathbf{A}$ is a symmetric tensor and $\mathbf{B}$ is an arbitrary tensor, show that

$$\mathbf{A} \cdot \mathbf{B} = \mathbf{A} \cdot \mathbf{B}^T = \tfrac{1}{2}\mathbf{A} \cdot (\mathbf{B} + \mathbf{B}^T)$$

40. If $\mathbf{A}$ is a skew tensor and $\mathbf{B}$ is an arbitrary tensor, show that

$$\mathbf{A} \cdot \mathbf{B} = -\mathbf{A} \cdot \mathbf{B}^T = \tfrac{1}{2}\mathbf{A} \cdot (\mathbf{B} - \mathbf{B}^T)$$

41. Prove that the representation of a second-order tensor as a sum of a symmetric tensor and a skew tensor is unique.

42. Prove the following:
   (i) If $\mathbf{A} \cdot \mathbf{B} = 0$ for every symmetric tensor $\mathbf{B}$, then the tensor $\mathbf{A}$ is skew.
   (ii) If $\mathbf{A} \cdot \mathbf{B} = 0$ for every skew tensor $\mathbf{B}$, then the tensor $\mathbf{A}$ is symmetric.

43. For every vector $\mathbf{a}$ and every tensor $\mathbf{A}$, show that
   (i) $\qquad\qquad\qquad\mathbf{a} \cdot (\text{skw } \mathbf{A})\mathbf{a} = 0$
   (ii) $\qquad\qquad\qquad\mathbf{a} \cdot (\text{sym } \mathbf{A})\mathbf{a} = \mathbf{a} \cdot \mathbf{Aa}$

44. Find the dual vector **ω** of the skew part of the tensor **A** whose matrix is

$$[\mathbf{A}] = \begin{bmatrix} 1 & -2 & 5 \\ 2 & 0 & -3 \\ -5 & 3 & 2 \end{bmatrix}$$

If $\mathbf{u} = 2\mathbf{e}_1 - 3\mathbf{e}_2 + \mathbf{e}_3$, verify that $(\text{skw } \mathbf{A})\mathbf{u} = \boldsymbol{\omega} \times \mathbf{u}$.

45. Find the skew tensor whose dual vector is $\boldsymbol{\omega} = \mathbf{e}_1 - 2\mathbf{e}_2 + 3\mathbf{e}_3$.

46. If a_i are components of a vector, show that

$$\begin{bmatrix} 0 & a_3 & -a_2 \\ -a_3 & 0 & a_1 \\ a_2 & -a_1 & 0 \end{bmatrix}$$

is the matrix of a tensor.

47. Let **a** and **b** be arbitrary vectors. Show that $(\mathbf{a} \otimes \mathbf{b} - \mathbf{b} \otimes \mathbf{a})$ is a skew tensor and that $\mathbf{b} \times \mathbf{a}$ is its dual vector. Deduce that $\frac{1}{2}(\mathbf{b} \times \mathbf{a})$ is the dual vector of $\text{skw}(\mathbf{a} \otimes \mathbf{b})$. (The tensor $(\mathbf{a} \otimes \mathbf{b} - \mathbf{b} \otimes \mathbf{a})$ is called the *exterior product* of **a** and **b**.)

48. Let **A** and **B** be tensors; prove the following

(i) $\quad\quad\quad\quad\quad\quad\quad\quad \text{tr } \mathbf{A} = \text{tr}(\text{sym } \mathbf{A})$

(ii) $\quad\quad\quad\quad\quad\quad\quad\quad \text{tr}(\mathbf{AB}) = \text{tr}(\mathbf{BA})$

(iii) If **A** is symmetric and **B** is skew, then $\text{tr}(\mathbf{AB}) = 0$

(iv) $\quad\quad\quad\quad\quad\quad\quad\quad \text{tr}(\mathbf{A}^T \mathbf{B}) = \mathbf{A} \cdot \mathbf{B}$

49. Let **A** and **B** be skew tensors and **u** and **v** be their respective dual vectors. Show that

(i) $\mathbf{AB} = \mathbf{v} \otimes \mathbf{u} - (\mathbf{u} \cdot \mathbf{v})\mathbf{I}$ $\quad$ (ii) $\text{tr}(\mathbf{AB}) = 2(\mathbf{u} \cdot \mathbf{v})$

50. Show that the equation $\mathbf{AB} = \mathbf{BA}$ holds for two skew tensors **A** and **B** if and only if they have the same dual vectors.

51. Prove the identities (2.11.12).

52. Show that a skew tensor is not invertible.

53. Show that $\mathbf{e}_k \otimes \mathbf{e}'_k$ is a proper orthogonal tensor.

54. Let **A*** be the adjugate of a tensor **A** (see Example 2.11.3). Prove the following:

(i) $\quad\quad\quad\quad\quad\quad (\mathbf{A}^T)^* = (\mathbf{A}^*)^T$

(ii) $\quad\quad\quad\quad\quad\quad (\mathbf{A}^*)^* = (\det \mathbf{A})\mathbf{A}$

(iii) $\quad\quad\quad\quad\quad\quad \text{tr } \mathbf{A}^* = II_\mathbf{A}$

(iv) $\quad\quad\quad\quad\quad\quad \mathbf{A}(\mathbf{a} \times \mathbf{A}^T \mathbf{b}) = (\mathbf{A}^*\mathbf{a}) \times \mathbf{b}$, for all vectors **a** and **b**

55. By employing the Cayley–Hamilton theorem, show that

(i) $\quad (\mathbf{A}^*)^T = \mathbf{A}^2 - I_A \mathbf{A} + II_A \mathbf{I}$

(ii) $\quad$ If $\mathbf{A}$ is invertible, $\mathbf{A}^{-1} = (III_\mathbf{A})^{-1}\{\mathbf{A}^2 - I_A \mathbf{A} + II_\mathbf{A}\mathbf{I}\}$

56. Obtain the following expressions for the principal invariants of a tensor $\mathbf{A}$.

$$I_\mathbf{A} = \tfrac{1}{2}\varepsilon_{ipq}\varepsilon_{jpq}a_{ij}$$
$$II_\mathbf{A} = \tfrac{1}{2}\varepsilon_{ijk}\varepsilon_{rsk}a_{ir}a_{js}$$
$$III_\mathbf{A} = \tfrac{1}{6}\varepsilon_{ijk}\varepsilon_{pqr}a_{ip}a_{jq}a_{kr}$$

57. For any tensor $\mathbf{A}$ and any invertible tensor $\mathbf{B}$, prove that

$$I_{\mathbf{BAB}^{-1}} = I_\mathbf{A}, \qquad II_{\mathbf{BAB}^{-1}} = II_\mathbf{A}, \qquad III_{\mathbf{BAB}^{-1}} = III_\mathbf{A}$$

58. For any tensor $\mathbf{A}$ and any vectors $\mathbf{u}$ and $\mathbf{v}$, prove that

$$\mathbf{A}^T(\mathbf{Au} \times \mathbf{Av}) = III_\mathbf{A}(\mathbf{u} \times \mathbf{v})$$

59. If $\mathbf{A}$ is a tensor, show that for any noncoplanar vectors $\mathbf{u}, \mathbf{v}, \mathbf{w}$

(i) $\quad [\mathbf{u}, \mathbf{v}, \mathbf{w}]I_\mathbf{A} = [\mathbf{Au}, \mathbf{v}, \mathbf{w}] + [\mathbf{u}, \mathbf{Av}, \mathbf{w}] + [\mathbf{u}, \mathbf{v}, \mathbf{Aw}]$

(ii) $\quad [\mathbf{u}, \mathbf{v}, \mathbf{w}]II_\mathbf{A} = [\mathbf{Au}, \mathbf{Av}, \mathbf{w}] + [\mathbf{Au}, \mathbf{v}, \mathbf{Aw}] + [\mathbf{u}, \mathbf{Av}, \mathbf{Aw}]$

(iii) $\quad [\mathbf{u}, \mathbf{v}, \mathbf{w}]III_\mathbf{A} = [\mathbf{Au}, \mathbf{Av}, \mathbf{Aw}]$

60. For any vectors $\mathbf{a}, \mathbf{b}, \mathbf{c}, \mathbf{u}, \mathbf{v}$, prove that

$$[\mathbf{a}, \mathbf{b}, \mathbf{c}]\,\mathrm{tr}(\mathbf{u} \otimes \mathbf{v}) = [(\mathbf{a}\cdot\mathbf{v})\mathbf{u}, \mathbf{b}, \mathbf{c}] + [\mathbf{a}, (\mathbf{b}\cdot\mathbf{v})\mathbf{u}, \mathbf{c}] + [\mathbf{a}, \mathbf{b}, (\mathbf{c}\cdot\mathbf{v})\mathbf{u}]$$
$$= (\mathbf{a}\cdot\mathbf{u})[\mathbf{u}, \mathbf{b}, \mathbf{c}] + (\mathbf{b}\cdot\mathbf{v})[\mathbf{a}, \mathbf{u}, \mathbf{c}] + (\mathbf{c}\cdot\mathbf{v})[\mathbf{a}, \mathbf{b}, \mathbf{u}]$$

61. Show that

$$III_\mathbf{A} = \tfrac{1}{3}\{\mathrm{tr}\,\mathbf{A}^3 - I_\mathbf{A}\,\mathrm{tr}\,\mathbf{A}^2 + II_\mathbf{A}\,\mathrm{tr}\,\mathbf{A}\}$$
$$= \tfrac{1}{3}\{\mathrm{tr}\,\mathbf{A}^3 - \tfrac{3}{2}(\mathrm{tr}\,\mathbf{A}^2)(\mathrm{tr}\,\mathbf{A}) + \tfrac{1}{2}(\mathrm{tr}\,\mathbf{A})^3\}$$

62. If $I_1 = \mathrm{tr}\,\mathbf{A}$, $I_2 = \mathrm{tr}\,\mathbf{A}^2$, $I_3 = \mathrm{tr}\,\mathbf{A}^3$, show that

$$\frac{\partial I_1}{\partial a_{ij}} = \delta_{ij}, \qquad \frac{\partial I_2}{\partial a_{ij}} = 2a_{ij}, \qquad \frac{\partial I_3}{\partial a_{ij}} = 3a_{ik}a_{kj}$$

63. If $\mathbf{A}$ is an invertible tensor, show that

$$I_{\mathbf{A}^{-1}} = \frac{II_\mathbf{A}}{III_\mathbf{A}}, \qquad II_{\mathbf{A}^{-1}} = \frac{I_\mathbf{A}}{III_\mathbf{A}}, \qquad III_{\mathbf{A}^{-1}} = \frac{1}{III_\mathbf{A}}$$

64. Show that

$$III_{\mathbf{A}^{(d)}} = III_\mathbf{A} - \tfrac{1}{3}I_\mathbf{A}II_\mathbf{A} + \tfrac{2}{27}I_\mathbf{A}^3$$

2.15 EXERCISES

65. Find the principal invariants of tensors whose matrices are given below:

(i) $\begin{bmatrix} 1 & 0 & 0 \\ 0 & 2 & 3 \\ 0 & 0 & 2 \end{bmatrix}$ (ii) $\begin{bmatrix} 1 & 0 & -1 \\ 2 & -1 & 0 \\ 0 & 1 & -2 \end{bmatrix}$

Hence find their eigenvalues.

66. For the tensors with the following matrices, find the eigenvalues and the principal directions:

(i) $\begin{bmatrix} 2 & 0 & 0 \\ 0 & 2 & 0 \\ 0 & 0 & 3 \end{bmatrix}$ (ii) $\begin{bmatrix} 3 & -1 & 0 \\ -1 & 3 & 0 \\ 0 & 0 & 1 \end{bmatrix}$

(iii) $\begin{bmatrix} 1 & 1 & 1 \\ 1 & 1 & 1 \\ 1 & 1 & 1 \end{bmatrix}$ (iv) $\begin{bmatrix} 1 & 0 & 2 \\ 0 & 1 & 0 \\ 2 & 0 & -2 \end{bmatrix}$

67. If Λ is an eigenvalue of a tensor **A**, show that for any vectors **a**, **b**, **c**,

$$[\mathbf{Aa} - \Lambda\mathbf{a}, \mathbf{Ab} - \Lambda\mathbf{b}, \mathbf{Ac} - \Lambda\mathbf{c}] = 0$$

68. For the tensor $\mathbf{a} \otimes \mathbf{b}$, show that $\mathbf{a} \cdot \mathbf{b}$ is an eigenvalue and $\hat{\mathbf{a}}$ is a corresponding eigenvector.

69. Let **A** be a symmetric tensor and **a** be a unit vector. Show that **a** is an eigenvector of **A** if and only if $\mathbf{A}(\mathbf{a} \otimes \mathbf{a}) = (\mathbf{a} \otimes \mathbf{a})\mathbf{A}$.

70. If **A** is a positive definite symmetric tensor, show that $\mathbf{A}^{-1}$ is also positive definite and symmetric and that $\sqrt{\mathbf{A}^{-1}} = (\sqrt{\mathbf{A}})^{-1}$.

71. Show that a symmetric tensor is positive definite if and only if its fundamental invariants are positive.

72. Find the square roots of the tensors whose matrices follow:

(i) $\begin{bmatrix} 1 & 0 & -1 \\ 0 & 1 & 0 \\ -1 & 0 & 0 \end{bmatrix}$ (ii) $\begin{bmatrix} 1 & 0 & -3 \\ 0 & 4 & 0 \\ -3 & 0 & 4 \end{bmatrix}$

73. Find the polar decompositions of the tensors whose matrices follow:

(i) $\begin{bmatrix} 1 & 0 & 0 \\ 2 & 1 & 0 \\ 0 & 0 & 1 \end{bmatrix}$ (ii) $\begin{bmatrix} 1 & 0 & 0 \\ 0 & 2 & 3 \\ 0 & 0 & 2 \end{bmatrix}$

(iii) $\begin{bmatrix} 2 & -1 & 0 \\ 2 & 1 & 0 \\ 0 & 0 & -1 \end{bmatrix}$ (iv) $\begin{bmatrix} 0 & 1 & 0 \\ 0 & 0 & 2 \\ -2 & 0 & 3 \end{bmatrix}$

74. In the polar decomposition $\mathbf{A} = \mathbf{QU} = \mathbf{VQ}$ of an invertible tensor $\mathbf{A}$, show that $\mathbf{U}$ and $\mathbf{V}$ have the same fundamental invariants.

75. Let $\mathbf{A} = \mathbf{QU} = \mathbf{VQ}$ be the polar decomposition of an invertible tensor $\mathbf{A}$. If Λ_i are the eigenvalues of $\mathbf{U}$ and $\mathbf{v}_i$ are the corresponding eigenvectors, show that $\mathbf{A}^{-1}$ can be represented in the form

$$\mathbf{A}^{-1} = \sum_{k=1}^{3} (\Lambda_k)^{-1} \mathbf{v}_k \otimes (\mathbf{Q}\mathbf{v}_k)$$

CHAPTER 3
CALCULUS OF TENSORS

3.1 INTRODUCTION

This chapter deals with a brief study of differential and integral calculus of vector and tensor functions. It is assumed that the reader is already familiar with calculus of several variables, particularly vector calculus.

3.2 SCALAR, VECTOR AND TENSOR FUNCTIONS

By a *vector* or a *tensor function*, we mean a vector or a tensor (respectively) whose components are real-valued functions of one or more real variables. Thus, if the components a_i of a vector **a** are real-valued functions of a (single) real variable t, we say that **a** is a vector function of t, denoted $\mathbf{a}(t)$. Similarly, if the components a_{ij} of a tensor **A** are real-valued functions of t, we say that **A** is a tensor function of t, denoted $\mathbf{A}(t)$. Also, if a scalar ϕ is a real-valued function of t, it is denoted as usual by $\phi(t)$. Generally, t is assumed to vary over a specified interval on the real line.

3 CALCULUS OF TENSORS

In studying scalar, vector and tensor functions, we assume that the components being dealt with are defined with respect to *fixed* sets of coordinate axes and that the components possess derivatives of any desired order.

Thus, if $a_i(t)$ are components of a vector function $\mathbf{a}(t)$, we assume that the nth derivatives $[d^n/dt^n]\{a_i(t)\}$ exist for any desired n. It is easy to see that $[d^n/dt^n]\{a_i(t)\}$ are also components of a vector. Because, if we set $c_i = [d^n/dt^n]\{a_i(t)\}$ in all coordinate systems, we find that

$$c_i' = \frac{d^n}{dt^n}\{a_i'(t)\} = \frac{d^n}{dt^n}\{\alpha_{ik} a_k(t)\},$$

using (2.3.5a);

$$= \alpha_{ik} \frac{d^n}{dt^n}\{a_k(t)\} = \alpha_{ik} c_k$$

The vector whose components are $[d^n/dt^n]\{a_i(t)\}$ is called the nth *derivative of* $\mathbf{a}$, denoted $d^n\mathbf{a}/dt^n$. In particular, the first derivative $d\mathbf{a}/dt$ is the vector whose components are da_i/dt; that is,

$$\left[\frac{d\mathbf{a}}{dt}\right]_i = \frac{da_i}{dt} = \frac{d}{dt}\{[\mathbf{a}]_i\} \qquad (3.2.1)$$

Obviously, if $\mathbf{a}$ is a constant vector, then its derivatives of all orders are equal to the zero vector.

Similarly, if $a_{ij}(t)$ are components of a tensor function $\mathbf{A}(t)$, we assume that the nth derivatives $[d^n/dt^n]\{a_{ij}(t)\}$ exist for any desired n. It may be verified that these derivatives are components of a tensor; this tensor is called the nth *derivative of* $\mathbf{A}$, denoted $d^n\mathbf{A}/dt^n$. In particular, the first derivative $d\mathbf{A}/dt$ is a tensor whose components are $[d/dt]\{a_{ij}(t)\}$; that is,

$$\left[\frac{d\mathbf{A}}{dt}\right]_{ij} = \frac{d}{dt}\{a_{ij}(t)\} = \frac{d}{dt}\{[\mathbf{A}]_{ij}\} \qquad (3.2.2)$$

Obviously, if $\mathbf{A}$ is a constant tensor, then its derivatives of all orders are equal to the zero tensor.

The following identities follow from the usual rules of differentiation:

$$\frac{d}{dt}(\mathbf{a} \pm \mathbf{b}) = \frac{d\mathbf{a}}{dt} \pm \frac{d\mathbf{b}}{dt} \qquad (3.2.3)$$

$$\frac{d}{dt}(\phi\mathbf{a}) = \frac{d\phi}{dt}\mathbf{a} + \phi\frac{d\mathbf{a}}{dt} \qquad (3.2.4)$$

$$\frac{d}{dt}(\mathbf{a} \cdot \mathbf{b}) = \mathbf{a} \cdot \frac{d\mathbf{b}}{dt} + \frac{d\mathbf{a}}{dt} \cdot \mathbf{b} \qquad (3.2.5)$$

3.2 SCALAR, VECTOR AND TENSOR FUNCTIONS

$$\frac{d}{dt}(\mathbf{a} \times \mathbf{b}) = \mathbf{a} \times \frac{d\mathbf{b}}{dt} + \frac{d\mathbf{a}}{dt} \times \mathbf{b} \tag{3.2.6}$$

$$\frac{d}{dt}(\mathbf{a} \otimes \mathbf{b}) = \mathbf{a} \otimes \frac{d\mathbf{b}}{dt} + \frac{d\mathbf{a}}{dt} \otimes \mathbf{b} \tag{3.2.7}$$

$$\frac{d}{dt}(\mathbf{A} \pm \mathbf{B}) = \frac{d\mathbf{A}}{dt} \pm \frac{d\mathbf{B}}{dt} \tag{3.2.8}$$

$$\frac{d}{dt}(\phi \mathbf{A}) = \frac{d\phi}{dt}\mathbf{A} + \phi \frac{d\mathbf{A}}{dt} \tag{3.2.9}$$

$$\frac{d}{dt}(\mathbf{A} \cdot \mathbf{B}) = \frac{d\mathbf{A}}{dt} \cdot \mathbf{B} + \mathbf{A} \cdot \frac{d\mathbf{B}}{dt} \tag{3.2.10}$$

$$\frac{d}{dt}(\mathbf{AB}) = \left(\frac{d\mathbf{A}}{dt}\right)\mathbf{B} + \mathbf{A}\left(\frac{d\mathbf{B}}{dt}\right) \tag{3.2.11}$$

$$\frac{d}{dt}(\mathbf{A}\mathbf{a}) = \left(\frac{d\mathbf{A}}{dt}\right)\mathbf{a} + \mathbf{A}\left(\frac{d\mathbf{a}}{dt}\right) \tag{3.2.12}$$

$$\frac{d}{dt}(\mathbf{A}^T) = \left(\frac{d\mathbf{A}}{dt}\right)^T \tag{3.2.13}$$

where ϕ is a real-valued (scalar) function of t, $\mathbf{a}$ and $\mathbf{b}$ are vector functions of t, and $\mathbf{A}$ and $\mathbf{B}$ are tensor functions of t.

Next, let us consider scalar, vector and tensor functions of more than one real variable. The independent real variables of our main interest in this text are the three coordinates x_i of a point $\mathbf{x}$. Generally, point $\mathbf{x}$ is assumed to vary over some region of three-dimensional space. A vector $\mathbf{a}$ whose components a_i are real-valued functions of x_i is called a *vector point function* or a *vector field*, denoted $\mathbf{a}(x_i)$ or $\mathbf{a}(\mathbf{x})$. Similarly, a tensor $\mathbf{A}$ whose components a_{ij} are real-valued functions of x_i is called a *tensor point function* or a *tensor field*, denoted $\mathbf{A}(x_i)$ or $\mathbf{A}(\mathbf{x})$. Also, a scalar ϕ that is a real-valued function of x_i is called a *scalar field*, denoted $\phi(x_i)$ or $\phi(\mathbf{x})$.

Scalars, vectors and tensors arising in physical situations are generally functions of both x_i and t, where t is the time variable. Such functions are analyzed by the combined use of properties of scalar, vector and tensor functions of t and those of scalar, vector and tensor fields.

EXAMPLE 3.2.1 Prove the identities (3.2.11) and (3.2.13).

Solution Let $a_{ij}(t)$ and $b_{ij}(t)$ be components of **A** and **B**, respectively. Then

$$\left[\frac{d}{dt}(\mathbf{AB})\right]_{ij} = \frac{d}{dt}\{[\mathbf{AB}]_{ij}\} = \frac{d}{dt}(a_{ik}b_{kj})$$

$$= \frac{da_{ik}}{dt}b_{kj} + a_{ik}\frac{db_{kj}}{dt}$$

$$= \left[\frac{d\mathbf{A}}{dt}\mathbf{B}\right]_{ij} + \left[\mathbf{A}\frac{d\mathbf{B}}{dt}\right]_{ij}$$

This proves the identity (3.2.11). Also,

$$\left[\frac{d}{dt}\mathbf{A}^T\right]_{ij} = \frac{d}{dt}\{[\mathbf{A}^T]_{ij}\} = \frac{d}{dt}\{[\mathbf{A}]_{ji}\}$$

$$= \left[\frac{d\mathbf{A}}{dt}\right]_{ji} = \left[\left(\frac{d\mathbf{A}}{dt}\right)^T\right]_{ij}$$

This proves the identity (3.2.13). ∎

EXAMPLE 3.2.2 If $\mathbf{Q}(t)$ is an orthogonal tensor, show that $(d\mathbf{Q}/dt)\mathbf{Q}^T$ is a skew tensor.

Solution Since **Q** is an orthogonal tensor, we have $\mathbf{QQ}^T = \mathbf{I}$. Differentiating both sides, we get

$$\left(\frac{d\mathbf{Q}}{dt}\right)\mathbf{Q}^T + \mathbf{Q}\left(\frac{d\mathbf{Q}^T}{dt}\right) = \mathbf{0}$$

so that

$$\left(\frac{d\mathbf{Q}}{dt}\right)\mathbf{Q}^T = -\mathbf{Q}\left(\frac{d\mathbf{Q}^T}{dt}\right) = -\mathbf{Q}\left(\frac{d\mathbf{Q}}{dt}\right)^T$$

Hence

$$\left\{\left(\frac{d\mathbf{Q}}{dt}\right)\mathbf{Q}^T\right\}^T = \left\{-\mathbf{Q}\left(\frac{d\mathbf{Q}}{dt}\right)^T\right\}^T = -\left(\frac{d\mathbf{Q}}{dt}\right)\mathbf{Q}^T$$

showing that the tensor $(d\mathbf{Q}/dt)\mathbf{Q}^T$ is skew. ∎

EXAMPLE 3.2.3 If $\mathbf{w}(t)$ is the dual vector of a skew tensor $\mathbf{W}(t)$, show that $d\mathbf{w}/dt$ is the dual vector of $d\mathbf{W}/dt$.

Solution The fact that $d\mathbf{W}/dt$ is skew when $\mathbf{W}$ is skew is obvious from (3.2.13). Let $\mathbf{a}$ be any vector (constant or function of t). Then

$$\left(\frac{d\mathbf{W}}{dt}\right)\mathbf{a} = \frac{d}{dt}(\mathbf{W}\mathbf{a}) - \mathbf{W}\left(\frac{d\mathbf{a}}{dt}\right)$$

$$= \frac{d}{dt}(\mathbf{w} \times \mathbf{a}) - \mathbf{w} \times \left(\frac{d\mathbf{a}}{dt}\right)$$

because $\mathbf{w}$ is the dual vector of $\mathbf{W}$;

$$= \frac{d\mathbf{w}}{dt} \times \mathbf{a}$$

Hence $d\mathbf{w}/dt$ is the dual vector of $d\mathbf{W}/dt$. ∎

3.3
COMMA NOTATION

Consider a real-valued function $f = f(x_i)$ that is a scalar, a component of a vector or a component of a tensor. As stated in Section 3.2, we assume that f has partial derivatives of all desired orders (w.r.t. x_i). We employ the following *comma notation*:

$$f_{,1} = \frac{\partial f}{\partial x_1}, \qquad f_{,2} = \frac{\partial f}{\partial x_2}, \qquad f_{,3} = \frac{\partial f}{\partial x_3}$$

or, briefly,

$$f_{,i} = \frac{\partial f}{\partial x_i}; \qquad i = 1, 2, 3 \tag{3.3.1}$$

Thus, for a specified value of $i (= 1, 2, 3)$ the *symbol* $(\ )_{,i}$ *denotes the partial differential operator* $(\partial/\partial x_i)(\ )$.

We will employ all the rules and conventions of the suffix notation to this comma notation—the comma not interfering with these rules and conventions. In particular, the summation convention will be employed across the comma.

Accordingly, the expression $i = 1, 2, 3$ can be suppressed in (3.3.1). Further, an expression like $(f_{,i}f_{,i})$ will stand for $\sum_{i=1}^{3}(f_{,i}f_{,i})$. Thus, for example, the differential df of f, which is given by

$$df = \frac{\partial f}{\partial x_1}dx_1 + \frac{\partial f}{\partial x_2}dx_2 + \frac{\partial f}{\partial x_3}dx_3$$

3 CALCULUS OF TENSORS

may be written in a compact form as follows:

$$df = f_{,k}\, dx_k \tag{3.3.2}$$

The comma notation can be extended to second-order partial derivatives also. We employ the following notations:

$$f_{,11} = \frac{\partial}{\partial x_1}\left(\frac{\partial f}{\partial x_1}\right) = \frac{\partial^2 f}{\partial x_1^2}, \qquad f_{,12} = \frac{\partial}{\partial x_2}\left(\frac{\partial f}{\partial x_1}\right) = \frac{\partial^2 f}{\partial x_2\, \partial x_1},$$

$$f_{,21} = \frac{\partial}{\partial x_1}\left(\frac{\partial f}{\partial x_2}\right) = \frac{\partial^2 f}{\partial x_1\, \partial x_2}$$

and so forth. In general,

$$f_{,ij} = \frac{\partial}{\partial x_j}\left(\frac{\partial f}{\partial x_i}\right) = \frac{\partial^2 f}{\partial x_j\, \partial x_i}, \qquad f_{,ji} = \frac{\partial}{\partial x_i}\left(\frac{\partial f}{\partial x_j}\right) = \frac{\partial^2 f}{\partial x_i\, \partial x_j}$$

Since the functions dealt with are assumed to be differentiable up to the desired order, the order of partial differentiation is taken to be interchangeable. That is,

$$f_{,ij} = f_{,ji} = \frac{\partial^2 f}{\partial x_i\, \partial x_j} = \frac{\partial^2 f}{\partial x_j\, \partial x_i} \tag{3.3.3}$$

Also,

$$f_{,ii} = f_{,11} + f_{,22} + f_{,33} = \frac{\partial^2 f}{\partial x_1^2} + \frac{\partial^2 f}{\partial x_2^2} + \frac{\partial^2 f}{\partial x_3^2}$$

The sum $(\partial^2 f/\partial x_1^2 + \partial^2 f/\partial x_2^2 + \partial^2 f/\partial x_3^2)$ is called the *Laplacian* of f, usually denoted $\nabla^2 f$. (The symbol ∇ is pronounced as *del* or *nabla*). Thus, we have

$$f_{,ii} = \nabla^2 f \tag{3.3.4}$$

the symbol $(\ )_{,ii}$ denoting the *Laplacian operator* $\nabla^2(\ )$.

The comma notation for third and higher order partial derivatives is analogous. Thus,

$$f_{,111} = \frac{\partial^3 f}{\partial x_1^3}, \qquad f_{,123} = \frac{\partial^3 f}{\partial x_3\, \partial x_2\, \partial x_1}, \qquad f_{,1123} = \frac{\partial^4 f}{\partial x_3\, \partial x_2\, \partial x_1^2}$$

and so forth.

EXAMPLE 3.3.1 Show that

$$\varepsilon_{imn} f_{,mn} = \varepsilon_{imn} f_{,nm} = 0 \tag{3.3.5}$$

3.3 COMMA NOTATION

Solution The result is obvious if we note that

$$\varepsilon_{imn} f_{,mn} = \varepsilon_{inm} f_{,nm} \quad \text{(interchange of dummies)}$$
$$= -\varepsilon_{imn} f_{,nm} \quad \text{(definition of the } \varepsilon \text{ symbol)}$$
$$= -\varepsilon_{imn} f_{,mn} \quad \text{(changing the order of differentiation)} \quad \blacksquare$$

EXAMPLE 3.3.2 Prove the following:

(i) $\quad x_{i,j} = \delta_{ij}; \quad x_{i,i} = 3$

(ii) $\quad (x_m x_n)_{,i} = \delta_{im} x_n + \delta_{in} x_m$

(iii) $\quad (x_m x_n)_{,ij} = \delta_{im} \delta_{jn} + \delta_{in} \delta_{jm}$

(iv) $\quad \nabla^2(x_m x_n) = 2\delta_{mn}$ (3.3.6)

(v) $\quad r_{,i} = (1/r) x_i, \quad r \neq 0$

(vi) $\quad (r^2)_{,ij} = 2\delta_{ij}$

(vii) $\quad \nabla^2(r^2) = 6$

Here $r^2 = x_i x_i$. (*This notation is used throughout.*)

Solution (i) We note that

$$x_{1,1} = \frac{\partial x_1}{\partial x_1} = 1$$

Similarly, $x_{2,2} = x_{3,3} = 1$. Also,

$$x_{1,2} = \frac{\partial x_1}{\partial x_2} = 0$$

Similarly, $x_{1,3} = x_{3,1} = x_{2,3} = x_{3,2} = x_{2,1} = 0$. Thus,

$$x_{i,j} = \begin{cases} 1 & \text{if } i \text{ and } j \text{ take the same value} \\ 0 & \text{if } i \text{ and } j \text{ take different values} \end{cases}$$
$$= \delta_{ij}$$

Consequently, $x_{i,i} = \delta_{ii} = 3$.

(ii) By the use of the product rule of differentiation, we get

$$(x_m x_n)_{,i} = x_{m,i} x_n + x_m x_{n,i} = \delta_{mi} x_n + x_m \delta_{ni}$$

by (i).

116   3   CALCULUS OF TENSORS

(iii) Hence
$$(x_m x_n)_{,ij} = [(x_m x_n)_{,i}]_{,j}$$
$$= (\delta_{mi} x_n)_{,j} + (x_m \delta_{ni})_{,j}$$
$$= \delta_{mi} x_{n,j} + x_{m,j} \delta_{ni}$$

(because the δ symbol is a constant)
$$= \delta_{mi} \delta_{nj} + \delta_{mj} \delta_{ni}$$

by (i);
$$= \delta_{im} \delta_{jn} + \delta_{in} \delta_{jm}$$

(iv) Consequently, by using (3.3.4) we obtain
$$\nabla^2(x_m x_n) = (x_m x_n)_{,ii} = \delta_{im} \delta_{in} + \delta_{im} \delta_{in} = 2\delta_{mn}$$

(v) We have $(r^2)_{,i} = 2rr_{,i}$. On the other hand, since $r^2 = x_m x_m$, we get, from the result (ii),
$$(r^2)_{,i} = (x_m x_m)_{,i} = 2x_m \delta_{mi} = 2x_i$$

Hence $r_{,i} = x_i/r$, if $r \neq 0$.

(vi) From the result (iii), we get
$$(r^2)_{,ij} = (x_m x_m)_{,ij} = 2\delta_{im} \delta_{jm} = 2\delta_{ij}$$

(vii) Consequently,
$$\nabla^2(r^2) = (r^2)_{,ii} = 2\delta_{ii} = 6 \quad \blacksquare$$

EXAMPLE 3.3.3   If $f_k = f_k(x_i)$, prove the following:

(i) $(x_k f_k)_{,i} = f_i + x_k f_{k,i}$

(ii) $(x_k f_k)_{,ij} = f_{i,j} + f_{j,i} + x_k f_{k,ij}$ (3.3.7)

(iii) $\nabla^2(x_k f_k) = 2f_{k,k} + x_k(\nabla^2 f_k)$

Solution   (i) By the use of the product rule of differentiation, we get
$$(x_k f_k)_{,i} = x_{k,i} f_k + x_k f_{k,i} = \delta_{ki} f_k + x_k f_{k,i} = f_i + x_k f_{k,i}$$

(ii) Hence
$$(x_k f_k)_{,ij} = (f_i + x_k f_{k,i})_{,j} = f_{i,j} + x_{k,j} f_{k,i} + x_k f_{k,ij}$$
$$= f_{i,j} + \delta_{kj} f_{k,i} + x_k f_{k,ij} = f_{i,j} + f_{j,i} + x_k f_{k,ij}$$

(iii) Consequently
$$\nabla^2(x_k f_k) = (x_k f_k)_{,ii} = 2f_{i,i} + x_k f_{k,ii} = 2f_{k,k} + x_k(\nabla^2 f_k) \quad \blacksquare$$

3.4
GRADIENT OF A SCALAR, DIVERGENCE AND CURL OF A VECTOR

In this section we consider three important entities that arise frequently in practical applications. The first one is a vector field associated with a scalar field and the other two are scalar and vector fields associated with a vector field.

3.4.1 GRADIENT OF A SCALAR FIELD

Consider a scalar field ϕ and put $a_i = \phi_{,i}$ in all coordinate systems. Then

$$a'_i = \frac{\partial \phi}{\partial x'_i}$$

By using the chain rule of differentiation, this can be rewritten as

$$a'_i = \frac{\partial \phi}{\partial x_k} \frac{\partial x_k}{\partial x'_i} = \phi_{,k} \frac{\partial x_k}{\partial x'_i}$$

From the transformation rule (2.2.7) we find that

$$\frac{\partial x_k}{\partial x'_i} = \alpha_{ik} \tag{3.4.1}$$

Hence

$$a'_i = \alpha_{ik} \phi_{,k} = \alpha_{ik} a_k$$

This transformation rule shows that $a_i = \phi_{,i}$ are components of a vector. This vector is evidently a function of x_i in general; it is called the *gradient* of ϕ, denoted grad ϕ, or $\nabla \phi$.

Thus, *if ϕ is a scalar field then $\nabla \phi$ is a vector field* with components given by

$$[\nabla \phi]_i = \phi_{,i} \tag{3.4.2}$$

It is obvious that the gradient of a constant scalar field is the zero vector. Also,

$$\nabla(\alpha \phi \pm \beta \psi) = \alpha \nabla \phi \pm \beta \nabla \psi \tag{3.4.3}$$

$$\nabla(\phi \psi) = \phi(\nabla \psi) + \psi(\nabla \phi) \tag{3.4.4}$$

for all scalar fields ϕ and ψ and scalar constants α and β.

By virtue of (3.4.2), expression (3.3.2) can be rewritten as

$$df = [\nabla f]_k \, dx_k = \nabla f \cdot d\mathbf{x} \tag{3.4.5}$$

3.4.2 DIRECTIONAL DERIVATIVE AND NORMAL DERIVATIVE

From geometry, we recall that an equation of the form $\phi(x_i) \equiv \phi(x_1, x_2, x_3) = constant$ represents a surface in three-dimensional space, and that a normal to the surface has direction ratios $(\partial\phi/\partial x_1, \partial\phi/\partial x_2, \partial\phi/\partial x_3)$. From (3.4.2) we note that these direction ratios are indeed the components of $\nabla\phi$. It therefore follows that *at any point* **x** *of the surface* $\phi(x_i) = constant$, *the vector* $\nabla\phi$ *is directed along a normal to the surface.* This is illustrated in Figure 3.1.

If **n** is the *unit vector* along $\nabla\phi$, then

$$\mathbf{n} = \frac{\nabla\phi}{|\nabla\phi|} \qquad (3.4.6)$$

Let **a** be a *unit vector* inclined at an angle θ to the direction of $\nabla\phi$. Then

$$\nabla\phi \cdot \mathbf{a} = |\nabla\phi|(\mathbf{n} \cdot \mathbf{a}) = |\nabla\phi| \cos\theta \qquad (3.4.7)$$

The scalar $\nabla\phi \cdot \mathbf{a}$, which represents the component of $\nabla\phi$ along **a**, is called the *directional derivative* of ϕ along **a**, usually denoted $\partial\phi/\partial\mathbf{a}$. Thus,

$$\frac{\partial\phi}{\partial\mathbf{a}} = \nabla\phi \cdot \mathbf{a} \qquad (3.4.8)$$

In particular, the directional derivative of ϕ along **n** is called the *normal derivative* of ϕ, denoted $\partial\phi/\partial\mathbf{n}$. Thus,

$$\frac{\partial\phi}{\partial\mathbf{n}} = \nabla\phi \cdot \mathbf{n} = |\nabla\phi| \qquad (3.4.9)$$

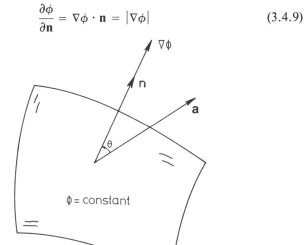

Figure 3.1. The vector $\nabla\phi$.

3.4 GRADIENT OF A SCALAR

From (3.4.7) and (3.4.8) it is evident that $|\partial\phi/\partial\mathbf{a}|$ is maximum when $\theta = 0$, that is, $\mathbf{a} = \mathbf{n}$, so that

$$\max\left|\frac{\partial\phi}{\partial\mathbf{a}}\right| = \frac{\partial\phi}{\partial\mathbf{n}} \quad (3.4.10)$$

Thus, *the normal derivative of ϕ is the maximum of all the directional derivatives of ϕ.*

From (3.4.6) and (3.4.9) we get the useful expression

$$\nabla\phi = |\nabla\phi|\mathbf{n} = \left(\frac{\partial\phi}{\partial\mathbf{n}}\right)\mathbf{n} \quad (3.4.11)$$

3.4.3 DIVERGENCE OF A VECTOR FIELD

Consider a vector field $\mathbf{u}$ and put $a = u_{k,k}$ in all coordinate systems. Then

$$a' = \frac{\partial u'_k}{\partial x'_k}$$

By using the chain rule of differentiation, this takes the form

$$a' = \frac{\partial u'_k}{\partial x_n}\frac{\partial x_n}{\partial x'_k}$$

Noting that $u'_k = \alpha_{km}u_m$ and using (3.4.1) this becomes

$$a' = (\alpha_{km}u_m)_{,n}\alpha_{kn} = \alpha_{km}\alpha_{kn}u_{m,n}$$
$$= \delta_{mn}u_{m,n} = u_{m,m} = a$$

This transformation rule shows that $a = u_{k,k}$ is a scalar (invariant). This scalar is evidently a function of x_i in general; it is called the *divergence* of $\mathbf{u}$, denoted div $\mathbf{u}$, or $\nabla \cdot \mathbf{u}$.

Thus, *if $\mathbf{u}$ is a vector field, then div $\mathbf{u}$ is a scalar field* given by

$$\text{div } \mathbf{u} = u_{k,k} = \{[\mathbf{u}]_k\}_{,k} \quad (3.4.12)$$

It is obvious that the divergence of a constant vector field is equal to 0.

If $\mathbf{u} = \nabla\phi$, then $u_i = \phi_{,i}$ and $u_{k,k} = \phi_{,kk} = \nabla^2\phi$, or div $\mathbf{u} = \nabla^2\phi$. Thus, we have the *identity*

$$\text{div}(\nabla\phi) = \nabla^2\phi \quad (3.4.13)$$

If $\mathbf{v} = \phi\mathbf{u}$, then $v_i = \phi u_i$ and div $\mathbf{v} = v_{k,k} = (\phi u_k)_{,k} = \phi_{,k}u_k + \phi u_{k,k} = \mathbf{u}\cdot\nabla\phi + \phi\text{ div }\mathbf{u}$. Thus we have another *identity*:

$$\text{div}(\phi\mathbf{u}) = \phi(\text{div }\mathbf{u}) + \nabla\phi\cdot\mathbf{u} \quad (3.4.14)$$

3.4.4 CURL OF A VECTOR FIELD

Consider again a vector field **u** and put $a_i = \varepsilon_{imn} u_{n,m}$ in all coordinate systems. Then

$$a'_i = \varepsilon'_{imn} \frac{\partial u'_n}{\partial x'_m}$$

Using the fact that ε_{ijk} are components of a third-order tensor and the chain rule of differentiation, we get

$$a'_i = \alpha_{ip}\alpha_{mq}\alpha_{nr}\varepsilon_{pqr} \frac{\partial u'_n}{\partial x_s} \frac{\partial x_s}{\partial x'_m}$$

Noting that $u'_n = \alpha_{nk} u_k$ and using (3.4.1) this becomes

$$a'_i = \alpha_{ip}\alpha_{mq}\alpha_{nr}\alpha_{ms}\alpha_{nk}\varepsilon_{pqr} u_{k,s}$$
$$= \alpha_{ip}\delta_{qs}\delta_{rk}\varepsilon_{pqr} u_{k,s} = \alpha_{ip}\varepsilon_{pqr} u_{r,q} = \alpha_{ip} a_p$$

This transformation rule shows that $a_i = \varepsilon_{imn} u_{n,m}$ are components of a vector. This vector is a function of x_i in general, called the *curl* of **u**, denoted curl **u** or $\nabla \times \mathbf{u}$.

Thus, *if* **u** *is a vector field, then* curl **u** *is also a vector field* with components given by

$$[\text{curl } \mathbf{u}]_i = \varepsilon_{imn} u_{n,m} = \varepsilon_{imn}\{[\mathbf{u}]_n\}_{,m} \tag{3.4.15}$$

It is obvious that the curl of a constant vector field is the zero vector.

If $\mathbf{u} = \nabla \phi$, then (3.4.15) gives $[\text{curl } \mathbf{u}]_i = \varepsilon_{imn}\phi_{,nm} = 0$, see (3.3.5). Thus, we have the *identity*

$$\text{curl } \nabla \phi = \mathbf{0} \tag{3.4.16}$$

From (3.4.12) and (3.4.15), we get another *identity*:

$$\text{div}(\text{curl } \mathbf{u}) = \{[\text{curl } \mathbf{u}]_k\}_{,k} = \varepsilon_{kmn} u_{n,mk} = 0 \tag{3.4.17}$$

If $\mathbf{v} = \phi \mathbf{u}$, then

$$[\text{curl } \mathbf{v}]_i = \varepsilon_{ijk} v_{k,j} = \varepsilon_{ijk}(\phi u_k)_{,j}$$
$$= \varepsilon_{ijk}\phi u_{k,j} + \varepsilon_{ijk}\phi_{,j} u_k$$
$$= \phi[\text{curl } \mathbf{u}]_i + [\nabla\phi \times \mathbf{u}]_i$$

Thus, we have one more *identity*:

$$\text{curl}(\phi \mathbf{u}) = \phi\, \text{curl } \mathbf{u} + \nabla\phi \times \mathbf{u} \tag{3.4.18}$$

3.4 GRADIENT OF A SCALAR 121

From the definitions of div **u** and curl **u**, it is easy to verify that div and curl are linear differential operators on vectors; that is,

$$\left.\begin{array}{l} \text{div}(\alpha\mathbf{u} + \beta\mathbf{v}) = \alpha\,\text{div}\,\mathbf{u} + \beta\,\text{div}\,\mathbf{v} \\ \text{curl}(\alpha\mathbf{u} + \beta\mathbf{v}) = \alpha\,\text{curl}\,\mathbf{u} + \beta\,\text{curl}\,\mathbf{v} \end{array}\right\} \quad (3.4.19)$$

for all vectors **u** and **v** and all scalar constants α and β.

3.4.5 THE OPERATOR $\mathbf{u} \cdot \nabla$

In computations involving scalars and vectors, we often deal with differential operators of the form $u_j(\partial/\partial x_j)(\ )$, where u_j are components of a vector **u**. We denote this *operator* by $\mathbf{u} \cdot \nabla$. That is,

$$\mathbf{u} \cdot \nabla = u_j \frac{\partial}{\partial x_j} \quad (3.4.20)$$

so that

$$(\mathbf{u} \cdot \nabla)\phi = u_j \frac{\partial \phi}{\partial x_j} = u_j \phi_{,j} = \mathbf{u} \cdot (\nabla \phi) \quad (3.4.21)$$

for a scalar field ϕ, and

$$(\mathbf{u} \cdot \nabla)\mathbf{v} = (\mathbf{u} \cdot \nabla)v_i \mathbf{e}_i = (u_j v_{i,j})\mathbf{e}_i \quad (3.4.22)$$

for a vector field **v**. From (3.4.22), we note that

$$[(\mathbf{u} \cdot \nabla)\mathbf{v}]_i = u_j v_{i,j} = [\mathbf{u}]_j\{[\mathbf{v}]_i\}_{,j} \quad (3.4.23)$$

3.4.6 SOME IDENTITIES

The following are some other well-known identities:

$$\text{div}(\mathbf{u} \times \mathbf{v}) = \mathbf{v} \cdot \text{curl}\,\mathbf{u} - \mathbf{u} \cdot \text{curl}\,\mathbf{v} \quad (3.4.24)$$

$$\text{curl}(\mathbf{u} \times \mathbf{v}) = (\text{div}\,\mathbf{v})\mathbf{u} - (\text{div}\,\mathbf{u})\mathbf{v} + (\mathbf{v} \cdot \nabla)\mathbf{u} - (\mathbf{u} \cdot \nabla)\mathbf{v} \quad (3.4.25)$$

$$\nabla(\mathbf{u} \cdot \mathbf{v}) = \mathbf{v} \times \text{curl}\,\mathbf{u} + \mathbf{u} \times \text{curl}\,\mathbf{v} + (\mathbf{v} \cdot \nabla)\mathbf{u} + (\mathbf{u} \cdot \nabla)\mathbf{v} \quad (3.4.26)$$

$$\text{curl curl}\,\mathbf{u} = \nabla\,\text{div}\,\mathbf{u} - \nabla^2 \mathbf{u} \quad (3.4.27)$$

Here, the vector $\nabla^2 \mathbf{u}$, called the *Laplacian of* **u**, is *defined* by

$$[\nabla^2 \mathbf{u}]_i = \nabla^2 u_i = \nabla^2 [\mathbf{u}]_i \quad (3.4.28)$$

Taking $\mathbf{v} = \mathbf{u}$ in (3.4.26), we get the following useful identity:

$$(\mathbf{u} \cdot \nabla)\mathbf{u} = \tfrac{1}{2}\nabla(u^2) - \mathbf{u} \times \text{curl}\,\mathbf{u} = \tfrac{1}{2}\nabla(u^2) + (\text{curl}\,\mathbf{u}) \times \mathbf{u} \quad (3.4.29)$$

where $u^2 = \mathbf{u} \cdot \mathbf{u}$.

122   3 CALCULUS OF TENSORS

EXAMPLE 3.4.1   If $\phi = \phi(x_i)$ and $f = f(\phi)$, show that

(i) $$\nabla \phi \times \nabla f = \mathbf{0}$$

(ii) $$\nabla \left\{ \int f(\phi) \, d\phi \right\} = f(\phi)(\nabla \phi)$$

Solution   (i) Since $f = f(\phi)$ and $\phi = \phi(x_i)$, we have by the chain rule of differentiation $f_{,k} = \{f'(\phi)\}\phi_{,k}$ so that $\nabla f = \{f'(\phi)\}(\nabla \phi)$. Hence

$$\nabla \phi \times \nabla f = \{f'(\phi)\}(\nabla \phi) \times (\nabla \phi) = \mathbf{0}$$

(ii) If we set $\int f(\phi) \, d\phi = F(\phi)$, then we get $F_{,i} = \{F'(\phi)\}\phi_{,i} = \{f(\phi)\}\phi_{,i}$ so that $\nabla F = \{f(\phi)\}(\nabla \phi)$.   ∎

EXAMPLE 3.4.2   Write down $\text{div}(\mathbf{u} \times \mathbf{v})$ in the suffix notation and hence prove the identity (3.4.24).

Solution   By the expression (1.7.9), we have $[\mathbf{u} \times \mathbf{v}]_k = \varepsilon_{kij} u_i v_j$. Hence (3.4.12) gives

$$\text{div}(\mathbf{u} \times \mathbf{v}) = \{[\mathbf{u} \times \mathbf{v}]_k\}_{,k} = \varepsilon_{kij}(u_i v_j)_{,k} \tag{3.4.30}$$

This is the expression for $\text{div}(\mathbf{u} \times \mathbf{v})$ in the suffix notation.

Carrying out the indicated differentiation in (3.4.30), we get

$$\text{div}(\mathbf{u} \times \mathbf{v}) = \varepsilon_{kij}(u_{i,k} v_j + u_i v_{j,k}) = (\varepsilon_{jki} u_{i,k})v_j - (\varepsilon_{ikj} v_{j,k})u_i$$

$$= [\text{curl } \mathbf{u}]_j v_j - [\text{curl } \mathbf{v}]_i u_i$$

(by (3.4.15))

$$= \mathbf{v} \cdot \text{curl } \mathbf{u} - \mathbf{u} \cdot \text{curl } \mathbf{v} \quad \blacksquare$$

EXAMPLE 3.4.3   Show that

$$[\mathbf{u} \times \text{curl } \mathbf{v}]_i = (v_{j,i} - v_{i,j})u_j \tag{3.4.31}$$

Deduce the identity (3.4.26).

Solution   By virtue of the expressions (1.7.9) and (3.4.15), we have

$$[\mathbf{u} \times \text{curl } \mathbf{v}]_i = \varepsilon_{ijk}[\mathbf{u}]_j[\text{curl } \mathbf{v}]_k = \varepsilon_{ijk}\varepsilon_{kmn} u_j v_{n,m}$$

Using the ε-δ identity, this becomes

$$[\mathbf{u} \times \text{curl } \mathbf{v}]_i = (\delta_{im}\delta_{jn} - \delta_{in}\delta_{jm})u_j v_{n,m} = (v_{j,i} - v_{i,j})u_j$$

This is the required result (3.4.31).

From (3.4.31) and (3.4.23), we get

$$[\mathbf{u} \times \text{curl } \mathbf{v} + (\mathbf{u} \cdot \nabla)\mathbf{v}]_i = u_j v_{j,i} \tag{3.4.32}$$

3.4 GRADIENT OF A SCALAR 123

Similarly
$$[\mathbf{v} \times \text{curl } \mathbf{u} + (\mathbf{v} \cdot \nabla)\mathbf{u}]_i = v_j u_{j,i}. \tag{3.4.33}$$

Expressions (3.4.32) and (3.4.33) yield

$$[\mathbf{u} \times \text{curl } \mathbf{v} + (\mathbf{u} \cdot \nabla)\mathbf{v} + \mathbf{v} \times \text{curl } \mathbf{u} + (\mathbf{v} \cdot \nabla)\mathbf{u}]_i$$
$$= u_j v_{j,i} + v_j u_{j,i} = (u_j v_j)_{,i} = (\mathbf{u} \cdot \mathbf{v})_{,i} = [\nabla(\mathbf{u} \cdot \mathbf{v})]_i$$

(by (3.4.2)). This proves the validity of the identity (3.4.26). ■

EXAMPLE 3.4.4 Prove the identity:
$$\nabla^2(\mathbf{x} \cdot \mathbf{v}) = 2 \text{ div } \mathbf{v} + \mathbf{x} \cdot \nabla^2 \mathbf{v} \tag{3.4.34}$$

Solution By use of the relation (iii) of Example 3.3.3 we have
$$\nabla^2(x_k v_k) = 2v_{k,k} + x_k \nabla^2 v_k$$

When written in the direct notation, this becomes the identity (3.4.34). ■

EXAMPLE 3.4.5 Write down the following equation in the direct notation:
$$\mu \nabla^2 u_i + (\lambda + \mu) u_{k,ki} + \rho_0 b_i = \rho_0 a_i \tag{3.4.35}$$

where λ, μ and ρ_0 are constants.

Solution We first recall that
$$\nabla^2 u_i = [\nabla^2 \mathbf{u}]_i \quad \text{and} \quad u_{k,ki} = (u_{k,k})_{,i} = [\nabla(\text{div } \mathbf{u})]_i$$

Hence the given equation is
$$\mu[\nabla^2 \mathbf{u}]_i + (\lambda + \mu)[\nabla(\text{div } \mathbf{u})]_i + \rho_0[\mathbf{b}]_i = \rho_0[\mathbf{a}]_i$$

Therefore, in the direct notation the equation reads
$$\mu \nabla^2 \mathbf{u} + (\lambda + \mu)\nabla(\text{div } \mathbf{u}) + \rho_0 \mathbf{b} = \rho_0 \mathbf{a} \quad \blacksquare \tag{3.4.35}'$$

EXAMPLE 3.4.6 Show that $\mathbf{v}$ is orthogonal to curl $\mathbf{v}$ if and only if $\mathbf{v} = \phi \nabla \psi$ for some scalar functions ϕ and ψ.

Solution First, suppose that $\mathbf{v} = \phi \nabla \psi$. Then by using identities (3.4.16) and (3.4.18), we get
$$\text{curl } \mathbf{v} = \nabla \phi \times \nabla \psi$$

Hence
$$\mathbf{v} \cdot \text{curl } \mathbf{v} = \phi \nabla \psi \cdot (\nabla \phi \times \nabla \psi) = 0$$

Thus, $\mathbf{v}$ is orthogonal to curl $\mathbf{v}$.

124   3 CALCULUS OF TENSORS

Next, suppose that **v** is orthogonal to curl **v**; that is, $\mathbf{v} \cdot \operatorname{curl} \mathbf{v} = 0$, or equivalently,

$$v_1(v_{3,2} - v_{2,3}) + v_2(v_{1,3} - v_{3,1}) + v_3(v_{2,1} - v_{1,2}) = 0$$

This implies that the total differential equation

$$v_1\, dx_1 + v_2\, dx_2 + v_3\, dx_3 = 0$$

is integrable. Hence, there exist functions $\mu \neq 0$ and ψ such that

$$\mu(v_1\, dx_1 + v_2\, dx_2 + v_3\, dx_3) = d\psi$$

Setting $\phi = 1/\mu$ and using (3.3.2), this can be rewritten as $v_k\, dx_k = (\phi\psi_{,k})\, dx_k$, which yields $v_i = \phi\psi_{,i}$ so that $\mathbf{v} = \phi\nabla\psi$, as required. ∎

EXAMPLE 3.4.7 A vector field **v** is said to be a *Beltrami field* if curl $\mathbf{v} = \omega\mathbf{v}$ for some scalar field ω. (Then ω is called the *abnormality factor* of **v**.) Show that for such a field,

$$\omega = \frac{1}{|\mathbf{v}|}\hat{\mathbf{v}} \cdot \operatorname{curl} \mathbf{v} \tag{3.4.36}$$

$$= \frac{\operatorname{curl} \mathbf{v} \cdot \operatorname{curl}\operatorname{curl} \mathbf{v}}{|\operatorname{curl} \mathbf{v}|^2} \tag{3.4.37}$$

If $\mathbf{v} = \nabla\phi + \psi(\nabla\chi)$, show that

$$\omega = \frac{\nabla\phi \cdot \nabla\psi \times \nabla\chi}{(\nabla\phi)^2 - \psi^2(\nabla\chi)^2} \tag{3.4.38}$$

Solution By data, curl $\mathbf{v} = \omega\mathbf{v}$. Hence

$$\hat{\mathbf{v}} \cdot \operatorname{curl} \mathbf{v} = \hat{\mathbf{v}} \cdot (\omega\mathbf{v}) = \omega|\mathbf{v}|$$

This is the required result (3.4.36). Also,

$$\operatorname{curl}\operatorname{curl} \mathbf{v} = \operatorname{curl}(\omega\mathbf{v}) = \omega \operatorname{curl} \mathbf{v} + \nabla\omega \times \mathbf{v}$$

by the identity (3.4.18). Hence

$$\operatorname{curl} \mathbf{v} \cdot \operatorname{curl}\operatorname{curl} \mathbf{v} = \omega|\operatorname{curl} \mathbf{v}|^2 + \operatorname{curl} \mathbf{v} \cdot (\nabla\omega \times \mathbf{v})$$

Since

$$\operatorname{curl} \mathbf{v} \cdot (\nabla\omega \times \mathbf{v}) = \omega\mathbf{v} \cdot (\nabla\omega \times \mathbf{v}) = 0$$

the required result (3.4.37) follows.

If $\mathbf{v} = \nabla\phi + \psi(\nabla\chi)$, we find, by using the identities (3.4.16) and (3.4.18), that

$$\operatorname{curl} \mathbf{v} = \nabla\psi \times \nabla\chi$$

Hence

$$\nabla\phi \cdot (\nabla\psi \times \nabla\chi) = (\nabla\phi - \psi\nabla\chi) \cdot (\nabla\psi \times \nabla\chi) = (\nabla\phi - \psi\nabla\chi) \cdot \text{curl } \mathbf{v}$$
$$= (\nabla\phi - \psi\nabla\chi) \cdot \omega\mathbf{v} = \omega(\nabla\phi - \psi\nabla\chi) \cdot (\nabla\phi + \psi\nabla\chi)$$
$$= \omega\{(\nabla\phi)^2 - \psi^2(\nabla\chi)^2\}$$

This is the required result (3.4.38). ∎

3.5
GRADIENT OF A VECTOR, DIVERGENCE AND CURL OF A TENSOR

In this section we introduce the gradient, divergence, curl and Laplacian operators frequently encountered in tensor calculus. These are extensions of and analogous to the operators defined in the preceding section.

3.5.1 GRADIENT OF A VECTOR FIELD

Consider a vector field $\mathbf{u}$ with components u_i and put $a_{ij} = u_{i,j}$ in all coordinate systems. Then

$$a'_{ij} = \frac{\partial u'_i}{\partial x'_j}$$

By using the chain rule of differentiation, this becomes

$$a'_{ij} = \frac{\partial u'_i}{\partial x_n} \frac{\partial x_n}{\partial x'_j}$$

Noting that $u'_i = \alpha_{im} u_m$ and using (3.4.1) this becomes

$$a'_{ij} = \alpha_{im} \alpha_{jn} u_{m,n} = \alpha_{im} \alpha_{jn} a_{mn}$$

This transformation rule shows that $a_{ij} = u_{i,j}$ are components of a (second-order) tensor. This tensor is a function of x_i in general and is called the *gradient of* $\mathbf{u}$, denoted grad $\mathbf{u}$ or $\nabla \mathbf{u}$.

Thus, *if* $\mathbf{u}$ *is a vector field then* $\nabla \mathbf{u}$ *is a tensor field* with components given by

$$[\nabla\mathbf{u}]_{ij} = u_{i,j} = \{[\mathbf{u}]_i\}_{,j} \qquad (3.5.1)$$

The transpose of this tensor is denoted by $\nabla\mathbf{u}^T$. Thus, we have

$$[\nabla\mathbf{u}^T]_{ij} = u_{j,i} = \{[\mathbf{u}]_j\}_{,i} \qquad (3.5.2)$$

Also, if **v** is a vector, then by virtue of (2.8.19), it follows that $\mathbf{v} \wedge \nabla \mathbf{u}$ is a tensor whose components are given by

$$[\mathbf{v} \wedge \nabla \mathbf{u}]_{ij} = \varepsilon_{irs} v_r u_{j,s} \tag{3.5.3}$$

It is obvious that the gradient of a constant vector field is the zero tensor of the second order. It is straightforward to verify that

(i) $\qquad\qquad\qquad \nabla \mathbf{x} = \mathbf{I} \tag{3.5.4}$

(ii) $\qquad\qquad \mathbf{I} \cdot \nabla \mathbf{u} = \nabla \mathbf{u} \cdot \mathbf{I} = \text{tr}(\nabla \mathbf{u}) = \text{div}\,\mathbf{u} \tag{3.5.5}$

(iii) $\qquad\qquad \text{tr}(\mathbf{v} \wedge \nabla \mathbf{u}) = \mathbf{v} \cdot \text{curl}\,\mathbf{u} \tag{3.5.6}$

If $\mathbf{u} = \nabla \phi$, then $\nabla \mathbf{u} = \nabla(\nabla \phi)$ is called the *second gradient* of ϕ, denoted $\nabla\nabla\phi$. Thus

$$[\nabla\nabla\phi]_{ij} = \phi_{,ij} \tag{3.5.7}$$

Evidently,

$$(\nabla\nabla\phi)^T = \nabla\nabla\phi \tag{3.5.8}$$

3.5.2 GRADIENT OF A TENSOR FIELD

Consider a tensor field **A** with components a_{ij} and put $c_{ijk} = a_{ij,k}$ in all coordinate systems. Then

$$c'_{ijk} = \frac{\partial a'_{ij}}{\partial x'_k}$$

By using the chain rule of differentiation, this becomes

$$c'_{ijk} = \frac{\partial}{\partial x_p}(a'_{ij}) \frac{\partial x_p}{\partial x'_k}$$

Noting that $a'_{ij} = \alpha_{im}\alpha_{jn}a_{mn}$ and using (3.4.1) this becomes

$$c'_{ijk} = \alpha_{im}\alpha_{jn}\alpha_{kp} a_{mn,p} = \alpha_{im}\alpha_{jn}\alpha_{kp} c_{mnp}$$

This transformation rule shows that $c_{ijk} = a_{ij,k}$ are components of a third-order tensor. This tensor is in general a function of x_i, called the *gradient of* **A**, denoted grad **A** or $\nabla\mathbf{A}$.

Thus, *if* **A** *is a (second-order) tensor field, then* $\nabla\mathbf{A}$ *is a third-order tensor field* with components given by

$$[\nabla\mathbf{A}]_{ijk} = a_{ij,k} = \{[\nabla\mathbf{A}]_{ij}\}_{,k} \tag{3.5.9}$$

Obviously, the gradient of a constant second-order tensor field is the zero tensor of order three.

3.5 GRADIENT OF A VECTOR

If $\mathbf{A} = \nabla \mathbf{u}$, then $\nabla \mathbf{A} = \nabla(\nabla \mathbf{u})$ is called the *second gradient* of $\mathbf{u}$, denoted $\nabla \nabla \mathbf{u}$. Thus

$$[\nabla \nabla \mathbf{u}]_{ijk} = u_{i,jk} = \{[\mathbf{u}]_i\}_{,jk} \qquad (3.5.10)$$

By taking contraction in (3.5.10) we find that $\nabla^2 u_i = u_{i,kk}$ are components of a vector. This vector is the *Laplacian* of $\mathbf{u}$, denoted $\nabla^2 \mathbf{u}$; see (3.4.28).

3.5.3 DIVERGENCE OF A TENSOR FIELD

If a_{ij} are components of a tensor field $\mathbf{A}$, it already has been seen that $a_{ij,k}$ are components of the tensor $\nabla \mathbf{A}$. Consequently, it follows (by contraction) that $a_{ij,j}$ are components of a vector field. This vector field is called the *divergence of* $\mathbf{A}$, denoted div $\mathbf{A}$.

Thus, *if $\mathbf{A}$ is a second-order tensor field, then* div $\mathbf{A}$ *is a vector field* with components given by

$$[\text{div } \mathbf{A}]_i = a_{ij,j} = \{[\mathbf{A}]_{ij}\}_{,j} \qquad (3.5.11)$$

Obviously, the divergence of a constant (second-order) tensor field is the zero vector.

From (3.5.11) it readily follows that div $\mathbf{A}^T$ is a vector field with components

$$[\text{div } \mathbf{A}^T]_i = a_{ji,j} = \{[\mathbf{A}]_{ji}\}_{,j} \qquad (3.5.12)$$

Setting $\mathbf{A} = \nabla \mathbf{u}$, we find from (3.5.11) and (3.5.12), that

$$[\text{div } \nabla \mathbf{u}]_i = u_{i,jj} = \nabla^2 u_i$$

so that

$$\text{div } \nabla \mathbf{u} = \nabla^2 \mathbf{u} \qquad (3.5.13)$$

and

$$[\text{div } \nabla \mathbf{u}^T]_i = u_{j,ij} = u_{j,ji} = [\nabla \text{ div } \mathbf{u}]_i$$

so that

$$\text{div } \nabla \mathbf{u}^T = \nabla(\text{div } \mathbf{u}) \qquad (3.5.14)$$

3.5.4 CURL OF A TENSOR FIELD

Consider again a tensor field $\mathbf{A}$ with components a_{ij}. Let us now put $c_{ij} = \varepsilon_{imn} a_{jn,m}$ in all coordinate systems. Then

$$c'_{ij} = \varepsilon'_{imn} \frac{\partial a'_{jn}}{\partial x'_m}$$

Using the fact that ε_{imn} are components of a third-order tensor and the chain rule of differentiation, this expression becomes

$$c'_{ij} = \alpha_{ip}\alpha_{mq}\alpha_{nr}\varepsilon_{pqr}\frac{\partial}{\partial x_s}(a'_{jn})\frac{\partial x_s}{\partial x'_m}$$

Noting that $a'_{jn} = \alpha_{jh}\alpha_{nk}a_{hk}$ and using (3.4.1), we get

$$c'_{ij} = \alpha_{ip}\alpha_{mq}\alpha_{nr}\alpha_{jh}\alpha_{nk}\alpha_{ms}\varepsilon_{pqr}a_{hk,s}$$
$$= \alpha_{ip}\alpha_{jh}\delta_{qs}\delta_{rk}\varepsilon_{pqr}a_{hk,s}$$
$$= \alpha_{ip}\alpha_{jh}\varepsilon_{pqr}a_{hr,q} = \alpha_{ip}\alpha_{jh}c_{ph}$$

This transformation rule shows that $c_{ij} = \varepsilon_{imn}a_{jn,m}$ are components of a second-order tensor. This tensor is a function of x_i in general and is called *curl of* **A**, denoted curl **A**.

Thus, *if* **A** *is a second-order tensor field, then curl* **A** *is also a second-order tensor field* with components given by

$$[\text{curl } \mathbf{A}]_{ij} = \varepsilon_{imn}a_{jn,m} = \varepsilon_{imn}\{[\mathbf{A}]_{jn}\}_{,m} \tag{3.5.15}$$

It is obvious that the curl of a constant tensor field is the zero tensor.

From (3.5.15) it readily follows that curl $\mathbf{A}^T$ and $(\text{curl } \mathbf{A})^T$ are tensors with components given by

$$[\text{curl } \mathbf{A}^T]_{ij} = \varepsilon_{imn}a_{nj,m} = \varepsilon_{imn}\{[\mathbf{A}]_{nj}\}_{,m} \tag{3.5.16}$$

$$[(\text{curl } \mathbf{A})^T]_{ij} = \varepsilon_{jmn}a_{in,m} = \varepsilon_{jmn}\{[\mathbf{A}]_{in}\}_{,m} \tag{3.5.17}$$

Evidently, $(\text{curl } \mathbf{A})^T \neq \text{curl } \mathbf{A}^T$, in general.

If $\mathbf{A} = \nabla\mathbf{u}$ then from (3.5.15) and (3.5.16), we find that

$$[\text{curl } \nabla\mathbf{u}]_{ij} = \varepsilon_{imn}u_{j,nm} = 0$$

so that

$$\text{curl } \nabla\mathbf{u} = \mathbf{0} \tag{3.5.18}$$

and

$$[\text{curl } \nabla\mathbf{u}^T]_{ij} = \varepsilon_{imn}u_{n,jm} = (\varepsilon_{imn}u_{n,m})_{,j}$$
$$= \{[\text{curl } \mathbf{u}]_i\}_{,j} = [\nabla \text{curl } \mathbf{u}]_{ij}$$

so that

$$\text{curl } \nabla\mathbf{u}^T = \nabla \text{curl } \mathbf{u} \tag{3.5.19}$$

3.5.5 LAPLACIAN OF A TENSOR FIELD

If a_{ij} are components of a tensor **A**, it has been noted that $c_{ijk} = a_{ij,k}$ are components of a third-order tensor $\nabla\mathbf{A}$. It is not hard to verify that $c_{ijk,m} = a_{ij,km}$ are components of a fourth-order tensor field. This tensor is

3.5 GRADIENT OF A VECTOR 129

called the *second-gradient of* **A**, denoted $\nabla\nabla\mathbf{A}$. Consequently, it follows (by contraction) that $c_{ijk,k} = a_{ij,kk}$ are components of a second-order tensor field. This tensor is called the *Laplacian of* **A**, denoted $\nabla^2\mathbf{A}$.

Thus, *if* **A** *is a second-order tensor field, then* $\nabla^2\mathbf{A}$ *is also a second-order tensor field* with components given by

$$[\nabla^2\mathbf{A}]_{ij} = \nabla^2(a_{ij}) = \nabla^2[\mathbf{A}]_{ij} \qquad (3.5.20)$$

Obviously, the Laplacian of a constant tensor field is the zero tensor. From (3.5.20) it readily follows that

$$(\nabla^2\mathbf{A})^T = \nabla^2\mathbf{A}^T \qquad (3.5.21)$$

It is easy to verify that ∇, *div*, *curl* and ∇^2 are linear differential operators in tensor calculus also; that is,

$$\nabla(\alpha\mathbf{u} + \beta\mathbf{v}) = \alpha\nabla\mathbf{u} + \beta\nabla\mathbf{v}$$

$$\nabla(\alpha\mathbf{A} + \beta\mathbf{B}) = \alpha\nabla\mathbf{A} + \beta\nabla\mathbf{B}$$

$$\text{div}(\alpha\mathbf{A} + \beta\mathbf{B}) = \alpha\,\text{div}\,\mathbf{A} + \beta\,\text{div}\,\mathbf{B} \qquad (3.5.22)$$

$$\text{curl}(\alpha\mathbf{A} + \beta\mathbf{B}) = \alpha\,\text{curl}\,\mathbf{A} + \beta\,\text{curl}\,\mathbf{B}$$

$$\nabla^2(\alpha\mathbf{A} + \beta\mathbf{B}) = \alpha\nabla^2\mathbf{A} + \beta\nabla^2\mathbf{B}$$

for any vectors **u**, **v**, any tensors **A** and **B** and any constant scalars α and β.

EXAMPLE 3.5.1 Let u_i, a_i and b_i be components of vectors **u**, **a** and **b**, and e_{ij}, w_{ij} and τ_{ij} be components of tensors **E**, **W** and **T**, respectively. Write down the following equations in the direct notation:

$$e_{ij} = \tfrac{1}{2}(u_{i,j} + u_{j,i}), \qquad w_{ij} = \tfrac{1}{2}(u_{i,j} - u_{j,i}), \qquad (3.5.23\text{a, b})$$

$$\tau_{ji,j} + \rho_0 b_i = \rho_0 a_i \qquad (3.5.24)$$

Solution By data and expressions (3.5.1) and (3.5.2), equations (3.5.23a) may be rewritten as

$$[\mathbf{E}]_{ij} = \tfrac{1}{2}[\nabla\mathbf{u} + \nabla\mathbf{u}^T]_{ij}$$

In the direct notation, these read

$$\mathbf{E} = \tfrac{1}{2}[\nabla\mathbf{u} + \nabla\mathbf{u}^T] = \text{sym}(\nabla\mathbf{u}) \qquad (3.5.23\text{a})'$$

Similarly, equations (3.5.23b) read as follows in the direct notation:

$$\mathbf{W} = \tfrac{1}{2}[\nabla\mathbf{u} - \nabla\mathbf{u}^T] = \text{skw}(\nabla\mathbf{u}) \qquad (3.5.23\text{b})'$$

By the data and expression (3.5.12), equations (3.5.24) may be rewritten as

$$[\text{div}\,\mathbf{T}^T]_i + \rho_0[\mathbf{b}]_i = \rho_0[\mathbf{a}]_i$$

130 3 CALCULUS OF TENSORS

In the direct notation, this reads

$$\text{div } \mathbf{T}^T + \rho_0 \mathbf{b} = \rho_0 \mathbf{a} \quad \blacksquare \tag{3.5.24}'$$

EXAMPLE 3.5.2 For the vector field $\mathbf{u} = \alpha(x_1^2 x_2 \mathbf{e}_1 + x_2^2 x_3 \mathbf{e}_2 + x_3^2 x_1 \mathbf{e}_3)$, where α is a constant, find (i) the gradient of $\mathbf{u}$, (ii) the divergences of $\nabla \mathbf{u}$ and $\nabla \mathbf{u}^T$ and (iii) the curl of $\nabla \mathbf{u}^T$.

Solution For the given vector field, we have

$$u_1 = \alpha x_1^2 x_2, \qquad u_2 = \alpha x_2^2 x_3, \qquad u_3 = \alpha x_3^2 x_1 \tag{3.5.25}$$

These give

$$[u_{i,j}] = \alpha \begin{pmatrix} 2x_1 x_2 & x_1^2 & 0 \\ 0 & 2x_2 x_3 & x_2^2 \\ x_3^2 & 0 & 2x_1 x_3 \end{pmatrix} \tag{3.5.26}$$

This is the matrix of $\nabla \mathbf{u}$.

From (3.5.26), we get

$$[u_{j,i}] = \alpha \begin{pmatrix} 2x_1 x_2 & 0 & x_3^2 \\ x_1^2 & 2x_2 x_3 & 0 \\ 0 & x_2^2 & 2x_1 x_3 \end{pmatrix} \tag{3.5.27}$$

This is the matrix of $\nabla \mathbf{u}^T$.

On using (3.5.25), expression (3.5.13) yields

$$[\text{div } \nabla \mathbf{u}]_1 = \nabla^2 u_1 = 2\alpha x_2, \qquad [\text{div } \nabla \mathbf{u}]_2 = \nabla^2 u_2 = 2\alpha x_3,$$

$$[\text{div } \nabla \mathbf{u}]_3 = \nabla^2 u_3 = 2\alpha x_1$$

Thus

$$\text{div } \nabla \mathbf{u} = \nabla^2 \mathbf{u} = 2\alpha(x_2 \mathbf{e}_1 + x_3 \mathbf{e}_2 + x_1 \mathbf{e}_3) \tag{3.5.28}$$

On using (3.5.25), expression (3.5.14) yields

$$[\text{div } \nabla \mathbf{u}^T]_1 = 2\alpha(x_2 + x_3), \qquad [\text{div } \nabla \mathbf{u}^T]_2 = 2\alpha(x_1 + x_3),$$

$$[\text{div } \nabla \mathbf{u}^T]_3 = 2\alpha(x_2 + x_1).$$

Thus

$$\text{div}(\nabla \mathbf{u}^T) = 2\alpha[(x_2 + x_3)\mathbf{e}_1 + (x_1 + x_3)\mathbf{e}_2 + (x_2 + x_1)\mathbf{e}_3] \tag{3.5.29}$$

It may be verified that this is equal to $\nabla \text{ div } \mathbf{u}$.

From (3.5.19) and (3.5.25), we find that

$$[\text{curl } \nabla \mathbf{u}^T]_{12} = -2x_2, \qquad [\text{curl } \nabla \mathbf{u}^T]_{23} = -2x_3, \qquad [\text{curl } \nabla \mathbf{u}^T]_{31} = 2x_1$$

and
$$[\text{curl } \nabla \mathbf{u}^T]_{ij} = 0$$
for other values of i, j. Hence
$$[\text{curl } \nabla \mathbf{u}^T] = \begin{pmatrix} 0 & -2x_2 & 0 \\ 0 & 0 & -2x_3 \\ -2x_1 & 0 & 0 \end{pmatrix} \quad (3.5.30)$$

This is the matrix of curl $\nabla \mathbf{u}^T$. It may be verified that curl $\nabla \mathbf{u}^T = \nabla$ curl $\mathbf{u}$. ∎

EXAMPLE 3.5.3 Let $\mathbf{u}$ be a given vector, and $\mathbf{A}$ be a given tensor. For any fixed vector $\mathbf{c}$ prove the following:

(i) $\quad\quad\quad\quad\quad\quad (\nabla \mathbf{u}^T)\mathbf{c} = \nabla(\mathbf{u} \cdot \mathbf{c}) \quad\quad\quad\quad\quad (3.5.31)$

(ii) $\quad\quad\quad\quad\quad (\text{div } \mathbf{A}) \cdot \mathbf{c} = \text{div}(\mathbf{A}^T \mathbf{c}) \quad\quad\quad\quad (3.5.32)$

(iii) $\quad\quad\quad\quad\quad (\text{curl } \mathbf{A})\mathbf{c} = \text{curl}(\mathbf{A}^T \mathbf{c}) \quad\quad\quad\quad (3.5.33)$

Solution Let a_{ij} be components of $\mathbf{A}$ and u_i and c_i be the components of $\mathbf{u}$ and $\mathbf{c}$, respectively. Then

(i) $\quad [(\nabla \mathbf{u}^T)\mathbf{c}]_i = [\nabla \mathbf{u}^T]_{ij}[\mathbf{c}]_j = u_{j,i} c_j = (u_j c_j)_{,i} = (\mathbf{u} \cdot \mathbf{c})_{,i} = [\nabla(\mathbf{u} \cdot \mathbf{c})]_i$

This proves (3.5.31).

(ii) $(\text{div } \mathbf{A}) \cdot \mathbf{c} = [\text{div } \mathbf{A}]_i [\mathbf{c}]_i = a_{ij,j} c_i = a_{ji,i} c_j = (a_{ji} c_j)_{,i} = ([\mathbf{A}^T \mathbf{c}]_i)_{,i}$
$\quad\quad\quad\quad\quad = \text{div } \mathbf{A}^T \mathbf{c}$

(iii) $[(\text{curl } \mathbf{A})\mathbf{c}]_i = [\text{curl } \mathbf{A}]_{ij}[\mathbf{c}]_j = (\varepsilon_{imn} a_{jn,m}) c_j = \varepsilon_{imn}(a_{jn} c_j)_{,m}$
$\quad\quad\quad\quad\quad = \varepsilon_{imn}([\mathbf{A}^T \mathbf{c}]_n)_{,m} = [\text{curl}(\mathbf{A}^T \mathbf{c})]_i$

This proves (3.5.33). ∎

EXAMPLE 3.5.4 Prove the following identities:

(i) $\quad\quad\quad\quad (\mathbf{u} \cdot \nabla)\mathbf{v} = (\nabla \mathbf{v})\mathbf{u} \quad\quad\quad\quad\quad (3.5.34)$

(ii) $\quad\quad\quad\quad \text{div}(\phi \mathbf{A}) = \phi \text{ div } \mathbf{A} + \mathbf{A}\nabla\phi \quad\quad\quad\quad (3.5.35)$

(iii) $\quad\quad\quad\quad \text{div}(\mathbf{A}\mathbf{u}) = \mathbf{u} \cdot \text{div } \mathbf{A}^T + \mathbf{A}^T \cdot \nabla \mathbf{u} \quad\quad (3.5.36)$

(iv) $\quad\quad\quad\quad \text{div}(\text{curl } \mathbf{A}) = \text{curl}(\text{div } \mathbf{A}^T) \quad\quad\quad (3.5.37)$

(v) $\quad\quad\quad\quad \text{div}(\text{curl } \mathbf{A})^T = \mathbf{0} \quad\quad\quad\quad\quad (3.5.38)$

132   **3 CALCULUS OF TENSORS**

(vi) $$\operatorname{tr}(\operatorname{curl} \mathbf{A}) = 0 \qquad (3.5.39)$$

if $\mathbf{A}$ is symmetric;

(vii) $$\operatorname{curl}(\operatorname{curl} \mathbf{A}^T) = (\operatorname{curl} \operatorname{curl} \mathbf{A})^T \qquad (3.5.40)$$

Solution   (i) $[(\nabla \mathbf{v})\mathbf{u}]_i = [\nabla \mathbf{v}]_{ij}[\mathbf{u}]_j = v_{i,j} u_j = [(\mathbf{u} \cdot \nabla)\mathbf{v}]_i$, see (3.4.23). This proves (3.5.34).

(ii) $[\operatorname{div}(\phi \mathbf{A})]_i = \{[\phi \mathbf{A}]_{ij}\}_{,j} = (\phi a_{ij})_{,j} = \phi a_{ij,j} + a_{ij}\phi_{,j}$. This proves (3.5.35).

(iii) $[\operatorname{div}(\mathbf{A}\mathbf{u})] = \{[\mathbf{A}\mathbf{u}]_i\}_{,i} = (a_{ij} u_j)_{,i}$

$$= a_{ij,i} u_j + a_{ij} u_{j,i} = a_{ji,j} u_i + a_{ji} u_{i,j}$$
$$= [\mathbf{u}]_i [\operatorname{div} \mathbf{A}^T]_i + [\mathbf{A}^T]_{ij} [\nabla \mathbf{u}]_{ij} = \mathbf{u} \cdot \operatorname{div} \mathbf{A}^T + \mathbf{A}^T \cdot (\nabla \mathbf{u})$$

(iv) $[\operatorname{div}(\operatorname{curl} \mathbf{A})]_i = \{[\operatorname{curl} \mathbf{A}]_{ij}\}_{,j}$

$$= \varepsilon_{imn} a_{jn,mj} = \varepsilon_{imn}(a_{jn,j})_{,m}$$
$$= \varepsilon_{imn}\{[\operatorname{div} \mathbf{A}^T]_n\}_{,m} = [\operatorname{curl}(\operatorname{div} \mathbf{A}^T)]_i$$

This proves (3.5.37).

(v) $[\operatorname{div}(\operatorname{curl} \mathbf{A})^T]_i = \{[(\operatorname{curl} \mathbf{A})^T]_{ij}\}_{,j} = \{[\operatorname{curl} \mathbf{A}]_{ji}\}_{,j} = \varepsilon_{jmn} a_{in,mj} = 0$. This proves (3.5.38).

(vi) $\operatorname{tr}(\operatorname{curl} \mathbf{A}) = [\operatorname{curl} \mathbf{A}]_{ii} = \varepsilon_{imn} a_{in,m}$. If $\mathbf{A}$ is symmetric, we have

$$\varepsilon_{imn} a_{in,m} = \varepsilon_{inm} a_{ni,m} = \varepsilon_{nmi} a_{in,m} = -\varepsilon_{imn} a_{in,m}$$

so that $\varepsilon_{imn} a_{in,m} = 0$. Hence $\operatorname{tr}(\operatorname{curl} \mathbf{A}) = 0$, if $\mathbf{A}$ is symmetric.

(vii) $[\operatorname{curl}(\operatorname{curl} \mathbf{A}^T)]_{ij} = \varepsilon_{imn}\{[\operatorname{curl} \mathbf{A}^T]_{jn}\}_{,m} = \varepsilon_{imn}(\varepsilon_{jpq}\{[\mathbf{A}^T]_{nq}\}_{,p})_{,m}$

$$= \varepsilon_{imn} \varepsilon_{jpq} a_{qn,pm} = \varepsilon_{jpq}(\varepsilon_{imn} a_{qn,m})_{,p}$$
$$= \varepsilon_{jpq}\{[\operatorname{curl} \mathbf{A}]_{iq}\}_{,p} = [\operatorname{curl}(\operatorname{curl} \mathbf{A})]_{ji}$$

This proves (3.5.40). ∎

EXAMPLE 3.5.5   Let $\mathbf{a}$ and $\mathbf{b}$ be constant vectors. For any vectors $\mathbf{u}$ and $\mathbf{v}$ and any tensor $\mathbf{A}$, prove the following:

(i) $\quad \operatorname{div}\{(\mathbf{u} \cdot \mathbf{a})\mathbf{A}^T \mathbf{b}\} = \mathbf{a} \cdot \{(\nabla \mathbf{u})\mathbf{A}^T + \mathbf{u} \otimes \operatorname{div} \mathbf{A}\}\mathbf{b} \qquad (3.5.41)$

(ii) $\quad \operatorname{curl}\{(\mathbf{u} \cdot \mathbf{a})\mathbf{A}^T \mathbf{b}\} \cdot \mathbf{v} = \mathbf{a} \cdot [(\mathbf{v} \wedge \nabla \mathbf{u})^T \mathbf{A}^T + \mathbf{u} \otimes (\operatorname{curl} \mathbf{A})^T \mathbf{v}]\mathbf{b} \qquad (3.5.42)$

Solution   Let a_i, b_i, u_i and v_i be components of $\mathbf{a}$, $\mathbf{b}$, $\mathbf{u}$ and $\mathbf{v}$, respectively, and a_{ij} be components of $\mathbf{A}$. Then,

(i) $\quad \operatorname{div}\{(\mathbf{u} \cdot \mathbf{a})\mathbf{A}^T \mathbf{b}\} = (u_k a_k a_{ji} b_j)_{,i} = a_k b_j (u_{k,i} a_{ji} + u_k a_{ji,i})$

$$= a_k b_j \{[(\nabla \mathbf{u})\mathbf{A}^T]_{kj} + [\mathbf{u} \otimes \operatorname{div} \mathbf{A}]_{kj}\}$$
$$= a_k [\{(\nabla \mathbf{u})\mathbf{A}^T + \mathbf{u} \otimes \operatorname{div} \mathbf{A}\}\mathbf{b}]_k$$
$$= \mathbf{a} \cdot \{(\nabla \mathbf{u})\mathbf{A}^T + \mathbf{u} \otimes \operatorname{div} \mathbf{A}\}\mathbf{b}$$

3.5 GRADIENT OF A VECTOR

(ii) $\text{curl}\{(\mathbf{u} \cdot \mathbf{a})\mathbf{A}^T\mathbf{b}\} \cdot \mathbf{v} = \varepsilon_{irs}(u_k a_k a_{js} b_j)_{,r} v_i$

$\quad = \varepsilon_{irs} a_k b_j v_i (u_{k,r} a_{js} + u_k a_{js,r})$

$\quad = a_k\{(\varepsilon_{sir} v_i u_{k,r}) a_{js} + u_k(\varepsilon_{irs} a_{js,r}) v_i\} b_j$

$\quad = a_k\{[\mathbf{v} \wedge \nabla\mathbf{u}]_{sk}[\mathbf{A}]_{js} + [\mathbf{u}]_k[\text{curl }\mathbf{A}]_{ij}[\mathbf{v}]_i\} b_j$

$\quad = a_k\{[(\mathbf{v} \wedge \nabla\mathbf{u})^T \mathbf{A}^T]_{kj} + [\mathbf{u} \otimes (\text{curl }\mathbf{A})^T \mathbf{v}]_{kj}\} b_j$

$\quad = \mathbf{a} \cdot \{(\mathbf{v} \wedge \nabla\mathbf{u})^T \mathbf{A}^T + \mathbf{u} \otimes (\text{curl }\mathbf{A})^T \mathbf{v}\} \mathbf{b}$ ∎

EXAMPLE 3.5.6 If **W** is a skew tensor and **w** is the dual vector thereof, show that

$$\text{curl }\mathbf{W} = (\text{div }\mathbf{w})\mathbf{I} - \nabla\mathbf{w} \quad (3.5.43)$$

Solution Let w_{ij} be components of **W** and w_i be components of **w**. Then w_{ij} and w_i are related as follows (see (2.10.4)): $w_{ij} = -\varepsilon_{ijk} w_k$. Hence

$\varepsilon_{imn} w_{jn,m} = -\varepsilon_{imn}\varepsilon_{jnp} w_{p,m} = \varepsilon_{imn}\varepsilon_{jpn} w_{p,m}$

$\quad = (\delta_{ij}\delta_{mp} - \delta_{ip}\delta_{mj}) w_{p,m} = \delta_{ij} w_{p,p} - w_{i,j}$

When written in direct notation, this becomes (3.5.43). ∎

EXAMPLE 3.5.7 Let **u** be a vector, $\mathbf{E} = \text{sym }\nabla\mathbf{u}$, $\mathbf{W} = \text{skw }\nabla\mathbf{u}$ and **w** be the dual vector of **W**. Prove the following:

(i) $\quad \mathbf{w} = \tfrac{1}{2}\text{curl }\mathbf{u} \quad (3.5.44)$

(ii) $\quad |\nabla\mathbf{u}|^2 = |\mathbf{E}|^2 + |\mathbf{W}|^2 = |\mathbf{E}|^2 + \tfrac{1}{2}|\text{curl }\mathbf{u}|^2$

$\quad = \text{div}(\nabla\mathbf{u}^T)\mathbf{u} - \mathbf{u}\cdot\nabla^2\mathbf{u}. \quad (3.5.45)$

(iii) $\quad \nabla\mathbf{u} \cdot \nabla\mathbf{u}^T = |\mathbf{E}|^2 - |\mathbf{W}|^2 = |\mathbf{E}|^2 - \tfrac{1}{2}|\text{curl }\mathbf{u}|^2$

$\quad = \text{div}(\mathbf{u}\cdot\nabla)\mathbf{u} - (\mathbf{u}\cdot\nabla)(\text{div }\mathbf{u})$

$\quad = \text{div}\{(\nabla\mathbf{u})\mathbf{u} - (\text{div }\mathbf{u})\mathbf{u}\} + (\text{div }\mathbf{u})^2 \quad (3.5.46)$

(iv) $\quad \mathbf{EW} + \mathbf{WE} = \text{skw}(\nabla\mathbf{u})^2 \quad (3.5.47)$

Solution (i) Let u_i, w_i, e_{ij} and w_{ij} be components of **u**, **w**, **E** and **W**, respectively. Then by use of (2.10.2) we get

$w_i = -\tfrac{1}{2}\varepsilon_{ipq} w_{pq}$

$\quad = -\tfrac{1}{4}\varepsilon_{ipq}(u_{p,q} - u_{q,p}) = \tfrac{1}{2}\varepsilon_{ipq} u_{q,p}$

When written in the direct notation, this becomes (3.5.44).

(ii) Consequently, by use of (2.10.8), we have
$$|\mathbf{W}|^2 = \tfrac{1}{2}|\text{curl }\mathbf{u}|^2 \tag{3.5.48}$$
Noting that $\nabla \mathbf{u} = \mathbf{E} + \mathbf{W}$, we find
$$|\nabla \mathbf{u}|^2 = \nabla \mathbf{u} \cdot \nabla \mathbf{u} = (\mathbf{E} + \mathbf{W}) \cdot (\mathbf{E} + \mathbf{W}) = |\mathbf{E}|^2 + |\mathbf{W}|^2 \tag{3.5.49}$$
because $\mathbf{E} \cdot \mathbf{W} = 0$. On the other hand,
$$\begin{aligned}|\nabla \mathbf{u}|^2 &= u_{i,j} u_{i,j} = (u_i u_{i,j})_{,j} - u_i u_{i,jj}\\ &= \{[(\nabla \mathbf{u}^T)\mathbf{u}]_j\}_{,j} - [\mathbf{u}]_i [\nabla^2 \mathbf{u}]_i\\ &= \text{div }(\nabla \mathbf{u}^T)\mathbf{u} - \mathbf{u} \cdot \nabla^2 \mathbf{u} \end{aligned} \tag{3.5.50}$$

Expressions (3.5.48), (3.5.49) and (3.5.50) constitute the relations (3.5.45).

(iii) Since $\nabla \mathbf{u}^T = \mathbf{E} - \mathbf{W}$, we get
$$\nabla \mathbf{u} \cdot \nabla \mathbf{u}^T = (\mathbf{E} + \mathbf{W}) \cdot (\mathbf{E} - \mathbf{W}) = |\mathbf{E}|^2 - |\mathbf{W}|^2 \tag{3.5.51}$$

On the other hand,
$$\begin{aligned}\nabla \mathbf{u} \cdot \nabla \mathbf{u}^T &= u_{i,j} u_{j,i} = (u_{i,j} u_j)_{,i} - u_j u_{i,ij}\\ &= \{[(\mathbf{u} \cdot \nabla)\mathbf{u}]_i\}_{,i} - (\mathbf{u} \cdot \nabla)(u_{i,i})\\ &= \text{div}\{(\mathbf{u} \cdot \nabla)\mathbf{u}\} - (\mathbf{u} \cdot \nabla)(\text{div }\mathbf{u}) \end{aligned} \tag{3.5.52}$$

Also,
$$\begin{aligned}\nabla \mathbf{u} \cdot \nabla \mathbf{u}^T &= u_{i,j} u_{j,i} = (u_{i,j} u_j - u_{k,k} u_i)_{,i} + (u_{k,k})^2\\ &= \{[(\nabla \mathbf{u})\mathbf{u}]_i - [(u_{k,k})\mathbf{u}]_i\}_{,i} + (u_{k,k})^2\\ &= \text{div}\{(\nabla \mathbf{u})\mathbf{u} - (\text{div }\mathbf{u})\mathbf{u}\} + (\text{div }\mathbf{u})^2 \end{aligned} \tag{3.5.53}$$

Expressions (3.5.48), (3.5.51), (3.5.52) and (3.5.53) give (3.5.46).

(iv) Next,
$$\begin{aligned}(\nabla \mathbf{u})^2 &= (\nabla \mathbf{u})(\nabla \mathbf{u}) = (\mathbf{E} + \mathbf{W})(\mathbf{E} + \mathbf{W})\\ &= \mathbf{E}^2 + \mathbf{W}^2 + \mathbf{EW} + \mathbf{WE}. \end{aligned} \tag{3.5.54}$$

Hence
$$[(\nabla \mathbf{u})^2]^T = \mathbf{E}^2 + \mathbf{W}^2 - (\mathbf{EW} + \mathbf{WE}) \tag{3.5.55}$$

Expressions (3.5.54) and (3.5.55) yield
$$(\nabla \mathbf{u})^2 - [(\nabla \mathbf{u})^2]^T = 2(\mathbf{EW} + \mathbf{WE}) \tag{3.5.56}$$

from which (3.5.47) is immediate. ∎

3.5 GRADIENT OF A VECTOR   135

EXAMPLE 3.5.8 Show that for any tensor $\mathbf{A}$,

$$\text{curl curl } \mathbf{A} = [\nabla^2(\text{tr } \mathbf{A}) - \text{div}(\text{div } \mathbf{A})]\mathbf{I} + \nabla \text{ div } \mathbf{A}^T$$
$$+ (\nabla \text{ div } \mathbf{A})^T - \nabla\nabla(\text{tr } \mathbf{A}) - \nabla^2 \mathbf{A}^T \qquad (3.5.57)$$

Deduce that, if $\mathbf{A}$ is symmetric, then the equation

$$\text{curl curl } \mathbf{A} = \mathbf{0} \qquad (3.5.58)$$

is equivalent to the equation

$$\nabla^2 \mathbf{A} + \nabla\nabla(\text{tr } \mathbf{A}) - \nabla(\text{div } \mathbf{A}) - (\nabla \text{ div } \mathbf{A})^T = \mathbf{0} \qquad (3.5.59)$$

Solution If a_{ij} are components of $\mathbf{A}$, we note that

$$[\text{curl curl } \mathbf{A}]_{ij} = \varepsilon_{irs}\varepsilon_{jmn}a_{sn,mr} \qquad (3.5.60)$$

By using the identity (1.7.22) this becomes

$$[\text{curl curl } \mathbf{A}]_{ij} = (a_{sn,mr}) \begin{vmatrix} \delta_{ij} & \delta_{im} & \delta_{in} \\ \delta_{rj} & \delta_{rm} & \delta_{rn} \\ \delta_{sj} & \delta_{sm} & \delta_{sn} \end{vmatrix} \qquad (3.5.61)$$

On expanding the determinant on the righthand side, this simplifies to

$$[\text{curl curl } \mathbf{A}]_{ij} = \delta_{ij}(a_{ss,mm} - a_{mr,mr}) + a_{jr,ir} + a_{mi,mj} - a_{ji,rr} - a_{ss,ij}$$
$$= \{\nabla^2(\text{tr } \mathbf{A}) - \text{div}(\text{div } \mathbf{A})\}\delta_{ij} + [(\nabla \text{ div } \mathbf{A})^T]_{ij}$$
$$+ [\nabla(\text{div } \mathbf{A}^T)]_{ij} - [\nabla^2 \mathbf{A}^T]_{ij} - [\nabla\nabla(\text{tr } \mathbf{A})]_{ij} \qquad (3.5.62)$$

This is precisely the identity (3.5.57).

If $\mathbf{A}$ is symmetric, the identity (3.5.57) reads as follows in the suffix notation:

$$\varepsilon_{irs}\varepsilon_{jmn}a_{sn,mr} = (a_{ss,mm} - a_{mr,mr})\delta_{ij} + a_{im,mj}$$
$$+ a_{jm,im} - a_{ij,mm} - a_{mm,ij} \qquad (3.5.63)$$

Now, suppose that (3.5.58) holds; then we have

$$\varepsilon_{irs}\varepsilon_{jmn}a_{sn,mr} = 0 \qquad (3.5.64)$$

Consequently, $\varepsilon_{irs}\varepsilon_{imn}a_{sn,mr} = 0$, which, on using the ε-δ identity, reduces to

$$a_{ss,mm} - a_{mr,mr} = 0 \qquad (3.5.65)$$

In view of (3.5.64) and (3.5.65), identity (3.5.63) yields

$$a_{ij,mm} + a_{mm,ij} - a_{im,mj} - a_{jm,im} = 0 \qquad (3.5.66)$$

136 3 CALCULUS OF TENSORS

When written in the direct notation, these equations become the tensor equation (3.5.59). Thus equation (3.5.58) implies equation (3.5.59).

Conversely, suppose that the tensor equation (3.5.59) holds. Then (3.5.66) hold. Taking the contraction of (3.5.66), we find that (3.5.65) holds. Since (3.5.65) and (3.5.66) hold, identity (3.5.63) yields (3.5.64). This means that the tensor equation (3.5.58) holds.

Thus, equations (3.5.58) and (3.5.59) imply each other; the equations are therefore equivalent. ∎

3.6 INTEGRAL THEOREMS FOR VECTORS

Here we state the divergence theorem and Stokes's theorem in vector integral calculus, with which the reader should be already familiar. Some consequences of these theorems are also summarized.

3.6.1 DIVERGENCE THEOREM

Let V be the volume of a three-dimensional region bounded by a closed regular surface S; see Figure 3.2. Then for a vector field $\mathbf{u}$ defined in V and on S,

$$\int_V (\operatorname{div} \mathbf{u})\, dV = \int_S (\mathbf{u} \cdot \mathbf{n})\, dS \qquad (3.6.1)$$

where $\mathbf{n}$ is the unit outward normal to S. In the suffix notation, the expression (3.6.1) reads:

$$\int_V u_{k,k}\, dV = \int_S u_k n_k\, dS \qquad (3.6.1)'$$

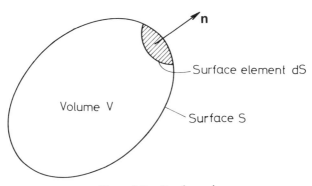

Figure 3.2. Regular region.

3.6 INTEGRAL THEOREMS FOR VECTORS

The following are some immediate consequences of expression (3.6.1):

(i) $\quad \int_V \nabla \phi \, dV = \int_S \phi \mathbf{n} \, dS,$ or $\quad \int_V \phi_{,k} \, dV = \int_S \phi n_k \, dS \quad (3.6.2)$

(ii) $\quad \int_V (\text{curl } \mathbf{u}) \, dV = \int_S (\mathbf{n} \times \mathbf{u}) \, dS$ or $\quad \int_V \varepsilon_{ijk} u_{k,j} \, dV = \int_S \varepsilon_{ijk} n_j u_k \, dS \quad (3.6.3)$

(iii) $\quad \int_V (\nabla^2 \phi) \, dV = \int_S (\mathbf{n} \cdot \nabla)\phi \, dS$ or $\quad \int_V \phi_{,kk} \, dV = \int_S n_k \phi_{,k} \, dS \quad (3.6.4)$

(iv) $\quad \int_V (\nabla^2 \mathbf{u}) \, dV = \int_S (\mathbf{n} \cdot \nabla)\mathbf{u} \, dS$ or $\quad \int_V u_{i,kk} \, dV = \int_S n_k u_{i,k} \, dS \quad (3.6.5)$

In (3.6.2) and (3.6.4), ϕ is a scalar field defined in V and on S.

3.6.2 SOLENOIDAL VECTORS

The surface integral $\int_S \mathbf{u} \cdot \mathbf{n} \, dS$ is usually called the *outward normal flux* or just *flux* of $\mathbf{u}$ across S. A vector is said to be *solenoidal* in a region if its flux across every closed regular surface in the region is 0. From the divergence theorem, it follows that $\mathbf{u}$ *is solenoidal in a simply connected region if and only if* div $\mathbf{u} = 0$ *in that region*.

A vector field whose divergence is exactly 0 is called a *divergence-free vector*. From the statement just made, it follows that a vector field is solenoidal in a simply connected region if and only if it is divergence free.

Since div(curl $\mathbf{u}$) = 0, see (3.4.17), therefore curl $\mathbf{u}$ is a divergence-free vector for every vector $\mathbf{u}$. It can be proven that every divergence-free vector $\mathbf{u}$ defined in a simply connected region can be represented as

$$\mathbf{u} = \text{curl } \mathbf{w} \quad (3.6.6)$$

where $\mathbf{w}$ is itself a divergence-free vector. Here $\mathbf{w}$ is called a *vector potential* of $\mathbf{u}$.

3.6.3 STOKES'S THEOREM

Let C be a simple closed curve in three-dimensional space and S be an open regular surface bounded by C; see Figure 3.3. Then for a vector field $\mathbf{u}$ defined on S as well as C,

$$\oint_C \mathbf{u} \cdot \mathbf{t} \, ds = \int_S (\text{curl } \mathbf{u}) \cdot \mathbf{n} \, dS \quad (3.6.7)$$

138   3 CALCULUS OF TENSORS

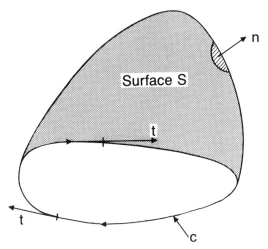

Figure 3.3. Open regular surface.

where **t** is a unit tangent vector to C that is assumed to be positively oriented relative to the unit normal **n** to S.

In the suffix notation, expression (3.6.7) reads

$$\oint_C u_i t_i \, ds = \int_S \varepsilon_{ijk} u_{k,j} n_i \, dS \qquad (3.6.7)'$$

If S is a closed surface, the lefthand side of (3.6.7) reduces to 0, then

$$\int_S (\text{curl } \mathbf{u}) \cdot \mathbf{n} \, dS = 0 \quad \text{or} \quad \int_S \varepsilon_{ijk} u_{k,j} n_i \, dS = 0 \qquad (3.6.8)$$

This expression also follows from the divergence theorem applied to curl **u**.

The particular case of (3.6.7), where C lies entirely in the $x_1 x_2$-plane and S is the part of the plane bounded by C, is important in its own right. In this case, expression (3.6.7) reduces to the form

$$\int_C (u_1 \, dx_1 + u_2 \, dx_2) = \int_S (u_{2,1} - u_{1,2}) \, dx_1 \, dx_2 \qquad (3.6.9)$$

where u_1 and u_2 are the x_1 and x_2-components of **u**. This particular case of Stokes's theorem is usually referred to as *Green's theorem in the plane*.

Following are some immediate consequences of the expression (3.6.7):

$$\text{(i)} \quad \oint_C \phi \mathbf{t} \, ds = \int_S \mathbf{n} \times \nabla \phi \, dS \quad \text{or} \quad \oint_C \phi t_i \, ds = \int_S \varepsilon_{ijk} n_j \phi_{,k} \, dS \qquad (3.6.10)$$

3.6 INTEGRAL THEOREMS FOR VECTORS

(ii) $\oint_C (\mathbf{u} \times \mathbf{t}) \, ds = \int_S [(\text{div } \mathbf{u})\mathbf{n} - (\nabla \mathbf{u})^T \mathbf{n}] \, dS$ or

$$\oint_C \varepsilon_{ijk} u_j t_k \, ds = \int_S [u_{k,k} n_i - u_{k,i} n_k] \, dS \qquad (3.6.11)$$

(iii) $\oint_C \text{curl } \mathbf{u} \cdot \mathbf{t} \, ds = \int_S \left[\frac{\partial}{\partial n} (\text{div } \mathbf{u}) - \mathbf{n} \cdot (\nabla^2 \mathbf{u}) \right] dS$ or

$$\oint_C \varepsilon_{ijk} u_{k,j} t_i \, ds = \int_S \left[\frac{\partial}{\partial n} (u_{k,k}) - n_i u_{i,kk} \right] dS \qquad (3.6.12)$$

In (3.6.10), ϕ is a scalar field defined on S as well as C.

Note: The *arc element* $\mathbf{t} \, ds$ is often denoted $d\mathbf{x}$. Consequently, we write $\int_C (\ ) \, d\mathbf{x}$ for $\int_C (\ ) \mathbf{t} \, ds$.

3.6.4 CONSERVATIVE AND IRROTATIONAL VECTORS

The line integral in the lefthand side of expression (3.6.7), namely, $\oint_C \mathbf{u} \cdot \mathbf{t} \, ds$, which represents the integral of $\mathbf{u} \cdot \mathbf{t}$ around C, is called the *circulation* of $\mathbf{u}$ round C. A vector $\mathbf{u}$ defined in a region is said to be *conservative* if its circulation round every simple closed curve in the region is 0, or equivalently, if the value of the line integral $\int_A^B \mathbf{u} \cdot \mathbf{t} \, ds$, defined in the region, depends only on the endpoints A and B and not on the curve joining A and B on which integration is carried out. Also, a vector is said to be *irrotational* (or *curl free*) in a region if its curl is exactly 0 in the region. From Stokes's theorem, it follows that, in a simply connected region, *a vector is conservative if and only if it is irrotational* in the region.

Since curl $\nabla \phi = 0$, see (3.4.16), therefore $\nabla \phi$ is an irrotational vector for every scalar field ϕ. It can be proven that every irrotational vector $\mathbf{u}$ defined in a simply connected region can be represented

$$\mathbf{u} = \nabla \phi \qquad (3.6.13)$$

Then ϕ is called a *scalar potential* of $\mathbf{u}$.

If a vector $\mathbf{u}$ is both divergence free and irrotational, then the identity (3.4.27) shows that $\nabla^2 \mathbf{u} = \mathbf{0}$; in this case we say that $\mathbf{u}$ is a *harmonic vector*.

3.6.5 HELMHOLTZ'S REPRESENTATION

It has been noted that a divergence-free vector has a representation as given by (3.6.6) and an irrotational vector has a representation as given by (3.6.13). A representation valid for a general vector, known as the *Helmholtz's representation* follows.

Consider a vector field **u**. Define the vector field **v** by

$$\mathbf{v}(\mathbf{x}) = -\frac{1}{4\pi} \int_V \frac{\mathbf{u}(\bar{\mathbf{x}})}{|\mathbf{x} - \bar{\mathbf{x}}|} dV \qquad (3.6.14)$$

where V is some volume in the region of definition of **u** and the integral is taken by varying $\bar{\mathbf{x}}$ over V, keeping **x** as a fixed point. It can be proven that $\nabla^2 \mathbf{v} = \mathbf{u}$. Then by using the identity (3.4.27) we get

$$\mathbf{u} = \nabla \phi + \text{curl } \mathbf{w} \qquad (3.6.15)$$

where

$$\phi = \text{div } \mathbf{v}, \qquad \mathbf{w} = -\text{curl } \mathbf{v} \qquad (3.6.16)$$

Thus, given a vector field **u**, there exist a scalar field ϕ and a vector field **w**, defined by (3.6.16) and (3.6.14), such that **u** has a representation as given by (3.6.15). This is the Helmholtz's representation. Note that the vector **w** present in this representation is divergence free.

EXAMPLE 3.6.1 Prove expressions (3.6.3) and (3.6.11).

Solution Let **a** be an arbitrary constant vector field. Employing the divergence theorem to the vector $\mathbf{a} \times \mathbf{u}$, we get

$$\int_V \text{div}(\mathbf{a} \times \mathbf{u}) \, dV = \int_S (\mathbf{a} \times \mathbf{u}) \cdot \mathbf{n} \, dS \qquad (3.6.17)$$

Using identities (3.4.24) and (1.4.18), expression (3.6.17) becomes

$$\mathbf{a} \cdot \left\{ \int_V \text{curl } \mathbf{u} \, dV - \int_S \mathbf{n} \times \mathbf{u} \, dS \right\} = 0$$

Since **a** is arbitrary, this yields (3.6.3).

On the other hand, if we employ Stokes's theorem to the vector $\mathbf{a} \times \mathbf{u}$, we get

$$\oint_C (\mathbf{a} \times \mathbf{u}) \cdot \mathbf{t} \, ds = \int_S \text{curl}(\mathbf{a} \times \mathbf{u}) \cdot \mathbf{n} \, dS \qquad (3.6.18)$$

Using identity (1.4.18) and

$$[\text{curl}(\mathbf{a} \times \mathbf{u})] \cdot \mathbf{n} = [(\text{div } \mathbf{u})\mathbf{a} - (\mathbf{a} \cdot \nabla)\mathbf{u}] \cdot \mathbf{n} = [(\text{div } \mathbf{u})\mathbf{a} - (\nabla \mathbf{u})\mathbf{a}] \cdot \mathbf{n}$$

$$= \mathbf{a} \cdot [(\text{div } \mathbf{u})\mathbf{n} - (\nabla \mathbf{u})^T \mathbf{n}]$$

which is obtained by using (3.4.25), (3.5.34) and (2.8.14), expression

(3.6.18) becomes

$$\mathbf{a} \cdot \left\{ \oint_C \mathbf{u} \times \mathbf{t}\, ds - \int_S (\operatorname{div} \mathbf{u} - \nabla \mathbf{u}^T)\mathbf{n}\, dS \right\} = 0$$

Since $\mathbf{a}$ is arbitrary, this yields (3.6.11). ∎

EXAMPLE 3.6.2 Show that

$$\mathbf{n} \times \operatorname{curl} \mathbf{u} = (\nabla \mathbf{u})^T \mathbf{n} - (\mathbf{n} \cdot \nabla)\mathbf{u} \qquad (3.6.19)$$

Hence, deduce the following alternative version of (3.6.11):

$$\oint_C \mathbf{u} \times \mathbf{t}\, ds = \int_S [(\operatorname{div} \mathbf{u})\mathbf{n} - (\mathbf{n} \cdot \nabla)\mathbf{u} - \mathbf{n} \times \operatorname{curl} \mathbf{u}]\, dS \qquad (3.6.20)$$

Solution By use of (3.4.31), (3.5.2) and (3.4.23), we get

$$[\mathbf{n} \times \operatorname{curl} \mathbf{u}]_i = (u_{k,i} - u_{i,k})n_k = [(\nabla \mathbf{u})^T \mathbf{n} - (\mathbf{n} \cdot \nabla)\mathbf{u}]_i$$

from which (3.6.19) is immediate.

Substituting for $(\nabla \mathbf{u})^T \mathbf{n}$ from the identity (3.6.19) in the expression (3.6.11), we readily get the expression (3.6.20). ∎

EXAMPLE 3.6.3 If $\mathbf{A}$ is a tensor field such that $\operatorname{div} \mathbf{A}^T = \mathbf{0}$ in a simply connected region, show that $\mathbf{A} = \operatorname{curl} \mathbf{W}$ for some tensor $\mathbf{W}$.

Solution As usual, let $\mathbf{e}_i$ be any one of the base vectors of a fixed coordinate system. Then

$$\operatorname{div}(\mathbf{A}\mathbf{e}_i) = \mathbf{e}_i \cdot \operatorname{div} \mathbf{A}^T$$

see (3.5.32);

$$= 0$$

by data. Hence, by virtue of (3.6.6), $\mathbf{A}\mathbf{e}_i$ can be represented as

$$\mathbf{A}\mathbf{e}_i = \operatorname{curl} \mathbf{w}_i \qquad (3.6.21)$$

for some divergence-free vector $\mathbf{w}_i$. Let $\mathbf{W} = \mathbf{e}_k \otimes \mathbf{w}_k$. Then

$$\mathbf{W}^T \mathbf{e}_i = (\mathbf{w}_k \otimes \mathbf{e}_k)\mathbf{e}_i = (\mathbf{e}_k \cdot \mathbf{e}_i)\mathbf{w}_k = \mathbf{w}_i. \qquad (3.6.22)$$

Hence (3.6.21) and (3.6.22) give

$$\mathbf{A}\mathbf{e}_i = \operatorname{curl} \mathbf{w}_i = \operatorname{curl}(\mathbf{W}^T \mathbf{e}_i) = (\operatorname{curl} \mathbf{W})\mathbf{e}_i$$

see (3.5.33). By virtue of the result proven in Example 2.7.2, it follows that $\mathbf{A} = \operatorname{curl} \mathbf{W}$. ∎

142   3 CALCULUS OF TENSORS

EXAMPLE 3.6.4 If $\mathbf{A}$ is a tensor field such that curl $\mathbf{A} = \mathbf{0}$ in a simply connected region, show that $\mathbf{A} = \nabla \mathbf{u}$ for some vector $\mathbf{u}$. If, further, tr $\mathbf{A} = 0$, show that $\mathbf{A} = \text{curl } \mathbf{W}$ for some skew tensor $\mathbf{W}$.

Solution   As usual, let $\mathbf{e}_i$ be any one of the base vectors of a fixed coordinate system. Then,

$$\text{curl}(\mathbf{A}^T \mathbf{e}_i) = (\text{curl } \mathbf{A}) \mathbf{e}_i$$

see (3.5.33);

$$= \mathbf{0}$$

by data. Hence by virtue of (3.6.13), there exists a scalar field ϕ_i such that

$$\mathbf{A}^T \mathbf{e}_i = \nabla \phi_i \tag{3.6.23}$$

Let $\mathbf{u} = \phi_k \mathbf{e}_k$. Then $\mathbf{u} \cdot \mathbf{e}_i = \phi_i$ so that, by using (3.5.31),

$$\nabla \phi_i = \nabla(\mathbf{u} \cdot \mathbf{e}_i) = (\nabla \mathbf{u}^T) \mathbf{e}_i \tag{3.6.24}$$

Hence (3.6.23) and (3.6.24) give $\mathbf{A}^T \mathbf{e}_i = (\nabla \mathbf{u}^T) \mathbf{e}_i$. By virtue of the result proven in Example 2.7.2, it follows that $\mathbf{A}^T = \nabla \mathbf{u}^T$ or $\mathbf{A} = \nabla \mathbf{u}$. Consequently, tr $\mathbf{A} = \text{tr } \nabla \mathbf{u} = \text{div } \mathbf{u}$. Hence, if tr $\mathbf{A} = 0$, then div $\mathbf{u} = 0$.

Let $\mathbf{W}$ be the skew tensor of which $-\mathbf{u}$ is the dual vector. Then, from (3.5.43), it follows that

$$\text{curl } \mathbf{W} = \text{div}(-\mathbf{u})\mathbf{I} - \nabla(-\mathbf{u})$$

Since div $\mathbf{u} = 0$, we get

$$\text{curl } \mathbf{W} = \nabla \mathbf{u} = \mathbf{A} \quad \blacksquare$$

EXAMPLE 3.6.5   Let $\mathbf{u}$ be a vector field defined on and inside a closed regular surface S enclosing a volume V, and $\mathbf{E} = \text{sym } \nabla \mathbf{u}$ and $\mathbf{W} = \text{skw } \nabla \mathbf{u}$. If $\mathbf{u} = \mathbf{0}$ on S, prove the following:

(i)
$$\int_V |\mathbf{W}|^2 \, dV \leq \int_V |\mathbf{E}|^2 \, dV \tag{3.6.25}$$

(ii) Korn's inequality:

$$\int_V |\nabla \mathbf{u}|^2 \, dV \leq 2 \int_V |\mathbf{E}|^2 \, dV \tag{3.6.26}$$

Solution   (i) From (3.5.46), we get

$$\int_V \{|\mathbf{E}|^2 - |\mathbf{W}|^2\} \, dV = \int_V \text{div}\{(\nabla \mathbf{u})\mathbf{u} - (\text{div } \mathbf{u})\mathbf{u}\} \, dV + \int_V (\text{div } \mathbf{u})^2 \, dV$$

Employing the divergence theorem to the first of the integrals on the righthand side, and recalling that $(\nabla \mathbf{u})\mathbf{u} = (\mathbf{u} \cdot \nabla)\mathbf{u}$, we get

$$\int_V \{|\mathbf{E}|^2 - |\mathbf{W}|^2\} \, dV = \int_S \{(\mathbf{u} \cdot \nabla)(\mathbf{u} \cdot \mathbf{n}) - (\text{div } \mathbf{u})(\mathbf{u} \cdot \mathbf{n})\} \, dS + \int_V (\text{div } \mathbf{u})^2 \, dV \quad (3.6.27)$$

Since $\mathbf{u} = \mathbf{0}$ on S by data, this yields

$$\int_V |\mathbf{E}|^2 \, dV - \int_V |\mathbf{W}|^2 \, dV = \int_V (\text{div } \mathbf{u})^2 \, dV \geq 0$$

from which (3.6.25) is immediate.
(ii) From (3.5.46) we also get

$$\int_V |\nabla \mathbf{u}|^2 \, dV = \int_V |\mathbf{E}|^2 \, dV + \int_V |\mathbf{W}|^2 \, dV \quad (3.6.28)$$

Using the inequality (3.6.25), expression (3.6.28) yields

$$\int_V |\nabla \mathbf{u}|^2 \, dV \leq \int_V |\mathbf{E}|^2 \, dV + \int_V |\mathbf{E}|^2 \, dV$$

which is (3.6.26). ∎

3.7
INTEGRAL THEOREMS FOR TENSORS

We now obtain some extensions to tensor fields of the divergence theorem and Stokes's theorem stated in Section 3.6. As in the case of a vector field, an integral of a given tensor field is defined as a tensor field whose components are the integrals of the components of the given field.

THEOREM 3.7.1 (Divergence Theorem for a Tensor) Let V be the volume of a three-dimensional region bounded by a closed regular surface S. Then for a tensor field $\mathbf{A}$ defined in V and on S,

$$\int_V \text{div } \mathbf{A} \, dV = \int_S \mathbf{A}\mathbf{n} \, dS \quad (3.7.1)$$

where $\mathbf{n}$ is, as usual, the unit outward normal to S.

Proof Let **c** be an arbitrary constant vector. Then

$$\mathbf{c} \cdot \int_V \text{div } \mathbf{A} \, dV = \int_V \mathbf{c} \cdot (\text{div } \mathbf{A}) \, dV$$

$$= \int_V \text{div}(\mathbf{A}^T \mathbf{c}) \, dV$$

see (3.5.32);

$$= \int_S (\mathbf{A}^T \mathbf{c}) \cdot \mathbf{n} \, dS = \int_S \mathbf{c} \cdot (\mathbf{A} \mathbf{n}) \, dS$$

so that

$$\mathbf{c} \cdot \left\{ \int_V \text{div } \mathbf{A} \, dV - \int_S \mathbf{A} \mathbf{n} \, dS \right\} = 0$$

Since **c** is an arbitrary vector, this expression yields the result (3.7.1). ∎

In the suffix notation, expression (3.7.1) reads

$$\int_V a_{ik,k} \, dV = \int_S a_{ik} n_k \, dS \qquad (3.7.1)'$$

Consequences

(i) If $\mathbf{A} = \phi \mathbf{B}$ (or $a_{ij} = \phi b_{ij}$) where ϕ is a scalar, then (3.7.1) and (3.7.1)' become, respectively,

$$\int_V \text{div}(\phi \mathbf{B}) \, dV = \int_S \phi \mathbf{B} \mathbf{n} \, dS \qquad (3.7.2)$$

$$\int_V (\phi b_{ik})_{,k} \, dV = \int_S \phi b_{ik} n_k \, dS \qquad (3.7.2)'$$

(ii) If $\mathbf{A} = \phi \mathbf{I}$, expression (3.7.1) becomes, on noting that $\text{div}(\phi \mathbf{I}) = \nabla \phi$, see (3.5.35),

$$\int_V \nabla \phi \, dV = \int_S \phi \mathbf{n} \, dS \qquad (3.7.3)$$

This is identical with expression (3.6.2).

THEOREM 3.7.2 Let V be the volume of a three-dimensional region bounded by a closed regular surface S. Then for a vector field **u** defined in

3.7 INTEGRAL THEOREMS FOR TENSORS

V and on S,

$$\int_V \nabla \mathbf{u} \, dV = \int_S \mathbf{u} \otimes \mathbf{n} \, dS \qquad (3.7.4)$$

where $\mathbf{n}$ is, as usual, the unit outward normal to S.

Proof Let $\mathbf{a}$ and $\mathbf{c}$ be arbitrary constant vectors. Then

$$\mathbf{c} \cdot \left\{ \int_V (\nabla \mathbf{u}) \, dV \right\} \mathbf{a} = \int_V \mathbf{c} \cdot (\nabla \mathbf{u}) \mathbf{a} \, dV = \int_V \mathbf{a} \cdot (\nabla \mathbf{u})^T \mathbf{c} \, dV$$

$$= \mathbf{a} \cdot \int_V \nabla(\mathbf{c} \cdot \mathbf{u}) \, dV = \mathbf{a} \cdot \int_S (\mathbf{c} \cdot \mathbf{u}) \mathbf{n} \, dS$$

by (3.7.3);

$$= \int_S \mathbf{a} \cdot (\mathbf{n} \otimes \mathbf{u}) \mathbf{c} \, dS = \int_S \mathbf{c} \cdot (\mathbf{n} \otimes \mathbf{u})^T \mathbf{a} \, dS$$

$$= \mathbf{c} \cdot \int_S (\mathbf{u} \otimes \mathbf{n}) \mathbf{a} \, dS = \mathbf{c} \cdot \left\{ \int_S \mathbf{u} \otimes \mathbf{n} \, dS \right\} \mathbf{a}$$

Since $\mathbf{c}$ and $\mathbf{a}$ are arbitrary, this expression yields the result (3.7.4). ■

In the suffix notation, (3.7.4) reads

$$\int_V u_{i,j} \, dV = \int_S u_i n_j \, dS \qquad (3.7.4)'$$

Note: If we take the trace on both sides of (3.7.4), we recover the divergence theorem (3.6.1).

THEOREM 3.7.3 (Stokes's Theorem for a Tensor) Let C be a simple closed curve in three-dimensional space and S be an open regular surface bounded by C. Then for a vector field defined on S as well as C,

$$\oint_C \mathbf{A} \mathbf{t} \, ds = \int_S (\text{curl } \mathbf{A})^T \mathbf{n} \, dS \qquad (3.7.5)$$

where $\mathbf{t}$ is the unit tangent to C, which is assumed to be positively oriented relative to the unit normal $\mathbf{n}$ to S.

Proof Let $\mathbf{a}$ be an arbitrary constant vector. Then

$$\mathbf{a} \cdot \int_S (\text{curl } \mathbf{A})^T \mathbf{n} \, dS = \int_S \mathbf{a} \cdot (\text{curl } \mathbf{A})^T \mathbf{n} \, dS = \int_S \mathbf{n} \cdot (\text{curl } \mathbf{A}) \mathbf{a} \, dS$$

$$= \int_S \mathbf{n} \cdot \text{curl}(\mathbf{A}^T \mathbf{a}) \, dS. \qquad (3.7.6)$$

146 3 CALCULUS OF TENSORS

By virtue of expression (3.6.7), we have

$$\int_S \mathbf{n} \cdot \text{curl}(\mathbf{A}^T\mathbf{a})\, dS = \int_C (\mathbf{A}^T\mathbf{a}) \cdot \mathbf{t}\, ds$$

$$= \int_C \mathbf{a} \cdot (\mathbf{At})\, ds = \mathbf{a} \cdot \int_C \mathbf{At}\, ds \qquad (3.7.7)$$

Since **a** is arbitrary, expressions (3.7.6) and (3.7.7) yield the result (3.7.5). ∎

In the suffix notation, expression (3.7.5) reads

$$\oint_C a_{ik} t_k\, ds = \int_S \varepsilon_{jrs} a_{is,r} n_j\, dS \qquad (3.7.5)'$$

Note: In the special case when S is closed, we have by (3.6.8),

$$\int_S \mathbf{n} \cdot \text{curl}(\mathbf{A}^T\mathbf{a})\, dS = 0 \qquad (3.7.8)$$

Then (3.7.5) gives

$$\int_S (\text{curl } \mathbf{A})^T \mathbf{n}\, dS = \mathbf{0} \quad \text{or} \quad \int_S \varepsilon_{jrs} a_{ir,s} n_j\, dS = 0 \qquad (3.7.9)$$

A Particular Case In the particular case when C lies in the $x_1 x_2$-plane and S is the part of this plane bounded by C, we have $n_1 = n_2 = t_3 = 0$ and $n_3 = 1$. Then expression (3.7.5)' takes the following form:

$$\oint_C (a_{i1}\, dx_1 + a_{i2}\, dx_2) = \int_S (a_{i2,1} - a_{i1,2})\, dS \qquad (3.7.10)$$

This is a generalization of the Green's theorem in the plane, given by (3.6.9).

THEOREM 3.7.4 Let C be a simple closed curve in three-dimensional space and S be an open regular surface bounded by C. Then for a vector field **u** defined on S as well as C,

$$\oint_C \mathbf{t} \otimes \mathbf{u}\, ds = \int_S (\mathbf{n} \wedge \nabla \mathbf{u})\, dS \qquad (3.7.11)$$

where **t** is the unit tangent to C, which is assumed to be positively oriented relative to the unit normal **n** to S.

3.7 INTEGRAL THEOREMS FOR TENSORS

Proof Let **a** be a constant vector. Then

$$\left\{\oint_C \mathbf{t} \otimes \mathbf{u}\, ds\right\}\mathbf{a} = \oint_C (\mathbf{t} \otimes \mathbf{u})\mathbf{a}\, ds = \oint_C (\mathbf{u} \cdot \mathbf{a})\mathbf{t}\, ds$$

$$= \oint_S \mathbf{n} \times \nabla(\mathbf{u} \cdot \mathbf{a})\, dS$$

by (3.6.10);

$$= \int_S \mathbf{n} \times (\nabla \mathbf{u})^T \mathbf{a}\, dS = \int_S (\mathbf{n} \wedge \nabla \mathbf{u})\mathbf{a}\, dS$$

by (2.8.17);

$$= \left\{\int_S (\mathbf{n} \wedge \nabla \mathbf{u})\, dS\right\}\mathbf{a}$$

Since **a** is arbitrary, expression (3.7.11) follows. ∎

In the suffix notation, expression (3.7.11) reads

$$\oint_C t_i u_j\, ds = \int_S \varepsilon_{irs} n_r u_{j,s}\, dS \qquad (3.7.11)'$$

If we take the trace of expression (3.7.11), we recover Stokes's theorem (3.6.7).

EXAMPLE 3.7.1 Let S be a regular surface enclosing a region of volume V. For a vector field **u** and a tensor field **A** defined on V and on S, show that

$$\int_V \{\mathbf{u} \otimes \operatorname{div} \mathbf{A} + (\nabla \mathbf{u})\mathbf{A}^T\}\, dV = \int_S \mathbf{u} \otimes (\mathbf{A}\mathbf{n})\, dS \qquad (3.7.12)$$

Hence deduce expressions (3.7.1) and (3.7.4).

Solution Let **a** and **b** be arbitrary constant vectors. Employing the divergence theorem (3.6.1) to the vector $(\mathbf{u} \cdot \mathbf{a})(\mathbf{A}^T \mathbf{b})$, we get

$$\int_V \operatorname{div}\{(\mathbf{u} \cdot \mathbf{a})(\mathbf{A}^T \mathbf{b})\}\, dV = \int_S \{(\mathbf{u} \cdot \mathbf{a})(\mathbf{A}^T \mathbf{b})\} \cdot \mathbf{n}\, dS \qquad (3.7.13)$$

Using identities (2.8.16) and (3.5.41), expression (3.7.13) becomes

$$\mathbf{a} \cdot \left\{\int_V [(\nabla \mathbf{u})\mathbf{A}^T + \mathbf{u} \otimes \operatorname{div} \mathbf{A}]\, dV\right\}\mathbf{b} = \mathbf{a} \cdot \left\{\int_S [\mathbf{u} \otimes \mathbf{A}\mathbf{n}]\, dS\right\}\mathbf{b}$$

Since **a** and **b** are arbitrary, this expression yields (3.7.12).

148 3 CALCULUS OF TENSORS

If **u** is a constant vector, then (3.7.12) becomes

$$\mathbf{u} \otimes \int_V \operatorname{div} \mathbf{A}\, dV = \mathbf{u} \otimes \int_S \mathbf{An}\, dS$$

Since this result is true for any **u**, we obtain (3.7.1). On the other hand, if **A** is the identity tensor **I**, then (3.7.12) becomes (3.7.4). ∎

EXAMPLE 3.7.2 Starting with expression (3.7.4) deduce expression (3.6.3).

Solution For any constant vector **c**, expression (3.7.4) yields

$$\int_V (\nabla \mathbf{u} - \nabla \mathbf{u}^T)\mathbf{c}\, dV = \int_S (\mathbf{u} \otimes \mathbf{n} - \mathbf{n} \otimes \mathbf{u})\mathbf{c}\, dS \quad (3.7.14)$$

By virtue of (3.5.44), the dual vector of $(\nabla \mathbf{u} - \nabla \mathbf{u}^T)$ is curl **u**. Hence

$$(\nabla \mathbf{u} - \nabla \mathbf{u}^T)\mathbf{c} = (\operatorname{curl} \mathbf{u}) \times \mathbf{c} \quad (3.7.15)$$

Also,

$$[\mathbf{u} \otimes \mathbf{n} - \mathbf{n} \otimes \mathbf{u}]\mathbf{c} = (\mathbf{n} \cdot \mathbf{c})\mathbf{u} - (\mathbf{u} \cdot \mathbf{c})\mathbf{n} = \mathbf{c} \times (\mathbf{u} \times \mathbf{n}) \quad (3.7.16)$$

Using (3.7.15) and (3.7.16), expression (3.7.14) becomes

$$\mathbf{c} \times \int_V \operatorname{curl} \mathbf{u}\, dV = \mathbf{c} \times \int_S (\mathbf{n} \times \mathbf{u})\, dS$$

Since **c** is arbitrary, expression (3.6.3) follows. ∎

EXAMPLE 3.7.3 Let S be a regular surface enclosing a region of volume V. For a tensor field **A** defined in V and on S, show that

$$\int_S \mathbf{u} \times (\mathbf{An})\, dS = \int_V (2\mathbf{w} + \mathbf{u} \times \operatorname{div} \mathbf{A})\, dV \quad (3.7.17)$$

Here **n** is the unit outward normal to S and **w** is the dual vector of the skew part of $\mathbf{A}(\nabla \mathbf{u})^T$. Deduce that

$$\int_S \mathbf{x} \times (\mathbf{An})\, dS = \int_V (2\boldsymbol{\xi} + \mathbf{x} \times \operatorname{div} \mathbf{A})\, dV \quad (3.7.18)$$

where $\boldsymbol{\xi}$ is the dual vector of skw **A**.

3.7 INTEGRAL THEOREMS FOR TENSORS 149

Solution Let **a** be an arbitrary constant vector. Then

$$\mathbf{a} \times \int_S \mathbf{u} \times (\mathbf{An}) \, dS = \int_S [\mathbf{a} \times \{\mathbf{u} \times (\mathbf{An})\}] \, dS$$

$$= \int_S \{\mathbf{u} \otimes (\mathbf{An}) - (\mathbf{An}) \otimes \mathbf{u}\}\mathbf{a} \, dS \quad (3.7.19)$$

From expression (3.7.12), we have

$$\int_S \{\mathbf{u} \otimes (\mathbf{An})\} \, dS = \int_V \{\mathbf{u} \otimes \operatorname{div} \mathbf{A} + (\nabla \mathbf{u})\mathbf{A}^T\} \, dV \quad (3.7.20)$$

so that (on taking the transpose)

$$\int_S \{(\mathbf{An}) \otimes \mathbf{u}\} \, dS = \int_V \{(\operatorname{div} \mathbf{A}) \otimes \mathbf{u} + \mathbf{A}(\nabla \mathbf{u})^T\} \, dV \quad (3.7.21)$$

Using (3.7.20) and (3.7.21) in (3.7.19), we get

$$\mathbf{a} \times \int_S \mathbf{u} \times (\mathbf{An}) \, dS$$

$$= \int_V [\{\mathbf{u} \otimes (\operatorname{div} \mathbf{A}) - (\operatorname{div} \mathbf{A}) \otimes \mathbf{u}\}\mathbf{a} - 2 \operatorname{skw}\{\mathbf{A}(\nabla \mathbf{u})^T\}\mathbf{a}] \, dV$$

$$= \mathbf{a} \times \int_S \{\mathbf{u} \times \operatorname{div} \mathbf{A} + 2\mathbf{w}\} \, dS$$

Since **a** is arbitrary, (3.7.17) follows.

If we set $\mathbf{u} = \mathbf{x}$ in (3.7.17) and recall that $\nabla \mathbf{x} = \mathbf{I}$, we readily get (3.7.18). ∎

EXAMPLE 3.7.4 Let S be a regular open surface bounded by a simple closed curve C. For a vector field **u** and a tensor field **A** defined on S as well as C, show that

$$\oint_C (\mathbf{u} \otimes \mathbf{At}) \, ds = \int_S [\mathbf{u} \otimes (\operatorname{curl} \mathbf{A})^T \mathbf{n} + (\mathbf{n} \wedge \nabla \mathbf{u})^T \mathbf{A}^T] \, dS \quad (3.7.22)$$

where **t** and **n** are as defined in expression (3.7.5). Hence deduce (3.7.5) and (3.7.11).

Solution Let **a** and **b** be arbitrary constant vectors. Employing the Stokes's theorem (3.6.7) to the vector $(\mathbf{u} \cdot \mathbf{a})(\mathbf{A}^T \mathbf{b})$, we get

$$\oint_C \{(\mathbf{u} \cdot \mathbf{a})(\mathbf{A}^T \mathbf{b})\} \cdot \mathbf{t} \, ds = \int_S \operatorname{curl}\{(\mathbf{u} \cdot \mathbf{a})(\mathbf{A}^T \mathbf{b})\} \cdot \mathbf{n} \, dS \quad (3.7.23)$$

150   3 CALCULUS OF TENSORS

Using the identities (2.8.16) and (3.5.42), expression (3.7.23) becomes

$$\mathbf{a} \cdot \left\{ \oint_C (\mathbf{u} \otimes \mathbf{At})\, ds \right\} \mathbf{b} = \mathbf{a} \cdot \left\{ \int_S [\mathbf{u} \otimes (\operatorname{curl} \mathbf{A})^T \mathbf{n} + (\mathbf{n} \wedge \nabla \mathbf{u})^T \mathbf{A}^T]\, dS \right\} \mathbf{b}$$

Since **a** and **b** are arbitrary, this expression yields (3.7.22).

If **u** is a constant vector, expression (3.7.22) becomes

$$\mathbf{u} \otimes \oint_C \mathbf{At}\, ds = \mathbf{u} \otimes \int_S (\operatorname{curl} \mathbf{A})^T \mathbf{n}\, dS$$

Since this result is true for any **u**, expression (3.7.5) follows.

On the other hand, if **A** is the unit tensor, expression (3.7.22) becomes

$$\oint_C (\mathbf{u} \otimes \mathbf{t})\, ds = \int_S (\mathbf{n} \wedge \nabla \mathbf{u})^T\, dS$$

from which (3.7.11) is immediate. ∎

EXAMPLE 3.7.5   Let S be a regular open surface bounded by a simple closed curve C. For a tensor field **A** defined on S and C, show that

$$\oint_C (\mathbf{u} \times \mathbf{At})\, ds = \int_S \{2\mathbf{w} + \mathbf{u} \times (\operatorname{curl} \mathbf{A})^T \mathbf{n}\}\, dS \qquad (3.7.24)$$

where **t** and **n** are as defined in (3.7.5) and **w** is the dual of the skew part of $\mathbf{A}(\mathbf{n} \wedge \nabla \mathbf{u})$. Hence deduce that

$$\oint_C (\mathbf{x} \times \mathbf{At})\, ds = \int_S \{2\xi + \mathbf{x} \times (\operatorname{curl} \mathbf{A})^T \mathbf{n}\}\, dS \qquad (3.7.25)$$

where ξ is the dual vector of $\operatorname{skw}(\mathbf{n} \wedge \mathbf{A})$.

Solution   Let **a** be an arbitrary constant vector. Then

$$\mathbf{a} \times \oint_C (\mathbf{u} \times \mathbf{At})\, ds = \oint_C \{\mathbf{u} \otimes (\mathbf{At}) - (\mathbf{At}) \otimes \mathbf{u}\} \mathbf{a}\, ds$$

Using (3.7.22) and its transpose, this expression takes the form

$$\mathbf{a} \times \oint_C (\mathbf{u} \times \mathbf{At})\, ds$$

$$= \int_S [\{\mathbf{u} \otimes (\operatorname{curl} \mathbf{A})^T \mathbf{n} - (\operatorname{curl} \mathbf{A})^T \mathbf{n} \otimes \mathbf{u}\}\mathbf{a} - 2\{\operatorname{skw} \mathbf{A}(\mathbf{n} \wedge \nabla \mathbf{u})\}\mathbf{a}]\, dS$$

$$= \mathbf{a} \times \int_S \{\mathbf{u} \times (\operatorname{curl} \mathbf{A})^T \mathbf{n} + 2\mathbf{w}\}\, dS$$

Since **a** is an arbitrary vector, result (3.7.24) follows.

Setting $\mathbf{u} = \mathbf{x}$ in (3.7.24) and recalling that $\nabla \mathbf{x} = \mathbf{I}$ and $\mathbf{A}(\mathbf{n} \wedge \mathbf{I}) = -(\mathbf{n} \wedge \mathbf{A})^T$, we get (3.7.25). ∎

3.8
EXERCISES

1. Verify the identities (3.2.3) to (3.2.7).

2. For any vector $\mathbf{u} = \mathbf{u}(t)$ with magnitude u, show that

$$\mathbf{u} \cdot \frac{d\mathbf{u}}{dt} = u \frac{du}{dt}$$

3. If $\mathbf{u}(t)$ is a unit vector, show that

$$\left| \mathbf{u} \times \frac{d\mathbf{u}}{dt} \right| = \left| \frac{d\mathbf{u}}{dt} \right|$$

4. If $\mathbf{u}(t)$ and $\mathbf{v}(t)$ are such that $d\mathbf{u}/dt = \mathbf{w} \times \mathbf{u}$ and $d\mathbf{v}/dt = \mathbf{w} \times \mathbf{v}$, where $\mathbf{w}$ is a constant vector, show that

$$\frac{d}{dt}(\mathbf{u} \times \mathbf{v}) = (\mathbf{u} \otimes \mathbf{v} - \mathbf{v} \otimes \mathbf{u})\mathbf{w}$$

5. Verify the identities (3.2.8) to (3.2.10) and (3.2.12).

6. If $\mathbf{Q}(t)$ is an orthogonal tensor, show that $\mathbf{Q}(d\mathbf{Q}/dt)^T$ and $\mathbf{Q}^T(d\mathbf{Q}/dt)$ are skew tensors.

7. If $\mathbf{A}(t)$ is an invertible tensor, show that

$$\frac{d}{dt}(\det \mathbf{A}) = (\det \mathbf{A}) \operatorname{tr}\left(\frac{d\mathbf{A}}{dt} \mathbf{A}^{-1}\right)$$

8. If $u_1 = 2x_1 x_2$, $u_2 = -x_2 x_3$, $u_3 = 3x_1 x_3$, find the matrices of the following:
(i) $u_{i,j}$ (ii) $u_{j,i}$
(iii) $(u_{i,j} + u_{j,i})$ (iv) $(u_{i,j} - u_{j,i})$

9. Verify the identities (3.4.3) and (3.4.4).

10. Show that the directional derivatives of ϕ along $\mathbf{e}_i$ are $\phi_{,i}$.

11. Find the directional derivative of $\phi = x_1 x_2 x_3$ along the tangent to the curve given by the parametric equations $x_1 = t$, $x_2 = t^2$, $x_3 = t^3$, at the points $(1, 1, 1)$ and $(-1, 1, -1)$.

12. Find the direction along which the directional derivative of the function $\phi = x_1 x_2 x_3$ is maximum at the point $(1, 1, 1)$. Also, find the maximum directional derivative.

152   3 CALCULUS OF TENSORS

13. If $\phi = x_1^n + x_2^n + x_3^n$, show that $\mathbf{x} \cdot \nabla\phi = n\phi$.

14. If $f = f(r)$, show that $\nabla f = \{f'(r)/r\}\mathbf{x}$, $r \neq 0$.

15. If $\mathbf{A}$ is a constant tensor, show that $\nabla(\mathbf{Ax} \cdot \mathbf{x}) = (\mathbf{A} + \mathbf{A}^T)\mathbf{x}$.

16. Verify the identities (3.4.19).

17. Show that
  (i) div $\mathbf{u} = \mathbf{e}_k \cdot \mathbf{u}_{,k}$ (ii) curl $\mathbf{u} = \mathbf{e}_k \times \mathbf{u}_{,k}$

18. Prove the identities (3.4.25) and (3.4.27) by using the suffix notation.

19. If $\mathbf{u} = \text{curl } \mathbf{v}$ and $\mathbf{v} = \text{curl } \mathbf{u}$, show that
  (i) div(curl $\mathbf{u} \times$ curl $\mathbf{v}$) $= u^2 - v^2$
  (ii) $\nabla^2 \mathbf{u} = -\mathbf{u}$, $\nabla^2 \mathbf{v} = -\mathbf{v}$

20. Show that $\nabla^2 |\mathbf{x}|^n = 0$ only if $n = 0$ or -1.

21. Find $\phi = \phi(r)$, given that div$(\phi\mathbf{x}) = 0$.

22. Find $\phi = \phi(r)$, given that $\nabla^2(r\phi) = 0$.

23. If $f = f(r)$, show that $\nabla^2\{f(r)\} = f''(r) + (2/r)f'(r)$, $r \neq 0$. Deduce that $\nabla^2(1/r) = 0$.

24. Show that $\nabla^2(\phi\psi) = \phi\nabla^2\psi + \psi\nabla^2\phi + 2\nabla\phi \cdot \nabla\psi$.

25. Show that $(\mathbf{u} \cdot \nabla)\mathbf{x} = \mathbf{u}$.

26. Show that curl $\mathbf{u} = \frac{1}{2}[\text{curl}\{(\text{curl } \mathbf{u}) \times \mathbf{x}\} - (\mathbf{x} \cdot \nabla) \text{ curl } \mathbf{u}]$.

27. Show that $r^n\mathbf{x}$ is curl free for any n, but divergence free only if $n = -3$.

28. If $\mathbf{c}$ is a constant vector, show that $f(r)\mathbf{c} \times \mathbf{x}$ is divergence free.

29. If $\mathbf{u} \cdot \text{curl } \mathbf{v} = \mathbf{v} \cdot \text{curl } \mathbf{u}$, show that $\mathbf{u} \times \mathbf{v}$ is divergence free and that

$$\text{curl curl}(\mathbf{u} \times \mathbf{v}) = \nabla^2(\mathbf{v} \times \mathbf{u})$$

30. Show that $\nabla(\mathbf{v} \cdot \nabla(1/r)) + \text{curl}(\mathbf{v} \times \nabla(1/r)) = \mathbf{0}$, $r \neq 0$.

31. If the curl of a Beltrami vector is again a Beltrami vector, show that the abnormality factor is independent of $\mathbf{x}$.

32. A vector $\mathbf{u}$ is said to be a *complex lamellar vector* if $\mathbf{u} \cdot \text{curl } \mathbf{u} = 0$. If $\mathbf{u}$ is a complex lamellar vector, show that for any scalar field ϕ, the vector $\phi\mathbf{u}$ is also a complex lamellar vector.

33. Show that $\nabla\mathbf{u} = u_{r,s}\mathbf{e}_r \otimes \mathbf{e}_s$.

34. If $\nabla\mathbf{u} = -(\nabla\mathbf{u})^T$, show that $\nabla\nabla\mathbf{u} = \mathbf{0}$.

35. Show that $(\mathbf{v} \wedge \nabla\mathbf{u})\mathbf{w} = \mathbf{v} \times (\nabla\mathbf{u})^T\mathbf{w}$.

36. Show that $\nabla(\phi\mathbf{u}) = \mathbf{u} \otimes \nabla\phi + \phi\nabla\mathbf{u}$.

3.8 EXERCISES

37. Verify the identities (3.5.4) to (3.5.6) and (3.5.22).

38. If $\mathbf{u} = x_1^2\mathbf{e}_1 + x_3^2\mathbf{e}_2 + x_2^2\mathbf{e}_3$, find $\nabla\mathbf{u}$ and $(\nabla\mathbf{u})\mathbf{u}$.

39. If $\mathbf{c}$ is a constant vector, show that $(\nabla^2\mathbf{A})\mathbf{c} = \nabla^2(\mathbf{Ac})$.

40. Prove the following identities:
  (i) $\text{div}(\text{div }\mathbf{A}) = \text{tr }\nabla(\text{div }\mathbf{A})$
  (ii) $\text{div}(\nabla\mathbf{u}^T) = \text{div}[(\text{div }\mathbf{u})\mathbf{I}]$
  (iii) $\text{div}(\phi\mathbf{I}) = \nabla\phi$
  (iv) $\text{div}(\mathbf{Au}) = \mathbf{u} \cdot \text{div }\mathbf{A}^T + \text{tr}[\mathbf{A}^T(\nabla\mathbf{u})]$
  (v) $\text{div}(\mathbf{u} \otimes \mathbf{v}) = (\text{div }\mathbf{v})\mathbf{u} + (\nabla\mathbf{u})\mathbf{v}$
  (vi) $\text{div}[\phi(\mathbf{Au})] = \phi\{(\mathbf{u} \cdot \text{div }\mathbf{A}^T) + (\mathbf{A}^T \cdot \nabla\mathbf{u})\} + \nabla\phi \cdot \mathbf{Au}$
  (vii) $\text{div}[(\nabla\mathbf{u})\mathbf{u}] = \nabla\mathbf{u} \cdot \nabla\mathbf{u}^T + \mathbf{u} \cdot \nabla(\text{div }\mathbf{u})$
  (viii) $\text{div}[(\nabla\mathbf{u})\mathbf{u} - (\text{div }\mathbf{u})\mathbf{u}] = \nabla\mathbf{u} \cdot \nabla\mathbf{u}^T - (\text{div }\mathbf{u})^2$
  (ix) $\text{div}[(\mathbf{u} \otimes \mathbf{v})\mathbf{w}] = (\text{div }\mathbf{u})(\mathbf{v} \cdot \mathbf{w}) + \mathbf{u} \cdot \nabla(\mathbf{v} \cdot \mathbf{w})$
  (x) $\text{curl}(\phi\mathbf{I}) = -[\text{curl}(\phi\mathbf{I})]^T$
  (xi) $\text{tr}[\nabla^2\mathbf{A} + \nabla\nabla(\text{tr }\mathbf{A})] = 2\nabla^2(\text{tr }\mathbf{A})$

41. Let f be a scalar field or a component of a vector or a tensor field defined in a region R. If $\int_V f\, dV = 0$ for every volume V in R, show that $f = 0$ in R.

42. Calculate the circulation of $\mathbf{u} = x_1^2\mathbf{e}_1 - x_1 x_2 \mathbf{e}_2$ round the ellipse $x_1 = a\cos\tau$, $x_2 = b\sin\tau$, $x_3 = 0$, $0 \le \tau \le 2\pi$.

43. By using the divergence theorem or Stokes's theorem, prove the expressions (3.6.2), (3.6.4), (3.6.5), (3.6.9), (3.6.10) and (3.6.12).

44. By using the divergence theorem, prove the following:

(i) $\int_S \phi \frac{\partial \psi}{\partial n} dS = \int_V (\phi\nabla^2\psi + \nabla\phi \cdot \nabla\psi)\, dV$

(ii) $\int_S \left(\phi \frac{\partial \psi}{\partial n} - \psi \frac{\partial \phi}{\partial n}\right) dS = \int_V (\phi\nabla^2\psi - \psi\nabla^2\phi)\, dV$

45. If $\mathbf{u}$ is everywhere normal to a closed regular surface S bounding V, show that

$$\int_V \text{curl }\mathbf{u}\, dV = \mathbf{0}$$

46. If $\mathbf{v}$ is as defined by (3.6.14), prove that $\nabla^2\mathbf{v} = \mathbf{u}$.

47. Verify that the vector $\mathbf{w}$ present in Helmholtz's representation (3.6.15) obeys the equation

$$\mathbf{w} = \frac{1}{4\pi}\int_V \frac{\text{curl }\mathbf{u}(\bar{\mathbf{x}})}{|\mathbf{x} - \bar{\mathbf{x}}|} dV$$

154   3 CALCULUS OF TENSORS

48. Deduce representations (3.6.6) and (3.6.13) from Helmholtz's representation (3.6.15).

49. By starting with the expression (3.7.17), deduce expression (3.6.3).

50. By starting with the expression (3.7.25), derive expression (3.6.12).

51. If V is the volume of a region bounded by a closed regular surface S, show that

$$\int_S \mathbf{x} \otimes \mathbf{n}\, dS = V\mathbf{I}$$

52. Prove that

$$\int_V a_{ij,i}\, dV = \int_S a_{ij} n_i\, dS$$

53. Let $\mathbf{u}$ be a vector field defined in a region of volume V enclosed by a regular closed surface S. Also, let $\mathbf{a}$ be a constant vector and $\phi = \phi(x_3)$ is a scalar function defined in V. If $\mathbf{u}$ is divergence free in V and $\mathbf{u} = \mathbf{0}$ on S, prove the following:

(i) $\displaystyle\int_V (\mathbf{u} \cdot \operatorname{curl}\operatorname{curl}\mathbf{u})\, dV = \int_V |\operatorname{curl}\mathbf{u}|^2\, dV$

(ii) $\displaystyle\int_V (\mathbf{u} \times \mathbf{a}) \cdot \nabla(\mathbf{a} \cdot \operatorname{curl}\mathbf{u})\, dV = -\int_V |\mathbf{a} \cdot \operatorname{curl}\mathbf{u}|^2\, dV$

(iii) $\displaystyle\int_V (\mathbf{u} \cdot \mathbf{a})\nabla^2(\mathbf{u} \cdot \mathbf{a})\, dV = -\int_V |\nabla(\mathbf{u} \cdot \mathbf{a})|^2\, dV$

(iv) $\displaystyle\int_V (\phi_{,3}\mathbf{u}) \cdot \nabla u_3\, dV = -\int_V \phi_{,33} u_3{}^2\, dV$

54. Let $\mathbf{h}$ be a vector field defined in a region of volume V enclosed by a regular closed surface S. If $\mathbf{h}$ is divergence free in V and $\mathbf{n} \times \operatorname{curl}\mathbf{h} = \mathbf{0}$ on S, show that

$$\int_V \mathbf{h} \cdot \operatorname{curl}\operatorname{curl}\mathbf{h}\, dV = \int_V |\operatorname{curl}\mathbf{h}|^2\, dV$$

55. Show that

$$\int_V [(\operatorname{curl}\mathbf{u}) \times \mathbf{v} + (\operatorname{curl}\mathbf{v}) \times \mathbf{u} + \mathbf{u}(\operatorname{div}\mathbf{v}) + \mathbf{v}(\operatorname{div}\mathbf{u})]\, dV$$

$$= \int_S [\{\mathbf{u} \otimes \mathbf{v} + \mathbf{v} \otimes \mathbf{u}\}\mathbf{n} - (\mathbf{u} \cdot \mathbf{v})\mathbf{n}]\, dS$$

CHAPTER 4
CONTINUUM HYPOTHESIS

4.1
INTRODUCTION

The subject of mechanics is concerned with the study of external effects, such as forces and heat inputs, on a physical object. If the object is a solid whose volume or shape or both generally change, the subject is called *solid mechanics*. If the object is a fluid (liquid or gas), the subject is referred to as *fluid mechanics*. Historically, both solid and fluid mechanics were developed almost simultaneously. The foundations of these subjects were laid during the latter half of the eighteenth and the first half of the nineteenth centuries by celebrated mathematicians including Leonhard Euler (1707–1783), Augustin Louis Cauchy (1789–1857), Siméon Denis Poisson (1781–1840), George Green (1793–1841) and George Stokes (1819–1903). An examination of these foundations reveals that the basic postulates and general principles upon which both solid and fluid mechanics are based are indeed the same. The mathematical equations that describe the physical laws, called the *field equations*, are common to both the subjects. But the two subjects are not identical. Solids and fluids have individual characteristic properties, and these properties are reflected in the study of the respective subjects. Equations that represent the characteristic properties of a material (or a class of materials) and distinguish one material from the other are called *constitutive equations*. Obviously, solids and fluids cannot have the same constitutive equations. Hence, although both solid and fluid mechanics share the same field equations, the two subjects differ in constitutive equations.

The study of solid and fluid mechanics through a unified approach constitutes the subject of *continuum mechanics*. This subject can be divided into three natural parts. The first part deals with the field equations common to both solid and fluid mechanics. The second part is concerned with the constitutive equations that are different for different types of solids and fluids. The third part deals with different branches of the subject whose governing equations are obtained by combining the field equations with appropriate constitutive equations.

The whole theory of continuum mechanics rests on a fundamental hypothesis known as the *continuum hypothesis*. The hypothesis is explained in the following sections of this chapter. The field equations of the theory are formulated on the basis of the fundamental laws of physics: the laws of conservation of mass, momentum and energy. These equations are derived in Chapter 8. The basic kinematic concepts of deformation, strain and motion as well as the concept of stress that pave the way for the derivation of field equations are presented in Chapters 5 to 7. Historically, the concepts of strain and stress were introduced and discussed by Cauchy during the years 1823–1827. A large part of the theories of strain and stress, as is known today, has stemmed from his pioneering work. The development of kinematics of motion and field equations are essentially due to Euler.

Constitutive equations are formulated in two different ways. One way is to postulate their basic forms on experimental grounds. Hooke's law, which describes the behavior of elastic solids, and the Newton's law of viscosity, which describes the behaviour of viscous fluids, are two standard examples of constitutive equations that were first formulated in this way. The other way is through a rigorous axiomatic approach. In this text, which is meant primarily for the beginners in solid and fluid mechanics, we refrain from following the axiomatic approach. We confine ourselves to dealing with only three important examples of constitutive equations applicable to perfectly elastic solids, perfectly nonviscous fluids and the so-called Newtonian viscous fluids. When combined with the field equations, these particular constitutive equations give rise to the governing equations of the linear elasticity theory and the theories of nonviscous and Newtonian viscous fluid flows. The equations of the linear elasticity theory and their immediate consequences and simple applications are presented in Chapter 9. Chapter 10 is concerned with the equations of nonviscous and Newtonian viscous

4.2
NOTION OF A CONTINUUM

It is a common knowledge that every physical object is made up of molecules, atoms and even smaller particles. These particles are not continuously distributed over the object; microscopic observations reveal the presence of gaps (empty spaces) between particles. While studying the external effects on physical objects, these gaps may or may not be taken into consideration depending on the hypothesis made. The study that takes account of the existence of gaps is called *microscopic study*. On the other hand, the study that ignores the gaps and treats a physical object as a continuous distribution of matter is called *macroscopic study*. The subjects of solid and fluid mechanics are concerned mainly with macroscopic study. Although microscopic study is supposed to yield more accurate results than macroscopic study, at least in principle, macroscopic description of mechanical behavior of materials is adequate and useful for common engineering applications.

The macroscopic viewpoint adopted in the study of solid and fluid mechanics leads to the notion of a *continuous medium*, or briefly a *continuum*. By a continuum, we mean a hypothetical physical object in which the matter is continuously distributed over the entire object. Solids and fluids whose mechanical behavior is studied from macroscopic point of view stand as two concrete examples of a continuum. Continuum mechanics deals with the study of deformation and motion of a continuum, and thereby provides a unified approach to solid and fluid mechanics.

4.3
CONFIGURATION OF A CONTINUUM

Consider a physical object $\mathcal{B}$ that occupies (fills) a region B in three-dimensional space, at a given instant of time t. Then every part of the object $\mathcal{B}$ has a *position* (or place or location) in the region B. If the object $\mathcal{B}$ is a continuum, that is if $\mathcal{B}$ is made up of matter that is continuously distributed (without gaps) over B, then every part of the region B is filled by some part of the continuum $\mathcal{B}$. In particular, every point P in the region B is the position of a part $\mathcal{P}$ of the continuum $\mathcal{B}$; such a part $\mathcal{P}$ is called a *particle*

158   **4 CONTINUUM HYPOTHESIS**

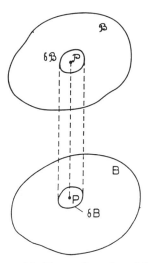

Figure 4.1. The configuration of $\mathcal{B}$.

of $\mathcal{B}$. Thus, a continuum $\mathcal{B}$ may be regarded as a connected (continuous) set of particles such that (i) every particle of $\mathcal{B}$ has a position in the region B occupied by $\mathcal{B}$, and (ii) every point P in the region B is the position of some particle $\mathcal{P}$ of the continuum $\mathcal{B}$. In other words, there exists a *one-to-one correspondence* between the particles of a continuum and the geometrical points of a region that the continuum occupies at any given instant of time; see Figure 4.1. The geometrical region that a continuum occupies at a given instant of time is called the *configuration* of the continuum *at that instant of time*.

It may be noted that the configuration of a continuum and the position of a particle of a continuum are defined with reference to a given instant of time. As a continuum moves, the position of every particle of the continuum and the configuration of the continuum change from one instant to another. Thus, a continuum has infinitely many configurations and every particle of the continuum has infinitely many positions in space. However, at any given instant of time, the position of every particle of a continuum and the configuration of the continuum are uniquely determined. At no instant can a particle have two distinct positions nor can two distinct particles have the same position.

The one-to-one correspondence between the particles of a continuum and the points of its configuration (at a given instant of time) enables us to speak of points, curves, surfaces and volumes in a continuum. A part of a continuum whose position is a geometrical point is referred to as a *material point* (or a *body point*). A material point is nothing but a particle, as defined earlier. A part of a continuum whose position is a geometrical curve is called a *material curve* or a *material arc*. The arc-length of a material curve at a given instant of time is defined to be identical to the arc-length of the geometrical curve that it occupies at that instant. A material arc of infinitesimally small length is referred to as a *material arc element*. A part of a continuum whose position is a geometrical surface is called a *material surface*. The surface area of a material surface at a given instant is defined to be identical to the surface area of the geometrical surface it occupies at that instant. A material surface of infinitesimally small area is referred to as a *material surface element*. A part (or the whole) of a continuum whose position is a three-dimensional geometrical region is called a *material body*. The position of a material body at a given instant of time is nothing but its configuration at that instant. The volume of a material body at a given instant of time is defined to be identical to the volume of its configuration at that instant. A material body of infinitesimally small volume is referred to as a *material body element, material volume element* or just a *material element*. For simplicity, material points, material arcs, material surfaces and material bodies are often referred to as *points, arcs, surfaces* and *bodies*. Unless stated to the contrary, all arcs, surfaces and bodies we consider are supposed to have finite lengths, finite areas and finite volumes, respectively.

4.4
MASS AND DENSITY

The one-to-one correspondence between the particles of a continuum and the points of its configuration allows us to study physical and kinematical quantities associated with a continuum through appropriate functions defined over its configuration. To illustrate this fact, let us consider the concepts of mass and density that every physical object is supposed to possess.

At an instant of time t, let B be the configuration of a material body $\mathcal{B}$ and P be the position of a particle $\mathcal{P}$ of $\mathcal{B}$. Consider an infinitesimal subregion δB of B containing point P, and let $\delta \mathcal{B}$ be the material element that occupies (fills) the subregion δB (see Figure 4.1). Because of the hypothesis of continuity in distribution of matter in $\mathcal{B}$, the subregion δB will not be empty, however small it may be. If δV is the volume of δB, then δV is the volume of $\delta \mathcal{B}$. Let δm be the mass of $\delta \mathcal{B}$. Like time and space, mass and volume are primitive quantities; these are postulated to be positive scalars. Thus, $\delta m > 0$; $\delta V > 0$. Consequently, the ratio $(\delta m/\delta V)$ is also a positive number. Since δB is not empty, however small it may be, we may think of the limit of $(\delta m/\delta V)$ as δV tends to 0 in the terminology of differential calculus. We suppose that this limit exists (finitely and uniquely) and denote it by ρ. That is,

$$\rho = \lim_{\delta V \to 0} \frac{\delta m}{\delta V} \qquad (4.4.1)$$

Thus, ρ is the limit of the ratio of the mass of the material element contained in an infinitesimal region around the point P to the volume of the region, as the volume tends to 0. The quantity ρ is called the *mass-density* (or just the *density*) of the material body at the point P. Note that the relation (4.4.1) defining ρ is consistent with the common definition of density and that the value of ρ depends on the point P chosen in the configuration B of the body $\mathcal{B}$. Hence, ρ is a function defined over B. Since the configuration B itself changes with time t, ρ is a function of t as well, in general.

While defining density through the relation (4.4.1), we have made the tacit assumption that mass is a differentiable function of volume. This means that the mass is assumed to be distributed over a region without sudden increase or decrease in any part of the region. In other words, no part of a material body is assumed to possess concentrated mass. Such assumptions are made throughout our discussions. In fact, every function considered in our analysis is supposed to be differentiable up to any desired order in the region of interest. Thus, for example, we suppose that ρ defined over the region B through the relation (4.4.1) is continuously differentiable over B. Consequently, it follows that if V denotes the volume of B then the relation (4.4.1) is equivalent to the relation

$$m = \int_V \rho \, dV \qquad (4.4.2)$$

where m is the *total mass* of the material body $\mathcal{B}$ filling the region B.

Total kinetic energy, total force, total linear momentum, total angular momentum, etc. of a material body are defined by relations analogous to (4.4.2). Total force on a material surface is defined in a similar way by considering a surface integral. The use of integral calculus over spatial variables distinguishes continuum mechanics from analytical mechanics dealing with discrete systems.

4.5
DESCRIPTIONS OF MOTION

Suppose we wish to study a motion of a material body $\mathcal{B}$ starting with some particular instant of time. It is convenient to take this instant of time as the origin in the time scale so that the instant corresponds to the *reference* or

4.5 DESCRIPTIONS OF MOTION   161

initial time $t = 0$. Let B_0 be the configuration of $\mathcal{B}$ at this instant of time. Then B_0 is called the *reference* or *initial configuration*. In view of the one-to-one correspondence between the particles of $\mathcal{B}$ and the points of B_0, every particle of $\mathcal{B}$ can be identified by its position in B_0. This means that if a typical point P_0 in B_0 is the position of a typical particle $\mathcal{P}$ of $\mathcal{B}$ at time $t = 0$, then $\mathcal{P}$ may be identified by the position vector $\mathbf{x}^0$ of the point P_0 with respect to some fixed origin O. Thus, the vector $\mathbf{x}^0$ serves as a label for the particle $\mathcal{P}$ (for all time), and when convenient, $\mathcal{P}$ itself may be called the particle $\mathbf{x}^0$.

As the body $\mathcal{B}$ moves out of the configuration B_0 at time $t = 0$, every particle of $\mathcal{B}$ moves with it. At any subsequent time $t > 0$, let P be the position of the particle $\mathcal{P}$ and let $\mathbf{x}$ be the position vector of the point P (with respect to the origin O); see Figure 4.2. Since the point P determines and is determined by $\mathbf{x}$, P is referred to as the point $\mathbf{x}$. Evidently, $\mathbf{x}$ depends on the particle $\mathcal{P}$, which is identified with $\mathbf{x}^0$, and t; and the motion being studied determines the type of dependence. Thus, a motion of $\mathcal{B}$ is described by an equation of the form

$$\mathbf{x} = \mathbf{x}(\mathbf{x}^0, t), \qquad t > 0 \tag{4.5.1}$$

subject to the consistency condition

$$\mathbf{x} = \mathbf{x}^0 \quad \text{for } t = 0 \tag{4.5.2}$$

The lefthand side of equation (4.5.1) indicates the position $\mathbf{x}$ that the particle $\mathbf{x}^0$ occupies at time $t > 0$, and the righthand side represents the function that determines this $\mathbf{x}$. The same symbol $\mathbf{x}$ is used to denote both the dependent variable and the function for simplicity in the notation.

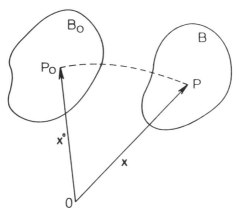

Figure 4.2. Initial and current positions of a particle.

Equation (4.5.1) determines the position **x** of every particle 𝒫 of ℬ at an instant of time $t > 0$. If we focus our attention on a specific particle, this equation determines the successive positions of that particle at different instants of time. On the other hand, at a specific instant of time the equation determines the positions of all particles of ℬ at that instant of time, and the totality of all these positions constitutes the configuration B of ℬ at that instant. Thus, for a given t, equation (4.5.1) defines a mapping from ℬ ≡ B_0 onto B that carries a particle $\mathbf{x}^0$ of ℬ to the point **x** in B. The point **x** is referred to as the *current* or *instantaneous position* of the particle $\mathbf{x}^0$. Also, the configuration B is referred to as the *current* or *instantaneous configuration* of ℬ.

Since no two distinct particles of ℬ can have the same position in any configuration and no two distinct points in a configuration can be positions of the same particle, equation (4.5.1) should determine not only a unique **x** for a given $\mathbf{x}^0$ and a given t, but also a unique $\mathbf{x}^0$ for a given **x** and a given t. In other words, the mapping defined by (4.5.1) should be a *one-one (invertible) mapping* so that (4.5.1) has a unique solution of the form

$$\mathbf{x}^0 = \mathbf{x}^0(\mathbf{x}, t), \qquad t > 0 \tag{4.5.3}$$

where the function $\mathbf{x}^0(\mathbf{x}, t)$ is the *inverse* of the function $\mathbf{x}(\mathbf{x}^0, t)$.

For any point **x** and for any instant of time $t > 0$, equation (4.5.3) specifies the particle $\mathbf{x}^0$ of which **x** is the position at that instant. If we focus our attention on a specific point **x**, equation (4.5.3) determines all those particles of ℬ that pass through that point at different instants of time $t > 0$. On the other hand, at a specific instant of time equation (4.5.3) specifies all particles that are positioned at different points of the current configuration B, and the totality of all these particles constitutes the body ℬ. Thus, for a given t, equation (4.5.3) defines a mapping from B onto ℬ ≡ B_0 that carries a point **x** of B to the particle $\mathbf{x}^0$ of ℬ. This mapping is the inverse of the mapping defined by (4.5.1).

Since the function $\mathbf{x}^0(\mathbf{x}, t)$ is the inverse of the function $\mathbf{x}(\mathbf{x}^0, t)$ it follows that the function $\mathbf{x}(\mathbf{x}^0, t)$ has to be the inverse of the function $\mathbf{x}^0(\mathbf{x}, t)$. This means that equation (4.5.1) can be recovered as a unique solution of equation (4.5.3), in principle. In other words, at any instant of time $t > 0$, equation (4.5.3) not only assigns a unique particle $\mathbf{x}^0$ to a given point **x** but also yields equation (4.5.1), which specifies a unique position **x** for a given particle $\mathbf{x}^0$. Thus, the motion described by equation (4.5.1) is described by equation (4.5.3) also. But the *ways* the two equations describe the motion are not identical; they are only equivalent. While equation (4.5.1) contains the particle $\mathbf{x}^0$ and time t as independent variables and specifies the position **x** of $\mathbf{x}^0$ for a given t, equation (4.5.3) contains the point **x** and time t as independent variables and specifies the particle $\mathbf{x}^0$ that occupies **x** for a given t. Thus, in the description of motion given by (4.5.1), attention is focused on a particle and we observe what is happening to the particle as it moves. This description is called the *material description*, and the independent variables $(\mathbf{x}^0, t)$ present in (4.5.1) are referred to as *material variables*. On the other hand, in the description of motion given by (4.5.3) attention is given to a point in space, and we study what is happening at the point as time passes. This description is called the *spatial description*, and the independent variables $(\mathbf{x}, t)$ present in (4.5.3) are referred to as *spatial variables*. Both material and spatial descriptions are useful in their own right. In solid mechanics, material description is more convenient than the spatial description. It is the other way round in fluid mechanics. Traditionally, material description is referred to as the *Lagrangian description* in honor of Lagrange, and the spatial description is referred to as the *Eulerian description* in honor of Euler. Both descriptions were given by Euler.

In Section 4.4, we mentioned that physical and kinematical quantities associated with a continuum are studied through appropriate functions defined over its configuration. If a quantity is represented by a function defined over the initial configuration B_0 of a body $\mathcal{B}$, equation (4.5.3) enables us to express that quantity as a function defined over a current configuration B of $\mathcal{B}$. On the other hand, if a quantity is represented by a function defined over a current configuration, equation (4.5.1) enables us to express that quantity as a function defined over the initial configuration. In this way, every quantity associated with a body $\mathcal{B}$ (or its motion) can be expressed in terms of the material variables $(\mathbf{x}^0, t)$ or in terms of the spatial variables $(\mathbf{x}, t)$. When a quantity is expressed in terms of $(\mathbf{x}^0, t)$, we say that it is in the *material* or *Lagrangian form* or *Lagrangian description*. When a quantity is expressed in terms of $(\mathbf{x}, t)$ we say that it is in *spatial* or *Eulerian form* or *Eulerian description*.

4.6
MATERIAL AND SPATIAL COORDINATES

We often deal with the component forms of equations (4.5.1) and (4.5.3). For this purpose, we set up at the origin O a fixed system of righthanded, rectangular Cartesian coordinate axes and consider the components of both $\mathbf{x}^0$ and $\mathbf{x}$ along these axes (see Figure 4.3). We denote the base vectors of the coordinate system by $\mathbf{e}_i$, as usual, and set

$$\mathbf{x}^0 \cdot \mathbf{e}_i = x_i^0, \qquad \mathbf{x} \cdot \mathbf{e}_i = x_i \tag{4.6.1}$$

so that x_i^0 are the components of $\mathbf{x}^0$ and x_i are the components of $\mathbf{x}$, along $\mathbf{e}_i$. To avoid possible confusion that may arise while dealing with x_i^0 and x_i

164   4 CONTINUUM HYPOTHESIS

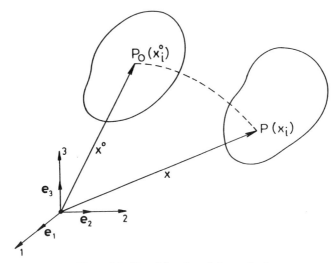

Figure 4.3. Material and spatial coordinates.

together, we refer to the coordinate axes as 1, 2, 3 axes, or briefly the i-axes. Thus the i-axis is the axis along which the base vector $\mathbf{e}_i$ is directed.

Then, equations (4.5.1) and (4.5.3) read respectively as follows in the component form:

$$x_i = x_i(x_k^0, t), \qquad t > 0 \qquad (4.6.2)$$

$$x_i^0 = x_i^0(x_k, t), \qquad t > 0 \qquad (4.6.3)$$

Also, the consistency condition (4.5.2) reads

$$x_i = x_i^0 \qquad \text{at } t = 0 \qquad (4.6.4)$$

Since $\mathbf{x}^0$ is the position vector of the point P_0 (which is the initial position of a particle $\mathcal{P}$), x_i^0 are the coordinates of P_0. These coordinates are termed *material coordinates*. Also, since $\mathbf{x}$ is the position vector of the point P (which is the current position of $\mathcal{P}$), x_i are the coordinates of P, called *spatial coordinates*. Often, material coordinates are referred to as *referential* or *initial coordinates*, and spatial coordinates are referred to as *instant* or *current coordinates*.

As a consequence of the one-to-one correspondence between the particles of a material body and the points of its configuration at any instant of time, it has been noted that the function $\mathbf{x}(\mathbf{x}^0, t)$ and $\mathbf{x}^0(\mathbf{x}, t)$ are inverses of each other and either equations (4.5.1) or (4.5.3) is a unique solution of the other. Therefore, equations (4.6.3) represent a unique solution of equations (4.6.2) and vice versa. An immediate implication of this is that the functions $x_i(x_k^0, t)$ and $x_i^0(x_k, t)$ possess continuous partial derivatives

4.6 MATERIAL AND SPATIAL COORDINATES

with respect to their arguments and that the *Jacobian J*, defined by

$$J = \det\left(\frac{\partial x_i}{\partial x_j^0}\right) \qquad (4.6.5)$$

is *nonzero* for any x_i^0 and any t of our interest. This amounts to saying that a motion described by (4.6.2) and (4.6.3) takes place smoothly and that it carries curves, surfaces and regions into curves, surfaces and regions, respectively. In particular, boundary surfaces get transformed to boundary surfaces only so that the particles that lie on a boundary surface of a material body in one configuration continue to remain on the boundary surface of the body in all configurations.

Another important consequence of the continuous differentiability of the functions $x_i(x_k^0, t)$ and $x_i^0(x_k, t)$ is that *the vanishing of an integral of a continuous function of spatial or material variables over an arbitrary volume implies the vanishing of the function at every point of the volume.* That is, if $\int_V f \, dV = 0$ for arbitrary volume V, then $f \equiv 0$ at every point of V. This result, referred to as the *localization theorem*, will be employed in Chapter 8 for deducing the local forms of balance equations from their integral forms.

With the view of making use of the conventions and advantages of the suffix notation in dealing with partial derivatives of the type present in (4.6.5), we will henceforth employ the *semicolon notation*:

$$\frac{\partial}{\partial x_j^0}(\ ) = (\ )_{;j}$$

reserving the usual comma notation for partial differentiation with respect to x_i. Then (4.6.5) can be rewritten as

$$J = \det(x_{i;j}) \qquad (4.6.6)$$

Since x_i are assumed to be continuously differentiable functions of x_i^0 and t, the partial derivatives $x_{i;j}$ are continuous in x_i^0 and t. Consequently, J is a continuous function of x_i^0 and t. The consistency conditions (4.6.4) require that $J = \det(x_{i;j}^0)$ at $t = 0$. Since $x_{i;j}^0 = \delta_{ij}$ and $\det(\delta_{ij}) = 1$, we get

$$J = 1 \quad \text{at } t = 0 \qquad (4.6.7)$$

Thus, (i) $J = 1$ for $t = 0$, and (ii) J is continuous and nonzero for $t > 0$. Therefore, J cannot be negative for any t; that is,

$$J > 0 \quad \text{for all } t \geq 0 \qquad (4.6.8)$$

The geometrical-physical meaning of this inequality will be discussed at a later stage.

4 CONTINUUM HYPOTHESIS

It is clear that every term defined, every assumption made and every inference drawn in this chapter stems from the hypothesis of continuity in distribution of matter over a region of space. The whole theory of continuum mechanics is based on this *continuum hypothesis*.

CHAPTER 5
DEFORMATION

5.1
INTRODUCTION

In Chapter 4, we explained what we mean by a material body and how its motion is described. The present chapter is concerned with the analysis of the geometrical changes that take place in a material body during its motion from one configuration to the other. The tensors which serve to measure these changes will be introduced and the related aspects will be considered in some detail.

5.2
DEFORMATION GRADIENT TENSOR

For analyzing the geometrical changes that take place in a material body $\mathcal{B}$, as it moves, we focus our attention on only two of its configurations: B_0 and B, where B_0 is the reference configuration and B is the current configuration at a specified (fixed) instant of time $t > 0$. We will refer to the transition of $\mathcal{B}$ from B_0 to B as a *deformation*. Also, B_0 will be referred to as the *initial*

168   5 DEFORMATION

or *undeformed configuration*, and B will be referred to as the *final* or *deformed configuration* (see Figure 5.1).

Consider a representative particle $\mathcal{P}$ of the body $\mathcal{B}$ and a material arc element $d\mathcal{C}$ issuing from $\mathcal{P}$. In the initial configuration B_0 this arc lies on some geometrical arc, say dC_0, issuing from the initial position P_0 of $\mathcal{P}$. Let $\mathbf{x}^0$ be the position vector of the point P_0 and $d\mathbf{x}^0$ be the vector representing the arc element dC_0 in both length and direction. Then the *initial direction* of $d\mathcal{C}$ is the direction of the vector $d\mathbf{x}^0$ and the *initial length* of $d\mathcal{C}$ is $ds_0 = |d\mathbf{x}^0|$.

When the body $\mathcal{B}$ moves over to the final configuration B, carrying with it the particle $\mathcal{P}$ and its neighboring particles, the material arc $d\mathcal{C}$ lies on some geometrical arc, say dC, in B that initiates from the position P of $\mathcal{P}$ in B. (Recall that, by the continuum hypothesis, curves transform to curves.) Let $\mathbf{x}$ be the position vector of the point P and $d\mathbf{x}$ be the vector representing the arc element dC both in direction and length. Then $ds = |d\mathbf{x}|$ is the *final length*, and the direction of $d\mathbf{x}$ is the *final direction* of $d\mathcal{C}$. We will call dC the *final location* of $d\mathcal{C}$. On the other hand, dC_0 will be called the *initial location* of $d\mathcal{C}$.

In the material description, a motion of $\mathcal{B}$ is described by equations of the form (4.6.2) in which x_i^0 and t are independent variables and x_i are dependent variables, x_i^0 and x_i being the components of $\mathbf{x}^0$ and $\mathbf{x}$, respectively. Therefore, the differentials dx_i can be expressed in terms of the differentials dx_i^0 through the following formula of differential calculus:

$$dx_i = \sum_{j=1}^{3} \frac{\partial x_i}{\partial x_j^0} dx_j^0, \qquad i = 1, 2, 3, \tag{5.2.1}$$

the time variable t being held fixed.

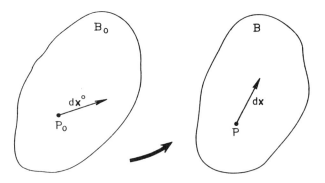

Figure 5.1. Initial and final configurations.

5.2 DEFORMATION GRADIENT TENSOR

Using the conventions of the suffix notation, and the semicolon notation defined in Section 4.6, expression (5.2.1) can be rewritten as

$$dx_i = x_{i;j}\, dx_j^0 \qquad (5.2.2)$$

We note that $(x_{i;j})$ is a 3×3 nonzero matrix related to x_i^0 system of coordinates. For an arbitrary vector $d\mathbf{x}^0$ with components dx_i^0, expression (5.2.2) shows that $x_{i;j}\, dx_j^0$ are components of a vector, $d\mathbf{x}$. Therefore, by quotient law (Section 2.4, property 7(ii)), it follows that $x_{i;j}$ are components of a nonzero tensor. This tensor is immediately recognized as the gradient of the vector $\mathbf{x}$, the gradient being taken with respect to x_i^0, called the *deformation gradient tensor in the material form*, or *the material deformation gradient*. We denote it by $\nabla^0 \mathbf{x}$ or $\mathbf{F}$, the symbol ∇^0 denoting the gradient operator with respect to x_i^0. Thus,

$$[\nabla^0 \mathbf{x}]_{ij} = [\mathbf{F}]_{ij} = x_{i;j} \qquad (5.2.3)$$

and

$$J = \det(x_{i;j}) = \det(\nabla^0 \mathbf{x}) = \det \mathbf{F} \qquad (5.2.4)$$

Consequently, (5.2.2) can be rewritten as follows in the direct notation

$$d\mathbf{x} = (\nabla^0 \mathbf{x})\, d\mathbf{x}^0 = \mathbf{F}\, d\mathbf{x}^0 \qquad (5.2.5)$$

This expression clearly exhibits that the vector $d\mathbf{x}$ is generated by the action of the tensor $\mathbf{F}$ on the vector $d\mathbf{x}^0$. In other words, the effects of the deformation on the material arc $d\mathcal{C}$ is represented by the tensor $\mathbf{F}$. Hence, (5.2.5) serves as a transformation rule satisfied by $d\mathcal{C}$ as it deforms from its initial location represented by the vector $d\mathbf{x}^0$ to the final location represented by the vector $d\mathbf{x}$. Since $d\mathbf{x}^0$ is a nonzero vector, we find from (5.2.5) that $d\mathbf{x}$ is also a nonzero vector. Thus, if a material arc has a nonzero length in one configuration it continues to have a nonzero length in every configuration. In other words, the tensor $\mathbf{F}$ transforms arcs of nonzero lengths to arcs of nonzero lengths. Relation (5.2.5) is the fundamental relation upon which the whole analysis of this chapter is based.

Let us consider two special cases of the relation (5.2.5).

5.2.1 TRANSLATION

Suppose $\mathbf{F} = \mathbf{I}$. In this (trivial) case, (5.2.5) reads $d\mathbf{x} = d\mathbf{x}^0$. This means that both the length and the orientation of $d\mathcal{C}$ are now preserved under the deformation. Such a deformation is called a *rigid-translation* or just a *translation*; see Figure 5.2.

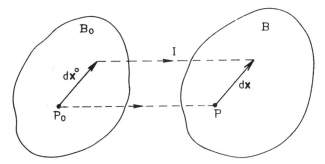

Figure 5.2. Translation.

5.2.2 ROTATION

Suppose $\mathbf{F}$ is an orthogonal tensor $\neq \mathbf{I}$. Then by virtue of the result proved in Example 2.8.3, we find that $|d\mathbf{x}| = |d\mathbf{x}^0|$. But, since $\mathbf{F} \neq \mathbf{I}$, $d\mathbf{x} \neq d\mathbf{x}^0$. Hence, in this case, the length of $d\mathcal{C}$ remains unchanged but its orientation undergoes a change. In other words, the effect of the deformation, now, is just to change the orientation of $d\mathcal{C}$ without changing its length. Such a deformation is called a *rigid rotation* or just a *rotation*; see Figure 5.3. It is to be noted that, under a rigid rotation, the arc $d\mathcal{C}$ may experience a translation as well. This is because, $\mathbf{F}$ always contains $\mathbf{I}$ as a factor; that is, $\mathbf{F} = \mathbf{FI} = \mathbf{IF}$.

In the general case, where $\mathbf{F}$ is neither $\mathbf{I}$ nor an orthogonal tensor, $d\mathcal{C}$ generally experiences a change in length as well as a change in orientation. An analysis of this general case requires the use of the polar decomposition of $\mathbf{F}$, and this will be taken up in detail in Section 5.3. Pending this analysis, we proceed to examine how a material surface element and a material volume element change under a deformation.

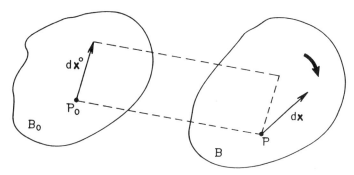

Figure 5.3. Rotation.

5.2.3 DEFORMATION OF A SURFACE ELEMENT

At the particle $\mathcal{P}$ consider another material arc element $d\bar{\mathcal{C}}$ so that $d\mathcal{C}$ and $d\bar{\mathcal{C}}$ together determine a material surface element $d\mathcal{S}$. Let $d\bar{\mathcal{C}}$ be represented by the vectors $d\bar{\mathbf{x}}^0$ and $d\bar{\mathbf{x}}$ in the configurations B_0 and B, respectively. Then, by virtue of the geometrical meaning of the cross product of vectors, the areas of $d\mathcal{S}$ in B_0 and B are respectively given by

$$dS_0 = |d\mathbf{x}^0 \times d\bar{\mathbf{x}}^0| \qquad (5.2.6a)$$

$$dS = |d\mathbf{x} \times d\bar{\mathbf{x}}| \qquad (5.2.6b)$$

Also, if $\mathbf{n}^0$ and $\mathbf{n}$ are unit normal vectors to $d\mathcal{S}$ in B_0 and B, respectively, such that the triads $(d\mathbf{x}^0, d\bar{\mathbf{x}}^0, \mathbf{n}^0)$ and $(d\mathbf{x}, d\bar{\mathbf{x}}, \mathbf{n})$ are righthanded (see Fig. 5.4), we have

$$d\mathbf{x}^0 \times d\bar{\mathbf{x}}^0 = (dS_0)\mathbf{n}^0 \qquad (5.2.7a)$$

$$d\mathbf{x} \times d\bar{\mathbf{x}} = (dS)\mathbf{n} \qquad (5.2.7b)$$

Since $d\mathcal{C}$ and $d\bar{\mathcal{C}}$ are of nonzero initial lengths, the initial area dS_0 of $d\mathcal{S}$ is not zero, provided $d\mathcal{C}$ and $d\bar{\mathcal{C}}$ are initially noncoincident. Under the deformation defined by (5.2.5), $d\mathcal{C}$ and $d\bar{\mathcal{C}}$ continue to be of nonzero lengths in every configuration. Hence, $d\mathbf{x}$ and $d\bar{\mathbf{x}}$, which represent these arcs in B, are nonzero vectors. Therefore, from (5.2.6b) we find that $dS = 0$ if and only if $d\mathbf{x}$ and $d\bar{\mathbf{x}}$ are collinear; that is, if and only if $d\mathcal{C}$ and $d\bar{\mathcal{C}}$ become coincident in B. If $d\mathcal{C}$ and $d\bar{\mathcal{C}}$ are initially noncoincident, the possibility of $d\mathcal{C}$ and $d\bar{\mathcal{C}}$ becoming coincident in B is ruled out, because two different

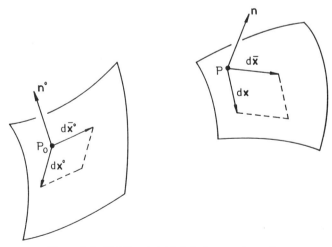

Figure 5.4. Righthanded triads in two configurations.

particles cannot occupy the same position in the same configuration. Hence $dS \neq 0$ whenever $dS_0 \neq 0$. Thus, if a material surface element has a nonzero area in one configuration, it continues to have a nonzero area in every configuration. In other words, $\mathbf{F}$ transforms surface elements of nonzero areas to surface elements of nonzero areas.

From (5.2.5), we find that

$$d\mathbf{x} \times d\bar{\mathbf{x}} = (\mathbf{F}\, d\mathbf{x}^0) \times (\mathbf{F}\, d\bar{\mathbf{x}}^0)$$

which on using (2.11.25) becomes

$$d\mathbf{x} \times d\bar{\mathbf{x}} = \mathbf{F}^*(d\mathbf{x}^0 \times d\bar{\mathbf{x}}^0) \tag{5.2.8}$$

where $\mathbf{F}^*$ is the cofactor of $\mathbf{F}$. By use of (5.2.7), expression (5.2.8) yields

$$(dS)\mathbf{n} = (dS_0)\mathbf{F}^*\mathbf{n}^0 \tag{5.2.9}$$

which gives

$$(dS)\mathbf{F}^T\mathbf{n} = (dS_0)\mathbf{F}^T\mathbf{F}^*\mathbf{n}^0 \tag{5.2.10}$$

Using (2.11.23) and (5.2.4), we note that

$$\mathbf{F}^T\mathbf{F}^* = J\mathbf{I} \tag{5.2.11}$$

so that (5.2.10) can be rewritten as

$$(dS)\mathbf{F}^T\mathbf{n} = J(dS_0)\mathbf{n}^0 \tag{5.2.12}$$

This expression explicitly exhibits how the oriented surface element $(dS)\mathbf{n}$ representing $d\mathcal{S}$ in the final configuration is related to the oriented surface element $(dS_0)\mathbf{n}^0$ representing the same $d\mathcal{S}$ in the initial configuration. In fact, (5.2.12) reveals an important characteristic property of the tensor $\mathbf{F}$. Recall that dS and dS_0 are nonzero areas. Also, $\mathbf{n}^0$ and therefore $\mathbf{n}$ are arbitrary vectors; hence $\mathbf{F}^T\mathbf{n}$ is a nonzero vector. Consequently, it follows from (5.2.12) that $J \neq 0$, which is precisely what we noted immediately after defining J through the relation (4.6.5). An immediate consequence of this important relation is that the tensor $\mathbf{F}$ is an *invertible tensor*; that is, the inverse $\mathbf{F}^{-1}$ of $\mathbf{F}$ exists.

From (5.2.5), we find that

$$d\mathbf{x}^0 = \mathbf{F}^{-1}\, d\mathbf{x} \tag{5.2.13}$$

which shows that the tensor $\mathbf{F}^{-1}$ carries the vector $d\mathbf{x}$ to the vector $d\mathbf{x}^0$. In other words, (5.2.13) defines the transformation that carries $d\mathcal{C}$ back to the initial location dC_0 from the final location dC.

5.2 DEFORMATION GRADIENT TENSOR

In the spatial description, x_i^0 are functions of x_i and t. Hence

$$dx_i^0 = \sum_{j=1}^{3} \frac{\partial x_i^0}{\partial x_j} dx_j = x_{i,j}^0 \, dx_j \tag{5.2.14}$$

if t is held fixed. Evidently, $x_{i,j}^0$ are the components of the gradient of $\mathbf{x}^0$, namely $\nabla \mathbf{x}^0$, the gradient being taken with respect to x_i. As such, (5.2.14) can be rewritten, in the direct notation, as

$$d\mathbf{x}^0 = (\nabla \mathbf{x}^0) \, d\mathbf{x}. \tag{5.2.15}$$

Comparing this with (5.2.13), we find that

$$\mathbf{F}^{-1} = \nabla \mathbf{x}^0; \qquad [\mathbf{F}^{-1}]_{ij} = x_{i,j}^0 \tag{5.2.16}$$

Consequently (since $\det \mathbf{A}^{-1} = (\det \mathbf{A})^{-1}$ for any invertible tensor $\mathbf{A}$),

$$\det \mathbf{F}^{-1} = \det(x_{i,j}^0) = \frac{1}{J} \tag{5.2.17}$$

The tensor $\mathbf{F}^{-1}$ is called the *deformation gradient in the spatial form*, or briefly the *spatial deformation gradient*.

Let us return to (5.2.12) and analyze another of its important implications. To this end, consider one more material arc element $d\bar{\bar{\mathcal{C}}}$ issuing from the particle $\mathcal{P}$ and initially directed along $\mathbf{n}^0$. Then $d\bar{\bar{\mathcal{C}}}$ is initially perpendicular to $d\mathcal{S}$. Let $d\bar{\mathbf{x}}^0$ and $d\bar{\mathbf{x}}$ be the vectors representing $d\bar{\bar{\mathcal{C}}}$ in the initial and final configurations respectively. Then, by use of (5.2.5), (5.2.12), (2.8.11) and (2.8.14), we find that

$$d\bar{\mathbf{x}} \cdot \mathbf{n} = (\mathbf{F} \, d\bar{\mathbf{x}}^0) \cdot \left\{ J\left(\frac{dS_0}{dS}\right)(\mathbf{F}^T)^{-1} \mathbf{n}^0 \right\}$$

$$= (\mathbf{F}^{-1} \mathbf{F} \, d\bar{\mathbf{x}}^0) \cdot J\left(\frac{dS_0}{dS}\right) \mathbf{n}^0$$

$$= J\left(\frac{dS_0}{dS}\right) d\bar{\mathbf{x}}^0 \cdot \mathbf{n}^0 \tag{5.2.18}$$

If $d\bar{s}_0$ and $d\bar{s}$ are the initial and final lengths of $d\bar{\bar{\mathcal{C}}}$ and if θ is the angle between $d\bar{\mathbf{x}}$ and $\mathbf{n}$ then $d\bar{\mathbf{x}}^0 \cdot \mathbf{n}^0 = d\bar{s}_0$ and $d\bar{\mathbf{x}} \cdot \mathbf{n} = (d\bar{s}) \cos \theta$, and (5.2.18) yields

$$\cos \theta = J\left(\frac{dS_0}{dS}\right)\left(\frac{d\bar{s}_0}{d\bar{s}}\right) \tag{5.2.19}$$

Evidently, $\theta \neq 0$ in general, the case $\theta = 0$ being not ruled out. This means that in the final configuration, $d\bar{\bar{\mathcal{C}}}$ is not necessarily perpendicular to $d\mathcal{S}$. In other words, a material arc that is perpendicular to $d\mathcal{S}$ in one configuration

174 5 DEFORMATION

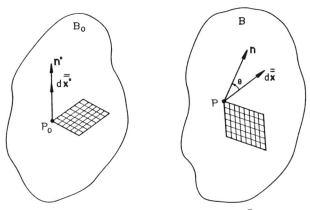

Figure 5.5. Initial and final orientations of $d\bar{\bar{\mathcal{C}}}$.

need not be perpendicular to dS in a subsequent configuration; see Figure 5.5.

A question now arises: can θ be equal to $\pi/2$? The answer is no, because θ can become equal to $\pi/2$ only if $d\bar{\bar{\mathcal{C}}}$ falls onto dS and becomes a part of dS in the final configuration. This is not possible, as there is no gap in dS to take in $d\bar{\bar{\mathcal{C}}}$. The next question is this: can θ be greater than $\pi/2$? The answer to this question is also in the negative, because θ, which starts from the value 0 in the initial configuration, can acquire a value greater than $\pi/2$ only after passing through the value $\pi/2$, which has already been ruled out. (Remember that we deal with continuous transformations!) Thus, θ must lie in the interval $0 \le \theta < \pi/2$. As such, the lefthand side of (5.2.19) is (always) positive. Since $dS_0, dS, d\bar{s}_0, d\bar{s}$ are all positive, it follows from (5.2.19) that $J > 0$, which is precisely the inequality (4.6.8) obtained by the assumption of continuity of J. This inequality asserts that the deformation preserves the sense of orientation of surface elements. Hence if dS is positively oriented in one configuration, it continues to be positively oriented in every configuration.

It is to be pointed out that the positive nature of J is a synthesis of the various postulates and assumptions made in the course of the preceding discussions. The geometrical meaning of J is obtained next.

5.2.4 DEFORMATION OF A VOLUME ELEMENT

Let us now consider the case where $d\bar{\bar{\mathcal{C}}}$ is not necessarily orthogonal to and not a part of dS in the initial configuration. Then, $d\mathcal{C}, d\bar{\mathcal{C}}$ and $d\bar{\bar{\mathcal{C}}}$ determine a material volume element, $d\mathcal{V}$. By virtue of the geometrical meaning of the scalar triple product, the volumes of $d\mathcal{V}$ in the initial and final

5.2 DEFORMATION GRADIENT TENSOR

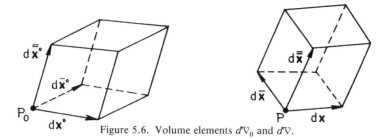

Figure 5.6. Volume elements $d\mathcal{V}_0$ and $d\mathcal{V}$.

configurations are respectively given by (see Figure 5.6)

$$dV_0 = |(d\mathbf{x}^0 \times d\bar{\mathbf{x}}^0) \cdot d\bar{\bar{\mathbf{x}}}^0|, \qquad dV = |(d\mathbf{x} \times d\bar{\mathbf{x}}) \cdot d\bar{\bar{\mathbf{x}}}|. \qquad (5.2.20)$$

Now, by using (5.2.8) and (5.2.5), we find that

$$(d\mathbf{x} \times d\bar{\mathbf{x}}) \cdot d\bar{\bar{\mathbf{x}}} = \mathbf{F}^* (d\mathbf{x}^0 \times d\bar{\mathbf{x}}^0) \cdot \mathbf{F}\, d\bar{\bar{\mathbf{x}}}^0$$

which, by use of (2.8.14) and (5.2.11), becomes

$$\begin{aligned}(d\mathbf{x} \times d\bar{\mathbf{x}}) \cdot d\bar{\bar{\mathbf{x}}} &= \mathbf{F}^T\mathbf{F}^*(d\mathbf{x}^0 \times d\bar{\mathbf{x}}^0) \cdot d\bar{\bar{\mathbf{x}}}^0 \\ &= J(d\mathbf{x}^0 \times d\bar{\mathbf{x}}^0) \cdot d\bar{\bar{\mathbf{x}}}^0 \end{aligned} \qquad (5.2.21)$$

Taking the absolute values on both sides of this expression and using (5.2.20) we get

$$dV = J\, dV_0 \qquad (5.2.22)$$

This expression explicitly exhibits how the final volume dV of $d\mathcal{V}$ is related to its initial volume dV_0. In fact, (5.2.22) reveals that dV is generally not equal to dV_0, the case $dV = dV_0$ not being ruled out. Further, J represents the ratio of the final volume of $d\mathcal{V}$ to its initial volume. More fundamentally, $dV \neq 0$ whenever $dV_0 \neq 0$. Thus, a material element having nonzero volume in one configuration continues to have nonzero volume in every configuration. This characteristic property of motion is a direct consequence of the invertibility of $\mathbf{F}$. Expression (5.2.22) is given by Euler.

5.2.5 ISOCHORIC DEFORMATION

From the Euler's formula (5.2.22) we note that a material volume element at the particle $\mathcal{P}$ does not suffer a change in volume under a deformation if and only if $J \equiv 1$. In such a case we say that $\mathcal{P}$ undergoes an *isochoric (volume-preserving) deformation*. If a deformation is such that *every* particle undergoes an isochoric deformation, we say that the deformation is an *isochoric deformation*. A material body is said to be *incompressible* if *every* deformation it undergoes is an isochoric deformation. Thus, a deformation is isochoric if and only if $J = 1$ at every particle in every configuration, and

176   5 DEFORMATION

a material body is incompressible if and only if every deformation it undergoes is an isochoric deformation.

EXAMPLE 5.2.1   For the deformation defined by equations

$$x_1 = x_1^0 + x_2^0, \quad x_2 = x_1^0 - 2x_2^0, \quad x_3 = x_1^0 + x_2^0 - x_3^0 \quad (5.2.23)$$

find $\mathbf{F}$, $\mathbf{F}^{-1}$ and J.

Solution   The given equations (5.2.23) express x_i in terms of x_i^0. Hence the description of deformation is in the material form. From these equations, we obtain

$$[\mathbf{F}] = [x_{i;j}] = \begin{bmatrix} 1 & 1 & 0 \\ 1 & -2 & 0 \\ 1 & 1 & -1 \end{bmatrix} \quad (5.2.24)$$

This is the matrix of $\mathbf{F}$ for the given deformation. From this matrix we find that

$$J = \det \mathbf{F} = \begin{vmatrix} 1 & 1 & 0 \\ 1 & -2 & 0 \\ 1 & 1 & -1 \end{vmatrix} = 3 \quad (5.2.25)$$

Solving equations (5.2.23) for x_i^0, we obtain

$$x_1^0 = \tfrac{1}{3}(2x_1 + x_2), \quad x_2^0 = \tfrac{1}{3}(x_1 - x_2), \quad x_3^0 = x_1 - x_3 \quad (5.2.26)$$

These equations describe the given deformation in the spatial form. From these equations we get

$$[\mathbf{F}^{-1}] = [x_{i,j}^0] \begin{bmatrix} 2/3 & 1/3 & 0 \\ 1/3 & -1/3 & 0 \\ 1 & 0 & -1 \end{bmatrix} \quad (5.2.27)$$

This is the matrix of $\mathbf{F}^{-1}$ for the given deformation.   ∎

EXAMPLE 5.2.2   For the deformation defined by the equations

$$x_1^0 = \tfrac{1}{2}(x_1^2 + x_2^2), \quad x_2^0 = \tan^{-1}(x_2/x_1), \quad x_1 \neq 0, \quad x_3^0 = x_3 \quad (5.2.28)$$

find $\mathbf{F}$ and $\mathbf{F}^{-1}$. Show that the deformation is an isochoric deformation.

5.2 DEFORMATION GRADIENT TENSOR

Solution From the equations (5.2.28), we observe that the given deformation is described in the spatial form. These equations give

$$[\mathbf{F}^{-1}] = [x^0_{i,j}] = \begin{bmatrix} x_1 & x_2 & 0 \\ -\dfrac{x_2}{R^2} & \dfrac{x_1}{R^2} & 0 \\ 0 & 0 & 1 \end{bmatrix} \quad (5.2.29)$$

where $R^2 = x_1^2 + x_2^2$, which is taken to be nonzero.

Solving equations (5.2.28), we obtain

$$x_1 = (2x_1^0)^{1/2} \cos x_2^0, \qquad x_2 = (2x_1^0)^{1/2} \sin x_2^0, \qquad x_3 = x_3^0 \quad (5.2.30)$$

These equations describe the given deformation in the material form. From these equations we obtain

$$[\mathbf{F}] = [x_{i;j}] = \begin{bmatrix} \dfrac{1}{\xi_0} \cos x_2^0 & -\xi_0 \sin x_2^0 & 0 \\ \dfrac{1}{\xi_0} \sin x_2^0 & \xi_0 \cos x_2^0 & 0 \\ 0 & 0 & 1 \end{bmatrix} \quad (5.2.31)$$

where $\xi_0 = (2x_1^0)^{1/2}$.

Thus, (5.2.31) gives the matrix of $\mathbf{F}$ and (5.2.29) gives the matrix of $\mathbf{F}^{-1}$ for the given deformation.

Also, from (5.2.31), we find that

$$J = \det(x_{i;j}) = 1 \qquad \text{for any } x_i^0$$

Hence the given deformation is an isochoric deformation. ∎

EXAMPLE 5.2.3 Prove the following:

(i) $$[\mathbf{F}^{-1}]_{ij} = x^0_{i,j} = \dfrac{1}{2J} \varepsilon_{irs} \varepsilon_{jpq} x_{p;r} x_{q;s} \quad (5.2.32)$$

(ii) $$[\mathbf{F}]_{ij} = x_{i;j} = \tfrac{1}{2} J \varepsilon_{irs} \varepsilon_{jpq} x^0_{p,r} x^0_{q,s} \quad (5.2.33)$$

Solution (i) From the identity (5.2.11), we get

$$\mathbf{F}^{-1} = \dfrac{1}{J} (\mathbf{F}^*)^T \quad (5.2.34)$$

178 5 DEFORMATION

Since $x_{i;j}$ are the components of **F**, the components of the cofactor **F*** of **F** are given by [see (2.11.22)]

$$[\mathbf{F}^*]_{ij} = \tfrac{1}{2}\varepsilon_{ipq}\varepsilon_{jrs}x_{p;r}x_{q;s} \tag{5.2.35}$$

Since $x^0_{i,j}$ are the components of $\mathbf{F}^{-1}$, we find from (5.2.34) and (5.2.35) that

$$x^0_{i,j} = [\mathbf{F}^{-1}]_{ij} = \frac{1}{J}[\mathbf{F}^*]_{ji}$$

$$= \frac{1}{2J}\varepsilon_{jpq}\varepsilon_{irs}x_{p;r}x_{q;s}$$

which is the relation (5.2.32).

(ii) From (5.2.11), we find

$$\mathbf{F} = J[(\mathbf{F}^{-1})^*]^T \tag{5.2.36}$$

Since $x^0_{i,j}$ are the components of $\mathbf{F}^{-1}$, the components of $(\mathbf{F}^{-1})^*$ are given by

$$[(\mathbf{F}^{-1})^*]_{ij} = \tfrac{1}{2}\varepsilon_{ipq}\varepsilon_{jrs}x^0_{p,r}x^0_{q,s} \tag{5.2.37}$$

Since $x_{i;j}$ are components of **F**, we find from (5.2.36) and (5.2.37),

$$x_{i;j} = [\mathbf{F}]_{ij} = J[(\mathbf{F}^{-1})^*]_{ji}$$

$$= \tfrac{1}{2}J\varepsilon_{jpq}\varepsilon_{irs}x^0_{p,r}x^0_{q,s}$$

which is the relation (5.2.33). ∎

EXAMPLE 5.2.4 Prove the following identities:

(i) $$[J(\mathbf{F}^{-1})^T]_{ij} = \frac{\partial J}{\partial(x_{i;j})} \tag{5.2.38}$$

(ii) $$\operatorname{div}\left(\frac{1}{J}\mathbf{F}^T\right) = \mathbf{0} \tag{5.2.39}$$

Solution (i) Since $J = \det(x_{i;j})$, we have, by virtue of the relation (1.7.26),

$$J = \tfrac{1}{6}\varepsilon_{imn}\varepsilon_{pqr}x_{i;p}x_{m;q}x_{n;r} \tag{5.2.40}$$

Differentiating this partially with respect to $(x_{i;j})$ and noting that

$$\frac{\partial(x_{i;p})}{\partial(x_{i;j})} = \delta_{pj} = \frac{\partial x^0_j}{\partial x^0_p} = \frac{\partial x^0_j}{\partial x_i}\frac{\partial x_i}{\partial x^0_p} = x^0_{j,i}x_{i;p}$$

5.2 DEFORMATION GRADIENT TENSOR

we obtain

$$\frac{\partial J}{\partial(x_{i;j})} = \frac{1}{6}\varepsilon_{imn}\varepsilon_{pqr}X_{m;q}X_{n;r}(x_{j,i}^0 x_{i;p})$$

which by use of (5.2.40) becomes

$$\frac{\partial J}{\partial(x_{i;j})} = Jx_{j,i}^0 \qquad (5.2.41)$$

This is precisely the identity (5.2.38).

(ii) Since

$$\left[\frac{1}{J}\mathbf{F}^T\right]_{ij} = \frac{1}{J}x_{j;i}$$

we have

$$\left[\operatorname{div}\left(\frac{1}{J}\mathbf{F}^T\right)\right]_i = \left(\frac{1}{J}x_{j;i}\right)_{,j} = \frac{\partial}{\partial x_j}\left(\frac{1}{J}\frac{\partial x_j}{\partial x_i^0}\right)$$

$$= \frac{1}{J}\frac{\partial}{\partial x_j}\left(\frac{\partial x_j}{\partial x_i^0}\right) + \frac{\partial}{\partial x_j}\left(\frac{1}{J}\right)\frac{\partial x_j}{\partial x_i^0}$$

$$= \frac{1}{J}\left[\frac{\partial}{\partial x_j}\left(\frac{\partial x_j}{\partial x_i^0}\right) - \frac{1}{J}\frac{\partial J}{\partial x_j}\frac{\partial x_j}{\partial x_i^0}\right] \qquad (5.2.42)$$

By using the chain rule of differentiation and the identity (5.2.41), we obtain

$$\frac{\partial J}{\partial x_j} = \frac{\partial J}{\partial(x_{r;s})}\frac{\partial}{\partial x_j}(x_{r;s})$$

$$= J\frac{\partial x_s^0}{\partial x_r}\frac{\partial}{\partial x_j}\left(\frac{\partial x_r}{\partial x_s^0}\right)$$

Therefore

$$\frac{1}{J}\frac{\partial J}{\partial x_j}\frac{\partial x_j}{\partial x_i^0} = \frac{\partial x_s^0}{\partial x_r}\frac{\partial}{\partial x_j}\left(\frac{\partial x_r}{\partial x_s^0}\right)\frac{\partial x_j}{\partial x_i^0}$$

$$= \frac{\partial x_s^0}{\partial x_r}\frac{\partial}{\partial x_i^0}\left(\frac{\partial x_r}{\partial x_s^0}\right) = \frac{\partial x_s^0}{\partial x_r}\frac{\partial}{\partial x_s^0}\left(\frac{\partial x_r}{\partial x_i^0}\right)$$

$$= \frac{\partial}{\partial x_r}\left(\frac{\partial x_r}{\partial x_i^0}\right)$$

In view of this relation, we find that the righthand side of (5.2.42) vanishes, and the identity (5.2.39) is proven. ∎

5.3
STRETCH AND ROTATION

We now proceed to analyze the effect of deformation defined by (5.2.5), namely,

$$d\mathbf{x} = \mathbf{F} \, d\mathbf{x}^0 \tag{5.3.1}$$

on the material arc element $d\mathcal{C}$ considered in Section 5.2, in the general case where $\mathbf{F}$ is neither the identity tensor nor an orthogonal tensor.

Let $\mathbf{a}^0$ and $\mathbf{a}$ be the *unit* vectors directed along $d\mathbf{x}^0$ and $d\mathbf{x}$, respectively (see Figure 5.7), so that

$$d\mathbf{x}^0 = (ds_0)\mathbf{a}^0 \tag{5.3.2a}$$

$$d\mathbf{x} = (ds)\mathbf{a} \tag{5.3.2b}$$

Using (5.3.2), expression (5.3.1) can be rewritten as

$$\eta \mathbf{a} = \mathbf{F}\mathbf{a}^0 \tag{5.3.3}$$

where

$$\eta = \frac{(ds)}{(ds_0)} \tag{5.3.4}$$

Evidently, η is a positive number representing the ratio of the final and initial lengths of $d\mathcal{C}$. It follows that $d\mathcal{C}$ does not change in length as it moves from initial location dC_0 to the final location dC if and only if $\eta = 1$, or equivalently $\mathbf{F}$ *preserves* the length of $\mathbf{a}^0$. Not every $\mathbf{F}$ can have this property. As such, $d\mathcal{C}$ generally changes in length as it moves from dC_0 to dC. This change is called an *extension* if $ds > ds_0$ (i.e., $\eta > 1$) and a *contraction* if $ds < ds_0$ (i.e., $\eta < 1$). The number η is called the *stretch* of $d\mathcal{C}$ or *stretch along* $\mathbf{a}^0$.

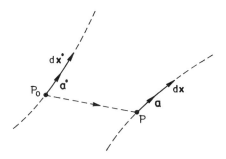

Figure 5.7. Vectors $\mathbf{a}^0$ and $\mathbf{a}$.

From (5.3.3), we find that $\mathbf{a}$ is aligned with $\mathbf{a}^0$ if and only if $\mathbf{F}\mathbf{a}^0 = \eta \mathbf{a}^0$, or equivalently $\mathbf{a}^0$ is an eigenvector of $\mathbf{F}$ with η as the corresponding eigenvalue. Not every $\mathbf{F}$ can have $\mathbf{a}^0$ as its eigenvector. Hence $\mathbf{a}$ is generally not aligned with $\mathbf{a}^0$. This means that the orientation of $d\mathcal{C}$ generally changes as it moves from dC_0 to dC.

Thus, under a deformation, $d\mathcal{C}$ generally changes in length as well as orientation. Later we show that this change can be interpreted as the net result of a translation, a rotation and proportional extensions or contractions in three mutually perpendicular directions.

5.3.1 DECOMPOSITION OF A DEFORMATION

In Section 5.2, it was noted that $\mathbf{F}$ is an invertible tensor. Hence, by the polar decomposition theorem (Theorem 2.14.4), $\mathbf{F}$ can be decomposed uniquely in the forms

$$\mathbf{F} = \mathbf{Q}\mathbf{U} \tag{5.3.5a}$$

$$\mathbf{F} = \mathbf{V}\mathbf{Q} \tag{5.3.5b}$$

where $\mathbf{Q}$ is an orthogonal tensor, and $\mathbf{U}$ and $\mathbf{V}$ are positive definite symmetric tensors such that

$$\mathbf{U}^2 = \mathbf{F}^T\mathbf{F} \tag{5.3.6a}$$

$$\mathbf{V}^2 = \mathbf{F}\mathbf{F}^T \tag{5.3.6b}$$

Using the right polar decomposition of $\mathbf{F}$, given by (5.3.5a), the relation (5.3.1) can be rewritten as

$$d\mathbf{x} = (\mathbf{Q}\mathbf{U})\, d\mathbf{x}^0 = \mathbf{Q}(\mathbf{U}d\mathbf{x}^0) \tag{5.3.7}$$

or, equivalently,

$$d\mathbf{y} = \mathbf{U}d\mathbf{x}^0 \tag{5.3.8a}$$

$$d\mathbf{x} = \mathbf{Q}d\mathbf{y} \tag{5.3.8b}$$

Thus, the transformation of $d\mathcal{C}$ from its initial location dC_0 to the final location dC can be decomposed into two parts. Under the first part of the transformation, given by (5.3.8a), $d\mathcal{C}$ is carried from dC_0 to an intermediate location, say dC', represented by the vector $d\mathbf{y}$, and in the second part given by (5.3.8b), it is carried from dC' to the final location dC. Note that the first part of the transformation is effected by the tensor $\mathbf{U}$ acting on $d\mathbf{x}^0$ and the second part is effected by the tensor $\mathbf{Q}$ acting on $d\mathbf{y}$.

5.3.2 TRIAXIAL STRETCH

Since **U** is positive definite and symmetric, it has at least one set of mutually orthogonal eigenvectors, say $\mathbf{p}_1, \mathbf{p}_2, \mathbf{p}_3$, which may be taken as the base vectors of a coordinate system, and three corresponding eigenvalues, say η_1, η_2, η_3, which are all positive; see sections 2.13 and 2.14. Hence **U** has the following spectral representation (see Example 2.13.5):

$$\mathbf{U} = \sum_{k=1}^{3} \eta_k (\mathbf{p}_k \otimes \mathbf{p}_k) \tag{5.3.9}$$

Substituting this expression into (5.3.8a) and using the identity (2.4.37) we get

$$d\mathbf{y} = \sum_{k=1}^{3} \eta_k (\mathbf{p}_k \cdot d\mathbf{x}^0) \mathbf{p}_k \tag{5.3.10}$$

If dy_i' and $dx_i^{0\prime}$ are the components of $d\mathbf{y}$ and $d\mathbf{x}^0$ along the base vectors $\mathbf{p}_i$ (that is, if $dy_i' = d\mathbf{y} \cdot \mathbf{p}_i$ and $dx_i^{0\prime} = d\mathbf{x}^0 \cdot \mathbf{p}_i$), then (5.3.10) yields

$$dy_1' = \eta_1 \, dx_1^{0\prime} \tag{5.3.11a}$$

$$dy_2' = \eta_2 \, dx_2^{0\prime} \tag{5.3.11b}$$

$$dy_3' = \eta_3 \, dx_3^{0\prime} \tag{5.3.11c}$$

We note that $dx_i^{0\prime}$ are the projections of the length of $d\mathcal{C}$ along the directions of $\mathbf{p}_i$ in the initial location dC_0 and dy_i' are the corresponding projections in the intermediate location dC'. Expressions (5.3.11) show that these two sets of projections are not the same. We find from (5.3.11a) that the projection of the initial length of $d\mathcal{C}$ in the direction of $\mathbf{p}_1$ gets multiplied by the factor η_1 as it moves from dC_0 to dC'. This means that $d\mathcal{C}$ experiences a stretch equal to η_1 in the direction of $\mathbf{p}_1$ as it moves from dC_0 to dC'. Similarly, (5.3.11b and c) show that $d\mathcal{C}$ generally experiences stretches equal to η_2 and η_3 along the directions of $\mathbf{p}_2$ and $\mathbf{p}_3$. (See Figure 5.8.) This experience of $d\mathcal{C}$ (as it moves from dC to dC') is referred to as *triaxial stretch* suffered by $d\mathcal{C}$. The numbers η_i are called the *principal stretches* and the directions of $\mathbf{p}_i$ (along which η_i are the stretches) are called the *principal directions of stretch*.

5.3.3 RIGID-BODY TRANSFORMATION

Let us now consider the second part of the transformation that carries $d\mathcal{C}$ from the intermediate location dC' to the final location dC. This part of the transformation, effected by **Q** acting on $d\mathbf{y}$, is described by the equation (5.3.8b). Since **Q** is an orthogonal tensor, we have $|d\mathbf{x}| = |d\mathbf{y}|$. Hence, **Q**

5.3 STRETCH AND ROTATION 183

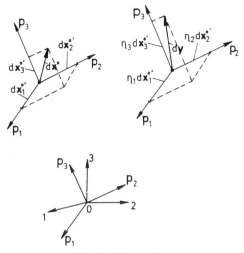

Figure 5.8. Triaxial stretch.

does not effect a change in the length of $d\mathbf{y}$; the only (possible) effect of $\mathbf{Q}$ on $d\mathbf{y}$ is to change the orientation of $d\mathbf{y}$. This means that, under the second part of the transformation, $d\mathcal{C}$ generally changes its orientation but retains the length it had acquired in the first part of the transformation. That is, $d\mathcal{C}$ acquires its final length in full in the first part of the transformation itself; in the second part of the transformation, it just gets tilted (rigidly) to the final orientation. Thus, the second part of the transformation is just a *pure (rigid) rotation*.

In the process of the full transformation from dC_0 to dC, $d\mathcal{C}$ may experience a *pure translation* as well. The translation may occur before the triaxial stretch, between the triaxial stretch and the rigid rotation, after the rigid rotation, or simultaneously with the triaxial stretch and rigid rotation. For definiteness in the interpretation, we suppose, without any loss of generality in the analysis, that the translation occurs after the rigid rotation. (This supposition amounts to treating $\mathbf{Q}(d\mathbf{y})$ as $\mathbf{I}(\mathbf{Q}d\mathbf{y})$ so that the identity tensor $\mathbf{I}$ represents the effect of translation.) Then, the second part of the transformation, effected by $\mathbf{Q}$, may be interpreted as a *rigid-body transformation* consisting of a pure translation and a pure rotation.

The preceding analysis shows that, during its transition from the initial location to the final location, $d\mathcal{C}$ may be thought of as being subjected to triaxial stretch followed by a rigid-body transformation (see Figure 5.9). And the net result is a change in length as well as a change in the orientation of $d\mathcal{C}$. This interpretation is based on the right polar decomposition of $\mathbf{F}$ given by (5.3.5a). A similar interpretation can be arrived at on the basis of

184   5 DEFORMATION

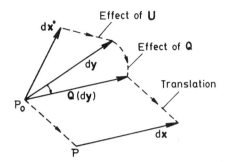

Figure 5.9. Stretch followed by rotation and translation.

the left polar decomposition given by (5.3.5b). In this case, triaxial stretch is effected by the tensor **V** and the rigid-body transformation is effected (as in the previous case) by **Q**. But here the rigid-body transformation precedes the triaxial stretch. Bearing in mind that **U** and **V** have the same eigenvalues but not necessarily the same eigenvectors (see Example 2.14.3), we may infer that, while the principal stretches remain the same in both the cases, the directions in which these occur are generally different in the two cases. The tensor **Q** that essentially effects the rotation (in both the cases) is referred to as the *rotation tensor*. The tensors **U** and **V** that effect triaxial stretches are known as the *right* and *left stretch tensors*, respectively.

In the special case when $\mathbf{U} = \mathbf{V} = \mathbf{I}$, we have $\mathbf{F} = \mathbf{Q}$ (by 5.3.5), and the deformation then consists of no triaxial stretch; $d\mathcal{C}$ experiences just a rigid-body transformation. Conversely, if $d\mathcal{C}$ experiences just a rigid-body transformation, then $\mathbf{U} = \mathbf{V} = \mathbf{I}$ so that $\mathbf{F} = \mathbf{Q}$. Thus the deformation given by (5.3.1) is a rigid-body transformation if and only if $\mathbf{F} = \mathbf{Q}$.

Since material arcs generally experience changes in their lengths and orientations (under a deformation) as seen in the preceding analysis, the relative distances between the particles of a material body are generally altered as the body moves from one configuration to the other. Consequently, the size and shape of the body generally change under deformation. Material bodies whose size or shape change under deformation are called *deformable materials*. Material bodies that are not deformable are called *rigid materials*. All real materials are deformable to a certain degree, and the concept of rigid material is just an abstract one. Rigid materials are of little interest in continuum mechanics.

EXAMPLE 5.3.1 For the deformation defined by equations (5.2.23) find (i) the direction after deformation of a material arc $d\mathcal{C}_1$ initially having direction ratios $1:1:1$ and (ii) the direction before deformation of a material arc $d\mathcal{C}_2$ finally having direction ratios $1:1:1$.

5.3 STRETCH AND ROTATION

Solution For the given deformation, [**F**] and [**F**$^{-1}$] are as given by (5.2.24) and (5.2.27).

For the arc $d\mathcal{C}_1$ we have

$$a_1^0 = a_2^0 = a_3^0 = \frac{1}{\sqrt{3}} \qquad (5.3.12)$$

Using (5.2.24) and (5.3.12) in (5.3.3) we find

$$\eta a_1 = \frac{2}{\sqrt{3}} \qquad \eta a_2 = -\frac{1}{\sqrt{3}} \qquad \eta a_3 = \frac{1}{\sqrt{3}} \qquad (5.3.13)$$

Hence, after deformation, $d\mathcal{C}_1$ has direction ratios $2:-1:1$.

For the arc $d\mathcal{C}_2$ we have

$$a_1 = a_2 = a_3 = \frac{1}{\sqrt{3}} \qquad (5.3.14)$$

Using (5.2.27) and (5.3.14) in (5.3.3) we get

$$a_1^0 = \eta \frac{1}{\sqrt{3}} \qquad a_2^0 = 0 \qquad a_3^0 = 0 \qquad (5.3.15)$$

Hence, before deformation, $d\mathcal{C}_2$ had direction ratios $1:0:0$. That is, initially $d\mathcal{C}_2$ laid along the 1 axis. ∎

EXAMPLE 5.3.2 Deformations defined by equations of the form

$$x_i = a_{ij}x_j^0 + c_i \qquad (5.3.16)$$

where a_{ij} and c_i are constants or at most functions of time are called *homogeneous deformations*. Show that, under a homogeneous deformation, plane elements transform to plane elements and straight lines transform to straight lines.

Solution From equations (5.3.16) we find that [**F**] = $[a_{ij}]$. Since **F** has to be invertible, $[a_{ij}]$ has to be a nonsingular matrix and equations (5.3.16) have to be solvable for x_i^0 to get equations of the form

$$x_i^0 = a_{ij}^{-1}(x_j - c_j) \qquad (5.3.17)$$

where a_{ij}^{-1} are elements of the matrix [**F**$^{-1}$]; that is,

$$[a_{ij}^{-1}] = [a_{ij}]^{-1} = [\mathbf{F}]^{-1} \qquad (5.3.18)$$

In the initial configuration, let a material surface element be a plane element with equation

$$\alpha_i x_i^0 + \beta = 0 \qquad (5.3.19)$$

186   5   DEFORMATION

where α and β are constants. Using (5.3.17), this equation becomes

$$\alpha_i a_{ij}^{-1}(x_j - c_j) + \beta = 0 \tag{5.3.20}$$

This is the equation of the material surface element in the deformed configuration. Being a linear equation in x_i, equation (5.3.20) represents a plane. As such, the surface element continues to be a plane element after deformation also.

Thus, under a homogeneous deformation, a plane element transforms to a plane element. Consequently, since a straight line may be thought of as the intersection of two planes, it follows that straight line elements transform to straight line elements under homogeneous deformations. ∎

EXAMPLE 5.3.3 (i) Show that a homogeneous deformation defined by equations (5.3.16) is a rigid-body transformation if and only if $[a_{ij}]$ is an orthogonal matrix.

(ii) Deduce that the deformation defined by the equations

$$x_1 = x_1^0 \cos\theta + x_2^0 \sin\theta$$
$$x_2 = -x_1^0 \sin\theta + x_2^0 \cos\theta \tag{5.3.21}$$
$$x_3 = x_3^0$$

where θ is a constant, is a rigid-body rotation.

Solution (i) We recall that a deformation is a rigid-body transformation if and only if $\mathbf{F} = \mathbf{Q}$, or equivalently $[\mathbf{F}]$ is an orthogonal matrix. Since $[\mathbf{F}] = [a_{ij}]$ for a homogeneous deformation defined by (5.3.16), the required result is immediate.

(ii) We observe that equations (5.3.21) are of the form (5.3.16). Therefore, the given deformation is a homogeneous deformation. From (5.3.21) we find that

$$[\mathbf{F}] = \begin{pmatrix} \cos\theta & \sin\theta & 0 \\ -\sin\theta & \cos\theta & 0 \\ 0 & 0 & 1 \end{pmatrix} \tag{5.3.22}$$

It is easy to verify that this matrix is orthogonal. Hence, $\mathbf{F}$ is an orthogonal tensor and the given deformation is a rigid-body rotation. ∎

EXAMPLE 5.3.4 For the deformation defined by the following equations, find $\mathbf{Q}$, $\mathbf{U}$ and $\mathbf{V}$:

$$x_1 = 2x_3^0 \quad x_2 = -x_1^0 \quad x_3 = -2x_2^0 - 3x_3^0 \tag{5.3.23}$$

Give a geometrical description of the deformation.

5.3 STRETCH AND ROTATION

Solution From equations (5.3.23), we get

$$[\mathbf{F}] = \begin{bmatrix} 0 & 0 & 2 \\ -1 & 0 & 0 \\ 0 & -2 & -3 \end{bmatrix} \qquad (5.3.24)$$

In Example 2.14.2 the method of obtaining the polar decomposition of a tensor has been illustrated. Using that method, we find that for the deformation gradient given by (5.3.24) the matrices of the right and left stretch tensors **U** and **V** and the matrix of the rotation tensor **Q** are as follows:

$$[\mathbf{U}] = \begin{bmatrix} -1 & 0 & 0 \\ 0 & 0 & -2 \\ 0 & -2 & -3 \end{bmatrix} \qquad (5.3.25a)$$

$$[\mathbf{V}] = \begin{bmatrix} 0 & 0 & 2 \\ 0 & -1 & 0 \\ 2 & 0 & -3 \end{bmatrix} \qquad (5.3.25b)$$

$$[\mathbf{Q}] = \begin{bmatrix} 0 & -1 & 0 \\ 1 & 0 & 0 \\ 0 & 0 & 1 \end{bmatrix} \qquad (5.3.25c)$$

From (5.3.25a), we find that the eigenvalues of **U** are

$$\eta_1 = -1, \quad \eta_2 = 1, \quad \eta_3 = -4 \qquad (5.3.26)$$

These are the principal stretches for the given deformation.

It can be verified that the eigenvectors of **U** are as follows:

$$\mathbf{p}_1 = \mathbf{e}_1, \quad \mathbf{p}_2 = -\frac{2}{\sqrt{5}}\mathbf{e}_2 + \frac{1}{\sqrt{5}}\mathbf{e}_3, \quad \mathbf{p}_3 = \frac{1}{\sqrt{5}}\mathbf{e}_2 + \frac{2}{\sqrt{5}}\mathbf{e}_3 \qquad (5.3.27)$$

These vectors represent the principal directions of stretch for the given deformation.

Comparing the matrix of **Q** given by (5.3.25c) with the transformation matrix (2.2.13), we find that **Q** represents a (rigid) rotation about the 3 axis through an angle $3\pi/2$ in the sense of a righthanded screw.

Thus, the given deformation may be thought of as the one in which a material arc experiences triaxial stretches -1, 1 and -4, respectively, along the directions of the vectors $\mathbf{p}_1$, $\mathbf{p}_2$ and $\mathbf{p}_3$ given by (5.3.27) followed by a

5.4
STRAIN TENSORS

In the preceding section it was shown that, under a deformation, a material arc generally experiences a change in length and a change in orientation. We now proceed to derive expressions that will enable us to compute these changes.

Let us start with expressions (5.3.3) and (5.3.4):

$$\eta \mathbf{a} = \mathbf{F}\mathbf{a}^0 \tag{5.4.1}$$

$$\eta = \frac{ds}{ds_0} \tag{5.4.2}$$

From (5.4.1), we find that

$$\eta^2 = \eta\mathbf{a} \cdot \eta\mathbf{a} = \mathbf{F}\mathbf{a}^0 \cdot \mathbf{F}\mathbf{a}^0 = \mathbf{a}^0 \cdot (\mathbf{F}^T\mathbf{F})\mathbf{a}^0 \tag{5.4.3}$$

by use of (2.8.14). If we set

$$\mathbf{C} = \mathbf{F}^T\mathbf{F}, \tag{5.4.4}$$

expression (5.4.3) becomes

$$\eta^2 = \mathbf{a}^0 \cdot \mathbf{C}\mathbf{a}^0 \tag{5.4.5}$$

From (5.4.4), (5.4.5) and (5.3.6a) we find that $\mathbf{C}$ is a *positive definite symmetric tensor* and that $\mathbf{C} = \mathbf{U}^2$. Expression (5.4.5) shows that the tensor $\mathbf{C}$ serves as a means of calculating η when $\mathbf{a}^0$ is known. This tensor was introduced by George Green in 1841 and is known as the *Green deformation tensor*.

For a rigid-body transformation, we have $\mathbf{F} = \mathbf{Q}$. Expression (5.4.4) then becomes $\mathbf{C} = \mathbf{Q}^T\mathbf{Q} = \mathbf{I}$. Thus, for a rigid-body transformation $\mathbf{C}$ is the identity tensor. The converse is also true; because when $\mathbf{C} = \mathbf{I}$, $\mathbf{F}$ becomes an orthogonal tensor that effects only a pure rotation possibly combined with a translation.

Recalling that the components of $\mathbf{F}$ are $x_{i;j}$, we find from (5.4.4) that the components c_{ij} of $\mathbf{C}$ are

$$c_{ij} = x_{k;i} x_{k;j} \tag{5.4.6}$$

5.4 STRAIN TENSORS

Consequently, (5.4.5) reads as follows in the suffix notation:

$$\eta^2 = c_{ij}a_i^0 a_j^0 = x_{k;i}x_{k;j}a_i^0 a_j^0 \tag{5.4.7}$$

For a deformation given in the material description, the numbers c_{ij} can be computed from (5.4.6). The stretch of any material arc whose initial orientation is known can then be determined by the use of (5.4.7).

5.4.1 NORMAL STRAIN

In many discussions, we will be interested in knowing the changes in the lengths and relative orientations of material arcs rather than their stretches. Here we introduce a tensor that enables us to compute these changes.

To this end, we first define a number e as follows:

$$e = \eta - 1 = \frac{ds - ds_0}{ds_0} \tag{5.4.8}$$

Evidently, e represents the change in length per unit initial length of a material arc; this number is referred to as the *normal strain* of the arc. We note that e is positive or negative depending on whether the element experiences an *extension* or a *contraction*. Also, $e = 0$ if and only if the arc retains its length (during a deformation).

From (5.4.5) and (5.4.8), we obtain

$$1 + e = (\eta^2)^{1/2} = (\mathbf{a}^0 \cdot \mathbf{C}\mathbf{a}^0)^{1/2}$$

$$= \{1 + \mathbf{a}^0 \cdot (\mathbf{C} - \mathbf{I})\mathbf{a}^0\}^{1/2} \tag{5.4.9}$$

If we set

$$\mathbf{G} = \tfrac{1}{2}(\mathbf{C} - \mathbf{I}) = \tfrac{1}{2}(\mathbf{F}^T\mathbf{F} - \mathbf{I}) \tag{5.4.10}$$

then (5.4.9) becomes

$$e = \{1 + 2\mathbf{a}^0 \cdot \mathbf{G}\mathbf{a}^0\}^{1/2} - 1 \tag{5.4.11}$$

From (5.4.10) we note that $\mathbf{G}$ is a *symmetric tensor* and that $\mathbf{G} = \mathbf{0}$ if and only if the deformation is a rigid-body transformation. Expression (5.4.11) shows that $\mathbf{G}$ serves as a means of calculating e for any known $\mathbf{a}^0$. The tensor $\mathbf{G}$ was introduced by Green in 1841 and by B. Saint-Venant in 1844, and it is known as the *Green strain tensor*.

By using (5.4.6) and (5.4.10), we obtain the following expression for the components g_{ij} of $\mathbf{G}$:

$$g_{ij} = \tfrac{1}{2}(c_{ij} - \delta_{ij}) = \tfrac{1}{2}(x_{k;i}x_{k;j} - \delta_{ij}) \tag{5.4.12}$$

190 5 DEFORMATION

Consequently, (5.4.11) reads as follows in the suffix notation:

$$e = \{1 + 2g_{ij}a_i^0 a_j^0\}^{1/2} - 1 \tag{5.4.13}$$

For a deformation given in the material description, the numbers g_{ij} can be computed from (5.4.12). The normal strain of any material arc whose initial orientation is known can then be determined from (5.4.13).

If e is so small that e^2 can be neglected, then (5.4.13) becomes

$$e \cong g_{ij}a_i^0 a_j^0 \tag{5.4.14}$$

Thus, to the first-order approximation, $g_{ij}a_i^0 a_j^0$ represents the normal strain of a material arc that was initially directed along the unit vector $\mathbf{a}^0$.

In the particular case when $\mathbf{a}^0$ is aligned with $\mathbf{e}_1$, we have $a_1^0 = 1$, $a_2^0 = a_3^0 = 0$. Expressions (5.4.7) and (5.4.14) then reduce to $\eta^2 = c_{11}$ and $e \approx g_{11}$. Thus, the component c_{11} of $\mathbf{C}$ represents the square of the stretch of a material arc that was initially directed along the 1 axis, and to the first-approximation, the component g_{11} of $\mathbf{G}$ represents the normal strain of this arc. The components c_{22} and c_{33} of $\mathbf{C}$ and g_{22} and g_{33} of $\mathbf{G}$ have similar geometrical meanings.

5.4.2 SHEAR STRAIN

In order to analyze the changes in the relative orientations of material arcs, let us consider another material arc $d\bar{\mathcal{C}}$ initiating from the particle $\mathcal{P}$ and directed along the unit vectors $\bar{\mathbf{a}}^0$ and $\bar{\mathbf{a}}$ in the initial and final configurations respectively. If $\bar{\eta}$ is the stretch and $\bar{e}$ is the normal strain of this arc, then by virtue of the relations (5.4.1), (5.4.7), (5.4.8) and (5.4.13), we have the following expressions:

$$\bar{\eta}\bar{\mathbf{a}} = \mathbf{F}\bar{\mathbf{a}}^0 \tag{5.4.1}'$$

$$\bar{\eta}^2 = c_{ij}\bar{a}_i^0 \bar{a}_j^0 \tag{5.4.7}'$$

$$\bar{\eta} = 1 + \bar{e} \tag{5.4.8}'$$

$$\bar{e} = \{1 + 2g_{ij}\bar{a}_i^0 \bar{a}_j^0\}^{1/2} - 1 \tag{5.4.13}'$$

If θ_0 is the angle between $d\mathcal{C}$ and $d\bar{\mathcal{C}}$ in the initial configuration and θ is the angle between them in the final configuration, we have (see Figure 5.10)

$$\cos \theta_0 = \mathbf{a}^0 \cdot \bar{\mathbf{a}}^0 = a_i^0 \bar{a}_i^0 \tag{5.4.15a}$$

$$\cos \theta = \mathbf{a} \cdot \bar{\mathbf{a}} = a_i \bar{a}_i \tag{5.4.15b}$$

5.4 STRAIN TENSORS

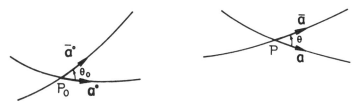

Figure 5.10. Initial and final orientations of material arcs.

From (5.4.15b), (5.4.1), (5.4.1)' and (5.4.4) we obtain

$$\eta\bar{\eta}\cos\theta = \eta\mathbf{a}\cdot\bar{\eta}\bar{\mathbf{a}}$$
$$= \mathbf{Fa}^0 \cdot \mathbf{F}\bar{\mathbf{a}}^0$$
$$= \mathbf{a}^0 \cdot \mathbf{F}^T\mathbf{F}\bar{\mathbf{a}}^0$$
$$= \mathbf{a}^0 \cdot \mathbf{C}\bar{\mathbf{a}}^0 = c_{ij}a_i^0\bar{a}_j^0 \quad (5.4.16)$$

Expressions (5.4.7), (5.4.7)' and (5.4.16) show that θ is completely determined by c_{ij}. Thus, the deformation components c_{ij} not only determine the stretch of a material arc but also the final angle between two material arcs whose initial orientations are known.

Using (5.4.12) and (5.4.15a), expression (5.4.16) can be rewritten in terms of g_{ij} as follows:

$$\eta\bar{\eta}\cos\theta - \cos\theta_0 = 2g_{ij}a_i^0\bar{a}_j^0 \quad (5.4.17)$$

Evidently, θ and θ_0 are different from one another in general. This means that the relative orientations of material arcs generally change under a deformation. If we set

$$2\gamma = \theta_0 - \theta \quad (5.4.18)$$

then γ represents one-half the *reduction* in the angle between the arcs $d\mathcal{C}$ and $d\bar{\mathcal{C}}$ due to deformation. The number γ is called the *shear strain* between these arcs.

If $\theta_0 = \pi/2$, expression (5.4.17) becomes, on using (5.4.8), (5.4.8)' and (5.4.18),

$$(1 + e)(1 + \bar{e})\sin 2\gamma = 2g_{ij}a_i^0\bar{a}_j^0 \quad (5.4.19)$$

Expressions (5.4.13), (5.4.13)' and (5.4.19) show that γ is completely determined by g_{ij}. Thus, the strain components g_{ij} not only determine the normal strain of a material arc but also the shear strain between two initially orthogonal arcs.

If γ, e and $\bar{e}$ are so small that nonlinear terms involving these can be neglected, then (5.4.19) reduces to

$$\gamma = g_{ij}a_i^0\bar{a}_j^0 \quad (5.4.20)$$

192   **5 DEFORMATION**

Thus, to the first-order approximation, $g_{ij}a_i^0\bar{a}_j^0$ represents the shear strain between the material arcs that were initially directed along two orthogonal unit vectors $\mathbf{a}^0$ and $\bar{\mathbf{a}}^0$.

In the particular case when $\mathbf{a}^0$ is aligned with $\mathbf{e}_1$ and $\bar{\mathbf{a}}^0$ is aligned with $\mathbf{e}_2$, we have $a_1^0 = 1$, $a_2^0 = a_3^0 = 0$ and $\bar{a}_2^0 = 1$, $\bar{a}_1^0 = \bar{a}_3^0 = 0$. Then, together with (5.4.7) and (5.4.7)′, expression (5.4.16) yields $c_{12} = \sqrt{c_{11}c_{22}}\cos\theta$, and (5.4.20) yields $g_{12} = \gamma$. Thus, c_{12} represents the product of the stretches suffered by the material arcs that were initially directed along the 1 and 2 axes and the cosine of their relative orientation in the final configuration. Also, to the first-order approximation, g_{12} represents the shear strain between these arcs. Other off-diagonal components of $\mathbf{C}$ and $\mathbf{G}$ have similar geometrical meanings.

Thus, the components of the two tensors $\mathbf{C}$ and $\mathbf{G}$, whose knowledge enables us to compute the changes in length and relative orientation of material arcs, have definite geometrical meanings. While the diagonal components of both the tensors are associated with the changes in length of material arcs, which were initially laying along the coordinate directions, the off-diagonal components are associated with the changes in their relative orientations.

5.4.3 EXPRESSIONS IN SPATIAL FORM

We observe that (5.4.5) and (5.4.13), which give the stretch and the normal strain of a material arc, and (5.4.16) and (5.4.17), which give the final angle and the change in the angle between two material arcs, are all expressed in terms of the *initial* orientations of the arcs. Here we deduce their counterparts expressed in terms of the *final* orientations of the arcs.

From (5.4.1), we obtain

$$\mathbf{a}^0 = \eta \mathbf{F}^{-1}\mathbf{a} \qquad (5.4.21)$$

from which it follows that

$$\frac{1}{\eta^2} = \frac{1}{\eta^2}\mathbf{a}^0 \cdot \mathbf{a}^0 = \mathbf{F}^{-1}\mathbf{a} \cdot \mathbf{F}^{-1}\mathbf{a}$$

$$= \mathbf{a} \cdot (\mathbf{F}^{-1})^T(\mathbf{F}^{-1}\mathbf{a})$$

$$= \mathbf{a} \cdot (\mathbf{F}\mathbf{F}^T)^{-1}\mathbf{a} \qquad (5.4.22)$$

If we set

$$\mathbf{B} = \mathbf{F}\mathbf{F}^T \qquad (5.4.23)$$

then (5.4.22) becomes

$$\frac{1}{\eta^2} = \mathbf{a} \cdot \mathbf{B}^{-1}\mathbf{a}. \qquad (5.4.24)$$

5.4 STRAIN TENSORS

From (5.4.23), (5.4.24) and (5.3.6b) we find that **B** is a *positive definite symmetric tensor* and that $\mathbf{B} = \mathbf{V}^2$. Expression (5.4.24) shows that the tensor $\mathbf{B}^{-1}$ serves as a means of calculating η when **a** is known. It can be easily shown that for a rigid-body transformation, $\mathbf{B} = \mathbf{B}^{-1} = \mathbf{I}$, and conversely. The tensor $\mathbf{B}^{-1}$ was introduced by Cauchy in 1827 and is known as the *Cauchy deformation tensor*. The tensor **B** was introduced by Finger in 1894 and is known as the *Finger deformation tensor*. However, **C** and **B** are commonly referred to as the *right* and *left Cauchy–Green deformation tensors* in view of their direct connection with the right and left polar decompositions of **F**.

Recalling that $x_{i;j}$ and $x^0_{i,j}$ are components of **F** and $\mathbf{F}^{-1}$, respectively, we find from (5.4.23) that the components b_{ij} of **B** and the components b_{ij}^{-1} of $\mathbf{B}^{-1}$ are given by

$$b_{ij} = x_{i;k} x_{j;k} \tag{5.4.25a}$$

$$b_{ij}^{-1} = x^0_{k,i} x^0_{k,j} \tag{5.4.25b}$$

Consequently, (5.4.24) reads as follows in the suffix notation:

$$\frac{1}{\eta^2} = b_{ij}^{-1} a_i a_j = x^0_{k,i} x^0_{k,j} a_i a_j \tag{5.4.26}$$

For a deformation given in spatial description, the numbers b_{ij}^{-1} can be computed from (5.4.25b). The stretch of any material arc whose final orientation is known can be determined by the use of (5.4.26).

From (5.4.8) and (5.4.24) we obtain

$$1 + e = (\eta^2)^{1/2} = (\mathbf{a} \cdot \mathbf{B}^{-1} \mathbf{a})^{-1/2}$$
$$= \{1 - \mathbf{a} \cdot (\mathbf{I} - \mathbf{B}^{-1}) \mathbf{a}\}^{-1/2} \tag{5.4.27}$$

Noting from (5.4.23) that

$$\mathbf{B}^{-1} = (\mathbf{F}\mathbf{F}^T)^{-1} = (\mathbf{F}^{-1})^T \mathbf{F}^{-1}$$

and setting

$$\mathbf{A} = \tfrac{1}{2}(\mathbf{I} - \mathbf{B}^{-1}) = \tfrac{1}{2}\{\mathbf{I} - (\mathbf{F}^{-1})^T \mathbf{F}^{-1}\} \tag{5.4.28}$$

we find from (5.4.27) that

$$e = (1 - 2\mathbf{a} \cdot \mathbf{A}\mathbf{a})^{-1/2} - 1 \tag{5.4.29}$$

From (5.4.28) we note that **A** is a *symmetric tensor* and that $\mathbf{A} = \mathbf{0}$ if and only if the deformation is a rigid-body transformation. Expression (5.4.29) shows that **A** serves as a means of computing e for any known **a**. The tensor **A** was introduced by E. Almansi in 1911 and by G. Hamel in 1912 and is known as *Almansi strain tensor*.

194 **5 DEFORMATION**

By using (5.4.25b) and (5.4.28) we obtain the following expression for the components a_{ij} of **A**:

$$a_{ij} = \tfrac{1}{2}(\delta_{ij} - b_{ij}^{-1}) = \tfrac{1}{2}(\delta_{ij} - x_{k,i}^0 x_{k,j}^0) \qquad (5.4.30)$$

Consequently, (5.4.29) reads as follows in the suffix notation:

$$e = (1 - 2a_{ij}a_i a_j)^{-1/2} - 1 \qquad (5.4.31)$$

For a deformation given in spatial description the numbers a_{ij} can be computed from (5.4.30). The normal strain of any material arc whose final orientation is known can be determined from (5.4.31).

If e is so small that e^2 can be neglected, (5.4.31) becomes

$$e \approx a_{ij}a_i a_j \qquad (5.4.32)$$

Thus, to the first-order approximation, $a_{ij}a_i a_j$ represents the normal strain of a material arc that is finally directed along the unit vector **a**.

Computations analogous to those employed in obtaining (5.4.16) and (5.4.17) lead to the following expressions:

$$\cos\theta_0 = \eta\bar{\eta}b_{ij}^{-1}a_i\bar{a}_j \qquad (5.4.33)$$

$$\cos\theta - \frac{1}{\eta\bar{\eta}}\cos\theta_0 = 2a_{ij}a_i\bar{a}_j \qquad (5.4.34)$$

When $\theta = \pi/2$, (5.4.34) becomes, on using (5.4.8), (5.4.8)' and (5.4.18),

$$(1 + e)^{-1}(1 + \bar{e})^{-1}\sin 2\gamma = 2a_{ij}a_i\bar{a}_j \qquad (5.4.35)$$

Expressions (5.4.31) and (5.4.35) show that γ is completely determined by a_{ij}. Thus, the strain components a_{ij} not only determine the normal strain of a material arc but also determine the shear strain between two finally orthogonal material arcs.

If γ, e and $\bar{e}$ are so small that nonlinear terms involving these can be ignored, then (5.4.35) reduces to the following form:

$$\gamma \approx a_{ij}a_i\bar{a}_j \qquad (5.4.36)$$

Thus, to the first-order approximation, $a_{ij}a_i\bar{a}_j$ represents the shear strain between the material arcs that are finally directed along two orthogonal unit vectors **a** and **ā**.

By using (5.4.32) and (5.4.36), it can be verified that, to the first-order approximation, a_{11} represents the normal strain of a material arc that is finally oriented along the 1 direction, and a_{12} represents the shear strain between two material arcs that are finally oriented along the 1 and 2 directions. Other components of **A** have similar geometrical meanings.

5.4 STRAIN TENSORS

Thus, the diagonal components of **A** are associated with the normal strains of material arcs that are finally directed along the coordinate directions and the off-diagonal components of **A** are associated with the shear strains between these arcs.

It is to be noted that the tensors **C** and **G** are defined in terms of the material deformation gradient **F**. As such, **C** and **G** are tensors associated with the material (Lagrangian) description of deformation. For this reason, **C** and **G** are often referred to as the *Lagrangian deformation tensor* and *Lagrangian strain tensor*, respectively. On the other hand, the tensors $\mathbf{B}^{-1}$ and **A** are defined in terms of the spatial deformation gradient $\mathbf{F}^{-1}$. As such, $\mathbf{B}^{-1}$ and **A** are tensors associated with the spatial (Eulerian) description of deformation. Therefore, $\mathbf{B}^{-1}$ and **A** are often referred to as the *Eulerian deformation tensor* and the *Eulerian strain tensor*, respectively.

EXAMPLE 5.4.1 For the deformation considered in Example 5.2.1, find the tensors **C** and **G**. Hence find (i) the stretch and the normal strain of a material arc $d\mathcal{C}$ whose initial direction ratios are $1:1:1$, and (ii) the change in the angle between this arc and the material arc $d\bar{\mathcal{C}}$ whose initial direction ratios are $-1:2:-1$.

Solution For the given deformation, [**F**] is given by (5.2.24), from which we find

$$[c_{ij}] = [\mathbf{C}] = [\mathbf{F}^T\mathbf{F}] = \begin{bmatrix} 3 & 0 & -1 \\ 0 & 6 & -1 \\ -1 & -1 & 1 \end{bmatrix} \quad (5.4.37)$$

Hence

$$[g_{ij}] = [\mathbf{G}] = \tfrac{1}{2}[\mathbf{C} - \mathbf{I}] = \begin{bmatrix} 1 & 0 & -\tfrac{1}{2} \\ 0 & \tfrac{5}{2} & -\tfrac{1}{2} \\ -\tfrac{1}{2} & -\tfrac{1}{2} & 0 \end{bmatrix} \quad (5.4.38)$$

For the given deformation, (5.4.37) and (5.4.38) give the matrices of the tensors **C** and **G**, respectively.

For the material arc $d\mathcal{C}$, we have

$$a_1^0 = a_2^0 = a_3^0 = \frac{1}{\sqrt{3}} \quad (5.4.39)$$

Using these a_i^0 and c_{ij} given by (5.4.37) in (5.4.7), we get $\eta = \sqrt{2} \approx 1.4142$. This is the stretch of the arc $d\mathcal{C}$. Its normal strain is $e = \eta - 1 \approx 0.4142$. Since $e > 0$, $d\mathcal{C}$ experiences an extension.

196 5 DEFORMATION

For the material arc $d\bar{\mathcal{C}}$, we have

$$\bar{a}_1^0 = -\frac{1}{\sqrt{6}}, \quad \bar{a}_2^0 = \frac{2}{\sqrt{6}}, \quad \bar{a}_3^0 = -\frac{1}{\sqrt{6}} \qquad (5.4.40)$$

Using these $\bar{a}_i^0$ and c_{ij} given by (5.4.37), we get the stretch of $d\bar{\mathcal{C}}$ as $\bar{\eta} = \sqrt{5} \approx 2.2360$ and the normal strain of $d\bar{\mathcal{C}}$ as $\bar{e} = \bar{\eta} - 1 \approx 1.2360$.

From (5.4.39) and (5.4.40), we find that $\mathbf{a}^0 \cdot \bar{\mathbf{a}}^0 = 0$. Therefore, $d\mathcal{C}$ and $d\bar{\mathcal{C}}$ are initially orthogonal. Using the values $\eta = \sqrt{2}$ and $\bar{\eta} = \sqrt{5}$ obtained previously and g_{ij} given by (5.4.38) in (5.4.17), we find that $\cos\theta \approx 0.6708$, or $\theta \approx 47.9°$. This is the final angle between $d\mathcal{C}$ and $d\bar{\mathcal{C}}$. Thus, the deformation causes a decrease of about 42.1° in the angle between $d\mathcal{C}$ and $d\bar{\mathcal{C}}$. ■

EXAMPLE 5.4.2 For the deformation considered in Example 5.2.2, find the tensor $\mathbf{B}^{-1}$. Show that one of the principal stretches is equal to 1 and that the squares of the other two principal stretches are the two roots of the quadratic equation

$$x^2 - \left(R^2 + \frac{1}{R^2}\right)x + 1 = 0 \qquad (5.4.41)$$

Deduce that $\det \mathbf{B}^{-1} = 1$.

Solution For the given deformation, $\mathbf{F}^{-1}$ is given by (5.2.29). Using this in (5.4.23) we obtain

$$[\mathbf{B}^{-1}] = [\mathbf{F}^{-1}]^T[\mathbf{F}^{-1}] = \begin{bmatrix} x_1^2 + \left(\dfrac{x_2^2}{R^4}\right) & \left(1 - \dfrac{1}{R^4}\right)x_1 x_2 & 0 \\ \left(1 - \dfrac{1}{R^4}\right)x_1 x_2 & x_2^2 + \left(\dfrac{x_1^2}{R^4}\right) & 0 \\ 0 & 0 & 1 \end{bmatrix} \qquad (5.4.42)$$

This determines the tensor $\mathbf{B}^{-1}$.

The principal stretches are, by definition, the eigenvalues of $\mathbf{V} = \sqrt{\mathbf{B}}$. Hence, if $\eta = \sqrt{\Lambda}$ is a principal stretch, then Λ is an eigenvalue of $\mathbf{B}$ so that $1/\Lambda$ is an eigenvalue of $\mathbf{B}^{-1}$. Thus, in the present case, $1/\Lambda$ must be a root of the following characteristic equation:

5.4 STRAIN TENSORS

$$\begin{vmatrix} x_1^2 + \left(\dfrac{x_2^2}{R^4}\right) - \dfrac{1}{\Lambda} & \left(1 - \dfrac{1}{R^4}\right)x_1 x_2 & 0 \\ \left(1 - \dfrac{1}{R^4}\right)x_1 x_2 & x_2^2 + \left(\dfrac{x_1^2}{R^4}\right) - \dfrac{1}{\Lambda} & 0 \\ 0 & 0 & 1 - \dfrac{1}{\Lambda} \end{vmatrix} = 0$$

Simplifying this equation, we find that Λ must satisfy the following equation:

$$(\Lambda - 1)\left[\Lambda^2 - \left(R^2 + \dfrac{1}{R^2}\right)\Lambda + 1\right] = 0 \qquad (5.4.43)$$

It is evident that one of the principal stretches is $\sqrt{\Lambda_1} = 1$ and that the squares of the other two principal stretches, Λ_2 and Λ_3, are the roots of the equation (5.4.41).

From equation (5.4.41) we note that $\Lambda_2 \Lambda_3 = 1$, so that $\Lambda_1 \Lambda_2 \Lambda_3 = 1$. Hence

$$\det(\mathbf{B}^{-1}) = (\Lambda_1 \Lambda_2 \Lambda_3)^{-1} = 1 \quad \blacksquare$$

EXAMPLE 5.4.3 Show that

$$(ds)^2 - (ds_0)^2 = 2g_{ij}\, dx_i^0\, dx_j^0 \qquad (5.4.44)$$

$$= 2a_{ij}\, dx_i\, dx_j \qquad (5.4.45)$$

Solution Recalling from (5.3.2) and (5.4.8) that

$$dx_i^0 = (ds_0)a_i^0, \qquad dx_i = (ds)a_i, \qquad 1 + e = \eta = \dfrac{ds}{ds_0} \qquad (5.4.46)$$

we find by use of (5.4.13) and (5.4.31) that

(i) $$(ds)^2 - (ds_0)^2 = (ds_0)^2\left[\left(\dfrac{ds}{ds_0}\right)^2 - 1\right]$$

$$= (ds_0)^2[(1 + e)^2 - 1]$$

$$= 2(ds_0)^2 g_{ij} a_i^0 a_j^0$$

$$= 2g_{ij}\, dx_i^0\, dx_j^0$$

and

(ii)
$$(ds)^2 - (ds_0)^2 = (ds)^2\left[1 - \left(\frac{ds_0}{ds}\right)^2\right]$$
$$= (ds)^2[1 - (1 + e)^{-2}]$$
$$= 2(ds)^2 a_{ij} a_i a_j$$
$$= 2a_{ij}\, dx_i\, dx_j$$

Thus, (5.4.44) and (5.4.45) are proven. ∎

Note: When g_{ij} are known, (5.4.44) gives the change in the square of the length of a material arc that was initially located at $\mathbf{x}^0$. When a_{ij} are known, (5.4.45) gives the change in the square of the length of a material arc that is finally located at $\mathbf{x}$. Some authors use (5.4.44) and (5.4.45) for defining the Lagrangian and Eulerian strain tensors, respectively.

EXAMPLE 5.4.4 Show that $\mathbf{B}$ and $\mathbf{C}$ have the same principal invariants and that

$$III_\mathbf{B} = III_\mathbf{C} = J^2, \qquad II_\mathbf{B} = II_\mathbf{C} = J^2(\mathrm{tr}\, \mathbf{C}^{-1}) \qquad (5.4.47)$$

Solution Recall that $\mathbf{U}$ and $\mathbf{V}$ have the same eigenvalues, which are the principal stretches, and $\mathbf{U}^2 = \mathbf{C}$ and $\mathbf{V}^2 = \mathbf{B}$. Hence, if η_1, η_2, η_3 are the principal stretches, then η_1^2, η_2^2 and η_3^2 are the eigenvalues of both $\mathbf{B}$ and $\mathbf{C}$, and η_1^{-2}, η_2^{-2} and η_3^{-2} are the eigenvalues of both $\mathbf{B}^{-1}$ and $\mathbf{C}^{-1}$. Consequently, by use of (2.13.14), we obtain

$$I_\mathbf{B} = I_\mathbf{C} = \eta_1^2 + \eta_2^2 + \eta_3^2$$
$$II_\mathbf{B} = II_\mathbf{C} = \eta_1^2\eta_2^2 + \eta_2^2\eta_3^2 + \eta_3^2\eta_1^2 \qquad (5.4.48)$$
$$III_\mathbf{B} = III_\mathbf{C} = \eta_1^2\eta_2^2\eta_3^2$$

which show that $\mathbf{B}$ and $\mathbf{C}$ have the same principal invariants. Also,

$$\mathrm{tr}\, \mathbf{C}^{-1} = I_{\mathbf{C}^{-1}} = \eta_1^{-2} + \eta_2^{-2} + \eta_3^{-2}$$
$$= \eta_1^{-2}\eta_2^{-2}\eta_3^{-2}(\eta_1^2\eta_2^2 + \eta_2^2\eta_3^2 + \eta_3^2\eta_1^2)$$
$$= \frac{1}{III_\mathbf{C}} II_\mathbf{C} \qquad (5.4.49)$$

Further,

$$III_\mathbf{C} = \det \mathbf{C} = \det(\mathbf{F}^T\mathbf{F}) = (\det \mathbf{F})^2 = J^2 \qquad (5.4.50)$$

Expressions (5.4.48), (5.4.49) and (5.4.50) yield (5.4.47). ∎

5.4 STRAIN TENSORS

EXAMPLE 5.4.5 Show that the stretch of a material arc element $d\mathcal{C}$ assumes an extreme value when $d\mathcal{C}$ initially lies along a principal direction of **C**.

Solution In terms of its initial orientation, the stretch η of $d\mathcal{C}$ is given by (5.4.7). We have to find a_i^0 for which η is an extremum. Since a_i^0 are components of a unit vector we have $a_k^0 a_k^0 = 1$. Thus, η^2 assumes extreme values when a_i^0 obey the equations

$$\frac{\partial}{\partial a_i^0}\{\eta^2 - \alpha(a_k^0 a_k^0 - 1)\} = 0 \tag{5.4.51}$$

where α is a Lagrangian multiplier.

Substituting for η^2 from (5.4.7) in (5.4.51) and noting that $(\partial a_i^0/\partial a_j^0) = \delta_{ij}$, we obtain the following conditions on a_i^0 under which η^2 is an extremum:

$$(c_{ij} - \alpha\delta_{ij})a_j^0 = 0 \tag{5.4.52}$$

These conditions are satisfied iff $\mathbf{a}^0$ is an eigenvector of **C**. Thus, η is an extremum when $d\mathcal{C}$ initially lies along a principal direction of **C**.

EXAMPLE 5.4.6 Consider a material surface element $d\mathcal{S}$ that is initially orthogonal to a unit vector $\mathbf{n}^0$. After deformation, suppose that it becomes orthogonal to a unit vector $\mathbf{n}$. If dS_0 and dS are the initial and final areas of $d\mathcal{S}$, prove that

$$\frac{(dS)^2}{(dS_0)^2} = J^2 \mathbf{n}^0 \cdot (\mathbf{C}^{-1}\mathbf{n}^0) = J^2\{\mathbf{n} \cdot \mathbf{Bn}\}^{-1} \tag{5.4.53}$$

Solution From (5.2.12), let us recall that $\mathbf{dS}_0$ and $d\mathbf{S}$ are related as follows:

$$(dS)\mathbf{n} = J(dS_0)(\mathbf{F}^T)^{-1}\mathbf{n}^0 \tag{5.4.54}$$

Since **n** is a unit vector, (5.4.54) yields

$$(dS)^2 = (dS)\mathbf{n} \cdot (dS)\mathbf{n}$$
$$= J^2(dS_0)^2\{(\mathbf{F}^{-1})^T\mathbf{n}^0 \cdot (\mathbf{F}^{-1})^T\mathbf{n}^0\}$$
$$= J^2(dS_0)^2\{\mathbf{n}^0 \cdot (\mathbf{F}^T\mathbf{F})^{-1}\mathbf{n}^0\} \tag{5.4.55}$$

which on using (5.4.4) becomes

$$\frac{(dS)^2}{(dS_0)^2} = J^2\mathbf{n}^0 \cdot (\mathbf{C}^{-1}\mathbf{n}^0) \tag{5.4.56}$$

200 5 DEFORMATION

By use of (5.4.54), we also get

$$J^2(dS_0)^2 = J(dS_0)\mathbf{n}^0 \cdot J(dS_0)\mathbf{n}^0 = (dS)^2(\mathbf{F}^T\mathbf{n}) \cdot (\mathbf{F}^T\mathbf{n})$$
$$= (dS)^2\mathbf{n} \cdot (\mathbf{FF}^T\mathbf{n}) \tag{5.4.57}$$

which on using (5.4.23) becomes

$$\frac{(dS)^2}{(dS_0)^2} = J^2(\mathbf{n} \cdot \mathbf{Bn})^{-1} \tag{5.4.58}$$

Expressions (5.4.56) and (5.4.58) prove (5.4.53). ∎

The results (5.4.54) and (5.4.53) are by E. J. Nanson (1878), known as *Nanson's formulae*.

5.5
STRAIN-DISPLACEMENT RELATIONS

As in Section 5.2, let P_0 be the initial position and P be the final position of a particle $\mathcal{P}$. Then the directed line segment $\overrightarrow{P_0P}$ is called the *displacement vector* of $\mathcal{P}$ from the initial configuration to the final configuration. We denote it by $\mathbf{u}$. If $\mathbf{x}^0$ is the position vector of P_0 and $\mathbf{x}$ that of P, then we have (see Figure 5.11)

$$\mathbf{u} = \overrightarrow{P_0P} = \mathbf{x} - \mathbf{x}^0 \tag{5.5.1}$$

In the material description, $\mathbf{x}$ is a function of $\mathbf{x}^0$ and t. Hence, in this description, $\mathbf{u}$ is a function of $\mathbf{x}^0$ and t so that (5.5.1) stands for

$$\mathbf{u}(\mathbf{x}^0, t) = \mathbf{x}(\mathbf{x}^0, t) - \mathbf{x}^0 \tag{5.5.2}$$

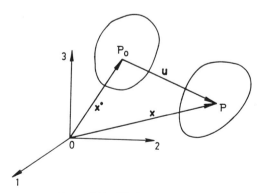

Figure 5.11. Displacement vector.

5.5 STRAIN-DISPLACEMENT RELATIONS

This expression specifies the displacement of the particle $\mathcal{P}$ at time t in terms of its initial position $\mathbf{x}^0$.

In the spatial description, $\mathbf{x}^0$ is a function of $\mathbf{x}$ and t. Hence, in this description, $\mathbf{u}$ is a function of $\mathbf{x}$ and t so that (5.5.1) now stands for

$$\mathbf{u}(\mathbf{x}, t) = \mathbf{x} - \mathbf{x}^0(\mathbf{x}, t) \tag{5.5.3}$$

This expression specifies the displacement of the particle $\mathcal{P}$ at time t in terms of its spatial position $\mathbf{x}$. In this case, $\mathbf{u}$ is referred to as the *displacement at the point* $\mathbf{x}$.

We now proceed to deduce relations connecting the deformation and strain tensors with the displacement vector. We find from (5.5.2) that

$$\nabla^0 \mathbf{u} = \nabla^0 \mathbf{x} - \nabla^0 \mathbf{x}^0 = \mathbf{F} - \mathbf{I}. \tag{5.5.4}$$

The tensor $\nabla^0 \mathbf{u}$ is referred to as the *displacement-gradient tensor in the material description*, or briefly the *material displacement-gradient*.

From (5.5.4) and (5.4.4) we get

$$\mathbf{C} = \mathbf{F}^T \mathbf{F} = (\nabla^0 \mathbf{u}^T + \mathbf{I})(\nabla^0 \mathbf{u} + \mathbf{I})$$
$$= \mathbf{I} + \nabla^0 \mathbf{u} + \nabla^0 \mathbf{u}^T + (\nabla^0 \mathbf{u}^T)(\nabla^0 \mathbf{u}) \tag{5.5.5}$$

Consequently, (5.4.10) yields

$$\mathbf{G} = \tfrac{1}{2}(\mathbf{C} - \mathbf{I}) = \tfrac{1}{2}[\nabla^0 \mathbf{u} + \nabla^0 \mathbf{u}^T + (\nabla^0 \mathbf{u}^T)(\nabla^0 \mathbf{u})] \tag{5.5.6}$$

This is called the *strain-displacement relation in the material description*.

In the suffix notation, (5.5.4) to (5.5.6) read respectively as follows:

$$u_{i;j} = x_{i;j} - \delta_{ij} \tag{5.5.4}'$$

$$c_{ij} = \delta_{ij} + u_{i;j} + u_{j;i} + u_{k;i} u_{k;j} \tag{5.5.5}'$$

$$g_{ij} = \tfrac{1}{2}(u_{i;j} + u_{j;i} + u_{k;i} u_{k;j}) \tag{5.5.6}'$$

When the *displacement components* u_i are known as functions of x_i^0, (5.5.4)' to (5.5.6)' may be employed to compute the components of $\nabla^0 \mathbf{u}$, $\mathbf{C}$ and $\mathbf{G}$, respectively.

From (5.5.3) we find that

$$\nabla \mathbf{u} = \nabla \mathbf{x} - \nabla \mathbf{x}^0 = \mathbf{I} - \mathbf{F}^{-1} \tag{5.5.7}$$

The tensor $\nabla \mathbf{u}$ is referred to as the *displacement-gradient tensor in the spatial description*, or briefly the *spatial displacement gradient*.

From (5.4.23) and (5.5.7) we obtain

$$\mathbf{B}^{-1} = (\mathbf{F}^{-1})^T \mathbf{F}^{-1} = (\mathbf{I} - \nabla \mathbf{u}^T)(\mathbf{I} - \nabla \mathbf{u})$$
$$= \mathbf{I} - \nabla \mathbf{u} - \nabla \mathbf{u}^T + (\nabla \mathbf{u}^T)(\nabla \mathbf{u}) \tag{5.5.8}$$

202   5 DEFORMATION

Consequently, (5.4.28) yields

$$\mathbf{A} = \tfrac{1}{2}(\mathbf{I} - \mathbf{B}^{-1}) = \tfrac{1}{2}[\nabla\mathbf{u} + \nabla\mathbf{u}^T - (\nabla\mathbf{u}^T)(\nabla\mathbf{u})] \quad (5.5.9)$$

This is called the *strain-displacement relation in the spatial description*.
In the suffix notation, (5.5.7) to (5.5.9) read respectively as follows:

$$u_{i,j} = \delta_{ij} - x^0_{i,j} \quad (5.5.7)'$$

$$b^{-1}_{ij} = \delta_{ij} - u_{i,j} - u_{j,i} + u_{k,i}u_{k,j} \quad (5.5.8)'$$

$$a_{ij} = \tfrac{1}{2}(u_{i,j} + u_{j,i} - u_{k,i}u_{k,j}) \quad (5.5.9)'$$

When u_i are known as functions of x_i, (5.5.7)' to (5.5.9)' may be employed to compute the components of $\nabla\mathbf{u}$, $\mathbf{B}^{-1}$ and $\mathbf{A}$, respectively.

EXAMPLE 5.5.1   In a certain deformation, the displacement components are found to be

$$u_1 = 3(x^0_1)^2 + x^0_2, \quad u_2 = 2(x^0_2)^2 + x^0_3, \quad u_3 = 4(x^0_3)^2 + x^0_1 \quad (5.5.10)$$

Find (i) the Green strain tensor at the origin O, and (ii) the stretch and the final length of a material arc of initial length ds_0 that was initially based at O and directed along the line making equal angles with the axes.

Solution   (i) From (5.5.10), we find that the matrix of the displacement-gradient $\nabla^0\mathbf{u}$ is

$$[\nabla^0\mathbf{u}] = [u_{i;j}] = \begin{bmatrix} 6x^0_1 & 1 & 0 \\ 0 & 4x^0_2 & 1 \\ 1 & 0 & 8x^0_3 \end{bmatrix} \quad (5.5.11)$$

By virtue of (5.5.6) and (5.5.6)' the matrix of the Green strain tensor $\mathbf{G}$ is given by

$$2[g_{ij}] = 2[\mathbf{G}] = [\nabla^0\mathbf{u}] + [\nabla^0\mathbf{u}]^T + [\nabla^0\mathbf{u}]^T[\nabla^0\mathbf{u}] \quad (5.5.12)$$

Substituting for $[\nabla^0\mathbf{u}]$ from (5.5.11) in (5.5.12), we obtain

$$2[g_{ij}] = 2[\mathbf{G}]$$

$$= \begin{bmatrix} 1 + 6x^0_1 + 36(x^0_1)^2 & 1 + 6x^0_1 & 1 + 8x^0_3 \\ 1 + 6x^0_1 & 1 + 8x^0_2 + 16(x^0_2)^2 & 1 + 4x^0_2 \\ 1 + 8x^0_3 & 1 + 4x^0_2 & 1 + 16x^0_3 + 64(x^0_3)^2 \end{bmatrix}$$

$$(5.5.13)$$

5.5 STRAIN-DISPLACEMENT RELATIONS

At the origin, this becomes

$$2[g_{ij}] = 2[\mathbf{G}] = \begin{bmatrix} 1 & 1 & 1 \\ 1 & 1 & 1 \\ 1 & 1 & 1 \end{bmatrix} \quad (5.5.14)$$

(ii) The stretch of a material arc whose initial orientation is given may be computed by using the following expression obtained from (5.4.2), (5.4.7), and (5.4.12):

$$\eta^2 = \frac{(ds)^2}{(ds_0)^2} = c_{ij} a_i^0 a_j^0 = (2g_{ij} + \delta_{ij}) a_i^0 a_j^0 \quad (5.5.15)$$

For the given arc, we have $a_1^0 = a_2^0 = a_3^0 = 1/\sqrt{3}$. Using these a_i^0 and substituting for g_{ij} from (5.5.14) in (5.5.15), we get the stretch of the arc as $\eta = 2$. Hence the final length of the arc is $ds = 2\, ds_0$. ∎

EXAMPLE 5.5.2 (i) Show that

$$J = \det(\mathbf{I} + \nabla^0 \mathbf{u}) = \{\det(\mathbf{I} - \nabla \mathbf{u})\}^{-1} \quad (5.5.16)$$

(ii) Deduce that the deformation for which the displacements (displacement components) are

$$u_1 = x_1 + x_2, \quad u_2 = x_1 - x_2, \quad u_3 = 2x_3 \quad (5.5.17)$$

is isochoric.

Solution (i) From (5.5.4) and (5.5.7), we have

$$\mathbf{F} = \nabla^0 \mathbf{u} + \mathbf{I} \quad (5.5.18a)$$

$$\mathbf{F}^{-1} = \mathbf{I} - \nabla \mathbf{u} \quad (5.5.18b)$$

Recalling that $J = \det \mathbf{F}$ and $(1/J) = \det \mathbf{F}^{-1}$, we readily get (5.5.16) from (5.5.18).

(ii) From (5.5.18b), we have

$$\frac{1}{J} = \det \mathbf{F}^{-1} = \det(\delta_{ij} - u_{i,j}) \quad (5.5.19)$$

Computing $u_{i,j}$ from (5.5.17) and substituting them in (5.5.19), we get

$$\frac{1}{J} = \begin{vmatrix} 0 & -1 & 0 \\ -1 & 2 & 0 \\ 0 & 0 & -1 \end{vmatrix} = 1$$

Hence, $J = 1$ so that the given deformation is isochoric. ∎

EXAMPLE 5.5.3 Prove the identities:

$$[\operatorname{curl} \mathbf{u}]_i = \frac{1}{J} \varepsilon_{jrs} X_{i;r} X_{j;s} \qquad (5.5.20)$$

$$= \frac{1}{J} \varepsilon_{kpq} x_{p;k} x_{q;i} \qquad (5.5.21)$$

Solution Since $u_i = x_i - x_i^0$, we get

$$[\operatorname{curl} \mathbf{u}]_i = \varepsilon_{ijk} u_{k,j} = \varepsilon_{ijk}(x_{k,j} - x_{k,j}^0)$$
$$= -\varepsilon_{ijk} x_{k,j}^0 = \varepsilon_{ijk} x_{j,k}^0 \qquad (5.5.22)$$

By using the identity (5.2.32), the relation (5.5.22) becomes

$$[\operatorname{curl} \mathbf{u}]_i = \frac{1}{2J} \varepsilon_{ijk} \varepsilon_{jrs} \varepsilon_{kpq} x_{p;r} x_{q;s} \qquad (5.5.23)$$

Employing the ε-δ identity (1.7.23) to the factor $\varepsilon_{ijk}\varepsilon_{kpq}$, expression (5.5.23) becomes

$$[\operatorname{curl} \mathbf{u}]_i = \frac{1}{2J} \varepsilon_{jrs}(\delta_{ip}\delta_{jq} - \delta_{iq}\delta_{jp}) x_{p;r} x_{q;s}$$

$$= \frac{1}{2J} \varepsilon_{jrs}(X_{i;r} X_{j;s} - X_{i;s} X_{j;r})$$

$$= \frac{1}{J} \varepsilon_{jrs} X_{i;r} X_{j;s}$$

which proves the identity (5.5.20).

On the other hand, if we employ the ε-δ identity to the factor $\varepsilon_{ijk}\varepsilon_{jrs}$, expression (5.5.23) becomes

$$[\operatorname{curl} \mathbf{u}]_i = \frac{1}{2J} \varepsilon_{kpq}(\delta_{kr}\delta_{is} - \delta_{ks}\delta_{ir}) x_{p;r} x_{q;s}$$

$$= \frac{1}{2J} \varepsilon_{kpq}(x_{p;k} x_{q;i} - x_{q;k} x_{p;i})$$

$$= \frac{1}{J} \varepsilon_{kpq} x_{p;k} x_{q;i}$$

which proves the identity (5.5.21). ■

5.6
INFINITESIMAL STRAIN TENSOR

In great many applications, particularly those concerned with common elastic materials, we deal with deformations wherein the *derivatives* of displacement components are so small that their squares and products can be ignored. Such deformations are referred to as *infinitesimal* or *small deformations*. A deformation that is not small is referred to as a *finite deformation*. In Sections 5.2 to 5.5 we dealt with finite deformations. From now on we will deal with infinitesimal deformations.

Using the fact that $u_i = u_i(x_k^0, t)$ in the material description and that $x_k^0 = x_k - u_k$, we get

$$u_{i,j} = u_{i;k}x_{k,j}^0 = u_{i;k}(x_k - u_k)_{,j} = (u_{i;j} - u_{i;k}u_{k,j}) \quad (5.6.1)$$

For small deformation, we neglect the *nonlinear terms* $u_{i;k}u_{k,j}$, so that (5.6.1) becomes

$$u_{i,j} = u_{i;j} \quad \text{or} \quad \nabla \mathbf{u} = \nabla^0 \mathbf{u} \quad (5.6.2)$$

Thus, for small deformation the material displacement gradient and the spatial displacement gradient are (almost) equal. Accordingly, in dealing with small deformation it is immaterial whether the displacement gradient is formed w.r.t. $\mathbf{x}^0$ or w.r.t. $\mathbf{x}$.

For small deformation, (5.5.5)′, (5.5.6)′, (5.5.8)′ and (5.5.9)′ yield

$$c_{ij} - \delta_{ij} = \delta_{ij} - b_{ij}^{-1} = 2g_{ij} = 2a_{ij} = u_{i;j} + u_{j;i} = u_{i,j} + u_{j,i} \quad (5.6.3)$$

or, in the direct notation,

$$\mathbf{C} - \mathbf{I} = \mathbf{I} - \mathbf{B}^{-1} = 2\mathbf{G} = 2\mathbf{A} = \nabla^0 \mathbf{u} + \nabla^0 \mathbf{u}^T = \nabla \mathbf{u} + \nabla \mathbf{u}^T \quad (5.6.3)'$$

From these expressions we find that the tensors **G** and **A** are identical. As such, for small deformation, the distinction between the Lagrangian and the Eulerian strain tensors disappears. Hence in dealing with small deformation it is immaterial whether the strain components are referred to the initial configuration or the final configuration.

In the case of small deformation, the tensor **G** (or **A**) is redesignated as **E** and referred to as the *infinitesimal* or *linear strain tensor*. Thus, from (5.6.3)′ we get the following *strain-displacement relation for small deformation*:

$$\mathbf{E} = \tfrac{1}{2}(\nabla \mathbf{u} + \nabla \mathbf{u}^T) = \tfrac{1}{2}(\nabla^0 \mathbf{u} + \nabla^0 \mathbf{u}^T) \quad (5.6.4)$$

Evidently, **E** is just the symmetric part of $\nabla \mathbf{u} = \nabla^0 \mathbf{u}$, and the symmetry of **E** is inherent in its definition itself. The tensor **E** was introduced by Cauchy in 1827 and is also referred to as the *Cauchy's strain tensor*. The components of **E**, called *infinitesimal strain components* or *Cauchy's strain components*, are denoted e_{ij}. Thus,

$$e_{ij} = g_{ij} = a_{ij} = \tfrac{1}{2}(u_{i,j} + u_{j,i}) = \tfrac{1}{2}(u_{i;j} + u_{j;i}) \qquad (5.6.4)'$$

Let us recall expression (5.4.13) for the normal strain e and rewrite it in terms of $u_{i;j}$ by using (5.6.3) to obtain

$$e = (1 + 2u_{i;j}a_i^0 a_j^0)^{1/2} - 1 \qquad (5.6.5)$$

Since $u_{i;j}$ are taken to be infinitesimally small, the first term in the righthand side of (5.6.5) can be expanded by using the binomial theorem. On carrying out this expansion, neglecting nonlinear terms involving $u_{i;j}$ and subsequently using (5.6.3) again we obtain

$$e = u_{i;j}a_i^0 a_j^0 = g_{ij}a_i^0 a_j^0 \qquad (5.6.6)$$

We find that this expression is identical with the expression (5.4.14), which was obtained by neglecting the squares and higher powers of e. It therefore follows that the small displacement-gradient approximation and the small strain approximation are of the same order. As such, e_{ij} have the same geometrical meanings as g_{ij} have under the small strain approximation. Thus, for example, e_{11} represents the normal strain of a material arc that initially lay along the 1 axis and e_{12} represents the shear strain between two material arcs that initially lay along the 1 and 2 axes. The numbers e_{11}, e_{22} and e_{33} are called *infinitesimal normal strains*, and $e_{12} = e_{21}$, $e_{13} = e_{31}$ and $e_{23} = e_{32}$ are called *infinitesimal shear strains*.

5.6.1 DILATATION

From (5.5.18a), we have

$$J = \det \mathbf{F} = \det(\delta_{ij} + u_{i;j}) = \begin{vmatrix} 1 + u_{1;1} & u_{1;2} & u_{1;3} \\ u_{2;1} & 1 + u_{2;2} & u_{2;3} \\ u_{3;1} & u_{3;2} & 1 + u_{3;3} \end{vmatrix} \qquad (5.6.7)$$

Expanding the determinant at the far righthand side of (5.6.7) and neglecting the nonlinear terms in displacement gradients, we obtain the following expression valid for small deformation:

$$J = 1 + u_{k;k} = 1 + e_{kk} \qquad (5.6.8)$$

By use of the Euler's formula (5.2.22), $dV = J\,dV_0$, expression (5.6.8) can

5.6 INFINITESIMAL STRAIN TENSOR

be rewritten as

$$\frac{dV - dV_0}{dV_0} = e_{kk} \tag{5.6.9}$$

Thus, e_{kk} represents the change in volume per unit initial volume of a material element undergoing small deformation. The number e_{kk} is called the *cubical dilatation* or just the *dilatation*.

We note from (5.6.4) and (5.6.4)' that

$$e_{kk} = \text{tr } \mathbf{E} = \text{div}^0 \mathbf{u} = \text{div } \mathbf{u} \tag{5.6.10}$$

where div^0 denotes the divergence w.r.t. the initial coordinates x_i^0. Thus, for small deformation, $\text{tr } \mathbf{E} = \text{div}^0 \mathbf{u} = \text{div } \mathbf{u}$ represents the change in volume per unit initial volume.

From (5.6.8) and (5.6.10), we infer that for isochoric small deformation $e_{kk} = \text{div}^0 \mathbf{u} = \text{div } \mathbf{u} = 0$ and for an incompressible continuum $e_{kk} = \text{div}^0 \mathbf{u} = \text{div } \mathbf{u} = 0$ for every small deformation.

EXAMPLE 5.6.1 For small deformation, prove the following:

(i)
$$\mathbf{C} \approx \mathbf{B} \approx \mathbf{I} + 2\mathbf{E} \tag{5.6.11a}$$
$$\mathbf{C}^{-1} \approx \mathbf{B}^{-1} \approx \mathbf{I} - 2\mathbf{E} \tag{5.6.11b}$$

(ii)
$$\mathbf{U} \approx \mathbf{V} \approx \mathbf{I} + \mathbf{E} \tag{5.6.12a}$$
$$\mathbf{U}^{-1} \approx \mathbf{V}^{-1} \approx \mathbf{I} - \mathbf{E} \tag{5.6.12b}$$

(iii)
$$\mathbf{Q} \approx \mathbf{F} - \mathbf{E} \tag{5.6.13a}$$
$$\mathbf{Q}^{-1} \approx \mathbf{F}^{-1} + \mathbf{E} \tag{5.6.13b}$$

Solution (i) From (5.6.3)' and (5.6.4), we get

$$\mathbf{C} = \mathbf{I} + 2\mathbf{E} \tag{5.6.14a}$$
$$\mathbf{B}^{-1} = \mathbf{I} - 2\mathbf{E} \tag{5.6.14b}$$

From (5.5.18) and (5.6.4), we get

$$\mathbf{B} = \mathbf{F}\mathbf{F}^T = (\nabla^0 \mathbf{u} + \mathbf{I})(\nabla^0 \mathbf{u}^T + \mathbf{I}) \approx \mathbf{I} + \nabla^0 \mathbf{u} + \nabla^0 \mathbf{u}^T = \mathbf{I} + 2\mathbf{E} \tag{5.6.15a}$$

and

$$\mathbf{C}^{-1} = (\mathbf{F}^T \mathbf{F})^{-1} = \mathbf{F}^{-1}(\mathbf{F}^{-1})^T = (\mathbf{I} - \nabla \mathbf{u})(\mathbf{I} - \nabla \mathbf{u}^T)$$
$$\approx \mathbf{I} - \nabla \mathbf{u} - \nabla \mathbf{u}^T = \mathbf{I} - 2\mathbf{E} \tag{5.6.15b}$$

Expressions (5.6.14) and (5.6.15) constitute the relations (5.6.11).

(ii) Expressions (5.6.11) may be rewritten as

$$\mathbf{C} \approx \mathbf{B} \approx (\mathbf{I} + 2\mathbf{E}) \approx (\mathbf{I} + \mathbf{E})(\mathbf{I} + \mathbf{E}) = (\mathbf{I} + \mathbf{E})^2 \quad (5.6.16a)$$

$$\mathbf{C}^{-1} \approx \mathbf{B}^{-1} \approx (\mathbf{I} - 2\mathbf{E}) \approx (\mathbf{I} - \mathbf{E})(\mathbf{I} - \mathbf{E}) = (\mathbf{I} - \mathbf{E})^2 \quad (5.6.16b)$$

Recalling that $\mathbf{U}^2 = \mathbf{C}$, $\mathbf{V}^2 = \mathbf{B}$, $(\mathbf{U}^{-1})^2 = \mathbf{C}^{-1}$ and $(\mathbf{V}^{-1})^2 = \mathbf{B}^{-1}$, we immediately get (5.6.12) from (5.6.16).

(iii) On using (5.5.18a) and (5.6.12b), expression (5.3.5a) yields

$$\mathbf{Q} = \mathbf{F}\mathbf{U}^{-1} = (\nabla^0 \mathbf{u} + \mathbf{I})(\mathbf{I} - \mathbf{E})$$

$$\approx \nabla^0 \mathbf{u} + \mathbf{I} - \mathbf{E} = \mathbf{F} - \mathbf{E}$$

which is (5.6.13a).

On using (5.5.18b) and (5.6.12a), expression (5.3.5b) yields

$$\mathbf{Q}^{-1} = \mathbf{F}^{-1}\mathbf{V} = (\mathbf{I} - \nabla \mathbf{u})(\mathbf{I} + \mathbf{E})$$

$$= \mathbf{I} - \nabla \mathbf{u} + \mathbf{E} = \mathbf{F}^{-1} + \mathbf{E}$$

which is (5.6.13b). ∎

EXAMPLE 5.6.2 Show that a small deformation for which $\mathbf{E} = \mathbf{0}$ represents a rigid-body transformation. Deduce that in this case the displacement can be represented in the form

$$\mathbf{u} = \boldsymbol{\omega} \times \mathbf{x} + \mathbf{c} \quad (5.6.17)$$

where $\mathbf{c}$ is a vector not dependent on $\mathbf{x}$ and $\boldsymbol{\omega}$ is a vector such that

$$(\nabla \mathbf{u})\mathbf{a} = \boldsymbol{\omega} \times \mathbf{a} \quad (5.6.18)$$

for every vector $\mathbf{a}$.

Show also that if two displacement vectors $\mathbf{u}$ and $\mathbf{u}'$ correspond to the same infinitesimal strain tensor, then $\mathbf{u} - \mathbf{u}'$ is a rigid displacement. (This result is known as *Kirchhoff's theorem*.)

Solution If $\mathbf{E} = \mathbf{0}$, then we find from (5.6.12a) that $\mathbf{U} = \mathbf{V} = \mathbf{I}$. Therefore, the deformation is a rigid-body transformation.

Further, when $\mathbf{E} = \mathbf{0}$, we find from (5.6.4) that $\nabla \mathbf{u}$ is a skew tensor, so that $u_{i,j} = -u_{j,i}$ from which it follows that

$$u_{i,jk} = u_{i,kj} = -u_{k,ij} = -u_{k,ji} = u_{j,ki} = u_{j,ik} = -u_{i,jk}$$

or

$$u_{i,jk} = 0 \quad (5.6.19)$$

5.6 INFINITESIMAL STRAIN TENSOR

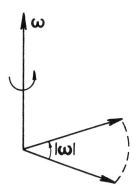

Figure 5.12. Rigid rotation.

The displacement components u_i must therefore be linear functions of x_i; that is

$$u_i = a_{ik} x_k + c_i \qquad (5.6.20)$$

where a_{ik} and c_i are independent of x_k. We easily note from (5.6.20) that $a_{ik} = u_{i,k}$; hence, (5.6.20) can be rewritten as

$$u_i = u_{i,k} x_k + c_i$$

or

$$\mathbf{u} = (\nabla \mathbf{u})\mathbf{x} + \mathbf{c} \qquad (5.6.21)$$

Since $\nabla \mathbf{u}$ is a skew tensor, there exists a unique dual vector $\boldsymbol{\omega}$ such that (5.6.18) holds for every vector $\mathbf{a}$. Using (5.6.18) in (5.6.21) we obtain (5.6.17).

We observe that $\mathbf{u}$ given by (5.6.17) consists of two parts: $\mathbf{c}$ and $\boldsymbol{\omega} \times \mathbf{x}$. The part $\mathbf{c}$ represents a rigid translation while the part $\boldsymbol{\omega} \times \mathbf{x}$ represents a rigid rotation through an angle $|\boldsymbol{\omega}|$ about the axis directed along $\boldsymbol{\omega}$ (see Figure 5.12).

If two displacement vectors $\mathbf{u}$ and $\mathbf{u}'$ correspond to the same infinitesimal strain tensor, then $\mathbf{u} - \mathbf{u}'$ corresponds to the zero infinitesimal strain. From the result just proven, it follows that $\mathbf{u} - \mathbf{u}'$ is a rigid displacement. ∎

EXAMPLE 5.6.3 Find **E** for the rigid rotation defined by equations (5.3.21) of Example 5.3.3.

Solution From (5.3.21) we obtain the following expressions for the displacement components associated with the rotation considered here.

$$\begin{aligned} u_1 &= x_1 - x_1^0 = x_1^0(\cos\theta - 1) + x_2^0 \sin\theta \\ u_2 &= x_2 - x_2^0 = -x_1^0 \sin\theta + x_2^0(\cos\theta - 1) \\ u_3 &= x_3 - x_3^0 = 0 \end{aligned} \qquad (5.6.22)$$

These give

$$[\nabla^0 \mathbf{u}] = \begin{bmatrix} \cos\theta - 1 & \sin\theta & 0 \\ -\sin\theta & \cos\theta - 1 & 0 \\ 0 & 0 & 0 \end{bmatrix} \quad (5.6.23)$$

Hence the strain-displacement relation (5.6.4) yields

$$[\mathbf{E}] = \tfrac{1}{2}[\nabla^0 \mathbf{u}] + \tfrac{1}{2}[\nabla^0 \mathbf{u}]^T = \begin{bmatrix} \cos\theta - 1 & 0 & 0 \\ 0 & \cos\theta - 1 & 0 \\ 0 & 0 & 0 \end{bmatrix} \quad (5.6.24)$$

This is the matrix of the infinitesimal strain tensor $\mathbf{E}$ for the given rigid rotation. Evidently, $\mathbf{E}$ is a nonzero tensor (for $\theta \neq 0$). ∎

Note: This example illustrates the important fact that, unlike the finite strain tensors $\mathbf{G}$ and $\mathbf{A}$, the infinitesimal strain tensor $\mathbf{E}$ need not be 0 for a rigid-body transformation. The reason for this is that $\mathbf{E}$ is not an exact measure of deformation.

EXAMPLE 5.6.4 Let V_0 be the initial volume and V be the final volume of a material volume $\mathcal{V}$ undergoing small deformation. If $f = f(x_i)$, show that

$$\int_V f \, dV = \int_{V_0} (1 + e_{kk}) f \, dV_0 \quad (5.6.25)$$

Hence deduce that the change in volume of $\mathcal{V}$ is given by

$$\delta\mathcal{V} = \int_V e_{kk} \, dV = \int_{V_0} e_{kk} \, dV_0 \quad (5.6.26)$$

Solution From the Euler's formula (5.2.22) let us recall that $dV = J \, dV_0$. Therefore

$$\int_V f \, dV = \int_{V_0} Jf \, dV_0 \quad (5.6.27)$$

Substituting for J from (5.6.8) in (5.6.27) we immediately obtain (5.6.25).
For $f = 1$, (5.6.25) becomes

$$\int_V dV = \int_{V_0} (1 + e_{kk}) \, dV_0$$

so that

$$\delta\mathcal{V} \equiv V - V_0 = \int_V dV - \int_{V_0} dV_0 = \int_{V_0} e_{kk}\, dV_0 \quad (5.6.28)$$

Taking $f = e_{kk}$ in (5.6.25), we get

$$\int_V e_{kk}\, dV = \int_{V_0} (1 + e_{kk}) e_{kk}\, dV_0 \approx \int_{V_0} e_{kk}\, dV_0 \quad (5.6.29)$$

which together with (5.6.28) yields (5.6.26). ∎

EXAMPLE 5.6.5 Show that for small deformation

$$\mathbf{n}\, dS = \mathbf{n}^0\, dS_0 \quad (5.6.30)$$

Solution From (5.2.12) let us recall that

$$\mathbf{F}^T \mathbf{n}\, dS = J\mathbf{n}^0\, dS_0 \quad (5.6.31)$$

Substituting for $\mathbf{F}$ from (5.5.4) and for J from (5.6.8) together with (5.6.10) in (5.6.31), we get

$$\mathbf{n}\, dS + (\nabla^0 \mathbf{u})^T \mathbf{n}\, dS = \mathbf{n}^0\, dS_0 + (\text{div}^0 \mathbf{u})\mathbf{n}^0\, dS_0 \quad (5.6.32)$$

Equating the terms of the same order in displacements on the two sides of (5.6.32), we obtain (5.6.30). ∎

5.7
INFINITESIMAL STRETCH AND ROTATION

In Section 5.3, it was seen that a transformation of a material arc element $d\mathcal{C}$ from its initial location $d\mathcal{C}_0$ to the final location $d\mathcal{C}$ under a deformation may be decomposed into two parts: the first part described by (5.3.8a), representing a triaxial stretch of the element; and the second part described by (5.3.8b), representing a rotation of the element. Let us now reexamine these two parts in the case of small deformation.

For small deformation, we recall from (5.6.12a) that

$$\mathbf{U} = \mathbf{E} + \mathbf{I} \quad (5.7.1)$$

Consequently, (5.3.8a) becomes

$$d\mathbf{y} = d\mathbf{x}^0 + \mathbf{E}\, d\mathbf{x}^0 \quad (5.7.2)$$

This equation shows that the first part of the transformation (which carries $d\mathcal{C}$ from its initial location $d\mathcal{C}_0$ to the intermediate location $d\mathcal{C}'$) is

212   **5 DEFORMATION**

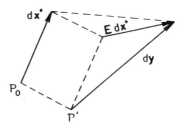

Figure 5.13. Translation preceding by stretch.

now made up of a (pure) translation and a stretch effected by **E**. The translation may follow or precede the stretch. Translation preceding by stretch is depicted in Figure 5.13.

By use of (5.7.1) and (5.3.9), we find that

$$\mathbf{E}\,d\mathbf{x}^0 = \sum_{k=1}^{3} \eta_k (\mathbf{p}_k \otimes \mathbf{p}_k)\,d\mathbf{x}^0 - d\mathbf{x}^0$$

$$= \sum_{k=1}^{3} e_k (\mathbf{p}_k \otimes \mathbf{p}_k)\,d\mathbf{x}^0$$

$$= \sum_{k=1}^{3} e_k (d\mathbf{x}^0 \cdot \mathbf{p}_k)\mathbf{p}_k \tag{5.7.3}$$

where

$$e_k = \eta_k - 1 \tag{5.7.4}$$

Expression (5.7.3) shows that the stretch effected by **E** is made up of three simple elongations directed along the principal directions of stretch with e_k as the corresponding normal strains.

Let us now look at the second part of the transformation (which carries $d\mathcal{C}$ from the intermediate location dC' to the final location dC) given by (5.3.8b). By use of (5.6.13a), (5.5.4) and (5.6.4), expression (5.3.8b) becomes

$$d\mathbf{x} = (\mathbf{F} - \mathbf{E})\,d\mathbf{y} = (\mathbf{I} + \nabla^0 \mathbf{u} - \mathbf{E})\,d\mathbf{y}$$

$$= d\mathbf{y} + \Omega\,d\mathbf{y} \tag{5.7.5}$$

where

$$\Omega = \tfrac{1}{2}(\nabla^0 \mathbf{u} - \nabla^0 \mathbf{u}^T) = \text{skw}\,\nabla^0 \mathbf{u} \tag{5.7.6}$$

Since Ω is a skew tensor, (5.7.5) may be rewritten as

$$d\mathbf{x} = d\mathbf{y} + \boldsymbol{\omega} \times d\mathbf{y} \tag{5.7.7}$$

where $\boldsymbol{\omega}$ is the dual vector of Ω. Expression (5.7.7) shows that the second part of the transformation of $d\mathcal{C}$ is now made up of a (pure) translation and

5.7 INFINITESIMAL STRETCH AND ROTATION 213

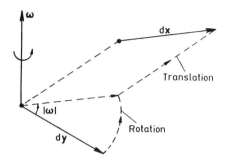

Figure 5.14. Translation following rotation.

a pure rotation through an angle $|\omega|$ about an axis directed along ω. The translation may follow or precede rotation. Translation following rotation is depicted in Figure 5.14.

From this analysis, it follows that, under small deformation, a material arc element $d\mathcal{C}$ may be thought as subjected to a pure translation and a triaxial elongation as described by (5.7.2) followed by another pure translation and a pure rotation as described by (5.7.7). This interpretation is illustrated in Figure 5.15.

It is to be noted that, in this analysis, the stretch part of the transformation is represented by the strain tensor **E**, which is the symmetric part of $\nabla^0 \mathbf{u}$, and the rotation part is represented by the tensor Ω, which is the skew part of $\nabla^0 \mathbf{u}$. The tensor Ω is called the *infinitesimal* or *linear rotation tensor*. The vector ω, which is the dual of Ω, is called the *infinitesimal* or *linear rotation vector*.

Since $\nabla^0 \mathbf{u} \approx \nabla \mathbf{u}$ for a small deformation, the expression (5.7.6) defining Ω may be replaced by the following expression:

$$\Omega = \text{skw } \nabla \mathbf{u} = \tfrac{1}{2}(\nabla \mathbf{u} - \nabla \mathbf{u}^T) \tag{5.7.8}$$

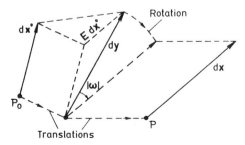

Figure 5.15. Translation and elongation followed by another translation and rotation.

214 5 DEFORMATION

Recalling that the dual vector of skw $\nabla \mathbf{u}$ is $\frac{1}{2}$ curl $\mathbf{u}$ (see Example 3.5.7), we obtain the following expression for the linear rotation vector:

$$\boldsymbol{\omega} = \tfrac{1}{2} \text{curl } \mathbf{u} \approx \tfrac{1}{2} \text{curl}^0 \mathbf{u} \qquad (5.7.9)$$

where curl^0 is the curl taken w.r.t. x_i^0.

If ω_{ij} and ω_i are the components of Ω and $\boldsymbol{\omega}$, respectively, then (5.7.6), (5.7.8) and (5.7.9) yield the following expressions for ω_{ij} and ω_i in terms of u_i:

$$\omega_{ij} = \tfrac{1}{2}(u_{i;j} - u_{j;i}) \approx \tfrac{1}{2}(u_{i,j} - u_{j,i}) \qquad (5.7.10)$$

$$\omega_i = \tfrac{1}{2}\varepsilon_{ijk} u_{k;j} \approx \tfrac{1}{2}\varepsilon_{ijk} u_{k,j} \qquad (5.7.11)$$

EXAMPLE 5.7.1 For the small deformation defined by

$$\mathbf{u} = \alpha x_1 x_2 (\mathbf{e}_1 + \mathbf{e}_2) + 2\alpha(x_1 + x_2) x_3 \mathbf{e}_3 \qquad (5.7.12)$$

where α is a constant, find the strain tensor, the rotation tensor and the rotation vector.

Solution From (5.7.12), we find that

$$[\nabla \mathbf{u}] = \alpha \begin{bmatrix} x_2 & x_1 & 0 \\ x_2 & x_1 & 0 \\ 2x_3 & 2x_3 & 2(x_1 + x_2) \end{bmatrix} \qquad (5.7.13)$$

Hence

$$[\mathbf{E}] = \tfrac{1}{2}[\nabla \mathbf{u}] + \tfrac{1}{2}[\nabla \mathbf{u}]^T = \alpha \begin{bmatrix} x_2 & \tfrac{1}{2}(x_1 + x_2) & x_3 \\ \tfrac{1}{2}(x_1 + x_2) & x_1 & x_3 \\ x_3 & x_3 & 2(x_1 + x_2) \end{bmatrix}$$
$$(5.7.14)$$

and

$$[\Omega] = \tfrac{1}{2}[\nabla \mathbf{u}] - \tfrac{1}{2}[\nabla \mathbf{u}]^T = \alpha \begin{bmatrix} 0 & \tfrac{1}{2}(x_1 - x_2) & -x_3 \\ -\tfrac{1}{2}(x_1 - x_2) & 0 & -x_3 \\ x_3 & x_3 & 0 \end{bmatrix} \qquad (5.7.15)$$

For the given deformation, relations (5.7.14) and (5.7.15) give the matrices of the strain and rotation tensors respectively. Also, the rotation vector, obtained from (5.7.12), is

$$\boldsymbol{\omega} = \tfrac{1}{2} \text{curl } \mathbf{u} = \alpha x_3(\mathbf{e}_1 - \mathbf{e}_2) + \tfrac{1}{2}\alpha(x_2 - x_1)\mathbf{e}_3 \quad \blacksquare \qquad (5.7.16)$$

5.7 INFINITESIMAL STRETCH AND ROTATION

EXAMPLE 5.7.2 Consider the displacement components

$$u_1 = a_1 x_1 + b_1 x_2 + c_1 x_3$$
$$u_2 = a_2 x_1 + b_2 x_2 + c_2 x_3 \quad (5.7.17)$$
$$u_3 = a_3 x_1 + b_3 x_2 + c_3 x_3$$

where a_i, b_i and c_i are constants. Show that these displacement components correspond to (i) a pure deformation if and only if

$$b_1 = a_2, \quad c_1 = a_3, \quad c_2 = b_3 \quad (5.7.18)$$

and (ii) a pure rotation if and only if

$$a_1 = b_2 = c_3 = 0, \quad b_1 = -a_2, \quad c_1 = -a_3, \quad c_2 = -b_3 \quad (5.7.19)$$

Find the rotation vector in the latter case.

Solution From (5.7.17), we find that

$$[\nabla \mathbf{u}] = \begin{bmatrix} a_1 & b_1 & c_1 \\ a_2 & b_2 & c_2 \\ a_3 & b_3 & c_3 \end{bmatrix} \quad (5.7.20)$$

so that

$$[\mathbf{E}] = \tfrac{1}{2} \begin{bmatrix} 2a_1 & b_1 + a_2 & c_1 + a_3 \\ a_2 + b_1 & 2b_2 & c_2 + b_3 \\ a_3 + c_1 & b_3 + c_2 & 2c_3 \end{bmatrix} \quad (5.7.21)$$

and

$$[\Omega] = \tfrac{1}{2} \begin{bmatrix} 0 & b_1 - a_2 & c_1 - a_3 \\ a_2 - b_1 & 0 & c_2 - b_3 \\ a_3 - c_1 & b_3 - c_2 & 0 \end{bmatrix} \quad (5.7.22)$$

Evidently, there occurs a pure deformation, that is $\Omega = \mathbf{0}$, if and only if (5.7.18) hold. Also, there occurs a pure rotation, that is $\mathbf{E} = \mathbf{0}$, if and only if (5.7.19) hold.

From (5.7.17), we find that

$$\text{curl } \mathbf{u} = (b_3 - c_2)\mathbf{e}_1 + (c_1 - a_3)\mathbf{e}_2 + (a_2 - b_1)\mathbf{e}_3 \quad (5.7.23)$$

In the case of pure rotation, where (5.7.19) hold, this yields

$$\boldsymbol{\omega} = \tfrac{1}{2} \text{curl } \mathbf{u} = (b_3 \mathbf{e}_1 + c_1 \mathbf{e}_2 + a_2 \mathbf{e}_3) \quad (5.7.24)$$

as the rotation vector. ∎

216 5 DEFORMATION

EXAMPLE 5.7.3 If $\mathbf{u}$ is the displacement vector at a point $\mathbf{x}$ and $\mathbf{u} + d\mathbf{u}$ is the displacement vector at a neighboring point $\mathbf{x} + d\mathbf{x}$, then $d\mathbf{u}$ is called the *relative displacement* in the neighborhood of $\mathbf{x}$. Show that $d\mathbf{u}$ is generally a superposition of a displacement due to stretch and a rigid-body displacement.

Solution From the definition of $d\mathbf{u}$, we note that

$$du_i = \frac{\partial u_i}{\partial x_j} dx_j = u_{i,j} dx_j$$

$$= e_{ij} dx_j + \omega_{ij} dx_j$$

or

$$d\mathbf{u} = \mathbf{E}\, d\mathbf{x} + \mathbf{\Omega}\, d\mathbf{x} \tag{5.7.25}$$

Evidently, $d\mathbf{u}$ is made up of two parts: $\mathbf{E}\, d\mathbf{x}$, which is effected by $\mathbf{E}$, and $\mathbf{\Omega}\, d\mathbf{x}$ which is effected by $\mathbf{\Omega}$. Since $\mathbf{E}$ causes a stretch and $\mathbf{\Omega}$ causes a pure rotation, it follows that $d\mathbf{u}$ is a superposition of a displacement caused by stretch and a rigid-body displacement. ∎

EXAMPLE 5.7.4 For small deformation, show that

$$\mathbf{Q} = \mathbf{I} + \mathbf{\Omega} \tag{5.7.26a}$$

$$\mathbf{Q}^{-1} = \mathbf{I} - \mathbf{\Omega} \tag{5.7.26b}$$

Solution Comparing (5.7.5) with (5.3.8b), we readily get (5.7.26a). By use of (5.6.13b), (5.5.7), (5.6.4) and (5.7.8) we obtain

$$\mathbf{Q}^{-1} = \mathbf{F}^{-1} + \mathbf{E} = (\mathbf{I} - \nabla\mathbf{u}) + \tfrac{1}{2}(\nabla\mathbf{u} + \nabla\mathbf{u}^T) = \mathbf{I} - \mathbf{\Omega}$$

which is (5.7.26b). ∎

EXAMPLE 5.7.5 Show that for small deformation the polar decompositions of $\mathbf{F}$ reduce to the additive decompositions of $\nabla^0 \mathbf{u}$ into symmetric and skew parts.

Solution Using the relations (5.5.18a), (5.6.12a) and (5.7.26a), the right polar decomposition of $\mathbf{F}$ given by (5.3.5a) becomes

$$\mathbf{I} + \nabla^0 \mathbf{u} = (\mathbf{I} + \mathbf{\Omega})(\mathbf{I} + \mathbf{E}) = \mathbf{I} + \mathbf{E} + \mathbf{\Omega}$$

or

$$\nabla^0 \mathbf{u} = \mathbf{E} + \mathbf{\Omega} \tag{5.7.27}$$

which is the additive decomposition of $\nabla^0 \mathbf{u}$ into symmetric and skew parts $\mathbf{E}$ and $\mathbf{\Omega}$.

On the other hand, using the relations (5.5.18b), (5.6.12b) and (5.7.26b), the left polar decomposition of $\mathbf{F}$ given by (5.3.5b) which is equivalent to $\mathbf{F}^{-1} = \mathbf{Q}^{-1}\mathbf{V}^{-1}$ becomes

$$\mathbf{I} - \nabla\mathbf{u} = (\mathbf{I} - \Omega)(\mathbf{I} - \mathbf{E}) = \mathbf{I} - \mathbf{E} - \Omega$$

or

$$\nabla\mathbf{u} = \mathbf{E} + \Omega \tag{5.7.28}$$

which is again the decomposition of $\nabla^0\mathbf{u} = \nabla\mathbf{u}$ into symmetric and skew parts. ∎

EXAMPLE 5.7.6 Show that, for small deformation,

$$\eta\mathbf{a} = \mathbf{a}^0 + \mathbf{E}\mathbf{a}^0 + \Omega\mathbf{a}^0 \tag{5.7.29}$$

Solution For small deformation, we have by (5.5.18a) and (5.7.27),

$$\mathbf{F} = \mathbf{I} + \mathbf{E} + \Omega \tag{5.7.30}$$

Substituting this into (5.4.1), we readily get (5.7.29). ∎

5.8
COMPATIBILITY CONDITIONS

The strain-displacement relations (5.6.4)′ enable us to find $\mathbf{E}$ for any given $\mathbf{u}$. But, for any given $\mathbf{E}$, it may not always be possible to find $\mathbf{u}$ from (5.6.4)′, because, for a given e_{ij}, (5.6.4)′ is a system of six partial differential equations for only three unknown functions u_i, and the system may not possess a solution. For example, suppose that e_{ij} are given as follows:

$$e_{11} = e_{22} = e_{33} = e_{12} = e_{13} = 0, \qquad e_{23} = x_2 x_3 \tag{5.8.1}$$

Then (5.6.4)′ yield the following partial differential equations:

$$\frac{\partial u_1}{\partial x_1} = 0, \qquad \frac{\partial u_2}{\partial x_2} = 0, \qquad \frac{\partial u_3}{\partial x_3} = 0$$

$$\frac{\partial u_1}{\partial x_2} + \frac{\partial u_2}{\partial x_1} = 0, \qquad \frac{\partial u_1}{\partial x_3} + \frac{\partial u_3}{\partial x_1} = 0 \tag{5.8.2}$$

$$\frac{\partial u_2}{\partial x_3} + \frac{\partial u_3}{\partial x_2} = 2x_2 x_3$$

Solutions of the second and the third of these equations are $u_2 = f(x_1, x_3)$, $u_3 = g(x_1, x_2)$, where f and g are arbitrary functions of their arguments.

Inserting these solutions into the last of the equations, we get

$$\frac{\partial f}{\partial x_3}(x_1, x_3) + \frac{\partial g}{\partial x_2}(x_1, x_2) - 2x_2 x_3 = 0 \tag{5.8.3}$$

Since the first or second term in this equation cannot have terms of the form $x_2 x_3$, the equation can never be satisfied. Thus, for the e_{ij} given by (5.8.1), equations (5.6.4)' do not yield a solution for u_i.

We now proceed to establish a necessary and sufficient condition on **E** that ensures the existence of **u** as a solution of (5.6.4)'.

THEOREM 5.8.1 For a given **E**, if the strain-displacement equation

$$\mathbf{E} = \tfrac{1}{2}(\nabla \mathbf{u} + \nabla \mathbf{u}^T) \tag{5.8.4}$$

has a solution for **u**, then **E** should satisfy the condition:

$$\operatorname{curl} \operatorname{curl} \mathbf{E} = \mathbf{0} \tag{5.8.5}$$

Conversely, if **E** satisfies the condition (5.8.5) in a simply connected region, then equation (5.8.4) possesses a solution for **u**. Furthermore, this solution is unique within a rigid-body displacement.

Proof First, suppose that for a given **E** there exists **u** such that equation (5.8.4) holds. Taking curl on both sides of this equation, we get

$$\operatorname{curl} \mathbf{E} = \tfrac{1}{2} \operatorname{curl} \nabla \mathbf{u} + \tfrac{1}{2} \operatorname{curl} \nabla \mathbf{u}^T \tag{5.8.6}$$

But, for any vector **a**,

$$\operatorname{curl} \nabla \mathbf{a} = \mathbf{0} \tag{5.8.7a}$$

$$\operatorname{curl} \nabla \mathbf{a}^T = \nabla \operatorname{curl} \mathbf{a} \tag{5.8.7b}$$

see identities (3.5.18) and (3.5.19). Hence (5.8.6) becomes

$$\operatorname{curl} \mathbf{E} = \tfrac{1}{2} \nabla (\operatorname{curl} \mathbf{u}) \tag{5.8.8}$$

Taking curl of this equation and using (5.8.7a), we get the condition (5.8.5).

Conversely, suppose that the condition (5.8.5) holds for a given **E** in a simply connected region. This condition may be written as

$$\operatorname{curl} \bar{\mathbf{E}} = \mathbf{0} \tag{5.8.9}$$

where

$$\bar{\mathbf{E}} = \operatorname{curl} \mathbf{E} \tag{5.8.10}$$

Then, since **E** is symmetric, the identity (3.5.39) yields

$$\operatorname{tr} \bar{\mathbf{E}} = \operatorname{tr}(\operatorname{curl} \mathbf{E}) = 0 \tag{5.8.11}$$

5.8 COMPATIBILITY CONDITIONS

Consequently, by virtue of the result proved in Example 3.6.4, there exists a skew tensor **W** such that

$$\bar{\mathbf{E}} = \text{curl } \mathbf{W} \tag{5.8.12}$$

Using (5.8.10) this becomes

$$\text{curl}(\mathbf{E} - \mathbf{W}) = \mathbf{0} \tag{5.8.13}$$

from which it follows that *there exists* a vector **u** such that

$$\mathbf{E} - \mathbf{W} = \nabla \mathbf{u} \tag{5.8.14}$$

see again Example 3.6.4. Equating the symmetric parts on both sides of (5.8.14) and bearing in mind that **E** is symmetric and **W** is skew, we find that **u** satisfies equation (5.8.4). Equation (5.8.4) thus possesses a solution.

Suppose **u**′ is also a solution of (5.8.4) for the same **E**. Then the strain tensor corresponding to $\mathbf{u} - \mathbf{u}'$ is 0 so that $\mathbf{u} - \mathbf{u}'$ is a rigid-body displacement (see Example 5.6.2). Thus, two solutions of (5.8.4), when they exist, differ at most by a rigid-body displacement.

This completes the proof of the theorem. ∎

The criterion (5.8.5) that ensures the possibility of obtaining **u** by solving the equation (5.8.4) (in a simply connected region) is known as the *integrability* or *compatibility condition* for **E**. A strain tensor **E** that satisfies this condition is referred to as a *possible* strain tensor.

By virtue of the result proved in Example 3.5.8, the compatibility condition (5.8.5) can be expressed in the following equivalent form:

$$\nabla^2 \mathbf{E} + \nabla\nabla(\text{tr } \mathbf{E}) - \nabla \text{ div } \mathbf{E} - (\nabla \text{ div } \mathbf{E})^T = \mathbf{0} \tag{5.8.15}$$

In the suffix notation, the conditions (5.8.5) and (5.8.15) read respectively as follows:

$$\varepsilon_{irs}\varepsilon_{jmn}e_{sn,rm} = 0 \tag{5.8.5}'$$

$$e_{ij,kk} + e_{kk,ij} - e_{ik,kj} - e_{jk,ki} = 0 \tag{5.8.15}'$$

When written in the unabridged form, these conditions yield the following six explicit and distinct conditions:

$$\begin{aligned} e_{11,22} + e_{22,11} &= 2e_{12,12} \\ e_{22,33} + e_{33,22} &= 2e_{23,23} \\ e_{33,11} + e_{11,33} &= 2e_{31,31} \\ e_{12,13} + e_{13,12} - e_{11,23} - e_{23,11} &= 0 \\ e_{23,21} + e_{21,23} - e_{22,31} - e_{31,22} &= 0 \\ e_{31,32} + e_{32,31} - e_{33,12} - e_{12,33} &= 0 \end{aligned} \tag{5.8.16}$$

These six conditions were first obtained by Saint-Venant in 1860 and are known as *Saint-Venant compatibility conditions* for the infinitesimal strain components.

The conditions (5.8.16) can also be extracted from the following set of conditions:

$$e_{ij,kl} + e_{kl,ij} - e_{ik,jl} - e_{jl,ik} = 0 \tag{5.8.17}$$

As such, conditions (5.8.17) are also often taken as an alternative version of the compatibility conditions (5.8.5)'. Note that the conditions (5.8.15)' are a contracted version of the conditions (5.8.17).

EXAMPLE 5.8.1 Show that the following are not possible strain components:

$$e_{11} = k(x_1^2 + x_2^2), \quad e_{22} = k(x_2^2 + x_3^2)$$
$$e_{12} = k'x_1x_2x_3, \quad e_{13} = e_{23} = e_{33} = 0 \tag{5.8.18}$$

Here k and k' are constants.

Solution The given numbers e_{ij} are possible strain components only if each of the six compatibility conditions (5.8.16) is satisfied. Substituting the given e_{ij} into the first of the conditions (5.8.16), we get

$$2k = 2k'x_3 \tag{5.8.19}$$

Since k and k' are constants, the condition (5.8.19) cannot be satisfied for $x_3 \neq 0$. For $x_3 = 0$, (5.8.19) gives $k = 0$; but then all the given e_{ij} vanish. Hence, the given e_{ij} are not possible strain components. ∎

EXAMPLE 5.8.2 (i) Find the compatibility condition for the strain tensor **E** if e_{11}, e_{22} and e_{12} are independent of x_3 and e_{13}, e_{23} and e_{33} are all 0.

(ii) Find the condition under which the following are possible strain components:

$$e_{11} = k(x_1^2 - x_2^2), \quad e_{12} = k'x_1x_2$$
$$e_{22} = kx_1x_2, \quad e_{13} = e_{23} = e_{33} = 0 \tag{5.8.20}$$

where k and k' are constants.

Solution (i) Since e_{11}, e_{22} and e_{12} are independent of x_3 and $e_{13} = e_{23} = e_{33} = 0$, we readily verify that all the conditions except the first one in (5.8.16) are automatically satisfied. Hence the only compatibility condition to be satisfied in this case is

$$e_{11,22} + e_{22,11} = 2e_{12,12} \tag{5.8.21}$$

5.8 COMPATIBILITY CONDITIONS

(ii) For e_{ij} given by (5.8.20), the only compatibility condition to be satisfied is (5.8.21). From (5.8.20) we find that

$$e_{11,22} = -2k, \qquad e_{22,11} = 0, \qquad e_{12,12} = k' \qquad (5.8.22)$$

Hence (5.8.21) is satisfied if and only if

$$k = -k' \qquad (5.8.23)$$

Thus, e_{ij} given by (5.8.20) are possible strain components if and only if $k = -k'$. ∎

Note: A deformation for which e_{11}, e_{22} and e_{33} are independent of x_3 and $e_{13} = e_{23} = e_{33} = 0$ is called a *plane deformation* parallel to the $x_1 x_2$ plane. Study of such deformations forms an important part of the elasticity theory.

EXAMPLE 5.8.3 When e_{ij} given by (5.8.20) are possible strain components, find the corresponding displacements, given that $u_3 = 0$.

Solution In Example 5.8.2, it was shown that e_{ij} given by (5.8.20) can yield displacements only when $k = -k'$. Substituting these e_{ij} with $k = -k'$ into the strain-displacement relations (5.6.4)′, we obtain the following partial differential equations:

$$u_{1,1} = k(x_1^2 - x_2^2), \qquad u_{2,2} = kx_1 x_2$$

$$u_{1,2} + u_{2,1} = -2kx_1 x_2 \qquad (5.8.24)$$

$$u_{2,3} + u_{3,2} = 0, \qquad u_{3,1} + u_{1,3} = 0, \qquad u_{3,3} = 0$$

Since $u_3 = 0$, the fourth and the fifth of equations (5.8.24) show that u_1 and u_2 are independent of x_3. Integration of the first two of the equations then yields

$$u_1 = \tfrac{1}{3}k(x_1^3 - 3x_1 x_2^2) + f(x_2) \qquad (5.8.25)$$

$$u_2 = \tfrac{1}{2}kx_1 x_2^2 + g(x_1) \qquad (5.8.26)$$

where f and g are arbitrary functions of x_2 and x_1, respectively. Inserting (5.8.25) and (5.8.26) into the third equation in (5.8.24), we get

$$g'(x_1) = \tfrac{1}{2}kx_2^2 - f'(x_2) \qquad (5.8.27)$$

The lefthand side of this condition is a function of x_1 while the righthand side is a function of x_2. As such, this condition is possible only if

$$g'(x_1) = c, \qquad \tfrac{1}{2}kx_2^2 - f'(x_2) = c \qquad (5.8.28)$$

222   5   DEFORMATION

where c is a constant. Consequently, we get
$$f(x_2) = \tfrac{1}{6}kx_2^3 - cx_2 + c_1, \qquad g(x_1) = cx_1 + c_2 \qquad (5.8.29)$$
where c_1 and c_2 are arbitrary constants. Expressions (5.8.25) and (5.8.26) now become
$$u_1 = \tfrac{1}{6}k(2x_1^3 - 6x_1 x_2^2 + x_2^3) - cx_2 + c_1 \qquad (5.8.30)$$
$$u_2 = \tfrac{1}{2}kx_1 x_2^2 + cx_1 + c_2 \qquad (5.8.31)$$
These are the displacement components associated with the given e_{ij} when the compatibility conditions are obeyed. ∎

EXAMPLE 5.8.4. For each of the following strain systems obtain the corresponding displacement components representing pure deformation, given that there is no displacement at the origin:

(i) $\; e_{11} = \dfrac{1}{E} N, \qquad e_{22} = e_{33} = -\dfrac{\nu}{E} N, \qquad e_{12} = e_{23} = e_{31} = 0 \quad (5.8.32)$

(ii) $\; e_{11} = e_{22} = e_{33} = e_{12} = 0, \qquad e_{13} = -\dfrac{M}{\mu \pi a b^3} x_2,$

$\qquad e_{23} = \dfrac{M}{\mu \pi a^3 b} x_1 \hfill (5.8.33)$

(iii) $\; e_{12} = e_{22} = -\dfrac{\nu \rho g}{E}(l - x_3), \qquad e_{33} = \dfrac{\rho g}{E}(l - x_3),$

$\qquad e_{11} = e_{13} = e_{23} = 0 \hfill (5.8.34)$

(iv) $\; e_{11} = e_{22} = \dfrac{\nu M}{EI} x_1, \qquad e_{33} = -\dfrac{M}{EI} x_1,$

$\qquad e_{12} = e_{23} = e_{31} = 0 \hfill (5.8.35)$

Here, E, N, ν, M, μ, π, a, b, l, ρ, g, I are constants.

Solution It may be verified that the strain systems (5.8.32) to (5.8.35) satisfy the compatibility conditions.

(i) Substituting for e_{ij} from (5.8.32) in the strain-displacement relations (5.6.4)', we obtain the following equations:

$$u_{1,1} = \dfrac{1}{E} N, \qquad u_{2,2} = u_{3,3} = -\dfrac{\nu}{E} N, \qquad (5.8.36)$$

$$u_{1,2} + u_{2,1} = 0, \qquad u_{2,3} + u_{3,2} = 0, \qquad u_{3,1} + u_{1,3} = 0 \quad (5.8.37)$$

5.8 COMPATIBILITY CONDITIONS

The righthand sides of equations (5.8.36) are constants; hence the solutions of these equations are linear functions of x_1, x_2, x_3 and are given by

$$u_1 = \frac{1}{E} N x_1 + a_1 x_2 + b_1 x_3 + c_1$$

$$u_2 = -\frac{v}{E} N x_2 + a_2 x_3 + b_2 x_1 + c_2 \qquad (5.8.38)$$

$$u_3 = -\frac{v}{E} N x_3 + a_3 x_1 + b_3 x_2 + c_3$$

where a_i, b_i, c_i are constants.

Using (5.8.38) in equations (5.8.37), we obtain the following relations connecting a_i and b_i:

$$a_1 = -b_2, \qquad a_2 = -b_3, \qquad a_3 = -b_1 \qquad (5.8.39)$$

Since there are no rigid-body rotations, we have Ω = skw $\nabla \mathbf{u} \equiv \mathbf{0}$ so that $u_{i,j} = u_{j,i}$. The solutions (5.8.38) therefore yield (see Example 5.7.2):

$$a_1 = b_2, \qquad a_2 = b_3, \qquad a_3 = b_1 \qquad (5.8.40)$$

The conditions (5.8.39) and (5.8.40) imply that a_i and b_i are 0. Also, since there is no displacement at the origin, (5.8.38) yield $c_i = 0$. Thus, (5.8.38) become

$$u_1 = \frac{1}{E} N x_1, \qquad u_2 = -\frac{v}{E} N x_2, \qquad u_3 = -\frac{v}{E} N x_3 \qquad (5.8.41)$$

These are the displacement components associated with the strain components given by (5.8.32).

(ii) Substituting for e_{ij} from (5.8.33) in the strain-displacement relations (5.6.4)', we obtain the following equations

$$u_{1,1} = 0, \qquad u_{2,2} = 0, \qquad u_{3,3} = 0, \qquad u_{1,2} + u_{2,1} = 0$$

$$u_{1,3} + u_{3,1} = -\frac{2M}{\mu \pi a b^3} x_2, \qquad u_{2,3} + u_{3,2} = \frac{2M}{\mu \pi a^3 b} x_1 \qquad (5.8.42)$$

The form of these equations suggests that u_1, u_2, u_3 are of the form

$$u_1 = a_1 x_2 x_3 + b_1 x_2 + c_1 x_3 + d_1$$

$$u_2 = a_2 x_3 x_1 + b_2 x_3 + c_2 x_1 + d_2 \qquad (5.8.43)$$

$$u_3 = a_3 x_1 x_2 + b_3 x_1 + c_3 x_2 + d_3$$

where a_i, b_i, c_i and d_i are constants. Substituting (5.8.43) into (5.8.42) we

224   5 DEFORMATION

find that

$$a_2 = -a_1 = \frac{M(a^2 + b^2)}{\mu\pi a^3 b^3}, \qquad a_3 = -\frac{M}{\mu\pi a^3 b^3}(a^2 - b^2) \quad (5.8.44)$$

$$b_1 = -c_2, \qquad b_2 = -c_3, \qquad b_3 = -c_1 \quad (5.8.45)$$

Since there is no rigid-body rotation, we have $u_{i,j} - u_{j,i} = 0$. In view of this, we find from (5.8.43) and (5.8.45) that b_i and c_i are all 0. Further, since there is no displacement at O, the constants d_i are all 0. Thus, the solutions (5.8.43) reduce, with the use of (5.8.44), to

$$u_1 = -\frac{M(a^2 + b^2)}{\mu\pi a^3 b^3} x_2 x_3$$

$$u_2 = \frac{M(a^2 + b^2)}{\mu\pi a^3 b^3} x_3 x_1 \quad (5.8.46)$$

$$u_3 = -\frac{M}{\mu\pi a^3 b^3}(a^2 - b^2) x_1 x_2$$

These are the displacement components associated with the strains given by (5.8.33).

(iii) Substituting for e_{ij} from (5.8.34) into the strain-displacement relations (5.6.4)', we obtain the following equations:

$$u_{1,1} = u_{2,2} = -\frac{\nu\rho g}{E}(l - x_3), \qquad u_{3,3} = \frac{\rho g}{E}(l - x_3) \quad (5.8.47)$$

$$u_{1,2} + u_{2,1} = 0, \qquad u_{2,3} + u_{3,2} = 0, \qquad u_{3,1} + u_{1,3} = 0 \quad (5.8.48)$$

Integrating the last of the equations in (5.8.47) gives

$$u_3 = \frac{\rho g}{E}\left(l - \frac{1}{2}x_3\right)x_3 + f_3(x_1, x_2) \quad (5.8.49)$$

where $f_3(x_1, x_2)$ is a function of x_1, x_2 only. Using (5.8.49) in the last two of equations (5.8.48) and integrating the resulting equations we obtain

$$u_1 = -x_3 f_{3,1} + f_1(x_1, x_2)$$
$$u_2 = -x_3 f_{3,2} + f_2(x_1, x_2) \quad (5.8.50)$$

where f_1 and f_2 are functions of x_1, x_2 only. Substituting (5.8.50) into the first two equations in (5.8.47) gives

$$f_{1,1} = f_{2,2} = -\frac{\nu\rho g l}{E}, \qquad f_{3,11} = f_{3,22} = -\frac{\nu\rho g}{E} \quad (5.8.51)$$

5.8 COMPATIBILITY CONDITIONS

Similarly, from (5.8.50) and first of (5.8.48) we obtain

$$f_{3,12} = 0, \quad f_{1,2} = -f_{2,1} \tag{5.8.52}$$

Further, (5.8.49) and the last two equations in (5.8.48) yield

$$f_{3,23} = 0, \quad f_{3,13} = 0 \tag{5.8.53}$$

The conditions (5.8.51) to (5.8.53) imply that f_i are of the form

$$f_1(x_1, x_2) = -\frac{\nu \rho g l}{E} x_1 + a_1 x_2 + c_1$$

$$f_2(x_1, x_2) = -\frac{\nu \rho g l}{E} x_2 - a_1 x_1 + c_2 \tag{5.8.54}$$

$$f_3(x_1, x_2) = -\frac{\nu \rho g}{2E}(x_1^2 + x_2^2) + a_3 x_1 + b_3 x_2 + c_3$$

where a_1, a_3, b_3 and c_i are constants. Using these in (5.8.49) and (5.8.50) we obtain

$$u_1 = -\frac{\nu \rho g}{E}(l - x_3)x_1 - a_3 x_3 + a_1 x_2 + c_1$$

$$u_2 = -\frac{\nu \rho g}{E}(l - x_3)x_2 - a_3 x_3 - a_1 x_1 + c_2 \tag{5.8.55}$$

$$u_3 = \frac{-\rho g}{2E}[(x_3 - 2l)x_3 + \nu(x_1^2 + x_2^2)] + a_3 x_1 + b_3 x_2 + c_3$$

Since there is no rigid-rotation, we have $u_{i,j} - u_{j,i} = 0$, and expressions (5.8.55) yield $a_3 = a_1 = b_3 = 0$. Also, since there is no displacement at the origin, $c_i = 0$. Consequently, (5.8.55) become

$$u_1 = -\frac{\nu \rho g}{E}(l - x_3)x_1$$

$$u_2 = -\frac{\nu \rho g}{E}(l - x_3)x_2 \tag{5.8.56}$$

$$u_3 = -\frac{\rho g}{2E}[(x_3 - 2l)x_3 + \nu(x_1^2 + x_2^2)]$$

These are the displacement components associated with the strains given by (5.8.34).

226   5 DEFORMATION

(iv) Substituting for e_{ij} from (5.8.35) into the strain-displacement relations (5.6.4)' we obtain the following equations:

$$u_{1,1} = u_{2,2} = \frac{\nu M}{EI}x_1, \qquad u_{3,3} = -\frac{M}{EI}x_1 \qquad (5.8.57)$$

$$u_{1,2} + u_{2,1} = 0, \qquad u_{2,3} + u_{3,2} = 0 \qquad u_{3,1} + u_{1,3} = 0 \quad (5.8.58)$$

From the third equation in (5.8.57), we get

$$u_3 = -\frac{M}{EI}x_1 x_3 + f_3(x_1, x_2) \qquad (5.8.59)$$

where f_3 is a function of x_1, x_2. The last two equations in (5.8.58) then give

$$u_1 = \frac{M}{2EI}x_3^2 - x_3 f_{3,1} + f_1(x_1, x_2) \qquad (5.8.60)$$

$$u_2 = -x_3 f_{1,2} + f_2(x_1, x_2) \qquad (5.8.61)$$

where f_1 and f_2 are functions of x_1, x_2. Substituting for u_1 and u_2 from (5.8.60) and (5.8.61) in the first two equations in (5.8.57), we obtain the following relations:

$$f_{3,11} = 0, \qquad f_{3,22} = 0 \qquad (5.8.62)$$

$$f_1 = \frac{\nu M}{2EI}x_1^2 + g_1(x_2), \qquad f_2 = \frac{\nu M}{2EI}x_1 x_2 + g_2(x_1) \qquad (5.8.63)$$

where g_1 and g_2 are functions of x_2 and x_1 respectively.

Using (5.8.60), (5.8.61) and (5.8.63) in the first equation of (5.8.58), we obtain the following relations:

$$f_{3,12} = 0 \qquad (5.8.64)$$

$$g_1 = -\frac{\nu M}{2EI}x_2^2 + a_1 x_2 + c_1, \qquad g_2 = -a_1 x_1 + c_2 \qquad (5.8.65)$$

where a_1, c_1, c_2 are constants.

From (5.8.62) and (5.8.64) we find that

$$f_3 = a_3 x_1 + b_3 x_2 + c_3 \qquad (5.8.66)$$

where a_3, b_3, c_3 are constants.

Substituting for f_i from (5.8.63), (5.8.65) and (5.8.66) in (5.8.59) to (5.8.61), we obtain

$$u_1 = \frac{M}{2EI}(x_3^2 + vx_1^2 - vx_2^2) + a_1 x_2 - a_3 x_3 + c_1$$

$$u_2 = \frac{Mv}{EI} x_1 x_2 - a_1 x_1 - b_3 x_3 + c_2 \quad (5.8.67)$$

$$u_3 = -\frac{M}{EI} x_3 x_1 + a_3 x_1 + b_3 x_2 + c_3$$

Since there is no rigid-body rotation, we have $u_{i,j} - u_{j,i} = 0$, and (5.8.67) yield $a_1 = a_3 = b_3 = 0$. Also, since there is no displacement at O, we get $c_i = 0$. Thus, (5.8.67) reduce to

$$u_1 = \frac{M}{2EI}(x_3^2 + vx_1^2 - vx_2^2), \quad u_2 = \frac{Mv}{EI} x_1 x_2, \quad u_3 = -\frac{M}{EI} x_3 x_1 \quad (5.8.68)$$

These are the displacement components associated with the strains given by (5.8.35). ∎

5.9 PRINCIPAL STRAINS

It has been seen that, under a deformation, a material arc generally changes in length and orientation. From a particle $\mathcal{P}$ there emerge infinitely many material arcs, and each of these arcs generally changes in length and orientation. We now consider the arcs that change only in length, under small deformation.

As in Section 5.3, let $\mathbf{a}^0$ and $\mathbf{a}$ be unit vectors representing the initial and final orientations of a material arc $d\mathcal{C}$ initiating from $\mathcal{P}$. In the case of small deformation, the stretch of $d\mathcal{C}$ is fully effected by the strain tensor $\mathbf{E}$ and the rotation is fully effected by the rotation tensor Ω. Hence, $d\mathcal{C}$ has no change in orientation if and only if $\Omega = \mathbf{0}$ and $\mathbf{a} = \mathbf{a}^0$. A necessary and sufficient condition for this may be obtained by setting $\Omega = \mathbf{0}$ and $\mathbf{a} = \mathbf{a}^0$ in (5.7.29). The condition so obtained is

$$\mathbf{E}\mathbf{a}^0 = (\eta - 1)\mathbf{a}^0 = e\mathbf{a}^0 \quad (5.9.1)$$

where $e = \eta - 1$ is the normal strain experienced by $d\mathcal{C}$. Evidently, this condition holds if and only if $\mathbf{a}^0$ is an eigenvector of $\mathbf{E}$ and e is the corresponding eigenvalue. Since $\mathbf{E}$ is a symmetric tensor, it has exactly three (not necessarily distinct) eigenvalues and at least three mutually orthogonal eigenvectors (by virtue of Theorems 2.13.2 and 2.13.4). Hence there exist at least three mutually orthogonal unit vectors and three normal strains for which condition (5.9.1) holds. This means that at least three mutually orthogonal material arcs at $\mathcal{P}$ change only in length. The normal strain of such an arc is called a *principal strain*, and its direction is called a *principal direction of strain* at $\mathcal{P}$. The material surface elements at $\mathcal{P}$ that are orthogonal to these principal directions are called the *principal planes of strain* at $\mathcal{P}$.

Thus, at a particle $\mathcal{P}$, there exist (i) at least three mutually orthogonal principal directions of strain, (ii) at least three mutually orthogonal principal planes of strain, and (iii) exactly three principal strains (which are not necessarily distinct).

Since the principal strains are the eigenvalues of $\mathbf{E}$, these are the roots of the characteristic equation

$$-\Lambda^3 + I_\mathbf{E}\Lambda^2 - II_\mathbf{E}\Lambda + III_\mathbf{E} = 0 \tag{5.9.2}$$

where $I_\mathbf{E}$, $II_\mathbf{E}$ and $III_\mathbf{E}$ are the fundamental invariants of the tensor $\mathbf{E}$, which by virtue of (2.11.1), (2.11.4) and (2.11.11) are given by

$$I_\mathbf{E} = e_{kk}, \qquad II_\mathbf{E} = \tfrac{1}{2}(e_{ii}e_{jj} - e_{ij}e_{ji}), \qquad III_\mathbf{E} = \det(e_{ij}) \tag{5.9.3}$$

These invariants are called the *small strain invariants*.

Once the three principal strains, say e_1, e_2, e_3, are found by solving the cubic (5.9.2), the corresponding principal directions of strain can be found by solving the vector equation (5.9.1) for $\mathbf{a}^0 = \mathbf{a}$ in the cases: $\Lambda = e_1$, $\Lambda = e_2$ and $\Lambda = e_3$. With respect to a set of axes chosen along the principal directions so determined—such axes are called *principal axes of strain*—the matrix of $\mathbf{E}$ (or briefly the *strain matrix*) is given as follows (by virtue of Theorem 2.13.5):

$$[e_{ij}] = \begin{bmatrix} e_1 & 0 & 0 \\ 0 & e_2 & 0 \\ 0 & 0 & e_3 \end{bmatrix} \tag{5.9.4}$$

In view of the geometrical meanings of e_{ij} noted in Section 5.6, it is verified from (5.9.4) that e_1, e_2, e_3 represent the normal strains of material arcs initially lying along the principal directions of strain. Since the shear strains are all 0 in the representation (5.9.4), it is also verified that these arcs remain in the same directions after deformation also.

Since e_1, e_2, e_3 are the roots of the cubic (5.9.2), the strain invariants have the following expressions in terms of the principal strains:

$$I_\mathbf{E} = e_1 + e_2 + e_3, \qquad II_\mathbf{E} = e_1 e_2 + e_2 e_3 + e_3 e_1, \qquad III_\mathbf{E} = e_1 e_2 e_3 \tag{5.9.5}$$

5.9 PRINCIPAL STRAINS

EXAMPLE 5.9.1 For the small deformation considered in Example 5.7.1, find the strain invariants, principal strains and the principal directions of strain at the point (1, 1, 0).

Solution For the deformation considered here, the strain matrix is given by (5.7.13). At the point (1, 1, 0), this matrix becomes

$$[E] = \begin{bmatrix} \alpha & \alpha & 0 \\ \alpha & \alpha & 0 \\ 0 & 0 & 4\alpha \end{bmatrix} \quad (5.9.6)$$

whose characteristic equation is found as

$$\Lambda(\Lambda - 2\alpha)(\Lambda - 4\alpha) = 0 \quad (5.9.7)$$

Hence the principal strains, which are the roots of equation (5.9.7), are

$$e_1 = 0, \quad e_2 = 2\alpha, \quad e_3 = 4\alpha \quad (5.9.8)$$

The strain invariants now follow from (5.9.5) as

$$I_E = 6\alpha, \quad II_E = 8\alpha^2, \quad III_E = 0 \quad (5.9.9)$$

Since $e_1 = 0$, the eigenvector **a** corresponding to e_1 is given by $\mathbf{Ea} = e_1\mathbf{a} = \mathbf{0}$, which on using (5.9.6) yields

$$a_1 + a_2 = 0, \quad a_1 + a_2 = 0, \quad 4a_3 = 0$$

so that

$$a_1 = -a_2 = \pm \frac{1}{\sqrt{2}}, \quad a_3 = 0 \quad (5.9.10)$$

Here we have used the fact that **a** is a unit vector.

Thus, an eigenvector associated with the principal strain $e_1 = 0$ is

$$\mathbf{a}_1 = \frac{1}{\sqrt{2}}(\mathbf{e}_1 - \mathbf{e}_2) \quad (5.9.11)$$

Similarly, the eigenvectors corresponding to $e_2 = 2\alpha$ and $e_3 = 4\alpha$ are obtained as

$$\mathbf{a}_2 = \frac{1}{\sqrt{2}}(\mathbf{e}_1 + \mathbf{e}_2), \quad \mathbf{a}_3 = \mathbf{e}_3 \quad (5.9.12)$$

The vectors $\mathbf{a}_1, \mathbf{a}_2, \mathbf{a}_3$ just determined specify the required principal directions. ∎

EXAMPLE 5.9.2 For small deformation, show that the normal strain of a material arc assumes an extremum value when the element lays along a principal direction of strain.

Solution In terms of its initial orientation, the normal strain of a material arc is given by

$$e \approx u_{i;j}a_i^0 a_j^0 = e_{ij}a_i^0 a_j^0 \tag{5.9.13}$$

see (5.6.6). We have to find a_i^0 for which e is an extremum. Since a_i^0 are components of a unit vector, we have $a_k^0 a_k^0 = 1$. Thus, e assumes extreme values when a_i^0 obey the equations

$$\frac{\partial}{\partial a_i^0}\{e - \alpha(a_k^0 a_k^0 - 1)\} = 0 \tag{5.9.14}$$

where α is a Lagrangian multiplier.

Substituting for e from (5.9.13), the conditions (5.9.14) become

$$(e_{ij} - \alpha\delta_{ij})a_j^0 = 0 \tag{5.9.15}$$

These conditions are satisfied if and only if $\mathbf{a}^0$ is along a principal direction of strain. Thus, e assumes an extreme value when the material arc lies along a principal direction of strain. ∎

Note: This result obtained is analogous to (and a consequence of) the result obtained in Example 5.4.5.

5.10

STRAIN DEVIATOR

In Section 2.12, we have seen that every second-order tensor can be decomposed uniquely into a spherical part and a deviator part. Employing this decomposition to the tensor $\mathbf{E}$, we get

$$\mathbf{E} = \tfrac{1}{3}(\operatorname{tr} \mathbf{E})\mathbf{I} + \mathbf{E}^{(d)} \tag{5.10.1}$$

We note that the trace of the spherical part of $\mathbf{E}$, which is equal to the trace of $\mathbf{E}$, represents dilatation. As such, the spherical part of $\mathbf{E}$ accounts for the volume change of a material element, under deformation. (In fact, the adjective *spherical part* is motivated by this interpretation.) Consequently, the deviator part $\mathbf{E}^{(d)}$ accounts for the change in shape of the element. The tensor $\mathbf{E}^{(d)}$ is called the (infinitesimal) *strain deviator tensor*. The components $e_{ij}^{(d)}$ of $\mathbf{E}^{(d)}$ are called the (infinitesimal) *strain deviator components* and the matrix $[e_{ij}^{(d)}]$ is called the *strain deviator matrix*.

In Example 2.13.7, it has been shown that a tensor and its deviator part have the same eigenvectors. It therefore follows that the principal directions of $\mathbf{E}^{(d)}$ coincide with those of $\mathbf{E}$. With respect to a set of principal axes, the strain deviator matrix is purely diagonal, the diagonal elements being the eigenvalues of $\mathbf{E}^{(d)}$. These eigenvalues, $e_1^{(d)}, e_2^{(d)}, e_3^{(d)}$, are called the *principal deviator strains*. By virtue of expressions (2.13.30), we note that $e_i^{(d)}$ are related to the principal strains e_i through the following relations:

$$e_1^{(d)} = \tfrac{1}{3}(2e_1 - e_2 - e_3),$$
$$e_2^{(d)} = \tfrac{1}{3}(2e_2 - e_3 - e_1), \qquad (5.10.2)$$
$$e_3^{(d)} = \tfrac{1}{3}(2e_3 - e_1 - e_2)$$

Since the deviator part of a tensor has zero trace, we have $I_{\mathbf{E}^{(d)}} = 0$. By virtue of expressions (2.13.31) and (2.13.33), the other two fundamental invariants of $\mathbf{E}^{(d)}$ are given as follows:

$$\begin{aligned} II_{\mathbf{E}^{(d)}} &= e_1^{(d)}e_2^{(d)} + e_2^{(d)}e_3^{(d)} + e_3^{(d)}e_1^{(d)} \\ &= -\tfrac{1}{6}\{(e_1 - e_2)^2 + (e_2 - e_3)^2 + (e_3 - e_1)^2\} \\ III_{\mathbf{E}^{(d)}} &= e_1^{(d)}e_2^{(d)}e_3^{(d)} \\ &= \tfrac{1}{27}(2e_1 - e_2 - e_3)(2e_2 - e_3 - e_1)(2e_3 - e_1 - e_2) \end{aligned} \qquad (5.10.3)$$

These are called the *strain-deviator invariants*.

EXAMPLE 5.10.1 For the small deformation considered in Example 5.7.1, find (i) the strain deviator matrix, (ii) principal deviator strains, and (iii) strain deviator invariants, at the point $(1, 1, 0)$.

Solution For the deformation considered here, the strain matrix at the point $(1, 1, 0)$ is given by (5.9.6). By use of (5.10.1), we find that the strain deviator matrix is given by

$$[\mathbf{E}^{(d)}] = [\mathbf{E}] - \tfrac{1}{3}(\operatorname{tr} \mathbf{E})[\mathbf{I}] \qquad (5.10.4)$$

which, on using (5.9.6), yields

$$[\mathbf{E}^{(d)}] = \begin{bmatrix} -\alpha & \alpha & 0 \\ \alpha & -\alpha & 0 \\ 0 & 0 & 2\alpha \end{bmatrix} \qquad (5.10.5)$$

Recalling that the principal strains are as given by (5.9.8) and using (5.10.2), we obtain the corresponding principal deviator strains as

$$e_1^{(d)} = -2\alpha, \qquad e_2^{(d)} = 0, \qquad e_3^{(d)} = \tfrac{4}{3}\alpha \qquad (5.10.6)$$

Using (5.10.6) in (5.10.3), we get the strain deviator invariants as

$$II_{\mathbf{E}^{(d)}} = -\tfrac{8}{3}\alpha^2, \qquad III_{\mathbf{E}^{(d)}} = 0 \qquad (5.10.7)$$

EXAMPLE 5.10.2 Show that

$$\frac{\partial}{\partial e_{ij}}\{II_{E^{(d)}}\} = -e_{ij}^{(d)} \qquad (5.10.8)$$

Solution By virtue of expression (2.12.11), we have

$$II_{E^{(d)}} = \tfrac{1}{2}[\tfrac{1}{3}e_{kk}^2 - e_{ij}e_{ji}] \qquad (5.10.9)$$

Hence

$$\frac{\partial}{\partial e_{ij}}[II_{E^{(d)}}] = [\tfrac{1}{3}\delta_{ij}e_{kk} - e_{ij}] = -e_{ij}^{(d)}$$

which is (5.10.8). ∎

5.11
EXERCISES

1. For the deformations defined by the following sets of equations, find the tensors F, F^{-1}, Q, U and V.

(i) $x_1 = \alpha x_1^0$, $\quad x_2 = \beta x_2^0$, $\quad x_3 = \beta x_3^0$

(ii) $x_1 = x_1^0$, $\quad x_2 = x_2^0 - \alpha x_3^0$, $\quad x_3 = x_3^0 + \alpha x_2^0$

(iii) $x_1 = \alpha x_1^0 + \beta x_2^0$, $\quad x_2 = -\alpha x_1^0 + \beta x_2^0$, $\quad x_3 = \gamma x_3^0$

(iv) $x_1 = x_1^0 - x_2^0 x_3^0$, $\quad x_2 = x_2^0 + x_1^0 x_3^0$, $\quad x_3 = x_3^0$

Here α, β and γ are positive constants. Which of these deformations are isochoric? Also, find the principal stretches in each case.

2. Give the geometrical descriptions of the deformations defined by the sets of equations (i), (ii) and (iv) of Exercise 1.

3. Deduce the relation (5.2.33) from the relation (5.2.32) and vice versa.

4. Prove the following identities:

(i) $\quad \varepsilon_{jmn} x_{j,k}^0 = \dfrac{1}{J} \varepsilon_{kpq} X_{p,m} X_{q,n}$

(ii) $\quad \varepsilon_{kpq} x_{j,k}^0 = \dfrac{1}{J} \varepsilon_{jmn} X_{p;m} X_{q;n}$

(iii) $\quad \varepsilon_{jmn} X_{j;k} = J \varepsilon_{kpq} x_{p,m}^0 x_{q,n}^0$

(iv) $\quad \varepsilon_{kpq} X_{j;k} = J \varepsilon_{jmn} x_{p,m}^0 x_{q,n}^0$

5.11 EXERCISES

5. Show that the deformation given by the equations
$$x_1 = x_1^0, \quad x_2 = x_2^0 + \alpha x_3^0, \quad x_3 = x_3^0 + \alpha x_2^0$$
where $\alpha (\neq 1)$ is a constant, is just a triaxial stretch. Deduce that the particles that lie on the circle
$$x_1^0 = 0, \quad (x_2^0)^2 + (x_3^0)^2 = (1 - \alpha^2)^{-1}$$
before deformation move over to the ellipse
$$x_1 = 0, \quad (1 + \alpha^2)x_2^2 - 4\alpha x_2 x_3 + (1 + \alpha^2)x_3^2 = 1 - \alpha^2$$

6. Under the deformation defined by the set of equations iv of Exercise 1 find the surface into which the cylinder $(x_1^0)^2 + (x_2^0)^2 = a^2$ deforms.

7. Show that, under a homogeneous deformation,
  (i) Parallel plane elements transform to parallel plane elements;
  (ii) Parallel straight line elements transform to parallel straight line elements;
  (iii) spherical surface elements transform to ellipsoidal surface elements.

8. Consider a homogeneous deformation defined by $x_i = a_{ij} x_j^0 + c_i$, where c_i and a_{ij} are constants. Show that, in this deformation, the particle that lies on a spherical surface of radius a in the current configuration initially lay on the surface of an ellipsoid. Show further that this ellipsoid is a sphere of radius b if and only if $a^2 a_{ki} a_{kj} = b^2 \delta_{ij}$.

9. For the deformations defined by the following sets of equations, find the tensors **C**, **B** and **G**:

  (i) $x_1 = x_1^0, \quad x_2 = x_2^0 + \alpha x_1^0, \quad x_3 = x_3^0$

  (ii) $x_1 = x_1^0, \quad x_2 = x_2^0 - \alpha x_3, \quad x_3 = x_3^0 + \alpha x_2^0$

  (iii) $x_1 = \sqrt{(2\alpha x_1^0 + \beta)}, \quad x_2 = \gamma x_2^0, \quad x_3 = \delta x_3^0$

  (iv) $x_1 = x_1^0 + x_2^0 \tan \alpha, \quad x_2 = x_2^0, \quad x_3 = x_3^0$

  (v) $x_1 = x_1^0 \cos(\alpha x_3^0) + x_2^0 \sin(\alpha x_3^0), \quad x_2 = -x_1^0 \sin(\alpha x_3^0) + x_2^0 \cos(\alpha x_3^0),$
  $x_3 = (1 + \alpha\beta)x_3^0$

Here α, β, γ and δ are nonzero constants.

10. For the deformations defined by the following sets of equations, find the tensors $\mathbf{B}^{-1}$ and **A**:

  (i) $x_1^2 = 2\alpha x_1^0 + \beta, \quad x_2 = \gamma x_2^0, \quad x_3 = \delta x_3^0$

  (ii) $x_1^0 = x_1 \cos(\alpha x_3) - x_2 \sin(\alpha x_3), \quad x_2^0 = x_1 \sin(\alpha x_3) + x_2 \cos(\alpha x_3),$
  $x_3^0 = (1 + \alpha\beta)x_3$

5 DEFORMATION

11. For the deformation defined by the equations

$$x_1 = x_1^0, \qquad x_2 = x_2^0 + \alpha x_1^0, \qquad x_3 = x_3^0$$

where α is a constant, show that

$$I_B = II_B = III_{B^{-1}} = 3 + \alpha^2, \qquad III_B = 1$$

12. Prove the following identities:

$$I_C = 3 + 2I_G$$

$$II_C = 3 + 4I_G + 4II_G$$

$$III_C = 1 + 2I_G + 4II_G + 8III_G$$

13. Show that a deformation is a rigid-body transformation if and only if $I_C = II_C = 3$ and $III_C = 1$.

14. For the deformation defined by the equations

$$x_1 = x_1^0 + \alpha x_2^0, \qquad x_2 = x_2^0 + \alpha x_1^0, \qquad x_3 = x_3^0$$

where α is a constant, find the normal strain of the material arc initially lying along the vector $\mathbf{e}_2 + \mathbf{e}_3$.

15. Show that under the deformation considered in the Exercise 11, material arcs initially lying along the 3 axis retain their lengths.

16. For the deformation defined by the equations

$$x_1 = x_1^0, \qquad x_2 = x_2^0, \qquad x_3 = x_3^0 + \alpha x_2^0$$

where α is a constant, find the material arcs initially lying parallel to the 23 plane, for which the normal strain is 0.

17. Derive the expressions (5.4.33) and (5.4.34).

18. Show that the final angle θ between two material arcs whose initial directions are specified by the unit vectors $\mathbf{a}^0$ and $\bar{\mathbf{a}}^0$ inclined at an angle θ_0 is given by

$$\cos \theta = \frac{2g_{ij} a_i^0 \bar{a}_j^0 + \cos \theta_0}{(1 + 2g_{ij} a_i^0 a_j^0)^{1/2}(1 + 2g_{ij} \bar{a}_i^0 \bar{a}_j^0)^{1/2}}$$

Obtain the counterpart of this expression in the spatial description.

19. Show that the angle between two material arcs initially represented by the unit vectors $\mathbf{a}^0$ and $\bar{\mathbf{a}}^0$ remains unchanged if and only if

$$\mathbf{a}^0 \cdot \mathbf{C}\bar{\mathbf{a}}^0 = (\mathbf{a}^0 \cdot \bar{\mathbf{a}}^0)\{(\mathbf{a}^0 \cdot \mathbf{C}\mathbf{a}^0)(\bar{\mathbf{a}}^0 \cdot \mathbf{C}\bar{\mathbf{a}}^0)\}^{1/2}$$

Express this condition in terms of the unit vectors $\mathbf{a}$ and $\bar{\mathbf{a}}$ along which the arcs lay in the deformed configuration.

20. Show that the deformation defined by the equations
$$x_1 = 2x_1^0 + x_2^0, \qquad x_2 = x_1^0 + 2x_2^0, \qquad x_3 = x_3^0$$
does not change the angle between the material arcs lying along $\mathbf{e}_1$ and $\mathbf{e}_2$ in the initial configuration.

21. For the deformation defined by the equations
$$x_1 = x_1^0 - x_2^0 + x_3^0, \qquad x_2 = x_2^0 - x_3^0 + x_1^0, \qquad x_3 = x_3^0 - x_1^0 + x_2^0$$
find (i) the stretch in the direction $\mathbf{a}^0 = (1/\sqrt{2})(\mathbf{e}_1 + \mathbf{e}_2)$, and (ii) the angle in the deformed configuration between the material arcs that were initially in the directions of $\mathbf{a}^0$ and $\bar{\mathbf{a}}^0 = \mathbf{e}_2$.

22. For the deformation defined by the equations
$$x_1 = 3x_1^0 + x_2^0, \qquad x_2 = x_2^0 + x_3^0, \qquad x_3 = 2x_3^0 - x_2^0$$
find (i) the normal strain of a material arc initially having direction ratios $1:1:1$, (ii) the initial direction ratios of the material arc that finally ends up in the 3 direction, and (iii) the change in the angle between the material arcs initially having direction ratios $1:0:0$ and $1:1:1$.

23. For the deformation considered in Exercise 22, find (i) the direction of the normal of the material surface element in the final configuration given that normal had the direction ratios $1:1:1$ in the initial configuration, (ii) the ratio of the area of this surface element in the final configuration to its area in the initial configuration.

24. For the deformation defined by
$$x_1^2 = 4x_1^0 \cos^2(x_2^0/2), \qquad x_2^2 = 4x_1^0 \sin^2(x_2^0/2), \qquad x_3 = x_3^0, \qquad x_1^0 > 0$$
show that the lengths of sides, angles and volume of the element $dx_1\, dx_2\, dx_3$ located initially at the point $(1, 1, 0)$ remain unchanged.

25. Show that a material surface element initially orthogonal to a unit vector $\mathbf{n}^0$ retains its area, under deformation, if and only if
$$\mathbf{n}^0 \cdot \mathbf{C}^{-1} \mathbf{n}^0 = \frac{1}{J^2}$$

26. Prove that
$$[\mathbf{C}^{-1}]_{ij} = \frac{2}{J} \frac{\partial J}{\partial c_{ij}} = \frac{1}{J} \frac{\partial J}{\partial g_{ij}}$$

27. Show that
$$\frac{dV}{dV_0} = (III_{\mathbf{B}^{-1}})^{1/2}$$

236   **5 DEFORMATION**

28. For the deformation defined by

$$x_1 = x_1(x_1^0, x_2^0), \qquad x_2 = x_2(x_1^0, x_2^0), \qquad x_3 = x_3^0$$

show that one of the principal stretches is 1. Show further that the deformation is isochoric if and only if the product of the other two principal stretches is 1.

29. If η_i are the principal stretches, show that $\frac{1}{2}(\eta_i^2 - 1)$ are the principal values of **G** and $\frac{1}{2}(1 - \eta_i^{-2})$ are the principal values of **A**.

30. Show that the stretch of a material arc assumes an extreme value when it finally lies along a principal direction of **B**.

31. For the deformations defined by the following sets of equations find the displacement components in material and spatial forms:

(i) $x_1 = x_1^0 t^2 + 2x_2^0 t + x_1^0$
     $x_2 = 2x_1^0 t^2 + x_2^0 t + x_2^0$
     $x_3 = \frac{1}{2}x_3^0 t + x_3^0$

(ii) $x_1 = x_1^0 e^t + x_3^0(e^t - 1)$,
     $x_2 = x_2^0 + x_3^0(e^t - e^{-t})$
     $x_3 = x_3^0$

32. For each of the following sets of displacement components, find the Lagrangian and Eulerian strain tensors:

(i) $u_1 = \alpha x_3^0$, $u_2 = -\alpha x_3^0$, $u_3 = \alpha(x_2^0 - x_1^0)$ (here α is a constant)

(ii) $u_1 = (x_1^0)^2 + (x_2^0)^2$, $u_2 = 3x_1^0 + 4(x_2^0)^2$, $u_3 = 2(x_1^0)^3 + 4x_3^0$

(iii) $u_1 = (x_1^0)^2 + 2(x_2^0)^2 x_3^0 + x_2^0 x_3^0$, $u_2 = x_1^0 x_2^0 + x_1^0 x_3^0 + 3(x_1^0)^2 x_3^0$
     $u_3 = 0$

If $\alpha \ll 1$, show that case (i) corresponds to a rigid-body transformation.

33. For the deformation defined by the displacement vector

$$\mathbf{u} = x_1^0 x_2^0 \mathbf{e}_1 + x_2^0 x_3^0 \mathbf{e}_2 + 2(x_1^0 + x_2^0)\mathbf{e}_3$$

find (i) the normal strain of the material arc initially having direction ratios $1:-1:1$, (ii) the normal strain of the material arc finally having direction ratios $1:-1:1$, (iii) the final angle between the material arcs which initially lay along $\mathbf{e}_2$ and $\mathbf{e}_3$.

34. Prove the following:

(i) $\mathbf{F} + \mathbf{F}^{-1} = 2\mathbf{I} + (\nabla \mathbf{u})(\nabla^0 \mathbf{u})$

(ii) $\mathbf{B} = \mathbf{I} + \nabla^0 \mathbf{u} + \nabla^0 \mathbf{u}^T + (\nabla^0 \mathbf{u})(\nabla^0 \mathbf{u}^T)$

35. Prove the following:

(i) $[\text{curl } \mathbf{u}]_i = \dfrac{1}{J}\varepsilon_{pqr} x_{p;i} x_{r;q} = \dfrac{1}{J}\varepsilon_{pqr} X_{i;p} x_{r;q}$

(ii) $\varepsilon_{pqr} x_{r;q} = J x_{i;p}^0 [\text{curl } \mathbf{u}]_i = J x_{p,i}^0 [\text{curl } \mathbf{u}]_i$

5.11 EXERCISES

36. For small deformation, show that

(i) $\quad \dfrac{1}{J} \approx (1 - \text{tr } \mathbf{E})$

(ii) $\quad I_C \approx 3 + 2 \text{ tr } \mathbf{E}, \qquad II_C \approx 3 + 4 \text{ tr } \mathbf{E}, \qquad III_C \approx 1 + 2 \text{ tr } \mathbf{E}$

(iii) $\quad \mathbf{E} = \frac{1}{2}(\mathbf{F} + \mathbf{F}^T) - \mathbf{I} = \mathbf{I} - \frac{1}{2}\{\mathbf{F}^{-1} + (\mathbf{F}^T)^{-1}\}$

37. For the small deformation defined by the equations

$$x_1 = x_1^0 + \alpha x_3^0, \qquad x_2 = x_2^0, \qquad x_3 = x_3^0 - \alpha x_1^0$$

where α is a small nonzero constant, find the dilatation. Deduce the condition under which the deformation is isochoric.

38. Find the constants α and β such that the small deformation defined by

$$u_1 = \alpha x_1 + 3x_2, \qquad u_2 = x_1 - \beta x_2, \qquad u_3 = 3x_3$$

is isochoric.

39. For small deformations defined by the following sets of displacements, find the strain tensor, rotation tensor and rotation vector.

(i) $\quad u_1 = -\alpha x_2^0 x_3^0, \qquad u_2 = \alpha x_1^0 x_2^0, \qquad u_3 = 0$

(ii) $\quad u_1 = \alpha x_1^0 x_2^0, \qquad u_2 = \alpha x_2^0 x_3^0, \qquad u_3 = 2\alpha(x_1^0 + x_2^0)x_3^0$

(iii) $\quad u_1 = \alpha^2(x_1 - x_3)^2, \qquad u_2 = \alpha^2(x_2 + x_3)^2, \qquad u_3 = -\alpha x_1 x_2$

(iv) $\quad u_1 = 3\alpha(x_1^2 x_2 + 2), \qquad u_2 = \alpha(x_2^2 + 6x_1 x_3), \qquad u_3 = 3\alpha(2x_3^2 + x_2 x_3)$

(v) $\quad u_1 = \dfrac{\alpha}{r^3} x_1 x_3, \qquad u_2 = \dfrac{\alpha}{r^3} x_2 x_3, \qquad u_3 = \dfrac{\alpha}{r^3}(x_3^2 + kr^2)$

Here α is a small constant, k is a constant and $r^2 = x_i x_i \neq 0$.

40. For the deformation corresponding to the case iii of Exercise 39, find (i) the normal strain at the point $(0, 2, -1)$ along the direction $8:-1:4$; and (ii) the change in the right angle between the material arcs at $(0, 2, -1)$ lying along the directions $8:-1:4$ and $4:4:-7$.

41. For the deformation corresponding to the case v of Exercise 39, find the deformed shape of a spherical surface $r = $ constant.

42. The linear strain tensor at a point is given by

$$[e_{ij}] = \begin{bmatrix} 1 & -3 & \sqrt{2} \\ -3 & 1 & -\sqrt{2} \\ \sqrt{2} & -\sqrt{2} & 4 \end{bmatrix}$$

Find the normal strain in the direction $1:-1:\sqrt{2}$ and the shear strain between the directions $1:-1:\sqrt{2}$ and $-1:1:\sqrt{2}$.

238 5 DEFORMATION

43. Find which of the following values of e_{ij} are possible linear strains. Compute the corresponding displacements in the appropriate cases.

(i) $e_{12} = x_1 x_2$, other $e_{ij} = 0$

(ii) $e_{11} = 2x_1$, $e_{12} = x_1 + 2x_2$, $e_{22} = 2x_1$, $e_{33} = 2x_3$,
$e_{12} = e_{23} = 0$

(iii) $e_{11} = \alpha(x_1^2 + x_2^2)$, $e_{22} = \alpha x_2^2$, $e_{12} = 2\alpha x_1 x_2$,
$e_{13} = e_{23} = e_{33} = 0$ (α = constant)

(iv) $e_{11} = \alpha x_3(x_1^2 + x_2^2)$, $e_{22} = \alpha x_2^2 x_3$, $e_{12} = 2\alpha x_1 x_2 x_3$,
$e_{13} = e_{23} = e_{33} = 0$ (α = constant)

(v) $[e_{ij}] = \begin{bmatrix} x_1 + x_2 & x_1 & x_2 \\ x_1 & x_2 + x_3 & x_3 \\ x_2 & x_3 & x_1 + x_3 \end{bmatrix}$

(vi) $[e_{ij}] = \begin{bmatrix} x_1^2 & x_2^2 & x_1 x_3 \\ x_2^2 & x_3 & x_3^2 \\ x_1 x_3 & x_3^2 & 0 \end{bmatrix}$

44. Obtain the conditions under which the following values of e_{ij} are possible linear strains:

(i) $e_{11} = \alpha x_2^2$, $e_{22} = \alpha x_1^2$, $e_{12} = \beta x_1 x_2$, $e_{13} = e_{23} = e_{33} = 0$

(ii) $e_{11} = \alpha x_3(x_1^2 + x_2^2)$, $e_{22} = \alpha x_1^2 x_3$, $e_{12} = \beta x_1 x_2 x_3$,
$e_{13} = e_{23} = e_{33} = 0$

(Here α and β are constants.)

45. Find the nature of the function f such that

$$e_{11} = \alpha f(x_2, x_3), \quad e_{22} = e_{33} = \beta f(x_2, x_3), \quad e_{12} = e_{23} = e_{33} = 0$$

where α and β are constants, is a possible system of strain.

46. If e_{13} and e_{23} are the only nonzero strain components and e_{13} and e_{23} are independent of x_3, show that the compatibility conditions may be reduced to the following single condition:

$$e_{13,2} - e_{23,1} = \text{constant}$$

47. If $e_{11} = e_{22} = e_{33} = e_{12} = 0$, $e_{13} = \phi_{,2}$ and $e_{23} = \phi_{,1}$, where ϕ is a function of x_1 and x_2, show that ϕ must satisfy the equation

$$\nabla^2 \phi = \text{constant}$$

5.11 EXERCISES

48. Derive the compatibility conditions (5.8.17) by differentiating the strain-displacement relations (5.6.4)'. Verify that the six conditions given in (5.8.16) are the only distinct conditions represented by (5.8.17).

49. For the small strain tensor considered in Exercise 42, find (i) the deviator strain components, (ii) the strain invariants, (iii) the principal strains and (iii) the direction of the maximum normal strain.

50. For a certain small deformation, the displacement gradient at a point is given by

$$[\nabla \mathbf{u}] = \begin{bmatrix} 4 & 1 & 4 \\ -1 & -4 & 0 \\ 0 & 2 & 6 \end{bmatrix}$$

Find (i) the strain components, (ii) the deviator strain components, (iii) the strain invariants, (iv) the principal strains and (v) the principal directions of strain.

CHAPTER 6
MOTION

6.1
INTRODUCTION

In the preceding chapter we studied some geometrical aspects of a transformation of a material body from the initial configuration to a current configuration with the time held fixed. Such a transformation was termed a *deformation*. We introduced and analyzed tensor fields that serve to measure the changes in length and relative orientation suffered by material arcs during a deformation. In the present chapter, we introduce and analyze tensors that serve to measure the time rates (or briefly, rates) of these changes during a motion, which may be viewed as a one-parameter sequence of deformations with time as the (continuous) parameter. The concepts of vorticity and circulation are also introduced and their kinematic aspects analyzed. Some important transport formulas are also proven.

6.2
MATERIAL AND LOCAL TIME DERIVATIVES

In the study of motion of a continuum, we deal with the time rates of changes of entities that vary from one continuum particle to the other. The displacement vector, introduced in Section 5.5, is one such entity; the velocity and the acceleration, to be introduced shortly, are also among such entities. These entities may be expressed as functions described in the material form or the spatial form, and the meaning of the time rate of their change depends on the nature of the description. This basic aspect is explained in the following paragraphs.

6.2.1 MATERIAL TIME DERIVATIVE

First, consider a real-valued function $f = f(\mathbf{x}^0, t)$ that represents a scalar or a component of a vector or tensor. As pointed out in Section 4.5, the point $\mathbf{x}^0$ determines a continuum particle $\mathcal{P}$ uniquely, namely the one located at $\mathbf{x}^0$ at $t = 0$, and this particle is referred to as the *particle* $\mathbf{x}^0$. With this notation, we interpret $f(\mathbf{x}^0, t)$ as the value of f experienced at time t by the particle $\mathbf{x}^0$. Also, the partial derivative of f with respect to t, with $\mathbf{x}^0$ held *fixed*, is interpreted as the time rate of change of f at the particle $\mathbf{x}^0$. This derivative is called the *particle* or *material time derivative of* f, denoted Df/Dt. Thus,

$$\frac{Df}{Dt} = \left(\frac{\partial f(\mathbf{x}^0, t)}{\partial t}\right)\bigg|_{\mathbf{x}^0} \tag{6.2.1}$$

where the subscript $\mathbf{x}^0$ accompanying the vertical bar indicates that $\mathbf{x}^0$ is held constant in the differentiation of f. Note that, like f, Df/Dt is a function of $\mathbf{x}^0$ and t by definition. In other words, Df/Dt given by (6.2.1) is a function in the *material form*.

6.2.2 LOCAL TIME DERIVATIVE

Next, consider a real-valued function $\phi = \phi(\mathbf{x}, t)$ that represents a scalar or a component of a vector or tensor. Since $\mathbf{x}$ is a point in the current configuration of a continuum, we interpret $\phi(\mathbf{x}, t)$ (which represents the value of ϕ at the point $\mathbf{x}$ at time t) as the value of ϕ experienced by the particle currently located at $\mathbf{x}$. Also, the partial derivative of ϕ with respect to t, with $\mathbf{x}$ held fixed, is interpreted as the time rate of change of ϕ at the particle currently located at $\mathbf{x}$. This derivative is called the *local time derivative* of ϕ, denoted by the usual partial derivative symbol $\partial \phi/\partial t$. Thus,

$$\frac{\partial \phi}{\partial t} = \left(\frac{\partial \phi(\mathbf{x}, t)}{\partial t}\right)\bigg|_{\mathbf{x}} \tag{6.2.2}$$

6.2 MATERIAL AND LOCAL TIME DERIVATIVES

Note that, like ϕ, $\partial\phi/\partial t$ is a function of $\mathbf{x}$ and t by definition. That is, $\partial\phi/\partial t$ given by (6.2.2) is a function in the *spatial form*.

The distinction between the material time derivative and the local time derivative should be emphasized. While both are partial derivatives with respect to t, the former is defined for a function of $\mathbf{x}^0$ and t whereas the latter is defined for a function of $\mathbf{x}$ and t. Physically, the local time derivative of a function represents the rate at which the function changes with time as seen by an observer currently (momentarily) stationed at a point, whereas the material time derivative represents the rate at which the function changes with time as seen by an observer stationed at a particle and moving with it. The material time derivative is therefore also called the *mobile time derivative* or the *derivative following a particle*. For brevity, the material time derivative will be referred to as the *material derivative* or *material rate*, and the local time derivative as the *local derivative* or *local rate*.

6.2.3 VELOCITY AND ACCELERATION

Since $\mathbf{x}$ is a function of $\mathbf{x}^0$ and t in the material description of motion, the material derivative of $\mathbf{x}$, namely $D\mathbf{x}/Dt$, can be defined; we denote this derivative by $\mathbf{v}$. Thus,

$$\mathbf{v} = \frac{D\mathbf{x}}{Dt} = \left(\frac{\partial \mathbf{x}}{\partial t}\right)\bigg|_{\mathbf{x}^0} \qquad (6.2.3)$$

By virtue of the meaning of the material derivative, it is evident that $\mathbf{v}$ represents the time rate of change of position, at time t, of the particle $\mathbf{x}^0$. This is called the *velocity* of the particle $\mathbf{x}^0$ at time t. If v_i are the components of $\mathbf{v}$, then (6.2.3) reads as follows in the component form:

$$v_i = \frac{Dx_i}{Dt} = \left(\frac{\partial x_i}{\partial t}\right)\bigg|_{x_k^0} \qquad (6.2.3)'$$

Consequently, v_i are called the *velocity components* of the particle $\mathbf{x}^0$ at time t.

In Section 5.5, the displacement vector $\mathbf{u}$ of the particle $\mathbf{x}^0$ has been defined by

$$\mathbf{u} = \mathbf{x} - \mathbf{x}^0 \qquad (6.2.4)$$

It has also been noted that $\mathbf{u}$ may be regarded as a function of $\mathbf{x}^0$ and t, or of $\mathbf{x}$ and t. Treating $\mathbf{u}$ as a function of $\mathbf{x}^0$ and t, we get from (6.2.3) and (6.2.4),

$$\mathbf{v} = \frac{\partial}{\partial t}(\mathbf{x}^0 + \mathbf{u})\bigg|_{\mathbf{x}^0} = \left(\frac{\partial \mathbf{u}}{\partial t}\right)\bigg|_{\mathbf{x}^0} = \frac{D\mathbf{u}}{Dt} \qquad (6.2.5)$$

Thus, the velocity of a particle at time t is precisely the rate of change of displacement of that particle at time t. Expression (6.2.5) reads as follows in the component form:

$$v_i = \frac{Du_i}{Dt} \tag{6.2.5}'$$

It may be pointed out that, in solid mechanics, the deformation and motion are generally described in terms of the displacement vector. In fluid mechanics, the motion is generally described in terms of the velocity vector. When a motion is described in terms of velocity, it is commonly referred to as a *flow*.

Since **v** is a function of $\mathbf{x}^0$ and t by definition, namely, (6.2.3), the material derivative of **v**, namely, $D\mathbf{v}/Dt$, can be defined. This derivative is called the *acceleration* of the particle $\mathbf{x}^0$ at time t. We often write $\dot{\mathbf{v}}$ for $D\mathbf{v}/Dt$. Thus, the acceleration of a particle at time t is the rate of change of velocity of that particle at time t. The components of the acceleration are denoted by Dv_i/Dt or $\dot{v}_i$.

It is to be emphasized that the velocity and acceleration are defined with reference to a particle and are basically functions of $\mathbf{x}^0$ and t. In the spatial description of motion, $\mathbf{x}^0$ is a function of $\mathbf{x}$ and t. Hence, like the displacement, velocity and acceleration can also be expressed as functions of $\mathbf{x}$ and t. When **v** is expressed as a function of $\mathbf{x}$ and t, $\mathbf{v}(\mathbf{x}, t)$ is referred to as the *instantaneous velocity at the point* $\mathbf{x}$. This actually means that $\mathbf{v}(\mathbf{x}, t)$ is the velocity at time t of the particle currently located at the point $\mathbf{x}$. Similar terminology is used in respect of acceleration also.

Next, we deduce a formula that will enable us to compute the instantaneous acceleration from the instantaneous velocity.

6.2.4 MATERIAL DERIVATIVE IN SPATIAL FORM

Consider again the function $\phi = \phi(\mathbf{x}, t)$ for which the local derivative was defined by (6.2.2). This function can be expressed as a function of x_i^0 and t as explicitly indicated in the following:

$$\phi = \phi(x_i, t) = \phi(x_i(x_k^0, t), t) \tag{6.2.6}$$

Consequently, the material derivative of ϕ can also be defined. By the chain rule of partial differentiation, we obtain from (6.2.6),

$$\left(\frac{\partial \phi}{\partial t}\right)\bigg|_{x_k^0} = \left(\frac{\partial \phi}{\partial t}\right)\bigg|_{x_i} + \left(\frac{\partial \phi}{\partial x_i}\right)\bigg|_{t} \left(\frac{\partial x_i}{\partial t}\right)\bigg|_{x_k^0} \tag{6.2.7}$$

6.2 MATERIAL AND LOCAL TIME DERIVATIVES

By virtue of expressions (6.2.1), (6.2.2) and (6.2.3)', we find that

$$\left(\frac{\partial \phi}{\partial t}\right)\bigg|_{x_k^0} = \frac{D\phi}{Dt}, \quad \left(\frac{\partial \phi}{\partial t}\right)\bigg|_{x_i} = \frac{\partial \phi}{\partial t}, \quad \left(\frac{\partial x_i}{\partial t}\right)\bigg|_{x_k^0} = v_i$$

Hence, denoting $(\partial\phi/\partial x_i)|_t$ as just $\partial\phi/\partial x_i = \phi_{,i}$, expression (6.2.7) can be rewritten as follows

$$\frac{D\phi}{Dt} = \frac{\partial \phi}{\partial t} + v_i \phi_{,i} = \frac{\partial \phi}{\partial t} + (\mathbf{v} \cdot \nabla)\phi \tag{6.2.8}$$

When $\mathbf{v}$ is known as a function of $\mathbf{x}$ and t, expression (6.2.8) enables us to compute $D\phi/Dt$ as a function of $\mathbf{x}$ and t. As such, (6.2.8) serves as a *formula for the material derivative in the spatial form*. Note that the first term on the righthand side of this formula, namely, $\partial\phi/\partial t$, represents the *local rate of change* of ϕ and the second term, namely, $v_i\phi_{,i} = (\mathbf{v} \cdot \nabla)\phi$, is the contribution due to the motion. The second term is referred to as the *convective rate of change* of ϕ.

It can be easily verified that the *material derivative operator*

$$\frac{D}{Dt} = \frac{\partial}{\partial t} + v_i(\ )_{,i} = \frac{\partial}{\partial t} + \mathbf{v} \cdot \nabla \tag{6.2.9}$$

which operates on functions represented in spatial form, satisfies all the rules of partial differentiation.

The concept of the material derivative and the formula (6.2.8) are attributed to Euler (1770) and Lagrange (1783).

6.2.5 ACCELERATION IN SPATIAL FORM

Taking $\phi = v_i$ in (6.2.8), we get the following expression for the acceleration:

$$\frac{Dv_i}{Dt} = \frac{\partial v_i}{\partial t} + v_k v_{i,k} \tag{6.2.10a}$$

or

$$\frac{D\mathbf{v}}{Dt} = \frac{\partial \mathbf{v}}{\partial t} + (\mathbf{v} \cdot \nabla)\mathbf{v} \tag{6.2.10b}$$

When $\mathbf{v}$ is known as a function of $\mathbf{x}$ and t, expression (6.2.10b) determines $D\mathbf{v}/Dt$ directly in terms of $\mathbf{x}$ and t; this expression therefore serves as a *formula for acceleration in the spatial form*.

By using the vector identity (3.4.29), formula (6.2.10b) can be put in the following useful form:

$$\frac{D\mathbf{v}}{Dt} = \frac{\partial \mathbf{v}}{\partial t} + \frac{1}{2}\nabla v^2 + (\text{curl } \mathbf{v}) \times \mathbf{v} \qquad (6.2.11)$$

From (6.2.10b) and (6.2.11), we note that the acceleration vector is made up of two parts: (i) the *local rate of change of velocity*, namely $\partial \mathbf{v}/\partial t$, and (ii) the *convective rate of change of velocity*, namely, $(\mathbf{v} \cdot \nabla)\mathbf{v} = \frac{1}{2}\nabla v^2 + (\text{curl } \mathbf{v}) \times \mathbf{v}$. Evidently, the second part is *quadratically nonlinear* in nature. Thus, the acceleration depends quadratically on the velocity field, and a given motion cannot be viewed as a superposition of two independent motions in general.

6.2.6 STEADY AND UNIFORM MOTIONS

We often deal with what are called *steady* and *uniform motions*. A motion is said to be *steady* if the velocity at each point is independent of time (in the spatial description) so that $\mathbf{v} = \mathbf{v}(\mathbf{x})$ and $\partial \mathbf{v}/\partial t = \mathbf{0}$. On the other hand, a motion is said to be *uniform* if the velocity at each instant is independent of position (in the spatial description) so that $\mathbf{v} = \mathbf{v}(t)$ and $\nabla \mathbf{v} = \mathbf{0}$. Thus,

(i) *for steady motion*

$$\frac{\partial \mathbf{v}}{\partial t} = \mathbf{0}; \qquad \frac{D\mathbf{v}}{Dt} = (\mathbf{v} \cdot \nabla)\mathbf{v} = \frac{1}{2}\nabla v^2 + (\text{curl } \mathbf{v}) \times \mathbf{v} \qquad (6.2.12)$$

and (ii) for *uniform motion*,

$$\nabla \mathbf{v} = \mathbf{0}; \qquad \frac{D\mathbf{v}}{Dt} = \frac{\partial \mathbf{v}}{\partial t} \qquad (6.2.13)$$

6.2.7 LINEAR MOTION

While analyzing infinitesimal deformations and motions for which the velocity is very small compared with other field functions, we *approximate* the material derivative operator D/Dt to the local derivative operator $\partial/\partial t$. In such a case, the motion is referred to as a *linear motion*. For such a motion,

$$\mathbf{v} = \frac{\partial \mathbf{u}}{\partial t}, \qquad \frac{D\mathbf{v}}{Dt} \approx \frac{\partial \mathbf{v}}{\partial t} \approx \frac{\partial^2 \mathbf{u}}{\partial t^2} \qquad (6.2.14)$$

EXAMPLE 6.2.1 For a continuum rotating like a rigid body about the origin, find the velocity and acceleration in the material and spatial forms.

6.2 MATERIAL AND LOCAL TIME DERIVATIVES

Solution In the material description, a motion is represented by an equation of the form

$$\mathbf{x} = \mathbf{x}(\mathbf{x}^0, t) \tag{6.2.15}$$

If this motion is a rigid rotation about the origin, we should have $|\mathbf{x}| = |\mathbf{x}^0|$. Then equation (6.2.15) takes the form

$$\mathbf{x} = \mathbf{Q}\mathbf{x}^0 \tag{6.2.16}$$

where $\mathbf{Q}$ is an orthogonal tensor depending solely on t.

By use of (6.2.3), we obtain from (6.2.16),

$$\mathbf{v} = \frac{D\mathbf{x}}{Dt} = \frac{d\mathbf{Q}}{dt}\mathbf{x}^0 \tag{6.2.17a}$$

$$\frac{D\mathbf{v}}{Dt} = \frac{d^2\mathbf{Q}}{dt^2}\mathbf{x}^0 \tag{6.2.17b}$$

Thus, for rigid rotation, (6.2.17) give the velocity and acceleration in the material form.

From (6.2.16) we get $\mathbf{x}^0 = \mathbf{Q}^{-1}\mathbf{x} = \mathbf{Q}^T\mathbf{x}$. Using this, (6.2.17) can be written as

$$\mathbf{v} = \frac{d\mathbf{Q}}{dt}\mathbf{Q}^T\mathbf{x} \tag{6.2.18a}$$

$$\frac{D\mathbf{v}}{Dt} = \frac{d^2\mathbf{Q}}{dt^2}\mathbf{Q}^T\mathbf{x} \tag{6.2.18b}$$

Thus, for rigid rotation, (6.2.18) give the velocity and acceleration in the spatial form.

Expressions (6.2.18) can be expressed in more useful forms as follows. From Example 3.2.2, let us recall that $\mathbf{W} = (d\mathbf{Q}/dt)\mathbf{Q}^T$ is a skew tensor. Hence, if $\mathbf{w}$ is the dual vector of $\mathbf{W}$ expression (6.2.18a) can be rewritten as

$$\mathbf{v} = \mathbf{w} \times \mathbf{x} \tag{6.2.19}$$

This is a well-known formula for rigid rotation about an axis through O with $\mathbf{w}$ as the *angular velocity vector*.

Further, we note that

$$\frac{d^2\mathbf{Q}}{dt^2}\mathbf{Q}^T = \frac{d}{dt}\left(\frac{d\mathbf{Q}}{dt}\mathbf{Q}^T\right) - \left(\frac{d\mathbf{Q}}{dt}\right)\left(\frac{d\mathbf{Q}^T}{dt}\right)$$

$$= \frac{d\mathbf{W}}{dt} - \left(\frac{d\mathbf{Q}}{dt}\mathbf{Q}^T\right)\left(\mathbf{Q}\frac{d\mathbf{Q}^T}{dt}\right)$$

$$= \frac{d\mathbf{W}}{dt} - \mathbf{W}\mathbf{W}^T = \frac{d\mathbf{W}}{dt} + \mathbf{W}^2$$

Hence expression (6.2.18b) can be rewritten as

$$\frac{D\mathbf{v}}{Dt} = \left(\frac{d\mathbf{W}}{dt} + \mathbf{W}^2\right)\mathbf{x}$$

$$= \frac{d\mathbf{w}}{dt} \times \mathbf{x} + \mathbf{w} \times (\mathbf{w} \times \mathbf{x}) \tag{6.2.20}$$

For rigid rotation, this gives the acceleration in spatial form, in terms of the angular velocity vector $\mathbf{w}$. ∎

EXAMPLE 6.2.2 For a material body in motion the displacement field is given as follows:

$$u_1 = 0, \quad u_2 = x_2 - \tfrac{1}{2}(x_2 + x_3)e^{-t} - \tfrac{1}{2}(x_2 - x_3)e^{t}$$
$$u_3 = x_3 - \tfrac{1}{2}(x_2 + x_3)e^{-t} + \tfrac{1}{2}(x_2 - x_3)e^{t} \tag{6.2.21}$$

Find the velocity and acceleration fields in the material and spatial forms.

Solution The given displacement field is in the spatial form. In order to find the velocity field, it is convenient to have the equations describing the motion rewritten in the material form. Since $u_i = x_i - x_i^0$, equations (6.2.21) may be rewritten as

$$x_1^0 = x_1$$
$$x_2^0 = \tfrac{1}{2}(x_2 + x_3)e^{-t} + \tfrac{1}{2}(x_2 - x_3)e^{t} \tag{6.2.22}$$
$$x_3^0 = \tfrac{1}{2}(x_2 + x_3)e^{-t} - \tfrac{1}{2}(x_2 - x_3)e^{t}$$

Solving these equations for x_1, x_2, x_3, we get

$$x_1 = x_1^0$$
$$x_2 = \tfrac{1}{2}(x_2^0 + x_3^0)e^{t} + \tfrac{1}{2}(x_2^0 - x_3^0)e^{-t} \tag{6.2.23}$$
$$x_3 = \tfrac{1}{2}(x_2^0 + x_3^0)e^{t} - \tfrac{1}{2}(x_2^0 - x_3^0)e^{-t}$$

These are the equations describing the given motion in the material form. From these we find that

$$v_1 = \frac{Dx_1}{Dt} = 0$$

$$v_2 = \frac{Dx_2}{Dt} = \frac{1}{2}(x_2^0 + x_3^0)e^{t} - \frac{1}{2}(x_2^0 - x_3^0)e^{-t} \tag{6.2.24}$$

$$v_3 = \frac{Dx_3}{Dt} = \frac{1}{2}(x_2^0 + x_3^0)e^{t} + \frac{1}{2}(x_2^0 - x_3^0)e^{-t}$$

6.2 MATERIAL AND LOCAL TIME DERIVATIVES

For the given motion, these are the velocity components in the material form. The corresponding acceleration components are

$$\frac{Dv_1}{Dt} = 0$$

$$\frac{Dv_2}{Dt} = \frac{1}{2}(x_2^0 + x_3^0)e^t + \frac{1}{2}(x_2^0 - x_3^0)e^{-t} \quad (6.2.25)$$

$$\frac{Dv_3}{Dt} = \frac{1}{2}(x_2^0 + x_3^0)e^t - \frac{1}{2}(x_2^0 - x_3^0)e^{-t}$$

These are also expressed in the material form.

To deduce the velocity and acceleration components in the spatial form, we substitute for x_i^0 from (6.2.22) in (6.2.24) and (6.2.25). Thus, we obtain

$$v_1 = 0, \quad v_2 = x_3, \quad v_3 = x_2 \quad (6.2.26)$$

$$\frac{Dv_1}{Dt} = 0, \quad \frac{Dv_2}{Dt} = x_2, \quad \frac{Dv_3}{Dt} = x_3 \quad (6.2.27)$$

For the given motion, (6.2.26) give the components of velocity and (6.2.27) give the components of acceleration in the spatial form. ∎

Note: Expressions (6.2.27) can also be deduced from (6.2.26) by making use of (6.2.10a), as follows. From (6.2.26) we get

$$\frac{\partial v_1}{\partial t} = \frac{\partial v_2}{\partial t} = \frac{\partial v_3}{\partial t} = 0$$

$$(\mathbf{v} \cdot \nabla)v_1 = 0, \quad (\mathbf{v} \cdot \nabla)v_2 = x_2, \quad (\mathbf{v} \cdot \nabla)v_3 = x_3 \quad (6.2.28)$$

Using these in (6.2.10a), we obtain expressions (6.2.27).

It may be noted that $\partial \mathbf{v}/\partial t = \mathbf{0}$ and $D\mathbf{v}/Dt \neq \mathbf{0}$. The motion is therefore steady and nonuniform.

EXAMPLE 6.2.3 For a certain motion of a continuum, the velocity is given by

$$v_1 = \frac{x_1}{1+t} \quad (6.2.29a)$$

$$v_2 = \frac{2x_2}{1+t} \quad (6.2.29b)$$

$$v_3 = \frac{3x_3}{1+t} \quad (6.2.29c)$$

Find the velocity in the material form, and the acceleration in both material and spatial forms.

Solution Since $v_i = Dx_i/Dt$, we get from equation (6.2.29a)

$$\frac{Dx_1}{Dt} = \frac{x_1}{1+t} \tag{6.2.30}$$

This is a first-order linear differential equation. Bearing in mind that $Dx_1/Dt = (\partial x_1/\partial t)_{x_k^0}$, a general solution of this equation is obtained as

$$x_1 = (1+t)f(x_k^0) \tag{6.2.31}$$

where $f(x_k^0)$ is an arbitrary function of x_k^0. Since $x_1 = x_1^0$ at $t = 0$, we get $f(x_k^0) = x_1^0$. Thus,

$$x_1 = x_1^0(1+t) \tag{6.2.32a}$$

Similarly, from equations (6.2.29b and c), we find that

$$x_2 = x_2^0(1+t)^2 \tag{6.2.32b}$$

$$x_3 = x_3^0(1+t)^3 \tag{6.2.32c}$$

Substituting for x_i from (6.2.32) in (6.2.29) we obtain

$$v_1 = x_1^0, \qquad v_2 = 2x_2^0(1+t), \qquad v_3 = 3x_3^0(1+t)^2 \tag{6.2.33}$$

For the given motion, these are the velocity components in the material form.

Expressions (6.2.33) yield the following expressions for the acceleration components in the material form:

$$\frac{Dv_1}{Dt} = 0, \qquad \frac{Dv_2}{Dt} = 2x_2^0, \qquad \frac{Dv_3}{Dt} = 6x_3^0(1+t) \tag{6.2.34}$$

If we substitute for x_2^0 and x_3^0 from (6.2.32b, c) into (6.2.34), we get

$$\frac{Dv_1}{Dt} = 0 \tag{6.2.35a}$$

$$\frac{Dv_2}{Dt} = \frac{2x_2}{(1+t)^2} \tag{6.2.35b}$$

$$\frac{Dv_3}{Dt} = \frac{6x_3}{(1+t)^2} \tag{6.2.35c}$$

For the given motion, (6.2.35) are the acceleration components in the spatial form. [These can also be obtained by employing the expression (6.2.10a) to the given relations (6.2.29)]. ∎

6.2 MATERIAL AND LOCAL TIME DERIVATIVES

EXAMPLE 6.2.4 If $\mathbf{v} = \nabla \phi$, show that

$$\frac{D\mathbf{v}}{Dt} = \nabla \left(\frac{\partial \phi}{\partial t} + \frac{1}{2} v^2 \right). \tag{6.2.36}$$

Solution Since $\mathbf{v} = \nabla \phi$, we have curl $\mathbf{v} = \mathbf{0}$. Consequently, (6.2.36) follows from (6.2.11). ∎

Note: The result just proven shows that if the velocity is the gradient of a potential then so is the acceleration.

EXAMPLE 6.2.5 At a point $\mathbf{x}$ in the current configuration of a continuum a certain physical quantity ϕ associated with the motion is given by

$$\phi = \frac{1}{r} e^{-at}, \qquad r \neq 0 \tag{6.2.37}$$

where $r = |\mathbf{x}|$ and a is a positive constant. Also, the velocity field describing the motion is given by

$$v_1 = x_1 x_2 t, \qquad v_2 = x_2^2 t, \qquad v_3 = x_2 x_3 t \tag{6.2.38}$$

Find the material rate of change of ϕ at the point $(0, a, 0)$ of the configuration.

Solution Since ϕ and $\mathbf{v}$ are given in the spatial forms, the formula (6.2.8) is convenient for the computation of $D\phi/Dt$.

From (6.2.37), we find that

$$\phi_{,i} = -\frac{1}{r^2} e^{-at} r_{,i} = -\frac{1}{r^3} x_i e^{-at} \tag{6.2.39}$$

Using (6.2.38), we now get

$$v_i \phi_{,i} = -\left(\frac{1}{r}\right) t x_2 e^{-at} \tag{6.2.40}$$

Also,

$$\frac{\partial \phi}{\partial t} = -\frac{a}{r} e^{-at} \tag{6.2.41}$$

Using (6.2.40) and (6.2.41), formula (6.2.8) gives

$$\frac{D\phi}{Dt} = \frac{\partial \phi}{\partial t} + v_i \phi_{,i} = -\frac{1}{r}(a + t x_2) e^{-at} \tag{6.2.42}$$

This is the material derivative of ϕ at the point $\mathbf{x}$ at time $t > 0$. At the point $(0, a, 0)$, we get

$$\frac{D\phi}{Dt} = -(1 + t) e^{-at} \quad ∎ \tag{6.2.43}$$

6 MOTION

EXAMPLE 6.2.6 Prove the following identities:

(i) $$\frac{D\mathbf{F}}{Dt} = (\nabla \mathbf{v})\mathbf{F} \qquad (6.2.44)$$

(ii) $$\frac{D\mathbf{F}^{-1}}{Dt} = -\mathbf{F}^{-1}(\nabla \mathbf{v}) \qquad (6.2.45)$$

(iii) $$\frac{D}{Dt}(d\mathbf{x}) = (\nabla \mathbf{v})(d\mathbf{x}) \qquad (6.2.46)$$

(iv) $$\frac{D}{Dt}(ds)^2 = 2\,d\mathbf{x} \cdot (\nabla \mathbf{v})(d\mathbf{x}) \qquad (6.2.47)$$

Solution (i) Since x_i and $x_{i;j} = [\mathbf{F}]_{ij}$ are functions of x_i^0 and t, we have

$$\frac{D}{Dt}(x_{i;j}) = \left.\frac{\partial}{\partial t}\left(\frac{\partial x_i}{\partial x_j^0}\right)\right|_{x_k^0} = \left.\frac{\partial}{\partial x_j^0}\left(\frac{\partial x_i}{\partial t}\right)\right|_{x_k^0} = \frac{\partial}{\partial x_j^0}\left(\frac{Dx_i}{Dt}\right) = v_{i;j} \qquad (6.2.48)$$

Treating v_i as functions of x_i and t, and using the chain rule of differentiation, we get $v_{i;j} = v_{i,k} x_{k;j}$. Hence (6.2.48) can be rewritten as

$$\frac{D}{Dt}(x_{i;j}) = v_{i,k} x_{k;j} \qquad (6.2.49)$$

This proves (6.2.44).

(ii) From (6.2.44), we readily get

$$\left(\frac{D\mathbf{F}}{Dt}\right)\mathbf{F}^{-1} = (\nabla \mathbf{v}) \qquad (6.2.50)$$

Also, we note that

$$\mathbf{0} = \frac{D\mathbf{I}}{Dt} = \frac{D}{Dt}(\mathbf{F}\mathbf{F}^{-1}) = \mathbf{F}\frac{D\mathbf{F}^{-1}}{Dt} + \left(\frac{D\mathbf{F}}{Dt}\right)\mathbf{F}^{-1}$$

Hence

$$-\mathbf{F}\frac{D\mathbf{F}^{-1}}{Dt} = \frac{D\mathbf{F}}{Dt}\mathbf{F}^{-1} = \nabla \mathbf{v}$$

by (6.2.50), from which (6.2.45) is immediate.

(iii) Recalling from relation (5.2.5) that $d\mathbf{x} = \mathbf{F}\,d\mathbf{x}^0$, we get

$$\frac{D}{Dt}(d\mathbf{x}) = \frac{D}{Dt}(\mathbf{F}\,d\mathbf{x}^0) = \left(\frac{D\mathbf{F}}{Dt}\right)d\mathbf{x}^0$$
$$= (\nabla \mathbf{v})\mathbf{F}\,d\mathbf{x}^0$$

by (6.2.44);

$$= (\nabla \mathbf{v})(d\mathbf{x})$$

6.2 MATERIAL AND LOCAL TIME DERIVATIVES 253

(iv) Since $(ds)^2 = d\mathbf{x} \cdot d\mathbf{x}$, we find

$$\frac{D}{Dt}(ds)^2 = \frac{D}{Dt}(d\mathbf{x} \cdot d\mathbf{x}) = 2\, d\mathbf{x} \cdot \frac{D}{Dt}(d\mathbf{x})$$

$$= 2\, d\mathbf{x} \cdot (\nabla \mathbf{v})(d\mathbf{x})$$

by (6.2.46). ∎

EXAMPLE 6.2.7 Prove the following *Euler's formulas*:

(i) $$\frac{DJ}{Dt} = J \operatorname{div} \mathbf{v} \tag{6.2.51}$$

(ii) $$\frac{D}{Dt}(\log J) = \operatorname{div} \mathbf{v} \tag{6.2.52}$$

Solution By definition, $J = \det \nabla^0 \mathbf{x} = \det(x_{i;j})$. Hence by use of expression (1.7.18), we have

$$J = \varepsilon_{pqr} x_{1;p} x_{2;q} x_{3;r} \tag{6.2.53}$$

Therefore,

$$\frac{DJ}{Dt} = \varepsilon_{pqr}\left\{\frac{D}{Dt}(x_{1;p})x_{2;q}x_{3;r} + \frac{D}{Dt}(x_{2;q})x_{1;p}x_{3;r} + \frac{D}{Dt}(x_{3;r})x_{1;p}x_{2;q}\right\} \tag{6.2.54}$$

From (6.2.49), we note that

$$\frac{D}{Dt}(x_{1;p}) = v_{1,k} x_{k;p}$$

Hence

$$\varepsilon_{pqr}\frac{D}{Dt}(x_{1;p})x_{2;q}x_{3;r} = \varepsilon_{pqr} v_{1,k} x_{k;p} x_{2;q} x_{3;r}$$

$$= v_{1,1}(\varepsilon_{pqr}x_{1;p}x_{2;q}x_{3;r}) + v_{1,2}(\varepsilon_{pqr}x_{2;p}x_{2;q}x_{3;r})$$

$$+ v_{1,3}(\varepsilon_{pqr}x_{3;p}x_{2;q}x_{3;r}) \tag{6.2.55}$$

With the aid of (1.7.18), we note that the parenthetical expressions in the last two terms of (6.2.55) represent determinants with two identical rows; these are therefore 0. Hence (6.2.53) and (6.2.55) yield

$$\varepsilon_{pqr}\frac{D}{Dt}(x_{1;p})x_{2;q}x_{3;r} = J v_{1,1} \tag{6.2.56a}$$

Similarly,

$$\varepsilon_{pqr}\frac{D}{Dt}(x_{2;q})x_{1;p}x_{3;r} = Jv_{2,2} \quad (6.2.56b)$$

$$\varepsilon_{pqr}\frac{D}{Dt}(x_{3;r})x_{1;p}x_{2;q} = Jv_{3,3} \quad (6.2.56c)$$

Putting (6.2.56) into (6.2.54), we obtain

$$\frac{DJ}{Dt} = Jv_{k,k} = J\,\text{div}\,\mathbf{v}$$

which is (6.2.51). Since

$$\frac{D}{Dt}(\log J) = \frac{1}{J}\frac{DJ}{Dt}$$

expression (6.2.52) follows from (6.2.51). ∎

EXAMPLE 6.2.8 Prove the following expressions for the material derivatives of the surface and volume elements in the current configuration:

(i) $$\frac{D}{Dt}\{(dS)\mathbf{n}\} = \{(\text{div}\,\mathbf{v}) - (\nabla\mathbf{v})^T\}\mathbf{n}(dS) \quad (6.2.57)$$

(ii) $$\frac{D}{Dt}(dV) = (\text{div}\,\mathbf{v})\,dV \quad (6.2.58)$$

Solution (i) Recall from relation (5.2.12) that

$$(dS)\mathbf{n} = J(\mathbf{F}^T)^{-1}(dS_0)\mathbf{n}^0 \quad (6.2.59)$$

This gives

$$\frac{D}{Dt}\{(dS)\mathbf{n}\} = \left\{\frac{DJ}{Dt}(\mathbf{F}^T)^{-1} + J\frac{D(\mathbf{F}^T)^{-1}}{Dt}\right\}(dS_0)\mathbf{n}^0 \quad (6.2.60)$$

Noting that

$$\frac{D(\mathbf{F}^T)^{-1}}{Dt} = \frac{D(\mathbf{F}^{-1})^T}{Dt} = \left(\frac{D\mathbf{F}^{-1}}{Dt}\right)^T$$

and using (6.2.45) and (6.2.51), expression (6.2.60) becomes

$$\frac{D}{Dt}\{(dS)\mathbf{n}\} = \{(\text{div}\,\mathbf{v}) - (\nabla\mathbf{v})^T\}J(\mathbf{F}^T)^{-1}(dS_0)\mathbf{n}^0$$

which, with the aid of (6.2.59) yields (6.2.57).

6.2 MATERIAL AND LOCAL TIME DERIVATIVES

(ii) Recalling the Euler's formula (5.2.22), namely, $dV = J(dV_0)$, we get

$$\frac{D}{Dt}(dV) = \frac{DJ}{Dt}(dV_0) = J(\text{div } \mathbf{v})\, dV_0$$

by (6.2.51);

$$= (\text{div } \mathbf{v})\, dV \quad \blacksquare$$

Note: Since dS_0 and dV_0 are surface and volume elements in the *initial* configuration, their material derivatives are 0. This fact has been used in the preceding computations.

EXAMPLE 6.2.9 Show that a surface represented by an equation of the form $F(\mathbf{x}, t) = 0$ is a material surface if and only if

$$\frac{DF}{Dt} = 0. \tag{6.2.61}$$

Deduce that the normal component of velocity at a point of a material surface $F(\mathbf{x}, t) = 0$ is given by

$$\mathbf{v} \cdot \mathbf{n} = -\frac{(\partial F/\partial t)}{|\nabla F|} \tag{6.2.62}$$

$\mathbf{n}$ being the positively oriented unit normal to the surface.

Solution First suppose that the surface given by the equation $F(\mathbf{x}, t) = 0$ is a material surface. Then the surface consists of the same set of particles for all time. Hence, its equation in the initial configuration should be $F(\mathbf{x}^0) = 0$. Consequently,

$$\frac{DF}{Dt} = \left(\frac{\partial F}{\partial t}\right)\bigg|_{\mathbf{x}^0} = 0$$

which is the required condition (6.2.61).

Conversely, suppose that the condition (6.2.61) holds. Then the value of the function $F(\mathbf{x}, t)$ computed for a particle currently positioned at $\mathbf{x}$ remains unchanged during the motion. In particular, those particles for which $F = 0$ at some time retain this value for all times. This means that the surface represented by the equation $F(\mathbf{x}, t) = 0$ contains the same set of particles for all time. Hence the surface is a material surface.

In the spatial description, the condition (6.2.61) reads, on using (6.2.8),

$$\frac{\partial F}{\partial t} + (\mathbf{v} \cdot \nabla)F = 0 \tag{6.2.63}$$

We note that $(\mathbf{v} \cdot \nabla)F = \mathbf{v} \cdot (\nabla F)$ and

$$\mathbf{n} = \frac{\nabla F}{|\nabla F|} \qquad (6.2.64)$$

Consequently, the relation (6.2.63) may be rewritten as

$$\frac{\partial F}{\partial t} + (\mathbf{v} \cdot \mathbf{n})|\nabla F| = 0 \qquad (6.2.65)$$

from which the expression (6.2.62) is immediate. ∎

Note: Expression (6.2.62) serves as a formula for the speed with which the surface under consideration advances in the direction normal to itself in space.

6.3
STRETCHING AND VORTICITY

In Section 5.4, we introduced the strain tensors that serve as measures of the changes in length and relative orientation of material arcs that generally occur when a material body moves from one configuration to another. We now introduce and analyze the tensors that serve as measures of the *time rates* of these changes. The tensor that serves as a means of computing the rate of rigid rotation will also be introduced.

Let us start with expression (6.2.46); namely,

$$\frac{D}{Dt}(d\mathbf{x}) = (\nabla \mathbf{v}) \, d\mathbf{x} \qquad (6.3.1)$$

Substituting $d\mathbf{x} = (ds)\mathbf{a}$ (which holds for a material arc $d\mathcal{C}$ currently directed along a unit vector $\mathbf{a}$) in (6.3.1) and carrying out the differentiation, we obtain

$$(ds)\frac{D\mathbf{a}}{Dt} + \mathbf{a}\frac{D}{Dt}(ds) = (ds)(\nabla \mathbf{v})\mathbf{a} \qquad (6.3.2)$$

Taking the scalar product with $\mathbf{a}$ on both sides of (6.3.2) and noting that $\mathbf{a} \cdot \mathbf{a} = 1$ and (therefore) $\mathbf{a} \cdot D\mathbf{a}/Dt = 0$, we get

$$\frac{1}{(ds)}\frac{D}{Dt}(ds) = \mathbf{a} \cdot (\nabla \mathbf{v})\mathbf{a} = \frac{1}{2}\mathbf{a} \cdot (\nabla \mathbf{v} + \nabla \mathbf{v}^T)\mathbf{a} \qquad (6.3.3)$$

6.3 STRETCHING AND VORTICITY

Setting
$$\mathbf{D} = \tfrac{1}{2}(\nabla \mathbf{v} + \nabla \mathbf{v}^T) = \text{sym } \nabla \mathbf{v} \tag{6.3.4}$$
expression (6.3.3) becomes
$$\frac{1}{(ds)} \frac{D}{Dt}(ds) = \mathbf{a} \cdot \mathbf{Da} \tag{6.3.5}$$

Evidently, $\mathbf{D}$ defined by (6.3.4) is a symmetric tensor completely determined by the *spatial velocity gradient* $\nabla \mathbf{v}$. Also, $(1/(ds))(D/Dt)(ds)$ represents the rate of change in length, per unit of current length, of the element $d\mathcal{C}$; this rate is called the *stretching rate* (or just *stretching*) *of $d\mathcal{C}$* or *stretching along* $\mathbf{a}$. Expression (6.3.5) shows that, when the current orientation of an arc element is known, the tensor $\mathbf{D}$ completely determines the stretching of the element. In a rigid motion $ds = ds_0$, and from expression (6.3.5) it follows that a necessary and sufficient condition for a motion to be rigid is that $\mathbf{D} = \mathbf{0}$. Thus, the tensor $\mathbf{D}$ serves as a measure of stretching; this tensor is called the *stretching tensor* or *the rate of deformation tensor* and was introduced by Euler in 1770.

Next we obtain the *geometrical interpretations* of the components of $\mathbf{D}$. In the component form, expressions (6.3.4) and (6.3.5) read

$$d_{ij} = \tfrac{1}{2}(v_{i,j} + v_{j,i}) \tag{6.3.4}'$$

$$\frac{1}{(ds)} \frac{D}{Dt}(ds) = d_{ij} a_i a_j \tag{6.3.5}'$$

By taking the particular case in which $\mathbf{a}$ is aligned with $\mathbf{e}_1$ we find from (6.3.5)' that d_{11} represents the stretching along the x_1-direction. The components d_{22} and d_{33} of $\mathbf{D}$ have similar geometrical meanings.

In order to obtain the meanings of the off-diagonal components of $\mathbf{D}$, let us consider another material arc $d\bar{\mathcal{C}}$, of current length $(d\bar{s})$, directed along a unit vector $\bar{\mathbf{a}}$. If $\theta(\neq 0)$ is the current angle between $d\bar{\mathcal{C}}$ and $d\mathcal{C}$, then $\cos \theta = \mathbf{a} \cdot \bar{\mathbf{a}}$, which yields, on material differentiation,

$$-(\sin \theta) \frac{D\theta}{Dt} = \mathbf{a} \cdot \frac{D\bar{\mathbf{a}}}{Dt} + \frac{D\mathbf{a}}{Dt} \cdot \bar{\mathbf{a}} \tag{6.3.6}$$

Noting from (6.3.2) and (6.3.5) that

$$\frac{D\mathbf{a}}{Dt} = (\nabla \mathbf{v})\mathbf{a} - (\mathbf{a} \cdot \mathbf{Da})\mathbf{a}$$

and, similarly,

$$\frac{D\bar{\mathbf{a}}}{Dt} = (\nabla \mathbf{v})\bar{\mathbf{a}} - (\bar{\mathbf{a}} \cdot \mathbf{D}\bar{\mathbf{a}})\bar{\mathbf{a}}$$

we obtain

$$\mathbf{a} \cdot \frac{D\bar{\mathbf{a}}}{Dt} + \bar{\mathbf{a}} \cdot \frac{D\mathbf{a}}{Dt} = \mathbf{a} \cdot (\nabla \mathbf{v})\bar{\mathbf{a}} + \bar{\mathbf{a}} \cdot (\nabla \mathbf{v})\mathbf{a} - (\mathbf{a} \cdot \bar{\mathbf{a}})(\mathbf{a} \cdot \mathbf{D}\mathbf{a} + \bar{\mathbf{a}} \cdot \mathbf{D}\bar{\mathbf{a}}) \quad (6.3.7)$$

Also, by use of (2.8.14) and (6.3.4) we get

$$\mathbf{a} \cdot (\nabla \mathbf{v})\bar{\mathbf{a}} + \bar{\mathbf{a}} \cdot (\nabla \mathbf{v})\mathbf{a} = \mathbf{a} \cdot (\nabla \mathbf{v} + \nabla \mathbf{v}^T)\bar{\mathbf{a}} = 2\mathbf{a} \cdot \mathbf{D}\bar{\mathbf{a}} \quad (6.3.8)$$

Using (6.3.7) and (6.3.8), expression (6.3.6) becomes

$$(\sin \theta) \frac{D\theta}{Dt} = (\mathbf{a} \cdot \bar{\mathbf{a}})(\mathbf{a} \cdot \mathbf{D}\mathbf{a} + \bar{\mathbf{a}} \cdot \mathbf{D}\bar{\mathbf{a}}) - 2\mathbf{a} \cdot \mathbf{D}\bar{\mathbf{a}} \quad (6.3.9)$$

This shows that **D** not only determines the stretchings of material arcs, but also the time rate of change in the angles between them.

If $\theta = \pi/2$, we have $\mathbf{a} \cdot \bar{\mathbf{a}} = 0$ and (6.3.9) reduces to the following simple form:

$$\mathbf{a} \cdot \mathbf{D}\bar{\mathbf{a}} = -\frac{1}{2} \frac{D\theta}{Dt} \quad (6.3.10)$$

This expression shows that $\mathbf{a} \cdot \mathbf{D}\bar{\mathbf{a}}$ represents one-half the rate of decrease in the angle suffered by two (currently) orthogonal material arcs. This rate is called the *shearing rate* or just the *shearing* between the arcs. Evidently, the arcs suffer no shearing if and only if the motion is rigid.

For the particular case in which **a** is aligned with $\mathbf{e}_1$ and $\bar{\mathbf{a}}$ is aligned with $\mathbf{e}_2$, we find from (6.3.10) that d_{12} *represents the shearing between material arcs currently lying along the* x_1 *and* x_2 directions. The other off-diagonal components of **D** have similar geometrical interpretations.

Thus, all the components of **D** have definite geometrical interpretations. While the diagonal components of **D** represent the stretching of material arcs currently lying along the coordinate directions, the off-diagonal components represent the shearing between such arcs. These geometrical interpretations are analogous to those of the components of the strain tensor **A** (which is also defined in the current configuration). But there is an important difference: whereas the meanings of the components of **D** are *exact*, the meanings of the components of **A** are just *approximations*.

Since **D** is a symmetric tensor, it follows from results proven in Section 2.13 that **D** possesses three eigenvalues and at least one set of three mutually orthogonal principal directions and that, with respect to the axes chosen along these principal directions, the matrix of **D** is purely diagonal, the diagonal elements being the eigenvalues of **D**. By virtue of the geometrical interpretations of the components of **D** obtained previously, it follows that the eigenvalues of **D** represent the stretchings along the principal directions of **D**. The eigenvalues of **D** are called the *principal stretchings* and the principal directions of **D** are called the *principal directions of stretching*. As in Section 5.3, it can be shown that the stretching of a general material arc is the net result of triaxial principal stretchings.

6.3.1 ISOCHORIC MOTION

From (6.3.4) we find that tr **D** = div **v**. Hence expression (6.2.58) yields

$$\text{tr } \mathbf{D} = \text{div } \mathbf{v} = \frac{1}{(dV)} \frac{D}{Dt}(dV) \qquad (6.3.11)$$

Evidently, tr **D** = (div **v**) *represents the rate of change in volume per unit volume of a material element* (as it moves with velocity **v**). Further, dV is constant during the motion if and only if div **v** = 0. From Euler's formula (6.2.51) we note that div **v** = 0 if and only if $DJ/Dt = 0$; that is, J remains unchanged during the motion. Initially, when $t = 0$, we have $J = 1$. Hence div **v** = 0 if and only if $J = 1$.

Thus, the *following statements are equivalent*:

1. dV = constant   (6.3.12a)

2.   div **v** = 0   (6.3.12b)

3. $J = 1$ (6.3.12c)

A motion is said to be an *isochoric (volume-preserving) motion* whenever dV = constant during the motion. It follows that a motion is isochoric if and only if div **v** = 0 or equivalently $J = 1$ during the motion. Note that this statement is consistent with that made in Section 5.2 in respect of an isochoric deformation.

In Section 5.2, it has been indicated that a continuum is said to be an *incompressible continuum* if and only if every deformation it undergoes is an isochoric deformation. Since a motion is a (one-parameter) family of deformations, it follows that a continuum is incompressible if and only if all its motions are isochoric motions. Consequently, it is inferred that *for every motion of an incompressible continuum each of the conditions in (6.3.12) holds*. The condition (6.3.12b), namely, div **v** = 0, plays a fundamental role in the study of motions of incompressible fluids. Some further aspects of the conditions (6.3.12) will be considered in Section 8.2.

6.3.2 ANALYSIS OF RELATIVE MOTION

We return to expression (6.3.1) and analyze it on slightly different lines. We note that $d\mathbf{x}$ may be regarded as the position vector of the point $(\mathbf{x} + d\mathbf{x})$

260 6 MOTION

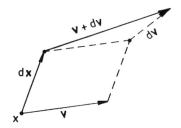

Figure 6.1. Relative velocity.

with respect to the point **x**. Consequently, $(D/Dt)(d\mathbf{x})$ may be interpreted as the velocity at the point $(\mathbf{x} + d\mathbf{x})$ relative to the point **x**. Let us denote this *relative velocity* by $d\mathbf{v}$; see Figure 6.1. Thus, (6.3.1) may be rewritten as

$$d\mathbf{v} = (\nabla \mathbf{v})(d\mathbf{x}) \qquad (6.3.13)$$

This expression shows that the velocity gradient $\nabla \mathbf{v}$, which is a tensor, serves as a means of computing $d\mathbf{v}$. Let us decompose $\nabla \mathbf{v}$ into symmetric and skew parts. It has already been noted that **D** is the symmetric part of $\nabla \mathbf{v}$ (by definition). Let **W** be the skew part of $\nabla \mathbf{v}$; that is,

$$\mathbf{W} = \tfrac{1}{2}(\nabla \mathbf{v} - \nabla \mathbf{v}^T) = \text{skw } \nabla \mathbf{v} \qquad (6.3.14)$$

Then expression (6.3.13) becomes

$$d\mathbf{v} = \mathbf{D}(d\mathbf{x}) + \mathbf{W}(d\mathbf{x}) \qquad (6.3.15)$$

Thus, $d\mathbf{v}$ is made up of two parts: (i) $d\mathbf{v}_1 = \mathbf{D}(d\mathbf{x})$, and (ii) $d\mathbf{v}_2 = \mathbf{W}(d\mathbf{x})$. By virtue of the geometrical significance of **D**, $d\mathbf{v}_1$ may be interpreted as the velocity caused by the stretching. To interpret $d\mathbf{v}_2$, let us look for the dual vector of **W**. Since $\mathbf{W} = \text{skw } \nabla \mathbf{v}$, we find from Example 3.5.7 that the dual vector of **W** is $\tfrac{1}{2}$ curl **v**. Hence

$$d\mathbf{v}_2 = \mathbf{W}(d\mathbf{x}) = (\tfrac{1}{2} \text{ curl } \mathbf{v}) \times (d\mathbf{x}) \qquad (6.3.16)$$

This expression is analogous to the expression (6.2.19) giving the velocity due to rigid rotation. Comparing these two expressions, we find that $d\mathbf{v}_2$ represents the velocity due to a *rigid rotation* about an axis through the point **x**, with *angular velocity* $\tfrac{1}{2}$ curl **v**. Thus, $d\mathbf{v}$ may be considered a superposition of the velocity caused by the stretching determined by **D** and the velocity due to the rigid rotation determined by **W**.

In view of its significance as described earlier, the tensor **W** is called the *spin* or *vorticity tensor*. Also, curl **v**, which is equal to twice the dual vector of **W**, is called the *spin* or *vorticity vector*. We denote this vector by **w**; that is,

$$\mathbf{w} = \text{curl } \mathbf{v} \tag{6.3.17}$$

The concept of vorticity was first introduced and analyzed by Cauchy in 1841.

6.3.3 IRROTATIONAL MOTION

In the particular case when **W** = **0** at **x**, $d\mathbf{v}$ is completely determined by **D** and there occurs no rigid rotation about **x**; in such a situation we say that the motion is *irrotational*. Thus, *a motion is irrotational at a point* **x** *if and only if the vorticity tensor* **W** *and (therefore) the vorticity vector* **w** *are 0 at* **x**

6.3.4 POTENTIAL FLOW

We often deal with a special type of motion wherein the velocity is the gradient of a scalar function; that is, $\mathbf{v} = \nabla \phi$ for some scalar function ϕ. Such a motion is called a *potential flow* and the function ϕ is called the *velocity potential* of the flow. Since curl $\nabla \phi \equiv \mathbf{0}$ by identity (3.4.16), it readily follows that *every potential flow is automatically an irrotational motion*. The converse is valid only if the region of interest is simply connected; the proof requires the use of the Stokes's theorem (3.6.7). Thus, *in a simply connected region, every irrotational motion is a potential flow*.

From the result of Example 6.2.4, we note that, in a potential flow, the acceleration is also the gradient of a scalar function.

Isochoric motions, irrotational motions and potential flows are of great importance, particularly in fluid mechanics.

6.3.5 COMPARISON WITH SMALL DEFORMATION

There is a striking resemblance between the set of expressions (6.3.4) and (6.3.14), through which **D** and **W** have been defined, and the set of expressions (5.6.4), (5.7.6) and (5.7.38), relating the strain tensor **E** and the rotation tensor Ω with the displacement vector occurring in *small* (infinitesimal) deformation. Although all of these expressions are linear, the linearity in the expressions (6.3.4) and (6.3.14) is *exact*, while that in (5.6.4), (5.7.6) and (5.7.38) is just *approximate*. In other words, while **D** and **W** measure the stretching and the angular velocity (due to rigid rotation) *exactly*, **E** and Ω measure the stretch and angular displacement (due to rigid rotation) only to the *first-order approximation*. In particular, it must be noted that, *whereas* **D** *is exactly* 0 *in a rigid-motion*, **E** *need not be* 0 *in such a motion*; see Example 5.6.3. However, because of the formal similarity

262　6 MOTION

between the expressions (6.3.4) and (5.6.4), all the results obtained in respect of **E** on the basis of (5.6.4) can be rewritten in respect of **D** with the words *displacement* and *strain tensor* replaced by the words *velocity* and *stretching tensor*.

From expression (5.6.4), we obtain

$$\frac{D\mathbf{E}}{Dt} = \frac{1}{2}\frac{D}{Dt}(\nabla^0\mathbf{u} + \nabla^0\mathbf{u}^T) = \frac{1}{2}\left\{\nabla^0\frac{D\mathbf{u}}{Dt} + \nabla^0\left(\frac{D\mathbf{u}}{Dt}\right)^T\right\}$$

$$= \tfrac{1}{2}(\nabla^0\mathbf{v} + \nabla^0\mathbf{v}^T)$$

using (6.2.5);

$$\approx \tfrac{1}{2}(\nabla\mathbf{v} + \nabla\mathbf{v}^T)$$

for small deformation;

$$= \mathbf{D}$$

by (6.3.4). Similarly, from (5.7.6), (5.7.38), (6.2.5) and (6.3.14), we obtain $D\Omega/Dt \cong \mathbf{W}$.

Thus, for *small deformation*, we have

$$\mathbf{D} \cong \frac{D\mathbf{E}}{Dt} \qquad (6.3.18a)$$

$$\mathbf{W} \cong \frac{D\Omega}{Dt} \qquad (6.3.18b)$$

These expressions show that, when the displacement gradient is infinitesimally small, **D** represents the rate of change of the strain tensor and **W** represents the rate of change of the rotation tensor. For this reason, **D** is often referred to as the *rate of strain* or *strain-rate tensor*, and **W** as the *rate of rotation* or *rotation-rate tensor*. When the deformation is not small, the strain tensor is either **G** or **A**. Then **D** does not represent the actual rate of change of **G** or **A**; see Example 6.3.3 later.

EXAMPLE 6.3.1 Prove the following identities:

(i) $\qquad (\mathbf{v}\cdot\nabla)\mathbf{v} = \mathbf{w}\times\mathbf{v} + \tfrac{1}{2}\nabla|\mathbf{v}|^2 \qquad (6.3.19)$

(ii) $\quad \mathrm{div}(\mathbf{v}\cdot\nabla)\mathbf{v} - (\mathbf{v}\cdot\nabla)(\mathrm{div}\,\mathbf{v}) = |\mathbf{D}|^2 - \tfrac{1}{2}|\mathbf{w}|^2 = \nabla\mathbf{v}\cdot\nabla\mathbf{v}^T \quad (6.3.20)$

(iii) $\qquad \mathrm{div}(\mathbf{v}\times\mathbf{w}) = \tfrac{1}{2}(\nabla^2|\mathbf{v}|^2 + |\mathbf{w}|^2)$

$$- \{|\mathbf{D}|^2 + (\mathbf{v}\cdot\nabla)(\mathrm{div}\,\mathbf{v})\} \quad (6.3.21)$$

Solution (i) By use of the vector identity (3.4.29) and expression (6.3.17) we readily obtain (6.3.19).

6.3 STRETCHING AND VORTICITY

(ii) Recalling that $\mathbf{D} = \text{sym } \nabla \mathbf{v}$ and that $\frac{1}{2}\mathbf{w} = \frac{1}{2}\text{curl }\mathbf{v}$ is the dual vector of $\mathbf{W} = \text{skw } \nabla \mathbf{v}$, we obtain (6.3.20) by changing the notation appropriately in (3.5.46).

(iii) Taking divergence of (6.3.19) and using (6.3.20), we arrive at (6.3.21). ∎

EXAMPLE 6.3.2 Prove the following identities:

(i)
$$\text{div}\left(\frac{D\mathbf{v}}{Dt}\right) = \frac{D}{Dt}(\text{div }\mathbf{v}) + |\mathbf{D}|^2 - \frac{1}{2}|\mathbf{w}|^2 \qquad (6.3.22)$$

(ii)
$$\text{curl}\left(\frac{D\mathbf{v}}{Dt}\right) = \frac{\partial \mathbf{w}}{\partial t} + \text{curl}(\mathbf{w} \times \mathbf{v}) \qquad (6.3.23)$$

$$= \frac{D\mathbf{w}}{Dt} + (\text{div }\mathbf{v})\mathbf{w} - (\mathbf{w} \cdot \nabla)\mathbf{v} \qquad (6.3.24)$$

(iii)
$$\frac{D}{Dt}(J\mathbf{F}^{-1}\mathbf{w}) = J\mathbf{F}^{-1}\text{curl}\left(\frac{D\mathbf{v}}{Dt}\right) \qquad (6.3.25)$$

Solution (i) By using (6.2.8), we get

$$\frac{D}{Dt}(v_{k,k}) = \frac{\partial}{\partial t}(v_{k,k}) + v_i v_{k,ki} = \left(\frac{\partial v_k}{\partial t}\right)_{,k} + (v_i v_{k,i})_{,k} - v_{i,k} v_{k,i}$$

$$= \left[\frac{\partial v_k}{\partial t} + v_i v_{k,i}\right]_{,k} - v_{i,k} v_{k,i} = \left(\frac{Dv_k}{Dt}\right)_{,k} - v_{i,k} v_{k,i}$$

or, in the direct notation,

$$\frac{D}{Dt}(\text{div }\mathbf{v}) = \text{div}\left(\frac{D\mathbf{v}}{Dt}\right) - \nabla \mathbf{v} \cdot \nabla \mathbf{v}^T \qquad (6.3.26)$$

Recalling from (6.3.20) that

$$\nabla \mathbf{v} \cdot \nabla \mathbf{v}^T = |\mathbf{D}|^2 - \tfrac{1}{2}|\mathbf{w}|^2$$

expression (6.3.26) readily yields (6.3.22).

(ii) Taking the curl of expression (6.2.11), we obtain (6.3.23). Since $\text{div }\mathbf{w} = \text{div curl }\mathbf{v} = \mathbf{0}$, the identity (3.4.25) gives

$$\text{curl}(\mathbf{w} \times \mathbf{v}) = (\text{div }\mathbf{v})\mathbf{w} + (\mathbf{v} \cdot \nabla)\mathbf{w} - (\mathbf{w} \cdot \nabla)\mathbf{v}$$

Using this in (6.3.23), we obtain

$$\text{curl}\left(\frac{D\mathbf{v}}{Dt}\right) = \frac{\partial \mathbf{w}}{\partial t} + (\mathbf{v} \cdot \nabla)\mathbf{w} + (\text{div }\mathbf{v})\mathbf{w} - (\mathbf{w} \cdot \nabla)\mathbf{v}$$

which in turn becomes (6.3.24) on using (6.2.8).

(iii) Operating both sides of the expression (6.3.24) by $J\mathbf{F}^{-1}$ and noting by use of (3.5.34) that $(\mathbf{w} \cdot \nabla)\mathbf{v} = (\nabla\mathbf{v})\mathbf{w}$, we obtain

$$J\mathbf{F}^{-1}\operatorname{curl}\left(\frac{D\mathbf{v}}{Dt}\right) = J\mathbf{F}^{-1}\frac{D\mathbf{w}}{Dt} + J(\operatorname{div} \mathbf{v})\mathbf{F}^{-1}\mathbf{w} - J\mathbf{F}^{-1}(\nabla\mathbf{v})\mathbf{w}$$

$$= \frac{D}{Dt}(J\mathbf{F}^{-1}\mathbf{w}) - \left\{\frac{DJ}{Dt} - J(\operatorname{div} \mathbf{v})\right\}\mathbf{F}^{-1}\mathbf{w}$$

$$- J\left\{\frac{D\mathbf{F}^{-1}}{Dt} + \mathbf{F}^{-1}(\nabla\mathbf{v})\right\}\mathbf{w} \tag{6.3.27}$$

Substituting for DJ/Dt and $D\mathbf{F}^{-1}/Dt$ from (6.2.51) and (6.2.45), respectively, in (6.3.27), we arrive at (6.3.25). ∎

EXAMPLE 6.3.3 Show that

$$\mathbf{D} = (\mathbf{F}^{-1})^T\left(\frac{D\mathbf{G}}{Dt}\right)\mathbf{F}^{-1} \tag{6.3.28}$$

$$= \frac{D\mathbf{A}}{Dt} + \{\mathbf{A}(\nabla\mathbf{v}) + (\nabla\mathbf{v})^T\mathbf{A}\} \tag{6.3.29}$$

Solution From (5.4.10) and (5.4.28), let us recall that

$$\mathbf{G} = \tfrac{1}{2}(\mathbf{F}^T\mathbf{F} - \mathbf{I}) \tag{6.3.30a}$$

$$\mathbf{A} = \tfrac{1}{2}\{\mathbf{I} - (\mathbf{F}^{-1})^T\mathbf{F}^{-1}\} \tag{6.3.30b}$$

Using expressions (6.2.44) and (6.3.4) we find from (6.3.30a) that

$$\frac{D\mathbf{G}}{Dt} = \frac{1}{2}[\{(\nabla\mathbf{v})\mathbf{F}\}^T\mathbf{F} + \mathbf{F}^T(\nabla\mathbf{v})\mathbf{F}]$$

$$= \frac{1}{2}[\mathbf{F}^T(\nabla\mathbf{v}^T + \nabla\mathbf{v})\mathbf{F}] = \mathbf{F}^T\mathbf{D}\mathbf{F}$$

from which (6.3.28) is immediate.

Using expressions (6.2.45) and (6.3.4), we obtain from (6.3.30b) that

$$\frac{D\mathbf{A}}{Dt} = \frac{1}{2}[(\nabla\mathbf{v}^T)(\mathbf{F}^{-1})^T\mathbf{F}^{-1} + (\mathbf{F}^{-1})^T\mathbf{F}^{-1}(\nabla\mathbf{v})]$$

$$= \frac{1}{2}[(\nabla\mathbf{v}^T)(\mathbf{I} - 2\mathbf{A}) + (\mathbf{I} - 2\mathbf{A})(\nabla\mathbf{v})]$$

$$= \mathbf{D} - (\nabla\mathbf{v}^T)\mathbf{A} - \mathbf{A}(\nabla\mathbf{v})$$

from which (6.3.29) follows. ∎

6.3 STRETCHING AND VORTICITY

Note: From expressions (6.3.28) and (6.3.29) it is evident that **D** does not generally represent the material rate of change of Lagrangian or Eulerian strain tensors.

EXAMPLE 6.3.4 For the flow defined by velocity components of the form $v_1 = v_1(x_1, x_2, t)$, $v_2 = v_2(x_1, x_2, t)$, $v_3 = 0$ (such a flow is called a *plane flow*), show that

$$\mathbf{DW} + \mathbf{WD} = (\text{div } \mathbf{v})\mathbf{W} \qquad (6.3.31)$$

Solution By using the identity (3.5.47), we note that

$$\mathbf{DW} + \mathbf{WD} = \text{skw}(\nabla \mathbf{v})^2$$

so that

$$[\mathbf{DW} + \mathbf{WD}]_{ij} = \tfrac{1}{2}(v_{i,k}v_{k,j} - v_{j,k}v_{k,i}) \qquad (6.3.32)$$

For the given v_i, this yields the following expression for the matrix of $\mathbf{DW} + \mathbf{WD}$:

$$[\mathbf{DW} + \mathbf{WD}] = (v_{1,1} + v_{2,2})\begin{bmatrix} 0 & \tfrac{1}{2}(v_{1,2} + v_{2,1}) & 0 \\ -\tfrac{1}{2}(v_{1,2} - v_{2,1}) & 0 & 0 \\ 0 & 0 & 0 \end{bmatrix}$$

$$= (\text{div } \mathbf{v})[\mathbf{W}]$$

This proves (6.3.31). ∎

EXAMPLE 6.3.5 Find the stretching and vorticity tensors for the flow described by the velocity components

$$v_1 = \frac{f(R)}{R}x_2, \qquad v_2 = -\frac{f(R)}{R}x_1, \qquad v_3 = 0 \qquad (6.3.33)$$

where $R = (x_1^2 + x_2^2)^{1/2} \neq 0$. Deduce that the motion is isochoric.
Further, show that the flow defined by

$$v_1 = -\frac{x_2}{R^2}, \qquad v_2 = \frac{x_1}{R^2}, \qquad v_3 = 0 \qquad (6.3.34)$$

is irrotational as well.

Solution For the velocity components given by (6.3.33), we find that

$$[\nabla \mathbf{v}] = \begin{bmatrix} \bar{\phi}x_1 x_2 & \bar{\phi}x_2^2 + \bar{\psi} & 0 \\ -(\bar{\psi} + \bar{\phi}x_1^2) & -\bar{\phi}x_1 x_2 & 0 \\ 0 & 0 & 0 \end{bmatrix} \qquad (6.3.35)$$

where

$$\bar{\phi} = \frac{1}{R^2}\left[f'(R) - \frac{1}{R}f(R)\right] \quad \text{and} \quad \bar{\psi} = \frac{1}{R}f(R)$$

Noting that $x_1 = R\cos\theta$, $x_2 = R\sin\theta$, where R and θ are plane polar coordinates, we obtain from (6.3.35) that

$$[\mathbf{D}] = \tfrac{1}{2}[\nabla\mathbf{v}] + \tfrac{1}{2}[\nabla\mathbf{v}]^T = \begin{bmatrix} \phi\sin 2\theta & -\phi\cos 2\theta & 0 \\ -\phi\cos 2\theta & -\phi\sin 2\theta & 0 \\ 0 & 0 & 0 \end{bmatrix} \quad (6.3.36)$$

and

$$[\mathbf{W}] = \tfrac{1}{2}[\nabla\mathbf{v}] - \tfrac{1}{2}[\nabla\mathbf{v}]^T = \begin{bmatrix} 0 & \psi & 0 \\ -\psi & 0 & 0 \\ 0 & 0 & 0 \end{bmatrix} \quad (6.3.37)$$

where

$$\phi = \tfrac{1}{2}R^2\bar{\phi}, \quad \psi = \tfrac{1}{2}[f'(R) + \bar{\psi}]$$

Expressions (6.3.36) and (6.3.37) give the matrices of $\mathbf{D}$ and $\mathbf{W}$ for the given motion. From (6.3.36), we note that tr $\mathbf{D} = 0$. Hence the motion is isochoric.

Expressions (6.3.34) correspond to the particular case $f(R) = -1/R$ of (6.3.33). In this case we find that $\psi = 0$. Consequently, $\mathbf{W} = \mathbf{0}$ by (6.3.37), and the motion is irrotational as well in addition to being isochoric. [This can be proven directly by showing that for v_i given by (6.3.34), div $\mathbf{v} = 0$ and curl $\mathbf{v} = \mathbf{0}$.] ∎

6.4
PATH LINES, STREAM LINES AND VORTEX LINES

6.4.1 PATH LINES

Let us recall the equation (4.5.1), namely

$$\mathbf{x} = \mathbf{x}(\mathbf{x}^0, t) \quad (6.4.1)$$

representing the motion of a material body in the Lagrangian description. As already pointed out, for a specified particle $\mathcal{P}$ this equation gives the successive positions $\mathbf{x}$ of $\mathcal{P}$ with increasing t. In other words, equation (6.4.1) represents the path in which the particle $\mathcal{P}$ moves during the motion of the continuum. This path, which is a space curve having its initial point as $\mathbf{x}^0$, is called the *path line* of the particle $\mathcal{P}$; see Figure 6.2.

6.4 PATH LINES, STREAM LINES AND VORTEX LINES

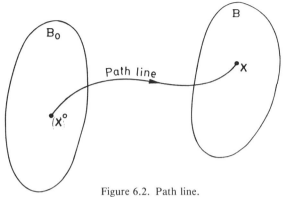

Figure 6.2. Path line.

For example, consider the motion described by the equations

$$x_1 = \sin(\alpha x_1^0 + \omega t)$$
$$x_2 = 1 - \cos(\alpha x_2^0 + \omega t) \qquad (6.4.2)$$
$$x_3 = 0$$

where α and ω are constants. For the particle $\mathcal{P}_0$ initially located at the origin, equations (6.4.2) become

$$x_1 = \sin \omega t, \qquad x_2 = 1 - \cos \omega t, \qquad x_3 = 0 \qquad (6.4.3)$$

These equations determine the location $\mathbf{x}$ of the particle $\mathcal{P}_0$ at any time t. Rewriting the equations we get

$$x_1^2 + (x_2 - 1)^2 = 1, \qquad x_3 = 0 \qquad \text{for all } t \qquad (6.4.4)$$

Therefore, the path line of the particle $\mathcal{P}_0$ is the circle $x_1^2 + (x_2 - 1)^2 = 1$ in the x_3 plane.

Reverting to the general discussion, suppose that, for a certain motion, the velocity is known at a point $\mathbf{x}$ of the current configuration. By the definition of $\mathbf{v}$, we have

$$\mathbf{v} = \frac{D\mathbf{x}}{Dt} \qquad (6.4.5)$$

Since $\mathbf{v}$ is known, equation (6.4.5) may be viewed as a differential equation for $\mathbf{x}$ and integrated under the (initial) condition $\mathbf{x} = \mathbf{x}^0$ at $t = 0$; the resulting solution will be in the form of (6.4.1). Thus, (6.4.5) serves as the *differential equation* for path lines.

6 MOTION

For example, consider the motion for which the velocity field is as follows:

$$v_1 = \frac{x_1}{1+t}, \quad v_2 = \frac{2x_2}{1+t}, \quad v_3 = \frac{3x_3}{1+t} \qquad (6.4.6)$$

For this field, equation (6.4.5) has been integrated in Example 6.2.3 and the following solution has been obtained; see (6.2.32):

$$x_1 = x_1^0(1+t), \quad x_2 = x_2^0(1+t)^2, \quad x_3 = x_3^0(1+t)^3 \qquad (6.4.7)$$

These are the equations for the path line of the particle whose current velocity is given by (6.4.6).

6.4.2 FIELD LINES

Consider a geometrical arc γ in the current configuration of a continuum. Let its equation be $\mathbf{x} = \mathbf{x}(\tau)$, where τ is a parameter. From three-dimensional geometry, we recall that the vector $d\mathbf{x}/d\tau$ is tangential to γ at the point $\mathbf{x}$ on it. Suppose $\mathbf{f}$ is a non-zero vector known at every point of the configuration. Then it follows that at any point $\mathbf{x}$ on γ the tangent to γ is directed along $\mathbf{f}$ if and only if

$$\mathbf{f} = \alpha \frac{d\mathbf{x}}{d\tau} \qquad (6.4.8)$$

for some positive real number α. Such an arc is called the *field line* of $\mathbf{f}$. Thus, *the field line of a vector $\mathbf{f}$ at a point $\mathbf{x}$ is an arc whose tangent at $\mathbf{x}$ is directed along $\mathbf{f}$*; see Figure 6.3.

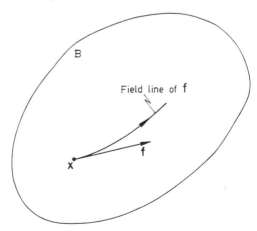

Figure 6.3. Field line.

6.4 PATH LINES, STREAM LINES AND VORTEX LINES

For a given **f**, the field line of **f** satisfies the differential equation (6.4.8) by definition. When integrated, this differential equation yields a solution of the form $\mathbf{x} = \mathbf{x}(\tau)$, which represents the field line of **f**. Equation (6.4.8) is therefore the *differential equation for the field line of* **f**. This equation can be rewritten in the following two forms:

$$f_i = \alpha \frac{dx_i}{d\tau} \tag{6.4.9}$$

$$\frac{dx_1}{f_1} = \frac{dx_2}{f_2} = \frac{dx_3}{f_3} \tag{6.4.10}$$

6.4.3 STREAM LINES AND VORTEX LINES

The two vectors that generally determine the nature of the motion of a continuum are the velocity and the vorticity. The field lines of our major interest here are therefore those of velocity and vorticity. The field lines of velocity are called the *stream lines* and those of vorticity are the *vortex lines*. Thus, a stream line is an arc in the current configuration such that the tangent to the arc at any of its points is directed along the velocity vector at that point. The meaning of a vortex line is analogous. The differential equations for a stream line and a vortex line are obtained by replacing **f** (or f_i) by **v** (or v_i) and **w** (or w_i), respectively, in the equations (6.4.8) to (6.4.10). It is important to note that, since the velocity and the vorticity vectors are uniquely determined at a point, there cannot be two distinct stream lines and two distinct vortex lines at a point. In other words, two stream lines cannot cross each other; neither can vortex lines. It is obvious that the stream lines do not make their appearance when there is no motion, and that the vortex lines are nonexistent in an irrotational motion.

From their definitions, it is evident that the path lines and stream lines are not the same. Whereas a path line is the trajectory of a specific particle as it moves from its initial location, a stream line is a curve tangential to the velocity of the particle at its current location (Figure 6.4). However, if the motion is steady, that is, if $\partial \mathbf{v}/\partial t = \mathbf{0}$, the path lines and the stream lines become *coincident*, because, when $\partial \mathbf{v}/\partial t = \mathbf{0}$, the differential equation for the path lines, namely, (6.4.5), yields the solution

$$\mathbf{x} = t\,\mathbf{v}(\mathbf{x}^0) + \mathbf{x}^0 \tag{6.4.11}$$

which evidently represents the curve tangential to **v**.

However, the converse is not true. To see this, let us consider the motion for which the velocity is given by (6.4.6). For this motion the path lines are given by (6.4.7), as already seen. Let us now find the stream lines. The

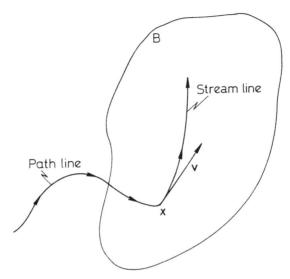

Figure 6.4. Stream line.

differential equation for these lines, namely

$$\frac{dx_1}{v_1} = \frac{dx_2}{v_2} = \frac{dx_3}{v_3} \tag{6.4.12}$$

are

$$\frac{dx_1}{x_1} = \frac{dx_2}{2x_2} = \frac{dx_3}{3x_3} = \frac{dt}{1+t} \tag{6.4.13}$$

The solution of these equations obeying the condition $\mathbf{x} = \mathbf{x}^0$ for $t = 0$ is given by

$$x_1 = x_0(1+t), \qquad x_2 = x_2^0(1+t)^2, \qquad x_3 = x_3^0(1+t)^3 \tag{6.4.14}$$

These equations are identical with equations (6.4.7). Hence, for the motion given by (6.4.6), the stream lines and the path lines are coincidental. But, we note from (6.4.6) that $\partial \mathbf{v}/\partial t \neq \mathbf{0}$; that is, the motion is not steady. Thus, the path lines and stream lines can be coincident in a nonsteady motion also. This analysis shows that whereas for a steady motion the path lines and the stream lines are necessarily coincident, the coincidence of these lines does not ensure that the motion is steady.

Returning to the motion for which the velocity is given by (6.4.6), we note that $\mathbf{w} = \mathbf{0}$ for this motion. As such, the motion is irrotational and (therefore) the vortex lines do not exist.

6.4.4 STREAM TUBES AND VORTEX TUBES

Consider a simple closed curve C in the current configuration and visualize the tube formed by the field lines of a vector $\mathbf{f}$, initiating from the points of C. This tube is called a *field tube of* $\mathbf{f}$ *at* C. If S is a section of the tube, then the integral $\int_S \mathbf{f} \cdot \mathbf{n}\, dS$ is called the *strength* of the tube at S. In particular, the field tube of $\mathbf{v}$ is called the *stream tube*, and the field tube of $\mathbf{w}$ is called the *vortex tube*; see Figure 6.5.

Consider two sections S_1 and S_2 of a field tube of $\mathbf{f}$ (Figure 6.6). Let V be the volume of the region bounded by S_1, S_2 and the curved surface S' of the tube. Then, by the divergence theorem (3.6.1), we obtain

$$\int_V \text{div } \mathbf{f}\, dV = \int_{S_1} \mathbf{f} \cdot \mathbf{n}\, dS + \int_{S_2} \mathbf{f} \cdot \mathbf{n}\, dS + \int_{S'} \mathbf{f} \cdot \mathbf{n}\, dS \quad (6.4.15)$$

where $\mathbf{n}$ is as usual the unit outward normal.

Since S' is formed by the field lines of $\mathbf{f}$, $\mathbf{f} \cdot \mathbf{n} = 0$ on S'. If the distance between S_1 and S_2 is so small that the tube can be considered straight, then $\mathbf{n}$ has opposite directions on S_1 and S_2. Consequently, expression (6.4.15) becomes

$$\int_V \text{div } \mathbf{f}\, dV = \pm \left\{ \int_{S_1} \mathbf{f} \cdot \mathbf{n}\, dS - \int_{S_2} \mathbf{f} \cdot \mathbf{n}\, dS \right\} \quad (6.4.16)$$

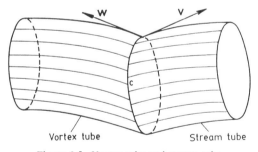

Figure 6.5. Vortex tube and stream tube.

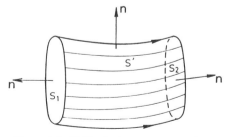

Figure 6.6. Field tube between two sections.

For $\mathbf{f} = \mathbf{w}$, the field tube of $\mathbf{f}$ is the vortex tube and in this case (6.4.16) yields, on using the fact that div $\mathbf{w} = 0$,

$$\int_{S_1} \mathbf{w} \cdot \mathbf{n}\, dS = \int_{S_2} \mathbf{w} \cdot \mathbf{n}\, dS \tag{6.4.17}$$

This expression shows that the strength of the vortex tube at S_1 is the same as that at S_2. In other words, *the strength of a vortex tube is constant*. This result is essentially by Helmholtz (1858).

For $\mathbf{f} = \mathbf{v}$, the field tube of $\mathbf{f}$ is the stream tube and in this case (6.4.16) yields

$$\int_B \operatorname{div} \mathbf{v}\, dV = \pm \left\{ \int_{S_1} \mathbf{v} \cdot \mathbf{n}\, dS - \int_{S_2} \mathbf{v} \cdot \mathbf{n}\, dS \right\}. \tag{6.4.18}$$

If the motion is isochoric, that is if div $\mathbf{v} = 0$, we find from (6.4.18) that

$$\int_{S_1} \mathbf{v} \cdot \mathbf{n}\, dS = \int_{S_2} \mathbf{v} \cdot \mathbf{n}\, dS \tag{6.4.19}$$

Thus, *in an isochoric motion, the strength of a stream tube is constant.*

For an incompressible continuum, every motion is isochoric. Hence, *for every motion of an incompressible continuum, the strength of a stream tube is constant*.

EXAMPLE 6.4.1 Show that the stream lines and the vortex lines are orthogonal if and only if $\mathbf{v} = \phi(\nabla \psi)$ for some scalar functions ϕ and ψ.

Solution We note that the stream lines and the vortex lines are orthogonal whenever $\mathbf{v}$ and curl $\mathbf{v}$ are orthogonal. From Example 3.4.6, we recall that this is possible if and only if $\mathbf{v} = \phi \nabla \psi$, for some ϕ and ψ. ∎

EXAMPLE 6.4.2 For the flow defined by the velocity field

$$\mathbf{v} = (1 + at)\mathbf{e}_1 + x_1 \mathbf{e}_2 \tag{6.4.20}$$

where a is a constant, find the path lines, stream lines and vortex lines.

Solution Recall that the path lines are governed by the differential equation (6.4.5). For the given motion, this equation yields

$$\frac{Dx_1}{Dt} = (1 + at) \tag{6.4.21a}$$

$$\frac{Dx_2}{Dt} = x_1 \tag{5.4.21b}$$

$$\frac{Dx_3}{Dt} = 0 \tag{6.4.21c}$$

Integration of the first of these equations gives, on using the fact that $x_1 = x_1^0$ for $t = 0$, the solution

$$x_1 = x_1^0 + (t + \tfrac{1}{2}at^2) \tag{6.4.22a}$$

Using this expression in equation (6.4.21b) and integrating the resulting equation under the condition $x_2 = x_2^0$ for $t = 0$, we get

$$x_2 = x_2^0 + x_1^0 t + \tfrac{1}{2}t^2 + \tfrac{1}{6}at^3 \tag{6.4.22b}$$

Equation (6.4.21c) gives

$$x_3 = x_3^0 \tag{6.4.22c}$$

For the given motion, (6.4.22) constitute the parametric equations for the path lines.

Next, recall that the stream lines are governed by the equations (6.4.12). For the given motion, these equations become

$$\frac{dx_1}{1+at} = \frac{dx_2}{x_1}, \quad x_3 = \text{constant} \tag{6.4.23}$$

These equations yield, on using the fact that $x_i = x_i^0$ for $t = 0$, the equations

$$x_1^2 = (x_1^0)^2 + 2[x_2(1+at) - x_2^0], \quad x_3 = x_3^0 \tag{6.4.24}$$

For the given motion, these are the equations of the stream lines.

From the given velocity field, we get $\mathbf{w} = \mathbf{e}_3$; therefore, the vortex lines are given by $d\mathbf{x}/d\tau = \alpha \mathbf{e}_3$, which yields, on using the conditions $x_i = x_i^0$ for $t = 0$, the equations

$$x_1 = x_1^0, \quad x_2 = x_2^0, \quad x_3 = x_3^0 + \tau\alpha \tag{6.4.25}$$

Evidently, vortex lines are straight lines in the x_3 direction. ∎

Note: For $a = 0$, equations (6.4.22) yield

$$x_1^2 = (x_1^0)^2 + 2(x_2 - x_2^0), \quad x_3 = x_3^0 \tag{6.4.26}$$

When $a = 0$, equations (6.4.24) also reduce to (6.4.26). Thus, for the given flow, the stream lines and the path lines coincide in the case of $a = 0$. This result is expected because for $a = 0$ the given flow is steady.

6.5
TRANSPORT FORMULAS

In Examples 6.2.6 and 6.2.8, the material derivatives of a material arc element, a material surface element and a material volume element located in the current configuration have been computed. We now generalize these expressions by deriving formulas for the material derivatives of integrals over material curves, material surfaces and material volumes located in the current configuration. These formulas, which are usually called the *transport formulas*, are of great utility in deriving the balance laws of mass, momentum and energy. The formulas are contained in the following.

THEOREM 6.5.1 Let ϕ denote a scalar, a component of a vector or a component of a tensor known in the *spatial form*. Then for a material curve $\mathcal{C}$, material surface $\mathcal{S}$ and a material body $\mathcal{B}$ occupying a curve C, a surface S and a volume V, respectively, in the current configuration, the following formulae hold:

(i) $$\frac{D}{Dt}\int_C \phi\,d\mathbf{x} = \int_C \left\{\frac{D\phi}{Dt} + \phi(\nabla\mathbf{v})\right\} d\mathbf{x} \tag{6.5.1}$$

(ii) $$\frac{D}{Dt}\int_S \phi\mathbf{n}\,dS = \int_S \left\{\left(\frac{D\phi}{Dt} + \phi\,\text{div}\,\mathbf{v}\right) - \phi(\nabla\mathbf{v})^T\right\}\mathbf{n}\,dS \tag{6.5.2}$$

(iii) $$\frac{D}{Dt}\int_V \phi\,dV = \int_V \left(\frac{D\phi}{Dt} + \phi\,\text{div}\,\mathbf{v}\right)dV \tag{6.5.3}$$

The general method of proving each of the formulas (6.5.1) to (6.5.3) is as follows. We first transfer the expression on the lefthand side of the formula to the initial configuration by changing the variables from x_i and t to x_i^0 and t. The range of integration then becomes a curve-surface-volume in the initial configuration. Since the material derivative is the partial derivative with respect to t, keeping x_i^0 fixed, the material differentiation can be carried out under the integral sign in the initial configuration. After this differentiation process is over, we transfer the expression back to the current configuration to arrive at the final result. The details follow.

Proof (i) Let C_0 be the initial location of the material curve $\mathcal{C}$ that is currently located on C. Then $d\mathbf{x}$ is related to the arc element $d\mathbf{x}^0$ of C_0 through the transformation $d\mathbf{x} = \mathbf{F}\,d\mathbf{x}^0$; see (5.3.1). Hence, the change of variables from x_i to x_i^0 yields

$$\int_C \phi(\mathbf{x},t)\,d\mathbf{x} = \int_{C_0} \phi\{\mathbf{x}(\mathbf{x}^0,t),t\}\mathbf{F}\,d\mathbf{x}^0 \tag{6.5.4}$$

Taking the material derivative on both sides of (6.5.4) and bearing in mind that C_0 is a curve in the initial configuration, we get

$$\frac{D}{Dt}\int_C \phi\,d\mathbf{x} = \int_{C_0} \frac{D}{Dt}(\phi\mathbf{F})\,d\mathbf{x}^0 = \int_{C_0}\left(\frac{D\phi}{Dt}\mathbf{F} + \phi\frac{D\mathbf{F}}{Dt}\right)d\mathbf{x}^0$$

6.5 TRANSPORT FORMULAS

By the use of (6.2.44), this becomes

$$\frac{D}{Dt} \int_C \phi \, d\mathbf{x} = \int_{C_0} \left\{ \frac{D\phi}{Dt} + \phi(\nabla \mathbf{v}) \right\} \mathbf{F} \, d\mathbf{x}^0 \tag{6.5.5}$$

Transferring the expression on the righthand side of (6.5.5) back to the current configuration with the use of (5.3.1), we obtain the formula (6.5.1).

(ii) Let S_0 be the initial location of the material surface S that is currently located on S. Then the oriented surface element $(\mathbf{n} \, dS)$ taken on S and the oriented surface element $(\mathbf{n}^0 \, dS_0)$ taken on S_0 are related to each other through the formula (5.2.12), namely, $(dS)\mathbf{n} = J(\mathbf{F}^T)^{-1}\mathbf{n}^0(dS_0)$. Hence

$$\int_S \phi \mathbf{n} \, dS = \int_{S_0} \phi J(\mathbf{F}^T)^{-1} \mathbf{n}^0 \, dS_0 \tag{6.5.6}$$

Taking the material derivative on both sides of (6.5.6) and bearing in mind that S_0 is a surface in the initial configuration, we get

$$\frac{D}{Dt} \int_S \phi \mathbf{n} \, dS = \int_{S_0} \frac{D}{Dt} \{\phi J(\mathbf{F}^{-1})^T\} \mathbf{n}^0 \, dS_0 \tag{6.5.7}$$

Carrying out the material differentiation in the righthand side of (6.5.7) and using the relations (6.2.45) and (6.2.51), we obtain

$$\frac{D}{Dt} \int_S \phi \mathbf{n} \, dS = \int_{S_0} \left\{ \frac{D\phi}{Dt} + \phi \, \text{div } \mathbf{v} - \phi(\nabla \mathbf{v})^T \right\} J(\mathbf{F}^{-1})^T \mathbf{n}^0 \, dS_0 \tag{6.5.8}$$

Transferring the expression on the righthand side of (6.5.8) back to the current configuration with the use of (5.2.12), we obtain the formula (6.5.2).

(iii) Let V_0 be the initial volume of the material body $\mathcal{B}$ that is currently occupying the volume V. Then the volume element dV of V is related to the volume element dV_0 of V_0 through the formula $dV = J \, dV_0$; see (5.2.22). Hence

$$\int_V \phi \, dV = \int_{V_0} \phi J \, dV_0 \tag{6.5.9}$$

Taking the material derivative on both sides of (6.5.9) and bearing in mind that V_0 is a volume in the initial configuration, we get

$$\frac{D}{Dt} \int_V \phi \, dV = \int_{V_0} \frac{D}{Dt} (\phi J) \, dV_0 = \int_{V_0} \left(\frac{D\phi}{Dt} J + \phi \frac{DJ}{Dt} \right) dV_0$$

By use of (6.2.51), this becomes

$$\frac{D}{Dt} \int_V \phi \, dV = \int_{V_0} \left(\frac{D\phi}{Dt} + \text{div } \mathbf{v} \right) J \, dV_0 \tag{6.5.10}$$

Transferring the expression on the righthand side back to the current configuration with the use of (5.2.22) we obtain the formula (6.5.3). This completes the proof of the theorem. ∎

6.5.1 REYNOLD'S TRANSPORT FORMULA

Here we obtain an alternative version of the formula (6.5.3). By use of the identity (3.4.14), the expression for the material derivative, given in (6.2.8), yields

$$\frac{D\phi}{Dt} + \phi \,\mathrm{div}\,\mathbf{v} = \frac{\partial \phi}{\partial t} + \mathrm{div}(\phi \mathbf{v}) \qquad (6.5.11)$$

Using this, the formula (6.5.3) becomes

$$\frac{D}{Dt}\int_V \phi \, dV = \int_V \left\{ \frac{\partial \phi}{\partial t} + \mathrm{div}(\phi \mathbf{v}) \right\} dV \qquad (6.5.12)$$

which on using the divergence theorem (3.6.1) reduces to

$$\frac{D}{Dt}\int_V \phi \, dV = \int_V \frac{\partial \phi}{\partial t} dV + \int_S \phi \mathbf{v} \cdot \mathbf{n} \, dS \qquad (6.5.13)$$

where S is the surface enclosing V.

Expression (6.5.13), which is an alternative version of (6.5.3), is known as the *Reynold's transport formula (theorem)*, after O. Reynold (1903).

This expression shows that the rate of change of the integral of ϕ over a material volume is made up of (i) the integral of the local rate of change of ϕ in the volume, and (ii) the flux of $\phi \mathbf{v}$ across the surface enclosing the volume.

Note: For simplicity in the terminology, the curve C, the surface S and the volume V, which are the current locations of a material curve $\mathcal{C}$, a material surface $\mathcal{S}$ and a material volume $\mathcal{V}$ (respectively) are themselves often referred to as a material curve, a material surface and a material volume (respectively).

EXAMPLE 6.5.1 Prove the following:

(i) $$\frac{D}{Dt}\int_C d\mathbf{x} = \int_C (\nabla \mathbf{v}) \, d\mathbf{x} \qquad (6.5.14)$$

(ii) $$\frac{D}{Dt}\int_S \mathbf{n} \, dS = \int_S \{\mathrm{div}\,\mathbf{v} - (\nabla \mathbf{v})^T\}\mathbf{n} \, dS \qquad (6.5.15)$$

(iii) $$\frac{D}{Dt}\int_V dV = \int_V (\mathrm{div}\,\mathbf{v}) \, dV \qquad (6.5.16)$$

6.5 TRANSPORT FORMULAS

Solution Taking $\phi = 1$ in the transport formulas (6.5.1) to (6.5.3), we readily obtain the results (6.5.14) to (6.5.16). ∎

Note: Expression (6.5.16) can be rewritten as

$$\frac{DV}{Dt} = \int_V \text{div } \mathbf{v} \, dV = \int_V (\text{tr } \mathbf{D}) \, dV \quad (6.5.17)$$

Thus, the rate of change of a material volume is equal to the integral of div $\mathbf{v}$ on that volume. Expression (6.5.17) is analogous to the expression (5.6.26) for the change in volume.

EXAMPLE 6.5.2 Let $\mathbf{f}$ be a vector function defined in the current configuration and C be a material curve in the configuration. Show that

$$\frac{D}{Dt} \int_C \mathbf{f} \otimes d\mathbf{x} = \int_C \left\{ \frac{D\mathbf{f}}{Dt} \otimes d\mathbf{x} + \mathbf{f} \otimes (\nabla \mathbf{v}) \, d\mathbf{x} \right\} \quad (6.5.18)$$

Deduce that

$$\frac{D}{Dt} \int_C \mathbf{f} \cdot d\mathbf{x} = \int_C \left\{ \frac{D\mathbf{f}}{Dt} \cdot d\mathbf{x} + (\nabla \mathbf{v})^T \mathbf{f} \cdot d\mathbf{x} \right\} \quad (6.5.19)$$

Solution From the transport formula (6.5.1), we have

$$\frac{D}{Dt} \int_C \phi \, dx_j = \int_C \left(\frac{D\phi}{Dt} dx_j + \phi v_{j,k} \, dx_k \right)$$

Setting $\phi = f_i$ in this expression, where f_i are the components of the given vector $\mathbf{f}$, we get

$$\frac{D}{Dt} \int_C f_i \, dx_j = \int_C \left(\frac{Df_i}{Dt} dx_j + f_i v_{j,k} \, dx_k \right)$$

which is precisely (6.5.18).

Taking the trace of expression (6.5.18) and using the fact that $\text{tr}\{\mathbf{a} \otimes (\mathbf{Ab})\} = \mathbf{a} \cdot \mathbf{Ab} = (\mathbf{A}^T \mathbf{a}) \cdot \mathbf{b}$, which follows by use of (2.4.35) and (2.8.14), we obtain (6.5.19). ∎

EXAMPLE 6.5.3 Let $\mathbf{f}$ be a vector function defined over the current configuration. For a material surface S in this configuration, show that

$$\frac{D}{Dt} \int_S \mathbf{f} \otimes \mathbf{n} \, dS = \int_S \left\{ \frac{D\mathbf{f}}{Dt} \otimes \mathbf{n} + (\text{div } \mathbf{v})\mathbf{f} \otimes \mathbf{n} - \mathbf{f} \otimes (\nabla \mathbf{v})^T \mathbf{n} \right\} dS \quad (6.5.20)$$

Deduce that

$$\frac{D}{Dt}\int_S \mathbf{f} \cdot \mathbf{n}\, dS = \int_S \left\{\frac{D\mathbf{f}}{Dt} + \mathbf{f}(\operatorname{div} \mathbf{v}) - (\mathbf{f} \cdot \nabla)\mathbf{v}\right\} \cdot \mathbf{n}\, dS \quad (6.5.21)$$

$$= \int_S \left\{\frac{\partial \mathbf{f}}{\partial t} + (\operatorname{div} \mathbf{f})\mathbf{v} + \operatorname{curl}(\mathbf{f} \times \mathbf{v})\right\} \cdot \mathbf{n}\, dS \quad (6.5.22)$$

Solution From the transport formula (6.5.2), we have

$$\frac{D}{Dt}\int_S \phi n_j\, dS = \int_S \left\{\frac{D\phi}{Dt} n_j + \phi v_{k,k} n_j - \phi v_{k,j} n_k\right\} dS$$

Setting $\phi = f_i$ in this expression, we get

$$\frac{D}{Dt}\int_S f_i n_j\, dS = \int_S \left\{\frac{Df_i}{Dt} n_j + f_i(v_{k,k}) n_j - f_i v_{k,j} n_k\right\} dS$$

This is precisely (6.5.20).

Taking the trace of expression (6.5.20) and using (2.4.35) and the fact that $\mathbf{f} \cdot (\nabla \mathbf{v})^T \mathbf{n} = (\nabla \mathbf{v})\mathbf{f} \cdot \mathbf{n} = (\mathbf{f} \cdot \nabla)\mathbf{v} \cdot \mathbf{n}$, which follows by use of (2.8.14) and (3.5.34), we obtain (6.5.21). Rewriting the righthand side of (6.5.21) by using the vector identity (3.4.25) and the result

$$\frac{D\mathbf{f}}{Dt} = \frac{\partial \mathbf{f}}{\partial t} + (\mathbf{v} \cdot \nabla)\mathbf{f}$$

we obtain (6.5.22). ∎

Note: It follows from (6.5.22) that *the flux of a vector* $\mathbf{f}$ *across every material surface is constant if and only if*

$$\frac{\partial \mathbf{f}}{\partial t} + (\operatorname{div} \mathbf{f})\mathbf{v} + \operatorname{curl}(\mathbf{f} \times \mathbf{v}) \equiv 0 \quad (6.5.23)$$

This is known as *Zorawski's criterion*.

6.6

CIRCULATION AND VORTICITY

In Section 3.6, the circulation of a vector field $\mathbf{f}$ round a simple closed curve C was defined as the line integral $\int_C \mathbf{f} \cdot \mathbf{t}\, ds$. We now study this concept when $\mathbf{f}$ is the velocity vector and C is a simple closed curve in the current configuration.

6.6 CIRCULATION AND VORTICITY

Let C be a simple closed curve (circuit) in the current configuration. Then the circulation of the velocity vector $\mathbf{v}$ around C will henceforth be referred to as the *circulation round C*, denoted I_C. Thus,

$$I_C = \int_C \mathbf{v} \cdot \mathbf{t}\, ds = \int_C \mathbf{v} \cdot d\mathbf{x} \qquad (6.6.1)$$

If C is a reducible curve, that is, if there is a regular open surface S bounded by C (which lies completely within the configuration), then by using Stokes's theorem (3.6.7), expression (6.6.1) can be rewritten as

$$I_C = \int_S \mathbf{w} \cdot \mathbf{n}\, dS \qquad (6.6.2)$$

This relation shows that the concept of circulation is closely related to the concept of vorticity; in fact I_C represents the flux of $\mathbf{w}$ across S. It is evident that $I_C = 0$ for every reducible curve C if and only if $\mathbf{w} \equiv \mathbf{0}$. In other words, *a motion is irrotational in a given region if and only if the circulation round every reducible circuit in the region is 0.*

6.6.1 MATERIAL DERIVATIVE OF CIRCULATION

Suppose that C is a material curve in the current configuration. Then DI_C/Dt can be computed with the aid of a transport formula. In fact, taking $\mathbf{f} = \mathbf{v}$ in (6.5.19) we readily get

$$\frac{DI_C}{Dt} = \int_C \left\{ \frac{D\mathbf{v}}{Dt} \cdot d\mathbf{x} + (\nabla \mathbf{v})^T \mathbf{v} \cdot d\mathbf{x} \right\} \qquad (6.6.3)$$

Since $(\nabla \mathbf{v})^T \mathbf{v} = \frac{1}{2} \nabla(\mathbf{v} \cdot \mathbf{v})$, we find that

$$\int_C (\nabla \mathbf{v})^T \mathbf{v} \cdot d\mathbf{x} = \frac{1}{2} \int_C \nabla(v^2) \cdot d\mathbf{x} = \frac{1}{2} \int_C dv^2 = 0$$

because C is a closed curve. Hence, expression (6.6.3) becomes

$$\frac{DI_C}{Dt} = \int_C \frac{D\mathbf{v}}{Dt} \cdot d\mathbf{x} \qquad (6.6.4)$$

This formula for the material derivative of circulation is given by Lord Kelvin (1869).

6.6.2 KELVIN'S CIRCULATION THEOREM

As a direct consequence of (6.6.4), we now prove the following famous result, known as the *Kelvin's circulation theorem*.

THEOREM 6.6.1 If the acceleration is the gradient of a potential, then the circulation round a material curve remains constant in time.

Proof If $D\mathbf{v}/Dt$ is the gradient of a potential, that is, if $D\mathbf{v}/Dt = \nabla\psi$ for some scalar function ψ, we have

$$\int_C \frac{D\mathbf{v}}{Dt} \cdot d\mathbf{x} = \int_C \nabla\psi \cdot d\mathbf{x} = \int_C d\psi = 0$$

Then expression (6.6.4) becomes

$$\frac{DI_C}{Dt} = 0 \quad \text{or} \quad I_C = \text{constant (w.r.t. } t) \tag{6.6.5}$$

This completes the proof the theorem. ∎

6.6.3 CIRCULATION-PRESERVING MOTION

A motion is said to be circulation preserving, if the circulation round every material curve remains constant in time. A necessary and sufficient condition for a motion to be circulation preserving follows immediately from the Kelvin's formula (6.6.4) and is given by the following theorem.

THEOREM 6.6.2 A motion is circulation preserving if and only if the acceleration is an irrotational vector.

Proof By employing Stokes's theorem (3.6.7) to (6.6.4), we obtain

$$\frac{DI_C}{Dt} = \int_S \text{curl}\left(\frac{D\mathbf{v}}{Dt}\right) \cdot \mathbf{n}\, dS \tag{6.6.6}$$

From this expression, it is evident that $DI_C/Dt = 0$ for every C if and only if $\text{curl}(D\mathbf{v}/Dt) \equiv 0$. This completes the proof. ∎

Note: In a potential flow, the acceleration is the gradient of a scalar function; see Example 6.2.4. Hence it follows from this theorem that *a potential flow is circulation preserving*.

COROLLARY For a circulation-preserving motion,

$$\frac{D}{Dt}(J\mathbf{F}^{-1}\mathbf{w}) = \mathbf{0} \tag{6.6.7}$$

Proof The result follows by setting $\text{curl}(D\mathbf{v}/Dt) = \mathbf{0}$ in (6.3.25).

6.6.4 CAUCHY'S VORTICITY EQUATION

Integrating equation (6.6.7) and noting that $J = 1$ and $\mathbf{F} = \mathbf{I}$ at $t = 0$, we get

$$J\mathbf{F}^{-1}\mathbf{w} = \mathbf{w}^0 \qquad (6.6.8)$$

where $\mathbf{w}^0 = \mathbf{w}^0(\mathbf{x}^0)$ is the vorticity vector at $t = 0$.

Equation (6.6.8) exhibits a simple relationship between the current vorticity and the initial vorticity, in a circulation-preserving motion. This remarkable equation is known as *Cauchy's vorticity equation*. We now give two important consequences.

6.6.5 PERMANENCE OF IRROTATIONAL MOTION

THEOREM 6.6.3 If a circulation-preserving motion is once irrotational, then it is irrotational throughout the history of the motion. (This is known as the *Cauchy–Lagrange theorem*.)

Proof From equation (6.6.8), we readily get

$$\mathbf{w} = \frac{1}{J}\mathbf{F}\mathbf{w}^0 \qquad (6.6.9)$$

It is evident that if $\mathbf{w}^0 = \mathbf{0}$, then $\mathbf{w} = \mathbf{0}$. Hence if the vorticity is initially 0, then it is 0 for all time. This completes the proof of the theorem. ∎

6.6.6 MOTION OF VORTEX LINES

THEOREM 6.6.4 In a circulation-preserving motion, the vortex lines are transported with the motion. This theorem is known as the *Helmholtz theorem*.

Proof Consider a vortex line through a point $\mathbf{x}^0$ in the initial configuration. Then the tangent to this line at $\mathbf{x}^0$ lays along $\mathbf{w}^0$ and therefore an arc element of this line is given by the equation

$$d\mathbf{x}^0 = \alpha \mathbf{w}^0 \qquad (6.6.10)$$

where α is a nonzero scalar. The corresponding arc $d\mathbf{x}$ in the current configuration is gotten by the use of (5.3.1) as

$$d\mathbf{x} = \alpha \mathbf{F}\mathbf{w}^0 \qquad (6.6.11)$$

By virtue of the vorticity equation (6.6.9), equation (6.6.11) becomes

$$d\mathbf{x} = (\alpha J)\mathbf{w} \qquad (6.6.12)$$

This equation shows that $d\mathbf{x}$ is the arc element of the vortex line at $\mathbf{x}$ (in the

EXAMPLE 6.6.1 For the flow defined by the velocity components

$$v_1 = -\frac{2\alpha}{b^2} x_2, \qquad v_2 = \frac{2\alpha}{a^2} x_1, \qquad v_3 = 0 \qquad (6.6.13)$$

where α, a and b are constants, find the circulation round the ellipse $(x_1^2/a^2) + (x_2^2/b^2) = 1$.

Solution For the given ellipse, the parametric equations are $x_1 = a \cos \theta$, $x_2 = b \sin \theta$, $0 \leq \theta \leq 2\pi$. Hence, if C denotes this ellipse, we obtain

$$I_C = \int_C \mathbf{v} \cdot d\mathbf{x} = \int_C (v_1 \, dx_1 + v_2 \, dx_2)$$

$$= \frac{2\alpha}{ab} \int_0^{2\pi} (a^2 \sin^2\theta + b^2 \cos^2\theta) \, d\theta$$

$$= \frac{2\alpha\pi}{ab} (a^2 + b^2) \qquad \blacksquare$$

EXAMPLE 6.6.2 Show that a motion is circulation preserving if and only if

$$\frac{\partial \mathbf{w}}{\partial t} = \text{curl}(\mathbf{v} \times \mathbf{w}) \qquad (6.6.14)$$

Solution From equation (6.6.2), we get

$$\frac{DI_C}{Dt} = \frac{D}{Dt} \int_S \mathbf{w} \cdot \mathbf{n} \, dS \qquad (6.6.15)$$

Using (6.5.22) and noting that div $\mathbf{w} = 0$, (6.6.15) becomes

$$\frac{DI_C}{Dt} = \int_S \left\{ \frac{\partial \mathbf{w}}{\partial t} + \text{curl}(\mathbf{w} \times \mathbf{v}) \right\} \cdot \mathbf{n} \, dS \qquad (6.6.16)$$

From this expression it is evident that I_C is constant for every C if and only if the integrand on the righthand side is 0 for every S. In other words, the motion is circulation preserving if and only if

$$\frac{\partial \mathbf{w}}{\partial t} + \text{curl}(\mathbf{w} \times \mathbf{v}) = \mathbf{0} \qquad (6.6.17)$$

which is (6.6.14). ∎

6.6 CIRCULATION AND VORTICITY

Note: From (6.6.15) and (6.6.16), we find that (6.6.17) is a necessary and sufficient condition for the flux of **w** across every material surface is constant. This result is known as *Helmholtz's third vorticity theorem*. Observe that the condition (6.6.17) is just a particular case of the Zorawski's criterion (6.5.23).

EXAMPLE 6.6.3 If ds_0 and ds are the arc lengths of a vortex line in the initial and current configurations of a circulation-preserving motion, show that

$$ds = \frac{|\mathbf{w}|}{|\mathbf{w}^0|} J ds_0 \qquad (6.6.18)$$

Deduce that

(i) $$\frac{D}{Dt}\left(\frac{ds}{J|\mathbf{w}|}\right) = 0 \qquad (6.6.19)$$

(ii) $$\frac{1}{ds}\frac{D}{Dt}(ds) = \frac{1}{|\mathbf{w}|}\frac{D}{Dt}|\mathbf{w}| + \operatorname{div} \mathbf{v} \qquad (6.6.20)$$

Solution From expression (6.6.10) and (6.6.12), we obtain

$$ds_0 = |d\mathbf{x}^0| = |\alpha||\mathbf{w}^0|, \qquad ds = |d\mathbf{x}| = |\alpha|J|\mathbf{w}|.$$

Eliminating $|\alpha|$ from these relations, we get (6.6.18).

From (6.6.18), we get

$$\frac{D}{Dt}\left(\frac{ds}{J|\mathbf{w}|}\right) = \frac{D}{Dt}\left(\frac{ds_0}{|\mathbf{w}_0|}\right) = 0$$

because both ds_0 and $|\mathbf{w}^0|$ correspond to the initial configuration. The relation (6.6.19) is thus proven.

Carrying out the differentiation in (6.6.19) we obtain

$$\frac{1}{ds}\frac{D}{Dt}(ds) = \frac{1}{J}\frac{DJ}{Dt} + \frac{1}{|\mathbf{w}|}\frac{D}{Dt}|\mathbf{w}|,$$

which, by use of (6.2.51), reduces to (6.6.20). ∎

Note: Expressions (6.6.18) and (6.6.20) serve as formulas for the stretch and stretching of vortex lines in circulation-preserving motions. The stretch is completely determined by the current and initial vorticity vectors, whereas the stretching is influenced by velocity in addition to vorticity. In an isochoric motion, the stretch of a vortex line is the ratio of the magnitude of the current vorticity to that of the initial vorticity, and its stretching is equal to ratio of the material rate of change of the magnitude of vorticity to the magnitude of vorticity, both evaluated in the current configuration.

6.7
EXERCISES

1. Find the velocity and acceleration in both material and spatial forms for the motions given by the following sets of equations:

(i) $x_1 = x_1^0(1 + \alpha^2 t^2)$, $\quad x_2 = x_2^0$, $\quad x_3 = x_3^0$

(ii) $x_1 = x_1^0 + x_2^0 t + x_3^0 t^2$, $\quad x_2 = x_2^0 + x_3^0 t + x_1^0 t^2$, $\quad x_3 = x_3^0 + x_1^0 t + x_2^0 t^2$

(iii) $x_1^0 = x_1 e^{-t} - x_3(1 - e^{-t})$, $\quad x_2^0 = x_2 - x_3(e^t - e^{-t})$, $\quad x_3^0 = x_3 e^{-t}$

(iv) $x_1^0 = x_1 \cos \alpha t - x_2 \sin \alpha t$, $\quad x_2^0 = x_1 \sin \alpha t + x_2 \cos \alpha t$, $\quad x_3^0 = x_3$,

(α is a constant)

2. For a certain motion, the velocity is given by $\mathbf{v} = \mathbf{x}/(1 + t)$. Show that $\mathbf{x} = \mathbf{x}^0(1 + t)$. Find the velocity in the material form and the acceleration in both material and spatial forms.

3. In a certain motion, the velocity is given by

$$v_1 = -k(x_1^3 + x_1 x_2^2)e^{-\alpha t}, \quad v_2 = k(x_1^2 x_2 + x_2^3)e^{-\alpha t}, \quad v_3 = 0$$

(k and α are positive constants). Find the acceleration of the particle currently located at the point $(0, 1, 0)$.

4. A motion for which the velocity is of the form $\mathbf{v} = \alpha(\mathbf{a} \otimes \mathbf{b})\mathbf{x}$, where α is a constant and $\mathbf{a}$ and $\mathbf{b}$ are fixed unit vectors, is called a *shear motion*. For this motion, show that

$$\frac{D\mathbf{v}}{Dt} = \frac{\partial \mathbf{v}}{\partial t}$$

5. For the motion given by the equations

$$x_1 = f(t)x_1^0, \quad x_2 = g(t)x_2^0, \quad x_3 = h(t)x_3^0$$

where $f(t)$, $g(t)$ and $h(t)$ are positive valued functions of t, find $\mathbf{v}$ and $D\mathbf{v}/Dt$ in both material and spatial forms. If $f(t) = e^{\alpha t}$, $g(t) = h(t) = 1$, where α is a nonzero constant, show that

$$\frac{\partial \mathbf{v}}{\partial t} = 0 \quad \text{but} \quad \frac{D\mathbf{v}}{Dt} \neq 0$$

6. Show that the motion defined by the velocity components $v_1 = -\alpha x_2$, $v_2 = \alpha x_1$, $v_3 = \beta$, where α and β are nonzero constants, is isochoric. Find the acceleration.

7. Show that the motion defined by the velocity components

$$v_1 = \frac{-2x_1 x_2 x_3}{R^4}, \quad v_2 = \frac{(x_1^2 - x_2^2)x_3}{R^4}, \quad v_3 = \frac{x_2}{R^2}, \quad \text{where } R^2 = x_1^2 + x_2^2 \neq 0$$

is irrotational. Find the acceleration.

6.7 EXERCISES

8. Show that the plane motion defined by the velocity components
$$v_1 = \psi_{,2}, \qquad v_2 = -\psi_{,1}, \qquad v_3 = 0$$
where $\psi = \psi(x_1, x_2)$ is a harmonic function, is irrotational. Find the acceleration components.

9. If in an isochoric motion, $\mathbf{w} = $ constant, show that $\nabla^2 \mathbf{v} = \mathbf{0}$.

10. For an isochoric and irrotational motion, show that $\nabla^2 |\mathbf{v}|^2 \geq 0$.

11. Find J for the motion defined by the velocity components
$$v_1 = x_1 - x_3, \qquad v_2 = x_3(e^t + e^{-t}), \qquad v_3 = 0$$

12. If $\partial \mathbf{v}/\partial t = \mathbf{0}$, show that $D^2 J/Dt^2 = J \operatorname{div}\{(\operatorname{div} \mathbf{v})\mathbf{v}\}$.

13. A certain motion is defined by the equations
$$x_1 = x_1^0 + at x_2^0, \qquad x_2 = x_2^0, \qquad x_3 = x_3^0$$
where a is a constant. Find the rate of change of the function $\phi = x_1^2 - x_2^2$ at the current position of the particle initially located at the origin.

14. At a point $\mathbf{x}$ in the current configuration of a continuum the temperature is given by $T = -e^{-t}(x_1 + 2x_2 - 3x_3)$. Also, the motion is defined by the velocity components $v_1 = x_1 e^{-t}$, $v_2 = -x_2 e^{-t}$, $v_3 = x_3 e^{-t}$. Find the material rate of change of T.

15. The temperature at a point $\mathbf{x}$ in a region filled by a material body is given by
$$T = T_0 e^{-\alpha t}(\sin \beta x_1)(\cos \beta x_2)$$
where T_0, α and β are constants. Also, the motion of the body is defined by the equations
$$x_1 = x_1^0 e^t + x_2^0 e^{-t}, \qquad x_2 = x_2^0 e^t + x_1^0 e^t, \qquad x_3 = x_3^0$$
Find the rate of change of T at the point $(1, 1, 1)$.

16. Suppose that an observer O moves with velocity $\mathbf{v} = \mathbf{e}_1 - 2\mathbf{e}_2 + \mathbf{e}_3$. Find the rate of change of the function $\phi = (x_1^2 - x_2^2)e^{-t}$ as seen by O, when moving past the point $(1, 0, 1)$.

17. For the velocity fields given in Exercises 2 and 3 find
   (i) The stretching tensor.
   (ii) The spin tensor and the vorticity vector.
   (iii) The invariants of the stretching tensor.
   (iv) The deviator of the stretching tensor.

18. For the velocity fields given in Exercises 2 and 3 find
   (i) The stretching of a material arc currently located in the direction $(1:1:1)$.
   (ii) The shearing between material arcs currently located in the directions $(1:1:1)$ and $(1:0:-1)$.
   (iii) The maximum stretchings.

19. For the motion defined by the velocity components

$$v_1 = f(R, t)x_2, \qquad v_2 = -f(R, t)x_1, \qquad v_3 = 0$$

where $R^2 = x_1^2 + x_2^2$, find the principal stretchings, the principal directions of stretching and the angular velocity vector.

20. By finding the most general integral of the equations $v_{i,j} + v_{j,i} = 0$, show that the stretching tensor is 0 if and only if the motion is rigid.

21. Show that the motion defined by the velocity $\mathbf{v} = \mathbf{c} \times \mathbf{x}$, where $\mathbf{c}$ is a constant vector, is a rigid motion.

22. Verify that the motion defined by the velocity components

$$v_1 = 3x_3, \qquad v_2 = -4x_3, \qquad v_3 = 4x_2 - 3x_1$$

is a rigid rotation. Find the spin tensor for the motion.

23. For a rigid motion show that

(i) $\operatorname{div}\left(\dfrac{D\mathbf{v}}{Dt}\right) = -\dfrac{1}{2}|\mathbf{w}|^2$

(ii) $\operatorname{curl}\left(\dfrac{D\mathbf{v}}{Dt}\right) = \dfrac{D\mathbf{w}}{Dt}$

(iii) $\mathbf{W} = \dfrac{D\mathbf{Q}}{Dt}\mathbf{Q}^T$

24. Check whether there exists a velocity field for which the tensor whose matrix follows is the stretching tensor:

$$\begin{bmatrix} x_2 + x_3 & x_1 - x_2 & x_3 \\ x_1 - x_2 & x_1 + x_3 & 3x_1 \\ x_3 & 3x_1 & x_1 + x_2 \end{bmatrix}$$

If so, find the velocity field.

25. For the shear motion defined in Exercise 4 with the angle between $\mathbf{a}$ and $\mathbf{b}$ equal to $\pi/2$, find the principal stretchings and the principal directions of stretchings.

26. Show that the stretching tensor $\mathbf{D}$ corresponds to a shear motion if and only if $\mathbf{D}$ is independent of $\mathbf{x}$ and $I_\mathbf{D} = III_\mathbf{D} = 0$.

27. Show that a motion is uniaxial if and only if $II_\mathbf{D} = III_\mathbf{D} = 0$.

28. Find a necessary and sufficient condition on $\mathbf{D}$ for a motion to be a plane motion.

29. Show that at each particle of a continuum there exists at least one material arc that is instantaneously stationary. If the velocity is of potential kind, show that there are at least three such material arcs, which are mutually orthogonal.

30. Show that
$$\text{curl}(\nabla \mathbf{v})\mathbf{v} = (\nabla \mathbf{v})\mathbf{w} - (\nabla \mathbf{w})\mathbf{v} - (\text{div } \mathbf{v})\mathbf{w}$$

31. Show that
$$\mathbf{D} \cdot \mathbf{D} = (\text{tr } \mathbf{D})^2 - 2II_\mathbf{D} = \nabla \mathbf{v} \cdot (\nabla \mathbf{v})^T + \tfrac{1}{2}\mathbf{w} \cdot \mathbf{w}$$
Deduce that $II_\mathbf{D} \leq 0$ for an isochoric motion.

32. If all the principal stretchings are equal, show that
$$(\det \mathbf{D})^2 = (\tfrac{1}{3} II_\mathbf{D})^3 = (\tfrac{1}{3} \text{tr } \mathbf{D})^6$$

33. Show that
$$\text{div} \frac{D\mathbf{v}}{Dt} = \frac{D}{Dt}(\text{tr } \mathbf{D}) + (\text{tr } \mathbf{D})^2 - 2II_\mathbf{D} - \tfrac{1}{2}|\mathbf{w}|^2$$

34. If $D\mathbf{v}/Dt = \nabla \phi$, show that ϕ satisfies the equations:

(i) $\nabla^2 \phi = \dfrac{D(\text{div } \mathbf{v})}{Dt} - \dfrac{1}{2}|\mathbf{w}|^2 + \mathbf{D} \cdot \mathbf{D}$

(ii) $\nabla^2 \left(\phi + \dfrac{1}{2} v^2 \right) = \dfrac{\partial}{\partial t}(\text{div } \mathbf{v}) - \text{div}(\mathbf{v} \times \mathbf{w})$

35. Show that
$$\frac{D}{Dt}(\log ds) = \frac{d\mathbf{x}}{ds} \cdot \mathbf{D}\left(\frac{d\mathbf{x}}{ds}\right)$$

36. Express the eigenvectors of $\mathbf{W}$ in terms of $\mathbf{w}$.

37. For a plane isochoric motion in which acceleration is the gradient of a potential, show that
$$\frac{D\mathbf{W}}{Dt} = 0$$

38. The tensor $\tilde{\mathbf{D}}$ is defined by
$$\tilde{\mathbf{D}} = \frac{D\mathbf{D}}{Dt} + \mathbf{D}(\nabla \mathbf{v}) + (\nabla \mathbf{v})^T \mathbf{D}$$
Show that
$$\frac{D^2}{Dt^2}(d\mathbf{x} \cdot d\bar{\mathbf{x}}) = 2 d\mathbf{x} \cdot \tilde{\mathbf{D}} \, d\bar{\mathbf{x}}$$

39. Show that
$$\nabla \left(\frac{D\mathbf{v}}{Dt} \right) = \frac{D}{Dt}(\nabla \mathbf{v}) + (\nabla \mathbf{v})^2$$
Deduce that
$$\nabla \left(\frac{D\mathbf{v}}{Dt} \right) \mathbf{F} = \frac{D^2 \mathbf{F}}{Dt^2}$$

40. For any vector **p**, show that

$$\frac{D\mathbf{p}}{Dt} = \mathbf{D}\mathbf{p} + \mathbf{W}\mathbf{p} - (\mathbf{p} \cdot \mathbf{D}\mathbf{p})\mathbf{p}$$

Deduce that if **p** is an eigenvector of **D**, then

$$\frac{D\mathbf{p}}{Dt} = \frac{1}{2}\mathbf{w} \times \mathbf{p}$$

41. Prove the following:

(i) $\dfrac{D\mathbf{C}}{Dt} = 2\mathbf{F}^T \mathbf{D} \mathbf{F}$

(ii) $\dfrac{D\mathbf{B}}{Dt} = (\nabla \mathbf{v})\mathbf{B} + \mathbf{B}(\nabla \mathbf{v})^T$

(iii) $\mathbf{W} = \dfrac{D\mathbf{Q}}{Dt}\mathbf{Q}^T + \dfrac{1}{2}\mathbf{Q}\left(\dfrac{D\mathbf{U}}{Dt}\mathbf{U}^{-1} - \mathbf{U}^{-1}\dfrac{D\mathbf{U}}{Dt}\right)\mathbf{Q}^T$

(iv) $\mathbf{D} = \dfrac{1}{2}\mathbf{Q}\left(\dfrac{D\mathbf{U}}{Dt}\mathbf{U}^{-1} + \mathbf{U}^{-1}\dfrac{D\mathbf{U}}{Dt}\right)\mathbf{Q}^T$

42. Prove the following:

(i) $\text{skw } \nabla\left(\dfrac{D\mathbf{v}}{Dt}\right) = \dfrac{D\mathbf{W}}{Dt} + \mathbf{D}\mathbf{W} + \mathbf{W}\mathbf{D}$

(ii) $\mathbf{F}^T\left\{\text{skw } \nabla\left(\dfrac{D\mathbf{v}}{Dt}\right)\right\}\mathbf{F} = \dfrac{D}{Dt}(\mathbf{F}^T \mathbf{W} \mathbf{F})$

(iii) $\text{skw } \nabla(D\mathbf{v}/Dt) = \mathbf{0}$ if and only if the flux of **w** through each material surface is constant.

43. Find the path lines and stream lines for the motion defined by the following velocity fields:

(i) $v_1 = \alpha x_2$, $\quad v_2 = -\alpha x_1$, $\quad v_3 = \beta$,
(α, β are constants).

(ii) $v_1 = \alpha x_1$, $\quad v_2 = -\alpha x_2$, $\quad v_3 = \beta R$
(α, β are constants and $R^2 = x_1^2 + x_2^2$)

(iii) $v_1 = \dfrac{x_1}{1+t}$, $\quad v_2 = x_2$, $\quad v_3 = 0$

(iv) $v_1 = \dfrac{x_1}{1+t}$, $\quad v_2 = -\dfrac{x_2}{1+t}$, $\quad v_3 = 0$

(v) $v_i = \dfrac{x_i}{1+t}$

44. For the motion considered in Exercise 3, show that the stream lines are circles.

45. Find the stream lines for the motion defined by the equations

$$x_1 = x_1^0 e^{-2\alpha t}, \qquad x_2 = x_2^0 e^{\alpha t}, \qquad x_3 = x_3^0 e^{\alpha t}$$

where α is a constant.

46. For the motion defined by the equation $\mathbf{x} = \mathbf{x}^0 + \tfrac{1}{2}\alpha t^2 \mathbf{a}$, where α is a real constant and $\mathbf{a}$ is a fixed unit vector, show that both the path lines and the stream lines are straight lines and that the motion is not steady.

47. For the motion defined by the equations

$$x_1 = a + \frac{1}{k} e^{-bk} \sin k(a+ct)$$

$$x_2 = -b - \frac{1}{k} e^{-bk} \cos k(a+ct)$$

$$x_3 = \text{constant}$$

where a, b, c and k are constants, show that the path lines are circles. Show that $(D/Dt)|\mathbf{v}| = 0$.

48. Show that the motion defined by the velocity components

$$v_1 = \alpha \frac{x_1}{R^2}, \qquad v_2 = -\alpha \frac{x_2}{R^2}, \qquad v_3 = 0$$

where α is a constant and $R^2 = x_1^2 + x_2^2 \neq 0$ is both isochoric and irrotational. Show further that the stream lines are circles and that $\phi = \alpha \tan^{-1}(x_1/x_2)$ is the velocity potential.

49. Show that in a motion where $\mathbf{v}/|\mathbf{v}|$ is independent of t, the stream lines and the path lines are coincident.

50. Show that for the motion considered in Exercise 48, the vortex lines are straight lines. Find their equations.

51. Find the stream lines and the vortex lines for the motion given by the velocity components

$$v_1 = \alpha \frac{x_1^2}{R^2 x_2}, \qquad v_2 = -\alpha \frac{x_2^2}{R^2 x_1}, \qquad v_3 = 0$$

where α is a constant and $R^2 = x_1^2 + x_2^2$.

52. A flow for which curl $\mathbf{v} = \omega \mathbf{v}$, where ω is a scalar field and is called a *Beltrami flow*. Show that for such a flow, the stream tubes (lines) and vortex tubes (lines) are coincident.

53. Let C be a closed curve on a vortex tube not encircling the tube. Show that

$$\int_C \mathbf{v} \cdot d\mathbf{x} = 0$$

54. If the cross-sectional area of a vortex tube is small, show that the product of the vorticity and the cross-sectional area is constant.

55. Let f be a scalar function defined in the current configuration. For a material curve C, show that

$$\frac{D}{Dt} \int_C f \, ds = \int_C \left(\frac{Df}{Dt} + f \mathbf{t} \cdot \mathbf{Dt} \right) ds$$

$\mathbf{t}$ being, as usual, the unit tangent to C.

56. Let $\mathbf{f}$ be a vector function defined in the current configuration. For a regular material surface S bounded by a curve C, show that

$$\frac{D}{Dt} \int_S \mathbf{f} \cdot \mathbf{n} \, dS = \int_S \left\{ \frac{\partial \mathbf{f}}{\partial t} + (\text{div } \mathbf{v}) \mathbf{f} \right\} \cdot \mathbf{n} \, dS + \int_C \mathbf{f} \times \mathbf{t} \, ds$$

57. Show that

$$\frac{\partial}{\partial t} \int_V \mathbf{w} \, dV = \int_S \mathbf{n} \times (\mathbf{v} \times \mathbf{w}) \, dS + \int_S \left(\mathbf{n} \times \frac{D\mathbf{v}}{Dt} \right) dS$$

Hence deduce that

$$\frac{D}{Dt} \int_V \mathbf{w} \, dV = \int_S \left\{ \mathbf{n} \times \frac{D\mathbf{v}}{Dt} + (\mathbf{w} \cdot \mathbf{n}) \mathbf{v} \right\} dS$$

58. Let $\mathbf{H}$ be a tensor function defined in the current configuration. For a material surface S, show that

$$\frac{D}{Dt} \int_S \mathbf{Hn} \, dS = \int_S \left\{ \frac{D\mathbf{H}}{Dt} + (\text{div } \mathbf{v}) \mathbf{H} - \mathbf{H}(\nabla \mathbf{v})^T \right\} \mathbf{n} \, dS$$

59. Let $\mathbf{f}$ be a vector function defined in the current configuration. For a material volume bounded by a regular closed surface S, show that

$$\frac{D}{Dt} \int_V \mathbf{f} \, dV = \int_V \left(\frac{D}{Dt} + \text{tr } \nabla \mathbf{v} \right) \mathbf{f} \, dV = \int_V \frac{\partial \mathbf{f}}{\partial t} \, dV + \int_S (\mathbf{v} \cdot \mathbf{n}) \mathbf{f} \, dS$$

60. Show that

$$\int_V \{2\mathbf{W} + (\text{div } \mathbf{v})\} \mathbf{v} \, dV = \int_S \left\{ (\mathbf{v} \cdot \mathbf{n}) \mathbf{v} - \frac{1}{2} v^2 \mathbf{n} \right\} dS$$

Deduce that for an isochoric motion with $\mathbf{v} = \mathbf{0}$ on S,

$$\int_V \mathbf{Wv} \, dV = \mathbf{0}$$

61. For the motion defined by the velocity components

$$v_1 = -\frac{\alpha x_1 + \beta x_2}{R^2}, \quad v_2 = \frac{\beta x_1 + \alpha x_2}{R^2}, \quad v_3 = 0$$

where $R^2 = x_1^2 + x_2^2 \neq 0$ and α and β are constants, find the circulation round the circle $x_1^2 + x_2^2 = c^2$.

62. For the motion described by the velocity components

$$v_1 = -\frac{\alpha x_2}{R^2}, \quad v_2 = \frac{\alpha x_1}{R^2}, \quad v_3 = 0$$

where α is a constant and $R^2 = x_1^2 + x_2^2 \neq 0$, show that the circulation round a circuit is 0 or $2\pi k$ according as the circuit encloses the origin or not.

63. Show that a motion defined by the equation $\mathbf{v} = \mathbf{A}\mathbf{x}$, where $\mathbf{A}$ is a constant tensor, is circulation preserving if and only if $(\nabla \mathbf{v})\mathbf{w} = (\text{div } \mathbf{v})\mathbf{w}$.

64. Let C_1 and C_2 be two circuits drawn on a vortex tube, each encircling the tube in the same direction. Show that the circulation round C_1 is equal to circulation round C_2. What does this circulation represent?

65. Show that, in a circulation-preserving motion, the vortex tubes move with the continuum.

66. Derive the expression

$$\text{curl}\left(\frac{D\mathbf{v}}{Dt}\right) = \frac{D\mathbf{w}}{Dt} + (\text{div } \mathbf{v})\mathbf{w} - (\mathbf{w} \cdot \nabla)\mathbf{v}$$

by computing DI_C/Dt from (6.6.2) with the aid of the transport formula (6.5.21), and comparing the resulting expression with (6.6.6).

CHAPTER 7
STRESS

7.1
INTRODUCTION

In Chapters 5 and 6, some kinematical aspects concerned with the deformation and motion of a continuum were discussed. Deformation and motion are generally caused by external forces that give rise to interactions between neighboring portions in the interior parts of a continuum. Such interactions are studied through the concept of stress. This chapter deals with the theory of stress.

7.2
BODY FORCES AND SURFACE FORCES

In continuum mechanics, two distinct types of forces are considered: body forces and surface forces. Body forces are forces that act on every element of a material and hence on the entire volume of the material. Gravitational force is an example of a body force. We postulate that the total body force acting on a material body $\mathcal{B}$ occupying a configuration B of volume V, at time t, is expressible in the form

$$\mathbf{f}^{(b)} = \int_V \rho \mathbf{b} \, dV \qquad (7.2.1)$$

where $\rho = \rho(\mathbf{x}, t)$ is the density at a point $\mathbf{x}$ of B and at time t and $\mathbf{b}$ is a vector with the physical dimension force per unit of mass. The vector $\mathbf{b}$ is referred to as the *body force per unit of mass* or just *body force*. Like ρ, $\mathbf{b}$ generally varies with $\mathbf{x}$ and t.

Surface forces (or contact forces) act on the surface of a material. This surface may be either a part or the whole of the boundary surface, if any, of the material or an imaginary surface visualized in the interior of the material. In the former case, surface forces are external forces that act on the boundary surface of the material. Wind forces and forces exerted by a liquid on a solid immersed in it are examples of such surface forces. In the latter case, surface forces are internal forces that arise from the action of one part of the material upon an adjacent part across the surface. For example, if we consider a heavy rod suspended vertically and visualize a horizontal cross section separating the rod into upper and lower parts, then the weight of the lower part of the rod acts as a surface force on the upper part across the cross section.

We postulate that the total surface force acting on an oriented surface S in the configuration B of a material body $\mathscr{B}$ is expressible in the form

$$\mathbf{f}^{(s)} = \int_S \mathbf{s} \, dS \qquad (7.2.2)$$

where $\mathbf{s}$ is a vector with physical dimension *force per unit of area*. Since $\mathbf{s}$ is the integrand on the righthand side of (7.2.2), $\mathbf{s}$ is a function of $\mathbf{x}$, in general, where $\mathbf{x}$ is a point of S. It may vary with time t as well. The vector $\mathbf{s}$ is called the *surface force per unit of area of S*, or the *stress vector* or *traction on S*.

For a chosen point $\mathbf{x}$ (in the current configuration of a material), let us consider an oriented surface element dS containing $\mathbf{x}$; see Figure 7.1. By virtue of (7.2.2), the total surface force acting on this surface element is $(dS)\mathbf{s}$. If we consider another oriented surface element, say dS_1, containing the same point $\mathbf{x}$, then the total surface force on this surface element is $(dS_1)\mathbf{s}$. Obviously, if dS and dS_1 have different areas, then the total surface forces acting on them are different. But if dS and dS_1 have the same areas but not necessarily the same orientation, then we cannot decisively say whether the total surface forces acting on them are the same or different, unless some definite postulate is made. The postulate made for this purpose is as follows: the stress vector $\mathbf{s}$ depends also on the orientation of the surface element upon which it acts; that is, $\mathbf{s}$ *is in general a function not*

7.2 BODY FORCES AND SURFACE FORCES

only of **x** *and t, but also of* **n**, where **n** is the *unit normal* to the surface element characterizing the orientation of the element: **s** = **s**(**x**, *t*, **n**). This postulate is by Cauchy and known as *Cauchy's stress postulate*. From this postulate, it follows that if dS and dS_1 have the same areas *and* the same orientation, then the total surface forces acting on them are the same and that, if dS and dS_1 have different orientations, the total surface forces acting on them are different.

When dS is taken on the boundary surface, if any, of a material, it is conventional to choose the unit normal **n** to dS as the exterior normal to the boundary surface. Then the stress vector **s** = **s**(**x**, *t*, **n**) is interpreted as the *external surface force* per unit of area on the boundary at time *t* and is usually referred to as *surface traction*.

When dS is considered in the interior of a material, **n** has two possible directions that are opposite to each other; we choose one of these directions. For a chosen **n**, the stress vector **s** = **s**(**x**, *t*, **n**) is interpreted as the *internal surface force* per unit area acting on dS at time *t* due to the action of that part of the material into which **n** is directed upon the adjacent part across dS. Consequently, **s**(**x**, *t*, −**n**) is the internal surface force per unit area acting on dS, at the same instant *t*, due to the action of that part of the material exterior to which **n** is directed upon the adjacent part across dS. Motivated by the Newton's third law of motion, we postulate that **s**(**x**, *t*, −**n**) is equal and opposite to **s**(**x**, *t*, **n**); that is,

$$\mathbf{s}(\mathbf{x}, t, -\mathbf{n}) = -\mathbf{s}(\mathbf{x}, t, \mathbf{n}) \quad (7.2.3)$$

For simplicity in the notation, we write **s**(**n**) for **s**(**x**, *t*, **n**). Then the relation (7.2.3) can be rewritten as

$$\mathbf{s}(-\mathbf{n}) = -\mathbf{s}(\mathbf{n}) \quad (7.2.4)$$

This relation is known as the *Cauchy's reciprocal relation*; see Figure 7.2.

While interpreting the stress vector acting on an internal surface element of a material, we have made the tacit assumption that the portions of a material lying on the two sides of a surface element interact with each other to produce a force across the surface. This assumption is well accepted in

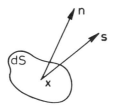

Figure 7.1. Stress vector.

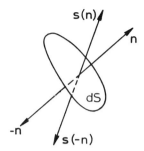

Figure 7.2. Cauchy's reciprocal rotation.

the theory of conventional solids and fluids. However, apart from a force, a couple may also exist across a surface. The inclusion of the effects of such couples in the theory gives rise to the concept of what is called *couple stress*. The effects of couple stresses are of interest only in special kinds of materials, called *polar materials*; these are not treated in this text.

The *resultant force* $\mathbf{f}^{(r)}$ acting on a material body $\mathcal{B}$ in a configuration B of volume V with boundary surface S is defined as the vector sum of the total body force acting on V and the total surface force acting on S; that is,

$$\mathbf{f}^{(r)} = \mathbf{f}^{(b)} + \mathbf{f}^{(s)} = \int_V \rho \mathbf{b} \, dV + \int_S \mathbf{s} \, dS \qquad (7.2.5)$$

Here, it is postulated that the internal forces across surfaces lying in the interior of V balance each other so that their resultant is zero.

7.3
STRESS COMPONENTS

Suppose we consider the particular case in which the surface element dS has $\mathbf{e}_1$ as the unit normal and denote the stress vector acting on this element by $\mathbf{s}^{(1)}$; that is, $\mathbf{s}^{(1)} = \mathbf{s}(\mathbf{e}_1)$. Resolving this vector along the coordinate axes, we get an expression of the form

$$\begin{aligned}\mathbf{s}^{(1)} &= \tau_{11}\mathbf{e}_1 + \tau_{12}\mathbf{e}_2 + \tau_{13}\mathbf{e}_3 \\ &= \tau_{1k}\mathbf{e}_k\end{aligned} \qquad (7.3.1)$$

where τ_{1k} are the components of $\mathbf{s}^{(1)}$ along the axes.

The stress vectors $\mathbf{s}^{(2)}$ and $\mathbf{s}^{(3)}$ are defined in an analogous way, and the following expressions are obtained:

$$\mathbf{s}^{(2)} = \tau_{2k}\mathbf{e}_k, \qquad \mathbf{s}^{(3)} = \tau_{3k}\mathbf{e}_k \qquad (7.3.2)$$

7.3 STRESS COMPONENTS

Expressions (7.3.1) and (7.3.2) can be put in the following condensed form:

$$\mathbf{s}^{(i)} = \tau_{ik} \mathbf{e}_k \qquad (7.3.3)$$

Taking dot product with $\mathbf{e}_j$ on both sides of (7.3.3) and noting that $\mathbf{e}_k \cdot \mathbf{e}_j = \delta_{kj}$ and $\tau_{ik} \delta_{kj} = \tau_{ij}$, we get

$$\mathbf{s}^{(i)} \cdot \mathbf{e}_j = \tau_{ij} \qquad (7.3.4)$$

Thus, for given i and j, τ_{ij} represents the jth component of $\mathbf{s}^{(i)}$, where $\mathbf{s}^{(i)}$ is the stress vector acting on a surface element having $\mathbf{e}_i$ as the unit normal. For example, τ_{23} represents the third component of the stress vector $\mathbf{s}^{(2)}$ (acting on a surface element having $\mathbf{e}_2$ as the unit normal). The nine real numbers τ_{ij} defined in this way are called the *stress components* at the point $\mathbf{x}$ and time t, in the x_i system. These components have dimensions: force/(length)^2. The matrix with τ_{ij} as elements, namely,

$$\tau_{ij} = \begin{bmatrix} \tau_{11} & \tau_{12} & \tau_{13} \\ \tau_{21} & \tau_{22} & \tau_{23} \\ \tau_{31} & \tau_{32} & \tau_{33} \end{bmatrix} \qquad (7.3.5)$$

is called the *stress matrix* at the point $\mathbf{x}$ and time t, in the x_i system. The three stress vectors $\mathbf{s}^{(i)}$ and the nine stress components τ_{ij} can be displayed pictorially as in Figure 7.3.

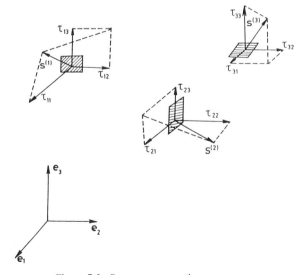

Figure 7.3. Stress vectors and components.

7.4
STRESS TENSOR

At a point $P(\mathbf{x})$ in the current configuration of a continuum, consider a "small" tetrahedron with three of its faces perpendicular to the coordinate axes and the fourth one perpendicular to an arbitrary unit vector $\mathbf{n}$ (see Figure 7.4). Let ΔA_1, ΔA_2 and ΔA_3 be the areas of the faces PQR, PSR and PQS perpendicular to the x_1, x_2 and x_3 axes, respectively, and ΔA be the area of the slant face QRS perpendicular to $\mathbf{n}$. Then

$$\Delta A_1 = (\Delta A) \cos(\mathbf{n}, x_1) = (\Delta A) n_1 \qquad (7.4.1a)$$

$$\Delta A_2 = (\Delta A) n_2 \qquad (7.4.1b)$$

$$\Delta A_3 = (\Delta A) n_3 \qquad (7.4.1c)$$

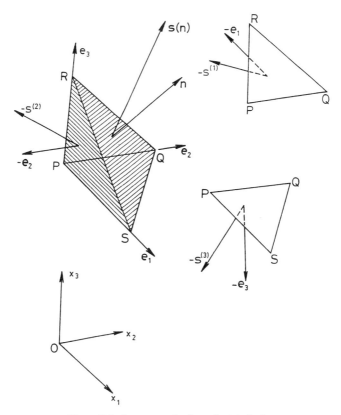

Figure 7.4. Stresses on the face of a tetrahedron.

7.4 STRESS TENSOR

where $\cos(\mathbf{n}, x_1)$ denotes the cosine of the angle between $\mathbf{n}$ and the positive x_1 axis and n_i are components of $\mathbf{n}$. Also, the volume of the tetrahedron is

$$\Delta V = \tfrac{1}{3} h (\Delta A) \tag{7.4.1d}$$

where h is the perpendicular distance of the point P from the slant face QRS.

We note that the plane element PQR is a part of the boundary surface of the material contained in the tetrahedron. As such, the unit normal to PQR, which is to be the exterior normal by convention, is $-\mathbf{e}_1$. The total surface force on this plane element is $(\Delta A_1)\{\mathbf{s}(-\mathbf{e}_1)\}$. In view of the reciprocal relation (7.2.4), this surface force is equal to $-(\Delta A_1)\{\mathbf{s}(\mathbf{e}_1)\}$ or $-(\Delta A_1)\mathbf{s}^{(1)}$. Similarly, the total surface forces on the plane elements PSR and PSQ are $-(\Delta A_2)\mathbf{s}^{(2)}$ and $-(\Delta A_3)\mathbf{s}^{(3)}$, respectively. Since $\mathbf{n}$ is the exterior normal to the tetrahedron on the plane element QRS, the total surface force on this plane element is $(\Delta A)\mathbf{s}(\mathbf{n})$. It is to be noted that while $\mathbf{s}^{(1)}$, $\mathbf{s}^{(2)}$ and $\mathbf{s}^{(3)}$ are stress vectors on plane elements passing through the point P, $\mathbf{s}(\mathbf{n})$ is the stress vector on a plane element that does not pass through P.

By virtue of the expression (7.2.5), the resultant force acting on the tetrahedron is given by

$$\mathbf{f}^{(r)} = (\Delta V)\rho \mathbf{b} + (\Delta A)\mathbf{s}(\mathbf{n}) - (\Delta A_1)\mathbf{s}^{(1)} - (\Delta A_2)\mathbf{s}^{(2)} - (\Delta A_3)\mathbf{s}^{(3)} \tag{7.4.2}$$

where ρ is the density and $\mathbf{b}$ is the body force both evaluated at the point P.

Let $\Delta m = \rho(\Delta V)$ be the mass of the material contained in the tetrahedron and $\mathbf{a}$ be the acceleration evaluated at P. Motivated by the Newton's second law of motion, we postulate that

$$\rho(\Delta V)\mathbf{a} = \mathbf{f}^{(r)} \tag{7.4.3}$$

Then, on taking note of (7.4.1) and (7.4.2), we get the equation

$$\mathbf{s}(\mathbf{n}) = n_1 \mathbf{s}^{(1)} + n_2 \mathbf{s}^{(2)} + n_3 \mathbf{s}^{(3)} + \tfrac{1}{3}\rho h(\mathbf{a} - \mathbf{b})$$

Now, with $\mathbf{n}$ and P fixed, let us take the limit as $h \to 0$. In this limit, the slant face QRS tends to a plane element normal to $\mathbf{n}$ at P. Then $\mathbf{s}(\mathbf{n})$ represents the stress vector at P on this element. Also, $\mathbf{s}^{(i)}$ are evaluated at P. Thus, in this limit,

$$\mathbf{s}(\mathbf{n}) = n_1 \mathbf{s}^{(1)} + n_2 \mathbf{s}^{(2)} + n_3 \mathbf{s}^{(3)} \tag{7.4.4}$$

This linear relation enables us to find $\mathbf{s}(\mathbf{n})$ for an arbitrary $\mathbf{n}$, when the stress vectors $\mathbf{s}^{(i)}$, $i = 1, 2, 3$, are known.

Substitution for $\mathbf{s}^{(i)}$ from (7.3.1) and (7.3.2) in (7.4.4) gives

$$\mathbf{s}(\mathbf{n}) = (\tau_{1j}\mathbf{e}_j)n_1 + (\tau_{2j}\mathbf{e}_j)n_2 + (\tau_{3j}\mathbf{e}_j)n_3$$
$$= \tau_{kj} n_k \mathbf{e}_j \tag{7.4.5}$$

Taking the scalar product with $\mathbf{e}_i$ on both sides of this expression, we get

$$s_i(\mathbf{n}) = \tau_{ki} n_k \qquad (7.4.6)$$

These relations connecting $s_i(\mathbf{n})$ and τ_{ij} show that, for any given $\mathbf{n}$, $s_i(\mathbf{n})$ can be determined when the stress components τ_{ij} are known.

Recall that the nine components τ_{ij} are defined w.r.t. a set of axes. Relations (7.4.6) show that for an arbitrary vector $\mathbf{n}$ with components n_i, $\tau_{ki} n_k$ are components of a vector, namely, $\mathbf{s}(\mathbf{n})$. From the quotient law proved in Example 2.8.6, it follows that τ_{ij} *are components of a second-order tensor*, called *Cauchy's stress tensor*, denoted $\mathbf{T}$. Then, by use of (2.7.6), we have

$$\mathbf{T} = \tau_{ij} \mathbf{e}_i \otimes \mathbf{e}_j \qquad (7.4.7)$$

The relations (7.4.6) can now be expressed in the direct tensor notation as follows:

$$\mathbf{s}(\mathbf{n}) = \mathbf{T}^T \mathbf{n} \qquad (7.4.8)$$

This relation connecting the stress vector $\mathbf{s}(\mathbf{n})$ and the stress tensor $\mathbf{T}$ is known as *Cauchy's law (hypothesis)*. The law can be expressed in terms of $\mathbf{T}$ (instead of $\mathbf{T}^T$) by interchanging the meanings of τ_{ij} and τ_{ji}. This is just a matter of notation and convention.

We note that, through a given point, there exist infinitely many surface elements. On every one of these elements we can define a stress vector. The totality of all these stress vectors is called the *state of stress* at the point. The relation (7.4.8) enables us to find the stress vector on any surface element at a point by knowing the stress tensor at the point. As such, the state of stress at a point is completely determined by the stress tensor at the point.

For computational purposes, it is convenient to have the equation (7.4.8) rewritten in the matrix notation. The resulting matrix equation has the explicit form

$$\begin{bmatrix} s_1 \\ s_2 \\ s_3 \end{bmatrix} = \begin{bmatrix} \tau_{11} & \tau_{21} & \tau_{31} \\ \tau_{12} & \tau_{22} & \tau_{32} \\ \tau_{13} & \tau_{23} & \tau_{33} \end{bmatrix} \begin{bmatrix} n_1 \\ n_2 \\ n_3 \end{bmatrix} \qquad (7.4.9)$$

where we have dropped $\mathbf{n}$ from the symbols $s_i(\mathbf{n})$ for simplicity in the notation.

It has to be pointed out that Cauchy's law (7.4.8), which asserts the existence of the stress tensor, is a direct consequence of the postulates (7.2.4) and (7.4.3). These two postulates are just two particular cases of a more general and fundamental postulate called the *law of balance of linear momentum*. This fundamental law will be introduced in Section 8.3.

7.4 STRESS TENSOR

EXAMPLE 7.4.1 The stress matrix at a point P in a material is given as follows:

$$[\tau_{ij}] = \begin{bmatrix} 3 & 1 & 4 \\ 1 & 2 & -5 \\ 4 & -5 & 0 \end{bmatrix}$$

Find (i) the stress vector on a plane element p through P and parallel to the plane $2x_1 + x_2 - x_3 = 1$, (ii) the magnitude of the stress vector and (iii) the angle that the stress vector makes with the normal to the plane.

Solution The plane element p on which the stress vector is required is parallel to the plane $2x_1 + x_2 - x_3 = 1$. Hence, the direction ratios of the normal to the plane are $(2, 1, -1)$ and the direction cosines are $(2/\sqrt{6}, 1/\sqrt{6}, -1/\sqrt{6})$. Thus, the components n_i of the unit normal to the plane element p are

$$n_1 = \frac{2}{\sqrt{6}}, \quad n_2 = \frac{1}{\sqrt{6}}, \quad n_3 = -\frac{1}{\sqrt{6}} \qquad (7.4.10)$$

Substituting for τ_{ij} from the given stress matrix and using (7.4.10) in Cauchy's law (7.4.9), we obtain

$$\begin{bmatrix} s_1 \\ s_2 \\ s_3 \end{bmatrix} = \begin{bmatrix} 3 & 1 & 4 \\ 1 & 2 & -5 \\ 4 & -5 & 0 \end{bmatrix} \begin{bmatrix} 2/\sqrt{6} \\ 1/\sqrt{6} \\ -1/\sqrt{6} \end{bmatrix}$$

This gives

$$s_1 = \sqrt{3/2}, \quad s_2 = 3\sqrt{3/2}, \quad s_3 = \sqrt{3/2} \qquad (7.4.11)$$

Thus, the required stress vector is

$$\mathbf{s} = \sqrt{3/2}\,(\mathbf{e}_1 + 3\mathbf{e}_2 + \mathbf{e}_3) \qquad (7.4.12)$$

and its magnitude is

$$|\mathbf{s}| = \sqrt{33/2} \qquad (7.4.13)$$

From (7.4.10) and (7.4.12), we get

$$\mathbf{s} \cdot \mathbf{n} = 2 \qquad (7.4.14)$$

Hence the angle θ between the directions of $\mathbf{s}$ and $\mathbf{n}$ is given by

$$\cos\theta = \frac{\mathbf{s} \cdot \mathbf{n}}{|\mathbf{s}||\mathbf{n}|} = \sqrt{8/33} \quad \blacksquare \qquad (7.4.15)$$

Note: Expression (7.4.15) illustrates the important fact that the stress vector on a surface element need not be along the normal to the surface.

EXAMPLE 7.4.2 The stress matrix at a point $P(x_i)$ in a material is given as follows:

$$[\tau_{ij}] = \begin{bmatrix} x_3 x_1 & x_3^2 & 0 \\ x_3^2 & 0 & -x_2 \\ 0 & -x_2 & 0 \end{bmatrix}$$

Find the stress vector at the point $Q(1, 0, -1)$ on the surface $x_2^2 + x_3^2 = x_1$.

Solution The stress vector is required on the surface $f(x_1, x_2, x_3) \equiv x_1 - x_2^2 - x_3^2 = 0$. We find that

$$\nabla f = \mathbf{e}_1 - 2x_2 \mathbf{e}_2 - 2x_3 \mathbf{e}_3; \quad |\nabla f| = (1 + 4x_2^2 + 4x_3^2)^{1/2} \quad (7.4.16)$$

At the point $Q(1, 0, -1)$, we get $\nabla f = \mathbf{e}_1 + 2\mathbf{e}_3$, $|\nabla f| = \sqrt{5}$. Hence the unit outward normal to the surface $f = 0$ at the point Q is

$$\mathbf{n} = \frac{\nabla f}{|\nabla f|} = \frac{1}{\sqrt{5}} (\mathbf{e}_1 + 2\mathbf{e}_3)$$

so that

$$n_1 = \frac{1}{\sqrt{5}}, \quad n_2 = 0, \quad n_3 = \frac{2}{\sqrt{5}} \quad (7.4.17)$$

At the point Q, the given stress matrix is

$$[\tau_{ij}] = \begin{bmatrix} -1 & 1 & 0 \\ 1 & 0 & 0 \\ 0 & 0 & 0 \end{bmatrix} \quad (7.4.18)$$

Substituting for n_i and τ_{ij} from (7.4.17) and (7.4.18) in Cauchy's law (7.4.9) and equating the corresponding elements, we get $s_1 = -1/\sqrt{5}$, $s_2 = 1/\sqrt{5}$, $s_3 = 0$. Hence, the required stress vector is

$$\mathbf{s} = -\frac{1}{\sqrt{5}} (\mathbf{e}_1 - \mathbf{e}_2) \quad \blacksquare$$

EXAMPLE 7.4.3 The state of stress at a point $\mathbf{x}$ is said to be *uniaxial* if $\mathbf{s(n)}$ is along a constant unit vector $\mathbf{a}$ for all $\mathbf{n}$. Show that such a state of stress arises if the stress tensor at $\mathbf{x}$ is of the form $\mathbf{T} = T(\mathbf{a} \otimes \mathbf{a})$, where T is a scalar. Interpret T.

Solution Suppose that
$$\mathbf{T} = T(\mathbf{a} \otimes \mathbf{a}) \tag{7.4.19}$$
where $\mathbf{a}$ is a constant unit vector and T is a scalar. Then for any vector $\mathbf{n}$,
$$\mathbf{T}^T\mathbf{n} = T(\mathbf{a} \otimes \mathbf{a})\mathbf{n} = T(\mathbf{a} \cdot \mathbf{n})\mathbf{a}$$
Using Cauchy's law (7.4.8), this becomes
$$\mathbf{s(n)} = T(\mathbf{a} \cdot \mathbf{n})\mathbf{a} \tag{7.4.20}$$
Evidently, $\mathbf{s(n)}$ is along $\mathbf{a}$ and the state of stress at a point $\mathbf{x}$ is uniaxial.

From (7.4.20), we get
$$\mathbf{s(a)} \cdot \mathbf{a} = T(\mathbf{a} \cdot \mathbf{a})(\mathbf{a} \cdot \mathbf{a}) = T \tag{7.4.21}$$
Since $\mathbf{s(a)}$ is the stress vector on a surface element with $\mathbf{a}$ as the unit normal, we find from (7.4.21) that T represents the component along $\mathbf{a}$ of the stress vector that acts on a plane perpendicular to $\mathbf{a}$. ∎

EXAMPLE 7.4.4 Show that $\mathbf{s(n)} = -p\mathbf{n}$ for all $\mathbf{n}$ if and only if $\tau_{ij} = -p\delta_{ij}$. (Such a state of stress occurs in fluids at rest and is called *hydrostatic stress*.)

Solution First suppose that
$$\mathbf{s(n)} = -p\mathbf{n} \tag{7.4.22}$$
for all $\mathbf{n}$. Then $\mathbf{s}^{(i)} \equiv \mathbf{s(e}_i) = -p\mathbf{e}_i$, and expression (7.3.4) yields
$$\tau_{ij} = \mathbf{s}^{(i)} \cdot \mathbf{e}_j = -p\mathbf{e}_i \cdot \mathbf{e}_j = -p\delta_{ij}. \tag{7.4.23}$$

Conversely, suppose that τ_{ij} are given by (7.4.23). Then $\mathbf{T} = -p\mathbf{I}$, and by using Cauchy's law we get (7.4.22).

EXAMPLE 7.4.5 Let $\mathbf{s(n)}$ and $\mathbf{s(n')}$ be stress vectors on two surface elements at a point $\mathbf{x}$. Prove that
$$\mathbf{n} \cdot \mathbf{s(n')} = \mathbf{n'} \cdot \mathbf{s(n)} \tag{7.4.24}$$
if and only if the stress tensor at $\mathbf{x}$ is symmetric.

Solution By Cauchy's law (7.4.8), we obtain
$$\mathbf{n'} \cdot \mathbf{s(n)} = \mathbf{n'} \cdot (\mathbf{T}^T\mathbf{n}); \quad \mathbf{n} \cdot \mathbf{s(n')} = \mathbf{n} \cdot (\mathbf{T}^T\mathbf{n'}). \tag{7.4.25}$$
Further, by (2.8.14) we note that
$$\mathbf{n} \cdot (\mathbf{T}^T\mathbf{n'}) = \mathbf{n'} \cdot (\mathbf{Tn}). \tag{7.4.26}$$
From (7.4.25) and (7.4.26), we find that (7.4.24) holds if and only if $\mathbf{T}^T = \mathbf{T}$. ∎

7.5
NORMAL AND SHEAR STRESSES

Example 7.4.1 illustrated that the stress vector $s(n)$ need not be collinear with n. Hence, $s(n)$ may be resolved along n and perpendicular to n, in general. Let

$$\sigma = s(n) \cdot n \qquad (7.5.1)$$

and

$$\tau = s(n) \cdot t \qquad (7.5.2)$$

where t is a unit vector perpendicular to n. Then σ is called the *normal stress* and τ is called the *shear stress* along t, on the surface element considered. The normal stress is said to be *tensile* if $\sigma > 0$ and *compressive* if $\sigma < 0$. If $\sigma = 0$, $s(n)$ acts tangential to the surface and is called a *pure tangential stress* or *pure shear stress*. When $s(n)$ acts along or opposite to n, the shear stress on the surface is 0 and $s(n)$ is (then) called *pure normal stress*. The pure normal stress acting opposite to n is called *pressure*.

In the particular case when $n = e_1$, expressions (7.5.1) and (7.3.4) yield

$$\sigma = s^{(1)} \cdot e_1 = \tau_{11}$$

Thus, τ_{11} is the normal stress on the surface element perpendicular to e_1. Similarly, τ_{22} and τ_{33} are normal stresses on surface elements perpendicular to e_2 and e_3 respectively. The stress components τ_{11}, τ_{22} and τ_{33} are therefore called *normal stresses*. Note that these stresses are the diagonal elements of the stress matrix.

Suppose we take $n = e_1$ and $t = e_2$. Then expressions (7.5.2) and (7.3.4) yield

$$\tau = s^{(1)} \cdot e_2 = \tau_{12}$$

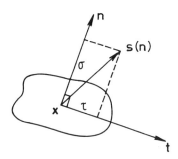

Figure 7.5. Normal and shear stresses.

7.5 NORMAL AND SHEAR STRESSES

Thus, τ_{12} is the shear stress along $\mathbf{e}_2$ on the surface element perpendicular to $\mathbf{e}_1$. Similar interpretation is given for τ_{21}, τ_{31}, τ_{13}, τ_{23} and τ_{32}. The stress components τ_{ij}, $i \neq j$ are therefore called the *shear stresses*. Note that these components are the nondiagonal elements of the stress matrix.

Using Cauchy's law (7.4.6), expression (7.5.1) yields

$$\sigma = s_i(\mathbf{n})n_i = \tau_{ki}n_k n_i$$
$$= \tau_{ik} n_i n_k \qquad (7.5.3)$$

on interchanging the dummy suffixes. Expression (7.5.3) enables us to determine the normal stress directly from the stress components τ_{ij}.

If $\mathbf{t}$ is chosen in the plane containing $\mathbf{n}$ and $\mathbf{s}(\mathbf{n})$, see Figure 7.5; then we have, in view of (7.5.1) and (7.5.2),

$$\mathbf{s}(\mathbf{n}) = \sigma \mathbf{n} + \tau \mathbf{t} \qquad (7.5.4)$$

This gives

$$|\mathbf{s}(\mathbf{n})|^2 = \sigma^2 + \tau^2 \qquad (7.5.5)$$

so that

$$\tau = |\{|\mathbf{s}(\mathbf{n})|^2 - \sigma^2\}|^{1/2} \qquad (7.5.6)$$

The expression in the righthand side of (7.5.6) is called the *magnitude of the shear stress*, or just the *shear stress*.

EXAMPLE 7.5.1 The stress matrix at a certain point in a material is given by

$$[\tau_{ij}] = \begin{bmatrix} 3 & 1 & 1 \\ 1 & 0 & 2 \\ 1 & 2 & 0 \end{bmatrix}$$

Find the normal stress and the shear stress on the octahedral plane element through the point. (Note: *Octahedral plane* is the plane whose normal makes equal angles with positive directions of the coordinate axes.)

Solution For the octahedral plane element, the normal $\mathbf{n}$ has direction ratios $(1, 1, 1)$. Hence the components of $\mathbf{n}$ are

$$n_1 = n_2 = n_3 = \frac{1}{\sqrt{3}} \qquad (7.5.7)$$

Substituting for τ_{ij} from the given matrix and for n_i from (7.5.7) in Cauchy's law (7.4.9), and equating the corresponding terms, we get

$$s_1 = \frac{5}{\sqrt{3}}, \qquad s_2 = \sqrt{3}, \qquad s_3 = \sqrt{3} \qquad (7.5.8)$$

as the components of the stress vector. The magnitude of this stress vector is

$$|\mathbf{s}(\mathbf{n})| = \sqrt{43/3} \qquad (7.5.9)$$

From (7.5.7) and (7.5.8), we get the normal stress acting on the given plane element as

$$\sigma = s_i n_i = 11/3 \qquad (7.5.10)$$

Since $\sigma > 0$, the normal stress on the plane is tensile.

Substituting for $|\mathbf{s}(\mathbf{n})|^2$ and σ^2 from (7.5.9) and (7.5.10) in the expression (7.5.6), we obtain the shear stress acting on the given plane element as $\tau = 2\sqrt{2}/3$. ∎

EXAMPLE 7.5.2 For the state of stress considered in Example 7.4.3, compute the normal stress and shear stress on an arbitrary plane element. Find the planes on which these stresses are maximum. Also, find the maximum values.

Solution For the state of stress considered, the stress vector on an arbitrary plane element with normal $\mathbf{n}$ is given by expression (7.4.20). If θ is the angle that $\mathbf{n}$ makes with $\mathbf{a}$, this expression can be rewritten as

$$\mathbf{s}(\mathbf{n}) = T \cos \theta \, \mathbf{a} \qquad (7.5.11)$$

so that

$$|\mathbf{s}(\mathbf{n})| = |T||\cos \theta| \qquad (7.5.12)$$

Hence the normal stress on the plane element considered is

$$\sigma = \mathbf{n} \cdot \mathbf{s}(\mathbf{n}) = T \cos^2 \theta \qquad (7.5.13)$$

Substituting for σ and $|\mathbf{s}(\mathbf{n})|$ from (7.5.13) and (7.5.12) in (7.5.6), we get the shear stress on the plane element as

$$\tau = |T^2 \cos^2 \theta - T^2 \cos^4 \theta|^{1/2} = \tfrac{1}{2}|T||\sin 2\theta| \qquad (7.5.14)$$

From (7.5.13) we see that σ is maximum for $\theta = 0$; this means that maximum normal stress occurs on a plane element perpendicular to $\mathbf{a}$. Also, $\sigma_{\max} = T$. From (7.5.14) we see that τ is maximum for $\theta = \pi/4$; this means that the maximum shear stress occurs on a plane inclined at an angle of 45° with $\mathbf{a}$. Also, $\tau_{\max} = \tfrac{1}{2}|T|$. ∎

7.6
PRINCIPAL STRESSES

It has been pointed out that the stress vector on a surface element need not be collinear with the normal to the surface element. We now proceed to investigate whether there exists any surface element at a given point on which the stress vector is collinear with the normal to the element.

7.6 PRINCIPAL STRESSES

We note that the stress vector $\mathbf{s(n)}$ is collinear with $\mathbf{n}$ if and only if

$$\mathbf{s(n)} = \Lambda \mathbf{n} \tag{7.6.1}$$

for some scalar Λ. Using Cauchy's law (7.4.8), the condition (7.6.1) becomes

$$\mathbf{T}^T \mathbf{n} = \Lambda \mathbf{n} \tag{7.6.2}$$

The condition (7.6.2) holds if and only if $\mathbf{n}$ is an *eigenvector* of $\mathbf{T}^T$ and Λ is the associated *eigenvalue*. Since every tensor has at least one eigenvector (Theorem 2.13.1), there exists at least one $\mathbf{n}$ such that the condition (7.6.2) holds. Thus, at any given point there *does* exist at least one surface element on which the stress vector is collinear with the normal. Such a surface element is called a *principal plane of stress* at the point. It follows that for a principal plane of stress, condition (7.6.1) holds and any surface element for which the condition (7.6.1) holds is a principal plane of stress. (Consequently, there is *no* shear stress on a principal plane of stress.) The direction of the normal to a principal plane of stress is called a *principal direction of stress*. The normal stress on a principal plane of stress is called a *principal stress*. From expressions (7.6.1) and (7.6.2), we find that a principal stress is nothing but an eigenvalue of the tensor $\mathbf{T}^T$ and a principal direction of stress is nothing but the direction of an eigenvector of $\mathbf{T}^T$.

In Section 2.13, it was shown that for a symmetric tensor there exist exactly three (not necessarily distinct) eigenvalues and at least three mutually orthogonal eigenvectors with the property that, if a system of axes is chosen along the orthogonal eigenvectors, then with respect to this system the matrix of the tensor is diagonal, the diagonal elements being the eigenvalues of the tensor. This property can be employed to the tensor $\mathbf{T}^T$, provided it is symmetric; that is $\mathbf{T}^T = \mathbf{T}$. In Section 8.4, we will indeed prove that $\mathbf{T} = \mathbf{T}^T$. Bearing this in mind, we restrict ourselves to the case $\mathbf{T} = \mathbf{T}^T$. Then, from what has been noted already, it follows that at every point of a material (i) there exist exactly three (not necessarily distinct) principal stresses, (ii) there exist at least three mutually orthogonal principal directions of stress; consequently there exist at least three mutually orthogonal principal planes of stress, and (iii) if a system of axes is chosen along the mutually orthogonal principal directions of stress, then the matrix of the stress tensor is purely diagonal, the diagonal elements being the principal stresses. A system of coordinate axes chosen along the principal directions of stress is referred to as *principal axes of stress*.

Since the principal stresses are the eigenvalues of $\mathbf{T}$, these are the roots of the characteristic equation

$$\det(\mathbf{T} - \Lambda \mathbf{I}) \equiv \det(\tau_{ij} - \Lambda \delta_{ij}) \equiv -\Lambda^3 + I_\mathbf{T} \Lambda^2 - II_\mathbf{T} \Lambda + III_\mathbf{T} = 0 \tag{7.6.3}$$

308 7 STRESS

where I_T, II_T and III_T are the fundamental invariants of the tensor **T**; that is

$$I_T = \tau_{kk}$$
$$II_T = \tfrac{1}{2}(\tau_{ii}\tau_{jj} - \tau_{ij}\tau_{ji}) \qquad (7.6.4)$$
$$III_T = \det(\tau_{ij})$$

The scalars I_T, II_T and III_T are called the *fundamental stress invariants*.

Once the three principal stresses, say τ_1, τ_2 and τ_3, are found by solving the cubic (7.6.3), the corresponding principal directions of stress can be found by solving the vector equation (7.6.2) for **n** in the cases: $\Lambda = \tau_1$, $\Lambda = \tau_2$ and $\Lambda = \tau_3$. With respect to the axes chosen along the principal directions so determined, the matrix of the stress tensor is as follows:

$$[\tau_{ij}] = \begin{bmatrix} \tau_1 & 0 & 0 \\ 0 & \tau_2 & 0 \\ 0 & 0 & \tau_3 \end{bmatrix} \qquad (7.6.5)$$

Since τ_1, τ_2, τ_3 are roots of the cubic (7.6.3), the fundamental stress invariants have the following expressions:

$$I_T = \tau_1 + \tau_2 + \tau_3$$
$$II_T = \tau_1\tau_2 + \tau_2\tau_3 + \tau_3\tau_1 \qquad (7.6.6)$$
$$III_T = \tau_1\tau_2\tau_3$$

EXAMPLE 7.6.1 The stress matrix at a point (x_i) is given by

$$[\tau_{ij}] = \begin{bmatrix} 0 & 0 & -ax_2 \\ 0 & 0 & bx_1 \\ -ax_2 & bx_1 & 0 \end{bmatrix}$$

where a is a constant. Show that the principal stresses at the point $(a, b, 0)$ are 0, $\pm\sqrt{2}\,ab$. Find the principal direction for which the principal stress is $\sqrt{2}\,ab$. Also, find the fundamental stress invariants.

Solution The principal stresses are the roots of the characteristic equation (7.6.3). Substituting for τ_{ij} from the given stress matrix with $x_1 = a$ and $x_2 = b$ in this equation we get the equation

$$\begin{vmatrix} -\Lambda & 0 & -ab \\ 0 & -\Lambda & ab \\ -ab & ab & -\Lambda \end{vmatrix} = 0$$

7.6 PRINCIPAL STRESSES

Expanding the determinant and solving the resulting equation, we get the three roots $\tau_1 = 0$, $\tau_2 = \sqrt{2}\,ab$, $\tau_3 = -\sqrt{2}\,ab$. These are the principal stresses.

To find the principal direction associated with the principal stress $\tau_2 = \sqrt{2}\,ab$, we have to solve equation (7.6.2) for **n** by using the given τ_{ij} with $x_1 = a$, $x_2 = b$ and taking $\Lambda = \tau_2 = \sqrt{2}\,ab$. Thus we obtain the following equations for n_i:

$$\sqrt{2}\,n_1 + n_3 = 0$$
$$\sqrt{2}\,n_2 - n_3 = 0$$
$$n_1 - n_2 + \sqrt{2}\,n_3 = 0$$

A solution of these equations subject to the constraint $n_i n_i = 1$ is

$$n_1 = -\frac{1}{2}, \quad n_2 = \frac{1}{2}, \quad n_3 = \frac{1}{\sqrt{2}}$$

Thus, the principal direction associated with the principal stress $\sqrt{2}\,ab$ has direction cosines $(-\frac{1}{2}, \frac{1}{2}, 1/\sqrt{2})$.

Substituting $\tau_1 = 0$, $\tau_2 = \sqrt{2}\,ab$ and $\tau_3 = -\sqrt{2}\,ab$ in expressions (7.6.6), we get

$$I_\mathrm{T} = 0, \quad II_\mathrm{T} = -2a^2b^2, \quad III_\mathrm{T} = 0$$

These are the fundamental stress invariants for the given state of stress at the point $(a, b, 0)$. ∎

EXAMPLE 7.6.2 Show that, as the orientation of a surface element at a point P varies, the normal stress on the surface element assumes an extreme value when the element is a principal plane of stress at P and that this extreme value is a principal stress.

Solution The expression for the normal stress is

$$\sigma = \tau_{ij} n_i n_j \tag{7.6.7}$$

see (7.5.3). We have to find n_i for which σ is an extremum. Since **n** is a unit vector, we have the restriction $n_k n_k = 1$. Hence, σ assumes extreme values when n_i obey the conditions

$$\frac{\partial}{\partial n_i}\{\sigma - \alpha(n_k n_k - 1)\} = 0 \tag{7.6.8}$$

where α is a Lagrangian multiplier. Substituting for σ from (7.6.7), the conditions (7.6.8) become

$$(\tau_{ij} - \alpha \delta_{ij})n_j = 0 \tag{7.6.9}$$

310   7 STRESS

These conditions are satisfied if and only if **n** is a principal direction of stress. Thus, σ assumes an extreme value on a principal plane of stress. For n_i given by (7.6.9), expression (7.6.7) yields $\sigma = \alpha$. From (7.6.9), we note that α is a principal stress. Thus, a principal stress is an extreme value of σ. ∎

EXAMPLE 7.6.3 Prove that, as the orientation of a surface element at a point P varies, the shear stress on the element assumes the maximum value when the normal to the element bisects the angle between the directions of the largest and the least principal stresses at P and that the maximum shear stress is equal to one-half the difference between these two principal stresses.

Solution It is convenient to choose the coordinate axes to coincide with the principal directions of stress at P. Then the stress matrix has the following diagonal form:

$$[\tau_{ij}] = \begin{bmatrix} \tau_1 & 0 & 0 \\ 0 & \tau_2 & 0 \\ 0 & 0 & \tau_3 \end{bmatrix} \quad (7.6.10)$$

where τ_1, τ_2, τ_3 are the principal stresses at P. Cauchy's law (7.4.6) then yields

$$s_1(\mathbf{n}) = \tau_1 n_1, \quad s_2(\mathbf{n}) = \tau_2 n_2, \quad s_3(\mathbf{n}) = \tau_3 n_3 \quad (7.6.11)$$

Hence the normal stress is

$$\sigma = s_i(\mathbf{n}) n_i = \tau_1 n_1^2 + \tau_2 n_2^2 + \tau_3 n_3^2 \quad (7.6.12)$$

From (7.6.11), we also get

$$|s(\mathbf{n})|^2 = \tau_1^2 n_1^2 + \tau_2^2 n_2^2 + \tau_3^2 n_3^2 \quad (7.6.13)$$

Substituting for σ and $|s(\mathbf{n})|$ from (7.6.12) and (7.6.13) in (7.5.6) we get

$$\tau^2 = \{\tau_1^2 n_1^2 + \tau_2^2 n_2^2 + \tau_3^2 n_3^2 - (\tau_1 n_1^2 + \tau_2 n_2^2 + \tau_3 n_3^2)^2\} \quad (7.6.14)$$

We have to find the maximum value of τ as n_i vary. Since $n_k n_k = 1$, the extreme values of τ^2 are found by solving the following three equations for n_i:

$$\frac{\partial}{\partial n_i}\{\tau^2 - \alpha(n_k n_k - 1)\} = 0 \quad (7.6.15)$$

7.6 PRINCIPAL STRESSES

where α is a Lagrangian multiplier. Using (7.6.14) and carrying out the differentiation, equations (7.6.15) give

$$\left.\begin{array}{l} \{(\tau_1 - 2\sigma)\tau_1 - \alpha\}n_1 = 0 \\ \{(\tau_2 - 2\sigma)\tau_2 - \alpha\}n_2 = 0 \\ \{(\tau_3 - 2\sigma)\tau_3 - \alpha\}n_3 = 0 \end{array}\right\} \qquad (7.6.16)$$

We verify that the following sets of values of n_i obey equations (7.6.16) as well as the constraint $n_k n_k = 1$:

(i) $\quad n_1 = \pm 1, \quad n_2 = 0, \quad n_3 = 0$

(ii) $\quad n_1 = 0, \quad n_2 = \pm 1, \quad n_3 = 0$

(iii) $\quad n_1 = 0, \quad n_2 = 0, \quad n_3 = \pm 1$

(iv) $\quad n_1 = n_2 = \pm \dfrac{1}{\sqrt{2}}, \quad n_3 = 0,$

(v) $\quad n_1 = n_3 = \pm \dfrac{1}{\sqrt{2}}, \quad n_2 = 0$

(vi) $\quad n_1 = 0, \quad n_2 = n_3 = \pm \dfrac{1}{\sqrt{2}}$

Thus τ^2 assumes extreme values for n_i given by relations (i) and (vi). The values of τ^2 corresponding to n_i given by (i) to (vi) may be obtained from (7.6.14). These values are

(a) $\tau^2 = 0,$ (b) $\tau^2 = 0,$ (c) $\tau^2 = 0$

(d) $\tau^2 = \tfrac{1}{4}(\tau_1 - \tau_2),$ (e) $\tau^2 = \tfrac{1}{4}(\tau_1 - \tau_3),$ (f) $\tau^2 = \tfrac{1}{4}(\tau_2 - \tau_3)$

We note that the value of τ^2 given by (a) to (c), namely, $\tau^2 = 0$, is the minimum value. The maximum value of τ^2 is one of the three values given by (d) to (f), depending on the values of τ_1, τ_2, τ_3. If τ_1 denotes the largest and τ_3 the least of τ_1, τ_2, τ_3, then the maximum value of τ is given by (e), the maximum value being $\tfrac{1}{2}(\tau_1 - \tau_3)$. Thus the maximum value of the shear stress is equal to one-half the difference between the largest and the smallest principal stresses. The values of n_i that correspond to this maximum value of τ are given by (v). We verify that these values of n_i correspond to the directions bisecting the angle between the x_1 and x_3 axes, which are along the principal directions of the largest and the smallest principal stresses. Thus, the normal to the surface element on which the maximum shear stress occurs bisects the angle between the directions of the largest and the least principal stresses. ∎

EXAMPLE 7.6.4 If the coordinate axes are chosen along the principal directions of stress at a point P, then the normal and shear stresses at P on the octahedral plane are called the *octahedral normal* and *shear stresses* at P. Show that these are given by

$$\sigma_{\text{oct}} = \tfrac{1}{3}(\tau_1 + \tau_2 + \tau_3) \tag{7.6.17}$$

$$\tau_{\text{oct}} = \tfrac{1}{3}\{(\tau_1 - \tau_2)^2 + (\tau_2 - \tau_3)^2 + (\tau_3 - \tau_1)^2\}^{1/2} \tag{7.6.18}$$

where τ_1, τ_2, τ_3 are the principal stresses at P.

Solution When the axes are along the principal directions of stress at P, the normal and shear stresses at P are given by (7.6.12) and (7.6.14).

For the octahedral plane, $n_1 = n_2 = n_3 = 1/\sqrt{3}$. With these values of n_i, (7.6.12) becomes (7.6.17), and (7.6.14) becomes

$$\tau_{\text{oct}}^2 = \tfrac{1}{9}\{3(\tau_1^2 + \tau_2^2 + \tau_3^2) - (\tau_1 + \tau_2 + \tau_3)^2\} \tag{7.6.19}$$

which can be simplified to obtain (7.6.18). ∎

7.7
STRESS DEVIATOR

In Section 2.12, we have seen that every second-order tensor can be decomposed into a spherical part and a deviator part. Employing this decomposition to the stress tensor $\mathbf{T}$, we get

$$\mathbf{T} = \tfrac{1}{3}(\text{tr } \mathbf{T})\mathbf{I} + \mathbf{T}^{(d)} \tag{7.7.1}$$

Setting

$$\bar{p} = -\tfrac{1}{3}(\text{tr } \mathbf{T}) = -\tfrac{1}{3}\tau_{kk} \tag{7.7.2}$$

(7.7.1) becomes

$$\mathbf{T} = -\bar{p}\mathbf{I} + \mathbf{T}^{(d)} \tag{7.7.3}$$

The number $\bar{p}$ is called the *mean pressure*. The deviator part $\mathbf{T}^{(d)}$, which is the excess of $\mathbf{T}$ over the spherical part $-\bar{p}\mathbf{I}$, is called the *stress deviator tensor*. It is evident that $\mathbf{T}^{(d)}$ is symmetric if $\mathbf{T}$ is symmetric. The components $\tau_{ij}^{(d)}$ of $\mathbf{T}^{(d)}$ are called the *stress deviator components* and the matrix $[\tau_{ij}^{(d)}]$ is called the *stress deviator matrix*.

Example 2.13.7 showed that a tensor and its deviator part have the same eigenvectors. It therefore follows that the principal directions of $\mathbf{T}^{(d)}$ coincide with those of $\mathbf{T}$ (which is assumed to be symmetric). With respect to a set of axes chosen along these principal directions, the stress deviator matrix is purely diagonal, the diagonal elements being the eigenvalues of

7.7 STRESS DEVIATOR

$\mathbf{T}^{(d)}$. These eigenvalues, say $\tau_1^{(d)}$, $\tau_2^{(d)}$, $\tau_3^{(d)}$, are called the *principal deviator stresses*. By virtue of expressions (2.13.30), these are related to the principal stresses τ_1, τ_2, τ_3 through the following relations:

$$\left.\begin{aligned} \tau_1^{(d)} &= \tfrac{1}{3}(2\tau_1 - \tau_2 - \tau_3) \\ \tau_2^{(d)} &= \tfrac{1}{3}(2\tau_2 - \tau_3 - \tau_1) \\ \tau_3^{(d)} &= \tfrac{1}{3}(2\tau_3 - \tau_1 - \tau_2) \end{aligned}\right\} \tag{7.7.4}$$

Since the deviator part of a tensor has its trace equal to 0, we have $I_{\mathbf{T}^{(d)}} = 0$. By virtue of expressions (2.13.14) and (2.13.31), the second and third fundamental invariants of $\mathbf{T}^{(d)}$ are given as follows:

$$\begin{aligned} II_{\mathbf{T}^{(d)}} &= \tau_1^{(d)}\tau_2^{(d)} + \tau_2^{(d)}\tau_3^{(d)} + \tau_3^{(d)}\tau_1^{(d)} \\ &= -\tfrac{1}{6}\{(\tau_1 - \tau_2)^2 + (\tau_2 - \tau_3)^2 + (\tau_3 - \tau_1)^2\} \end{aligned} \tag{7.7.5}$$

$$\begin{aligned} III_{\mathbf{T}^{(d)}} &= \tau_1^{(d)}\tau_2^{(d)}\tau_3^{(d)} \\ &= \tfrac{1}{27}(2\tau_1 - \tau_2 - \tau_3)(2\tau_2 - \tau_3 - \tau_1)(2\tau_3 - \tau_2 - \tau_3) \end{aligned} \tag{7.7.6}$$

The scalars $II_{\mathbf{T}^{(d)}}$ and $III_{\mathbf{T}^{(d)}}$ are called the *stress-deviator invariants*.

EXAMPLE 7.7.1 For the state of stress given by the stress matrix

$$[\tau_{ij}] = \begin{bmatrix} 1 & 0 & 0 \\ 0 & 3 & -1 \\ 0 & -1 & 3 \end{bmatrix}$$

find (i) the stress deviator matrix, (ii) principal deviator stresses, and (iii) stress deviator invariants.

Solution (i) From the given stress components, we get $\tau_{kk} = 7$. Hence, from expression (7.7.1), we obtain the stress deviator components as follows:

$$\tau_{11}^{(d)} = \tau_{11} - \tfrac{1}{3}\tau_{kk} = -\tfrac{4}{3}; \qquad \tau_{12}^{(d)} = \tau_{21}^{(d)} = 0$$

$$\tau_{22}^{(d)} = \tau_{22} - \tfrac{1}{3}\tau_{kk} = \tfrac{2}{3}; \qquad \tau_{23}^{(d)} = \tau_{32}^{(d)} = -1$$

$$\tau_{33}^{(d)} = \tau_{33} - \tfrac{1}{3}\tau_{kk} = \tfrac{2}{3}; \qquad \tau_{13}^{(d)} = \tau_{31}^{(d)} = 0$$

Thus, the stress deviator matrix is

$$[\tau_{ij}^{(d)}] = \begin{bmatrix} -\tfrac{4}{3} & 0 & 0 \\ 0 & \tfrac{2}{3} & -1 \\ 0 & -1 & \tfrac{2}{3} \end{bmatrix}$$

(ii) The characteristic equation of the given matrix is

$$\begin{vmatrix} 1 - \Lambda & 0 & 0 \\ 0 & 3 - \Lambda & -1 \\ 0 & -1 & 3 - \Lambda \end{vmatrix} = 0$$

Expanding the determinant and solving the resulting cubic equation we get the three roots as $\tau_1 = 1$, $\tau_2 = 2$, $\tau_3 = 4$. These are the principal stresses for the given state of stress. The corresponding principal deviator stresses may be computed by using expressions (7.7.4). Thus, we obtain

$$\tau_1^{(d)} = -\tfrac{4}{3}, \qquad \tau_2^{(d)} = -\tfrac{1}{3}, \qquad \tau_3^{(d)} = \tfrac{5}{3}$$

These may also be computed directly from the characteristic equation of the matrix $[\tau_{ij}^{(d)}]$.

(iii) Expressions (7.7.5) and (7.7.6) now yield the stress deviator invariants

$$II_{\mathbf{T}^{(d)}} = -\tfrac{21}{9}, \qquad III_{\mathbf{T}^{(d)}} = \tfrac{20}{27}$$ ∎

EXAMPLE 7.7.2 Show that the octahedral shear stress τ_{oct} is given in terms of the principal deviator stresses by

$$\tau_{oct}^2 = -\tfrac{2}{3}\{\tau_1^{(d)}\tau_2^{(d)} + \tau_2^{(d)}\tau_3^{(d)} + \tau_3^{(d)}\tau_1^{(d)}\} \qquad (7.7.7)$$

$$= -\tfrac{2}{3} II_{\mathbf{T}^{(d)}} \qquad (7.7.8)$$

Solution From (7.6.18), we have

$$\tau_{oct}^2 = \tfrac{1}{9}\{(\tau_1 - \tau_2)^2 + (\tau_2 - \tau_3)^2 + (\tau_3 - \tau_1)^2\} \qquad (7.7.9)$$

From expressions (7.7.5), we get

$$(\tau_1 - \tau_2)^2 + (\tau_2 - \tau_3)^2 + (\tau_3 - \tau_1)^2$$
$$= -6 II_{\mathbf{T}^{(d)}} = -6\{\tau_1^{(d)}\tau_2^{(d)} + \tau_2^{(d)}\tau_3^{(d)} + \tau_3^{(d)}\tau_1^{(d)}\} \qquad (7.7.10)$$

Using (7.7.10) in (7.7.9), we obtain (7.7.7) and (7.7.8). ∎

7.8

BOUNDARY CONDITION FOR THE STRESS TENSOR

It was noted that Cauchy's law (7.4.8) connecting $\mathbf{s}(\mathbf{n})$ and $\mathbf{T}$ holds at every point of the material and at every instant of time. As such, it holds also at points of the boundary surface, if any, on the material. Thus, if we take a

7.8 BOUNDARY CONDITION FOR THE STRESS TENSOR

point P on the boundary surface of the material, we should have

$$\mathbf{T}^T\mathbf{n} = \mathbf{s}(\mathbf{n}) \tag{7.8.1}$$

at P. For a boundary point, $\mathbf{n}$ is a known vector and by convention is taken as the exterior normal, and $\mathbf{s}(\mathbf{n})$ represents the surface force per unit of area caused by the external surface forces. Hence (7.8.1) serves as a boundary condition for $\mathbf{T}$ (for known $\mathbf{n}$ and known $\mathbf{s}(\mathbf{n})$).

The boundary surface of a material is said to be a *stress-free* (or *traction-free*) *surface* or just a *free surface* if the stress vector acting on it is 0. For such a boundary surface, the boundary condition (7.8.1) becomes

$$\mathbf{T}^T\mathbf{n} = \mathbf{0} \tag{7.8.2}$$

EXAMPLE 7.8.1 Obtain the boundary conditions for the stress components in the case when the boundary of the surface is subjected to a pressure p. Specialize these conditions for a material filling the region $x_1 < 0$ with the plane $x_1 = 0$ as the boundary.

Solution By definition, pressure is the pure normal stress acting opposite to the normal of a surface (see Sec. 7.5). Hence, if $p(>0)$ is the magnitude of pressure, we have

$$\mathbf{s}(\mathbf{n}) = -p\mathbf{n} \tag{7.8.3}$$

By using (7.8.1), this can be rewritten as

$$\mathbf{T}^T\mathbf{n} = -p\mathbf{n} \quad \text{or} \quad \tau_{ji}n_j = -pn_i \tag{7.8.4}$$

Thus, if a boundary surface is acted upon by a pressure of magnitude p, (7.8.4) are the conditions to be satisfied by the stresses τ_{ij} on the boundary, with n_i as the components of the outward unit normal.

If the material fills the space $x_1 < 0$ with $x_1 = 0$ as the boundary, the outward normal to the boundary is $\mathbf{n} = \mathbf{e}_1$; see Figure 7.6. The conditions (7.8.4) then yield the following conditions to be satisfied on $x_1 = 0$:

$$\tau_{11} = -p, \quad \tau_{21} = 0, \quad \tau_{31} = 0 \quad \blacksquare \tag{7.8.5}$$

Note: If the boundary $x_1 = 0$ is stress free, conditions (7.8.5) become $\tau_{11} = \tau_{21} = \tau_{31} = 0$ on $x_1 = 0$.

EXAMPLE 7.8.2 Let P be a point on a free surface S. Show that the stress vector on every other surface element at P acts tangential to S.

Solution Let $\mathbf{n}$ be the unit normal to the surface S at the point P. Since S is a free surface, we have $\mathbf{s}(\mathbf{n}) = \mathbf{0}$. Let S' be any other surface element at P with unit normal $\mathbf{n}'$. Then the stress vector on S' is $\mathbf{s}(\mathbf{n}')$. Since the stress

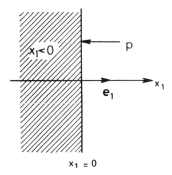

Figure 7.6. Example 7.8.1.

tensor is assumed to be symmetric, we have from (7.4.24), $\mathbf{n} \cdot \mathbf{s}(\mathbf{n}') = \mathbf{n}' \cdot \mathbf{s}(\mathbf{n})$. Since $\mathbf{s}(\mathbf{n}) = \mathbf{0}$, it follows that $\mathbf{s}(\mathbf{n}')$ is orthogonal to $\mathbf{n}$. That is, $\mathbf{s}(\mathbf{n}')$ is tangential to S. ∎

7.9
PIOLA-KIRCHHOFF STRESS TENSORS

In the preceding sections, we analyzed the concept of stress by considering points and surface elements in the current configuration of the material. The stress vector **s** on a surface element dS at a point in the current configuration was defined as the surface force per unit of area of that element. In some discussions, particularly in nonlinear elasticity, it is often more convenient to use the idea of a surface force on dS measured per unit of area of the corresponding element dS_0 in the initial configuration. Here we introduce tensors that will enable us to compute such a surface force. Like the Green strain tensor, these tensors are in Lagrangian description.

If dS and dS_0 are the corresponding elements in the current and initial configurations, respectively, we find from (5.2.12) that

$$(dS)\mathbf{n} = J(dS_0)(\mathbf{F}^T)^{-1}\mathbf{n}^0 \qquad (7.9.1)$$

where $\mathbf{n}$ and $\mathbf{n}^0$ are the unit vectors to dS and dS_0, respectively; see Figure 7.7.

By virtue of the relations (7.4.8), the total surface force acting on dS is

$$(dS)\mathbf{s} = (dS)(\mathbf{T}^T\mathbf{n}) = \mathbf{T}^T(dS)\mathbf{n} \qquad (7.9.2)$$

Substituting for $(dS)\mathbf{n}$ from (7.9.1), this becomes

$$(dS)\mathbf{s} = J(dS_0)(\mathbf{F}^{-1}\mathbf{T})^T\mathbf{n}^0 \qquad (7.9.3)$$

7.9 PIOLA-KIRCHHOFF STRESS TENSORS

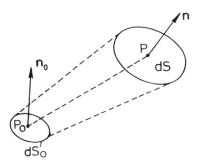

Figure 7.7. Surface elements dS_0 and dS.

Setting
$$\mathbf{T}^0 = J\mathbf{F}^{-1}\mathbf{T} \tag{7.9.4}$$
we get
$$(dS)\mathbf{s} = (dS_0)(\mathbf{T}^0)^T \mathbf{n}^0 \tag{7.9.5}$$

This expression is analogous to expression (7.9.2). It is evident that $(\mathbf{T}^0)^T\mathbf{n}^0$ represents the surface force on dS measured per unit area of dS_0; we denote this vector by $\mathbf{s}^0$. Thus,
$$(dS)\mathbf{s} = (dS_0)\mathbf{s}^0 \tag{7.9.6}$$
where
$$\mathbf{s}^0 = (\mathbf{T}^0)^T \mathbf{n}^0 \tag{7.9.7}$$

When the tensor $\mathbf{T}^0$ is known, expression (7.9.7) enables us to find $\mathbf{s}^0$ for any given $\mathbf{n}^0$. The vector $\mathbf{s}^0$ is called the *nominal stress vector* and the tensor $\mathbf{T}^0$ is called the *nominal stress tensor*. From (7.9.4) we note that $\mathbf{T}^0$ may not be symmetric even when $\mathbf{T}$ is symmetric. The transpose of $\mathbf{T}^0$, namely, $(\mathbf{T}^0)^T$, is called the *first Piola-Kirchhoff stress tensor*, after Piola (1833) and Kirchhoff (1852). Some authors refer to $\mathbf{T}^0$ as the Poila-Kirchhoff stress tensor and $(\mathbf{T}^0)^T$ as the nominal stress tensor.

In components, expression (7.9.7) reads
$$s_i^0 = \tau_{ji}^0 n_j^0 \tag{7.9.7'}$$

To interpret the meanings of the components τ_{ij}^0 of $\mathbf{T}^0$, let us take $\mathbf{n}^0$ to be a base vector, say, $\mathbf{e}_1$. Then $\mathbf{s}^0$ represents the surface force per unit of initial area of the surface element that initially had $\mathbf{e}_1$ as its unit normal. The expressions (7.9.7)' now become $s_i^0 = \tau_{1i}^0$. Thus τ_{1i}^0 represents the ith component of the surface force per unit initial area of the surface element which initially had $\mathbf{e}_1$ as its unit normal. The components τ_{2i}^0 and τ_{3i}^0 have analogous meanings. Thus, in general, τ_{ij}^0 represents the jth component of the surface force per unit of initial area on the surface element that initially had its normal in the ith-direction.

Setting
$$\mathbf{S} = \mathbf{T}^0(\mathbf{F}^{-1})^T \tag{7.9.8}$$

expression (7.9.7) becomes
$$\mathbf{s}^0 = \mathbf{F}\mathbf{S}^T\mathbf{n}^0 \tag{7.9.9}$$

This expression is often used as an alternative to the expression (7.9.7). The tensor $\mathbf{S}$ is called the *second Piola-Kirchhoff stress tensor*. Unlike the tensors $\mathbf{T}$ and $\mathbf{T}^0$, the tensor $\mathbf{S}$ does not admit a direct physical interpretation.

The following expressions follow from (7.9.4) and (7.9.8):

$$\left. \begin{array}{c} \mathbf{T} = \dfrac{1}{J}\mathbf{F}\mathbf{T}^0 = \dfrac{1}{J}\mathbf{F}\mathbf{S}\mathbf{F}^T \\ \mathbf{S} = J\mathbf{F}^{-1}\mathbf{T}(\mathbf{F}^{-1})^T \end{array} \right\} \tag{7.9.10}$$

From these expressions, we find that $\mathbf{S}$ is symmetric whenever $\mathbf{T}$ is symmetric and vice versa.

7.10
EXERCISES

1. The stress matrix at a point P is given by

$$[\tau_{ij}] = \begin{bmatrix} 1 & 0 & -1 \\ 0 & -2 & 0 \\ -1 & 0 & 2 \end{bmatrix}$$

Find the stress vector at P on a plane element whose normal is $\mathbf{e}_2$. Also, find its magnitude.

2. The stress matrix at a point P is given by

$$[\tau_{ij}] = \begin{bmatrix} 2 & -1 & 3 \\ -1 & 4 & 0 \\ 3 & 0 & -1 \end{bmatrix}$$

Find the stress vector $\mathbf{s}$ at P on a plane element with normal $\mathbf{n} = 2(\mathbf{e}_1 + \mathbf{e}_2) + \mathbf{e}_3$. Also, find (i) the angle that $\mathbf{s}$ makes with $\mathbf{n}$, and (ii) magnitude of $\mathbf{s}$.

3. The stress matrix at a point P is given by

$$[\tau_{ij}] = \begin{bmatrix} 1 & 1 & 0 \\ 1 & -1 & 0 \\ 0 & 0 & 1 \end{bmatrix}$$

Find the stress vector at P acting on the plane element parallel to the plane $x_1 + x_2 + 2x_3 = 0$. Also find the normal stress and the shear stress on the element.

4. The stress matrix at a point P is given by

$$[\tau_{ij}] = \begin{bmatrix} 1 & a & b \\ a & 1 & c \\ b & c & 1 \end{bmatrix}$$

where a, b, c are constants. Find a, b, c so that the stress vector at P on the octahedral plane is 0.

5. At a point P, the stress matrix is given by

$$[\tau_{ij}] = \begin{bmatrix} 1 & 0 & -1 \\ 0 & -2 & 0 \\ -1 & 0 & 2 \end{bmatrix}$$

Find a plane through P parallel to the x_3 axis on which the normal stress is 0.

6. At a point P, the stress matrix is as follows:

$$[\tau_{ij}] = \begin{bmatrix} 1 & 0 & -2 \\ 0 & -1 & 2 \\ -2 & 2 & 0 \end{bmatrix}$$

Find $\mathbf{n}$ such that $s_1(\mathbf{n}) = s_2(\mathbf{n}) = 0$ at P. Also, determine $s_3(\mathbf{n})$ at P.

7. The stress matrix at a point P is given by

$$[\tau_{ij}] = \begin{bmatrix} 0 & 1 & 2 \\ 1 & a & 1 \\ 2 & 1 & 1 \end{bmatrix}$$

where a is a constant. Find a so that the stress vector on some plane at P is 0. Also, specify this plane.

8. The stress matrix at a point (x_i) is given by

$$[\tau_{ij}] = \begin{bmatrix} 3x_1 x_2 & 5x_2^2 & 0 \\ 5x_2^2 & 0 & 2x_3 \\ 0 & 2x_3 & 0 \end{bmatrix}$$

Find the stress vector at a point $P(1, -1/2, \sqrt{3}/2)$ on the plane that is tangential to the cylindrical surface $x_2^2 + x_3^2 = 1$. Find also the normal and shear stresses.

9. The stress components at a point P with plane polar coordinates (R, θ) are

$$\tau_{11} = -\frac{2a}{R}\cos^3\theta, \qquad \tau_{22} = -\frac{a}{R}\sin\theta\sin 2\theta$$

$$\tau_{12} = \tau_{21} = -\frac{a}{R}\cos\theta\sin 2\theta, \qquad R \neq 0$$

$$\tau_{13} = \tau_{31} = \tau_{23} = \tau_{32} = \tau_{33} = 0$$

where a is a constant. Find the stress vector at P on the surface $R =$ constant. Also, find the normal and shear stresses on this surface. Evaluate all these at the point $(1, \pi/4)$.

10. At a point, the stress vectors on three planes are obtained as follows:

$$\mathbf{s(n)} = \mathbf{e}_1 + 2\mathbf{e}_2 + 3\mathbf{e}_3 \quad \text{for} \quad \mathbf{n} = -\mathbf{e}_1$$

$$\mathbf{s(n)} = 2\sqrt{3}\,\mathbf{e}_1 + 2\sqrt{3}\,\mathbf{e}_2 \quad \text{for} \quad \mathbf{n} = \frac{1}{\sqrt{3}}(\mathbf{e}_1 + \mathbf{e}_2 + \mathbf{e}_3)$$

$$\mathbf{s(n)} = 2(\mathbf{e}_1 + \mathbf{e}_2 + \mathbf{e}_3) \quad \text{for} \quad \mathbf{n} = \mathbf{e}_2$$

Find the stress matrix at the point.

11. By employing Cauchy's law, show directly (without appealing to a quotient law) that τ_{ij} are components of a tensor.

12. Let ΔS_1 and ΔS_2 be two arbitrary plane elements at a point P such that the stress vector on ΔS_1 lies in ΔS_2. Find a necessary and sufficient condition under which the stress vector on ΔS_2 lies in ΔS_1.

13. Let $\mathbf{s(n)}$ and $\mathbf{s(n')}$ be the stress vectors at a point on two plane elements with normals $\mathbf{n}$ and $\mathbf{n'}$. Find the direction of the stress vector on the plane containing $\mathbf{s(n)}$ and $\mathbf{s(n')}$, given that the stress tensor is symmetric.

14. Show that the sum of the squares of the magnitudes of stress vectors at a point on planes parallel to the coordinate planes is invariant under coordinate transformations.

15. The state of stress at a point P is such that the stress vector on a particular surface element ΔS is purely normal and the stress vector on every surface element perpendicular to ΔS is 0. Show that the state of stress is uniaxial.

16. If the state of stress is uniaxial, show that the stress tensor is of the form $\mathbf{T} = T(\mathbf{a} \otimes \mathbf{a})$, where $\mathbf{a}$ is a constant unit vector and T is a scalar (given that $\mathbf{T}$ is symmetric).

7.10 EXERCISES

17. The state of stress at a point P is said to be a *pure shear* of magnitude T relative to unit vectors **a** and **b** if **a** and **b** are orthogonal and if a surface element with normal **a** experiences a stress of magnitude T in the direction of **b** and a surface element with normal **b** experiences a stress of magnitude T in the direction of **a**. For such a state of stress, find the stress vector on an arbitrary surface element, and the stress tensor.

18. The stress matrix at a point is given by

$$[\tau_{ij}] = \begin{bmatrix} 11 & 2 & 8 \\ 2 & 2 & -10 \\ 8 & -10 & 5 \end{bmatrix}$$

Find the principal stresses and principal directions of stress at the point.

19. The stress matrix at a point is given as follows:

$$[\tau_{ij}] = \begin{bmatrix} 1 & 0 & 2 \\ 0 & 1 & 0 \\ 2 & 0 & -2 \end{bmatrix}$$

Find the principal stresses. Show that the principal directions that correspond to the largest and the smallest principal stresses both lie in the $x_3 x_1$ plane.

20. Find the principal stresses and the principal directions of stress at a point where the stress matrix is as follows:

$$[\tau_{ij}] = \begin{bmatrix} a & a & a \\ a & a & a \\ a & a & a \end{bmatrix}$$

21. Find the fundamental stress invariants for the state of stress defined by the stress matrix

$$[\tau_{ij}] = \begin{bmatrix} 0 & 1 & 2 \\ 1 & 0 & 2 \\ 2 & 2 & 0 \end{bmatrix}$$

22. Find the principal stresses at a point P, given that the stress matrix at P is

$$[\tau_{ij}] = \begin{bmatrix} 3 & 1 & 1 \\ 1 & 0 & 2 \\ 1 & 2 & 0 \end{bmatrix}$$

Hence, find the maximum of all the shear stresses acting on plane elements through P. Also, determine the element on which the maximum shear stress occurs.

7 STRESS

23. The stress matrix at a point is given by

$$[\tau_{ij}] = \begin{bmatrix} a & 0 & 0 \\ 0 & b & 0 \\ 0 & 0 & c \end{bmatrix}$$

where $a > b > c$ are constants. Find the maximum shear stress and the plane element on which it acts.

24. At a point P, the principal stresses are $\tau_1 = 4$, $\tau_2 = 1$, $\tau_3 = -2$. Find the stress vector, the normal stress and the shear stress on the octahedral plane at P.

25. The stress matrix at a point (x_i) is

$$[\tau_{ij}] = \begin{bmatrix} 0 & 0 & ax_2 \\ 0 & 0 & -ax_1 \\ ax_2 & -ax_1 & 0 \end{bmatrix}$$

where a is a constant. Find the principal stresses and the principal directions of stress at the point $P(1, 2, 4)$. Also, compute the stress invariants, octahedral normal stress and octahedral shear stress at P.

26. At a point the principal stresses are $\tau_1 = 2$, $\tau_2 = a$, $\tau_3 = 1$, where a is a constant. Find a so that the octahedral shear stress is the maximum shear stress.

27. At a point P, the principal stresses τ_1, τ_2, τ_3 are such that $\tau_2 = \frac{1}{2}(\tau_1 + \tau_3)$. Find the plane element at P on which the normal stress is τ_2 and shear stress is $\frac{1}{4}|\tau_1 - \tau_3|$.

28. Prove the following:

(i) $\tau^2 = (\tau_1 - \tau_2)^2 n_1^2 n_2^2 + (\tau_2 - \tau_3)^2 n_2^2 n_3^2 + (\tau_3 - \tau_1)^2 n_3^2 n_1^2$

(ii) $\tau_{oct}^2 = \frac{1}{9}(2I_T^2 - 6II_T)$

29. Suppose that the x_i system is coincident with the principal directions of stress at a point P and let τ_1, τ_2 and τ_3 be the principal stresses at P. If τ'_{ij} are the stress components w.r.t. the x'_i system also set up at P, show that

$$\tau'_{ij} = \sum_{k=1}^{3} \alpha_{ip} \alpha_{jp} \tau_k$$

30. Find the principal stresses for the state of stress that is a pure shear of magnitude T relative to $\mathbf{e}_1$ and $\mathbf{e}_2$, as defined in Exercise 17.

31. The stress components at a point are such that $\tau_{ij} = \tau_{ji}$ with $\tau_{13} = \tau_{23} = 0$ and $\tau_{33} = \frac{1}{2}(\tau_{11} + \tau_{22})$. Show that the maximum shear stress at the point is given by

$$\tau_{max} = \frac{1}{2}[(\tau_{11} - \tau_{22})^2 + 4\tau_{12}^2]^{1/2}$$

32. If a state of stress is such that $\tau_{31} = \tau_{32} = \tau_{33} \equiv 0$ and $\tau_{ij} = \tau_{ji}$, show that (i) the x_3 axis is a principal direction of stress and the corresponding principal stress is 0, and (ii) the principal directions in the $x_1 x_2$ plane are inclined at an angle θ with the x_1 and x_2 axes, respectively, where

$$\tan 2\theta = \frac{2\tau_{12}}{\tau_{11} - \tau_{22}}$$

and the corresponding principal stresses are

$$\tfrac{1}{2}(\tau_{11} + \tau_{22}) \pm [\tfrac{1}{4}(\tau_{11} - \tau_{22})^2 + \tau_{12}^2]^{1/2}$$

33. The stress components at a point P are given by $\tau_{ij} = \alpha(a_i b_j + a_j b_i)$, where a_i and b_i are components of constant unit vectors and α is a constant. Find the principal stresses and principal directions of stress at P.

34. Given the stress matrix

$$[\tau_{ij}] = \begin{bmatrix} 1 & -3 & \sqrt{2} \\ -3 & 1 & -\sqrt{2} \\ \sqrt{2} & -\sqrt{2} & 4 \end{bmatrix}$$

find (i) the stress deviator matrix, (ii) the principal deviator stresses and (iii) the stress deviator invariants.

35. The stresses in a circular cylinder of radius a with x_3 axis as the axis are given by

$$\tau_{13} = \tau_{31} = \mu\alpha x_1, \qquad \tau_{23} = \tau_{32} = -\mu\alpha x_2$$

$$\tau_{11} = \tau_{22} = \tau_{33} = \tau_{12} = \tau_{21} = 0$$

where μ and α are constants. Show that the boundary surface of the cylinder is stress free.

36. The stresses acting at a point (x_i) in a rectangular block bounded by the surfaces $x_1 = \pm a$, $x_2 = \pm a$ and $x_3 = \pm b$ are given as follows:

$$\tau_{11} = -k(x_1^2 - x_2^2), \qquad \tau_{22} = k(x_1^2 - x_2^2)$$

$$\tau_{12} = \tau_{21} = 2ax_1x_2, \qquad \tau_{13} = \tau_{31} = \tau_{23} = \tau_{32} = \tau_{33} = 0$$

Find the total force on each face of the block.

37. The stress components in the half-space $x_3 > 0$ are given by $\tau_{ij} = (ax_ix_j/r^5)x_3$, $r \neq 0$, $a > 0$. Find the total force on the surface of the hemisphere $r = a$, $x_3 > 0$.

38. A long circular cylinder with its axis along the x_3 axis is subjected to a pressure p on its surface. Write down the boundary conditions in terms of the stress components τ_{ij}.

39. A sphere with center at the origin is subjected to a pressure p on its surface. Write down the boundary conditions in terms of the stress components τ_{ij}.

40. Prove the following relations connecting the components of **T**, **T**^{0} and **S**:

(i) $\tau_{ij} = \dfrac{1}{J} x_{i;k} \tau^0_{kj} = \dfrac{1}{J} x_{i;k} x_{j;m} S_{km}$

(ii) $\tau^0_{ij} = J x^0_{i,k} \tau_{kj} = S_{ik} x_{j;k}$

(iii) $S_{ij} = J x^0_{i,k} x^0_{j,m} \tau_{km} = \tau^0_{ik} x^0_{j,k}$

41. Show that for small deformations, $s_{ij} \approx \tau_{ij}$.

CHAPTER 8

FUNDAMENTAL LAWS OF CONTINUUM MECHANICS

8.1
INTRODUCTION

Continuum mechanics is essentially based upon the following four fundamental mechanical principles, commonly known as *conservation laws* or *balance laws*: (i) the law of conservation of mass, (ii) the law of balance of linear momentum, (iii) the law of balance of angular momentum and (iv) the law of balance of energy. These laws are postulated in the form of equations involving certain integrals; such equations give rise to the so-called *field equations* that should hold at every point of a continuum and for all time. The important feature of the field equations is that these equations are applicable to all continua (solids, liquids, gases) regardless of their internal physical structure. This chapter is devoted primarily to the derivation of the field equations with some of their immediate consequences.

It will be seen that the field equations are inadequate to determine all the unknown functions involved in the theory. These equations are therefore to be supplemented by certain additional basic equations. The additional equations appear in the form of relations representing the intrinsic physical properties of the continuum under study. Such additional equations are called *constitutive equations*. Unlike the field equations, the constitutive equations distinguish one class of materials from the other. A brief discussion on the need for constitutive equations is given at the end of the chapter. Consideration of the general theory of constitutive equations falls beyond the scope of this book.

8.2
CONSERVATION OF MASS

In Section 4.4, the mass m of a material body $\mathcal{B}$ in a configuration B of volume V, at time t was expressed as

$$m = \int_V \rho \, dV \tag{8.2.1}$$

where $\rho = \rho(\mathbf{x}, t)$ is the density at a point $\mathbf{x}$ of B and at time t. We now postulate that mass m remains unchanged during the motion of $\mathcal{B}$. This postulate is called the *law (principle) of the conservation of mass*. As a mathematical expression, this law reads

$$\frac{Dm}{Dt} = 0 \tag{8.2.2}$$

Substituting for m from (8.2.1) in (8.2.2) and using the transport formula (6.5.3), we obtain

$$\int_V \left(\frac{D\rho}{Dt} + \rho \operatorname{div} \mathbf{v} \right) dV = 0$$

This equation is supposed to hold for any volume V; therefore, by invoking the localization theorem (see Section 4.6) it follows that

$$\frac{D\rho}{Dt} + \rho \operatorname{div} \mathbf{v} = 0 \tag{8.2.3}$$

at every point of V and for all t.

Equation (8.2.3) represents the law of conservation of mass in the *spatial (Eulerian) form*. This equation, which is a first-order partial differential equation in the functions ρ and $\mathbf{v}$, is one of the fundamental equations of continuum mechanics. Traditionally this equation is called the *equation of continuity*, in which the word *continuity* is used in the sense of constancy (of mass). The equation was first obtained by Euler in 1757; a particular case of the equation was given earlier, in 1752, by d'Alembert.

8.2 CONSERVATION OF MASS

The following equivalent forms of equation (8.2.3) are often useful:

$$\frac{D}{Dt}(\log \rho) + \text{div } \mathbf{v} = 0 \tag{8.2.4a}$$

$$\rho \frac{D}{Dt}\left(\frac{1}{\rho}\right) = \text{div } \mathbf{v} \tag{8.2.4b}$$

Using (6.2.9) and (3.4.14), equation (8.2.3) can be put in the following alternative forms also:

$$\frac{\partial \rho}{\partial t} + \mathbf{v} \cdot \nabla \rho + \rho \text{ div } \mathbf{v} = 0 \tag{8.2.5a}$$

$$\frac{\partial \rho}{\partial t} + \text{div}(\rho \mathbf{v}) = 0 \tag{8.2.5b}$$

8.2.1 EQUATION OF CONTINUITY IN MATERIAL FORM

Multiplying equation (8.2.3) by J and using the Euler's formula (6.2.51), namely,

$$\frac{DJ}{Dt} = J \text{ div } \mathbf{v} \tag{8.2.6}$$

we obtain

$$\frac{D}{Dt}(\rho J) = 0 \tag{8.2.7}$$

In the initial configuration where $t = 0$ and $\mathbf{x} = \mathbf{x}^0$, we have $J = 1$. Let ρ_0 denote the *initial density* $\rho(\mathbf{x}^0, 0)$. Then, integration of equation (8.2.7) yields

$$\rho J = \rho_0 \tag{8.2.8}$$

This is the equation of continuity in the *material (Lagrangian) form*. This equation is also given by Euler.

8.2.2 CASE OF INCOMPRESSIBLE CONTINUUM

From equation (8.2.3) we note that div $\mathbf{v} = 0$ if and only if $D\rho/Dt = 0$. But $D\rho/Dt = 0$ holds if and only if ρ remains unchanged during the motion; that is, $\rho = \rho_0$ for all time. Equation (8.2.8) shows that $\rho = \rho_0$ holds if and only if $J = 1$.

Thus, *the following four conditions are all equivalent to one another.*

$$\text{div } \mathbf{v} = 0 \tag{8.2.9a}$$

$$\frac{D\rho}{Dt} = 0 \tag{8.2.9b}$$

$$\rho = \rho_0 \tag{8.2.9c}$$

$$J = 1 \tag{8.2.9d}$$

In Section 6.3, it was noted that either of the conditions div $\mathbf{v} = 0$ or $J = 1$ is a necessary and sufficient condition for a motion to be isochoric. Since the four conditions (8.2.9) are all equivalent, it follows that each of these four conditions is a necessary and sufficient condition for a motion to be isochoric.

In Section 6.3, it was also pointed out that an incompressible continuum is a continuum for which every motion is isochoric. Accordingly, *each of the conditions* (8.2.9) *holds for every motion of an incompressible continuum*. Any of these four conditions can therefore be employed as the equation of continuity for studying motions of such continua.

8.2.3 AN IMPORTANT CONSEQUENCE

Let $\phi = \phi(\mathbf{x}, t)$ be an arbitrary scalar, vector or tensor function defined over the configuration B. Then by using the transport formula (6.5.3) we obtain

$$\frac{D}{Dt} \int_V (\rho\phi)\, dV = \int_V \left[\frac{D}{Dt}(\rho\phi) + \rho\phi \text{ div } \mathbf{v} \right] dV$$

$$= \int_V \left[\rho \frac{D\phi}{Dt} + \phi\left(\frac{D\rho}{Dt} + \rho \text{ div } \mathbf{v}\right) \right] dV$$

Using (8.2.3), this reduces to

$$\frac{D}{Dt} \int_V (\rho\phi)\, dV = \int_V \rho \frac{D\phi}{Dt}\, dV \tag{8.2.10}$$

This consequence of the equation of continuity is of great utility in our further discussions.

EXAMPLE 8.2.1 Obtain equation (8.2.8) directly by using the law of conservation of mass. Hence deduce equation (8.2.3).

8.2 CONSERVATION OF MASS

Solution If V_0 is the volume of the initial configuration B_0 of a body $\mathcal{B}$, the mass of $\mathcal{B}$ evaluated in this configuration is

$$m_0 = \int_{V_0} \rho(\mathbf{x}^0, 0)\, dV_0 = \int_{V_0} \rho_0\, dV_0 \tag{8.2.11}$$

The law of conservation of mass states that $m = m_0$. Hence, (8.2.1) and (8.2.11) give

$$\int_{V_0} \rho_0\, dV_0 = \int_V \rho(\mathbf{x}, t)\, dV \tag{8.2.12}$$

This is often referred to as the *integral* (or *global*) *form of the law of conservation of mass*.

Changing the variable of integration from $\mathbf{x}$ to $\mathbf{x}^0$ in the integral on the righthand side of (8.2.12) and bearing in mind that $dV = J\, dV_0$, we get

$$\int_{V_0} (\rho_0 - \rho J)\, dV_0 = 0$$

Since this holds for any volume V_0, it follows that $\rho_0 = \rho J$, which is (8.2.8). Since $\rho_0 = \rho(\mathbf{x}^0, 0)$ depends only on $\mathbf{x}^0$, equaton (8.2.8) gives

$$\frac{D}{Dt}(\rho J) = 0$$

that is,

$$\rho \frac{DJ}{Dt} + J \frac{D\rho}{Dt} = 0$$

On using the Euler's formula (8.2.6), this becomes

$$J\left(\frac{D\rho}{Dt} + \rho \,\text{div}\, \mathbf{v}\right) = 0$$

This leads to equation (8.2.3) since $J \neq 0$. ∎

EXAMPLE 8.2.2 For the motion of a continuum given by the equations

$$x_1 = x_1^0 + \alpha t x_3^0$$
$$x_2 = x_2^0 + \alpha t x_3^0$$
$$x_3 = x_3^0 - \alpha t (x_1^0 + x_2^0)$$

find the density in the current configuration in terms of the density in the reference configuration.

Solution For the given motion, we find that $J = 1 + 2\alpha^2 t^2$. The equation of continuity in the Lagrangian form, namely, (8.2.8), gives the current density as

$$\rho = \frac{1}{J}\rho_0 = \frac{\rho_0}{1 + 2\alpha^2 t^2} \quad \blacksquare$$

EXAMPLE 8.2.3 For a certain flow of a continuum the velocity field is given by $v_i = x_i/(1 + t)$. Show that the density at time t is $\rho = \rho_0/(1 + t)^3$. Deduce that $\rho x_1 x_2 x_3 = \rho_0 x_1^0 x_2^0 x_3^0$.

Solution For the given velocity field, we find that

$$\text{div } \mathbf{v} = v_{k,k} = \frac{3}{1 + t}$$

Hence the equation of continuity (8.2.4a) yields

$$\frac{D}{Dt}(\log \rho) = -\frac{3}{1 + t}$$

Integration of this equation gives

$$\log \rho = -3 \log(1 + t) + A \tag{8.2.13}$$

where A is, in general, a function of x_i^0.

For $t = 0$, we have $\rho = \rho_0$. Hence (8.2.13) gives $A = \log \rho_0$. Putting this into (8.2.13), we obtain

$$\rho = \frac{\rho_0}{(1 + t)^3} \tag{8.2.14}$$

The given velocity field can be rewritten as

$$\frac{Dx_i}{Dt} = \frac{x_i}{1 + t}$$

Solving this differential equation, we obtain

$$x_i = x_i^0 (1 + t) \tag{8.2.15}$$

Expressions (8.2.14) and (8.2.15) yield $\rho x_1 x_2 x_3 = \rho_0 x_1^0 x_2^0 x_3^0$. $\blacksquare$

EXAMPLE 8.2.4 (i) Show that in a moving continuum, the density, the velocity and the vorticity are connected through the equation:

$$\frac{D}{Dt}\left(\frac{\mathbf{w}}{\rho}\right) = \left(\frac{\mathbf{w}}{\rho} \cdot \nabla\right)\mathbf{v} + \frac{1}{\rho}\text{curl}\frac{D\mathbf{v}}{Dt} \tag{8.2.16}$$

This is known as *Beltrami's vorticity equation*.

(ii) Show that Cauchy's vorticity equation (6.6.8) can be expressed in the form:

$$\frac{\mathbf{w}}{\rho} = \mathbf{F}\frac{\mathbf{w}^0}{\rho_0} \tag{8.2.17}$$

Solution (i) In Example 6.3.2, we have obtained the identity:

$$\operatorname{curl}\frac{D\mathbf{v}}{Dt} = \frac{D\mathbf{w}}{Dt} + (\operatorname{div}\mathbf{v})\mathbf{w} - (\mathbf{w}\cdot\nabla)\mathbf{v} \tag{8.2.18}$$

Now,

$$\frac{D}{Dt}\left(\frac{\mathbf{w}}{\rho}\right) = \frac{1}{\rho}\frac{D\mathbf{w}}{Dt} - \frac{1}{\rho^2}\frac{D\rho}{Dt}\mathbf{w} = \frac{1}{\rho}\left[\frac{D\mathbf{w}}{Dt} + (\operatorname{div}\mathbf{v})\mathbf{w}\right] \tag{8.2.19}$$

by the equation of continuity (8.2.3).

Substitution for $(D\mathbf{w}/Dt) + (\operatorname{div}\mathbf{v})\mathbf{w}$ from (8.2.19) in (8.2.18) yields equation (8.2.16). For an incompressible material, equation (8.2.16) becomes

$$\frac{D\mathbf{w}}{Dt} = (\mathbf{w}\cdot\nabla)\mathbf{v} + \operatorname{curl}\frac{D\mathbf{v}}{Dt} \tag{8.2.20}$$

(ii) Substituting for J from (8.2.8) in Cauchy's vorticity equation (6.6.8) we obtain (8.2.17). ∎

8.3
BALANCE OF LINEAR MOMENTUM

For a material body $\mathcal{B}$ occupying a configuration B of volume V, at time t, we define the *linear momentum* $\mathbf{p}$ by

$$\mathbf{p} = \int_V \rho\mathbf{v}\,dV \tag{8.3.1}$$

As stated in Section 7.1, we consider two types of forces acting on $\mathcal{B}$: body forces and surface forces. The result of these forces in the configuration B is given by (7.2.5), namely

$$\mathbf{f}^{(r)} = \int_V \rho\mathbf{b}\,dV + \int_S \mathbf{s}\,dS \tag{8.3.2}$$

where $\mathbf{b}$ is the force per unit mass and $\mathbf{s}$ is the surface force per unit area of S, where S is the boundary surface enclosing V.

We postulate that the material time rate of change of $\mathbf{p}$ is equal to $\mathbf{f}^{(r)}$. This postulate, consistent with the Newton's second law of motion, is

332   8 FUNDAMENTAL LAWS OF CONTINUUM MECHANICS

known as the *law (principle) of balance of linear momentum* for a continuum. As a mathematical expression, this law reads

$$\frac{D\mathbf{p}}{Dt} = \mathbf{f}^{(r)} \tag{8.3.3}$$

Using (8.3.1) and (8.3.2), equation (8.3.3) becomes

$$\frac{D}{Dt}\int_V \rho \mathbf{v}\, dV = \int_V \rho \mathbf{b}\, dV + \int_S \mathbf{s}\, dS \tag{8.3.4}$$

By virtue of (8.2.10), we have

$$\frac{D}{Dt}\int_V \rho \mathbf{v}\, dV = \int_V \rho \frac{D\mathbf{v}}{Dt}\, dV \tag{8.3.5}$$

Cauchy's law connecting the stress vector and the stress tensor, namely, (7.4.8), and the divergence theorem given by (3.7.1) yield

$$\int_S \mathbf{s}\, dS = \int_S \mathbf{T}^T \mathbf{n}\, dS = \int_V \operatorname{div} \mathbf{T}^T\, dV \tag{8.3.6}$$

Substitution of (8.3.5) and (8.3.6) into (8.3.4) gives

$$\int_V \left\{\rho \frac{D\mathbf{v}}{Dt} - \rho\mathbf{b} - \operatorname{div} \mathbf{T}^T\right\} dV = 0$$

Since V is an arbitrary volume, the integrand must vanish identically (by the localization theorem). Thus

$$\operatorname{div} \mathbf{T}^T + \rho \mathbf{b} = \rho \frac{D\mathbf{v}}{Dt} \tag{8.3.7}$$

at every point of V and for all t.

Equation (8.3.7) expresses the law of balance of linear momentum in terms of Cauchy's stress tensor. This equation, first obtained by Cauchy in 1827, is known as *Cauchy's equation of motion*.

In components, (8.3.7) reads

$$\tau_{ji,j} + \rho b_i = \rho \frac{Dv_i}{Dt} \tag{8.3.7}'$$

For a continuum at rest (static equilibrium) or in motion with constant velocity, we have $\dot{\mathbf{v}} = \mathbf{0}$; then equation (8.3.7) becomes

$$\operatorname{div} \mathbf{T}^T + \rho \mathbf{b} = \mathbf{0} \tag{8.3.8}$$

8.3 BALANCE OF LINEAR MOMENTUM

This equation is referred to as *Cauchy's equation of equilibrium*. In components, the equation reads

$$\tau_{ji,j} + \rho b_i = 0 \qquad (8.3.8)'$$

Recall that Cauchy's stress tensor **T** has been defined in the current configuration. As such, the equation of motion (8.3.7) and the equation of equilibrium (8.3.8) are in the *spatial form*. Next we deduce the corresponding equations in the material form.

8.3.1 EQUATION OF MOTION IN MATERIAL FORM

In view of relations (5.2.22), (7.9.5) and (8.2.8), expression (8.3.4) may be rewritten, with reference to the initial configuration, as

$$\frac{D}{Dt} \int_{V_0} \rho_0 \mathbf{v} \, dV_0 = \int_{V_0} \rho_0 \mathbf{b} \, dV_0 + \int_{S_0} (\mathbf{T}^0)^T \mathbf{n}^0 \, dS_0 \qquad (8.3.9)$$

Employing the divergence theorem given by (3.7.1) to the surface integral in (8.3.9) and noting that V (and therefore V_0) is an arbitrary volume, we obtain the following equation of motion expressed in terms of the first Piola–Kirchhoff stress tensor, in the material form:

$$\text{div}^0 (\mathbf{T}^0)^T + \rho_0 \mathbf{b} = \rho_0 \frac{D\mathbf{v}}{Dt} \qquad (8.3.10)$$

We can rewrite equation (8.3.10) in terms of the second Piola–Kirchhoff stress tensor **S**. From expression (7.9.8), we get $\mathbf{T}^0 = \mathbf{S}\mathbf{F}^T$. Substituting this into (8.3.10), we obtain

$$\text{div}^0 (\mathbf{F}\mathbf{S}^T) + \rho_0 \mathbf{b} = \rho_0 \frac{D\mathbf{v}}{Dt} \qquad (8.3.11)$$

From expression (5.4.4), we have $\mathbf{F} = \mathbf{I} + \nabla^0 \mathbf{u}$. Using this and expression (6.2.5)', equation (8.3.11) can be expressed as

$$\text{div}^0 \{(\mathbf{I} + \nabla^0 \mathbf{u})\mathbf{S}^T\} + \rho_0 \mathbf{b} = \rho_0 \frac{D^2 \mathbf{u}}{Dt^2} \qquad (8.3.12)$$

Thus, (8.3.11) and (8.3.12) serve as alternative versions of the equation of motion (8.3.10). In nonlinear elasticity, equations (8.3.10) to (8.3.12) are more convenient than Cauchy's equation (8.3.7). Equations (8.3.10) and (8.3.11) are by Piola (1833).

334　8 FUNDAMENTAL LAWS OF CONTINUUM MECHANICS

In components, equations (8.3.10) to (8.3.12) read, respectively, as

$$\tau^0_{ji;j} + \rho_0 b_i = \rho_0 \frac{Dv_i}{Dt} \qquad (8.3.10)'$$

$$\{x_{i;k} s_{jk}\}_{;j} + \rho_0 b_i = \rho_0 \frac{Dv_i}{Dt} \qquad (8.3.11)'$$

$$\{(\delta_{ik} + u_{i;k}) s_{jk}\}_{;j} + \rho_0 b_i = \rho_0 \frac{D^2 u_i}{Dt^2} \qquad (8.3.12)'$$

The equilibrium counterparts of equations (8.3.10) to (8.3.12) and their corresponding component forms can be obtained by setting $D\mathbf{v}/Dt$ (or Dv_i/Dt) equal to 0 in these equations.

EXAMPLE 8.3.1 The stress components in a continuum in equilibrium are given by

$$\tau_{11} = x_1^2, \qquad \tau_{22} = x_2^2, \qquad \tau_{33} = x_1^2 + x_2^2$$

$$\tau_{12} = \tau_{21} = 2x_1 x_2, \qquad \tau_{23} = \tau_{32} = \tau_{31} = \tau_{13} = 0$$

Find the body force that must be acting on the continuum.

Solution Since the continuum is in equilibrium, the given stress components must obey Cauchy's equations of equilibrium (8.3.8)'. When written in the expanded form, these equations read

$$\left.\begin{array}{l} \dfrac{\partial \tau_{11}}{\partial x_1} + \dfrac{\partial \tau_{21}}{\partial x_2} + \dfrac{\partial \tau_{31}}{\partial x_3} + \rho b_1 = 0 \\[6pt] \dfrac{\partial \tau_{12}}{\partial x_1} + \dfrac{\partial \tau_{22}}{\partial x_2} + \dfrac{\partial \tau_{32}}{\partial x_3} + \rho b_2 = 0 \\[6pt] \dfrac{\partial \tau_{13}}{\partial x_1} + \dfrac{\partial \tau_{23}}{\partial x_2} + \dfrac{\partial \tau_{33}}{\partial x_3} + \rho b_3 = 0 \end{array}\right\} \qquad (8.3.13)$$

Substituting the given expressions for the stress components in these equations, we have

$$4x_1 + \rho b_1 = 0, \qquad 4x_2 + \rho b_2 = 0, \qquad b_3 = 0$$

Thus, the body force that must act on the material is given by

$$\rho \mathbf{b} = -4(x_1 \mathbf{e}_1 + x_2 \mathbf{e}_2) \qquad \blacksquare$$

EXAMPLE 8.3.2 At a point of a continuum in equilibrium under zero body force, the stress matrix is given as follows:

$$[\tau_{ij}] = \begin{bmatrix} x_1 + x_2 & f(x_1, x_2) & 0 \\ f(x_1, x_2) & x_1 - 2x_2 & 0 \\ 0 & 0 & x_2 \end{bmatrix}$$

Find $f(x_1, x_2)$, given that on the plane $x_1 = 1$, the stress vector is

$$\mathbf{s} = (1 + x_2)\mathbf{e}_1 - (x_2 - 5)\mathbf{e}_2$$

Solution Since the continuum is in equilibrium under zero body force, Cauchy's equations of equilibrium (8.3.13) are to be satisfied with $b_i = 0$. Substituting for the stress components from the given stress matrix in these equations, we find that the last equation is identically satisfied and that the first two equations yield

$$\frac{\partial f}{\partial x_2} = -1, \qquad \frac{\partial f}{\partial x_1} = 2$$

Since f depends only on x_1 and x_2, these equations show that f must be of the form

$$f(x_1, x_2) = 2x_1 - x_2 + A \qquad (8.3.14)$$

where A is a constant.

For the plane $x_1 = $ constant, $\mathbf{n}$ is given by $n_1 = 1$, $n_2 = n_3 = 0$, and Cauchy's law (7.4.6) gives

$$s_i = \tau_{ji} n_j = \tau_{1i}$$

For the given stress matrix, this yields

$$s_2 = \tau_{12} = f(x_1, x_2)$$

But, it is given that

$$s_2 = -(x_2 - 5) \quad \text{on} \quad x_1 = 1$$

Hence

$$f(x_1, x_2) = -(x_2 - 5) \quad \text{on} \quad x_1 = 1 \qquad (8.3.15)$$

But, from (8.3.14) we get

$$f(x_1, x_2) = A - x_2 + 2, \quad \text{for} \quad x_1 = 1 \qquad (8.3.16)$$

Relations (8.3.15) and (8.3.16) give $A = 3$. Putting this value of A back into (8.3.14), we get

$$f(x_1, x_2) = 2x_1 - x_2 + 3 \quad \blacksquare$$

EXAMPLE 8.3.3 The stress tensor in a continuum in equilibrium under zero body force is given by $\mathbf{T} = T(\mathbf{y} \otimes \mathbf{y})$, where $T = T(\mathbf{x})$ and $\mathbf{y}$ is the unit vector along $\mathbf{x}$. Show that

$$\mathbf{x} \cdot \nabla T + 2T = 0 \qquad (8.3.17)$$

Find T in the case when $T = T(r)$, where $r^2 = \mathbf{x} \cdot \mathbf{x}$.

Solution The components of the given stress tensor are $\tau_{ij} = T y_i y_j$. Since $\mathbf{y}$ is the unit vector along $\mathbf{x}$, we have $y_i = x_i/r$. Hence $\tau_{ij} = (T x_i x_j)/r^2$, so that

$$\tau_{ij,j} = T_{,j} \frac{x_i x_j}{r^2} + 2T \frac{x_i}{r^2} \qquad (8.3.18)$$

The continuum is in equilibrium under zero body force, so we have $\tau_{ji,j} = 0$, by Cauchy's equations (8.3.8)'. We note that the given $\mathbf{T}$ is symmetric. Hence it follows from (8.3.18) that

$$T_{,j} x_j + 2T = 0$$

which is the required result (8.3.17).

If $T = T(r)$, then $\nabla T = (dT/dr)(\mathbf{x}/r)$. Hence (8.3.17) becomes

$$\frac{dT}{dr} + \frac{2}{r} T = 0$$

Solving this differential equation, we find that $T = A/r^2$, where A is an arbitrary constant. ∎

8.4
BALANCE OF ANGULAR MOMENTUM

For a material body $\mathcal{B}$ occupying a configuration B of volume V, at time t, the angular momentum about the origin is defined by

$$\mathbf{h}_0 = \int_V \{\mathbf{x} \times (\rho \mathbf{v})\} \, dV \qquad (8.4.1)$$

Also, the resultant moment, about the origin, of the external forces acting on the body $\mathcal{B}$ in the configuration B is defined by

$$\mathbf{g}_0 = \int_V \{\mathbf{x} \times (\rho \mathbf{b})\} \, dV + \int_S (\mathbf{x} \times \mathbf{s}) \, dS \qquad (8.4.2)$$

where S is as usual the surface enclosing V.

8.4 BALANCE OF ANGULAR MOMENTUM

We postulate that the material time rate of change of $\mathbf{h}_0$ is equal to $\mathbf{g}_0$. This postulate is known as the *law (principle) of balance of angular momentum* for a continuum. As a mathematical expression, this law reads

$$\frac{D\mathbf{h}_0}{Dt} = \mathbf{g}_0 \qquad (8.4.3)$$

On using expressions (8.4.1) and (8.4.2), equation (8.4.3) becomes

$$\frac{D}{Dt}\int_V \{\mathbf{x} \times (\rho\mathbf{v})\}\, dV = \int_V \{\mathbf{x} \times (\rho\mathbf{b})\}\, dV + \int_S (\mathbf{x} \times \mathbf{s})\, dS \qquad (8.4.4)$$

By using (8.2.10) and recalling that $D\mathbf{x}/Dt = \mathbf{v}$, we get

$$\frac{D}{Dt}\int_V (\mathbf{x} \times \rho\mathbf{v})\, dV = \frac{D}{Dt}\int_V \rho(\mathbf{x} \times \mathbf{v})\, dV = \int_V \rho\frac{D}{Dt}(\mathbf{x} \times \mathbf{v})\, dV$$

$$= \int_V \rho\mathbf{x} \times \frac{D\mathbf{v}}{Dt} \qquad (8.4.5)$$

By using Cauchy's law (7.4.8) and (3.7.18), we get

$$\int_S \mathbf{x} \times \mathbf{s}\, dS = \int_S \mathbf{x} \times \mathbf{T}^T\mathbf{n}\, dS$$

$$= \int_V \{2\boldsymbol{\xi} + (\mathbf{x} \times \operatorname{div} \mathbf{T}^T)\}\, dV \qquad (8.4.6)$$

where $\boldsymbol{\xi}$ is the dual vector of skw $\mathbf{T}^T$.

Substituting (8.4.5) and (8.4.6) in (8.4.4), we get

$$\int_V \left\{\mathbf{x} \times \left(\rho\frac{D\mathbf{v}}{Dt} - \rho\mathbf{b} - \operatorname{div}\mathbf{T}^T\right) - 2\boldsymbol{\xi}\right\} dV = \mathbf{0} \qquad (8.4.7)$$

Using Cauchy's equation (8.3.7) and the localization theorem, equation (8.4.7) yields $\boldsymbol{\xi} = \mathbf{0}$. Consequently, we have skw $\mathbf{T}^T = \mathbf{0}$, or

$$\mathbf{T} = \mathbf{T}^T \qquad (8.4.8)$$

at every point of the continuum and for all time.

Thus, the law of balance of angular momentum, as represented by equation (8.4.3), leads to the interesting and important conclusion that *the stress tensor is symmetric*. The symmetry relation (8.4.8) is also by Cauchy (1827).

As a direct consequence of the expression (8.4.8) we find from (7.9.10) that

$$\mathbf{F}\mathbf{T}^0 = (\mathbf{T}^0)^T\mathbf{F}^T; \qquad \mathbf{S} = \mathbf{S}^T \qquad (8.4.9)$$

338 8 FUNDAMENTAL LAWS OF CONTINUUM MECHANICS

In recent years, the principle of angular momentum represented by an equation more general than (8.4.3) has also been considered. In this generalized representation, the vector $\mathbf{g}_0$ includes the possible contributions of the moments of internal forces developed within the continuum. Then the stress tensor turns out to be nonsymmetric and one deals with what is known as a polar continuum not considered in this text; see remarks following the Cauchy's reciprocal relation (7.2.4).

8.5
GENERAL SOLUTIONS OF THE EQUATION OF EQUILIBRIUM

In many static problems, the body force per unit of volume, namely, $\rho\mathbf{b}$ (rather than the body force per unit mass, namely, $\mathbf{b}$) is taken as a known function. Then, the equilibrium equations (8.3.8)' contain τ_{ij} as the only unknown functions. These equations are three in number and, therefore, are generally *not* adequate to find all of τ_{ij}. However, the equations can be solved to determine τ_{ij} in terms of some auxiliary functions, known as *stress functions*. One such general solution is obtained next. Some other solutions that follow from this solution are also presented.

For ready reference, let us write down the equations (8.3.8) and (8.4.8):

$$\text{div } \mathbf{T}^T + \rho\mathbf{b} = \mathbf{0} \tag{8.5.1}$$

$$\mathbf{T} = \mathbf{T}^T \tag{8.5.2}$$

A general solution of these equations is provided by the following theorem.

THEOREM 8.5.1 Let $\mathbf{A}$ be an arbitrary symmetric tensor, and $\mathbf{h}$ an arbitrary vector obeying the equation

$$\nabla^2 \mathbf{h} = -\rho\mathbf{b} \tag{8.5.3}$$

Then
$$\mathbf{T} = \text{curl curl } \mathbf{A} + \nabla\mathbf{h} + \nabla\mathbf{h}^T - (\text{div } \mathbf{h})\mathbf{I} \tag{8.5.4}$$

is a solution of equations (8.5.1) and (8.5.2). Furthermore, this solution is complete in the sense that every solution of equations (8.5.1) and (8.5.2) admits a representation of the form (8.5.4).

Proof If $\mathbf{A}$ is symmetric, the identity (3.5.40) yields

$$(\text{curl curl } \mathbf{A})^T = \text{curl curl } \mathbf{A}$$

showing that curl curl $\mathbf{A}$ is a symmetric tensor. Also, $\mathbf{I}$ and $(\nabla\mathbf{h} + \nabla\mathbf{h}^T)$ are symmetric tensors. Consequently, $\mathbf{T}$ given by (8.5.4) is a symmetric tensor (being a sum of symmetric tensors). Thus, equation (8.5.2) is satisfied.

8.5 GENERAL SOLUTIONS OF THE EQUATION OF EQUILIBRIUM

Using identities (3.5.13), (3.5.14), (3.5.35), (3.5.37) and (3.5.38), we have

$$\left.\begin{array}{c} \text{div curl curl } \mathbf{A} = \text{curl div}(\text{curl } \mathbf{A})^T = \mathbf{0} \\ \text{div } \nabla \mathbf{h} = \nabla^2 \mathbf{h}, \quad \text{div } \nabla \mathbf{h}^T = \nabla(\text{div } \mathbf{h}), \quad \text{div}(\text{div } \mathbf{h})\mathbf{I} = \nabla(\text{div } \mathbf{h}) \end{array}\right\} \quad (8.5.5)$$

Using these relations, we find from expression (8.5.4) that

$$\text{div } \mathbf{T} = \nabla^2 \mathbf{h} \quad (8.5.6)$$

If **h** obeys equation (8.5.3), relation (8.5.6) becomes

$$\text{div } \mathbf{T} + \rho \mathbf{b} = \mathbf{0}$$

showing that equation (8.5.1) is satisfied.

Thus, if **A** is a symmetric tensor and **h** is a vector obeying equation (8.5.3), then **T** given by (8.5.4) is a solution of equations (8.5.1) and (8.5.2).

Next, we prove that the solution (8.5.4) is complete. Suppose **T** is a solution of equations (8.5.1) and (8.5.2). Then we have

$$\mathbf{T} = \mathbf{T}^T, \quad \text{div } \mathbf{T} + \rho \mathbf{b} = \mathbf{0} \quad (8.5.7\text{ab})$$

Let $\bar{\mathbf{T}}$ be a tensor such that

$$\nabla^2 \bar{\mathbf{T}} = -\mathbf{T} \quad (8.5.8)$$

An equation of the form (8.5.8) is known to have a solution; therefore, the existence of $\bar{\mathbf{T}}$ is assured. Equation (8.5.7a) implies that $\bar{\mathbf{T}}$ is symmetric.
Let

$$\mathbf{A} = \bar{\mathbf{T}} - (\text{tr } \bar{\mathbf{T}})\mathbf{I} \quad (8.5.9)$$

Then **A** is also symmetric. The identity (3.5.57) now implies that

$$\text{curl curl } \mathbf{A} = \nabla \text{div } \bar{\mathbf{T}} + (\nabla \text{div } \bar{\mathbf{T}})^T - \text{div}(\text{div } \bar{\mathbf{T}})\mathbf{I} - \nabla^2 \bar{\mathbf{T}} \quad (8.5.10)$$

where we have used the results

$$\text{tr } \mathbf{A} = -2 \text{ tr } \bar{\mathbf{T}}, \quad \text{div } \mathbf{A} = \text{div } \bar{\mathbf{T}} - \nabla(\text{tr } \bar{\mathbf{T}})$$

Setting $\mathbf{h} = -\text{div } \bar{\mathbf{T}}$ and using (8.5.8), the relation (8.5.10) can be rewritten as

$$\mathbf{T} = \text{curl curl } \mathbf{A} + (\nabla \mathbf{h} + \nabla \mathbf{h}^T) - (\text{div } \mathbf{h})\mathbf{I} \quad (8.5.11)$$

This expression is in the form (8.5.4). We have already noted that **A** is symmetric. Since **T** obeys equation (8.5.7b) we verify from (8.5.11), with the aid of the identities (8.5.5), that **h** obeys equation (8.5.3).

Thus, if **T** is any solution of equations (8.5.1) and (8.5.2), then **T** can be expressed in the form (8.5.4), where **A** is a symmetric tensor and **h** is a vector obeying equation (8.5.3). In other words, the solution of equations (8.5.1) and (8.5.2) given by (8.5.4) is complete. The theorem is proven. ■

8 FUNDAMENTAL LAWS OF CONTINUUM MECHANICS

In the absence of the body force ($\mathbf{b} = \mathbf{0}$), $\mathbf{h}$ turns out to be a *harmonic vector*. This case of the solution (8.5.4), first obtained by H. Schaefer in 1953, is known as the *Schaefer's solution*. If $\mathbf{h}$ is set equal to $\mathbf{0}$ (with $\mathbf{b} = \mathbf{0}$), the solution (8.5.4) becomes

$$\mathbf{T} = \operatorname{curl} \operatorname{curl} \mathbf{A} \qquad (8.5.12)$$

This is known as the *Beltrami's solution*, after Beltrami (1892).

In the suffix notation, the solution (8.5.4) reads:

$$\tau_{ij} = \varepsilon_{imn}\varepsilon_{jpq}a_{mp,nq} + h_{i,j} + h_{j,i} - h_{k,k}\delta_{ij} \qquad (8.5.13)$$

where

$$a_{ij} = a_{ji} \quad \text{and} \quad h_{i,kk} = -\rho b_i \qquad (8.5.14)$$

8.5.1 CASE OF CONSERVATIVE BODY FORCE

In many practical problems, the body force is conservative; that is, $\rho\mathbf{b} = -\nabla\chi$ for some scalar function χ. Then, using the identity $\nabla\chi = \operatorname{div}(\chi\mathbf{I})$, equation (8.5.1) can be written in the form

$$\operatorname{div} \mathbf{T}_0^T = \mathbf{0} \qquad (8.5.15)$$

where

$$\mathbf{T}_0 = \mathbf{T} - \chi\mathbf{I} \qquad (8.5.16)$$

Also, equation (8.5.2) can be rewritten as

$$\mathbf{T}_0 = \mathbf{T}_0^T \qquad (8.5.17)$$

By virtue of Theorem 8.5.1, it follows that a complete solution for equations (8.5.15) and (8.5.17) is given by

$$\mathbf{T}_0 = \operatorname{curl} \operatorname{curl} \mathbf{A} + \nabla\mathbf{h} + \nabla\mathbf{h}^T - (\operatorname{div} \mathbf{h})\mathbf{I}$$

so that

$$\mathbf{T} = \operatorname{curl} \operatorname{curl} \mathbf{A} + \nabla\mathbf{h} + \nabla\mathbf{h}^T + (\chi - \operatorname{div} \mathbf{h})\mathbf{I} \qquad (8.5.18)$$

where $\mathbf{A}$ is a symmetric tensor and $\mathbf{h}$ is a harmonic vector.

Equations (8.5.15) to (8.5.17) are together equivalent to the following three equations for τ_{ij}:

$$\tau_{ij,j} - \chi_{,i} = 0 \qquad (8.5.19)$$

Also, the solution (8.5.18) yields:

$$\tau_{ij} = \varepsilon_{imn}\varepsilon_{jpq}a_{mp,nq} + (h_{i,j} + h_{j,i}) + (\chi - h_{k,k})\delta_{ij} \qquad (8.5.20)$$

where

$$a_{ij} = a_{ji} \quad \text{and} \quad h_{i,kk} = 0 \qquad (8.5.21)$$

8.5 GENERAL SOLUTIONS OF THE EQUATION OF EQUILIBRIUM

The explicit expressions that follow from (8.5.20) and (8.5.21) are

$$\tau_{11} = \chi + a_{22,33} + a_{33,22} - 2a_{23,23} - h_{2,2} - h_{3,3} + h_{1,1}$$
$$\tau_{22} = \chi + a_{33,11} + a_{11,33} - 2a_{31,31} - h_{3,3} - h_{1,1} + h_{2,2}$$
$$\tau_{33} = \chi + a_{11,22} + a_{22,11} - 2a_{12,12} - h_{1,1} - h_{2,2} + h_{3,3}$$
$$\tau_{12} = \tau_{21} = a_{23,31} + a_{31,23} - a_{33,12} - a_{12,33} + (h_{1,2} + h_{2,1}) \quad (8.5.22)$$
$$\tau_{23} = \tau_{32} = a_{31,12} + a_{12,31} - a_{11,23} - a_{23,11} + (h_{2,3} + h_{3,2})$$
$$\tau_{31} = \tau_{13} = a_{12,23} + a_{23,12} - a_{22,31} - a_{31,22} + (h_{3,1} + h_{1,3})$$

These expressions constitute a general and complete solution of equations (8.5.19).

8.5.2 PARTICULAR CASES

Several special solutions of equations (8.5.19) can be deduced from the expressions (8.5.22) by making special choices of a_{ij} and h_i.

CASE i Suppose

$$[a_{ij}] = \begin{bmatrix} \phi & 0 & 0 \\ 0 & \psi & 0 \\ 0 & 0 & \xi \end{bmatrix} \quad \text{and} \quad h_i = 0$$

Then expressions (8.5.22) become

$$\tau_{11} = \chi + \psi_{,33} + \xi_{,22}$$
$$\tau_{22} = \chi + \xi_{,11} + \phi_{,33}$$
$$\tau_{33} = \chi + \phi_{,22} + \psi_{,11}$$
$$\tau_{21} = \tau_{12} = -\xi_{,12} \quad (8.5.23)$$
$$\tau_{32} = \tau_{23} = -\phi_{,23}$$
$$\tau_{13} = \tau_{31} = -\psi_{,31}$$

The solution of equations (8.5.19) as represented by (8.5.23) is known as the *Maxwell's solution*, after Maxwell (1868). The function ϕ, ψ and ξ are referred to as *Maxwell's stress functions*.

CASE ii Suppose

$$[a_{ij}] = \begin{bmatrix} 0 & \phi^* & \psi^* \\ \phi^* & 0 & \xi^* \\ \psi^* & \xi^* & 0 \end{bmatrix} \quad \text{and} \quad h_i = 0$$

8 FUNDAMENTAL LAWS OF CONTINUUM MECHANICS

Then expressions (8.5.22) become

$$\tau_{11} = \chi - 2\xi^*_{,23}$$
$$\tau_{22} = \chi - 2\psi^*_{,31}$$
$$\tau_{33} = \chi - 2\phi^*_{,12}$$
$$\tau_{12} = \tau_{21} = \xi^*_{,31} + \psi^*_{,23} - \phi^*_{,33} \qquad (8.5.24)$$
$$\tau_{23} = \tau_{32} = \psi^*_{,12} + \phi^*_{,31} - \xi^*_{,11}$$
$$\tau_{31} = \tau_{13} = \phi^*_{,23} + \xi^*_{,12} - \psi^*_{,22}$$

The solution of equations (8.5.19) as represented by (8.5.24) is known as the *Morera's solution*, after Morera (1892). The functions ϕ^*, ψ^*, ξ^* are referred to as *Morera's stress functions*.

CASE iii Suppose

$$[a_{ij}] = \begin{bmatrix} v\xi & 0 & 0 \\ 0 & v\xi & 0 \\ 0 & 0 & \xi \end{bmatrix} \quad \text{and} \quad h_i = 0$$

where v is a constant and $\xi = \xi(x_1, x_2)$.

Then expressions (8.5.22) become

$$\tau_{11} = \chi + \xi_{,22}$$
$$\tau_{22} = \chi + \xi_{,11}$$
$$\tau_{33} = \chi + v(\xi_{,22} + \xi_{,11}) \qquad (8.5.25)$$
$$\tau_{12} = \tau_{21} = -\xi_{,12}$$
$$\tau_{23} = \tau_{32} = \tau_{13} = \tau_{31} = 0$$

The solution of equations (8.5.19) as represented by (8.5.25) is known as *Airy's solution*, after Airy (1863). The function ξ is referred to as *Airy's stress function*. This solution plays a fundamental role in the study of what are called *plane strain problems* in the theory of elasticity.

CASE iv Suppose

$$[a_{ij}] = \begin{bmatrix} 0 & 0 & 0 \\ 0 & 0 & 0 \\ 0 & 0 & \xi \end{bmatrix} \quad \text{and} \quad h_i = 0$$

8.5 GENERAL SOLUTIONS OF THE EQUATION OF EQUILIBRIUM

Then expressions (8.5.22) become

$$\tau_{11} = \chi + \xi_{,22}$$
$$\tau_{22} = \chi + \xi_{,11} \qquad (8.5.26)$$
$$\tau_{12} = \tau_{21} = -\xi_{,12}$$
$$\tau_{13} = \tau_{31} = \tau_{23} = \tau_{32} = \tau_{13} = \tau_{33} = 0$$

The solution of equations (8.5.19) as represented by (8.5.26) is also known as *Airy's solution*. This solution plays a fundamental role in the study of what are called *plane stress problems* in the theory of elasticity.

Note that solutions (8.5.25) and (8.5.26) are just particular cases of Maxwell's solution (8.5.23).

CASE v Suppose

$$[a_{ij}] = \begin{bmatrix} 0 & 0 & 0 \\ 0 & 0 & \xi^* \\ 0 & \xi^* & 0 \end{bmatrix} \quad \text{and} \quad h_i = 0$$

where $\xi^* = \xi^*(x_1, x_2)$.

Then expressions (8.5.22) become

$$\tau_{11} = \tau_{22} = \tau_{33} = \tau_{12} = 0$$
$$\tau_{23} = -\mu\alpha\Omega_{,1} \qquad (8.5.27)$$
$$\tau_{31} = \mu\alpha\Omega_{,2}$$

where

$$\Omega = \frac{1}{\mu\alpha}\xi^*_{,1}, \qquad \mu\alpha \neq 0$$

The solution of equations (8.5.19) as represented by (8.5.27) is known as *Prandtl's solution*, after L. Prandtl (1903). The function Ω is known as *Prandtl's stress function*. This solution is particularly useful in solving what is known as the *torsion problem* in the elasticity theory.

Note that the solution (8.5.27) is just a particular case of Morera's solution (8.5.24).

8.6
BALANCE OF ENERGY

The *kinetic energy* K of a material body $\mathcal{B}$ occupying a configuration B of volume V, at time t, is defined by

$$K = \frac{1}{2} \int_V \rho(\mathbf{v} \cdot \mathbf{v}) \, dV \tag{8.6.1}$$

Also, the *rate of work* or *power* of external forces acting on $\mathcal{B}$ in B is defined by

$$P = \int_V \rho \mathbf{b} \cdot \mathbf{v} \, dV + \int_S \mathbf{s} \cdot \mathbf{v} \, dS \tag{8.6.2}$$

where S is as usual the surface enclosing V.

If the continuum is a heat conducting material and if there is a temperature difference between the interior of S and its exterior, then it is a physical phenomenon that heat flows across S. In fact, if the temperature is higher in the interior of S, heat flows out of V across S and if the temperature is lower in the interior of S, heat flows into V across S. We postulate that the "amount" Q of heat flowing out of V across S per unit of time is representable as

$$Q = \int_S \mathbf{q} \cdot \mathbf{n} \, dS \tag{8.6.3}$$

where $\mathbf{n}$ is as usual the unit outward normal to S, and $\mathbf{q}$ is known as the *heat flux vector* (see Figure 8.1). It follows that the amount of heat that flows into V across S per unit of time is given by $-Q$.

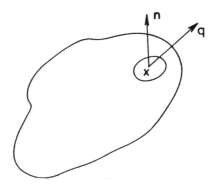

Figure 8.1. Heat flux.

8.6 BALANCE OF ENERGY

If any heat is generated within V because of the presence of heat sources, this heat contributes to the temperature difference across S. We postulate that the amount of heat H generated within V per unit of time is representable as

$$H = \int_V \rho h \, dV \tag{8.6.4}$$

where h is referred to as the *heat supply* or *strength of the (internal) heat source*. It is obvious that the net amount of heat contained in V per unit time is $H - Q$. We suppose that, apart from the heat flow, there is no other energy flow (like that due to chemical energy or electromagnetic energy) in the material.

We make a hypothesis that in addition to the kinetic energy the continuum possesses another type of energy called the *internal energy* and that the total energy of a body of the continuum is the sum of the kinetic energy and the internal energy. Like the concepts of time, mass, force, heat, etc., the concept of internal energy is also primitive. The internal energy E possessed by a body $\mathcal{B}$ in the configuration B is expressed as

$$E = \int_V \rho \varepsilon \, dV \tag{8.6.5}$$

where ε is known as the *internal energy per unit of mass* or the *specific internal energy*.

After having introduced the quantities K, P, Q, H and E, we now postulate a relation connecting these quantities as follows:

$$\frac{D}{Dt}(K + E) = P + (H - Q) \tag{8.6.6}$$

In words, this postulate states that the material time rate of change of the total energy of a body $\mathcal{B}$ is equal to the sum of the rate of work done by external forces acting on the volume and the boundary surface of $\mathcal{B}$ and the rate of the net amount of heat contained in the volume. This is known as the *law (principle) of balance of energy* or *the first law of thermodynamics*.

Substituting for K, E, P, Q and H from expressions (8.6.1), (8.6.5), (8.6.2), (8.6.3) and (8.6.4) in the equation (8.6.6), we obtain the following *integral form of the law of balance of energy*:

$$\frac{D}{Dt}\int_V \rho\left\{\frac{1}{2}(\mathbf{v} \cdot \mathbf{v}) + \varepsilon\right\} dV = \int_V \rho \mathbf{b} \cdot \mathbf{v} \, dV + \int_S \mathbf{s} \cdot \mathbf{v} \, dS + \int_V \rho h \, dV$$

$$- \int_S \mathbf{q} \cdot \mathbf{n} \, dS \tag{8.6.7}$$

By using (8.2.10), we note that

$$\frac{D}{Dt}\int_V \rho\left\{\frac{1}{2}(\mathbf{v}\cdot\mathbf{v})+\varepsilon\right\}dV = \int_V \rho\frac{D}{Dt}\left\{\frac{1}{2}(\mathbf{v}\cdot\mathbf{v})+\varepsilon\right\}dV$$

$$= \int_V \rho\left\{\mathbf{v}\cdot\frac{D\mathbf{v}}{Dt}+\frac{D\varepsilon}{Dt}\right\}dV \qquad (8.6.8)$$

By using (7.4.8), (2.8.14), (3.6.1) and (3.5.36), we get

$$\int_S \mathbf{s}\cdot\mathbf{v}\,dS = \int_S \mathbf{T}^T\mathbf{n}\cdot\mathbf{v}\,dS = \int_S \mathbf{n}\cdot\mathbf{Tv}\,dS$$

$$= \int_V \operatorname{div}(\mathbf{Tv})\,dV$$

$$= \int_V (\mathbf{v}\cdot\operatorname{div}\mathbf{T}^T + \mathbf{T}^T\cdot\nabla\mathbf{v})\,dV \qquad (8.6.9)$$

Also, by (3.6.1), we have

$$\int_S \mathbf{q}\cdot\mathbf{n}\,dS = \int_V \operatorname{div}\mathbf{q}\,dV \qquad (8.6.10)$$

With the aid of (8.6.8) to (8.6.10), expression (8.6.7) becomes

$$\int_V\left[\rho\frac{D\varepsilon}{Dt} + \mathbf{v}\cdot\left\{\frac{D\mathbf{v}}{Dt} - \operatorname{div}\mathbf{T}^T - \rho\mathbf{b}\right\} - \mathbf{T}^T\cdot\nabla\mathbf{v} + \operatorname{div}\mathbf{q} - \rho h\right]dV = 0 \qquad (8.6.11)$$

Using Cauchy's equation (8.3.7) and noting that V is an arbitrary volume, equation (8.6.11) yields

$$\rho\frac{D\varepsilon}{Dt} - \mathbf{T}^T\cdot\nabla\mathbf{v} + \operatorname{div}\mathbf{q} - \rho h = 0 \qquad (8.6.12)$$

Using the symmetry of $\mathbf{T}$ and noting then that $\mathbf{T}\cdot\nabla\mathbf{v} = \mathbf{T}\cdot\mathbf{D}$, equation (8.6.12) can be rewritten as

$$\rho\frac{D\varepsilon}{Dt} = \mathbf{T}\cdot\mathbf{D} - \operatorname{div}\mathbf{q} + \rho h \qquad (8.6.13)$$

This equation represents the law of balance of energy. This is a first-order partial differential equation that should hold at every point of the continuum and for all time. The equation is by Kirchhoff (1894) and usually referred to as the *energy equation*. The term $\mathbf{T}\cdot\nabla\mathbf{v} = \mathbf{T}\cdot\mathbf{D}$ is interpreted as the *rate of working of the stress*, called the *stress power*. Equation (8.6.13) is in the *spatial form*.

8.6.1 EQUATION OF ENERGY IN MATERIAL FORM

We recall the law of balance of energy given by equation (8.6.7) and rewrite it in the initial configuration. The resulting equation is

$$\frac{D}{Dt}\int_{V_0} \rho_0 \left\{\frac{1}{2}(\mathbf{v}\cdot\mathbf{v}) + \varepsilon\right\} dV_0 = \int_{V_0} \rho_0 \mathbf{b}\cdot\mathbf{v}\, dV_0 + \int_{S_0} \mathbf{v}\cdot\mathbf{T}^T\{J(\mathbf{F}^T)^{-1}\mathbf{n}^0\}\, dS_0$$
$$+ \int_{V_0} \rho_0 h\, dV - \int_{S_0} \mathbf{q}\cdot J(\mathbf{F}^T)^{-1}\mathbf{n}^0\, dS_0 \quad (8.6.14)$$

Here we have used (5.2.12), (5.2.22), (7.9.1) and (7.9.3) and the equation of continuity (8.2.8).

From expression (7.9.4), we note that $\mathbf{T}^T J(\mathbf{F}^T)^{-1} = (\mathbf{T}^0)^T$. Hence

$$\int_{S_0} \mathbf{v}\cdot\mathbf{T}^T\{J(\mathbf{F}^T)^{-1}\mathbf{n}^0\mathbf{v}\}\, dS_0 = \int_{S_0} (\mathbf{T}^0)^T \mathbf{n}^0 \cdot \mathbf{v}\, dS_0 = \int_{S_0} \mathbf{n}^0 \cdot (\mathbf{T}^0 \mathbf{v})\, dS_0$$
$$= \int_{V_0} \mathrm{div}^0(\mathbf{T}^0 \mathbf{v})\, dV_0$$
$$= \int_{V_0} \{\mathbf{v}\cdot\mathrm{div}^0(\mathbf{T}^0)^T + (\mathbf{T}^0)^T \cdot (\nabla^0 \mathbf{v})\}\, dV_0 \quad (8.6.15)$$

Here we have used the divergence theorem (3.7.1) and (3.5.36).

Setting

$$\mathbf{q}^0 = J\mathbf{F}^{-1}\mathbf{q} \quad (8.6.16)$$

the last of the integrals in (8.6.14) can be rewritten as

$$\int_{S_0} \mathbf{q}\cdot J(\mathbf{F}^T)^{-1}\mathbf{n}^0\, dS_0 = \int_{S_0} \mathbf{q}^0 \cdot \mathbf{n}^0\, dS_0 \quad (8.6.17)$$

Evidently, $\mathbf{q}^0$ represents heat flux measured per unit area in the *initial* configuration.

By the use of the divergence theorem, expression (8.6.17) becomes

$$\int_{S_0} \mathbf{q}\cdot J(\mathbf{F}^T)^{-1}\mathbf{n}^0\, dS_0 = \int_{V_0} \mathrm{div}^0 \mathbf{q}^0\, dV_0 \quad (8.6.18)$$

With the aid of expressions (8.6.15) and (8.6.18), expression (8.6.14) can be rewritten as

$$\int_{V_0} \left[\rho_0 \frac{D\varepsilon}{Dt} + \mathbf{v}\cdot\left\{\rho_0 \frac{D\mathbf{v}}{Dt} - \mathrm{div}^0(\mathbf{T}^0)^T - \rho_0 \mathbf{b}\right\}\right.$$
$$\left. - (\mathbf{T}^0)^T \cdot (\nabla^0 \mathbf{v}) + \mathrm{div}^0 \mathbf{q}^0 - \rho_0 h\right] dV_0 = 0 \quad (8.6.19)$$

Using the equation of motion (8.3.10) and noting that V (and therefore V_0) is an arbitrary volume, equation (8.6.19) yields

$$\rho_0 \frac{D\varepsilon}{Dt} = (\mathbf{T}^0)^T \cdot (\nabla^0 \mathbf{v}) - \text{div}^0 \mathbf{q}^0 + \rho_0 h \tag{8.6.20}$$

This is the equation of energy in the *material form*.

EXAMPLE 8.6.1 Prove the following formula for kinetic energy:

$$K = \int_V [\rho\{\mathbf{x} \cdot \mathbf{w} \times \mathbf{v} + (\mathbf{x} \cdot \mathbf{v}) \, \text{div } \mathbf{v}\}$$

$$+ \frac{1}{2}\{(\mathbf{x} \times \mathbf{v}) \cdot (\mathbf{v} \times \nabla\rho) + (\mathbf{x} \cdot \mathbf{v})(\mathbf{v} \cdot \nabla\rho)\}] \, dV$$

$$- \frac{1}{2}\int_S \rho[(\mathbf{x} \times \mathbf{v}) \cdot (\mathbf{v} \times \mathbf{n}) + (\mathbf{x} \cdot \mathbf{v})(\mathbf{v} \cdot \mathbf{n})] \, dS \tag{8.6.21}$$

Deduce that for a material of constant density,

$$K = \rho\left[\int_V (\mathbf{x} \cdot \mathbf{w} \times \mathbf{v}) \, dV + \int_S \left\{\frac{1}{2}v^2\mathbf{x} - (\mathbf{x} \cdot \mathbf{v})\mathbf{v}\right\} \cdot \mathbf{n} \, dS\right] \tag{8.6.22}$$

This relation is known as the *Lamb–Thomson formula*.

Solution By using identities (3.4.4), (3.4.14), (3.4.22) and (3.4.26), we find that

$$\text{div}(\rho v^2 \mathbf{x}) = 3\rho v^2 + \mathbf{x} \cdot (\rho \nabla v^2 + v^2 \nabla \rho) \tag{8.6.23}$$

$$\text{div}\{(\mathbf{x} \cdot \mathbf{v})(\rho\mathbf{v})\} = (\mathbf{x} \cdot \mathbf{v})(\rho \, \text{div } \mathbf{v} + \nabla\rho \cdot \mathbf{v}) + \rho v^2 + \rho\mathbf{x} \cdot \mathbf{w} \times \mathbf{v}$$

$$+ \tfrac{1}{2}\rho\mathbf{x} \cdot \nabla v^2 \tag{8.6.24}$$

These relations yield

$$\text{div}[\tfrac{1}{2}(\rho v^2 \mathbf{x}) - (\mathbf{x} \cdot \mathbf{v})(\rho\mathbf{v})] = \tfrac{1}{2}\rho v^2 + \tfrac{1}{2}v^2\mathbf{x} \cdot \nabla\rho - (\mathbf{x} \cdot \mathbf{v})(\rho \, \text{div } \mathbf{v} + \mathbf{v} \cdot \nabla\rho)$$

$$- \rho\mathbf{x} \cdot (\mathbf{w} \times \mathbf{v}) \tag{8.6.25}$$

Integrating both sides over the volume V and using the expression (8.6.1) and the divergence theorem, we get

$$K = \int_S \rho\left[\frac{1}{2}v^2(\mathbf{x} \cdot \mathbf{n}) - (\mathbf{x} \cdot \mathbf{v})(\mathbf{v} \cdot \mathbf{n})\right] dS$$

$$+ \int_V \left[(\mathbf{x} \cdot \mathbf{v})(\mathbf{v} \cdot \nabla\rho) - \frac{1}{2}v^2(\mathbf{x} \cdot \nabla\rho) + \rho\{(\mathbf{x} \cdot \mathbf{v}) \, \text{div } \mathbf{v} + \mathbf{x} \cdot \mathbf{w} \times \mathbf{v}\}\right] dV$$

$$\tag{8.6.26}$$

By use of the vector identity (1.7.51), expression (8.6.26) yields (8.6.21).

8.6 BALANCE OF ENERGY

If ρ is constant, we have div $\mathbf{v} = 0$, by the equation of continuity. Expression (8.6.26) then becomes (8.6.22). ■

EXAMPLE 8.6.2 Prove the following:

(i) $$\frac{DK}{Dt} = \int_V (\rho \mathbf{b} \cdot \mathbf{v} - \mathbf{T} \cdot \mathbf{D}) \, dV + \int_S \mathbf{s} \cdot \mathbf{v} \, dS \qquad (8.6.27)$$

(ii) $$\frac{DE}{Dt} = \int_V (\rho h + \mathbf{T} \cdot \mathbf{D}) \, dV - \int_S \mathbf{q} \cdot \mathbf{n} \, dS \qquad (8.6.28)$$

Equation (8.6.27) is known as the *equation of balance of mechanical energy*, and (8.6.28) is known as the *equation of balance of thermal energy*.

Solution (i) Using (8.2.10), we find from equation (8.6.1) that

$$\frac{DK}{Dt} = \int_V \rho \frac{D}{Dt}\left(\frac{1}{2} \mathbf{v} \cdot \mathbf{v}\right) dV = \int_V \rho \mathbf{v} \cdot \frac{D\mathbf{v}}{Dt} \, dV \qquad (8.6.29)$$

With the aid of (8.3.7), (8.6.9) and the symmetry of $\mathbf{T}$, (8.6.29) becomes

$$\frac{DK}{Dt} = \int_V (\text{div } \mathbf{T} + \rho \mathbf{b}) \cdot \mathbf{v} \, dV$$

$$= \int_S \mathbf{s} \cdot \mathbf{v} \, dS + \int_V (\rho \mathbf{b} \cdot \mathbf{v} - \mathbf{T} \cdot \mathbf{D}) \, dV$$

This is the equation (8.6.27).

(ii) In view of (8.6.1) and (8.6.5), the equation of energy (8.6.7) can be rewritten as

$$\frac{D}{Dt}(K + E) = \int_V \rho(\mathbf{b} \cdot \mathbf{v} + h) \, dV + \int_S (\mathbf{s} \cdot \mathbf{v} - \mathbf{q} \cdot \mathbf{n}) \, dS \qquad (8.6.30)$$

Subtracting equation (8.6.27) from equation (8.6.30) we get equation (8.6.28). ■

Note: Equation (8.6.28) can also be deduced from the energy equation (8.6.13) and vice versa. Indeed, equation (8.6.13) is just the local form of equation (8.6.28).

Because of the presence of the common term $\int_V \mathbf{T} \cdot \mathbf{D} \, dV$, equations (8.6.27) and (8.6.28) are interlinked. This means that, in general, the motion and the thermal state of a material influence one another.

EXAMPLE 8.6.3 In the absence of body force and a heat supply, show that the material rate of total energy in a material body contained in a

350   **8 FUNDAMENTAL LAWS OF CONTINUUM MECHANICS**

volume V is equal to the flux of the vector $(\mathbf{Tv} - \mathbf{q})$ across the surface S enclosing V.

Solution   When $\mathbf{b} = \mathbf{0}$ and $h = 0$, the energy equation (8.6.7) becomes

$$\frac{D}{Dt}(K + E) = \int_S (\mathbf{s} \cdot \mathbf{v} - \mathbf{q} \cdot \mathbf{n})\, dS = \int_S (\mathbf{Tv} - \mathbf{q}) \cdot \mathbf{n}\, dS \quad (8.6.31)$$

because $\mathbf{s} \cdot \mathbf{v} = \mathbf{T}^T \mathbf{n} \cdot \mathbf{v} = \mathbf{Tv} \cdot \mathbf{n}$. This proves the result. ∎

Note: The vector $(\mathbf{Tv} - \mathbf{q})$ is referred to as the *energy flux vector*.

EXAMPLE 8.6.4   Show that

$$\mathbf{T} \cdot \mathbf{D} = \frac{\rho}{\rho_0}(\mathbf{T}^0)^T \cdot \frac{D\mathbf{F}}{Dt} = \frac{\rho}{\rho_0} \mathbf{S} \cdot \frac{D\mathbf{G}}{Dt} \quad (8.6.32)$$

Solution   We have

$$\mathbf{T} \cdot \mathbf{D} = \operatorname{tr}\{\mathbf{T}(\nabla \mathbf{v})\} = \operatorname{tr}\{(\nabla \mathbf{v})\mathbf{T}\}$$

$$= \operatorname{tr}\left\{(\nabla \mathbf{v})\frac{1}{J}\mathbf{F}\mathbf{T}^0\right\}$$

using (7.9.4);

$$= \frac{1}{J}\operatorname{tr}\left(\frac{D\mathbf{F}}{Dt}\mathbf{T}^0\right)$$

by (6.2.44);

$$= \frac{\rho}{\rho_0}(\mathbf{T}^0)^T \cdot \frac{D\mathbf{F}}{Dt} \quad (8.6.33)$$

Also,

$$\operatorname{tr}\left(\frac{D\mathbf{F}}{Dt}\mathbf{T}^0\right) = \operatorname{tr}\left(\mathbf{T}^0\frac{D\mathbf{F}}{Dt}\right) = \operatorname{tr}\left(\mathbf{S}\mathbf{F}^T\frac{D\mathbf{F}}{Dt}\right)$$

using (7.9.8);

$$= \mathbf{S} \cdot \left(\mathbf{F}^T \frac{D\mathbf{F}}{Dt}\right)$$

since $\mathbf{S} = \mathbf{S}^T$;

$$= \mathbf{S} \cdot \operatorname{sym}\left(\mathbf{F}^T \frac{D\mathbf{F}}{Dt}\right) + \frac{1}{2}\mathbf{S} \cdot \frac{D}{Dt}(\mathbf{F}^T\mathbf{F}) \quad (8.6.34)$$

Using (5.4.3) and (5.4.8), this becomes

$$\operatorname{tr}\left(\frac{D\mathbf{F}}{Dt}\mathbf{T}^0\right) = \frac{1}{2}\mathbf{S} \cdot \frac{D\mathbf{C}}{Dt} = \mathbf{S} \cdot \frac{D\mathbf{G}}{Dt} \quad (8.6.35)$$

Expressions (8.6.33) and (8.6.35) together yield (8.6.32). ∎

8.7 ENTROPY INEQUALITY 351

EXAMPLE 8.6.5 Show that equation (8.6.20) can be put in the following alternative forms:

$$\rho_0 \frac{D\varepsilon}{Dt} = \text{tr}\left\{\mathbf{T}^0 \frac{D\mathbf{F}}{Dt}\right\} - \text{div}^0 \mathbf{q}^0 + \rho_0 h \qquad (8.6.36)$$

$$\rho_0 \frac{D\varepsilon}{Dt} = \mathbf{S} \cdot \frac{D\mathbf{G}}{Dt} - \text{div}^0 \mathbf{q}^0 + \rho_0 h \qquad (8.6.37)$$

Solution We first note that

$$(\mathbf{T}^0)^T \cdot (\nabla^0 \mathbf{v}) = (\mathbf{T}^0)^T \cdot \frac{D}{Dt}(\nabla^0 \mathbf{x}) = (\mathbf{T}^0)^T \cdot \frac{D\mathbf{F}}{Dt} = \text{tr}\left\{\mathbf{T}^0 \left(\frac{D\mathbf{F}}{Dt}\right)\right\}$$

Using (8.6.35), this becomes

$$(\mathbf{T}^0)^T \cdot (\nabla^0 \mathbf{v}) = \frac{1}{2} \mathbf{S} \cdot \frac{D\mathbf{C}}{Dt} = \mathbf{S} \cdot \frac{D\mathbf{G}}{Dt} \qquad (8.6.38)$$

Using (8.6.38) in equation (8.6.20), we obtain equations (8.6.36) and (8.6.37). ∎

8.7
ENTROPY INEQUALITY

In the preceding section, we postulated the law of balance of energy on the assumption that a material body possesses an internal energy. In this section we postulate what is called an *entropy inequality* on the assumption that a material body possesses *entropy*, a primitive quantity that increases or decreases accordingly as heat is supplied or withdrawn from the body. We suppose that the entropy $\tilde{H}$ of a body $\mathcal{B}$ is expressible as

$$\tilde{H} = \int_V \rho \eta \, dV \qquad (8.7.1)$$

where η is known as the *entropy per unit of mass* or the *specific entropy*.

Since the entropy is associated with the heat content of a body, it is directly related to the temperature, which is a measure of the degree of hotness (or coldness) of a body. In what is called the *absolute scale* (or the *Kelvin scale*), the temperature of a body is *always positive*. We denote the absolute temperature at a point of a body $\mathcal{B}$ by T. Then

$$\tilde{Q} = \int_V \frac{\rho h}{T} dV - \int_S \frac{\mathbf{q}}{T} \cdot \mathbf{n} \, dS \qquad (8.7.2)$$

8 FUNDAMENTAL LAWS OF CONTINUUM MECHANICS

is called the *rate of entropy input* into the body ℬ in its current configuration. The integrand in the volume integral in (8.7.2), namely, $(\rho h/T)$, is called the *entropy source* and the integrand in the surface integral, namely, $(\mathbf{q}/T)$, is called the *entropy flow*.

With $\tilde{H}$ and $\tilde{Q}$ defined by (8.7.1) and (8.7.2), we now postulate that

$$\frac{D\tilde{H}}{Dt} \geq \tilde{Q} \tag{8.7.3}$$

In words, this postulate states that the material time rate of change of entropy of a material body is no less than the rate of entropy input into the body. This is known as the *law (principle) of entropy* or *the second law of thermodynamics*.

Substituting for $\tilde{H}$ and $\tilde{Q}$ from (8.7.1) and (8.7.2) in (8.7.3) we obtain the following *integral form of the law of entropy*:

$$\frac{D}{Dt}\int_V \rho\eta \, dV - \int_V \frac{\rho h}{T} dV + \int_S \frac{\mathbf{q}}{T} \cdot \mathbf{n} \, dS \geq 0 \tag{8.7.4}$$

Rewriting the first term of this inequality by use of (8.2.1) and the last term by use of the divergence theorem (3.6.1) and noting that V is an arbitrary volume, we obtain the following inequality representing the law of entropy in the *local form*:

$$\rho \frac{D\eta}{Dt} + \operatorname{div}\left(\frac{\mathbf{q}}{T}\right) - \frac{\rho h}{T} \geq 0 \tag{8.7.5}$$

This is known as the *Clausius–Duhem inequality*, after R. Clausius (1854) and P. Duhem (1901).

Using the fact that

$$\operatorname{div}\left(\frac{\mathbf{q}}{T}\right) = \frac{1}{T}\operatorname{div}\mathbf{q} - \frac{1}{T^2}(\nabla T)\cdot\mathbf{q} \tag{8.7.6}$$

which follows by the use of (3.4.14) and the definition of $\nabla\phi$, the inequality (8.7.5) can be put in the following useful form:

$$\rho T \frac{D\eta}{Dt} - \rho h + \operatorname{div}\mathbf{q} - \frac{1}{T}(\nabla T)\cdot\mathbf{q} \geq 0 \tag{8.7.7}$$

An alternative version of this inequality, obtained by the use of the energy equation (8.6.13), reads

$$\rho T \frac{D\eta}{Dt} - \rho \frac{D\varepsilon}{Dt} + \mathbf{T}\cdot\mathbf{D} - \frac{1}{T}(\nabla T)\cdot\mathbf{q} \geq 0 \tag{8.7.8}$$

8.7 ENTROPY INEQUALITY

As mentioned earlier, it is a physical phenomenon that heat flows from a region of higher temperature to a region of lower temperature. In other words, heat does not flow against a temperature gradient. This means that the vector fields $\mathbf{q}$ and ∇T act in opposite directions. That is,

$$(\nabla T) \cdot \mathbf{q} \leq 0 \tag{8.7.9}$$

This is known as the *classical heat conduction inequality*. The equality in (8.7.9) holds only when $\nabla T = 0$.

In view of (8.7.9), the inequality (8.7.7) implies

$$\rho T \frac{D\eta}{Dt} + \text{div } \mathbf{q} - \rho h \geq 0 \tag{8.7.10}$$

This is known as the *Clausius-Planck inequality*, after Clausius (1854) and M. Planck (1887).

An alternative version of (8.7.10), obtained by the use of (8.7.8) and (8.7.9), is

$$\rho T \frac{D\eta}{Dt} - \rho \frac{D\varepsilon}{Dt} + \mathbf{T} \cdot \mathbf{D} \geq 0 \tag{8.7.11}$$

Returning to the inequality (8.7.5), we note that the inequality is in the spatial form. Its counterpart in the material form is obtained next.

8.7.1 ENTROPY INEQUALITY IN MATERIAL FORM

Let us recall the integral form of the entropy law as given by (8.7.4) and rewrite it in the initial configuration of $\mathcal{B}$ by use of the relations (5.2.12), (5.2.22), (8.2.8) and (8.6.16). The resulting inequality is

$$\int_{V_0} \rho_0 \frac{D\eta}{Dt} dV_0 - \int_{V_0} \rho_0 \frac{h}{T} dV_0 + \int_{S_0} \frac{1}{T} \mathbf{q}^0 \cdot \mathbf{n}^0 dS_0 \geq 0 \tag{8.7.12}$$

Employing the divergence theorem to the surface integral in (8.7.12) and noting that V_0 is an arbitrary volume, we obtain

$$\rho_0 \frac{D\eta}{Dt} - \rho_0 \frac{h}{T} + \text{div}^0 \left(\frac{\mathbf{q}^0}{T} \right) \geq 0 \tag{8.7.13}$$

This is the Clausius-Duhem inequality in the *material form*. Observe that this inequality is *formally* identical with its spatial counterpart (8.7.5).

The material versions of inequalities (8.7.7) to (8.7.11) can be written by changing ∇ and div to ∇^0 and div^0, ρ and $\mathbf{q}$ to ρ_0 and $\mathbf{q}^0$ and $\mathbf{T} \cdot \mathbf{D}$ to $(\mathbf{T}^0)^T \cdot \nabla^0 \mathbf{v}$ in (8.7.7) to (8.7.11).

The inequalities (8.7.5) and (8.7.13) are useful in studying restrictions on the thermomechanical behavior of material models. In recent years, entropy

354 8 FUNDAMENTAL LAWS OF CONTINUUM MECHANICS

inequalities that are more general than these inequalities have been proposed; their discussion falls beyond the scope of this text.

EXAMPLE 8.7.1 The *specific free energy* (or the *Helmholtz free energy*) ψ is defined by

$$\psi = \varepsilon - \eta T \qquad (8.7.14)$$

Show that

(i) $$\rho \frac{D\psi}{Dt} = \mathbf{T} \cdot \mathbf{D} - \text{div } \mathbf{q} + \rho h - \rho \frac{D}{Dt}(\eta T) \qquad (8.7.15)$$

(ii) $$\rho \frac{D\psi}{Dt} + \rho \eta \frac{DT}{Dt} - \mathbf{T} \cdot \mathbf{D} + \frac{1}{T}(\nabla T) \cdot \mathbf{q} \leq 0 \qquad (8.7.16)$$

Solution Substituting for ε from (8.7.14) in the energy equation (8.6.13), we obtain equation (8.7.15). Similarly, elimination of ε from (8.7.14) and (8.7.8) yields the inequality (8.7.16). ∎

EXAMPLE 8.7.2 Assuming that the heat-flux vector $\mathbf{q}$ is related to the temperature gradient ∇T through the relation

$$\mathbf{q} = -\mathbf{K}(\nabla T) \qquad (8.7.17)$$

show that $\mathbf{K}$ is a positive definite tensor.

Solution Substituting for $\mathbf{q}$ from (8.7.17) into the classical heat conduction inequality (8.7.9), we find

$$(\nabla T) \cdot \mathbf{K}(\nabla T) \geq 0 \qquad (8.7.18)$$

Since this inequality holds for arbitrary ∇T and the equality in (8.7.9) is valid only if $\nabla T = 0$, it follows that $\mathbf{K}$ is a positive definite tensor. ∎

Note: The relation (8.7.17) is known as the *Fourier–Duhamel law of heat conduction*, after J. Fourier (1822) and J. M. C. Duhamel (1832). This law is directly motivated by experimental observations at normal laboratory conditions and is generally employed to study thermal states of solids, liquids and gases. The tensor $\mathbf{K}$ is called the *thermal conductivity tensor*; its components k_{ij} are found to be either constants or functions of T. Experimental observations also show that $k_{ij} = k_{ji}$; that is, $\mathbf{K}$ is a *symmetric tensor*.

A material is said to be *thermally isotropic* if $\mathbf{K}$ is an isotropic tensor; in such a material there is no preferred direction of heat flow. By virtue of

Theorem 2.5.1, it follows that, for a thermally isotropic material,

$$\mathbf{K} = k\mathbf{I} \qquad (8.7.19)$$

where k is a scalar. The positive definiteness of $\mathbf{K}$ implies that $k > 0$.

By use of (8.7.19) in (8.7.17), we obtain

$$\mathbf{q} = -k\,\nabla T \qquad (8.7.20)$$

This is known as *Fourier's law of heat conduction*. The positive scalar k is called the *coefficient of thermal conductivity*.

In recent years, heat conduction laws that are more general than (8.7.17) have been proposed and used. These generalizations, called *non-Fourier heat conduction laws*, are relevant only in specialized applications (not considered in this text).

8.8
CONSTITUTIVE EQUATIONS

In the earlier sections of this chapter we have developed the following equations.

Equation of continuity:

$$\frac{D\rho}{Dt} + \rho\,\text{div}\,\mathbf{v} = 0 \qquad \text{(spatial form)} \qquad (8.8.1a)$$

$$\rho_0 = \rho J \qquad \text{(material form)} \qquad (8.8.1b)$$

Equations of motion:

$$\left.\begin{array}{c} \text{div}\,\mathbf{T}^T + \rho\mathbf{b} = \rho\dfrac{D\mathbf{v}}{Dt} \\[4pt] \mathbf{T} = \mathbf{T}^T \end{array}\right\} \qquad \text{(spatial form)} \qquad (8.8.2a)$$

$$\left.\begin{array}{c} \text{div}^0\{(\mathbf{I} + \nabla^0\mathbf{u})\mathbf{S}^T\} + \rho_0\mathbf{b} = \rho_0\dfrac{D^2\mathbf{u}}{Dt^2} \\[4pt] \mathbf{S} = \mathbf{S}^T \end{array}\right\} \qquad \text{(material form)} \qquad (8.8.2b)$$

Equation of energy:

$$\rho\frac{D\varepsilon}{Dt} = \mathbf{T}\cdot\nabla\mathbf{v} - \text{div}\,\mathbf{q} + \rho h \qquad \text{(spatial form)} \qquad (8.8.3a)$$

$$\rho_0\frac{D\varepsilon}{Dt} = (\mathbf{I} + \nabla^0\mathbf{u})\mathbf{S}^T\cdot(\nabla^0\mathbf{v}) - \text{div}^0\mathbf{q}^0 + \rho_0 h \qquad \text{(material form)} \qquad (8.8.3b)$$

These are the *field equations* representing the laws of conservation of mass, balance of momenta and balance of energy, in the local form. These are all partial differential equations and hold at every point of a continuum and for all time. The initial density ρ_0, the body force **b** and the heat supply h are taken to be *known functions*.

It is readily seen that a total of 8 scalar equations are embodied in the field equations while a total of 17 unknown field functions are involved in these equations, the unknown functions being the density ρ, the three velocity components v_i (or the three displacement components u_i), the nine stress components τ_{ij} (or τ_{ij}^0 or s_{ij}), the three heat-flux components q_i (or q_i^0) and the specific internal energy ε. Hence, these equations are inadequate to determine all the field functions, and we need some more basic equations to close the system.

Recall that the field equations (8.8.1) to (8.8.3) have been developed for a general continuum without focusing on any particular material. As such, these equations are valid for all continua irrespective of their internal constitution. The equations developed in Chapters 5 to 7 also fall into this category. This means that the equations developed so far hold equally well to all solids, liquids and gases. But solids, liquids and gases have several individual characteristic properties. We should therefore have appropriate additional basic equations that reflect these properties. Equations that represent the characteristic property of a material (or a class of materials) and distinguish one material from the other are called *constitutive equations*. In other words, constitutive equations are the equations describing the relations between observable effects and the internal constitution of matter. Thus, whereas the equations developed so far are based purely on geometrical-dynamical principles and hold for all continua, the constitutive equations are aimed at portraying the intrinsic physical properties that are different for different continua. In the preceding paragraph, the need for having more basic equations of the theory has been observed. The constitutive equations adequately fill this need.

The constitutive equations are usually postulated directly from experimental data or by judiciously developed mathematical generalizations of experimental data. However, in recent years, these equations are also being developed by employing rigorous axiomatic approach based upon what are known as the *principles of consistency, coordinate invariance and material frame indifference*. The constitutive equations generally consist of (i) a law of heat conduction, (ii) a caloric equation of state and (iii) a material law. The law of heat conduction relates the heat-flux vector with temperature gradient and is usually taken as given by (8.7.20). Caloric equation of state is an equation determining the specific internal energy in terms of temperature and other field functions. This equation is again a postulate based upon experimental grounds, and the form of the equation depends on the particular material under study. A material law usually consists of a set of equations, each equation specifying a stress component in terms of other field functions. The form of these equations again depends on the particular material under study.

Sometimes, in the development of a theory, the number of field functions is increased. In such cases, apart from the constitutive equations described previously, some additional constitutive equations are postulated. The kinetic equations of state, as represented by the Boyle's law for a "perfect gas," is an example of such an additional constitutive equation. In many situations, constitutive equations will be required to obey certain restrictions. The *entropy inequality* (8.7.5), which holds for all continua (solids, liquids and gases), is an example of such a restriction.

The field equations (8.8.1) to (8.8.3) together with relevant equations obtained in Chapters 5 to 7 as well as appropriate constitutive equations generally serve as an adequate set of basic equations for determining all the field functions. But since the constitutive equations necessarily differ from one material to the other, we cannot have (even if we desire so) a common set of such basic equations valid for all continua. We are thus lead to the study of different branches of continuum mechanics separately, each branch dealing with one set of constitutive equations. The branch of continuum mechanics in which constitutive equations valid for solids are utilized is known as *solid mechanics*. The branch in which constitutive equations valid for fluids are utilized is known as *fluid mechanics*. The subjects of elasticity, thermoelasticity, viscoelasticity, plasticity, hydrodynamics, gas dynamics and so on, emerge as further subdivisions of continuum mechanics.

It is again emphasized that all the equations and results developed in Chapters 5 to 7 and in the earlier sections of this chapter hold for all (nonpolar) continua and therefore remain valid in all branches of continuum mechanics. The constitutive equations distinguish branches of continuum mechanics.

8.9
EXERCISES

1. For the velocity and the density fields given here, check whether the equation of continuity is satisfied.

(i) $\mathbf{v} = x_1 \mathbf{e}_1, \quad \rho = \rho_0 e^{-t}$

(ii) $\mathbf{v} = \dfrac{x_1}{1 + t} \mathbf{e}_1, \quad \rho = \dfrac{\rho_0}{1 + t}$

2. Given the velocity $\mathbf{v} = t(x_2 \mathbf{e}_2 + x_3 \mathbf{e}_3)$, find the density $\rho = \rho(t)$ such that the equation of continuity is satisfied.

3. Given the velocity $\mathbf{v} = \{x_3/(1+t)\mathbf{e}_3\}$, find the density $\rho = \rho(x_3)$ such that the equation of continuity is satisfied.

4. If the current density of an incompressible continuum is $\rho = kx_1$, where k is a constant, find the nature of velocity $\mathbf{v}$ such that $v_2 = 0$.

5. Show that
$$\mathbf{v} = \frac{x_2 \mathbf{e}_1 - x_1 \mathbf{e}_2}{x_1^2 + x_2^2}$$
can be velocity of an incompressible continuum.

6. Find k such that
$$v_1 = kx_3(x_2 - 2)^2, \quad v_2 = -x_1 x_2, \quad v_3 = kx_1 x_3$$
may be velocity components of an incompressible continuum.

7. Show that $v_i = ax_i/r^3$, where a is a constant and $r^2 = x_k x_k \neq 0$ is a possible velocity field for an incompressible continuum.

8. For an irrotational motion of a continuum with velocity $\mathbf{v} = \nabla \phi$, show that the equation of continuity is
$$\frac{D\rho}{Dt} + \rho \nabla^2 \phi = 0$$
Deduce that if the continuum is incompressible, ϕ is a harmonic function.

9. A deformation of a continuum is defined by
$$x_1 = (1+a)x_1^0 + bx_2^0, \quad x_2 = bx_1^0 + (1+a)x_2^0, \quad x_3 = x_3^0$$
where a and b are constants such that $a > b - 1$. Show that
$$\rho = \rho_0 \{(1+a)^2 - b^2\}^{-1}$$

10. For any scalar function f, show that
$$\rho \frac{Df}{Dt} = \frac{\partial}{\partial t}(\rho f) + \text{div}(\rho f \mathbf{v})$$

11. Show that
$$\rho \frac{D\mathbf{v}}{Dt} + \frac{\partial}{\partial t}(\rho \mathbf{v}) + \text{div}(\rho \mathbf{v} \otimes \mathbf{v})$$
Write down this expression in the suffix form.

12. For any scalar function f, obtain the following generalized vorticity equation:
$$\frac{D}{Dt}\left(\frac{\mathbf{w}}{\rho} \cdot \nabla f\right) = \frac{\mathbf{w}}{\rho} \cdot \nabla \frac{Df}{Dt} + \frac{1}{\rho} \text{curl} \frac{D\mathbf{v}}{Dt} \cdot \nabla f$$

13. If $\nabla(\text{div } \mathbf{v}) = \mathbf{0}$, show that

$$\rho = \rho_0 \exp\left\{-\int_0^t \text{div } \mathbf{v} \, dt\right\}$$

14. Verify that each of the following stress systems obeys Cauchy's equations of equilibrium:

(i) $\tau_{11} = \tau_{22} = \tau_{12} = \tau_{23} = \tau_{13} = 0$, $\tau_{33} = \rho g x_3$, where ρ and g are constants, with $\mathbf{b} = -g\mathbf{e}_3$.

(ii) $\tau_{11} = \tau_{22} = \tau_{33} = \tau_{12} = 0$, $\tau_{23} = \mu\alpha x_1$, $\tau_{13} = -\mu\alpha x_2$, where μ and α are constants, with $\mathbf{b} = \mathbf{0}$.

(iii) $\tau_{11} = \tau_{22} = \tau_{33} = \tau_{12} = 0$, $\tau_{13} = \mu\alpha(\partial\phi/\partial x_1 - x_2)$, $\tau_{23} = \mu\alpha(\partial\phi/\partial x_2 + x_1)$, where μ and α are constants, and $\phi = \phi(x_1, x_2)$ is a harmonic function, with $\mathbf{b} = \mathbf{0}$.

(iv) $\tau_{11} = x_2 + v(x_1^2 - x_2^2)$, $\tau_{22} = x_1^2 + v(x_2^2 - x_1^2)$
$\tau_{33} = v(x_1^2 + x_2^2)$, $\tau_{12} = -2vx_1 x_2$, $\tau_{23} = \tau_{31} = 0$
where v is a constant, with $\mathbf{b} = \mathbf{0}$.

15. The stress field in a continuum in equilibrium is given by

$$[\tau_{ij}] = \begin{bmatrix} x_2^2 x_3 & x_2(1 - x_3^2) & 0 \\ x_2(1 - x_3^2) & x_3^3 - 3x_3 & 0 \\ 0 & 0 & 2x_1^3 \end{bmatrix}$$

Find the body force that must be acting.

16. Repeat Exercise 15 for the stress matrix

$$[\tau_{ij}] = \begin{bmatrix} x_1^2 & 2x_1 x_2 & 0 \\ 2x_1 x_2 & x_2^2 & 0 \\ 0 & 0 & 2(x_1^2 + x_2^2) \end{bmatrix}$$

17. The stress tensor in a continuum in equilibrium with zero body force is such that $\tau_{i3} = 0$. Show that

$$\tau_{11,11} + \tau_{22,22} + 2\tau_{12,12} = 0$$

18. The stress tensor in a continuum in equilibrium with zero body force is given by $\mathbf{T} = T\mathbf{a} \otimes \mathbf{a}$, where $T = T(\mathbf{x})$ and $\mathbf{a}$ is a constant unit vector. Show that ∇T must be everywhere perpendicular to $\mathbf{a}$.

19. Find the function $\phi = \phi(r)$ such that the stresses

$$\tau_{ij} = \lambda \delta_{ij}\{3\phi(r) + r\phi'(r)\} + 2\mu\left\{\phi(r)\delta_{ij} + \frac{1}{r}\phi'(r)x_i x_j\right\}$$

obey the equations of equilibrium under zero body force. Here λ and μ are constants and $r^2 = x_i x_i \neq 0$.

360   **8 FUNDAMENTAL LAWS OF CONTINUUM MECHANICS**

20. In a continuum in equilibrium, the stresses are given by $\tau_{ij} = \alpha x_3 x_i x_j / r^5$, where α is a constant and $r^2 = x_i x_i \neq 0$. Find the body force that acts on the continuum.

21. Using the result of Exercise 11, show that Cauchy's equation of motion can be written as

$$\frac{\partial}{\partial t}(\rho \mathbf{v}) = \text{div}[\mathbf{T} - \rho \mathbf{v} \otimes \mathbf{v}] + \rho \mathbf{b}$$

Write down this equation in the suffix form.

22. For a function $f = f(\mathbf{x}, t)$, prove that

$$\int_S f \mathbf{T} \mathbf{n} \, dS = \int_V [\mathbf{T}(\nabla f) + \rho f(\dot{\mathbf{v}} - \mathbf{b})] \, dV$$

23. For a function $\mathbf{f} = \mathbf{f}(\mathbf{x}, t)$, prove that

$$\int_V \{(\nabla \mathbf{f})\mathbf{T} + \rho \mathbf{f} \otimes (\dot{\mathbf{v}} - \mathbf{b})\} \, dV = \int_S \mathbf{f} \otimes \mathbf{s} \, dS$$

Deduct that

(i) $\displaystyle\int_V \mathbf{T} \, dV = \int_S \mathbf{x} \otimes \mathbf{s} \, dS + \int_V \rho \mathbf{x} \otimes (\mathbf{b} - \dot{\mathbf{v}}) \, dV$

(ii) $\displaystyle\int_V \mathbf{T} \cdot (\nabla \mathbf{u}) \, dV = \int_S \mathbf{s} \cdot \mathbf{u} \, dS + \int_V \rho (\mathbf{b} - \dot{\mathbf{v}}) \cdot \mathbf{u} \, dV$

24. Deduce equations (7.2.4) and (7.4.3) as particular cases of equation (8.3.4).

25. Deduce equations (8.3.10) and (8.3.11) from equation (8.3.7).

26. Show that

$$\int_S [\mathbf{x} \times \mathbf{s}]_i \, dS = \int_V \varepsilon_{ijk} \{\tau_{jk} + x_j \tau_{mk,m}\} \, dV$$

Using this expression and the suffix form of (8.4.5), show that equation (8.4.4) yields $\varepsilon_{ijk} \tau_{jk} = 0$. Hence deduce the relation (8.4.8).

27. The center of mass $\mathbf{x}_c$ of a body $\mathcal{B}$ of mass m is defined by

$$\mathbf{x}_c = \frac{1}{m} \int_V \rho \mathbf{x} \, dV$$

By employing the law of balance of linear momentum, show that

$$m \frac{D^2 \mathbf{x}_c}{Dt^2} = \rho \mathbf{b}$$

28. Show that in the law of balance of angular momentum the origin may be taken at the mass center with the velocity interpreted as that relative to the mass center.

29. Assuming that the amount Q of heat flowing out of a volume V across its boundary surface S per unit time is representable as $Q = \int_S q(\mathbf{x}, \mathbf{n})\, dS$, show that $q = \mathbf{q} \cdot \mathbf{n}$.

30. For a continuum rotating like a rigid body with angular velocity $\mathbf{w}$, show that the kinetic energy is given by

$$K = \frac{1}{2}\int_V \rho\{r^2 \mathbf{I} - \mathbf{x} \otimes \mathbf{x}\} \cdot (\mathbf{w} \otimes \mathbf{w})\, dV$$

31. For a material in which the stress tensor is of the form $\mathbf{T} = -p\mathbf{I}$, show that the stress power is given by

$$\mathbf{T} \cdot \mathbf{D} = \frac{p}{\rho}\frac{D\rho}{Dt}$$

If, further, the material is incompressible, deduce that the equation of mechanical and thermal energy balances become uncoupled.

32. Prove that

$$\int_V \mathbf{T} \cdot \mathbf{D}\, dV = \int_{V_0} (\mathbf{T}^0)^T \cdot \frac{D\mathbf{F}}{Dt}\, dV_0 = \int_{V_0} \mathbf{S} \cdot \frac{D\mathbf{G}}{Dt}\, dV_0$$

33. Prove the following:

(i) $\dfrac{DK}{Dt} = \displaystyle\int_{V_0} \left\{\rho_0 \mathbf{b} \cdot \mathbf{v} - \mathbf{S} \cdot \frac{D\mathbf{G}}{Dt}\right\} dV_0 + \int_{S_0} \mathbf{s}^0 \cdot \mathbf{n}^0\, dS_0$

(ii) $\dfrac{DE}{Dt} = \displaystyle\int_{V_0} \left\{\rho_0 h + \mathbf{S} \cdot \frac{D\mathbf{G}}{Dt}\right\} dV_0 - \int_{S_0} \mathbf{q}^0 \cdot \mathbf{n}^0\, dS_0$

34. Deduce equation (8.6.20) from equation (8.6.13).

35. Show that the energy equation (8.6.13) can be expressed in the following form:

$$\mathrm{div}[(\mathbf{T}^T \mathbf{v}) - \mathbf{q}] + \rho \mathbf{v} \cdot \mathbf{b} + \rho h = \rho \frac{\partial \bar{\varepsilon}}{\partial t} + \rho(\nabla \bar{\varepsilon}) \cdot \mathbf{v} + \mathrm{div}(\rho \bar{\varepsilon})$$

where

$$\bar{\varepsilon} = \varepsilon + \tfrac{1}{2}v^2$$

36. If T is uniform at each time, show that the Clausius–Duhem inequality (8.7.5) reduces to the Clausius–Planck inequality (8.7.10).

37. If $\delta = \mathbf{T} \cdot \mathbf{D} - \rho(D\psi/Dt + \eta(DT/Dt))$, where ψ is the specific free energy defined by (8.7.14), show that the Clausius–Planck inequality is given by $\delta \geq 0$ and that the Clausius–Duhamel inequality is given by $T\delta \geq \nabla T \cdot \mathbf{q}$. (Here, δ is called *internal dissipation*.)

38. If δ defined in Exercise 37 is exactly 0, show that the Clausius–Duhem inequality reduces to the classical heat conduction inequality (8.7.9).

39. Show that the material version of the energy equation (8.7.15) is

$$\rho_0 \frac{D\psi}{Dt} = -\rho_0 \frac{D}{Dt}(\eta T) + (\mathbf{T}^0)^T \cdot \nabla^0 \mathbf{v} - \text{div}^0 \mathbf{q}^0 + \rho_0 h$$

40. Show that the material version of the inequality (8.7.16) is

$$\rho_0 \left(\frac{D\psi}{Dt} + \eta \frac{DT}{Dt} \right) - (\mathbf{T}^0)^T \cdot \nabla^0 \mathbf{v} + \frac{1}{T} \mathbf{q}^0 \cdot \nabla^0 T \leq 0$$

CHAPTER 9
EQUATIONS OF LINEAR ELASTICITY

9.1
INTRODUCTION

In this chapter, we consider one of the most important branches of continuum mechanics: the classical theory of elasticity. This theory deals with a class of continua called *linear elastic solids*. An elastic solid is a deformable continuum that possesses the property of recovering its original configuration when forces causing deformation are removed. An elastic solid that undergoes only an infinitesimal deformation and for which the governing material law is linear is called a *linear elastic solid*. From experimental observations it is known that, under normal loadings, many structural materials such as metals, concrete, wood and rocks behave as linear elastic solids. The classical theory of elasticity serves as an excellent model for studying the mechanical behavior of a wide variety of such solid materials.

The classical elasticity theory is an essential part of solid mechanics and its scope is vast. We restrict ourselves to the derivation of the governing equations of the theory and some of its immediate consequences. Some simple and standard applications are also presented.

9.2
GENERALIZED HOOKE'S LAW

As just mentioned, an elastic solid is a deformable continuum that recovers its original configuration when forces causing deformation are removed. In order to find a material law that portrays this characteristic behavior of elastic solids undergoing infinitesimal deformation, we start with the following set of linear relations:

$$\tau_{11} = c_{1111}e_{11} + c_{1112}e_{12} + \cdots + c_{1133}e_{33}$$
$$\tau_{12} = c_{1211}e_{11} + c_{1212}e_{12} + \cdots + c_{1233}e_{33} \quad (9.2.1)$$
$$\cdots$$
$$\tau_{33} = c_{3311}e_{11} + c_{3312}e_{12} + \cdots + c_{3333}e_{33}$$

Here, τ_{ij} are, as usual, the components of the stress tensor **T** arising due to the presence of external forces, e_{ij} are the components of Cauchy's strain tensor **E** describing the infinitesimal deformation caused by these forces, and $c_{1111}, c_{1112}, \ldots, c_{3333}$ are 81 scalar coefficients that depend on the physical properties of the solid and are independent of the strain components e_{ij}. We suppose that relations (9.2.1) hold at every point of the continuum and at every instant of time and are solvable for e_{ij} in terms of τ_{ij}. Then it follows that τ_{ij} are all 0 whenever e_{ij} are all 0, and that e_{ij} are all 0 whenever τ_{ij} are all 0. That is, τ_{ij} and e_{ij} are homogeneous, linear functions of each other. Physically, this means that the material for which the linear relations (9.2.1) hold deforms in the presence of stresses and it recovers its undeformed configuration when the stresses are removed. Further, since e_{ij} are components of Cauchy's strain tensor, the deformation is infinitesimal. Thus, the linear relations (9.2.1) describe the characteristic property of an elastic solid undergoing infinitesimal deformation and constitute a material law for such an elastic solid. An elastic solid for which the linear relations (9.2.1) constitute *the* material law is referred to as a *linear* (or *linearly*) *elastic solid*.

A particular case of relations (9.2.1) is

$$\tau_{11} = Ee_{11} \quad (9.2.2)$$

If the coefficient E is a constant, then the relation (9.2.2) states that for a linear elastic solid the normal stress in the x_1 direction is directly proportional to the normal strain in the same direction. In effect, this relation was first enunciated, on experimental grounds, by Robert Hooke in 1678 and is known as *Hooke's law*. The material law described by relations (9.2.1) is a natural generalization of Hooke's law, referred to as the *generalized Hooke's law*. This (generalized) law is attributed to Cauchy (1822).

9.2 GENERALIZED HOOKE'S LAW

The nine relations in (9.2.1) can be expressed in a concise form as

$$\tau_{ij} = c_{ijkl} e_{kl} \qquad (9.2.3)$$

Since τ_{ij} and e_{ij} are components of second-order tensors, using a quotient law proved in Example 2.4.7, it follows from (9.2.3) that c_{ijkl} are components of a fourth-order tensor. This tensor characterizes the mechanical properties of the material and is called the *elasticity tensor*. The 81 components c_{ijkl} of this tensor are called *elastic moduli*. Because of the symmetry of τ_{ij} and e_{ij}, not more than 36 of these 81 elastic moduli can be independent. Since e_{ij} are dimensionless quantities, it follows from (9.2.3) that the elastic moduli have the same dimensions as the stresses (force/length^2).

By hypothesis, relations (9.2.3) should hold at all points x_i of the material and at all time t. The stresses τ_{ij} and the strains e_{ij} vary in general with x_i and t. Likewise, c_{ijkl} may also vary with x_i and t. If, however, c_{ijkl} are independent of x_i and t, we say that the material is (elastically) *homogeneous*. Thus, for a homogeneous elastic solid, the elastic moduli are constants so that the mechanical properties remain the same throughout the solid for all time.

The tensor equation (9.2.3) represents the generalized Hooke's law in the x_i system. When written in the x_i' system, the law would read

$$\tau_{ij}' = c_{ijkl}' e_{kl}' \qquad (9.2.4)$$

where τ_{ij}' are components of **T**, e_{ij}' are components of **E** and c_{ijkl}' are components of the elasticity tensor, in the x_i' system. In general, τ_{ij}' and e_{ij}' are different from τ_{ij} and e_{ij}. Likewise, c_{ijkl}' are generally different from c_{ijkl}. If, however, c_{ijkl}' are the same as c_{ijkl}, we say that the material is (elastically) *isotropic*. Thus, for an isotropic elastic solid, $c_{ijkl}' = c_{ijkl}$; that is, the elasticity tensor is isotropic, which means to say that the mechanical properties of the solid are independent of the orientation of the coordinate axes.

In Section 2.6 it has been shown that every isotropic tensor of order 4 can be represented in the form given by (2.6.1). Hence, for an isotropic elastic solid, c_{ijkl} are of the form

$$c_{ijkl} = \alpha \delta_{ij} \delta_{kl} + \beta \delta_{ik} \delta_{jl} + \gamma \delta_{il} \delta_{jk} \qquad (9.2.5)$$

where α, β and γ are scalars. Substituting for c_{ijkl} from (9.2.5) in (9.2.3) and noting that $e_{ij} = e_{ji}$ we obtain (see Example 2.6.1),

$$\tau_{ij} = \alpha \delta_{ij} e_{kk} + (\beta + \gamma) e_{ij}$$

which on redesignating α as λ and $\beta + \gamma$ as 2μ yields

$$\tau_{ij} = \lambda \delta_{ij} e_{kk} + 2\mu e_{ij} \qquad (9.2.6)$$

These relations represent the generalized Hooke's law for a linear, isotropic elastic solid. Evidently, the law now involves just *two* independent elastic moduli λ and μ. These moduli are constants if the solid is also homogeneous.

The generalized Hooke's law as given by (9.2.6) was proposed by Cauchy in 1822. The symbols λ and μ were introduced later by Gabriel Lamé in 1852; these are called the *Lamé moduli*.

From relations (9.2.6), we get

$$\tau_{kk} = (3\lambda + 2\mu)e_{kk} \qquad (9.2.7)$$

Substituting for e_{kk} from (9.2.7) back into (9.2.6) and solving the resulting equations for e_{ij} we obtain

$$e_{ij} = \frac{1}{2\mu}\left[\tau_{ij} - \frac{\lambda}{3\lambda + 2\mu}\delta_{ij}\tau_{kk}\right] \qquad (9.2.8)$$

provided $\mu \neq 0$, $3\lambda + 2\mu \neq 0$; see Example 1.6.2.

Relations (9.2.8) and (9.2.6) are equivalent to one another. While (9.2.6) express τ_{ij} in terms of e_{ij}, (9.2.8) express e_{ij} in terms of τ_{ij}.

In the direct notation, relations (9.2.6) and (9.2.8) read, respectively, as follows:

$$\mathbf{T} = \lambda(\operatorname{tr} \mathbf{E})\mathbf{I} + 2\mu\mathbf{E} \qquad (9.2.6)'$$

$$\mathbf{E} = \frac{1}{2\mu}\left[\mathbf{T} - \frac{\lambda}{3\lambda + 2\mu}(\operatorname{tr} \mathbf{T})\mathbf{I}\right] \qquad (9.2.8)'$$

Relations (9.2.6) can also be rewritten in the form of a matrix equation, as follows:

$$\begin{bmatrix} \tau_{11} \\ \tau_{22} \\ \tau_{33} \\ \tau_{12} \\ \tau_{23} \\ \tau_{31} \end{bmatrix} = \begin{bmatrix} \lambda + 2\mu & \lambda & \lambda & 0 & 0 & 0 \\ \lambda & \lambda + 2\mu & \lambda & 0 & 0 & 0 \\ \lambda & \lambda & \lambda + 2\mu & 0 & 0 & 0 \\ 0 & 0 & 0 & 2\mu & 0 & 0 \\ 0 & 0 & 0 & 0 & 2\mu & 0 \\ 0 & 0 & 0 & 0 & 0 & 2\mu \end{bmatrix} \begin{bmatrix} e_{11} \\ e_{22} \\ e_{33} \\ e_{12} \\ e_{23} \\ e_{31} \end{bmatrix} \qquad (9.2.6)''$$

The matrix equation equivalent to relations (9.2.8) can be written similarly.

It should be emphasized that the relations (9.2.3) and their special case (9.2.6) are postulated to be valid only for *linear* elastic solids. For most solid materials such relations hold until the stresses reach a limit called the *proportional limit*. When the stresses exceed this limit, the deformation becomes nonlinear but the elastic behavior of the material continues to exist

9.2 GENERALIZED HOOKE'S LAW

until the stresses reach what is called the *elastic limit* of the material. For studying nonlinear deformations below the elastic limit, relations (9.2.3) are to be replaced by more general (nonlinear) functional relations. The theory is then called *nonlinear elasticity*. Consideration of this theory falls beyond the scope of this book.

Henceforth, we will be concerned with linear elastic solids that are *both homogeneous and isotropic*. As such, relations (9.2.6) or their equivalents (9.2.8) will represent the generalized Hooke's law for our further discussion of the elasticity theory in this text. These relations will be referred to as *stress-strain relations* or simply *Hooke's law*. Also, a homogeneous and isotropic linear elastic solid will be referred to simply as an *elastic body*.

EXAMPLE 9.2.1 Show that for a linear isotropic elastic solid, the principal directions of stress and the principal directions of strain coincide.

Solution Let $\mathbf{a}$ be a unit vector along a principal direction of strain at a point of the solid. Then $\mathbf{a}$ is an eigenvector of the strain tensor $\mathbf{E}$ at the point so that $\mathbf{E}\mathbf{a} = \Lambda \mathbf{a}$ for some scalar Λ. The Hooke's law (9.2.6)' then yields

$$\mathbf{T}\mathbf{a} = [\lambda(\operatorname{tr} \mathbf{E}) + 2\mu\Lambda]\mathbf{a} \qquad (9.2.9)$$

This relation shows that the vector $\mathbf{T}\mathbf{a}$ is collinear with $\mathbf{a}$; therefore, $\mathbf{a}$ is an eigenvector of $\mathbf{T}$ as well. Hence, $\mathbf{a}$ is along a principal direction of stress also. ∎

EXAMPLE 9.2.2 Show that the stress-strain relation (9.2.6)' is equivalent to the following relations taken together:

$$\operatorname{tr} \mathbf{T} = (3\lambda + 2\mu) \operatorname{tr} \mathbf{E} \qquad (9.2.10a)$$

$$\mathbf{T}^{(d)} = 2\mu \mathbf{E}^{(d)} \qquad (9.2.10b)$$

Solution From (7.7.1) and (5.10.1), we recall that

$$\mathbf{T}^{(d)} = \mathbf{T} - \tfrac{1}{3}(\operatorname{tr} \mathbf{T})\mathbf{I} \qquad (9.2.11a)$$

$$\mathbf{E}^{(d)} = \mathbf{E} - \tfrac{1}{3}(\operatorname{tr} \mathbf{E})\mathbf{I} \qquad (9.2.11b)$$

From (9.2.6)' we find that (9.2.10a) holds and that

$$\mathbf{T}^{(d)} = \{\lambda(\operatorname{tr} \mathbf{E})\mathbf{I} + 2\mu \mathbf{E}\} - \tfrac{1}{3}(\operatorname{tr} \mathbf{T})\mathbf{I}$$

using (9.2.11a);

$$= 2\mu\{\mathbf{E} - \tfrac{1}{3}(\operatorname{tr} \mathbf{E})\mathbf{I}\}$$

using (9.2.10a);

$$= 2\mu \mathbf{E}^{(d)}$$

by (9.2.11b). Thus, relation (9.2.6)' implies the relations (9.2.10).

Conversely, with the aid of (9.2.11), relations (9.2.10) give

$$\begin{aligned} \mathbf{T} &= \mathbf{T}^{(d)} + \tfrac{1}{3}(\operatorname{tr} \mathbf{T})\mathbf{I} \\ &= 2\mu\{\mathbf{E} - \tfrac{1}{3}(\operatorname{tr} \mathbf{E})\mathbf{I}\} + \tfrac{1}{3}(3\lambda + 2\mu)(\operatorname{tr} \mathbf{E})\mathbf{I} \\ &= \lambda(\operatorname{tr} \mathbf{E})\mathbf{I} + 2\mu\mathbf{E} \end{aligned}$$

which is the relation (9.2.6)'. ∎

EXAMPLE 9.2.3 In a vertical elastic beam deforming under its own weight (acting in the x_3 direction) the strain components are found to be

$$e_{11} = e_{22} = -\frac{\lambda}{2(\lambda + \mu)} a(b - x_3)$$

$$e_{33} = a(b - x_3), \qquad e_{12} = e_{23} = e_{31} = 0$$

where a and b are constants. Find the stress components.

Solution To find the stress components from the strain components, it is convenient to employ the matrix equation (9.2.6)″. Substituting the given strain components into this equation, and equating the corresponding terms, we obtain the required stress components as follows:

$$\tau_{11} = \tau_{22} = \tau_{12} = \tau_{23} = \tau_{13} = 0$$

$$\tau_{33} = \frac{\mu(3\lambda + 2\mu)}{\lambda + \mu} a(b - x_3) \qquad \blacksquare$$

9.3
PHYSICAL MEANINGS OF ELASTIC MODULI

With the view of obtaining the physical meanings of the elastic moduli appearing in the Hooke's law (9.2.6), we consider the following particular cases.

CASE i Suppose the stress tensor has only one nonzero component τ_{11}. Such a stress system occurs in a beam placed along the x_1-axis and subjected to a longitudinal stress (Figure 9.1). Then, the relations (9.2.8) yield the following expressions for the strain components:

$$e_{11} = \frac{\lambda + \mu}{\mu(3\lambda + 2\mu)} \tau_{11}, \qquad e_{22} = e_{33} = -\frac{\lambda}{2\mu(3\lambda + 2\mu)} \tau_{11}, \qquad (9.3.1)$$

$$e_{12} = e_{23} = e_{31} = 0$$

9.3 PHYSICAL MEANINGS OF ELASTIC MODULI

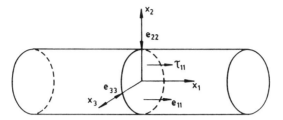

Figure 9.1. Case i.

If we set

$$E = \frac{\mu(3\lambda + 2\mu)}{\lambda + \mu} \qquad (9.3.2a)$$

$$\nu = \frac{\lambda}{2(\lambda + \mu)} \qquad (9.3.2b)$$

then the relations (9.3.1) yield

$$\frac{\tau_{11}}{e_{11}} = E \qquad (9.3.3)$$

$$\frac{e_{22}}{e_{11}} = \frac{e_{33}}{e_{11}} = -\nu \qquad (9.3.4)$$

Expression (9.3.3) is indeed the original version of the Hooke's law, given by (9.2.2). Experiments conducted on most naturally occurring elastic materials show that a tensile longitudinal stress produces a longitudinal extension together with a contraction in transverse directions. Accordingly, for $\tau_{11} > 0$ we take $e_{11} > 0$, $e_{22} < 0$, $e_{33} < 0$. It then follows from (9.3.3) and (9.3.4) that $E > 0$, $\nu > 0$.

From (9.3.3), we note that the constant E represents the ratio of the longitudinal stress to the corresponding longitudinal strain. This constant is referred to as *Young's modulus*, after Thomas Young, who gave a discussion of the elasticity theory in 1807.

From equation (9.3.4), we get

$$\nu = \left|\frac{e_{22}}{e_{11}}\right| = \left|\frac{e_{33}}{e_{11}}\right| \qquad (9.3.5)$$

The constant ν thus represents the numerical value of the ratio of the contraction in a transverse direction to the corresponding extension in the longitudinal direction. This ratio was introduced by Simon D. Poisson in 1829; it is known as *Poisson's ratio*.

370 9 EQUATIONS OF LINEAR ELASTICITY

Relations (9.3.2) can be solved for λ and μ in terms of E and v; the resulting relations are

$$\lambda = \frac{Ev}{(1 + v)(1 - 2v)} \qquad (9.3.6a)$$

$$\mu = \frac{E}{2(1 + v)} \qquad (9.3.6b)$$

Since $E > 0$ and $v > 0$, we have $\mu > 0$.

The following consequences of the relations (9.3.6) are frequently needed:

$$\lambda + 2\mu = \frac{E(1 - v)}{(1 + v)(1 - 2v)}, \qquad \frac{\lambda + \mu}{\mu} = \frac{1}{1 - 2v}$$

$$\frac{\lambda + 2\mu}{\mu} = \frac{2(1 - v)}{1 - 2v}, \qquad \frac{\lambda}{\lambda + 2\mu} = \frac{v}{1 - v} \qquad (9.3.7)$$

CASE ii Suppose the stress tensor is such that $\tau_{ij} = -p\delta_{ij}$. Such a stress system occurs when every point of a material is subjected to an all-around pressure p, as illustrated in Figure 9.2. Then we have $\tau_{kk} = -3p$ and (9.2.7) yields

$$e_{kk} = -\frac{3p}{3\lambda + 2\mu} \qquad (9.3.8)$$

If we set

$$K = \lambda + \tfrac{2}{3}\mu \qquad (9.3.9)$$

it follows from (9.3.8) that

$$K = -\frac{p}{e_{kk}} \qquad (9.3.10)$$

Experiments conducted on most naturally occurring elastic materials show that an all-around pressure tends to reduce the volume of the material; that

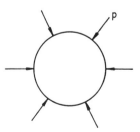

Figure 9.2. Material with all-around pressure.

9.3 PHYSICAL MEANINGS OF ELASTIC MODULI

is, if $p > 0$, then $e_{kk} < 0$. Consequently, it follows from (9.3.10) that $K > 0$. Relation (9.3.10) also shows that the constant K represents the numerical value of the ratio of the compressive stress to the dilatation. This constant is called the *modulus of compression* or the *bulk modulus*. This was introduced by George G. Stokes in 1845.

Substitution for λ and μ from (9.3.6) into (9.3.9) yields

$$K = \frac{E}{3(1 - 2\nu)} \qquad (9.3.11)$$

Since $K > 0$, it follows from (9.3.11) that $\nu \leq \frac{1}{2}$. The relation (9.3.6a) then yields $\lambda > 0$. Thus, the two Lamé constants λ and μ are both positive.

From (9.2.6)″ we note that

$$2\mu = \frac{\tau_{12}}{e_{12}} = \frac{\tau_{13}}{e_{13}} = \frac{\tau_{23}}{e_{23}} \qquad (9.3.12)$$

The constant 2μ is thus the ratio of a shear stress component to the corresponding shear strain component; it is therefore related to the rigidity of the material. For this reason, μ is called the *modulus of rigidity* or the *shear modulus*. The other Lamé constant λ has no direct physical meaning.

The symbols λ, μ, E, ν and K are the five basic elastic constants that appear in the theory of elasticity. As seen earlier, all of these are taken to be positive on the basis of experimental observations, and barring λ, all have definite physical meanings. (The positiveness of the elastic constants can be proven from thermodynamical considerations also.) In view of the interrelationships that exist between these constants, only two of these are independent. Some of these interrelationships are summarized in Table 9.1. The constants μ, E, ν and K that have definite physical meanings are often called *engineering elastic constants*. For a typical copper material, the approximate values of these engineering elastic constants, expressed in units of 10^{10} N/m² are $\mu = 4.8$, $E = 12.98$, $\nu = 0.343$, $K = 13.78$.

Using (9.3.9) and (9.3.11), the relation (9.2.7) can be written as

$$e_{kk} = \frac{1}{3K} \tau_{kk} = \frac{1 - 2\nu}{E} \tau_{kk} \qquad (9.3.13)$$

From (9.3.13) we find that $e_{kk} \equiv 0$ if and only if $\nu = \frac{1}{2}$, provided of course E and τ_{kk} remain finite. For $\nu \to \frac{1}{2}$, the relations (9.3.6) and (9.3.11) yield $\lambda \to \infty$, $K \to \infty$, $\mu = \frac{1}{3}E$. This limiting case corresponds to what is called an *incompressible elastic body*. (Recall from Section 5.6 that, for an incompressible continuum, $e_{kk} = 0$.) Unless stated to the contrary, we will be concerned with elastic bodies that are *not* incompressible; accordingly, we take $\nu < \frac{1}{2}$ so that λ and K remain finite, in general.

9 EQUATIONS OF LINEAR ELASTICITY

Table 9.1

Elastic Constant	In terms of							
	(λ, μ)	(λ, ν)	(λ, K)	(μ, E)	(μ, ν)	(μ, K)	(E, ν)	(E, K)	(ν, K)
$\lambda =$	λ	λ	λ	$\dfrac{\mu(E - 2\mu)}{3\mu - E}$	$\dfrac{2\mu\nu}{1 - 2\nu}$	$K - \dfrac{2}{3}\mu$	$\dfrac{E\nu}{(1 + \nu)(1 - 2\nu)}$	$\dfrac{3K(3K - E)}{9K - E}$	$\dfrac{3K\nu}{1 + \nu}$
$\mu =$	μ	$\dfrac{\lambda(1 - 2\nu)}{2\nu}$	$\dfrac{3}{2}(K - \lambda)$	μ	μ	μ	$\dfrac{E}{2(1 + \nu)}$	$\dfrac{3KE}{9K - E}$	$\dfrac{3K(1 - 2\nu)}{2(1 + \nu)}$
$E =$	$\dfrac{\mu(3\lambda + 2\mu)}{\lambda + \mu}$	$\dfrac{\lambda(1 + \nu)(1 - 2\nu)}{\nu}$	$\dfrac{9K(K - \lambda)}{3K - \lambda}$	E	$2\mu(1 + \nu)$	$\dfrac{9K\mu}{3K + \mu}$	E	E	$3K(1 - 2\nu)$
$\nu =$	$\dfrac{\lambda}{2(\lambda + \mu)}$	ν	$\dfrac{\lambda}{3K - \lambda}$	$\dfrac{E}{2\mu} - 1$	ν	$\dfrac{3K - 2\mu}{2(3K + \mu)}$	ν	$\dfrac{3K - E}{6K}$	ν
$K =$	$\lambda + \dfrac{2}{3}\mu$	$\dfrac{\lambda(1 + \nu)}{3\nu}$	K	$\dfrac{\mu E}{3(3\mu - E)}$	$\dfrac{2\mu(1 + \nu)}{3(1 - 2\nu)}$	K	$\dfrac{E}{3(1 - 2\nu)}$	K	K

9.3 PHYSICAL MEANINGS OF ELASTIC MODULI

It is often convenient to have relations (9.2.6) and (9.2.8) expressed in terms of the constants E and v. Substituting for λ and μ from (9.3.6) in (9.2.6), we obtain

$$\tau_{ij} = \frac{E}{1+v}\left[e_{ij} + \frac{v}{1-2v}\delta_{ij}e_{kk}\right] \qquad (9.3.14)$$

Similarly, with the aid of (9.3.6), the relations (9.2.8) can be rewritten as

$$e_{ij} = \frac{1+v}{E}\tau_{ij} - \frac{v}{E}\delta_{ij}\tau_{kk} \qquad (9.3.15)$$

In the direct notation, relations (9.3.14) and (9.13.15) read as follows:

$$\mathbf{T} = \frac{E}{1+v}\left[\mathbf{E} + \frac{v}{1-2v}(\text{tr }\mathbf{E})\mathbf{I}\right] \qquad (9.3.14)'$$

$$\mathbf{E} = \frac{1+v}{E}\mathbf{T} - \frac{v}{E}(\text{tr }\mathbf{T})\mathbf{I} \qquad (9.3.15)'$$

Relations (9.13.14) and (9.13.15) can also be rewritten in the form of matrix equations. For example, relations (9.13.15) read as follows:

$$\begin{bmatrix} e_{11} \\ e_{22} \\ e_{33} \\ e_{12} \\ e_{23} \\ e_{31} \end{bmatrix} = \begin{bmatrix} 1/E & -v/E & -v/E & 0 & 0 & 0 \\ -v/E & 1/E & -v/E & 0 & 0 & 0 \\ -v/E & -v/E & 1/E & 0 & 0 & 0 \\ 0 & 0 & 0 & (1+v)/E & 0 & 0 \\ 0 & 0 & 0 & 0 & (1+v)/E & 0 \\ 0 & 0 & 0 & 0 & 0 & (1+v)/E \end{bmatrix} \begin{bmatrix} \tau_{11} \\ \tau_{22} \\ \tau_{33} \\ \tau_{12} \\ \tau_{23} \\ \tau_{31} \end{bmatrix}$$

$$(9.3.15)''$$

EXAMPLE 9.3.1 If

$$W = \tfrac{1}{2}[\lambda e_{kk}^2 + 2\mu e_{ij}e_{ij}] \qquad (9.3.16)$$

prove the following:

(i) $\quad \dfrac{\partial W}{\partial e_{ij}} = \tau_{ij} \qquad (9.3.17)$

(ii) $\quad W = \tfrac{1}{2}\tau_{ij}e_{ij} \qquad (9.3.18)$

(iii) $\quad W$ is a *scalar invariant* $\qquad (9.3.19)$

(iv) $\quad W \geq 0 \quad$ and $\quad W = 0 \quad$ if and only if $e_{ij} = 0$

(v) $\quad \dfrac{\partial W}{\partial \tau_{ij}} = e_{ij} \qquad (9.3.20)$

374   9   EQUATIONS OF LINEAR ELASTICITY

(The function W is called the *strain-energy function*; its physical meaning is given in Section 9.4.)

Solution   (i) From (9.3.16), we note that W is a function of e_{ij}. Partial differentiation of this function with respect to e_{ij} gives

$$\frac{\partial W}{\partial e_{ij}} = \frac{1}{2}\left[2\lambda e_{kk}\frac{\partial e_{kk}}{\partial e_{ij}} + 4\mu e_{ij}\right] \qquad (9.3.21)$$

Since $e_{kk} = \delta_{ij}e_{ij}$, it follows that $\partial e_{kk}/\partial e_{ij} = \delta_{ij}$. Using this fact and Hooke's law (9.2.6) in (9.3.21) we immediately get (9.3.17).

(ii) Noting that $e_{kk}^2 = e_{kk}e_{ii} = e_{kk}e_{ij}\delta_{ij}$, (9.3.16) may be rewritten as

$$W = \tfrac{1}{2}[\lambda e_{kk}\delta_{ij} + 2\mu e_{ij}]e_{ij}$$

which yields (9.3.18), on using (9.2.6).

(iii) Since τ_{ij} and e_{ij} are components of tensors $\mathbf{T}$ and $\mathbf{E}$, respectively, $W = \tfrac{1}{2}\tau_{ij}e_{ij} = \tfrac{1}{2}\mathbf{T}\cdot\mathbf{E}$ is a scalar invariant.

(iv) Since $\lambda > 0$, $\mu > 0$, and $(e_{kk})^2$ and $e_{ij}e_{ij}$ are nonnegative, it follows from (9.3.16) that $W \geq 0$ and that $W \equiv 0$ if and only if $e_{kk} = 0$ and $e_{ij} = 0$. Since $e_{ij} = 0$ automatically implies that $e_{kk} = 0$, therefore $W \equiv 0$ holds if and only if $e_{ij} = 0$.

(v) Substituting for e_{ij} in terms of τ_{ij} from (9.3.15) in (9.3.18) we obtain the following expression for W expressed in terms of τ_{ij}:

$$W = \frac{1}{2}\left[\frac{1+\nu}{E}\tau_{ij}\tau_{ij} - \frac{\nu}{E}\tau_{kk}^2\right] \qquad (9.3.22)$$

Differentiating this expression w.r.t. τ_{ij}, we get

$$\frac{\partial W}{\partial \tau_{ij}} = \frac{1+\nu}{E}\tau_{ij} - \frac{\nu}{E}\tau_{kk}\frac{\partial \tau_{kk}}{\partial \tau_{ij}} \qquad (9.3.23)$$

Noting that $\partial \tau_{kk}/\partial \tau_{ij} = \delta_{ij}$ and using (9.3.15) in (9.3.23) we obtain (9.3.20). ∎

EXAMPLE 9.3.2   Show that, for an incompressible elastic body, the stress–strain relation can be written as

$$\mathbf{T} = -\bar{p}\mathbf{I} + 2\mu\mathbf{E} \qquad (9.3.24)$$

where $\bar{p} = -\tfrac{1}{3}(\operatorname{tr}\mathbf{T})$ is the mean compressive normal stress.

9.3 PHYSICAL MEANINGS OF ELASTIC MODULI

Solution For an incompressible elastic body, we have $v = \frac{1}{2}$ and $\mu = \frac{1}{3}E$. In this case, relation (9.3.15)' becomes

$$\mathbf{E} = \frac{1}{2\mu}\mathbf{T} - \frac{1}{6\mu}(\text{tr }\mathbf{T})\mathbf{I} \qquad (9.3.25)$$

so that

$$\mathbf{T} = \tfrac{1}{3}(\text{tr }\mathbf{T})\mathbf{I} + 2\mu\mathbf{E} \qquad (9.3.26)$$

Setting $\bar{p} = -\frac{1}{3}(\text{tr }\mathbf{T})$ in (9.3.26), we get (9.3.24). ∎

EXAMPLE 9.3.3 In an elastic beam placed along the x_3 axis and bent by a couple about the x_2 axis, the stresses are found to be

$$\tau_{33} = -\frac{E}{R}x_1, \qquad \tau_{11} = \tau_{22} = \tau_{12} = \tau_{23} = \tau_{31} = 0$$

where R is a constant. Find the corresponding strains.

Solution It is convenient to employ the matrix equation (9.3.15)″ to find the strain components from the stress components. Substituting the given stress components in (9.3.15)″ and equating the corresponding terms, we obtain the required strain components as

$$e_{11} = e_{22} = \frac{v}{R}x_1, \qquad e_{33} = -\frac{1}{R}x_1, \qquad e_{12} = e_{23} = e_{31} = 0 \qquad \blacksquare$$

EXAMPLE 9.3.4 A beam placed along the x_1 axis and subjected to a longitudinal stress τ_{11} at every point is so constrained that e_{22} and τ_{33} are 0 at every point. Show that

$$\tau_{22} = v\tau_{11} \qquad (9.3.27a)$$

$$e_{11} = \frac{1 - v^2}{E}\tau_{11} \qquad (9.3.27b)$$

$$e_{33} = -\frac{v(1 + v)}{E}\tau_{11} \qquad (9.3.27c)$$

Solution From (9.3.15)″, we get

$$e_{22} = \frac{1}{E}\tau_{22} - \frac{v}{E}(\tau_{11} + \tau_{33})$$

Since $e_{22} = 0$ and $\tau_{33} = 0$, this immediately yields (9.3.27a).

Using (9.3.27a) and the fact that $\tau_{33} = 0$ in the expressions for e_{11} and e_{33} contained in (9.3.15)″, we get (9.3.27b, c). ∎

376 9 EQUATIONS OF LINEAR ELASTICITY

9.4
GOVERNING EQUATIONS

In the preceding chapter we have obtained the following balance equations that are valid for all continua.
Equation of balance of mass:

$$\frac{D\rho}{Dt} + \rho \operatorname{div} \mathbf{v} = 0 \qquad (9.4.1)$$

Equations of balance of momentum:

$$\operatorname{div} \mathbf{T}^T + \rho \mathbf{b} = \rho \frac{D\mathbf{v}}{Dt} \qquad (9.4.2a)$$

$$\mathbf{T}^T = \mathbf{T} \qquad (9.4.2b)$$

Equation of balance of energy:

$$\rho \frac{D\varepsilon}{Dt} = \mathbf{T} \cdot \nabla \mathbf{v} - \operatorname{div} \mathbf{q} + \rho h \qquad (9.4.3)$$

Since the theory of elasticity is a branch of continuum mechanics (in which the continuum is an elastic solid), the field equations (9.4.1)–(9.4.3) are automatically valid in this theory.

The classical elasticity theory is based on the hypothesis that the deformation is infinitesimal. As such, the linear strain-displacement relation (5.6.4) holds in this theory; that is, we have

$$\text{Strain-displacement relation: } \mathbf{E} = \tfrac{1}{2}(\nabla \mathbf{u} + \nabla \mathbf{u}^T) \qquad (9.4.4)$$

The material law valid for homogeneous and isotropic linear elastic solids is given by (9.2.6)′; that is, we have

$$\textit{Material law: } \mathbf{T} = \lambda (\operatorname{tr} \mathbf{E})\mathbf{I} + 2\mu \mathbf{E} \qquad (9.4.5)$$

In Section 6.2, it was noted that for an infinitesimal deformation, $D/Dt \approx \partial/\partial t$ so that

$$\mathbf{v} \approx \frac{\partial \mathbf{u}}{\partial t} \qquad (9.4.6a)$$

$$\operatorname{div} \mathbf{v} \approx \frac{\partial}{\partial t}(\operatorname{div} \mathbf{u}) \qquad (9.4.6b)$$

$$\frac{D\mathbf{v}}{Dt} \approx \frac{\partial^2 \mathbf{u}}{\partial t^2} \qquad (9.4.6c)$$

9.4 GOVERNING EQUATIONS 377

Using (9.4.6b), equation (9.4.1) becomes

$$\frac{1}{\rho}\frac{\partial \rho}{\partial t} + \frac{\partial}{\partial t}(\text{div } \mathbf{u}) = 0 \qquad (9.4.7)$$

The general solution of this partial differential equation is

$$\rho = \rho_0 \exp(-\text{div } \mathbf{u}) \qquad (9.4.8)$$

Thus, under the linear approximation, the continuity equation (9.4.1) has been completely solved for ρ in terms of $\mathbf{u}$.

Using (9.4.8), equation (9.4.2a) can be rewritten as

$$\left(\frac{1}{\rho_0}\text{div } \mathbf{T}^T\right)\exp(\text{div } \mathbf{u}) + \mathbf{b} = \frac{D\mathbf{v}}{Dt} \qquad (9.4.9)$$

Simplifying the first term of this equation with the aid of MacLaurin's expansion applied to $\exp(\text{div } \mathbf{u})$ and neglecting products of stresses and displacements in the expansion and subsequently using (9.4.6c), we arrive at the following linearized form of Cauchy's equation of motion (9.4.2a):

$$\text{div } \mathbf{T}^T + \mathbf{f} = \rho_0 \frac{\partial^2 \mathbf{u}}{\partial t^2} \qquad (9.4.10)$$

where $\mathbf{f} = \rho_0 \mathbf{b}$ is the *body force per unit of undeformed volume*.

Since $\mathbf{E}$ is a symmetric tensor, we find from (9.4.5) that $\mathbf{T}$ is also a symmetric tensor. Thus, the symmetry of $\mathbf{T}$ is an intrinsic physical property of an isotropic linear elastic solid. Consequently, the balance equation (9.4.2b) is trivially satisfied, and equation (9.4.10) becomes

$$\text{div } \mathbf{T} + \mathbf{f} = \rho_0 \frac{\partial^2 \mathbf{u}}{\partial t^2} \qquad (9.4.11)$$

The classical elasticity theory is an isothermal theory (that is, the deformation is assumed to take place at a constant temperature). Therefore, the last two terms on the righthand side of the energy equation (9.4.3) may be set equal to 0 in this theory. Then, under the approximations given by (9.4.6), this equation becomes

$$\frac{\partial \varepsilon}{\partial t} \approx \frac{1}{\rho_0}\mathbf{T} \cdot \nabla \mathbf{v} \approx \frac{1}{\rho_0}\mathbf{T} \cdot \nabla\left(\frac{\partial \mathbf{u}}{\partial t}\right) \approx \frac{1}{\rho_0}\mathbf{T} \cdot \frac{\partial \mathbf{E}}{\partial t} \qquad (9.4.12)$$

Here we have used the relation (9.4.4), the symmetry of $\mathbf{T}$, the relation (2.9.8) and the approximation $\nabla \approx \nabla^0$ valid for small deformation.

9 EQUATIONS OF LINEAR ELASTICITY

Substituting for **T** from Hooke's law (9.4.5) in (9.4.12), we get

$$\rho_0 \frac{\partial \varepsilon}{\partial t} = \lambda(\text{tr } \mathbf{E}) \frac{\partial}{\partial t}(\text{tr } \mathbf{E}) + 2\mu \mathbf{E} \cdot \frac{\partial \mathbf{E}}{\partial t}$$

$$= \frac{1}{2} \frac{\partial}{\partial t} [\lambda(\text{tr } \mathbf{E})^2 + 2\mu \mathbf{E} \cdot \mathbf{E}] = \frac{\partial W}{\partial t} \quad (9.4.13)$$

where

$$W = \tfrac{1}{2} \{\lambda(\text{tr } \mathbf{E})^2 + 2\mu \mathbf{E} \cdot \mathbf{E}\} \quad (9.4.14)$$

Obviously, if $\varepsilon = 0$ in the reference configuration, a general solution of the equation (9.4.13), is

$$\rho_0 \varepsilon = W \quad (9.4.15)$$

Thus, the energy equation (9.4.3) is solved.

We note that, when thermal effects are absent, the only mechanism available for storing internal energy in a body is through deformation. As such, the solution (9.4.15), with W defined as a positive-definite function of **E** through the relation (9.4.14), is meaningful on physical grounds also. By virtue of this solution, *the function W may be interpreted as the internal energy of an elastic body per unit of undeformed volume*. The internal energy of an elastic body is called the *strain-energy*. The function W therefore represents the strain-energy per unit of undeformed volume, usually referred to as the *strain-energy function* or *the elastic potential*.

Note that (9.4.14) is the same as (9.3.16). As such, the elastic potential W has properties given by (9.3.17) through (9.3.20). The function W was first introduced through equations of the form (9.3.17) by George Green in 1839. Expressions (9.3.20) are by A. Castigliano (1875). Expression (9.3.18) is attributed to B. P. E. Clapeyron (1799–1864).

In the preceding paragraphs, the equation of conservation of mass and the equation of energy have been completely solved and the momentum equation (9.4.2a) has been linearized. Also, it is found that the momentum equation (9.4.2b) is identically satisfied. Thus, the *linearized momentum equation* as given by (9.4.11) is the only *field equation* to be taken account of in the study of the classical elasticity theory.

Fifteen individual equations are embodied in the set of equations (9.4.4), (9.4.5) and (9.4.11) and 15 unknowns also are present in the set, the unknowns being 3 displacement components u_i, 6 strain components e_{ij} and 6 stress components τ_{ij}. *The set of equations (9.4.4), (9.4.5) and (9.4.11) therefore serves as a complete set of governing equations for the linear theory of elasticity of homogeneous and isotropic solids*. We note that equations (9.4.4) and (9.4.11) hold for all continua undergoing linear deformation and equation (9.4.5) is the material law portraying the characteristic behavior of homogeneous and isotropic linear elastic solids. Since these equations form a complete set of governing equations, no more constitutive equations are needed. In other words, the material law (9.4.5) is the only constitutive equation needed for the theory under consideration.

In a problem in linear elasticity the displacement vector **u** is generally extracted by integrating (9.4.4). Hence, the following compatibility condition obeyed by **E**, namely, (5.8.15), is also required:

$$\nabla^2 \mathbf{E} + \nabla\nabla(\text{tr } \mathbf{E}) - \nabla \text{ div } \mathbf{E} - (\nabla \text{ div } \mathbf{E})^T = \mathbf{0} \quad (9.4.16)$$

Recall that this compatibility condition is just a consequence of the strain–displacement relation (9.4.4); as such, (9.4.16) serves just as a supplementary condition rather than a basic equation in its own right.

We observe that the governing equations (9.4.4), (9.4.5) and (9.4.11) are expressed in the spatial form. Let us now look at the corresponding equations in the material form.

9.4.1 EQUATIONS IN MATERIAL FORM

In Section 5.6, it has been noted that **E** could be regarded as a function of $\mathbf{x}^0$ as well and that the strain–displacement relation (5.6.4) has the following material form:

$$\mathbf{E} = \tfrac{1}{2}(\nabla^0 \mathbf{u} + \nabla^0 \mathbf{u}^T) \quad (9.4.17)$$

By using (5.5.4), (5.5.7), (5.6.8) and (7.9.10), we find that in the *linear case* $\mathbf{T} \approx \mathbf{T}^0$. Consequently, (8.3.10) which is the equation of motion in material form becomes

$$\text{div}^0 \mathbf{T} + \mathbf{f} = \rho_0 \frac{\partial^2 \mathbf{u}}{\partial t^2} \quad (9.4.18)$$

Here, we have used the symmetry of **T** and the approximations given by (9.4.6). Since **T** is now regarded as a function of $\mathbf{x}^0$ and t, **f** and **u** are also required to be regarded so.

Equations (9.4.17) and (9.4.18) are the counterparts of equations (9.4.4) and (9.4.11) in the Lagrangian description. We observe that these two sets of equations are identical in *structure*. In other words, (9.4.4) and (9.4.11) retain their form in the Lagrangian description also. Further, since the stress–strain relation (9.4.5) is an algebraic equation, it also retains its form in the Lagrangian description. As such, equations (9.4.4), (9.4.5) and (9.4.11) could be employed both in the Eulerian description and in the Lagrangian description, with x_i regarded as the current coordinates in the Eulerian description and as the initial coordinates in the Lagrangian description. The usual practice in the elastic theory is to employ the Lagrangian description. Accordingly, throughout the remaining part of this chapter, equations (9.4.4), (9.4.5) and (9.4.11) are employed with x_i regarded as the *initial* coordinates. Consequently, the region V over which x_i vary is regarded as the *initial configuration* of the body. On the same count, we will write ρ for ρ_0. This arrangement is of course a matter of notation.

EXAMPLE 9.4.1 Show that the change in volume of an elastic body of volume V, bounded by a closed surface S, is given by

$$\delta V = \frac{1-2\nu}{E}\left[\int_V \left(f_i - \rho\frac{\partial^2 u_i}{\partial t^2}\right)x_i\, dV + \int_S s_i x_i\, dS\right] \quad (9.4.19)$$

Solution We first recall that the change in volume of a continuum undergoing small deformation is given by (5.6.26):

$$\delta V = \int_V e_{kk}\, dV$$

Substituting for e_{kk} from (9.3.13) in this expression, we get

$$\delta V = \int_V \frac{1-2\nu}{E}\tau_{kk}\, dV$$

$$= \frac{1-2\nu}{E}\int_V \tau_{ij}\delta_{ij}\, dV = \frac{1-2\nu}{E}\int_V \tau_{ij}x_{i,j}\, dV$$

$$= \frac{1-2\nu}{E}\left[\int_V (\tau_{ij}x_i)_{,j}\, dV - \int_V \tau_{ij,j}x_i\, dV\right] \quad (9.4.20)$$

Employing the divergence theorem (3.6.1)' to the first integral on the righthand side of (9.4.20) and using Cauchy's equation (9.4.11) in the second integral, we get

$$\delta V = \frac{1-2\nu}{E}\left[\int_S \tau_{ij}n_j x_i\, dS + \int_V \left(f_i - \rho\frac{\partial^2 u_i}{\partial t^2}\right)x_i\, dV\right] \quad (9.4.21)$$

The result (9.4.19) is now immediate if we note that $\tau_{ij}n_j = s_i$. ∎

Note: In the absence of inertial effects, (9.4.19) reduces to

$$\delta V = \frac{1-2\nu}{E}\left[\int_S s_i x_i\, dS + \int_V f_i x_i\, dV\right] \quad (9.4.22)$$

EXAMPLE 9.4.2 If an elastic body of volume V, bounded by a closed surface S, is in equilibrium in a deformed state under a given system of body forces f_i and surface forces s_i, prove that the total strain energy is equal to one-half the work that would be done by the external forces acting through the displacements u_i from the undeformed state to the state of equilibrium. (This result is known as *Clapeyron's theorem*.)

Solution We recall that the elastic potential W represents the strain energy per unit volume. Hence the total strain energy U contained in a volume V is

$$U = \int_V W\,dV \qquad (9.4.23)$$

Using (9.3.18) and the fact that

$$\tau_{ij} e_{ij} = \tfrac{1}{2}\tau_{ij}(u_{i,j} + u_{j,i}) = \tau_{ij} u_{i,j}$$

expression (9.4.23) becomes

$$\begin{aligned} U &= \frac{1}{2}\int_V \tau_{ij} u_{i,j}\,dV \\ &= \frac{1}{2}\left[\int_V (\tau_{ij} u_i)_{,j}\,dV - \int_V \tau_{ij,j} u_i\,dV\right] \end{aligned} \qquad (9.4.24)$$

Employing the divergence theorem to the first of the integrals in the righthand side of (9.4.24) and using Cauchy's law $s_i = \tau_{ij} n_j$ and Cauchy's equations of equilibrium $\tau_{ij,j} + f_i = 0$, which follow by equating the righthand side of (9.4.11) to 0, we obtain

$$U = \frac{1}{2}\left[\int_S s_i u_i\,dS + \int_V f_i u_i\,dV\right] \qquad (9.4.25)$$

We note that $s_i u_i$ represents the work done by the surface force **s** per unit of area with displacement **u**. As such $\int_S s_i u_i\,dS$ represents the total work done by surface forces acting on S. Similarly, $\int_V f_i u_i\,dV$ represents the total work done by the body forces acting in V. Thus, the righthand side of (9.4.25) represents one-half the total work done by body forces as well as surface forces. This proves the required result. ∎

EXAMPLE 9.4.3 Let $\tau_{ij}^{(1)}$ be the stresses corresponding to the strains $e_{ij}^{(1)}$ and $\tau_{ij}^{(2)}$ be the stresses corresponding to the strains $e_{ij}^{(2)}$, in an elastic body. Prove that

$$\tau_{ij}^{(1)} e_{ij}^{(2)} = \tau_{ij}^{(2)} e_{ij}^{(1)} \qquad (9.4.26)$$

If $\tau_{ij}^{(1)}$ and $\tau_{ij}^{(2)}$ obey the equations of motion under the body forces $f_i^{(1)}$ and $f_i^{(2)}$, respectively, and if $u_i^{(1)}$, $u_i^{(2)}$ and $s_i^{(1)}$, $s_i^{(2)}$ are the corresponding

9 EQUATIONS OF LINEAR ELASTICITY

displacements and stress vector components, deduce that

$$\int_S s_i^{(1)} u_i^{(2)} \, dS + \int_V \left(f_i^{(1)} - \rho \frac{\partial^2 u_i^{(1)}}{\partial t^2} \right) u_i^{(2)} \, dV$$

$$= \int_S s_i^{(2)} u_i^{(1)} \, dS + \int_V \left(f_i^{(2)} - \rho \frac{\partial^2 u_i^{(2)}}{\partial t^2} \right) u_i^{(1)} \, dV \quad (9.4.27)$$

Interpret this result in the absence of inertia effects.

Solution By virtue of the Hooke's law (9.4.5), we have

$$\tau_{ij}^{(1)} = \lambda \delta_{ij} e_{kk}^{(1)} + 2\mu e_{ij}^{(1)}; \quad \tau_{ij}^{(2)} = \lambda \delta_{ij} e_{kk}^{(2)} + 2\mu e_{ij}^{(2)}$$

These give

$$\tau_{ij}^{(1)} e_{ij}^{(2)} = \lambda e_{ii}^{(2)} e_{kk}^{(1)} + 2\mu e_{ij}^{(1)} e_{ij}^{(2)}$$

$$= (\lambda e_{kk}^{(2)} \delta_{ij} + 2\mu e_{ij}^{(2)}) e_{ij}^{(1)} = \tau_{ij}^{(2)} e_{ij}^{(1)}$$

which is (9.4.26).

Using the symmetry of the stress components, the strain-displacement relations, the divergence theorem, the Cauchy's law and the Cauchy's equations of motion, we find that

$$\int_V \tau_{ij}^{(1)} e_{ij}^{(2)} \, dV = \int_V \tau_{ij}^{(1)} u_{i,j}^{(2)} \, dV$$

$$= \int_V (\tau_{ij}^{(1)} u_i^{(2)})_{,j} \, dV - \int_V \tau_{ij,j}^{(1)} u_i^{(2)} \, dV$$

$$= \int_S \tau_{ij}^{(1)} n_j u_i^{(2)} \, dS + \int_V \left(f_i^{(1)} - \rho \frac{\partial^2 u_i^{(1)}}{\partial t^2} \right) u_i^{(2)} \, dV$$

$$= \int_S s_i^{(1)} u_i^{(2)} \, dS + \int_V \left(f_i^{(1)} - \rho \frac{\partial^2 u_i^{(1)}}{\partial t^2} \right) u_i^{(2)} \, dV \quad (9.4.28)$$

Similarly,

$$\int_V \tau_{ij}^{(2)} e_{ij}^{(1)} \, dV = \int_S s_i^{(2)} u_i^{(1)} \, dS + \int_V \left(f_i^{(2)} - \rho \frac{\partial^2 u_i^{(2)}}{\partial t^2} \right) u_i^{(1)} \, dV \quad (9.4.29)$$

By the result (9.4.26), the lefthand sides of (9.4.28) and (9.4.29) are equal. Hence their righthand sides are also equal. This proves (9.4.27).

In the absence of inertia, (9.4.27) becomes

$$\int_S s_i^{(1)} u_i^{(2)} \, dS + \int_V f_i^{(1)} u_i^{(2)} \, dV = \int_S s_i^{(2)} u_i^{(1)} \, dS + \int_V f_i^{(2)} u_i^{(1)} \, dV \quad (9.4.30)$$

The lefthand side of (9.4.30) represents the total work that would be done by the force system $\{s_i^{(1)}, f_i^{(1)}\}$ in acting through the displacements $u_i^{(2)}$ while the righthand side represents the total work that would be done by the force system $\{s_i^{(2)}, f_i^{(2)}\}$ in acting through the displacements $u_i^{(1)}$. These two works are equal, according to (9.4.30).

Thus, (9.4.30) has the following interpretation: if an elastic body is subjected to two systems of body and surface forces producing two equilibrium states, then the work that would be done by the first system of forces in acting through the displacements of the second system is equal to the work that would be done by the second system of forces in acting through the displacements of the first system. (This result is known as the *reciprocal theorem of Betti and Rayleigh*.) ∎

9.5
BOUNDARY VALUE PROBLEMS

The most general problem of the elasticity theory is to determine the distribution of stresses and strains as well as displacements at all points of a body and at all time when certain boundary conditions and certain initial conditions are specified. It has been noted that, in the linear elasticity theory, the displacements, strains and stresses are governed by equations (9.4.4), (9.4.5) and (9.4.11). Accordingly, solving a problem in linear elasticity generally amounts to solving these equations for **u**, **E** and **T** in terms of **x** and t under certain specified boundary conditions and initial conditions, the body force **f** being assumed to be known beforehand. For a body occupying a region V with boundary S, the *boundary conditions* specified are usually of one of the following three kinds:

(i) The displacement vector is specified at every point of S and for all time $t \geq 0$; that is,

$$\mathbf{u} = \mathbf{u}^* \quad \text{on} \quad S \quad \text{for} \quad t \geq 0 \qquad (9.5.1)$$

where $\mathbf{u}^*$ is a known function.

(ii) The stress vector is specified at every point of S and for all time $t \geq 0$; that is,

$$\mathbf{Tn} = \mathbf{s}^* \quad \text{on} \quad S \quad \text{for} \quad t \geq 0 \qquad (9.5.2)$$

where $\mathbf{s}^*$ is a known function. (Here we have used Cauchy's law $\mathbf{s} = \mathbf{Tn}$.)

(iii) The displacement vector is specified at every point of a part S_u of S for all time $t \geq 0$ and the stress vector is specified at every point on the remaining part $S_\tau = S - S_u$ for all time $t \geq 0$; that is,

$$\left.\begin{array}{l} \mathbf{u} = \mathbf{u}^* \quad \text{on} \quad S_u \\ \mathbf{Tn} = \mathbf{s}^* \quad \text{on} \quad S_\tau \end{array}\right\} \quad \text{for} \quad t \geq 0 \qquad (9.5.3)$$

9 EQUATIONS OF LINEAR ELASTICITY

It has been assumed that at time $t = 0$, the body is in an undeformed state; that is,

$$\mathbf{u} = \mathbf{0} \quad \text{in} \quad V \quad \text{at} \quad t = 0 \tag{9.5.4}$$

This is an *initial condition*. In addition, the following initial condition is also employed:

$$\frac{\partial \mathbf{u}}{\partial t} = \mathbf{v}^* \quad \text{in} \quad V \quad \text{at} \quad t = 0 \tag{9.5.5}$$

where $\mathbf{v}^*$ is a prescribed function. This condition implies that the velocity is specified at every point of the body at time $t = 0$.

The problem of solving equations (9.4.4), (9.4.5) and (9.4.11) under the initial conditions (9.5.4) and (9.5.5) and one of the boundary conditions (9.5.1) to (9.5.3) to determine $\mathbf{u}$, $\mathbf{E}$ and $\mathbf{T}$ at every point $\mathbf{x}$ of V and for $t > 0$ is known as a *boundary value problem in elasticity*. A set $\{\mathbf{u}, \mathbf{E}, \mathbf{T}\}$ so determined, if it exists, is called a *solution* of the problem. When the boundary condition is of the form (9.5.1), the problem is referred to as the *displacement boundary value problem*; when the boundary condition is of the form (9.5.2), the problem is referred to as the *traction* (or *stress*) *boundary value problem*; and when the boundary condition is of the form (9.5.3), the problem is referred to as *mixed boundary value problem*. The three problems are together called the *fundamental boundary value problems*. The boundary conditions valid for all the three problems can be written down in the form of (9.5.3). For the displacement problem, $S = S_u$ and S_τ is empty; for the traction problem $S = S_\tau$ and S_u is empty, and for the mixed problem $S_u \neq S \neq S_\tau$.

In many applications, we consider problems in which inertia can be neglected. For such problems, equation (9.4.11) becomes the equilibrium equation:

$$\text{div } \mathbf{T} + \mathbf{f} = \mathbf{0} \tag{9.5.6}$$

Then, the boundary conditions become time independent and the initial conditions (9.5.4) and (9.5.5) are redundant. The problems governed by equations (9.4.4), (9.4.5) and (9.5.6) and the boundary conditions (9.5.3) are referred to as *boundary value problems in elastostatics*.

On the other hand, time-dependent problems governed by equations (9.4.4), (9.4.5) and (9.4.11), the initial conditions (9.5.4) and (9.5.5) and the boundary conditions (9.5.3) are referred to as *boundary value problems in elastodynamics*.

When a boundary-value problem is considered, natural questions arise: does a solution exist? and if so, is it possible to find more than one solution? A discussion of the first question involves great mathematical difficulties and is beyond the scope of this book. However the second question can be answered (in the negative) in a straightforward way.

9.6
UNIQUENESS OF SOLUTION (STATIC CASE)

First, we consider the case of elastostatic problems. For these problems, the governing equations are (9.4.4), (9.4.5) and (9.5.6); namely,

$$\mathbf{E} = \tfrac{1}{2}(\nabla \mathbf{u} + \nabla \mathbf{u}^T) \tag{9.6.1}$$

$$\mathbf{T} = \lambda(\operatorname{tr} \mathbf{E})\mathbf{I} + 2\mu \mathbf{E} \tag{9.6.2}$$

$$\operatorname{div} \mathbf{T} + \mathbf{f} = \mathbf{0} \tag{9.6.3}$$

Also, the boundary conditions are (9.5.3), namely

$$\mathbf{u} = \mathbf{u}^* \quad \text{on} \quad S_u, \qquad \mathbf{Tn} = \mathbf{s}^* \quad \text{on} \quad S_\tau \tag{9.6.4}$$

A solution of an elastostatic problem is a set $\{\mathbf{u}, \mathbf{E}, \mathbf{T}\}$ where $\mathbf{u}$, $\mathbf{E}$ and $\mathbf{T}$ satisfy equations (9.6.1) to (9.6.3) in V and the boundary conditions (9.6.4) on S, the body force $\mathbf{f}$ being taken to be a known function. Next we prove that such a solution is unique.

9.6.1 UNIQUENESS THEOREM IN ELASTOSTATICS

THEOREM 9.6.1 The solution of an elastostatic problem governed by equations (9.6.1) to (9.6.3) and the boundary conditions (9.6.4) is unique within a rigid body displacement.

Proof Suppose $\{\mathbf{u}^{(1)}, \mathbf{E}^{(1)}, \mathbf{T}^{(1)}\}$ and $\{\mathbf{u}^{(2)}, \mathbf{E}^{(2)}, \mathbf{T}^{(2)}\}$ are two solutions of an elastostatic problem. Then equations (9.6.1)–(9.6.3) and the boundary conditions (9.6.4) are satisfied by both the solutions. Thus,

$$\left. \begin{array}{l} \operatorname{div} \mathbf{T}^{(1)} + \mathbf{f} = \mathbf{0} \\ \operatorname{div} \mathbf{T}^{(2)} + \mathbf{f} = \mathbf{0} \end{array} \right\} \quad \text{in} \quad V \tag{9.6.5}$$

$$\left. \begin{array}{l} \mathbf{T}^{(1)} = \lambda(\operatorname{tr} \mathbf{E}^{(1)})\mathbf{I} + 2\mu \mathbf{E}^{(1)} \\ \mathbf{T}^{(2)} = \lambda(\operatorname{tr} \mathbf{E}^{(2)})\mathbf{I} + 2\mu \mathbf{E}^{(2)} \end{array} \right\} \quad \text{in} \quad V \tag{9.6.6}$$

$$\left. \begin{array}{l} \mathbf{E}^{(1)} = \tfrac{1}{2}(\nabla \mathbf{u}^{(1)} + \nabla \mathbf{u}^{(1)T}) \\ \mathbf{E}^{(2)} = \tfrac{1}{2}(\nabla \mathbf{u}^{(2)} + \nabla \mathbf{u}^{(2)T}) \end{array} \right\} \quad \text{in} \quad V \tag{9.6.7}$$

and

$$\left. \begin{array}{ll} \mathbf{u}^{(1)} = \mathbf{u}^*, & \mathbf{u}^{(2)} = \mathbf{u}^* \quad \text{on} \quad S_u \\ \mathbf{T}^{(1)}\mathbf{n} = \mathbf{s}^*, & \mathbf{T}^{(2)}\mathbf{n} = \mathbf{s}^* \quad \text{on} \quad S_\tau \end{array} \right\} \tag{9.6.8}$$

Suppose we set
$$\bar{\mathbf{u}} = \mathbf{u}^{(1)} - \mathbf{u}^{(2)}, \quad \bar{\mathbf{E}} = \mathbf{E}^{(1)} - \mathbf{E}^{(2)}, \quad \bar{\mathbf{T}} = \mathbf{T}^{(1)} - \mathbf{T}^{(2)} \quad (9.6.9)$$

Then (9.6.5) to (9.6.8) yield

$$\left. \begin{array}{l} \operatorname{div} \bar{\mathbf{T}} = \mathbf{0} \\ \bar{\mathbf{T}} = \lambda (\operatorname{tr} \bar{\mathbf{E}})\mathbf{I} + 2\mu \bar{\mathbf{E}} \\ \bar{\mathbf{E}} = \tfrac{1}{2}(\nabla \bar{\mathbf{u}} + \nabla \bar{\mathbf{u}}^T) \end{array} \right\} \quad \text{in} \quad V \quad \begin{array}{l}(9.6.10a)\\(9.6.10b)\\(9.6.10c)\end{array}$$

and

$$\bar{\mathbf{u}} = \mathbf{0} \quad \text{on} \quad S_u, \quad \bar{\mathbf{T}}\mathbf{n} = \mathbf{0} \quad \text{on} \quad S_\tau \quad (9.6.11)$$

Let

$$I = \int_V \bar{\mathbf{T}} \cdot \bar{\mathbf{E}} \, dV \quad (9.6.12)$$

Simplifying the integrand in (9.6.12) by using (9.6.10b) we obtain

$$I = \int_V \{\lambda (\operatorname{tr} \bar{\mathbf{E}})^2 + 2\mu \bar{\mathbf{E}} \cdot \bar{\mathbf{E}}\} \, dV \quad (9.6.13)$$

If $\bar{u}_i$, $\bar{e}_{ij}$ and $\bar{\tau}_{ij}$ are components of $\bar{\mathbf{u}}$, $\bar{\mathbf{E}}$ and $\bar{\mathbf{T}}$, respectively, we note that $\bar{\mathbf{T}} \cdot \bar{\mathbf{E}} = \bar{\tau}_{ij}\bar{e}_{ij} = \bar{\tau}_{ij}\bar{u}_{i,j}$. Expression (9.6.12) then yields

$$\begin{aligned} I &= \int_V \bar{\tau}_{ij}\bar{u}_{i,j} \, dV \\ &= \int_V \{(\bar{\tau}_{ij}\bar{u}_i)_{,j} - \bar{\tau}_{ij,j}\bar{u}_i\} \, dV \\ &= \int_S \bar{\tau}_{ij}\bar{u}_i n_j \, dS - \int_V \bar{\tau}_{ij,j}\bar{u}_i \, dV \end{aligned} \quad (9.6.14)$$

In (9.6.14), the surface integral vanishes because of the boundary conditions (9.6.11) and the volume integral vanishes because of equation (9.6.10a). Accordingly, we have $I \equiv 0$. Expression (9.6.13) therefore yields

$$\int_V \{\lambda (\operatorname{tr} \bar{\mathbf{E}})^2 + 2\mu \bar{\mathbf{E}} \cdot \bar{\mathbf{E}}\} \, dV \equiv 0 \quad (9.6.15)$$

Since $\lambda > 0$, $\mu > 0$, and $(\operatorname{tr} \bar{\mathbf{E}})^2$ and $\bar{\mathbf{E}} \cdot \bar{\mathbf{E}} = \bar{e}_{ij}\bar{e}_{ij}$ are nonnegative, the integrand in equation (9.6.15) must be identically 0. Consequently,

$$\operatorname{tr} \bar{\mathbf{E}} \equiv 0, \quad \bar{\mathbf{E}} \equiv \mathbf{0} \quad \text{in} \quad V \quad (9.6.16)$$

Equation (9.6.10b) then yields

$$\bar{\mathbf{T}} \equiv \mathbf{0} \quad \text{in} \quad V \quad (9.6.17)$$

Since $\bar{\mathbf{E}} \equiv \mathbf{0}$, the corresponding displacement $\bar{\mathbf{u}}$ represents a rigid-body displacement; see Example 5.6.2. The relations (9.6.9), (9.6.16) and (9.6.17) together imply that the two solutions $\{\mathbf{u}^{(1)}, \mathbf{E}^{(1)}, \mathbf{T}^{(1)}\}$ and $\{\mathbf{u}^{(2)}, \mathbf{E}^{(2)}, \mathbf{T}^{(2)}\}$ are identical within a rigid body displacement. The theorem is thus proved. ∎

This theorem was first established by Kirchhoff in 1859.

The crucial point in the proof of the theorem is the nonnegativeness of the integrand in (9.6.13). In view of expression (9.4.14), this integrand is nothing but twice the strain-energy function associated with the strain $\bar{\mathbf{E}}$. As such, the uniqueness of solution just proven is a direct consequence of the positive definiteness of the strain-energy function.

9.7
UNIQUENESS OF SOLUTION (DYNAMIC CASE)

We now consider the case of elastodynamic problems. For these problems, the governing equations are (9.4.4), (9.4.5) and (9.4.11), namely,

$$\text{div } \mathbf{T} + \mathbf{f} = \rho \frac{\partial^2 \mathbf{u}}{\partial t^2} \tag{9.7.1}$$

$$\mathbf{T} = \lambda (\text{tr } \mathbf{E})\mathbf{I} + 2\mu \mathbf{E} \tag{9.7.2}$$

$$\mathbf{E} = \tfrac{1}{2}(\nabla \mathbf{u} + \nabla \mathbf{u}^T) \tag{9.7.3}$$

the initial conditions are (9.5.4) and (9.5.5), namely,

$$\mathbf{u} = \mathbf{0}, \quad \frac{\partial \mathbf{u}}{\partial t} = \mathbf{v}^* \quad \text{for } t = 0 \tag{9.7.4}$$

and the boundary conditions are (9.5.3), namely,

$$\left. \begin{array}{l} \mathbf{n} = \mathbf{u}^* \quad \text{on } S_u \\ \mathbf{Tn} = \mathbf{s}^* \quad \text{on } S_\tau \end{array} \right\} \quad \text{for } t \geq 0 \tag{9.7.5}$$

A solution of an elastodynamic problem is a set $\{\mathbf{u}, \mathbf{E}, \mathbf{T}\}$ where $\mathbf{u}, \mathbf{E}, \mathbf{T}$ satisfy (i) equations (9.7.1) to (9.7.3) in V for $t > 0$, with $\mathbf{f}$ as known function, (ii) the initial conditions (9.7.4) in V for $t = 0$, and (iii) the boundary conditions (9.7.5) on S for $t \geq 0$. Next we prove that such a solution is unique.

9.7.1 UNIQUENESS THEOREM IN ELASTODYNAMICS

THEOREM 9.7.1 An elastodynamic problem governed by equations (9.7.1) to (9.7.3), the initial conditions (9.7.4) and the boundary conditions (9.7.5) cannot have more than one solution.

Proof Suppose $\{\mathbf{u}^{(1)}, \mathbf{E}^{(1)}, \mathbf{T}^{(1)}\}$ and $\{\mathbf{u}^{(2)}, \mathbf{E}^{(2)}, \mathbf{T}^{(2)}\}$ are two solutions of an elastodynamic problem. Then equations (9.7.1) to (9.7.3), the initial conditions (9.7.4) and the boundary conditions (9.7.5) are satisfied by both solutions. Accordingly, if we set

$$\bar{\mathbf{u}} = \mathbf{u}^{(1)} - \mathbf{u}^{(2)}, \quad \bar{\mathbf{E}} = \mathbf{E}^{(1)} - \mathbf{E}^{(2)}, \quad \bar{\mathbf{T}} = \mathbf{T}^{(1)} - \mathbf{T}^{(2)} \quad (9.7.6)$$

then $\bar{\mathbf{u}}, \bar{\mathbf{E}}, \bar{\mathbf{T}}$ satisfy the following equations and boundary conditions:

$$\operatorname{div} \bar{\mathbf{T}} = \rho \frac{\partial^2 \bar{\mathbf{u}}}{\partial t^2} \quad \text{in} \quad V \quad (9.7.7a)$$

$$\bar{\mathbf{T}} = \lambda (\operatorname{tr} \bar{\mathbf{E}}) \mathbf{I} + 2\mu \bar{\mathbf{E}} \quad \text{in} \quad V \quad (9.7.7b)$$

$$\bar{\mathbf{E}} = \tfrac{1}{2}(\nabla \bar{\mathbf{u}} + \nabla \bar{\mathbf{u}}^T) \quad \text{in} \quad V \quad (9.7.7c)$$

$$\bar{\mathbf{u}} = 0 \quad \text{on} \quad S_u, \quad \bar{\mathbf{T}}\mathbf{n} = 0 \quad \text{on} \quad S_\tau \quad (9.7.8)$$

Relations (9.7.7) and (9.7.8) must hold for $t \geq 0$.

Also, we obtain the following initial conditions from (9.7.4):

$$\bar{\mathbf{u}} = 0 \quad (9.7.9a)$$

$$\frac{\partial \bar{\mathbf{u}}}{\partial t} = 0 \quad (9.7.9b)$$

in V for $t = 0$. Let

$$N = \int_V \left\{ \bar{\mathbf{T}} \cdot \bar{\mathbf{E}} + \rho \frac{\partial \bar{\mathbf{u}}}{\partial t} \cdot \frac{\partial \bar{\mathbf{u}}}{\partial t} \right\} dV \quad (9.7.10)$$

Substituting for $\bar{\mathbf{T}}$ from (9.7.7b) in (9.7.10), we obtain

$$N = \int_V \left\{ \lambda (\operatorname{tr} \bar{\mathbf{E}})^2 + 2\mu \bar{\mathbf{E}} \cdot \bar{\mathbf{E}} + \rho \frac{\partial \bar{\mathbf{u}}}{\partial t} \cdot \frac{\partial \bar{\mathbf{u}}}{\partial t} \right\} dV$$

$$= \int_V \left\{ \lambda (\bar{e}_{kk})^2 + 2\mu \bar{e}_{ij} \bar{e}_{ij} + \rho \frac{\partial \bar{u}_i}{\partial t} \frac{\partial \bar{u}_i}{\partial t} \right\} dV \quad (9.7.11)$$

where $\bar{u}_i, \bar{e}_{ij}$ and $\bar{\tau}_{ij}$ are components of $\bar{\mathbf{u}}, \bar{\mathbf{E}}$ and $\bar{\mathbf{T}}$, respectively. Then

$$\frac{DN}{Dt} = 2 \int_V \left\{ \left(\lambda \bar{e}_{kk} \frac{\partial \bar{e}_{ii}}{\partial t} + 2\mu \bar{e}_{ij} \frac{\partial \bar{e}_{ij}}{\partial t} \right) + \rho \frac{\partial \bar{u}_i}{\partial t} \frac{\partial^2 \bar{u}_i}{\partial t^2} \right\} dV \quad (9.7.12)$$

9.7 UNIQUENESS OF SOLUTION (DYNAMIC CASE)

In obtaining this expression, we have used the approximation $D/Dt \approx \partial/\partial t$ and the fact that V is regarded as the initial configuration.

By use of (9.7.7b), the symmetry of $\bar{\mathbf{T}}$ and (9.7.7c), we find that

$$\lambda \bar{e}_{ii} \frac{\partial \bar{e}_{kk}}{\partial t} + 2\mu \bar{e}_{ij} \frac{\partial \bar{e}_{ij}}{\partial t} = \bar{\tau}_{ij} \left(\frac{\partial \bar{u}_i}{\partial t} \right)_{,j} \quad (9.7.13)$$

Using this expression, (9.7.12) becomes

$$\frac{DN}{Dt} = 2 \int_V \left\{ \bar{\tau}_{ij} \left(\frac{\partial \bar{u}_i}{\partial t} \right)_{,j} + \rho \frac{\partial \bar{u}_i}{\partial t} \frac{\partial^2 \bar{u}_i}{\partial t^2} \right\} dV$$

$$= 2 \int_V \left\{ \left(\bar{\tau}_{ij} \frac{\partial \bar{u}_i}{\partial t} \right)_{,j} - \bar{\tau}_{ij,j} \frac{\partial \bar{u}_i}{\partial t} + \rho \frac{\partial \bar{u}_i}{\partial t} \frac{\partial^2 \bar{u}_i}{\partial t^2} \right\} dV \quad (9.7.14)$$

Using the divergence theorem and (9.7.7a), expression (9.7.14) yields

$$\frac{DN}{Dt} = 2 \int_S \bar{\tau}_{ij} \frac{\partial \bar{u}_i}{\partial t} n_j \, dS \quad (9.7.15)$$

The boundary conditions (9.7.8) imply that the righthand side of (9.7.15) is 0. Consequently, we obtain

$$N = N_0 \quad (9.7.16)$$

where N_0 is independent of time t.

By the initial condition (9.7.9a), we have $\bar{u}_i = 0$ at $t = 0$. From (9.7.7c) it then follows that $\bar{e}_{ij} = 0$ at $t = 0$. We also have $\partial \bar{u}_i / \partial t = 0$ at $t = 0$ by condition (9.7.9b). Consequently, (9.7.10) implies that $N = 0$ for $t = 0$. It then follows from (9.7.16) that $N \equiv 0$ for all time $t \geq 0$. Hence, from (9.7.11) we get

$$\int_V \left\{ \lambda (\text{tr } \bar{\mathbf{E}})^2 + 2\mu \bar{\mathbf{E}} \cdot \bar{\mathbf{E}} + \rho \frac{\partial \bar{\mathbf{u}}}{\partial t} \cdot \frac{\partial \bar{\mathbf{u}}}{\partial t} \right\} dV \equiv 0 \quad (9.7.17)$$

Since $\lambda > 0$, $\mu > 0$, $\rho > 0$, and $(\text{tr } \bar{\mathbf{E}})^2$, $\bar{\mathbf{E}} \cdot \bar{\mathbf{E}}$ and $\partial \bar{\mathbf{u}}/\partial t \cdot \partial \bar{\mathbf{u}}/\partial t$ are nonnegative, it follows that the integrand in (9.7.17) is exactly 0; consequently,

$$\text{tr } \bar{\mathbf{E}} \equiv 0 \quad (9.7.18a)$$

$$\bar{\mathbf{E}} \equiv \mathbf{0} \quad (9.7.18b)$$

$$\frac{\partial \bar{\mathbf{u}}}{\partial t} \equiv \mathbf{0} \quad (9.7.18c)$$

in V for all $t \geq 0$. Since $\bar{\mathbf{u}} = \mathbf{0}$ in V at $t = 0$, (9.7.18c) yields

$$\bar{\mathbf{u}} \equiv \mathbf{0} \quad (9.7.19)$$

in V for all $t \geq 0$. The relation (9.7.18b) together with (7.7.7b) yields

$$\bar{\mathbf{T}} = \mathbf{0} \qquad (9.7.20)$$

in V for all $t \geq 0$.

Relations (9.7.6), (9.7.18b), (9.7.19) and (9.7.20) together show that the two solutions $\{\mathbf{u}^{(1)}, \mathbf{E}^{(1)}, \mathbf{T}^{(1)}\}$ and $\{\mathbf{u}^{(2)}, \mathbf{E}^{(2)}, \mathbf{T}^{(2)}\}$ are identical. The theorem is thus proven. ∎

This theorem was first established by Franz E. Neumann in 1855.

The crucial point in the proof of the theorem is the nonnegative character of the integrand in (9.7.10). This integrand is nothing but twice the sum of the strain-energy function associated with $\bar{\mathbf{E}}$ and the kinetic energy function associated with $\bar{\mathbf{u}}$. As such, the uniqueness of the solution just proven is a direct consequence of the positive definiteness of the strain-energy function and the nonnegativeness of the kinetic energy function.

Having answered the question of uniqueness of solutions in the affirmative (in both static and dynamic cases), our next aim is to look for ways of solving the boundary value problems, assuming of course that the solutions do exist.

Recall that solving a boundary-value problem means the determination of fifteen unknowns u_i, e_{ij} and τ_{ij} by solving the fifteen governing equations embodied in (9.4.4), (9.4.5) and (9.4.11) under appropriate initial and boundary conditions. But finding a solution of fifteen equations for fifteen unknowns is a formidable task. It is therefore convenient to have fewer governing equations with fewer unknowns. Since the boundary conditions are usually specified in terms of displacement or stress, it is convenient to express the governing equations entirely in terms of displacement or stress. This is done in the following sections.

9.8
NAVIER'S EQUATION

We now proceed to express the governing equations (9.4.4), (9.4.5) and (9.4.11) entirely in terms of the displacement vector.

Equations (9.4.4), (9.4.5) and (9.4.11) are

$$\mathbf{E} = \tfrac{1}{2}(\nabla \mathbf{u} + \nabla \mathbf{u}^T) \qquad (9.8.1)$$

$$\operatorname{div} \mathbf{T} + \mathbf{f} = \rho \frac{\partial^2 \mathbf{u}}{\partial t^2} \qquad (9.8.2)$$

$$\mathbf{T} = \lambda (\operatorname{tr} \mathbf{E})\mathbf{I} + 2\mu \mathbf{E} \qquad (9.8.3)$$

9.8 NAVIER'S EQUATION

Substituting for **E** from (9.8.1) in (9.8.3), we get the *stress-displacement relation*:

$$\mathbf{T} = \lambda(\text{div } \mathbf{u})\mathbf{I} + \mu(\nabla\mathbf{u} + \nabla\mathbf{u}^T) \tag{9.8.4}$$

Using the identities (3.5.35), (3.5.13) and (3.5.14), we find from (9.8.4) that

$$\text{div } \mathbf{T} = \mu\nabla^2\mathbf{u} + (\lambda + \mu)\nabla(\text{div } \mathbf{u}) \tag{9.8.5}$$

Substitution of (9.8.5) into (9.8.2) yields the following governing equation expressed entirely in terms of the displacement vector:

$$\mu\nabla^2\mathbf{u} + (\lambda + \mu)\nabla \text{ div } \mathbf{u} + \mathbf{f} = \rho\frac{\partial^2\mathbf{u}}{\partial t^2} \tag{9.8.6}$$

The suffix form of this equation is

$$\mu\nabla^2 u_i + (\lambda + \mu)u_{k,ki} + f_i = \rho\frac{\partial^2 u_i}{\partial t^2} \tag{9.8.6}'$$

Equation (9.8.6) forms a synthesis of the analysis of strain, analysis of stress and the stress–strain relation. This fundamental partial differential equation of the elasticity theory is known as *Navier's equation of motion*, after Navier (1821).

Navier's equation (9.8.6) can be written in several different forms. By using the vector identity (3.4.27), this equation can be put in the following two alternative forms:

$$(\lambda + 2\mu)\nabla^2\mathbf{u} + (\lambda + \mu)\text{ curl curl } \mathbf{u} + \mathbf{f} = \rho\frac{\partial^2\mathbf{u}}{\partial t^2} \tag{9.8.7}$$

$$(\lambda + 2\mu)\nabla \text{ div } \mathbf{u} - \mu \text{ curl curl } \mathbf{u} + \mathbf{f} = \rho\frac{\partial^2\mathbf{u}}{\partial t^2} \tag{9.8.8}$$

On using the relation [see (9.3.7)]

$$\frac{\lambda + \mu}{\mu} = \frac{1}{1 - 2\nu} \tag{9.8.9}$$

we can write equation (9.8.6) in one more form as

$$\Box_2\mathbf{u} + \frac{1}{1 - 2\nu}\nabla \text{ div } \mathbf{u} + \frac{1}{\mu}\mathbf{f} = \mathbf{0} \tag{9.8.10}$$

where

$$\Box_2 = \nabla^2 - \frac{1}{c_2^2}\frac{\partial^2}{\partial t^2} \tag{9.8.11a}$$

$$c_2^2 = \frac{\mu}{\rho} \tag{9.8.11b}$$

9.8.1 STATIC CASE

In the static case, equation (9.8.6) becomes

$$\mu\nabla^2\mathbf{u} + (\lambda + \mu)\nabla\,\mathrm{div}\,\mathbf{u} + \mathbf{f} = \mathbf{0} \qquad (9.8.12)$$

which reads as follows in the suffix notation:

$$\mu\nabla^2 u_i + (\lambda + \mu)u_{k,ki} + f_i = 0 \qquad (9.8.12)'$$

Equation (9.8.12) is known as *Navier's equation of equilibrium*.

Like (9.8.6), equation (9.8.12) can also be rewritten in several different forms. Of these, the following is a standard one:

$$\nabla^2\mathbf{u} + \frac{1}{1-2\nu}\nabla(\mathrm{div}\,\mathbf{u}) + \frac{1}{\mu}\mathbf{f} = \mathbf{0} \qquad (9.8.13)$$

Note that this is the static counterpart of equation (9.8.10).

EXAMPLE 9.8.1 If

$$\square_1 = \nabla^2 - \frac{1}{c_1^2}\frac{\partial^2}{\partial t^2} \qquad (9.8.14a)$$

$$c_1^2 = \frac{\lambda + 2\mu}{\rho} \qquad (9.8.14b)$$

prove the following:

(i) $$c_1^2 = \frac{E(1-\nu)}{(1+\nu)(1-2\nu)\rho} \qquad (9.8.15a)$$

$$(c_2^2/c_1^2) = \frac{E}{2\rho(1+\nu)} \qquad (9.8.15b)$$

(ii) $$c_1^2 = \frac{\mu}{\lambda + 2\mu} = \frac{1-2\nu}{2(1-\nu)} \qquad (9.8.16)$$

(iii) $$c_1^2\square_1 - c_2^2\square_2 = (c_1^2 - c_2^2)\nabla^2 \qquad (9.8.17)$$

(iv) $$2(1-\nu)\square_1 - (1-2\nu)\square_2 = \nabla^2 \qquad (9.8.18)$$

(v) $$\square_1 - \square_2 = \frac{\rho(1+\nu)}{E(1-\nu)}\frac{\partial^2}{\partial t^2} \qquad (9.8.19)$$

Solution (i) Substituting for $(\lambda + 2\mu)$ from (9.3.7) in (9.8.14b), we get (9.8.15a). Substitution for μ from (9.3.6b) in (9.8.11b) yields (9.8.15b).

(ii) Using (9.8.11b), (9.8.14b) and (9.8.15) we readily obtain (9.8.16).

(iii) Expressions (9.8.14a) and (9.8.11a) yield (9.8.17).

9.8 NAVIER'S EQUATION   393

(iv) Substituting for c_1^2 and c_2^2 from (9.8.15) in (9.8.17), and simplifying the resulting expression, we obtain (9.8.18).

(v) Expression (9.18.19) follows from (9.8.11a), (9.8.14a) and (9.8.15). ∎

EXAMPLE 9.8.2 Prove the following:

(i) $$\Box_1(\text{div } \mathbf{u}) = -\frac{1}{\rho c_1^2} \text{div } \mathbf{f} \qquad (9.8.20)$$

(ii) $$\Box_2(\text{curl } \mathbf{u}) = -\frac{1}{\rho c_2^2} \text{curl } \mathbf{f} \qquad (9.8.21)$$

(iii) $$\Box_1 \Box_2 \mathbf{u} = -\frac{1}{\rho c_2^2}\left[\Box_1 - \frac{1}{2(1-\nu)}\nabla \text{ div}\right]\mathbf{f} \qquad (9.8.22)$$

Solution (i) Taking the divergence of Navier's equation (9.8.6) with the aid of the identity (3.4.13), we obtain the equation

$$(\lambda + 2\mu)\nabla^2(\text{div } \mathbf{u}) + \text{div } \mathbf{f} = \rho \frac{\partial^2}{\partial t^2}(\text{div } \mathbf{u}) \qquad (9.8.23)$$

Rearranging the terms in this equation with the aid of (9.8.14), we get (9.8.20).

(ii) Taking the curl of equation (9.8.6) with the aid of the identity (3.4.16), we obtain the equation

$$\mu\nabla^2(\text{curl } \mathbf{u}) + \text{curl } \mathbf{f} = \rho \frac{\partial^2}{\partial t^2}(\text{curl } \mathbf{u}) \qquad (9.8.24)$$

Rearranging the terms in this equation with the aid of (9.8.11), we arrive at (9.8.21).

(iii) Operating the equation (9.8.10) throughout by $\Box_1$ and substituting for $\Box_1(\text{div } \mathbf{u})$ from (9.8.20) in the resulting equation, we get

$$\Box_1\Box_2(\text{div } \mathbf{u}) - \frac{1}{\rho c_1^2(1-2\nu)}\nabla(\text{div } \mathbf{f}) + \frac{1}{\mu}\Box_1\mathbf{f} = \mathbf{0} \qquad (9.8.25)$$

Substituting for μ from (9.3.6b) in (9.8.25) and simplifying the resulting expression with the use of (9.8.16), we obtain (9.8.22). ∎

Note: If $\mathbf{f}$ is constant, the preceding results become $\Box_1(\text{div } \mathbf{u}) = 0$, $\Box_2(\text{curl } \mathbf{u}) = \mathbf{0}$ and $\Box_1\Box_2\mathbf{u} = \mathbf{0}$. If, further, the deformation is static, we get $\nabla^2(\text{div } \mathbf{u}) = 0$, $\nabla^2(\text{curl } \mathbf{u}) = \mathbf{0}$, $\nabla^2\nabla^2\mathbf{u} = \mathbf{0}$.

394 9 EQUATIONS OF LINEAR ELASTICITY

EXAMPLE 9.8.3 Show that the stress vector on a surface in an elastic body is given by

$$\mathbf{s} = \lambda(\text{div } \mathbf{u})\mathbf{n} + 2\mu(\mathbf{n} \cdot \nabla)\mathbf{u} + \mu \mathbf{n} \times \text{curl } \mathbf{u} \tag{9.8.26}$$

Solution Substituting for $\mathbf{T}$ from the stress-displacement relation (9.8.4) into Cauchy law: $\mathbf{s} = \mathbf{Tn}$ and using the identities (3.5.34) and (3.6.19), we obtain (9.8.26). ∎

EXAMPLE 9.8.4 Show that for an incompressible elastic body,

$$\mathbf{T}^{(d)} = \mu(\nabla \mathbf{u} + \nabla \mathbf{u}^T) \tag{9.8.27}$$

Deduce that

$$\Box_2 \mathbf{u} + \frac{1}{E}\{\nabla(\text{tr } \mathbf{T}) + 3\mathbf{f}\} = \mathbf{0} \tag{9.8.28}$$

Solution For an incompressible elastic body, the stress–strain relation is given by (9.3.26). Substituting for $\mathbf{E}$ from the strain-displacement relation (9.8.1) in (9.3.26) and using the definition of $\mathbf{T}^{(d)}$ we readily obtain (9.8.27).

Taking the divergence throughout in (9.3.26), using (9.8.1) and the identities (3.5.35), (3.5.13), (3.5.14) and bearing in mind that div $\mathbf{u} = 0$ and $\mu = E/3$, we obtain

$$\text{div } \mathbf{T} = \tfrac{1}{3}[\nabla(\text{tr } \mathbf{T}) + E\nabla^2 \mathbf{u}] \tag{9.8.29}$$

Substitution of this expression into Cauchy's equation of motion (9.8.2) yields equation (9.8.28). ∎

EXAMPLE 9.8.5 When an elastic body undergoes spherically symmetric deformation, the displacement vector is of the form

$$\mathbf{u} = u(r)\mathbf{e}_r, \qquad r \neq 0 \tag{9.8.30}$$

where $\mathbf{e}_r$ is the unit vector along the radial direction. For such a displacement, compute (i) the corresponding stress components, (ii) the normal stress on a spherical surface $r = $ constant and (iii) the normal stress on a radial plane. Also, determine $u(r)$ so that Navier's equation of equilibrium with zero body force is satisfied.

Solution (i) The expression (9.8.30) can be rewritten as

$$\mathbf{u} = u(r)\frac{1}{r}\mathbf{x} = \phi(r)\mathbf{x} \tag{9.8.31}$$

where

$$\phi(r) = \frac{1}{r}u(r) \tag{9.8.32}$$

9.8 NAVIER'S EQUATION

From (9.8.31), we find that $u_i = \phi(r)x_i$ so that

$$u_{i,j} = \phi(r)\delta_{ij} + \phi'(r)\left(\frac{1}{r}x_j\right)x_i = u_{j,i} \qquad (9.8.33)$$

Therefore

$$u_{k,k} = 3\phi(r) + r\phi'(r) \qquad (9.8.34)$$

Substituting for $u_{i,j}$ and $u_{j,i}$ from (9.8.33) and for $u_{k,k}$ from (9.8.34) in the stress-displacement relation (9.8.4) rewritten in the suffix notation and using (9.8.32), we obtain the following expressions for the stresses associated with the given displacement field:

$$\tau_{ij} = 2\left\{(\lambda + \mu)\delta_{ij} - 2\mu\frac{1}{r}x_ix_j\right\}\frac{1}{r}u(r) + \left\{\lambda\delta_{ij} + 2\mu\frac{1}{r^2}x_ix_j\right\}u'(r) \qquad (9.8.35)$$

(ii) For a spherical surface r = constant, we have $\mathbf{n} = \mathbf{e}_r$ so that $n_i = x_i/r$. Therefore, by (7.5.3), the normal stress σ_r on this surface is given by $\sigma_r = \tau_{ij}n_in_j = (\tau_{ij}x_ix_j)/r^2$. Using (9.8.35), we obtain

$$\sigma_r = 2\lambda\frac{1}{r}u(r) + (\lambda + 2\mu)u'(r) \qquad (9.8.36)$$

This normal stress is called the *radial stress*; see Figure 9.3.

(iii) If $\bar{\mathbf{n}}$ is the unit normal to a radial plane, we have $\bar{\mathbf{n}} \cdot \mathbf{e}_r = 0$, and the normal stress σ_h on the plane is given by $\sigma_h = \tau_{ij}\bar{n}_i\bar{n}_j$. By use of (9.8.35), we obtain the following expression for σ_h:

$$\sigma_h = 2(\lambda + \mu)\frac{1}{r}u(r) + \lambda u'(r) \qquad (9.8.37)$$

This normal stress is called the *peripheral stress* or the *hoop stress*; see Figure 9.3.

(iv) Finally, to determine $u(r)$, we return to (9.8.33) and (9.8.34) and find from these expressions that

$$u_{i,ij} = u_{k,ki} = \left\{\phi''(r) + \frac{4}{r}\phi'(r)\right\}x_i \qquad (9.8.38)$$

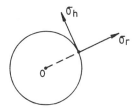

Figure 9.3. Radial stress and hoop stress.

396   9 EQUATIONS OF LINEAR ELASTICITY

Substituting these into Navier's equation of equilibrium (9.8.12)′ with $f_i = 0$, we find that this equation is satisfied if $\phi(r)$ obeys the following differential equation:

$$\frac{d^2\phi}{dr^2} + \frac{4}{r}\frac{d\phi}{dr} = 0 \qquad (9.8.39)$$

The general solution of this equation is

$$\phi(r) = \frac{A}{r^3} + B \qquad (9.8.40)$$

where A and B are arbitrary constants. Expression (9.8.32) then yields

$$u(r) = \frac{A}{r^2} + Br \qquad (9.8.41)$$

Thus, when $u(r)$ is as given by (9.8.41), **u** given by (9.8.30) obeys Navier's equation of equilibrium with zero body force. ∎

EXAMPLE 9.8.6   When an elastic body undergoes a deformation symmetric about the x_3 axis, the displacement vector is of the form

$$\mathbf{u} = u(R)\mathbf{e}_R, \qquad R \neq 0 \qquad (9.8.42)$$

where $R^2 = x_1^2 + x_2^2$ and $\mathbf{e}_R$ is the unit vector along the radial direction in the cylindrical polar coordinate system with x_3 axis as axis. For such a displacement, compute (i) the corresponding stress components, (ii) the normal stress on a cylindrical surface $R = $ constant and (iii) the normal stress on a plane containing the x_3 axis (such a plane is called an *axial plane*). Also, determine $u(R)$ such that Navier's equation of equilibrium with zero body force is satisfied.

Solution   (i) The expression (9.8.42) can be rewritten as

$$u_i = u(R)\frac{1}{R}x_i = \psi(R)x_i, \qquad i,j = 1,2; \qquad u_3 = 0 \qquad (9.8.43)$$

Then

$$\left.\begin{array}{l} u_{i,j} = \psi(R)\delta_{ij} + \dfrac{1}{R}\psi'(R)x_i x_j = u_{j,i} \quad \text{for} \quad i,j = 1,2 \\[4pt] u_{3,j} \equiv 0 \end{array}\right\} \qquad (9.8.44)$$

so that

$$u_{k,k} = 2\psi(R) + R\psi'(R) \qquad (9.8.45)$$

Substituting (9.8.44) and (9.8.45) into the stress-displacement relation (9.8.4) rewritten in the suffix notation and using the fact that $\psi(R) = (1/R)u(R)$, we obtain the following expressions for stresses associated with

9.8 NAVIER'S EQUATION

the given displacement field:

$$\begin{aligned}\tau_{ij} &= \left\{(\lambda + 2\mu)\delta_{ij} - 2\mu \frac{1}{R^2} x_i x_j\right\} \frac{1}{R} u(R) \\ &+ \left\{\lambda \delta_{ij} + 2\mu \frac{1}{R^2} x_i x_j\right\} u'(R), \quad i, j = 1, 2 \\ \tau_{33} &= \lambda \left[\frac{1}{R} u(R) + u'(R)\right] \\ \tau_{31} &= \tau_{32} = 0 \end{aligned} \quad (9.8.46)$$

(ii) For a cylindrical surface R = constant, we have $\mathbf{n} = \mathbf{e}_R$ so that $n_i = x_i/R$ for $i = 1, 2$ and $n_3 = 0$. Therefore, by (7.5.3), the normal stress σ_R on this surface is $\sigma_R = \tau_{ij} n_i n_j = (\tau_{ij} x_i x_j)/R^2$; $i, j = 1, 2$. Using (9.8.46), we find that

$$\sigma_R = \lambda \frac{1}{R} u(R) + (\lambda + 2\mu) u'(R) \quad (9.8.47)$$

(iii) If $\bar{\mathbf{n}}$ is the unit outward normal to an axial plane, we have $\bar{\mathbf{n}} \cdot \mathbf{e}_R = 0$, and the normal stress σ_h on the plane is given by $\sigma_h = \tau_{ij} \bar{n}_i \bar{n}_j$. Using (9.8.46), we obtain the following expression for σ_h:

$$\sigma_h = (\lambda + 2\mu) \frac{1}{R} u(R) + \lambda u'(R) \quad (9.8.48)$$

(iv) Finally, to determine $u(R)$, we return to (9.8.44) and (9.8.45) and find from these expressions that

$$\begin{aligned} u_{i,jj} = u_{k,ki} &= \left\{3 \frac{d\psi}{dR} + R \frac{d^2\psi}{dR^2}\right\} \frac{x_i}{R} \\ &= \left\{\frac{d}{dR}\left[\frac{1}{R} \frac{d}{dR}(R^2 \psi)\right]\right\} \frac{x_i}{R} \\ &= \left\{\frac{d}{dR}\left[\frac{1}{R} \frac{d}{dR}(Ru)\right]\right\} \frac{x_i}{R} \end{aligned} \quad (9.8.49)$$

Substituting these into Navier's equation of equilibrium (9.8.12)' with $f_i = 0$, we find that the equation is satisfied if u obeys the following differential equation:

$$\frac{d}{dR}\left[\frac{1}{R} \frac{d}{dR}(Ru)\right] = 0 \quad (9.8.50)$$

Integration of this equation yields

$$u(R) = AR + \frac{B}{R} \qquad (9.8.51)$$

where A and B are arbitrary constants.

Thus, when $u(R)$ is as given by (9.8.51), **u** given by (9.8.42) obeys the Navier's equation of equilibrium with zero body force. ∎

9.9
DISPLACEMENT FORMULATION

Having the governing equations expressed entirely in terms of the displacement vector **u**, as shown in the preceding section, the boundary-value problems of elastodynamics can now be described entirely in terms of **u**.

For a mixed problem the boundary conditions are given (9.5.3). By using the stress-displacement relation (9.8.4), these boundary conditions can be rewritten as

$$\left.\begin{array}{r}\mathbf{u} = \mathbf{u}^* \quad \text{on} \quad S_u \\ [\lambda(\text{div }\mathbf{u})\mathbf{I} + \mu(\nabla\mathbf{u} + \nabla\mathbf{u}^T)]\mathbf{n} = \mathbf{s}^* \quad \text{on} \quad S_\tau\end{array}\right\} \quad \text{for} \quad t \geq 0 \quad (9.9.1)$$

The Navier's equation (9.8.6), the initial conditions (9.5.4) and (9.5.5) as well as the boundary conditions (9.9.1), all expressed entirely in terms of **u**, constitute what is known as the *displacement formulation* of a mixed boundary-value problem in elastodynamics. In this formulation, the displacement **u** is determined by solving the equation (9.8.6) subject to conditions (9.5.4), (9.5.5) and (9.9.1). Once this problem is solved, the strain **E** and the stress **T** can be determined by using the strain-displacement relation (9.8.1) and Hooke's law (9.8.3). It could be said that almost all problems in elastodynamics are solved in this way. In the static case, the same procedure is adopted except that the initial conditions (9.5.4) and (9.5.5) are not required.

By virtue of the uniqueness theorems proven in Sections 9.6 and 9.7, the displacement boundary-value problem has a unique solution in the static as well as dynamic cases. However, the uniqueness of solution could be proven directly by using the displacement formulation. Here we prove one such uniqueness theorem in the dynamic case. The static case is left for the reader.

THEOREM 9.9.1 An elastodynamic problem governed by equations (9.8.6), the initial conditions (9.5.4) and (9.5.5) and the boundary conditions (9.9.1) with $S = S_u$ cannot have more than one solution.

9.9 DISPLACEMENT FORMULATION

Proof Suppose $\mathbf{u}^{(1)}$ and $\mathbf{u}^{(2)}$ are two solutions of one and the same problem obeying (9.8.6), (9.5.4), (9.5.5) and (9.9.1) for $S = S_u$, and let $\bar{\mathbf{u}} = \mathbf{u}^{(1)} - \mathbf{u}^{(2)}$. Then $\bar{u}_i$ satisfy the following equations and the initial and boundary conditions:

$$\mu \bar{u}_{i,jj} + (\lambda + \mu)\bar{u}_{k,ki} = \rho \frac{\partial^2 \bar{u}_i}{\partial t^2} \tag{9.9.2}$$

in V for $t \geq 0$;

$$\bar{u}_i = 0, \quad \frac{\partial \bar{u}_i}{\partial t} = 0 \tag{9.9.3}$$

in V for $t = 0$;

$$\bar{u}_i = 0 \tag{9.9.4}$$

on S for $t \geq 0$. It is sufficient to show that $\bar{u}_i \equiv 0$ in V for all $t \geq 0$.

Multiplying (9.9.2) by $\dot{\bar{u}}_i = \partial \bar{u}_i / \partial t$ and integrating the resulting equation over V and setting $\ddot{\bar{u}}_i = \partial^2 \bar{u}_i / \partial t^2$, we get

$$\int_V [\mu \bar{u}_{i,jj} \dot{\bar{u}}_i + (\lambda + \mu)\bar{u}_{k,ki}\dot{\bar{u}}_i - \rho \dot{\bar{u}}_i \ddot{\bar{u}}_i]\, dV = 0$$

This can be rewritten in the following form

$$\int_V [\mu \bar{u}_{i,j} \dot{\bar{u}}_i + (\lambda + \mu)\bar{u}_{k,k} \dot{\bar{u}}_j]_{,j}\, dV$$

$$- \int_V [\mu \bar{u}_{i,j} \dot{\bar{u}}_{i,j} + (\lambda + \mu)\bar{u}_{k,k} \dot{\bar{u}}_{i,i} + \rho \dot{\bar{u}}_i \ddot{\bar{u}}_i]\, dV = 0 \tag{9.9.5}$$

Employing the divergence theorem to the first of the integrals in (9.9.5) and noting that an over dot denotes partial derivative w.r.t. t, we obtain the equation

$$\int_S [\mu \bar{u}_{i,j} \dot{\bar{u}}_i + (\lambda + \mu)\bar{u}_{k,k} \dot{\bar{u}}_j]n_j\, dS$$

$$- \frac{1}{2}\frac{\partial}{\partial t}\int_V [\mu(\bar{u}_{i,j})(\bar{u}_{i,j}) + (\lambda + \mu)(\bar{u}_{k,k})^2 + \rho \dot{\bar{u}}_i \dot{\bar{u}}_i]\, dV = 0 \tag{9.9.6}$$

In view of the boundary conditions (9.9.4), the integrand under the surface integral in (9.9.6) vanishes identically, and we obtain

$$\int_V \left[\mu(\bar{u}_{i,j})(\bar{u}_{i,j}) + (\lambda + \mu)(\bar{u}_{k,k})^2 + \rho \dot{\bar{u}}_i \dot{\bar{u}}_i\right] dV = K_0 \tag{9.9.7}$$

for $t \geq 0$, where K_0 is independent of t. At $t = 0$, the integrand in the lefthand side of (9.9.7) vanishes by virtue of the initial conditions (9.9.3);

hence, $K_0 = 0$. Further, since $\lambda, \mu, \rho > 0$, the integrand in equation (9.9.7) is a sum of nonnegative terms, and so each term must vanish separately. Thus, we arrive at the following set of equations:

$$\int_V (\bar{u}_{i,j}\bar{u}_{i,j})\, dV \equiv 0, \qquad \int_V (\bar{u}_{k,k})^2\, dV \equiv 0, \qquad \int_V \dot{\bar{u}}_i \dot{\bar{u}}_i\, dV \equiv 0$$

for $t \geq 0$. Since the integrand in each of these integrals is nonnegative, we find

$$\bar{u}_{i,j} \equiv 0, \qquad \bar{u}_{k,k} \equiv 0, \qquad \dot{\bar{u}}_i \equiv 0 \qquad (9.9.8)$$

in V for all $t \geq 0$. The last result in (9.9.8) implies that $\bar{u}_i = u_i^0$ in V for $t \geq 0$, where u_i^0 are independent of t. In view of the initial conditions on $\bar{u}_i$, given in (9.9.3), it follows that $u_i^0 = 0$. Thus, $\bar{u}_i \equiv 0$ in V for all $t \geq 0$, and the proof is complete. ∎

9.10
STRESS FORMULATION

We now proceed to recast the governing equations of elasticity entirely in terms of the stress tensor **T**. As in the derivation of Navier's equation, let us start with the strain-displacement relation (9.4.4); that is,

$$\mathbf{E} = \tfrac{1}{2}(\nabla \mathbf{u} + \nabla \mathbf{u}^T) \qquad (9.10.1)$$

and the linearized Cauchy equation (9.4.10); that is,

$$\text{div } \mathbf{T} + \mathbf{f} = \rho \ddot{\mathbf{u}} \qquad (9.10.2)$$

but take Hooke's law in the form (9.3.15)'; that is,

$$\mathbf{E} = \frac{1+\nu}{E}\mathbf{T} - \frac{\nu}{E}(\text{tr } \mathbf{T})\mathbf{I} \qquad (9.10.3)$$

From equation (9.10.2), we obtain

$$\nabla(\text{div } \mathbf{T}) + \nabla(\text{div } \mathbf{T})^T + (\nabla \mathbf{f} + \nabla \mathbf{f}^T) = \rho(\nabla \ddot{\mathbf{u}} + \nabla \ddot{\mathbf{u}}^T) \qquad (9.10.4)$$

From (9.10.1) and (9.10.3), we find that

$$\nabla \ddot{\mathbf{u}} + \nabla \ddot{\mathbf{u}}^T = 2\left[\frac{1+\nu}{E}\ddot{\mathbf{T}} - \frac{\nu}{E}(\text{tr } \ddot{\mathbf{T}})\mathbf{I}\right]$$

By using this in the righthand side of (9.10.4), we obtain the equation

$$\nabla(\text{div } \mathbf{T}) + \nabla(\text{div } \mathbf{T})^T - \frac{2\rho(1+\nu)}{E}\left[\ddot{\mathbf{T}} - \frac{\nu}{1+\nu}\mathbf{I}(\text{tr } \ddot{\mathbf{T}})\right] + \nabla \mathbf{f} + \nabla \mathbf{f}^T = 0 \qquad (9.10.5)$$

9.10 STRESS FORMULATION

In the suffix notation, this equation reads

$$\tau_{ik,kj} + \tau_{jk,ki} - \frac{2\rho(1+v)}{E}\left[\ddot{\tau}_{ij} - \frac{v}{1+v}\delta_{ij}\ddot{\tau}_{kk}\right] + f_{i,j} + f_{j,i} = 0 \qquad (9.10.5)'$$

The partial differential equation (9.10.5) expressed entirely in terms of the stress tensor (**f** being regarded as known) is by Valcovici (1951) and referred to as the *stress equation of motion*. This serves as a counterpart of Navier's equation (9.8.6), which is often referred to as the *displacement equation of motion*.

Since (9.10.1) to (9.10.3) hold at every point of the volume V of the body and for all $t \geq 0$, equation (9.10.5) also holds in V for all $t \geq 0$. To this equation, we append the initial conditions

$$\mathbf{T} = \mathbf{T}_0, \qquad \dot{\mathbf{T}} = \mathbf{T}_1 \qquad (9.10.6)$$

in V at $t = 0$, and the boundary condition

$$\mathbf{T} = \mathbf{T}^0 \qquad (9.10.7)$$

on S for $t \geq 0$, where $\mathbf{T}_0$, $\mathbf{T}_1$ and $\mathbf{T}^0$ are specified functions.

Equation (9.10.5) together with the conditions (9.10.6) and (9.10.7) constitute a boundary value problem in the *stress formulation* of elastodynamics. Once this problem is solved, the strain and the displacement can be determined by using Hooke's law (9.10.3) and the strain-displacement relation (9.10.1). The uniqueness of the solution of this problem is proven next.

9.10.1 UNIQUENESS OF SOLUTION

THEOREM 9.10.1 The solution of the problem governed by equation (9.10.5) with the conditions (9.10.6) and (9.10.7) is unique.

Proof For given $\mathbf{f}$, $\mathbf{T}_0$, $\mathbf{T}_1$ and $\mathbf{T}^0$, let $\mathbf{T}^{(1)}$ and $\mathbf{T}^{(2)}$ be two solutions of the problem governed by (9.10.5)–(9.10.7), and let $\bar{\mathbf{T}} = \mathbf{T}^{(1)} - \mathbf{T}^{(2)}$. Then the components $\bar{\tau}_{ij}$ of $\bar{\mathbf{T}}$ obey the following equations and conditions:

$$\bar{\tau}_{ik,kj} + \bar{\tau}_{jk,ki} - \frac{2\rho(1+v)}{E}\left(\ddot{\bar{\tau}}_{ij} - \frac{v}{1+v}\delta_{ij}\ddot{\bar{\tau}}_{kk}\right) = 0 \qquad (9.10.5)''$$

in V for $t \geq 0$;

$$\bar{\tau}_{ij} = 0, \qquad \dot{\bar{\tau}}_{ij} = 0 \qquad (9.10.6)'$$

in V for $t = 0$;

$$\bar{\tau}_{ij} = 0 \qquad (9.10.7)'$$

on S for $t \geq 0$. It is sufficient to show that $\bar{\tau}_{ij} \equiv 0$ in V for $t \geq 0$.

Multiplying equation (9.10.5)″ by $\dot{\bar{\tau}}_{ij}$ and making use of the relations

$$\bar{\tau}_{ik,kj}\dot{\bar{\tau}}_{ij} = (\bar{\tau}_{ik,k}\dot{\bar{\tau}}_{ij})_{,j} - \bar{\tau}_{ik,k}\dot{\bar{\tau}}_{ij,j}$$
$$\bar{\tau}_{jk,ki}\dot{\bar{\tau}}_{ij} = \bar{\tau}_{ik,kj}\dot{\bar{\tau}}_{ij} \tag{9.10.8}$$

which are readily verified, we obtain the equation

$$[(\bar{\tau}_{ik,k}\dot{\bar{\tau}}_{ij})_{,j} - \bar{\tau}_{ik,k}\dot{\bar{\tau}}_{ij,j}] - \frac{\rho(1+\nu)}{E}\left[\dot{\bar{\tau}}_{ij}\ddot{\bar{\tau}}_{ij} - \frac{\nu}{1+\nu}\dot{\bar{\tau}}_{kk}\ddot{\bar{\tau}}_{ii}\right] = 0 \tag{9.10.9}$$

Integrating this equation over V and using the divergence theorem and the boundary conditions (9.10.7)′, we get

$$\frac{\partial}{\partial t}\int_V\left[(\bar{\tau}_{ik,k}\bar{\tau}_{ij,j}) + \frac{\rho(1+\nu)}{E}\left\{\dot{\bar{\tau}}_{ij}\dot{\bar{\tau}}_{ij} - \frac{\nu}{1+\nu}(\dot{\bar{\tau}}_{kk})^2\right\}\right]dV = 0 \tag{9.10.10}$$

Integrating this equation w.r.t. t, and using the identity

$$\dot{\bar{\tau}}_{ij}\dot{\bar{\tau}}_{ij} - \frac{\nu}{1+\nu}(\dot{\bar{\tau}}_{kk})^2 = \left(\dot{\bar{\tau}}_{ij} - \frac{1}{3}\delta_{ij}\dot{\bar{\tau}}_{kk}\right)\left(\dot{\bar{\tau}}_{ij} - \frac{1}{3}\delta_{ij}\dot{\bar{\tau}}_{pp}\right) + \frac{1}{3}\frac{1-2\nu}{1+\nu}(\dot{\bar{\tau}}_{kk})^2 \tag{9.10.11}$$

which may readily be verified, we obtain the equation

$$\int_V \bar{\tau}_{ik,k}\bar{\tau}_{ij,j}\,dV + \frac{\rho(1+\nu)}{E}\int_V\left(\dot{\bar{\tau}}_{ij} - \frac{1}{3}\delta_{ij}\dot{\bar{\tau}}_{kk}\right)\left(\dot{\bar{\tau}}_{ij} - \frac{1}{3}\delta_{ij}\dot{\bar{\tau}}_{mm}\right)dV$$
$$+ \frac{\rho(1-2\nu)}{3E}\int_V(\dot{\bar{\tau}}_{kk})^2\,dV = F \tag{9.10.12}$$

where F is independent of t. In view of the initial conditions (9.10.6)′, we find that $F \equiv 0$. Since $\rho > 0$, $E > 0$ and $0 < \nu \le \frac{1}{2}$, it follows that

$$\bar{\tau}_{ik,k} \equiv 0, \qquad \dot{\bar{\tau}}_{ij} - \frac{1}{3}\delta_{ij}\dot{\bar{\tau}}_{kk} \equiv 0, \qquad \dot{\bar{\tau}}_{kk} \equiv 0 \tag{9.10.13}$$

in V. The last two relations in (9.10.13) give $\dot{\bar{\tau}}_{ij} = 0$; that is, $\bar{\tau}_{ij}$ are independent of t. By virtue of the initial conditions (9.10.6)′, it follows that $\bar{\tau}_{ij} \equiv 0$ in V for all $t \ge 0$. This completes the proof. ∎

The uniqueness Theorem 9.10.1 and its proof were first given by J. Ignaczak in 1963.

9.11
BELTRAMI–MICHELL EQUATION

Another important governing differential equation of elasticity that is also expressed completely in terms of the stress tensor can be deduced by making use of the compatibility condition (9.4.16).

9.11 BELTRAMI-MICHELL EQUATION

Substituting for **E** from Hooke's law (9.10.3) into the compatibility condition (9.4.16), we obtain the equation

$$\nabla^2 \mathbf{T} - (\nabla \operatorname{div} \mathbf{T}) - (\nabla \operatorname{div} \mathbf{T})^T - \frac{v}{1+v}\nabla^2(\operatorname{tr} \mathbf{T})\mathbf{I} + \frac{1}{1+v}\nabla\nabla(\operatorname{tr} \mathbf{T}) = \mathbf{0} \tag{9.11.1}$$

Substitution for div **T** from Cauchy's equation (9.10.2) in (9.11.1) and simplification of the resulting equation by use of (9.10.1) and (9.10.3) yields the following equation expressed entirely in terms of **T**:

$$\Box_2 \mathbf{T} - \frac{v}{1+v}\Box_2(\operatorname{tr} \mathbf{T})\mathbf{I} + \frac{1}{1+v}\nabla\nabla(\operatorname{tr} \mathbf{T}) + \nabla\mathbf{f} + \nabla\mathbf{f}^T = \mathbf{0} \tag{9.11.2}$$

This tensor equation can be further simplified as described next.

Taking the trace of equation (9.11.2), we get

$$\frac{1}{1+v}[(1-2v)\Box_2 + \nabla^2](\operatorname{tr} \mathbf{T}) + 2\operatorname{div}\mathbf{f} = 0 \tag{9.11.3}$$

By use of (9.8.18), equation (9.11.3) becomes

$$\Box_1(\operatorname{tr} \mathbf{T}) = -\frac{1+v}{1-v}\operatorname{div}\mathbf{f} \tag{9.11.4}$$

so that

$$\Box_2(\operatorname{tr} \mathbf{T}) = -\left\{\frac{1+v}{1-v}\operatorname{div}\mathbf{f} + (\Box_1 - \Box_2)(\operatorname{tr} \mathbf{T})\right\} \tag{9.11.5}$$

Substituting this back into the second term in (9.11.2), we obtain

$$\Box_2 \mathbf{T} + \frac{1}{1+v}\{\nabla\nabla + v(\Box_1 - \Box_2)\mathbf{I}\}(\operatorname{tr} \mathbf{T}) + \frac{v}{1-v}(\operatorname{div}\mathbf{f})\mathbf{I} + \nabla\mathbf{f} + \nabla\mathbf{f}^T = \mathbf{0} \tag{9.11.6}$$

This is the desired governing differential equation expressed entirely in terms of the stress tensor. In the suffix notation, this equation reads

$$\Box_2 \tau_{ij} + \frac{1}{1+v}\left\{\frac{\partial^2}{\partial x_i \partial x_j} + v(\Box_1 - \Box_2)\delta_{ij}\right\}\tau_{kk}$$
$$+ \frac{v}{1-v}\delta_{ij}f_{k,k} + f_{i,j} + f_{j,i} = 0 \tag{9.11.6}'$$

In the static case, equations (9.11.6) and (9.11.6)' become

$$\nabla^2 \mathbf{T} + \frac{1}{1+v}\nabla\nabla(\operatorname{tr} \mathbf{T}) + \frac{v}{1-v}(\operatorname{div}\mathbf{f})\mathbf{I} + \nabla\mathbf{f} + \nabla\mathbf{f}^T = \mathbf{0} \tag{9.11.7}$$

$$\nabla^2 \tau_{ij} + \frac{1}{1+v}\tau_{kk,ij} + \frac{v}{1-v}\delta_{ij}f_{k,k} + f_{i,j} + f_{j,i} = 0 \tag{9.11.7}'$$

9 EQUATIONS OF LINEAR ELASTICITY

The six equations embodied in (9.11.7)' were obtained by Donati in 1894 and by Michell in 1900. Earlier, in 1892, Beltrami had obtained the equations for the case of zero body forces. The dynamic counterpart of these equations, namely, (9.11.6)', were given by Iacovache in 1950 for the case $v = \frac{1}{2}$ and by Valcovici in 1951 for the general case. Equations (9.11.6) to (9.11.7)' are referred to as the *Beltrami–Michell equations*.

Notice that equation (9.11.6) has been derived by starting with the compatibility condition (9.4.16) for the strain tensor **E**. For this reason, equation (9.11.6) is often referred to as the *equation of compatibility for stress*.

In Section 8.5, it was pointed out that Cauchy's equations of equilibrium, (8.3.8)', which are three in number, are inadequate to determine all of the six stress components τ_{ij}. Together with the compatibility equations (9.11.7)', of which there are six, Cauchy's equations (8.3.8)' serve as a closed system of governing equations for τ_{ij} in elastostatic problems.

EXAMPLE 9.11.1 Find whether the following stress system can be a solution of an elastostatic problem in the absence of body forces:

$$\tau_{11} = x_2 x_3, \quad \tau_{22} = x_3 x_1, \quad \tau_{12} = x_3^2, \quad \tau_{13} = \tau_{23} = \tau_{33} = 0$$

Solution In order that the given stress system can be a solution of an elastostatic problem in the absence of body forces, the following equations are to be satisfied:

(i) Cauchy's equations of equilibrium (8.3.8)' with $f_i = 0$; that is,

$$\tau_{11,1} + \tau_{12,2} + \tau_{13,3} = 0$$
$$\tau_{21,1} + \tau_{22,2} + \tau_{23,3} = 0 \quad\quad (9.11.8)$$
$$\tau_{31,1} + \tau_{32,2} + \tau_{33,3} = 0$$

(ii) Beltrami–Michell equations (9.11.7)' with $f_i = 0$; that is,

$$\nabla^2 \tau_{11} + \frac{1}{1+v}(\tau_{11} + \tau_{22} + \tau_{33})_{,11} = 0$$

$$\nabla^2 \tau_{22} + \frac{1}{1+v}(\tau_{11} + \tau_{22} + \tau_{33})_{,22} = 0$$

$$\nabla^2 \tau_{33} + \frac{1}{1+v}(\tau_{11} + \tau_{22} + \tau_{33})_{,33} = 0$$

$$\nabla^2 \tau_{12} + \frac{1}{1+v}(\tau_{11} + \tau_{22} + \tau_{33})_{,12} = 0 \quad\quad (9.11.9)$$

$$\nabla^2 \tau_{13} + \frac{1}{1+v}(\tau_{11} + \tau_{22} + \tau_{33})_{,13} = 0$$

$$\nabla^2 \tau_{23} + \frac{1}{1+v}(\tau_{11} + \tau_{22} + \tau_{33})_{,23} = 0$$

9.11 BELTRAMI-MICHELL EQUATION

It is easy to check that all the equations in (9.11.8) and all *except* the fourth one in (9.11.9) are satisfied by the given stress system. Since the given stress system *does not* satisfy the Beltrami-Michell equations fully, it cannot form a solution of an elastostatic problem. ∎

Note: This example illustrates the important fact that a stress system may not be a solution of an elasticity problem even though it satisfies Cauchy's equilibrium equations.

EXAMPLE 9.11.2 Deduce the Beltrami-Michell equation starting with Navier's equation.

Solution Let us start with the Navier's equation as given by (9.8.10). From this we get the equation

$$\Box_2 \nabla \mathbf{u} + \frac{1}{1 - 2\nu} \nabla \nabla (\text{div } \mathbf{u}) + \frac{1}{\mu} \nabla \mathbf{f} = \mathbf{0}$$

Adding this equation to its transpose and subsequently using the strain-displacement relation (9.8.1), we obtain the following equation of motion expressed entirely in terms of **E**:

$$\Box_2 \mathbf{E} + \frac{1}{1 - 2\nu} \nabla \nabla (\text{tr } \mathbf{E}) + \frac{1}{2\mu} (\nabla \mathbf{f} + \nabla \mathbf{f}^T) = \mathbf{0} \qquad (9.11.10)$$

Substituting for **E** from Hooke's law (9.10.3) in (9.11.10) and recalling that $2\mu(1 + \nu) = E$, see Table 9.1, we arrive at equation (9.11.2), from which the Beltrami-Michell equation follows. ∎

EXAMPLE 9.11.3 If **u** is a solution of Navier's equation, show that the corresponding **T** is automatically a solution of the Beltrami-Michell equation.

Solution By use of relations (9.3.14)', (9.3.13), (9.8.1) as well as identity (9.8.19), the Beltrami-Michell equation (9.11.6) can be decomposed as follows:

$$\frac{E}{1 + \nu} \Box_2 \left[\frac{1}{2} (\nabla \mathbf{u} + \nabla \mathbf{u}^T) + \frac{\nu}{1 - 2\nu} (\text{div } \mathbf{u}) \mathbf{I} \right]$$
$$+ \frac{E}{(1 + \nu)(1 - 2\nu)} [\nabla \nabla (\text{div } \mathbf{u}) + \nu (\Box_1 - \Box_2)(\text{div } \mathbf{u}) \mathbf{I}]$$
$$+ \frac{\nu}{1 - \nu} (\text{div } \mathbf{f}) \mathbf{I} + \nabla \mathbf{f}^T = \mathbf{0} \qquad (9.11.11)$$

Multiplying this equation by $2(1 + v)/E = 1/\mu$ and rearranging the terms, we get

$$\nabla\left[\square_2\mathbf{u} + \frac{1}{1-2v}\nabla(\text{div }\mathbf{u}) + \frac{1}{\mu}\mathbf{f}\right] + \nabla\left[\square_2\mathbf{u} + \frac{1}{1-2v}\nabla\text{ div }\mathbf{u} + \frac{1}{\mu}\mathbf{f}\right]^T$$
$$+ \frac{2v}{1-2v}\left[\square_1\text{ div }\mathbf{u} + \frac{(1-2v)}{2\mu(1-v)}\text{div }\mathbf{f}\right]\mathbf{I} = 0 \qquad (9.11.12)$$

By use of the identity (9.8.18), it is easy to check that

$$\square_1(\text{div }\mathbf{u}) + \frac{(1-2v)}{2\mu(1-v)}\text{div }\mathbf{f}$$
$$= \frac{1-2v}{2(1-v)}\text{div}\left[\square_2\mathbf{u} + \frac{1}{1-2v}\nabla(\text{div }\mathbf{u}) + \frac{1}{\mu}\mathbf{f}\right] \qquad (9.11.13)$$

In view of the identity (9.11.13), equation (9.11.12) can be rewritten as

$$\nabla\left[\square_2\mathbf{u} + \frac{1}{1-2v}\nabla\text{ div }\mathbf{u} + \frac{1}{\mu}\mathbf{f}\right] + \nabla\left[\square_2\mathbf{u} + \frac{1}{1-2v}\nabla\text{ div }\mathbf{u} + \frac{1}{\mu}\mathbf{f}\right]^T$$
$$+ \frac{v}{1-v}\text{div}\left[\square_2\mathbf{u} + \frac{1}{1-2v}\nabla\text{ div }\mathbf{u} + \frac{1}{\mu}\mathbf{f}\right]\mathbf{I} = 0 \qquad (9.11.14)$$

If $\mathbf{u}$ is a solution of Navier's equation, each term on the lefthand side of (9.11.14) is 0 and the equation is identically satisfied. Thus, if $\mathbf{u}$ is a solution of Navier's equation, the Beltrami–Michell equation is automatically satisfied. ∎

Note: If $\mathbf{T}$ obeys the Beltrami–Michell equation, then it can be similarly proved that the *corresponding* displacement $\mathbf{u}$ determined by using (9.10.3), and (9.10.1) is a solution of Navier's equation.

EXAMPLE 9.11.4 Show that

$$\square_1\square_2\mathbf{T} = \frac{1}{1-v}[(\nabla\nabla - v\square_2\mathbf{I})\text{ div }\mathbf{f} - (1-v)\square_1(\nabla\mathbf{f} + \nabla\mathbf{f}^T)] \qquad (9.11.15)$$

Solution Operating (9.11.2) by $\square_1$ and using (9.11.4), we readily get (9.11.15). ∎

Note: If $\mathbf{f}$ = constant, it follows from (9.11.3) and (9.11.15) that $\square_1(\text{tr }\mathbf{T}) = 0$ and $\square_1\square_2\mathbf{T} = \mathbf{0}$.

9.12
SOME STATIC PROBLEMS

In the preceding sections we formulated the governing equations of the linear elasticity theory in terms of displacements and stresses. We now employ these equations to solve some *simple* problems of practical interest. Static problems are considered in this section; the next section deals with dynamical problems. The four problems that follow are solved using the stress formulation, and problems 5 and 6 are solved using the displacement formulation.

9.12.1 AXIAL EXTENSION OF A BEAM

Consider an elastic beam of uniform cross section bounded by a cylindrical surface and by a pair of planes normal to this surface. The cylindrical surface is usually referred to as the *lateral surface* of the beam and the planes are the *bases* (or *end faces*). Suppose that the beam is in equilibrium under a uniform normal stress N acting on the bases. The lateral surface is stress free and the body force is neglected. The problem is to compute the stresses, strains and displacements at an arbitrary point of the beam.

It is convenient to choose the x_1 axis along the line of centroids of cross sections of the beam with the origin lying on one of the bases. If l is the initial length of the beam, the other base lies on the plane $x_1 = l$ before deformation; see Figure 9.4. Then, by hypothesis, the following boundary condition hold:

$$\left. \begin{array}{ll} \tau_{11} = N; \quad \tau_{12} = \tau_{13} = 0 \quad \text{for} \quad x_1 = 0, l \\ \tau_{ij} n_j = 0 \quad \text{on the lateral surface} \end{array} \right\} \quad (9.12.1)$$

From conditions (9.12.1) we note that the stresses are prescribed at every point of the boundary surface of the beam; the problem considered is therefore a traction boundary-value problem.

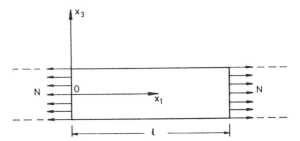

Figure 9.4. Axial extension of a beam.

Bearing in mind that $n_1 = 0$ at every point of the lateral surface, we verify that the following stress system obeys the conditions (9.12.1):

$$\tau_{11} = N, \qquad \tau_{12} = \tau_{13} = \tau_{22} = \tau_{23} = \tau_{33} = 0 \qquad (9.12.2)$$

Thus, if we take (9.12.2) as the expressions for the stress system at *any point* of the beam, then the boundary conditions of the problem are identically satisfied. In Section 9.11, it was pointed out that Cauchy's equilibrium equations (8.3.8)′ and the Beltrami–Michell compatibility equations (9.11.7)′ together serve as a set of governing equations for stresses in an elastostatic problem. Noting that N is a constant and $\mathbf{f} = \mathbf{0}$ by hypothesis, we check that the stress system (9.12.2) readily satisfies all these governing equations.

The strains e_{ij} associated with the stresses given by (9.12.2) can be obtained by substituting for τ_{ij} from (9.12.2) in the matrix equation (9.3.15)″. Thus, we obtain

$$\left. \begin{array}{l} e_{11} = \dfrac{1}{E} N; \quad e_{22} = e_{33} = -\dfrac{\nu}{E} N \\[6pt] e_{12} = e_{23} = e_{33} = 0 \end{array} \right\} \qquad (9.12.3)$$

The displacements u_i associated with these strains follow from the suffix form of the strain–displacement relation (9.4.4). Thus, we obtain (see Example 5.8.4):

$$u_1 = \frac{1}{E} N x_1, \qquad u_2 = -\frac{\nu}{E} N x_2, \qquad u_3 = -\frac{\nu}{E} N x_3 \qquad (9.12.4)$$

Expressions (9.12.2), (9.12.3) and (9.12.4) give the stresses, strains and displacements that occur at an arbitrary point $\mathbf{x}$ of the beam. By virtue of the uniqueness theorem proved in Section 9.6, these expressions constitute the only possible solution of the problem. It is important to note that this solution is completely independent of the length and the geometrical form of the cross section of the beam. As such, the solution is valid for a beam of any length and any cross section. Expressions (9.12.4) show that (i) the displacements u_1 and u_2 have the same values in all sections of the beam, and (ii) plane sections $x_1 = $ constant deform into plane sections $x_1 + u_1 = ((E + N)/E)x_1 = $ constant.

EXAMPLE 9.12.1 In the problem just considered, find the normal and shear stresses on an oblique section of the beam. Also, find their extreme values and the sections on which these occur.

9.12 SOME STATIC PROBLEMS

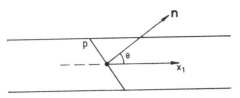

Figure 9.5. Example 9.12.1.

Solution If **n** denotes the unit normal to an oblique plane section p of the beam (Figure 9.5), the components of the stress vector on this section are obtained, by use of Cauchy's law (7.4.9) and the relations (9.12.2), as

$$s_1(\mathbf{n}) = Nn_1, \qquad s_2(\mathbf{n}) = s_3(\mathbf{n}) = 0 \qquad (9.12.5)$$

Using these expressions in (7.5.1) and (7.5.6), the normal stress σ and the shear stress τ acting on the plane p are found to be

$$\sigma = Nn_1^2, \qquad \tau = (1 - n_1^2)^{1/2} N n_1 \qquad (9.12.6)$$

If θ is the angle that **n** makes with the axis of the beam (x_1 axis), then $n_1 = \cos \theta$ and the expressions (9.12.6) can be put in the following alternative form:

$$\sigma = N \cos^2 \theta, \qquad \tau = \tfrac{1}{2} N |\sin 2\theta| \qquad (9.12.7)$$

From these expressions it is evident that σ is maximum for $\theta = 0$ and minimum for $\theta = 90°$. Also, τ is maximum for $\theta = 45°$ and minimum for $\theta = 0$. Thus, the normal stress is maximum on the right cross sections of the beam and minimum on sections parallel to the axis. The shear stress is maximum on sections inclined at an angle of 45° with the axis of the beam and minimum on the right sections. The extreme values of normal and shear stresses are given by

$$\sigma_{max} = N, \qquad \tau_{max} = \tfrac{1}{2}N; \qquad \sigma_{min} = 0, \qquad \tau_{min} = 0 \qquad \blacksquare \qquad (9.12.8)$$

REMARKS The matrix of the stresses given by (9.12.2) is purely diagonal. Hence, the coordinate axes chosen for the discussion of the problem are actually the principal axes of stress, and the principal stresses are $\tau_{11} = N$, $\tau_{22} = 0$ and $\tau_{33} = 0$. Since N and 0 are the maximum and minimum values of σ that occur when **n** is along the x_1 and x_2 axes, respectively, the principal stresses τ_{11} and τ_{22} are the maximum and the minimum normal stresses. We also observe that the maximum value of τ occurs on the plane that bisects the angle between the planes on which σ is maximum and minimum, and that its magnitude is equal to one half the difference between the maximum and minimum values of σ. These facts illustrate the results proven in Examples 7.6.2 and 7.6.3.

9.12.2 EXTENSION OF A BEAM BY ITS OWN WEIGHT

Consider an elastic beam of uniform cross section suspended vertically by fixing the centroid of the upper base. Suppose that a stress system that balances the weight of the beam acts on this base and that the lower base and the lateral surface of the beam are stress free. Suppose that there is no external force other than the force due to gravity and that the beam is in equilibrium. The problem is to find the stresses, strains and displacements at an arbitrary point of the beam.

For convenience, we choose the centroid of the upper base as the origin of the coordinate system with the x_3 axis directed vertically downward along the line of centroids of cross sections (Figure 9.6). Then the gravitational force per unit mass is given by $\mathbf{b} = g\mathbf{e}_3$, where g is the usual gravity factor. Consequently, we have

$$\mathbf{f} = \rho\mathbf{b} = \rho g\mathbf{e}_3 \tag{9.12.9}$$

If l is the undeformed length of the beam, then the weight of a column of unit cross sectional area of the beam is $\rho l g$. This weight acts along the x_3 axis and the stress at the upper base, $x_3 = 0$, balances this weight. Hence, the following conditions are to hold:

$$\tau_{33} = \rho l g, \qquad \tau_{31} = \tau_{32} = 0 \tag{9.12.10}$$

at $x_3 = 0$. Since the lower base and the lateral surface are stress-free, the following conditions are also to be satisfied:

$$\left.\begin{array}{l} \tau_{31} = \tau_{32} = \tau_{33} = 0 \quad \text{at} \quad x_3 = l \\ \tau_{ij}n_j = 0 \quad \text{on the lateral surface} \end{array}\right\} \tag{9.12.11}$$

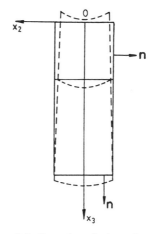

Figure 9.6. Extension of a beam by gravity.

9.12 SOME STATIC PROBLEMS

From conditions (9.12.10) and (9.12.11), we note that the stresses are prescribed at every point of the boundary surface of the beam; as such, this problem is a traction boundary-value problem.

Bearing in mind that $n_3 = 0$ at every point of the lateral surface, we verify that all the conditions contained in (9.12.10) and (9.12.11) are satisfied by the following stress system:

$$\tau_{11} = \tau_{22} = \tau_{12} = \tau_{13} = \tau_{23} = 0$$
$$\tau_{33} = \rho g(l - x_3)$$
(9.12.12)

That is, if we take (9.12.12) as the expressions for the stress system at *any point* of the beam, then all the boundary conditions of the problem are satisfied. It is easy to verify that this stress system obeys also Cauchy's equilibrium equations (8.3.8)' and the Beltrami–Michell equations (9.11.7)', with **f** given by (9.12.9).

The strains e_{ij} associated with the stresses τ_{ij} given by (9.12.12) can be obtained by using the matrix equation (9.3.15)''. Thus, we get

$$e_{11} = e_{22} = -\frac{\nu \rho g}{E}(l - x_3)$$

$$e_{33} = \frac{\rho g}{E}(l - x_3)$$
(9.12.13)

$$e_{12} = e_{13} = e_{23} = 0$$

The displacements u_i associated with these strains follow from the suffix form of the strain-displacement relation (9.4.4). Thus, we obtain (see Example 5.8.4, equation (5.8.56)):

$$u_1 = -\frac{\nu \rho g}{E} x_1(l - x_3)$$

$$u_2 = -\frac{\nu \rho g}{E} x_2(l - x_3)$$
(9.12.14)

$$u_3 = -\frac{\rho g}{2E}[x_3(x_3 - 2l) + \nu(x_1^2 + x_2^2)]$$

It is trivial to verify that expressions (9.12.14) meet the condition that the centroid of the upper base positioned at the origin is held fixed.

Relations (9.12.12), (9.12.13) and (9.12.14) give the stresses, strains and the displacements that occur at an arbitrary point **x** of the beam. By virtue of the uniqueness theorem proven in Section 9.6, these relations constitute the only possible solution of the problem. We note that the solution is valid for all geometrical forms of the sections of the beam.

412   9 EQUATIONS OF LINEAR ELASTICITY

From the results (9.12.14) we find that the points *on the axis* of the beam are displaced only in the vertical direction and that the displacement u_3 obeys the law:

$$u_3 = \frac{\rho g}{E} x_3 \left(l - \frac{1}{2} x_3 \right) \quad (9.12.15)$$

From equations (9.12.14) we also note that different points of a given section $x_3 = c$ of the beam have different vertical displacements. As such, the plane sections $x_3 = $ constant do not deform into plane sections. In fact, we find from the last equation in (9.12.14) that the deformed shape of such a section is a paraboloid of revolution having the x_3 axis as the axis.

From (9.12.13), we find that the lateral contractions e_{11} and e_{22} in a cross section vary directly with the distance of the section from the lower base of the beam. As such, the maximum contraction occurs at the upper base of the beam, the maximum contraction being $(\nu \rho g l)/E$. Also, no contraction occurs at the lower base. Consequently, vertical line elements of the beam do not remain vertical after deformation. The deformed shape of the beam is indicated by the dotted lines in Figure 9.6.

EXAMPLE 9.12.2   Show that the vertical displacement at the centroid of the lower base of a beam stretched by its own weight w is the same as that produced by the load $\frac{1}{2}w$ applied at the base.

Solution   With reference to the axes shown in Figure 9.6, the coordinates of the centroid of the lower base of the beam are $(0, 0, l)$. Result (9.12.15) gives the vertical displacement at this point as

$$(u_3)_{x_3 = l} = \frac{\rho g}{2E} l^2 \quad (9.12.16)$$

If A is the area of cross section of the beam, the weight w of the beam is given by $w = \rho l A g$. Using this, (9.12.16) may be rewritten as

$$(u_3)_{x_3 = l} = \frac{wl}{2AE} \quad (9.12.17)$$

On the other hand, if $\frac{1}{2}w$ is the load (total surface force) acting on the lower base ($x_3 = l$) of the beam, then the normal stress acting on this base is

$$(\tau_{33})_{x_3 = l} = \frac{w}{2A} \quad (9.12.18)$$

Hooke's law then gives

$$(e_{33})_{x_3 = l} = \frac{1}{E} (\tau_{33})_{x_3 = l} = \frac{w}{2AE} \quad (9.12.19)$$

9.12 SOME STATIC PROBLEMS

Consequently, the strain-displacement relation yields

$$(u_3)_{x_3 = l} = \frac{wl}{2AE} \qquad (9.12.20)$$

Relations (9.12.17) and (9.12.20) are identical, and the required result is proved. ∎

9.12.3 BENDING OF A BEAM BY TERMINAL COUPLES

Consider an elastic beam of uniform cross section and of length l. Suppose that a couple of moment **m** about a line perpendicular to the beam is applied at one of the bases and an opposite couple of moment $-\mathbf{m}$ is applied at the other base (so that the beam is in equilibrium). The lateral surface of the beam is stress free and the body forces are neglected. The problem is to find the stresses, strains and displacements that occur at an arbitrary point of the beam due to the bending it experiences because of the applied end couples. This kind of bending of a beam is called *simple bending* or *pure bending*.

Let us choose the centroid of the base of the beam on which the couple $-\mathbf{m}$ acts as the origin of the coordinate system, with the x_3 axis initially directed along the line of centroids of cross sections of the beam. Also, the x_2 axis is chosen along the axis of the couple **m** and the x_1 axis is chosen such that the coordinate system is righthanded (Figure 9.7).

Due to the bending caused by the applied couples, the longitudinal elements of the beam experience elongation or contraction. Consequently, the stress vector **s** acting at a point on a cross section of the beam produces a moment $\mathbf{x} \times \mathbf{s}$ per unit of area. The total moment on the cross section is therefore $\int_A (\mathbf{x} \times \mathbf{s})\,dA$, where A is the area of cross section. If the cross section is the base $x_3 = l$, then this moment should be equal to the moment of the applied couple; that is,

$$\int_A (\mathbf{x} \times \mathbf{s})\,dA = \mathbf{m} \qquad (9.12.21)$$

We note that on the base $x_3 = l$, we have $\mathbf{s} = \mathbf{s}^{(3)}$ and $\mathbf{m} = M\mathbf{e}_2$, where $M = |\mathbf{m}|$. Hence (9.12.21) yields, on equating the corresponding

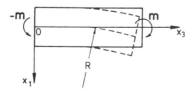

Figure 9.7. Simple bending.

components,

$$\left.\begin{aligned} \int_A (x_2 \tau_{33} - l\tau_{32})\, dA &= 0 \\ \int_A (l\tau_{31} - x_1 \tau_{33})\, dA &= M \\ \int_A (x_1 \tau_{32} - x_2 \tau_{31})\, dA &= 0 \end{aligned}\right\} \quad \text{on} \quad x_3 = l \quad (9.12.22)$$

On the base $x_3 = 0$, the moment of the applied couple is $-M\mathbf{e}_2$ and the stress vector is $\mathbf{s} = -\mathbf{s}^{(3)}$. Hence, we should have

$$\int_A (\mathbf{x} \times \mathbf{s}^{(3)})\, dA = M\mathbf{e}_2 \quad (9.12.23)$$

on $x_3 = 0$.

Equating the corresponding components in (9.12.23) we get

$$\left.\begin{aligned} \int_A x_2 \tau_{33}\, dA &= 0 \\ \int_A x_1 \tau_{33}\, dA &= -M \\ \int_A (x_1 \tau_{32} - x_2 \tau_{31})\, dA &= 0 \end{aligned}\right\} \quad \text{on} \quad x_3 = 0 \quad (9.12.24)$$

The conditions (9.12.22) and (9.12.24) are to be satisfied by the stresses at the bases $x_3 = l$ and $x_3 = 0$, respectively. It is easy to check that the stress system

$$\tau_{11} = \tau_{22} = \tau_{12} = \tau_{13} = \tau_{23} = 0, \qquad \tau_{33} = ax_1 \quad (9.12.25)$$

where a is a nonzero constant, meets all of the conditions in (9.12.22) and (9.12.24) provided that

$$\int_A x_1 x_2\, dA = 0 \quad (9.12.26a)$$

$$\int_A x_1^2\, dA = -\frac{M}{a} \quad (9.12.26b)$$

We note that the integral in (9.12.26b) represents the *moment of intertia* of a section about the x_2 axis, denoted I. Then (9.12.26b) is satisfied if we take $a = -(M/I)$. Further, the integral in (9.12.26a) represents the *product of intertia* of a section about the x_2 and x_3 axes. If the axes are assumed to

9.12 SOME STATIC PROBLEMS

be along the *principal axes of intertia*, then this product is identically 0, and (9.12.26a) is identically satisfied. Thus, with $a = -(M/I)$, the stress system (9.12.25) satisfies all of the conditions in (9.12.22) and (9.12.24).

The lateral surface of the beam has been assumed to be stress free. Hence the conditions $\tau_{ij}n_j = 0$ are to hold at every point of this surface. Since $n_3 = 0$ at all these points, we readily find that the stress system (9.12.25) meets these conditions as well.

Bearing in mind that $\mathbf{f} = \mathbf{0}$, we check that the stress system (9.12.25) also obeys Cauchy's equilibrium equations (8.3.8)' and the Beltrami–Michell compatibility equations (9.11.7)'.

Thus, the stress system

$$\tau_{11} = \tau_{22} = \tau_{12} = \tau_{13} = \tau_{23} = 0$$
$$\tau_{33} = -\frac{M}{I}x_1$$
(9.12.27)

satisfies all the governing equations and all the boundary conditions relevant to the problem.

The strains e_{ij} associated with the stresses τ_{ij} given by (9.12.27) follow by using Hooke's law (9.3.15)''. Thus, we get

$$e_{11} = e_{22} = \frac{\nu M}{EI}x_1, \qquad e_{33} = -\frac{M}{EI}x_1$$
$$e_{12} = e_{23} = e_{13} = 0$$
(9.12.28)

The corresponding displacements follow by the use of strain-displacement relation (9.4.4). Thus, (see example 5.8.4)

$$u_1 = \frac{M}{2EI}\{x_3^2 + \nu(x_1^2 - x_2^2)\}$$
$$u_2 = \frac{M\nu}{EI}x_1 x_2$$
(9.12.29)
$$u_3 = -\frac{M}{EI}x_1 x_3$$

Relations (9.12.27), (9.12.28) and (9.12.29) give the stresses, strains and displacements that occur at an arbitrary point $\mathbf{x}$ of the beam. By virtue of the uniqueness of solutions, these relations constitute the only possible solution of the problem. We note that this solution is valid for all geometrical forms of the sections of the beam provided that the moment of the applied couple is directed along a principal axis of inertia of the section.

From (9.12.28), we find that longitudinal elements of the beam extend or contract depending on whether $x_1 > 0$ or $x_1 < 0$. For $x_1 = 0$, all the strains are 0; as such the elements initially lying on the $x_2 x_3$ plane do not change in length. The $x_2 x_3$ plane is therefore referred to as the *neutral plane* of the beam. From (9.12.29), we find that $u_2 = 0$ for elements initially lying in the $x_3 x_1$ plane; as such, these elements continue to remain in the same plane after deformation. The $x_1 x_3$ plane is therefore referred to as the *plane of bending*. We note that the line element initially lying along the x_3 axis lies on the neutral plane as well as on the plane of bending; this element is called the *central line* of the beam. From (9.12.29) we find that the points of the central line are displaced completely in the x_1 direction and that the displacement obeys the equation

$$u_1 = \frac{M}{2EI} x_3^2 \qquad (9.12.30)$$

We note that the curve defined by the equation (9.12.30) is a parabola in the $x_1 x_3$ plane having the x_1 axis as the axis. Thus, the central line of the beam deforms into a parabola given by the equation (9.12.30). The curvature R of this parabola is given by

$$\frac{1}{R} = \frac{d^2 u_1 / dx_3^2}{[1 + (du_1/dx_3)^2]^{3/2}} \approx \frac{d^2 u_1}{dx_3^2} = \frac{M}{EI} \qquad (9.12.31)$$

where the nonlinear term $(du_1/dx_3)^2$ is neglected because deformation is small.

The result (9.12.31) is of great importance in the theory of bending of beams, usually referred to as the *Bernoulli–Euler law*.

Next, consider a particle initially located on the section $x_3 = c$. Let x_i' be the coordinates of the location of this particle after deformation, so that $x_i' = x_i + (u_i)_{x_3 = c}$. Then, we find from (9.12.29) and (9.12.31) that

$$x_3' = c - \frac{1}{R} c x_1 = c\left\{\left(1 - \frac{1}{R} x_1'\right) + \frac{1}{R}(u_1)_{x_3 = c}\right\} \approx c\left(1 - \frac{1}{R} x_1'\right) \qquad (9.12.32)$$

Equation (9.12.32) represents a plane in the deformed configuration of the beam, and it is easy to verify that this plane is normal to the deformed central line. Thus, planes normal to the central line before deformation are transferred to planes normal to the deformed central line. This result is also of importance in the theory of bending of beams.

EXAMPLE 9.12.3 Suppose that the beam considered in the above problem is rectangular with $x_1 = \pm h$ and $x_2 = \pm b$ as the boundary lines of a cross section. Find the deformed shape of the cross section.

9.12 SOME STATIC PROBLEMS 417

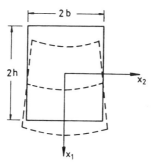

Figure 9.8. Example 9.12.3.

Solution Figure 9.8 shows a typical section of the beam for which x_3 is a constant, say, c. Due to deformation, suppose the particle initially located at the point (x_1, x_2) of this section gets displaced to the point (x'_1, x'_2), where $x'_1 = x_1 + u_1$ and $x'_2 = x_2 + u_2$. Then (9.12.29) and (9.12.31) give

$$\left. \begin{aligned} x'_1 &= x_1 + \frac{1}{2R}\{c^2 + v(x_1^2 - x_2^2)\} \\ x'_2 &= x_2\left(1 + \frac{v}{R}x_1\right) \end{aligned} \right\} \quad (9.12.33)$$

For the particle initially located on the lines $x_2 = \pm b$, equations (9.12.33) give

$$\begin{aligned} x'_2 &= \pm b\left(1 + \frac{v}{R}x_1\right) \\ &= \pm b\left[1 + \frac{v}{R}\left\{x'_1 - \frac{1}{2R}(c^2 + vx_1^2 - vx_2^2)\right\}\right] \\ &\approx \pm b\left(1 + \frac{v}{R}x'_1\right) \end{aligned} \quad (9.12.34)$$

Thus, the lines $x_2 = \pm b$ transform into lines given by equations (9.12.34), which make angles $\pm\tan^{-1}(bv/R)$ with the vertical.

For the particle initially located on the lines $x_1 = \pm h$, equations (9.12.33) give

$$\begin{aligned} x'_1 &= \pm h + \frac{1}{2R}\{c^2 + v(h^2 - x_2^2)\} \\ &= \pm h + \frac{1}{2R}\left[c^2 + vh^2 - v(x'_2)^2\left(1 + \frac{v}{R}x_1\right)^{-1}\right] \\ &\approx \pm h + \frac{1}{2R}\{c^2 + vh^2 - v(x'_2)^2\} \end{aligned} \quad (9.12.35)$$

Thus, the lines $x_1 = \pm h$ transform into curves given by equations (9.12.35), which represent parabolas whose axes are directed opposite to the x_1 axis. It is easy to verify that the curvature of both these parabolas is approximately v/R.

Thus, the boundary lines $x_1 = \pm h$ and $x_2 = \pm b$ of a cross section transform, respectively, to the oblique lines given by (9.12.34) and the parabolas given by (9.12.35). The deformed shape of the section is shown by dotted lines in Figure 9.8. ∎

9.12.4 TORSION OF ELLIPTIC AND CIRCULAR BEAMS

Consider an elastic beam of length l and uniform elliptic or circular cross section. Suppose that a couple (of moment) **m** about the axis of the beam is applied at one of the bases and an opposite couple $-\mathbf{m}$ is applied at the other base. The lateral surface is stress free and body forces are neglected. The problem is to compute the displacements, strains and stresses developed in the beam because of the *twist* (or *torsion*) it experiences due to the applied couples; see Figure 9.9.

Let us choose the coordinate axes such that the origin is the centroid of the base subjected to the couple $-\mathbf{m}$ and the x_3 axis is along the axis of the beam.

If **s** is the stress vector at a point of a cross section, then the total moment of the surface force acting on the cross section is $\int_A \mathbf{x} \times \mathbf{s}\, dA$. If this section is the base $x_3 = 0$, we have $\mathbf{s} = -\mathbf{s}^{(3)}$ and the total moment is equal to $-\mathbf{m}$; thus,

$$\int_A \mathbf{x} \times (-\mathbf{s}^{(3)})\, dA = -\mathbf{m} \tag{9.12.36}$$

for $x_3 = 0$. On the other hand, if the section is the base $x_3 = l$ we have $\mathbf{s} = \mathbf{s}^{(3)}$ and the total moment is equal to **m**; thus,

$$\int_A \mathbf{x} \times \mathbf{s}^{(3)}\, dA = \mathbf{m} \tag{9.12.37}$$

for $x_3 = l$.

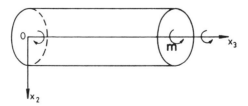

Figure 9.9. Torsion.

9.12 SOME STATIC PROBLEMS

By hypothesis, the applied couples act about the axis of the beam; hence, we may take $\mathbf{m} = M\mathbf{e}_3$, where $M = |\mathbf{m}|$. Then (9.12.36) and (9.12.37) yield the following set of conditions to be satisfied by the stress components:

$$\int_A x_2 \tau_{33}\, dA = 0, \quad \int_A x_1 \tau_{33}\, dA = 0, \quad \int_A (x_1 \tau_{32} - x_2 \tau_{31})\, dA = M$$

$$\int_A (x_2 \tau_{33} - l\tau_{32})\, dA = 0, \quad \int_A (l\tau_{31} - x_1 \tau_{33})\, dA = 0 \tag{9.12.38}$$

Since the beam is twisted by the applied couples, the cross sections $x_3 = $ constant experience shear strains and hence shear stresses. Therefore, τ_{31} and τ_{32}, are *not* 0. If we take all the remaining stress components to be identically 0, that is, if

$$\tau_{11} = \tau_{22} = \tau_{33} = \tau_{12} = 0 \tag{9.12.39}$$

then the conditions (9.12.38) become

$$\int_A \tau_{31}\, dA = 0, \quad \int_A \tau_{32}\, dA = 0, \quad \int_A (x_1 \tau_{32} - x_2 \tau_{31})\, dA = M \tag{9.12.40}$$

By hypothesis, the lateral surface of the beam is stress free. Hence $\tau_{ij} n_j = 0$ on this surface, for which $n_3 = 0$. Suppose that the beam is elliptic with the boundary of a cross section given by the equation:

$$\frac{x_1^2}{a^2} + \frac{x_2^2}{b^2} = 1 \tag{9.12.41}$$

(The circular beam corresponds to the case $b = a$.) Then the slope of the normal $\mathbf{n}$ to the boundary curve (9.12.41) (which lies parallel to the $x_1 x_2$ plane) is

$$-\frac{dx_1}{dx_2} = \frac{a^2 x_2}{b^2 x_1}$$

Hence, the components n_1 and n_2 of $\mathbf{n}$ are proportional to $b^2 x_1$ and $a^2 x_2$, respectively. Consequently, the conditions $\tau_{ij} n_j = 0$ to be satisfied on the lateral surface reduce to the single condition

$$b^2 \tau_{31} x_1 + a^2 \tau_{32} x_2 = 0 \tag{9.12.42}$$

where we have made use of (9.12.39).

Since the body force is neglected, Cauchy's equilibrium equations $\tau_{ij,j} = 0$ must be satisfied at every point of the beam. In view of (9.12.39), these equations take the following explicit forms:

$$\tau_{13,3} = 0, \quad \tau_{23,3} = 0, \quad \tau_{31,1} + \tau_{32,2} = 0 \tag{9.12.43}$$

9 EQUATIONS OF LINEAR ELASTICITY

The simplest solution of these equations that meets the condition (9.12.42) is

$$\tau_{31} = cx_2, \qquad \tau_{32} = -\frac{b^2}{a^2}cx_1 \qquad (9.12.44)$$

where c is a nonzero constant.

Using (9.12.44), the conditions (9.12.40) become

$$\int_A x_2\, dA = 0, \qquad \int_A x_1\, dA = 0, \qquad \int_A (b^2 x_1^2 + a^2 x_2^2)\, dA = -\frac{Ma^2}{c} \qquad (9.12.45)$$

Bearing in mind that A is the cross-sectional area bounded by the ellipse (9.12.41), we note that the integrals $\int_A x_1^2\, dA$ and $\int_A x_2^2\, dA$ denote the moments of inertia I_2 and I_1 of the section about the x_2 and x_1 axes, respectively, with $I_1 = \pi a b^3/4$ and $I_2 = \pi a^3 b/4$. As such, the last condition in (9.12.45) becomes

$$c = -\frac{2M}{\pi ab^3} \qquad (9.12.46)$$

Further, the first two conditions in (9.12.45) are identically satisfied. Hence, if c is chosen as given by (9.12.46), then Cauchy's equilibrium equations as well as all the boundary conditions of the problem are satisfied by the stress system given by (9.12.39) and (9.12.44). It is easy to verify that this system also satisfies the Beltrami–Michell equations.

Thus, a system of stresses that meets all the requirements of the problem is

$$\tau_{11} = \tau_{22} = \tau_{33} = \tau_{12} = 0$$
$$\tau_{31} = -\frac{2M}{\pi ab^3} x_2, \qquad \tau_{32} = \frac{2M}{\pi a^3 b} x_1 \qquad (9.12.47)$$

The corresponding system of strains follow from Hooke's law (9.3.15)″. Thus, we get

$$e_{11} = e_{22} = e_{33} = e_{12} = 0$$
$$e_{13} = -\frac{M}{\mu \pi ab^3} x_2, \qquad e_{32} = \frac{M}{\mu \pi a^3 b} x_1 \qquad (9.12.48)$$

The displacement components associated with these strains follow from the strain-displacement relation (9.4.4). Thus we obtain (see equation (5.8.46)),

$$u_1 = -\alpha x_2 x_3, \qquad u_2 = \alpha x_1 x_3, \qquad u_3 = -\alpha\left(\frac{a^2 - b^2}{a^2 + b^2}\right) x_1 x_2 \qquad (9.12.49)$$

where

$$\alpha = \frac{a^2 + b^2}{a^3 b^3} \cdot \frac{M}{\mu\pi} > 0 \qquad (9.12.50)$$

Expressions (9.12.49) show that in the case of an elliptic beam ($b \neq a$) there does occur a longitudinal displacement u_3, in addition to the transverse displacements u_1, u_2. Since u_3 depends on x_1 and x_2 and is independent of x_3, cross sections are warped according to the law given by the last relation in (9.12.49) and the warping is the same in all sections. We find that the lines of warping are rectangular hyperbolas and that (because of warping) the parts of a cross section for which x_1 and x_2 have the same signs get pulled toward the fixed base whereas the parts for which x_1 and x_2 have opposite signs get pushed away from this base.

If x'_1 and x'_2 are the coordinates of the new location P' of a particle initially located at the point $P = (x_1, x_2)$ in a cross section, we find by the use of (9.12.49) that

$$(x'_1)^2 + (x'_2)^2 = (1 + \alpha^2 x_3^2)(x_1^2 + x_2^2) \approx x_1^2 + x_2^2 \qquad (9.12.51)$$

Thus, the particle experiences a pure rotation about the axis of the beam; if θ is the angle of rotation, we find that (see Figure 9.10)

$$\begin{aligned} u_1 &= x'_1 - x_1 = r\cos(\gamma + \theta) - r\cos\gamma \approx -x_2 \theta \\ u_2 &= x'_2 - x_2 = r\sin(\gamma + \theta) - r\sin\gamma \approx x_1 \theta \end{aligned} \qquad (9.12.52)$$

Comparing these results with those given in (9.12.49), we find that $\theta = \alpha x_3$. Thus, the angle of rotation θ in a section is proportional to the distance of the section from the fixed base. The constant of proportionality, namely, α, represents the *angle of twist* (rotation) per unit length. The quantity (M/α), which represents the moment of the couple required to produce a unit angle of twist per unit of length, is called the *torsional rigidity* or the *torsional stiffness* of the beam, usually denoted D. From

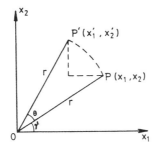

Figure 9.10. Rotation in a cross section.

422   9   EQUATIONS OF LINEAR ELASTICITY

(9.12.50), we find that

$$D = \frac{\mu \pi a^3 b^3}{a^2 + b^2} > 0 \tag{9.12.53}$$

From (9.12.47), we find that the stress vector at a point P of a cross section of the beam is given by

$$\mathbf{s} = \tau_{31}\mathbf{e}_1 + \tau_{32}\mathbf{e}_2 = \frac{2M}{\pi a^3 b^3}(-a^2 x_2 \mathbf{e}_1 + b^2 x_1 \mathbf{e}_2) \tag{9.12.54}$$

If we set $x_1 = ka \cos \gamma$, $x_2 = kb \sin \gamma$, then (9.12.54) becomes

$$\mathbf{s} = \frac{2Mk}{\pi a^2 b^2}(-\mathbf{e}_1 a \sin \gamma + \mathbf{e}_2 b \cos \gamma) \tag{9.12.55}$$

This is the stress vector at a point P ($ak \cos \gamma$, $bk \sin \gamma$) of the section. It follows immediately that the point P lies on the ellipse

$$\frac{x_1^2}{a^2} + \frac{x_2^2}{b^2} = k^2 \leq 1 \tag{9.12.56}$$

whose normal at P is along the vector

$$\bar{\mathbf{n}} = \left(\mathbf{e}_1 \frac{1}{a} \cos \gamma + \mathbf{e}_2 \frac{1}{b} \sin \gamma\right) \tag{9.12.57}$$

From (9.4.55) and (9.4.57), we readily find that $\mathbf{s} \cdot \bar{\mathbf{n}} = 0$, so that $\mathbf{s}$ is along the tangent to the ellipse (9.12.56). Thus, at a point P of a cross section, the stress vector is tangential to the ellipse that passes through the point and is concentric with and similar to the boundary curve of the section. In other words, the family of ellipses determined by (9.12.56) for different values of k are the *lines (curves) of shear stress* for the elliptic beam considered.

From (9.12.55), we find that the magnitude of $\mathbf{s}$ at P is given by

$$|\mathbf{s}| = \frac{2Mk}{\pi a^2 b^2}\{b^2 + (a^2 - b^2)\sin^2\gamma\}^{1/2} \tag{9.12.58}$$

If a and b are the lengths of the semi-major axis and the semi-minor axis, respectively, of the boundary curve (9.12.41), we find from (9.12.58) that $|\mathbf{s}|$ is maximum when k is maximum (that is $k = 1$) and $\gamma = \pi/2$ or $3\pi/2$. Thus, the magnitude of shear stress assumes a maximum value at the endpoints of the minor axis of the boundary curve, and the maximum value is given by

$$\max|\mathbf{s}| = \frac{2M}{\pi a b^2} \tag{9.12.59}$$

9.12 SOME STATIC PROBLEMS

For a *circular beam* ($b = a$), we find from (9.12.49) that $u_3 \equiv 0$. Accordingly, the cross sections of the beam remain undisplaced from their original position, and the beam experiences no warping along its length. There occur only transverse displacements u_1 and u_2 at a point, and these displacements vary from one cross section to the other. From (9.12.53) and (9.12.59) we find the following expressions for torsional rigidity and maximum shear stress for a circular beam of radius a:

$$D = \frac{\mu\pi}{2} a^4 \qquad (9.12.60)$$

$$\max|s| = \frac{2M}{\pi a^3} \qquad (9.12.61)$$

We note from (9.12.58) that for a circular beam there is no specific point on the boundary at which $|s|$ is maximum; $\max|s|$ given by (9.12.61) occurs at every point of the boundary. Further, the lines of shear stress are circles concentric with the boundary.

It may be pointed out that the torsion problem for a circular beam was first solved by Coloumb in 1787. The theory of torsion for noncircular beams was later developed by Saint-Venant in 1855. The Saint-Venant's theory of torsion is one of the most celebrated works in the mathematical theory of elasticity.

EXAMPLE 9.12.4 Show that the torsional rigidity of an elliptic beam with cross sectional area A and polar moment of intertia I_e is given by

$$D_e = \frac{\mu A^4}{4\pi^2 I_e} \qquad (9.12.62a)$$

If I_c is the polar moment of intertia of a circular beam having the same cross-sectional area as the previous elliptic beam and D_c is its torsional rigidity, show that

$$D_e = (I_c/I_e)D_c \qquad (9.12.62b)$$

Solution For a beam whose cross section is bounded by the ellipse (9.12.41), the area of cross section is $A = \pi ab$ and the polar moment of inertia is

$$I_e = I_1 + I_2 = \frac{\pi}{4} ab(a^2 + b^2) \qquad (9.12.63)$$

Hence, (9.12.53) may be rewritten as

$$D = D_e = \mu \frac{\pi a^3 b^3}{a^2 + b^2} \frac{4\pi^3 ab}{4\pi^2 (\pi ab)} = \mu \frac{A^4}{4\pi^2 I_e}$$

which is (9.12.62a). For a circular beam ($b = a$), this becomes

$$D = D_c = \frac{\mu A^4}{4\pi^2 I_c} \tag{9.12.64}$$

where

$$I_c = (I_e)_{b=a} = \frac{\pi}{2} a^4 \tag{9.12.65}$$

is the polar moment of inertia of a circular beam and A is its cross-sectional area.

For a given A, we readily find from (9.12.62a), (9.12.63), (9.12.64) and (9.12.65) that

$$D_e = \frac{\mu}{4\pi^2 I_e} \frac{4\pi^2 I_c D_c}{\mu} = \left(\frac{I_c}{I_e}\right) D_c$$

which is (9.12.62b). ∎

Note: From (9.12.63) and (9.12.65), we find that $I_c \leq I_e$. Consequently, from (9.12.62b) it follows that $D_e \leq D_c$. Thus, a circular beam has greater torsional stiffness than an elliptic beam with same cross-sectional area.

9.12.5 CYLINDRICAL TUBE UNDER PRESSURE

Consider a long, straight elastic circular tube of inner radius a and outer radius b. Suppose that a uniform pressure p acts on the inner surface and a uniform pressure $q(\neq p)$ acts on the outer surface. Body forces are ignored. The problem is to determine the displacements and stresses caused at an arbitrary point of the tube.

We note that the pressures p and q act completely along the radial direction of the circular boundary surfaces. Also, since the tube is long, the effects of surface forces acting on the ends of the tube (on the deformation in a cross section) can be neglected. Further, the body forces are ignored. In view of these facts, we naturally assume that (i) the particles get displaced completely along the radial direction, (ii) the amount of displacement depends solely on the distance of the particle from the central line (axis) of the tube and (iii) the same kind of deformation occurs in every section of the tube.

Let us choose the coordinate axes such that the $x_1 x_2$ plane coincides with a section of the tube and the x_3 axis lies along the axis of the tube. Then, in view of these assumptions, the displacement field at any point of the tube may be taken as

$$\mathbf{u} = u(R)\mathbf{e}_R \tag{9.12.66}$$

9.12 SOME STATIC PROBLEMS

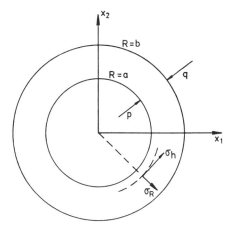

Figure 9.11. Pressures on a cylindrical tube.

where $\mathbf{e}_R$ is the unit vector along the radial vector $x_1\mathbf{e}_1 + x_2\mathbf{e}_2$, $R^2 = x_1^2 + x_2^2$ and $u = u(R)$ is a function to be determined.

We note that the displacement field given by (9.12.66) is parallel to the x_1x_2 plane and independent of x_3. A problem in which this kind of displacement arises is called a *plane strain* problem. Such problems, together with the so-called plane stress problems, belong to the area of *plane elastostatics*.

Returning to the problem under consideration, we note that the boundary surfaces of the tube are $R = a$ and $R = b$, where a and b are the inner and outer radii of the tube, and that these surfaces are subjected to uniform pressures p and q; see Figure 9.11. Hence, if σ_R is the normal stress on a cylindrical surface concentric with the tube, the boundary conditions for the problem are

$$\sigma_R = \begin{cases} -p & \text{for} \quad R = a \\ -q & \text{for} \quad R = b \end{cases} \qquad (9.12.67)$$

In order that $\mathbf{u}$ taken as in (9.12.66) may be a solution of the problem, it must obey Navier's equation of equilibrium with zero body force and the boundary conditions (9.12.67). In Example 9.8.6, it was shown that $\mathbf{u}$ as given by (9.12.66) obeys Navier's equation of equilibrium with zero body force if $u(R)$ is as given by (9.8.51); that is,

$$u = AR + \frac{B}{R} \qquad (9.12.68)$$

where A and B are arbitrary constants. These constants are now to be determined by using the boundary conditions (9.12.67). In Example 9.8.6,

9 EQUATIONS OF LINEAR ELASTICITY

σ_R has been obtained in terms of $u(R)$ as given by (9.8.47); that is,

$$\sigma_R = \lambda \frac{u}{R} + (\lambda + 2\mu) \frac{du}{dR} \qquad (9.12.69)$$

Substituting for u from (9.12.68) in (9.12.69) we get

$$\sigma_R = 2(\lambda + \mu)A - \frac{2\mu}{R^2} B \qquad (9.12.70)$$

The boundary conditions (9.12.67) now yield the following simultaneous equations for A and B:

$$\begin{aligned}(\lambda + \mu)A - \frac{\mu}{a^2} B &= -\frac{1}{2}p \\ (\lambda + \mu)A - \frac{\mu}{b^2} B &= -\frac{1}{2}q\end{aligned} \qquad (9.12.71)$$

Solving these equations, we get

$$(\lambda + \mu)A = \frac{a^2 p - b^2 q}{2(b^2 - a^2)}, \qquad \mu B = \frac{a^2 b^2 (p - q)}{2(b^2 - a^2)} \qquad (9.12.72)$$

With these values of A and B, (9.12.68) and (9.12.66) give the displacement vector

$$\mathbf{u} = \left\{ \frac{a^2 p - b^2 q}{2(\lambda + \mu)(b^2 - a^2)} R + \frac{a^2 b^2 (p - q)}{2\mu(b^2 - a^2)} \frac{1}{R} \right\} \mathbf{e}_R \qquad (9.12.73)$$

which satisfies the governing differential equation as well as the boundary conditions of the problem. By virtue of the uniqueness of solution, this $\mathbf{u}$ is unique. Thus, (9.12.73) gives the displacement that occurs at a point distant R from the axis of the tube.

Using (9.12.72) in (9.12.70) we obtain

$$\sigma_R = \frac{a^2 p - b^2 q}{b^2 - a^2} - \frac{a^2 b^2 (p - q)}{b^2 - a^2} \frac{1}{R^2} \qquad (9.12.74)$$

This is the expression for the normal stress acting on a cylindrical surface $R = $ constant in the tube.

The normal stress σ_h acting on an axial plane can be computed by use of expression (9.8.48) together with (9.12.68) and (9.12.72). Thus, we obtain

$$\sigma_h = \frac{a^2 p - b^2 q}{b^2 - a^2} + \frac{a^2 b^2 (p - q)}{b^2 - a^2} \frac{1}{R^2} \qquad (9.12.75)$$

9.12 SOME STATIC PROBLEMS

Let us now analyze the nature of σ_R and σ_h in the particular case where the external pressure q is absent. In this case, (9.12.74) and (9.12.75) become

$$\sigma_R = \frac{a^2 p}{b^2 - a^2}\left[1 - \frac{b^2}{R^2}\right] \quad (9.12.76)$$

$$\sigma_h = \frac{a^2 p}{b^2 - a^2}\left[1 + \frac{b^2}{R^2}\right] \quad (9.12.77)$$

Since $a < b$ and $R \leq b$, we readily find that $\sigma_R \leq 0$ and $\sigma_h > 0$. This means that every point of the tube experiences a compressive radial stress and a tensile peripheral stress. Also, the peripheral stress σ_h is maximum on the inner surface of the tube, with

$$\max \sigma_h = \frac{p(a^2 + b^2)}{b^2 - a^2} \quad (9.12.78)$$

If the thickness $(b - a)$ of the tube is very small, (9.12.78) becomes

$$\max \sigma_h \approx \frac{pb}{b - a} \quad (9.12.79)$$

Thus, for a given pressure p, max σ_h varies inversely as the thickness of the tube.

Two other particular cases of the problem are also of interest. If $b \to \infty$, the tube becomes an *unbounded body with a cylindrical cavity*. Then (9.12.73) becomes

$$\mathbf{u} = \left\{\frac{a^2 p}{2\mu R} - \frac{q}{2}\left(\frac{R}{\lambda + \mu} + \frac{a^2}{\mu R}\right)\right\}\mathbf{e}_R \quad (9.12.80)$$

It is evident that for $\mathbf{u}$ to have finite magnitude for *all* R, q should be 0. Then we get

$$\mathbf{u} = \frac{a^2 p}{2\mu R}\mathbf{e}_R \quad (9.12.81)$$

The corresponding expressions for σ_R and σ_h follow from (9.12.76) and (9.12.77) with $b \to \infty$; the results are

$$\sigma_R = -p\frac{a^2}{R^2}, \quad \sigma_h = p\frac{a^2}{R^2} \quad (9.12.82)$$

We find that these σ_R and σ_h have the same magnitude at every point of the body. Also, as $R \to \infty$, $\mathbf{u}$, σ_R and σ_h all tend to 0.

In the case when $a \to 0$ with b remaining finite, the tube becomes a *full cylinder* (with no hole). Then (9.12.73), (9.12.74) and (9.12.75) become (taking p to be bounded),

$$\mathbf{u} = -\frac{q}{2(\lambda + \mu)} R \mathbf{e}_R \qquad (9.12.83)$$

$$\sigma_R = \sigma_h = -q$$

We find that σ_R and σ_h are now everywhere equal to the applied pressure. Also, the magnitude of displacement varies directly with R and is maximum on the surface $R = b$.

9.12.6 SPHERICAL SHELL UNDER PRESSURE

Consider an elastic spherical shell of inner radius a and outer radius b. Suppose that a uniform pressure p acts on the inner surface and a uniform pressure $q(\neq p)$ acts on the outer surface. Body forces are ignored. The problem is to determine the displacements and stresses caused at an arbitrary point of the shell.

We note that the pressures p and q act completely along the radial direction from the center of the shell. Further, body forces are ignored. Hence, we assume that (i) the particles get displaced completely along the radial direction, and (ii) the amount of displacement depends solely on the distance of the particle from the center of the shell. Then, if we choose the coordinate axes with the origin at the center of the shell, the displacement vector at any point of the shell may be taken as

$$\mathbf{u} = u(r) \mathbf{e}_r \qquad (9.12.84)$$

where $\mathbf{e}_r$ is the unit vector along the radial direction and $u(r)$ is a function to be determined.

The surfaces $r = a$ and $r = b$ are subjected to uniform pressures p and q, respectively, where a and b are the inner and outer radii of the shell; see Figure 9.12. Hence, if σ_r is the normal stress on a spherical surface r = constant in the shell, the boundary conditions for the problem are

$$\sigma_r = \begin{cases} -p & \text{for } r = a \\ -q & \text{for } r = b \end{cases} \qquad (9.12.85)$$

In order that $\mathbf{u}$ taken as in (9.12.84) may be a solution of the problem, it must satisfy Navier's equation of equilibrium with zero body force and the boundary conditions (9.12.85). In Example 9.8.5, it was shown that $\mathbf{u}$ given by (9.12.84) obeys Navier's equation of equilibrium with zero body force if

9.12 SOME STATIC PROBLEMS

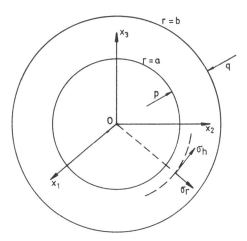

Figure 9.12. Spherical shell under pressure.

$u(r)$ is as given by (9.8.41); that is,

$$u(r) = \frac{A}{r^2} + Br \qquad (9.12.86)$$

where A and B are arbitrary constants. These constants are now to be determined by using the boundary conditions (9.12.85). In Example 9.8.5, σ_r was obtained in terms of $u(r)$ as given by (9.8.36); that is,

$$\sigma_r = 2\lambda \frac{1}{r} u(r) + (\lambda + 2\mu) \frac{du}{dr} \qquad (9.12.87)$$

Substituting for u from (9.12.86) in (9.12.87), we get

$$\sigma_r = -4\mu \frac{1}{r^3} A + (3\lambda + 2\mu)B \qquad (9.12.88)$$

The boundary conditions (9.12.85) now yield the following simultaneous equations for A and B:

$$\left. \begin{array}{r} 4\mu \dfrac{1}{a^3} A - (3\lambda + 2\mu)B = p \\[1em] 4\mu \dfrac{1}{b^3} A - (3\lambda + 2\mu)B = q \end{array} \right\} \qquad (9.12.89)$$

Solving these equations, we get

$$4\mu A = \frac{a^3 b^3 (p - q)}{b^3 - a^3}, \qquad (3\lambda + 2\mu)B = \frac{pa^3 - qb^3}{b^3 - a^3} \qquad (9.12.90)$$

With these values of A and B, (9.12.84) and (9.12.86) give the displacement vector

$$\mathbf{u} = \frac{1}{b^3 - a^3}\left[\frac{a^3b^3(p-q)}{4\mu r^2} + \frac{(pa^3 - qb^3)r}{(3\lambda + 2\mu)}\right]\mathbf{e}_r \quad (9.12.91)$$

which satisfies the governing differential equation as well as the boundary conditions of the problem. By virtue of the uniqueness of solution, this $\mathbf{u}$ is unique. Thus, (9.12.91) gives the displacement that occurs at a point distant r from the center of the shell.

By use of (9.12.90) in (9.12.88) we obtain

$$\sigma_r = \frac{1}{b^3 - a^3}\left[(pa^3 - qb^3) - \frac{a^3b^3}{r^3}(p-q)\right] \quad (9.12.92)$$

This is the expression for the normal stress acting on a spherical surface r = constant in the shell.

The normal stress σ_h acting on a radial plane can be computed with the use of expressions (9.3.37), (9.12.86) and (9.12.90). Thus, we obtain

$$\sigma_h = \frac{1}{b^3 - a^3}\left[\frac{a^3b^3(p-q)}{2r^3} + (pa^3 - qb^3)\right] \quad (9.12.93)$$

Let us now analyze the nature of σ_r and σ_h in the particular case where the external pressure q is absent. In this case, (9.12.92) and (9.12.93) become

$$\sigma_r = \frac{pa^3}{b^3 - a^3}\left[1 - \frac{b^3}{r^3}\right] \quad (9.12.94)$$

$$\sigma_h = \frac{pa^3}{b^3 - a^3}\left[1 + \frac{b^3}{2r^3}\right] \quad (9.12.95)$$

Since $a < b$ and $r \leq b$, we readily find that $\sigma_r \leq 0$ and $\sigma_h > 0$. This means that every point of the shell experiences a compressive radial stress and a tensile peripheral stress. Also, the peripheral stress σ_h is maximum on the inner surface of the shell, with

$$\max \sigma_h = \frac{p(2a^3 + b^3)}{2(b^3 - a^3)} \quad (9.12.96)$$

If the thickness $(b - a)$ of the shell is very small, (9.12.96) becomes

$$\max \sigma_h \approx \frac{pb}{2(b - a)} \quad (9.12.97)$$

Thus, for a given pressure, max σ_h varies inversely as the thickness of the shell.

Two other particular cases of the problem are also of interest. If $b \to \infty$, the shell becomes an *unbounded body with a spherical cavity*. Then (9.12.91) becomes

$$\mathbf{u} = \left[\frac{a^3(p-q)}{4\mu r^2} - \frac{qr}{3\lambda + 2\mu} \right] \mathbf{e}_r \qquad (9.12.98)$$

It is evident that, for $\mathbf{u}$ to have finite magnitude for all r, q should be 0. Then we get

$$\mathbf{u} = \frac{a^3 p}{4\mu r^2} \mathbf{e}_r \qquad (9.12.99)$$

The corresponding values for σ_r and σ_h follow from (9.12.94) and (9.12.95) with $b \to \infty$; they are

$$\sigma_r = -\frac{pa^3}{r^3}, \qquad \sigma_h = \frac{pa^3}{r^3} \qquad (9.12.100)$$

We find that these σ_r and σ_h have the same magnitude at every point of the body. Also, as $r \to \infty$, $\mathbf{u}$, σ_r and σ_h all tend to 0.

When $a \to 0$ with b remaining finite, the shell becomes a *full sphere* (with no hole). Then (9.12.91), (9.12.92) and (9.12.93) become (taking p to be bounded)

$$\mathbf{u} = -\frac{qr}{3\lambda + 2\mu} \mathbf{e}_r \qquad (9.12.101)$$

$$\sigma_r = \sigma_h = -q$$

We find that σ_r and σ_h are now everywhere equal to the applied pressure. Also, the magnitude of displacement varies directly with r and is maximum on the surface $r = b$.

The problems on the cylindrical tube and the spherical shell subjected to unequal pressures were first considered by Lamé in 1852. These problems are referred to as *Lamé's pressure-vessel problems*.

9.13
ELASTIC WAVES

As indicated earlier, in the displacement formulation, a dynamical problem in elasticity is governed by Navier's equation (9.8.6). A simple inspection of this equation reveals that it is a *hyperbolic-type* partial differential equation of the second order. A solution of this equation represents a *wave motion*. In the stress formulation of an elastodynamic problem, the stress equation of motion (9.10.5) or the Beltrami–Michell equation (9.11.6) serves as the governing equation; these are also hyperbolic partial differential equations. Therefore, the solutions of these equations also represent wave motions. Accordingly, every motion of an elastic body occurs in the form of a wave (called an *elastic wave*), and every dynamical problem in elasticity deals with the study of one or another type of elastic waves. Some simple elastic wave propagation problems are discussed in this section.

9.13.1 STRESS WAVES IN A SEMI-INFINITE BEAM

Consider a thin semi-infinite elastic beam that is initially at rest in an undeformed state. Suppose that at time $t = 0^+$ a time-dependent pressure $p(t)$ is applied to the end of the beam along the length, and this pressure is maintained for all subsequent times; see Figure 9.13. The body forces are ignored. The problem is to determine the stress and displacement that occur at an arbitrary point of the beam at any subsequent time t.

Let us choose the x_1 axis along the axis of the beam with the origin at the initial position of the end where the pressure is applied. Then the boundary condition to be satisfied is

$$\tau_{11} = -p(t) \tag{9.13.1}$$

for $x_1 = 0$, $t > 0$.

It is assumed that the beam is *thin*. Also, the load that causes deformation acts along the length of the beam. Hence, we suppose that at any point x_1 of the beam there occurs only the longitudinal stress τ_{11} and that this is a function of x_1 and t. Then the stress equation of motion (9.10.5) yields the following equation for τ_{11}:

$$\alpha^2 \frac{\partial^2 \tau_{11}}{\partial x_1^2} = \frac{\partial^2 \tau_{11}}{\partial t^2} \tag{9.13.2}$$

where

$$\alpha^2 = \frac{E}{\rho} \tag{9.13.3}$$

We have ignored the body forces.

Equation (9.13.2) is the governing partial differential equation for the problem. We note that this is a *one-dimensional wave equation* with α representing the *speed of propagation*. We therefore infer that the

Figure 9.13. Stress on a beam.

distribution of the stress τ_{11} occurs in the form of a wave propagating with speed $\alpha = \sqrt{E/\rho}$ along the length of the beam. Such a wave is called a *stress wave*.

Since the beam is initially at rest in an undeformed state, the following initial conditions hold:

$$\mathbf{u} = \frac{\partial \mathbf{u}}{\partial t} = \mathbf{0} \tag{9.13.4}$$

at $t = 0$ and $x_1 \geq 0$. Consequently, we get, on using the stress-displacement relation (9.8.4), the following conditions for τ_{11}:

$$\tau_{11} = \frac{\partial \tau_{11}}{\partial t} = 0 \tag{9.13.5}$$

at $t = 0$ and $x_1 \geq 0$.

Thus, (9.13.2) is the governing wave equation, (9.13.5) are the initial conditions and (9.13.1) is the boundary condition for the problem. Note that all of these are expressed in terms of the stress component τ_{11}. This is the stress formulation of the problem.

To solve the problem, we change the independent variables from x_1 and t to $\xi = t - (x_1/\alpha)$ and $\eta = t + (x_1/\alpha)$. Equation (9.13.2) then becomes

$$\frac{\partial^2 \tau_{11}}{\partial \xi \, \partial \eta} = 0 \tag{9.13.6}$$

Integration of this equation yields the following general solution for τ_{11}:

$$\tau_{11} = f(\xi) + g(\eta) \tag{9.13.7}$$

where $f(\xi)$ and $g(\eta)$ are arbitrary functions. This solution is usually called *D'Alembert's solution*.

Using (9.13.7), the initial conditions (9.13.5) take the form

$$\left. \begin{array}{l} f(-x_1/\alpha) + g(x_1/\alpha) = 0 \\ f'(-x_1/\alpha) + g'(x_1/\alpha) = 0 \end{array} \right\} \tag{9.13.8}$$

for $x_1 \geq 0$. These conditions are satisfied if we choose $f(\xi)$ and $g(\eta)$ such that

$$\left. \begin{array}{ll} g(\eta) = A & \text{for all } \eta \\ f(\xi) = -A & \text{for } \xi \leq 0 \end{array} \right\} \tag{9.13.9}$$

where A is an arbitrary constant. Then the solution (9.13.7) becomes

$$\tau_{11} = \begin{cases} 0 & \text{for } \xi \leq 0 \\ A + f(\xi) & \text{for } \xi > 0 \end{cases}$$

or

$$\tau_{11} = \begin{cases} 0 & \text{for } t \le x_1/\alpha \\ F\left(t - \dfrac{x_1}{\alpha}\right) & \text{for } t > x_1/\alpha \end{cases} \qquad (9.13.10)$$

where $F(t - x_1/\alpha) = A + f(t - x_1/\alpha)$ is an arbitrary function of $(t - x_1/\alpha)$. The boundary condition (9.13.1) is satisfied if we choose $F(t - x_1/\alpha) = -p(t - x_1/\alpha)$. Thus, a solution for τ_{11} that satisfies the governing equation (9.13.2) and the conditions (9.13.5), and (9.13.1) is

$$\tau_{11}(x_1, t) = \begin{cases} 0 & \text{for } t \le x_1/\alpha \\ -p\left(t - \dfrac{x_1}{\alpha}\right) & \text{for } t > x_1/\alpha \end{cases} \qquad (9.13.11)$$

From this solution, we note that at any chosen point $\bar{x}_1$ of the beam, no stress occurs until the time $\bar{t} = \bar{x}_1/\alpha$, and a time-dependent compressive stress $-p(t - \bar{x}_1/\alpha)$ occurs thereafter. This stress is due to the stress wave that starts from the end $x_1 = 0$ at $t = 0$ and arrives at the point $\bar{x}_1$ at time $\bar{t} = \bar{x}_1/\alpha$. The speed of the wave is $\alpha = \sqrt{E/\rho}$ as already indicated.

The longitudinal displacement u_1 associated with τ_{11} given by (9.13.11) can be computed by using Hooke's law (9.3.3) and the strain-displacement relation $e_{11} = u_{1,1}$. Thus we get

$$u_1(x_1, t) = \begin{cases} 0 & \text{for } t \le x_1/\alpha \\ \dfrac{\alpha}{E} \displaystyle\int_0^{t-x_1/\alpha} p(t_0)\, dt_0 & \text{for } t > x_1/\alpha \end{cases} \qquad (9.13.12)$$

This result shows that the displacement u_1 caused by the stress wave at a point $\bar{x}_1$ at time $t > \bar{x}_1/\alpha$ is directly proportional to the area under the pressure curve over the time interval $(0, t - \bar{x}_1/\alpha)$, the constant of proportionality being $\alpha/E = (E\rho)^{-1/2}$.

From (9.13.12), we get

$$\frac{\partial u_1}{\partial t}(x_1, t) = \frac{\alpha}{E} p\left(t - \frac{x_1}{\alpha}\right) \qquad (9.13.13)$$

for $t > x_1/\alpha$. This gives the speed at a point x_1 of the beam at time $t (> x_1/\alpha)$. From (9.13.11), (9.13.13) and (9.13.3) we get, for $t > x/\alpha$,

$$\tau_{11} \bigg/ \left(\frac{\partial u_1}{\partial t}\right) = -\frac{E}{\alpha} = -\sqrt{E\rho} \qquad (9.13.14)$$

9.13 ELASTIC WAVES

We note that the quantity $\tau_{11}/(\partial u_1/\partial t)$ represents the stress required to generate a speed of unit magnitude in the beam, called the *mechanical impedance* or *wave resistance* factor of the beam. From (9.13.14), we observe that this quantity is a constant (equal to $-\sqrt{E\rho}$) depending on the physical property of the beam.

9.13.2 PLANE WAVES IN A HALF-SPACE

Consider a semi-infinite elastic body (half-space) with plane boundary, which is initially at rest in an undeformed state. Suppose that at time $t = 0^+$ a time-dependent pressure $p(t)$ is applied to the boundary, and this pressure is maintained for all subsequent times. Body forces are ignored. The problem is to determine the displacement and stress that occur at an arbitrary point of the body at any subsequent time t.

Let us choose the axes such that the origin lies on the boundary of the half-space in its initial position with the x_1 axis perpendicular to the boundary and directed into the half-space. The geometrical space filled by the body is then $x_1 > 0$ with the plane $x_1 = 0$ as its boundary; see Figure 9.14. It is assumed that the deformation of the half-space is caused by the pressure $p(t)$ acting on the plane $x_1 = 0$ along the x_1 axis for $t > 0$. Hence the boundary condition for the problem is

$$\tau_{11} = -p(t) \qquad (9.13.15)$$

at $x_1 = 0$ for $t > 0$.

In view of this boundary condition, we assume that the displacement due to deformation is directed along the x_1 axis and that it depends only on x_1,

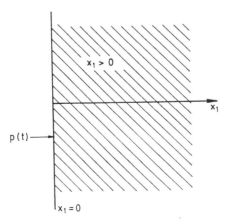

Figure 9.14. Half-space $x_1 > 0$.

436   9 EQUATIONS OF LINEAR ELASTICITY

apart from t. Thus, we seek a solution in the form

$$\mathbf{u} = u_1(x_1, t)\mathbf{e}_1 \tag{9.13.16}$$

for $x_1 \geq 0$, $t > 0$; where $u_1(x_1, t)$ is to be determined.

Since the body is initially at rest in an undeformed state, the following initial conditions hold:

$$u_1 = \frac{\partial u_1}{\partial t} = 0 \tag{9.13.17}$$

for $x_1 \geq 0$ at $t = 0$. By using (9.13.16) in the stress-displacement relation (9.8.4), we get

$$\tau_{11} = (\lambda + 2\mu)\frac{\partial u_1}{\partial x_1}, \quad \tau_{22} = \tau_{33} = \lambda \frac{\partial u_1}{\partial x_1} \tag{9.13.18}$$

$$\tau_{12} = \tau_{23} = \tau_{31} = 0$$

Using the first of these expressions, the condition (9.13.15) can be rewritten as

$$\frac{\partial u_1}{\partial x_1}(x_1, t) = -\frac{1}{\lambda + 2\mu}p(t) \tag{9.13.19}$$

for $x_1 = 0$, $t > 0$.

Thus, (9.13.17) and (9.13.19), expressed in terms of u_1, are the initial and boundary conditions for the problem.

Since the problem is being formulated in terms of the displacement, the governing equation is Navier's equation (9.8.6). In view of (9.13.16), Navier's equation becomes

$$c_1^2 \frac{\partial^2 u_1}{\partial x_1^2} = \frac{\partial^2 u_1}{\partial t^2} \tag{9.13.20}$$

where $c_1 = [(\lambda + 2\mu)/\rho]^{1/2}$ as usual. We have ignored the body forces.

We note that (9.13.20) is the one-dimensional wave equation with c_1 as the speed of propagation. Thus, under the assumptions made, the distribution of the displacement in the half-space occurs in the form of a plane wave propagating in the x_1 direction with speed c_1.

As in the case of equation (9.13.2), we find the D'Alembert's solution for equation (9.13.20) as

$$u_1 = f\left(t - \frac{x_1}{c_1}\right) + g\left(t + \frac{x_1}{c_1}\right) \tag{9.13.21}$$

for $x_1 \geq 0$, $t \geq 0$, where the functions f and g are to be determined. Using

9.13 ELASTIC WAVES

(9.13.21) in the initial conditions (9.13.17) we get the equations

$$\left.\begin{array}{r}f\left(-\dfrac{x_1}{c_1}\right) + g\left(\dfrac{x_1}{c_1}\right) = 0 \\[6pt] f'\left(-\dfrac{x_1}{c_1}\right) + g'\left(\dfrac{x_1}{c_1}\right) = 0\end{array}\right\}$$

for $x_1 \geq 0$, which are satisfied if we choose f and g as follows;

$$\left.\begin{array}{l}g\left(t + \dfrac{x_1}{c_1}\right) = A \quad \text{for} \quad t > 0, \\[6pt] f\left(t - \dfrac{x_1}{c_1}\right) = -A \quad \text{for} \quad t \leq \dfrac{x_1}{c_1}\end{array}\right\} \quad x_1 \geq 0 \qquad (9.13.22)$$

where A is an arbitrary constant. Then (9.13.21) reduces to the following form:

$$u_1(x_1, t) = \begin{cases} 0 & \text{for} \quad t \leq \dfrac{x_1}{c_1} \\[6pt] F\left(t - \dfrac{x_1}{c_1}\right) & \text{for} \quad t > \dfrac{x_1}{c_1} \end{cases} \qquad (9.13.23)$$

where $F(t - x_1/c_1) = A + f(t - x_1/c_1)$ is an arbitrary function. Hence

$$\frac{\partial u_1}{\partial x_1}(x_1, t) = \begin{cases} 0 & \text{for} \quad t \leq \dfrac{x_1}{c_1} \\[6pt] \left(-\dfrac{1}{c_1}\right) F'\left(t - \dfrac{x_1}{c_1}\right) & \text{for} \quad t > \dfrac{x_1}{c_1} \end{cases} \qquad (9.13.24)$$

Using this, the boundary condition (9.13.19) becomes

$$\left(-\frac{1}{c_1}\right) F'(t) = -\frac{1}{\lambda + 2\mu} p(t)$$

for $t > 0$. This is satisfied if we take

$$F\left(t - \frac{x_1}{c_1}\right) = \frac{c_1}{(\lambda + 2\mu)} \int_0^{t - x_1/c_1} p(t_0) \, dt_0 \qquad (9.13.25)$$

Putting this expression for $F(t - x_1/c_1)$ back into (9.13.23), we get the following solution for u_1 that meets all the requirements of the problem:

$$u_1(x_1, t) = \begin{cases} 0 & \text{for } t \leq \dfrac{x_1}{c_1} \\ \dfrac{c_1}{(\lambda + 2\mu)} \displaystyle\int_0^{t-x_1/c_1} p(t_0)\, dt_0 & \text{for } t > \dfrac{x_1}{c_1} \end{cases} \qquad (9.13.26)$$

The solution (9.13.26) shows that at any chosen point $\bar{x}_1$ of the half-space no displacement occurs until time $\bar{t} = \bar{x}_1/c_1$ and a longitudinal displacement occurs thereafter. The magnitude of this displacement is directly proportional to the area under the pressure curve over the time interval $(0, t - \bar{x}_1/c_1)$, the constant of proportionality being $[(\lambda + 2\mu)\rho]^{-1/2}$.

For $t > x_1/c_1$, (9.13.18) and (9.13.26) yield the following expressions for the normal stresses developed in the half-space:

$$\tau_{11} = -p\left(t - \frac{x_1}{c_1}\right)$$

$$\tau_{22} = \tau_{33} = -\left(\frac{\lambda}{\lambda + 2\mu}\right) p\left(t - \frac{x_1}{c_1}\right) \qquad (9.13.27)$$

We note that all these stresses are compressive in nature.

From (9.13.26), we also find that, for $t > x_1/c_1$,

$$\frac{\partial u_1}{\partial t}(x_1, t) = \frac{c_1}{(\lambda + 2\mu)} p\left(t - \frac{x_1}{c_1}\right) \qquad (9.13.28)$$

This gives the speed at a point x_1 and at time $t > x_1/c_1$. From (9.13.27) and (9.13.28), we get the following expression for the wave resistance of the half-space:

$$\tau_{11} \bigg/ \left(\frac{\partial u_1}{\partial t}\right) = -\frac{\lambda + 2\mu}{c_1} = -\sqrt{(\lambda + 2\mu)\rho} \qquad (9.13.29)$$

REMARK The two problems just discussed are analogous to one another. While the first was formulated in terms of stress, the second was formulated in terms of displacement.

9.13.3 PLANE WAVES IN AN UNBOUNDED ELASTIC BODY

Suppose that an unbounded elastic body experiences a displacement field that depends only on x_1 and t. The body forces are neglected. The problem is (i) to find the cause for the occurrence of such a displacement, and (ii) to analyze the associated physical phenomenon.

When $\mathbf{u} = \mathbf{u}(x_1, t)$, Navier's equation of motion (9.8.6) yields the following three equations if the body forces are neglected.

$$c_1^2 \frac{\partial^2 u_1}{\partial x_1^2} = \frac{\partial^2 u_1}{\partial t^2} \qquad (9.13.30\text{a})$$

$$c_2^2 \frac{\partial^2 u_2}{\partial x_1^2} = \frac{\partial^2 u_2}{\partial t^2} \qquad (9.13.30\text{b})$$

$$c_2^2 \frac{\partial^2 u_3}{\partial x_1^2} = \frac{\partial^2 u_3}{\partial t^2} \qquad (9.13.30\text{c})$$

where

$$c_1^2 = (\lambda + 2\mu)/\rho, \quad c_2^2 = \mu/\rho \qquad (9.13.31)$$

Therefore, in order that the displacement is of the form $\mathbf{u} = \mathbf{u}(x_1, t)$, the components u_i of $\mathbf{u}$ must obey the equations (9.13.30). We observe that each of the three equations in (9.13.30) is a one-dimensional wave equation governing a plane wave propagating in the x_1 direction. We note that u_1 is governed by the wave equation associated with the speed c_1, and u_2 and u_3 are both governed by the wave equation associated with the speed c_2. Thus, a displacement field of the form $\mathbf{u} = \mathbf{u}(x_1, t)$ occurs due to the propagation of three plane waves, of which one moves with speed c_1 and the other two move with speed c_2. Since the body is unbounded, there are no boundary effects.

We note that u_1 is the component of $\mathbf{u}$ taken along the x_1 direction, which is the direction of propagation. Hence, the wave associated with u_1 (which propagates with speed c_1) is called the *longitudinal wave*. When this wave is *not* absent (that is, when $u_1 \neq 0$) we find that div $\mathbf{u} \neq 0$; as such, the wave causes a change in volume of the material elements in the body and is therefore also referred to as the *dilatational wave*. If $u_2 = u_3 = 0$, the deformation is caused fully by this wave, and we find that curl $\mathbf{u} = 0$; hence the motion is *irrotational* so that no change occurs in the shape of material elements.

Next, we note that u_2 and u_3 are the components of $\mathbf{u}$ taken along directions perpendicular to the direction of propagation. Hence the waves associated with u_2 and u_3 (which propagate with speed c_2) are called the *transverse waves*. When either of these waves is *not* absent we find that curl $\mathbf{u} \neq 0$; as such, these waves produce a change in shape of the material elements and are therefore also referred to as *shear waves*. If $u_1 = 0$, the deformation is caused fully by these waves, and we find that div $\mathbf{u} = 0$; hence the motion is *isochoric* so that no change occurs in the volume of material elements.

Expressions (9.13.31) show that $c_1 > c_2$. As such, the dilatational wave propagates faster than the shear waves. The dilatational wave is therefore

440 9 EQUATIONS OF LINEAR ELASTICITY

also called the *primary wave* or the *P-wave*, and the shear waves the *secondary waves* or the *S-waves*. It is customary to take the x_3 axis in the vertical direction and the x_2 axis in a horizontal direction. Then, the S-wave with displacement u_2 is referred to as the *secondary horizontal wave* or the *SH-wave*, and the S-wave with displacement u_3 is referred to as the *secondary vertical wave* or the *SV-wave*. For a typical metal like copper, the speeds of primary and secondary waves are estimated as $c_1 = 4.36 \times 10^3$ m/s and $c_2 = 2.13 \times 10^3$ m/s, respectively.

We note for a plane wave propagating in the x_1 direction, a function $\phi(x_1, t)$ associated with the wave is a solution of an equation in the form

$$c_2^2 \frac{\partial^2 \phi}{\partial x_1^2} = \frac{\partial^2 \phi}{\partial t^2} \tag{9.13.32}$$

A particular form of ϕ that is periodic in both x_1 and t is

$$\phi = A \exp\{i(\omega t - kx_1)\} \tag{9.13.33}$$

where A, ω and k are real constants. Evidently, the period of ϕ in t is $2\pi/\omega$; this is referred to as the *period* of the wave. The reciprocal of the period, namely, $\omega/2\pi$, is called the *frequency* of the wave; the number ω is often referred to as the *angular frequency*. Also, the period of ϕ in x_1 is $2\pi/k$, referred to as the *wavelength*. The reciprocal of the wavelength, namely, $k/2\pi$, is called the *wave number*. The number ω/k is called the *phase speed* of the wave. The maximum value A of ϕ is called the *wave amplitude*.

Substitution of (9.13.33) into (9.13.32) yields $c^2 = \omega^2/k^2$. Accordingly, a solution of (9.13.32) that corresponds to a wave propagating with frequency $\omega/2\pi$ and speed c is

$$\phi = A \exp\left\{i\omega\left(t - \frac{x_1}{c}\right)\right\}$$

$$= A\left[\cos \omega\left(t - \frac{x_1}{c_1}\right) + i \sin \omega\left(t - \frac{x_1}{c_1}\right)\right] \tag{9.13.34}$$

This solution is just a particular form of the D'Alembert's solution of the wave equation. A wave associated with this type of solution is referred to as a *plane harmonic wave*.

In view of these remarks, we may note that when the P-wave, governed by equation (9.13.30a), is a harmonic wave, the associated displacement is

$$u_1 = A_1 \exp\left\{i\omega\left(t - \frac{x_1}{c_1}\right)\right\} \tag{9.13.35}$$

9.13 ELASTIC WAVES   441

where A_1 is a constant. The corresponding expressions for S-waves are

$$(u_2, u_3) = (A_2, A_3) \exp\left\{i\omega\left(t - \frac{x_1}{c_2}\right)\right\} \qquad (9.13.36)$$

where A_2, A_3 are constants. Consequently, the displacement field $\mathbf{u}(x_1, t)$ determined by (9.13.35) and (9.13.36) corresponds to a plane harmonic wave propagating with frequency $\omega/2\pi$.

9.13.4 RAYLEIGH WAVES

Consider an elastic half-space with a stress-free plane boundary in which a plane harmonic wave whose amplitude diminishes with the distance from the boundary propagates in a direction parallel to the boundary. The body forces are ignored. The problem is to find the speed of propagation of the wave.

Let us choose the axes such that (i) the half-space initially fills the region $x_3 > 0$ with $x_3 = 0$ as the boundary, and (ii) the x_1 axis is along the direction of wave propagation; see Figure 9.15. Then, if $\omega/2\pi$ is the frequency and c is the speed of propagation, we seek the displacement associated with the wave in the form

$$\mathbf{u} = \mathbf{a} e^{-\alpha x_3} \exp\left\{i\omega\left(t - \frac{x_1}{c}\right)\right\} \qquad (9.13.37)$$

where $\mathbf{a}$ is a constant vector and α is a positive number, so that the amplitude $\mathbf{a} e^{-\alpha x_3}$ diminishes with increasing x_3. The vector $\mathbf{a}$ and the speed of propagation c are to be determined by using the fact that $\mathbf{u}$ must be a solution of Navier's equation of motion and that the boundary $x_3 = 0$ is stress free.

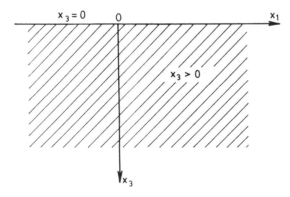

Figure 9.15. Half-space $x_3 > 0$.

9 EQUATIONS OF LINEAR ELASTICITY

From (9.13.37) we obtain the following expressions for the components of $(\nabla \text{ div } \mathbf{u})$ and $\Box_2 \mathbf{u}$:

$$[\nabla \text{ div } \mathbf{u}]_1 = -\left(\frac{\omega^2}{c^2} a_1 - i\alpha \frac{\omega}{c} a_3\right) \exp\left\{i\omega\left(t - \frac{x_1}{c}\right)\right\}$$

$$[\nabla \text{ div } \mathbf{u}]_2 = 0$$

$$[\nabla \text{ div } \mathbf{u}]_3 = \left(i\alpha \frac{\omega}{c} a_1 + \alpha^2 a_3\right) \exp\left\{i\omega\left(t - \frac{x_1}{c}\right)\right\}$$

$$[\Box_2 \mathbf{u}]_k = \left(\nabla^2 - \frac{1}{c_2^2}\frac{\partial^2}{\partial t^2}\right)[\mathbf{u}]_k = \left(\alpha^2 - \frac{\omega^2}{c^2} + \frac{\omega^2}{c_2^2}\right) a_k \exp\left\{i\omega\left(t - \frac{x_1}{c}\right)\right\}$$

Substituting these expressions into the component form of Navier's equation (9.8.10) with $\mathbf{f} = \mathbf{0}$ and using (9.8.16), we get the following three equations for the three components a_i of $\mathbf{a}$:

$$\left[c_2^2 \alpha^2 + \omega^2\left(1 - \frac{c_1^2}{c^2}\right)\right] a_1 + i\alpha(c_1^2 - c_2^2)\frac{\omega}{c} a_3 = 0 \qquad (9.13.38)$$

$$\left[c_2^2 \alpha^2 + \omega^2\left(1 - \frac{c_2^2}{c^2}\right)\right] a_2 = 0 \qquad (9.13.39)$$

$$(c_1^2 - c_2^2) i\alpha \frac{\omega}{c} a_1 + \left[c_1^2 \alpha^2 + \omega^2\left(1 - \frac{c_2^2}{c^2}\right)\right] a_3 = 0 \qquad (9.13.40)$$

We find that (9.13.39) is automatically satisfied if we take $a_2 = 0$. Then, for $\mathbf{u}$ to be nonzero, a_1 and a_3 cannot both be 0; that is, equations (9.13.38) and (9.13.40) should have a nontrivial solution. A necessary and sufficient condition for this is

$$\begin{vmatrix} c_2^2 \alpha^2 + \omega^2\left(1 - \frac{c_1^2}{c^2}\right) & (c_1^2 - c_2^2) i\alpha \frac{\omega}{c} \\ (c_1^2 - c_2^2) i\alpha \frac{\omega}{c} & c_1^2 \alpha^2 + \omega^2\left(1 - \frac{c_2^2}{c^2}\right) \end{vmatrix} = 0 \qquad (9.13.41)$$

which simplifies to

$$\left[\alpha^2 - \frac{\omega^2}{c^2}\left(1 - \frac{c^2}{c_1^2}\right)\right]\left[\alpha^2 - \frac{\omega^2}{c^2}\left(1 - \frac{c^2}{c_2^2}\right)\right] = 0 \qquad (9.13.41)'$$

9.13 ELASTIC WAVES

Thus, the two possible values of α^2 which make a_1 and a_3 both not 0 are

$$\alpha_1^2 = \frac{\omega^2}{c^2}\left(1 - \frac{c^2}{c_1^2}\right), \quad \alpha_2^2 = \frac{\omega^2}{c^2}\left(1 - \frac{c^2}{c_2^2}\right) \quad (9.13.42)$$

For $\alpha = \alpha_1$, expression (9.13.40) yields $(a_3/a_1) = -(c/\omega)ia_1$, so that we may take $a_3 = \alpha_1 A$ and $a_1 = (\omega/c)iA$, where A is an arbitrary real constant. Thus, for $\alpha = \alpha_1$, $a_1 = (\omega/c)iA$, $a_2 = 0$, $a_3 = \alpha_1 A$, expression (9.13.37) satisfies Navier's equation (9.8.10) with $\mathbf{f} = \mathbf{0}$. In other words

$$\mathbf{u} = \left(i\frac{\omega}{c}\mathbf{e}_1 + \alpha_1 \mathbf{e}_3\right) A e^{-\alpha_1 x_3} \exp\left\{i\omega\left(t - \frac{x_1}{c}\right)\right\} \quad (9.13.43)$$

is a solution of the equation (9.8.10) with $\mathbf{f} = \mathbf{0}$.

For $\alpha = \alpha_2$, (9.13.38) yields $(a_3 c/\omega) = -(ia_1/\alpha_2)$, so that we may take $a_3 = (\omega/c)B$ and $a_1 = i\alpha_2 B$, where B is an arbitrary real constant. Thus, for $\alpha = \alpha_2$, $a_1 = i\alpha_2 B$, $a_2 = 0$, $a_3 = (\omega/c)B$, expression (9.13.37) satisfies Navier's equation (9.8.10) with $\mathbf{f} = \mathbf{0}$. Accordingly,

$$\mathbf{u} = \left(i\alpha_2 \mathbf{e}_1 + \frac{\omega}{c}\mathbf{e}_3\right) B e^{-\alpha_2 x_3} \exp\left\{i\omega\left(t - \frac{x_1}{c}\right)\right\} \quad (9.13.44)$$

is another solution of the equation (9.8.10) with $\mathbf{f} = \mathbf{0}$.

A general solution for $\mathbf{u}$ associated with the wave under consideration is obtained by superposing the two solutions (9.13.43) and (9.13.44). We thus obtain the following solutions for the components of $\mathbf{u}$:

$$\begin{aligned} u_1 &= i\left(\frac{\omega}{c} A e^{-\alpha_1 x_3} + \alpha_2 B e^{-\alpha_2 x_3}\right)\exp\left\{i\omega\left(t - \frac{x_1}{c}\right)\right\} \\ u_2 &= 0 \qquad (9.13.45) \\ u_3 &= \left(\alpha_1 A e^{-\alpha_1 x_3} + \frac{\omega}{c} B e^{-\alpha_2 x_3}\right)\exp\left\{i\omega\left(t - \frac{x_1}{c}\right)\right\} \end{aligned}$$

The stresses τ_{ij} associated with these displacement components follow by the use of the stress-displacement relation (9.8.4). We find that

$$\begin{aligned} \tau_{31} &= -i\mu\frac{\omega}{c}\left\{2\alpha_1 A e^{-\alpha_1 x_3} + \left(2 - \frac{c^2}{c_2^2}\right)\frac{\omega}{c} B e^{-\alpha_2 x_3}\right\} \\ \tau_{32} &= 0 \qquad (9.13.46) \\ \tau_{33} &= -\mu\frac{\omega}{c}\left\{\left(2 - \frac{c^2}{c_2^2}\right)\frac{\omega}{c} A e^{-\alpha_1 x_3} + 2\alpha_2 B e^{-\alpha_2 x_3}\right\} \end{aligned}$$

where we have suppressed the factor $\exp\{i\omega(t - x_1/c)\}$ for brevity.

9 EQUATIONS OF LINEAR ELASTICITY

Since the boundary $x_3 = 0$ is stress free, we have $\tau_{31} = \tau_{32} = \tau_{33} = 0$ for $x_3 = 0$. By use of (9.13.46), these boundary conditions yield the following equations:

$$2\alpha_1 A + \left(2 - \frac{c^2}{c_2^2}\right)\frac{\omega}{c} B = 0$$

$$\left(2 - \frac{c^2}{c_2^2}\right)\frac{\omega}{c} A + 2\alpha_2 B = 0 \qquad (9.13.47)$$

From (9.13.45) we note that for **u** to be nonzero, A and B cannot vanish together. Hence the two equations for A and B given by (9.13.47) should have a nontrivial solution. A necessary and sufficient condition for this is

$$\begin{vmatrix} 2\alpha_1 & \left(2 - \dfrac{c^2}{c_2^2}\right)\dfrac{\omega}{c} \\ \left(2 - \dfrac{c^2}{c_2^2}\right)\dfrac{\omega}{c} & 2\alpha_2 \end{vmatrix} = 0 \qquad (9.13.48)$$

With the aid of expressions (9.13.42), this condition reduces to

$$\left(2 - \frac{c^2}{c_2^2}\right)^2 = 4\left(1 - \frac{c^2}{c_1^2}\right)^{1/2}\left(1 - \frac{c^2}{c_2^2}\right)^{1/2} \qquad (9.13.49)$$

Thus, under the assumptions made, this is a necessary and sufficient condition for the solution of the type (9.13.45) to exist. This condition contains c as the only unknown; as such the condition serves as the equation for the determination of c. Since α_1 and α_2 are to be positive, we find from expressions (9.13.42) that c has to be less than c_2, which is less than c_1. Of course, c has to be positive as well. Thus, a possible speed of the wave being considered is a root of the equation (9.13.49) that lies in the interval $(0, c_2)$.

Equation (9.13.49) can be simplified to the form

$$p^3 - 8p^2 + (24 - 16q^2)p - 16(1 - q^2) = 0 \qquad (9.13.50)$$

where

$$p = \frac{c^2}{c_2^2} \quad \text{and} \quad q = \frac{c_2}{c_1} < 1 \qquad (9.13.51)$$

If we denote the left side of (9.13.50) by $f(p)$, it is easily seen that $f(0) = -16(1 - q^2) < 0$ and $f(1) = 1 > 0$. Hence, equation (9.13.50) has one or three real roots for p between 0 and 1. We verify that $f''(p) = 6p - 16$ does not vanish for p lying between 0 and 1. Hence (9.13.50) has exactly one root for p between 0 and 1. Equivalently, equation (9.13.49) has exactly one root, say c_R, for c between 0 and c_2. Hence, the wave being considered does exist and propagates with only one possible speed c_R.

We note that equation (9.13.49) determining c_R, contains no term involving ω. As such, the speed c_R of the wave is independent of the frequency. Further, since $q^2 = (c_2^2/c_1^2) = (1 - 2v)/2(1 - v)$ [see (9.8.16)], c_R depends entirely on the shear wave speed c_2 and Poisson's ratio. For a material with $v = 0.25$ (such as rock), equation (9.13.50) becomes

$$3p^3 - 24p^2 + 56p - 32 = 0 \qquad (9.13.52)$$

whose roots are $p = 4, 2 \pm 2/\sqrt{3}$. Of these roots, only $p = 2 - 2/\sqrt{3}$ is less than 1, and this yields $c_R = 0.9194 c_2$ as the speed of the wave.

REMARK A wave in a semi-infinite body whose amplitude diminishes with the distance from the boundary surface of the body is called a *surface wave* or *Rayleigh wave*. Studies on such waves are of practical importance in areas like seismology and geophysics. Lord Rayleigh first initiated the study of surface waves by discussing the problem just considered in 1885.

It may be noted that the displacement vector associated with a Rayleigh wave is made up of two components: one along the direction of propagation and the other perpendicular to the boundary surface; see (9.13.45). Hence (if the boundary surface is taken as a horizontal surface), a Rayleigh wave is a coupled combination of the *P*-wave and the *SV*-wave.

EXAMPLE 9.13.1 Show that a surface wave of the type just considered cannot exist if $u_3 = 0$ and $\tau_{31} = 0$ on the boundary $x_3 = 0$. (Such a boundary is called a *rigid, lubricated boundary*.)

Solution In view of the last relation in (9.13.45), the boundary condition $u_3 = 0$ for $x_3 = 0$ becomes

$$\alpha_1 A + \frac{\omega}{c} B = 0 \qquad (9.13.53)$$

Also, the boundary condition $\tau_{31} = 0$ for $x_3 = 0$ is given by the first relation in (9.13.47), namely,

$$2\alpha_1 A + \left(2 - \frac{c^2}{c_2^2}\right) \frac{\omega}{c} B = 0 \qquad (9.13.54)$$

For the waves to exist, A and B involved in (9.13.53) and (9.13.54) cannot both be 0. A criterion for this is

$$\begin{vmatrix} \alpha_1 & \dfrac{\omega}{c} \\ 2\alpha_1 & \left(2 - \dfrac{c^2}{c_2^2}\right)\dfrac{\omega}{c} \end{vmatrix} = 0$$

which simplifies to
$$\alpha_1 \omega c = 0 \tag{9.13.55}$$

This condition is satisfied only if $\alpha_1 = 0$, because ω and c are not 0 when there is a wave propagation. But, for a wave of the surface type, $\alpha_1 > 0$. Hence, the condition (9.13.55) cannot be satisfied by a surface wave; such a wave therefore cannot exist under the assumed conditions. ∎

EXAMPLE 9.13.2 Show that an SH-wave of the surface type cannot exist in a half-space with a stress-free horizontal boundary.

Solution Suppose that the half-space initially fills the region $x_3 > 0$ with $x_3 = 0$ as the horizontal boundary. Since the boundary is stress free, we have $\tau_{31} = \tau_{32} = \tau_{33} = 0$ for $x_3 = 0$.

If there is a pure SH-wave of surface type propagating in the x_1 direction, the corresponding displacement vector is given by $\mathbf{u} = u_2 \mathbf{e}_2$, where

$$u_2 = A e^{-\alpha x_3} \exp\{i\omega(t - x_1/c)\} \tag{9.13.56}$$

$\alpha > 0$ and A is a nonzero constant. Also, $\omega/2\pi$ is the frequency and c is the speed of propagation.

For $\mathbf{u} = u_2 \mathbf{e}_2$, with u_2 given by (9.13.56), the stress-displacement relation (9.8.4) gives $\tau_{31} = \tau_{33} \equiv 0$ and

$$\tau_{32} = -\mu\alpha A e^{-\alpha x_3} \exp\{i\omega(t - x_1/c)\} \tag{9.13.57}$$

Hence the only boundary condition to be satisfied on $x_3 = 0$ is $\tau_{32} = 0$, which on using (9.13.57) becomes $\mu\alpha A = 0$. This condition cannot be satisfied, because all of μ, α and A are nonzero. Hence the wave considered is not a possible one. ∎

9.13.5 LOVE WAVES

Consider an elastic half-space with horizontal plane boundary, covered by an elastic layer of uniform thickness. Suppose that (i) the layer and the half-space have different densities and different shear moduli, (ii) there is a welded contact between the layer and the half-space and (iii) the horizontal boundary of the composite structure is stress free. The problem is to investigate the possibility of the propagation of an SH-wave of the surface type in this composite structure. Body forces are ignored.

Let us choose the axes such that the half-space initially fills the region $x_3 > 0$ and the layer the space $-H \leq x_3 \leq 0$, so that the horizontal plane boundary of the composite structure is $x_3 = -H$; see Figure 9.16. Also, let the x_1 axis be chosen along the direction of propagation. Then for an

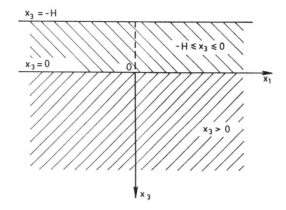

Figure 9.16. Half-space covered by a layer.

SH-wave of surface type we take $\mathbf{u} = u_2 \mathbf{e}_2$, where

$$u_2 = Ae^{-\alpha x_3} \exp[i\omega(t - x_1/c)] \qquad (9.13.58)$$

in $0 < x_3 < \infty$; and

$$u_2 = f(x_3) \exp[i\omega(t - x_1/c)] \qquad (9.13.59)$$

in $-H \le x_3 \le 0$. Then $\omega/2\pi$ is the frequency and c is the speed of propagation of the wave (if it exists). The unknowns A, $\alpha(>0)$ and $f(x_3)$ are to be determined by using the governing equation for $\mathbf{u}$ and the boundary conditions.

When $\mathbf{u} = u_2(x_1, x_3, t)\mathbf{e}_2$, the governing equation for $\mathbf{u}$, namely, Navier's equation (9.8.6), takes the following form in the absence of body forces

$$\mu(u_{2,11} + u_{2,33}) = \rho \frac{\partial^2 u_2}{\partial t^2} \qquad (9.13.60)$$

Let us first consider the case where $0 < x_3 < \infty$. Substitution for u_2 from (9.13.58) into (9.13.60) yields the relation

$$\alpha^2 = \frac{\omega^2}{c^2}\left(1 - \frac{c^2}{c_2^2}\right) \qquad (9.13.61)$$

Since α has to be greater than 0, c has to be less than c_2.

Next, let us consider the case where $-H \le x_3 \le 0$. If $\bar{\rho}$ is the density and $\bar{\mu}$ is the shear modulus of the layer, then the governing equation for u_2 in $-H \le x_3 \le 0$ is the equation (9.13.60) with ρ and μ replaced by $\bar{\rho}$ and $\bar{\mu}$. Substitution for u_2 from (9.13.59) in this governing equation yields the equation

$$f''(x_3) - \beta^2 f(x_3) = 0 \qquad (9.13.62)$$

448   9   EQUATIONS OF LINEAR ELASTICITY

where

$$\beta^2 = \frac{\omega^2}{c^2}\left(1 - \frac{c^2}{\bar{c}_2^2}\right) \tag{9.13.63a}$$

$$\bar{c}_2^2 = \frac{\bar{\mu}}{\bar{\rho}} \tag{9.13.63b}$$

A general solution of the differential equation (9.13.62) is

$$f(x_3) \equiv B \cosh \beta x_3 + C \sinh \beta x_3 \tag{9.13.64}$$

where B and C are arbitrary constants.

Since there is a welded contact between the layer and the half-space, the displacement and the stresses must be continuous across the interface $x_3 = 0$. That is

$$(u_2, \tau_{31}, \tau_{32}, \tau_{33})|_{\text{layer}} = (u_2, \tau_{31}, \tau_{32}, \tau_{33})|_{\text{half-space}} \tag{9.13.65}$$

for $x_3 = 0$.

Also, since the boundary of the composite structure, $x_3 = -H$, is stress free, we have

$$(\tau_{31}, \tau_{32}, \tau_{33})|_{\text{layer}} = (0, 0, 0) \tag{9.13.66}$$

for $x_3 = -H$.

Since $\mathbf{u} = u_2(x_1, x_3, t)\mathbf{e}_2$ we find from the stress-displacement relation (9.8.4) that

$$\tau_{31} = \tau_{33} \equiv 0 \quad \text{in the layer and the half-space} \tag{9.13.67}$$

and

$$\tau_{32} = \begin{cases} \mu u_{2,3} & \text{in the half-space} \\ \bar{\mu} u_{2,3} & \text{in the layer} \end{cases} \tag{9.13.68}$$

By use of (9.13.58), (9.13.59) and (9.13.64) in (9.13.68) we obtain

$$\tau_{32} = -\mu \alpha A e^{-\alpha x_3} \exp[i\omega(t - x_1/c)] \quad \text{in the half-space} \tag{9.13.69}$$

$$\tau_{32} = \bar{\mu}\beta[B \sinh \beta x_3 + C \cosh \beta x_3] \exp[i\omega(t - x_1/c)] \quad \text{in the layer} \tag{9.13.70}$$

Using (9.13.58), (9.13.59), (9.13.64), (9.13.67) and (9.13.70), we obtain from the interfacial conditions (9.13.65) and the boundary conditions (9.13.66) the following three relations to be satisfied by A, B and C:

$$A - B = 0$$
$$\mu \alpha A + \bar{\mu}\beta C = 0 \tag{9.13.71}$$
$$B \sinh \beta H - C \cosh \beta H = 0$$

9.13 ELASTIC WAVES

For the waves under consideration to exist, A, B, C cannot all be 0; that is, the system (9.13.71) should have a nontrivial solution. A necessary and sufficient condition for this is

$$\begin{vmatrix} 1 & -1 & 0 \\ \mu\alpha & 0 & \bar{\mu}\beta \\ 0 & \sinh\beta H & -\cosh\beta H \end{vmatrix} = 0$$

which simplifies to

$$\mu\alpha = -\bar{\mu}\beta \tanh \beta H \qquad (9.13.72)$$

We note that μ and $\bar{\mu}$ are positive constants, and $\alpha > 0$ for a surface-type wave. From (9.13.63a), we find that β is either real or purely imaginary according as $c \le \bar{c}_2$ or $c > \bar{c}_2$. When β is real (positive, negative or 0), $\beta(\tanh \beta H)$ is nonnegative. Hence, if β is real, the condition (9.13.72) *cannot* be satisfied. As such, in this case, $A = B = C = 0$ and, consequently, $\mathbf{u} = \mathbf{0}$ in the half-space and in the layer. Therefore, the wave propagation considered is *not* a possible one.

Let us now consider the other case, where β is purely imaginary. If we set $\beta = i\eta$, where η is a nonzero real number, the condition (9.13.72) becomes

$$\eta(\tan \eta H) = \frac{\mu\alpha}{\bar{\mu}} \qquad (9.13.73)$$

Since $\eta \ne 0$, both sides of this equation are positive. Also, $\tan \eta H$ takes every value between 0 and ∞; hence, $\eta(\tan \eta H)$ certainly takes the value $\mu\alpha/\bar{\mu}$ at least once. Hence, in this case the condition (9.13.72) is possible. Setting $\beta = i\eta$ in (9.13.63a), we get

$$\eta^2 = \frac{\omega^2}{c^2}\left(\frac{c^2}{\bar{c}_2^2} - 1\right) \qquad (9.13.74)$$

Since η is real and nonzero, c has to be necessarily greater than $\bar{c}_2$. Substituting for α and η from (9.13.61) and (9.13.74) into (9.13.73), we get the following equation governing the speed c:

$$\tan\left\{\frac{\omega}{c}\left(\frac{c^2}{\bar{c}_2^2} - 1\right)^{1/2} H\right\} = \frac{\mu(1 - c^2/c_2^2)^{1/2}}{\bar{\mu}(c^2/\bar{c}_2^2 - 1)^{1/2}} \qquad (9.13.75)$$

This is a transcendental equation that yields infinitely many roots for c. However, c has to be less than c_2 and greater than $\bar{c}_2$. Thus, the possible speeds of the wave considered are precisely the roots of the equation (9.13.75) that lie in the interval $(\bar{c}_2, c_2)$. If the layer and the half-space are such that $c_2 \le \bar{c}_2$, the wave considered is not a possible one.

450   **9 EQUATIONS OF LINEAR ELASTICITY**

It is important to observe that the equation (9.13.75) contains ω. Therefore, the values of c determined from this equation do depend on ω. This means that the speed of propagation depends on the frequency. That is, the wave is dispersive.

In the limiting case when the layer is absent, we have $\mu = \bar{\mu}$ and $\rho = \bar{\rho}$. Then the equation (9.13.75) leads to the impossible condition $0 = -1$. Hence, in this case, the wave considered cannot exist. This result is in agreement with that proven in Example 9.13.2.

Note: The problem discussed above is by A. E. H. Love (1911), and the wave considered in the problem is known as a *Love wave*. Like Rayleigh waves, Love waves are also of practical importance, particularly in seismology and geophysics.

9.14
EXERCISES

1. Write down the relations (9.2.8) in the matrix form.

2. Prove the identities (9.3.6), (9.3.7) and those contained in Table 9.1.

3. Write down the relations (9.3.14) in the matrix form.

4. Prove the following relations connecting the stress invariants and the strain invariants:
$$I_T = (3\lambda + 2\mu)I_E$$
$$II_T = \lambda(3\lambda + 4\mu)I_E^2 + 4\mu^2 II_E$$
$$III_T = \lambda^2(\lambda + 2\mu)I_E^3 + 4\lambda\mu^2 I_E II_E + 8\mu^3 III_E$$

5. Obtain the following expressions for the strain-energy function W:
$$W = \mu e_{ij}e_{ij} + \frac{1}{6}(3K - 2\mu)e_{kk}^2 = \frac{1}{2}K(e_{kk})^2 + \mu e_{ij}^{(d)}e_{ij}^{(d)}$$
$$= \frac{1}{6}\tau_{kk}e_{mm} + \frac{1}{2}\tau_{ij}^{(d)}e_{ij}^{(d)} = \frac{1}{18K}\tau_{kk}^2 + \frac{1}{4\mu}\tau_{ij}^{(d)}e_{ij}^{(d)}$$
$$= \frac{1}{4\mu}\left[\tau_{ij}\tau_{ij} - \frac{\lambda}{3\lambda + 2\mu}(\tau_{kk})^2\right] = \frac{1}{4\mu}\tau_{ij}\tau_{ij} + \frac{1}{6}\left[\frac{1}{3K} - \frac{1}{2\mu}\right]\tau_{kk}^2$$
$$= \left(\frac{1}{2}\lambda + \mu\right)I_E^2 - 2\mu II_E = \mu\left[\frac{\nu}{1 - 2\nu}I_E^2 + e_{ij}e_{ij}\right]$$
$$= \frac{1}{2E}[I_T^2 - 2(1 + \nu)II_T] = \frac{1}{18K}I_T^2 + \frac{1}{6\mu}(I_T^2 - 3I_T)$$
$$= W_1 + W_2$$

where

$$W_1 = \frac{1}{2}Ke_{kk}^2 = \frac{1}{18K}\tau_{kk}^2$$

$$W_2 = \frac{1}{3}\mu[(e_{11} - e_{22})^2 + (e_{22} - e_{33})^2 + (e_{33} - e_{11})^2 + 6(e_{12}^2 + e_{23}^2 + e_{31}^2)]$$

$$= \frac{1}{12\mu}[(\tau_{11} - \tau_{22})^2 + (\tau_{22} - \tau_{33})^2 + (\tau_{33} - \tau_{11})^2 + 6(\tau_{12}^2 + \tau_{23}^2 + \tau_{31}^2)]$$

6. Evaluate W for the stress field

$$\tau_{11} = \tau_{22} = \tau_{33} = \tau_{12} = 0, \qquad \tau_{13} = -\mu\alpha x_2, \qquad \tau_{23} = \mu\alpha x_1$$

where α is a constant.

7. If $\tau_{13} = \tau_{23} = \tau_{33} = 0$, show that

$$e_{11} + e_{22} = \frac{1-\nu}{E}(\tau_{11} + \tau_{22}) \quad \text{and} \quad e_{22} - e_{11} + 2ie_{12} = \frac{1+\nu}{E}(\tau_{22} - \tau_{11} + 2i\tau_{12})$$

Deduce that

$$W = \frac{1-\nu}{4E}(\tau_{11} + \tau_{22})^2 + \frac{1+\nu}{4E}|\tau_{22} - \tau_{11} + 2i\tau_{12}|^2$$

8. Find the stresses associated with the following displacement fields:

(i) $u_1 = 0, \qquad u_2 = 0, \qquad u_3 = k\theta$

(ii) $u_1 = kx_2x_3, \qquad u_2 = kx_3x_1, \qquad u_3 = kx_1x_2$

(iii) $u_1 = kx_2x_3, \qquad u_2 = kx_3x_1, \qquad u_3 = k(x_1^2 - x_2^2)$

(iv) $u_1 = -\frac{k}{2}\{x_3^2 + \nu(x_1^2 - x_2^2)\}, \qquad u_2 = k\nu x_1 x_3, \qquad u_3 = kx_3x_1$

where k is a constant and $\theta = \tan^{-1}(x_2/x_1)$; $x_1 \neq 0$.

9. For an elastic body subjected to the stress system $\tau_{ij} = -p\delta_{ij}$, show that the displacements are $u_i = (-p/3K)x_i$.

10. When an elastic solid is immersed in a fluid whose density is the same as that of the solid, the stresses in the solid are found to be $\tau_{11} = \tau_{22} = \tau_{33} = -p + \rho g x_3$, where p is the pressure at the level $x_3 = 0$ and g is the constant gravity factor. Compute the corresponding strains and displacements.

11. A rod placed along the x_1 axis and subjected to a longitudinal stress τ_{11} is so constrained that there is no lateral contraction. Show that

$$\tau_{11} = \frac{(1-v)E}{(1+v)(1-2v)} e_{11}$$

12. Derive the solution (9.4.8) for the equation (9.4.7).

13. Show that the relation (9.8.26) can be put in the following component form:

$$s_i = \frac{E}{1+v}\left[\frac{\partial u_i}{\partial n} - \omega_{ij}n_j + \frac{v}{1-2v}u_{k,k}n_i\right]$$

where

$$\frac{\partial}{\partial n} = \mathbf{n}\cdot\nabla \quad \text{and} \quad \omega_{ij} = \frac{1}{2}(u_{i,j} - u_{j,i})$$

14. For an incompressible elastic body, show that

$$\mathbf{s} = -\bar{p}\mathbf{n} + 2\mu\frac{\partial \mathbf{u}}{\partial n} + \mu(\mathbf{n}\times\operatorname{curl}\mathbf{u})$$

15. If $\mathbf{f}$ is such that div $\mathbf{f} = 0$ and curl $\mathbf{f} = \mathbf{0}$, prove the following in the static case:
   (i) $\mathbf{u}$, $\mathbf{E}$ and $\mathbf{T}$ are biharmonic functions.
   (ii) div $\mathbf{u}$, curl $\mathbf{u}$, tr $\mathbf{E}$ and tr $\mathbf{T}$ are harmonic functions.

16. Verify the following identities:

(i) $\dfrac{c_1^2 - c_2^2}{c_2^2} = \dfrac{1}{1-2v}$ (ii) $\dfrac{c_1^2 - c_2^2}{c_1^2} = \dfrac{1}{2(1-v)}$

17. If u_i are solutions of Navier's equations of equilibrium under zero body forces, show that so are $u_{i,k}$.

18. Show that, in the absence of body forces, Navier's equation of equilibrium may be written as follows

$$\nabla^2\left[\mathbf{u} + \frac{1}{2}\left(1 + \frac{\lambda}{\mu}\right)(\operatorname{div}\mathbf{u})\mathbf{x}\right] = \mathbf{0}$$

19. Let

$$\mathbf{u}_0 = \mu\mathbf{u} + \tfrac{1}{2}(\lambda + \mu)(\operatorname{div}\mathbf{u})\mathbf{x}$$

Show that in the static case, $\mathbf{u}_0$ satisfies the equation

$$\nabla^2\mathbf{u}_0 = -\mathbf{f} - \frac{\lambda + \mu}{2(\lambda + 2\mu)}(\operatorname{div}\mathbf{f})\mathbf{x}$$

20. For an incompressible elastic body in equilibrium, show that
   (i) $\mu\operatorname{curl}\operatorname{curl}\mathbf{u} + \nabla\bar{p} = \mathbf{f}$
   (ii) $\nabla^2\mathbf{T} - 2\nabla\nabla\bar{p} + (\nabla\mathbf{f} + \nabla\mathbf{f}^T) + (\operatorname{div}\mathbf{f})\mathbf{I} = \mathbf{0}$

9.14 EXERCISES 453

21. Show that the displacements

$$u_1 = \frac{Ax_1}{r(r+x_3)}, \quad u_2 = \frac{Ax_2}{r(r+x_3)}, \quad u_3 = \frac{A}{r},$$

where A is a constant, satisfy the Navier's equations of equilibrium in the absence of body forces. Find the corresponding stresses. Also compute the stress vector on the surface $r =$ constant.

22. In an elastic body in equilibrium under the body force $\mathbf{f} = ax_1x_2\mathbf{e}_3$, where a is a constant, the displacements are of the form

$$u_1 = Ax_1^2 x_2 x_3, \quad u_2 = Bx_1 x_2^2 x_3, \quad u_3 = Cx_1 x_2 x_3^2$$

where A, B, C are constants. Find A, B, C. Evaluate the corresponding stresses.

23. Verify that the following are solutions of Navier's equation of equilibrium in the absence of body forces:

(i) $\mathbf{u} = \nabla \phi, \quad \nabla^2 \phi = 0$

(ii) $\mathbf{u} = (\mathbf{a} \cdot \mathbf{x})\nabla g - (3 - 4\nu)g\nabla(\mathbf{a} \cdot \mathbf{x}), \quad \nabla^2 g = 0, \quad \mathbf{a} = $ constant

(iii) $\mathbf{u} = \nabla h \times \nabla(\mathbf{a} \cdot \mathbf{x}), \quad \nabla^2 h = 0, \quad \mathbf{a} = $ constant

(iv) $\mathbf{u} = \dfrac{\partial}{\partial x_1}(\mathbf{h}) - \dfrac{x_1}{3-4\nu}\nabla(\text{div }\mathbf{h}), \quad \nabla^2 \mathbf{h} = 0$

In each case find the corresponding stresses.

24. Find the functions ϕ and $\mathbf{h}$ such that $\mathbf{u} = \mathbf{h} + r^2(\nabla \phi)$ is a solution of Navier's equation of equilibrium under zero body force.

25. If ϕ and $\mathbf{h}$ are harmonic functions such that

$$(5 - 4\nu)\phi + \mathbf{x} \cdot \nabla \phi + \text{div }\mathbf{h} = \text{constant}$$

show that $\mathbf{u} = \mathbf{h} + \phi \mathbf{x}$ is a solution of Navier's equation of equilibrium under zero body force. What happens to this solution in the case $\nu = \frac{1}{2}$?

26. If ϕ and $\mathbf{h}$ are harmonic functions such that

$$(3 - 4\nu)\frac{\partial \phi}{\partial x_1} + \text{div }\mathbf{h} = \text{constant}$$

show that $\mathbf{u} = \mathbf{h} + x_1(\nabla \phi)$ is a solution of Navier's equation of equilibrium in the absence of body force. What happens to this solution for $\nu = \frac{1}{2}$?

27. If ϕ and $\mathbf{h}$ are such that $\nabla^2 \phi = \text{div }\mathbf{h}$ and $\nabla^2 \mathbf{h} = \mathbf{0}$, show that

$$\mathbf{u} = \mathbf{h} - \frac{1}{2(1-\nu)}\nabla \phi$$

is a solution of Navier's equation of equilibrium in the absence of body force. Prove that this solution is complete.

454   **9 EQUATIONS OF LINEAR ELASTICITY**

28. If $\mathbf{h}$ and $\mathbf{h}^*$ are such that $\nabla^2 \mathbf{h} = \text{curl } \mathbf{h}^*$ and $\nabla^2 \mathbf{h}^* = 0$, show that

$$\mathbf{u} = \mathbf{h}^* - \frac{1}{1-2\nu} \text{curl } \mathbf{h}$$

is a solution of Navier's equation of equilibrium in the absence of body force. Prove that this solution is complete.

29. If p_0 and p are harmonic functions and $\mathbf{c}$ is a constant unit vector, show that

$$\mathbf{u} = p\mathbf{c} - \frac{1}{4(1-\nu)} \nabla\{p_0 + (\mathbf{x} \cdot \mathbf{c})p\}$$

is a solution of Navier's equation of equilibrium in the absence of body force. (This solution is known as *Boussinesq's solution*).

30. Let $\mathbf{p}$ be an arbitrary vector function obeying the equation $\nabla^2 \mathbf{p} = -(1/\mu)\mathbf{f}$ and p_0 be a scalar function obeying the equation $\nabla^2 p_0 = (1/\mu)\mathbf{x} \cdot \mathbf{f}$. Show that $\mathbf{u}$ given by

$$\mathbf{u} = \mathbf{p} - \frac{1}{4(1-\nu)} \nabla(p_0 + \mathbf{x} \cdot \mathbf{p})$$

is a complete solution of Navier's equation of equilibrium. (This solution is known as the *Papkovitch-Neuber solution*. The functions $\mathbf{p}$ and p_0 are called *Papkovitch-Neuber potentials*).

31. If $\mathbf{f} = $ constant, show that in the equilibrium case the Papkovitch-Neuber potentials may be expressed in the form

$$p_0 = p_0^* + \frac{1}{6\mu}(\mathbf{x}^* \cdot \mathbf{f}), \qquad \mathbf{p} = \mathbf{p}^* - \frac{r^2}{6\mu}\mathbf{f}$$

where p_0^* and $\mathbf{p}^*$ are harmonic functions and

$$\mathbf{x}^* = x_1^3 \mathbf{e}_1 + x_2^3 \mathbf{e}_2 + x_3^3 \mathbf{e}_3$$

32. Let $\mathbf{g}$ be an arbitrary vector function obeying the equation $\nabla^2 \nabla^2 \mathbf{g} = -(1/\mu)\mathbf{f}$. Show that

$$\mathbf{u} = \nabla^2 \mathbf{g} - \frac{1}{2(1-\nu)} \nabla(\text{div } \mathbf{g})$$

is a complete solution of Navier's equation of equilibrium. (This solution is known as the *Galerkin solution*. The vector $\mathbf{g}$ is called the *Galerkin vector*.)

33. Given that

$$\mathbf{g} = \frac{A}{r}(x_2 \mathbf{e}_1 - x_1 \mathbf{e}_2)$$

where A is a constant, is the Galerkin vector in the equilibrium case in the absence of body forces, compute the corresponding displacements and stresses.

34. In the equilibrium case, the Galerkin vector is given by

$$\mathbf{g} = ar^2\mathbf{e}_1 + bx_3^4\mathbf{e}_2$$

where a and b are constants. Find the body force that must be acting. Compute the corresponding displacements and stresses.

35. Repeat the Exercise 34 in the following cases:

(i) $\mathbf{g} = r^2\mathbf{e}_3$ (ii) $\mathbf{g} = x_3^4\mathbf{e}_3$ (iii) $\mathbf{g} = r_2(x_2\mathbf{e}_1 - x_1\mathbf{e}_2)$

36. Suppose the body force $\mathbf{f}$ has magnitude f and acts in the x_3 direction. If a scalar function g obeys the equation $\nabla^4 g = -(1/\mu)f$, show that

$$\mathbf{u} = (\nabla^2 g)\mathbf{e}_3 - \frac{1}{2(1-\nu)}\nabla(g_{,3})$$

is a solution of Navier's equation of equilibrium. This solution is known as the *Love's solution*. The function g is called the *Love's strain function*.

37. Deduce the Papkovitch–Neuber solution of Exercise 30 from the Galerkin solution of Exercise 32 and vice versa.

38. Using Navier's equation of motion, show that e_{kk} satisfies the equation

$$\square_1 e_{kk} = -\frac{1-2\nu}{2\mu(1-\nu)}\,\text{div } \mathbf{f}$$

39. Using Navier's equation, show that $\omega_{ij} = \frac{1}{2}(u_{i,j} - u_{j,i})$ satisfies the equation

$$\nabla^2 \omega_{ij} = -\frac{1}{2\mu}(\bar{f}_{i,j} - \bar{f}_{j,i})$$

where $\bar{f}_i = f_i - \ddot{u}_i$.

40. Find c such that

$$u_1 = A \sin\frac{2\pi}{l}(x_1 \pm ct), \qquad u_2 = u_3 = 0$$

where A and l are constants, may satisfy Navier's equation of motion in the absence of body forces.

41. Find $f(r)$ such that $u_k = f(r)x_k \exp(i\omega t)$ may be a solution of Navier's equation of motion in the absence of body forces.

42. Let $\mathbf{g}$ be an arbitrary vector function obeying the equation $\square_2 \square_1 \mathbf{g} = -(1/\mu)\mathbf{f}$. Show that

$$\mathbf{u} = \square_1 \mathbf{g} - \frac{1}{2(1-\nu)}\nabla(\text{div }\mathbf{g})$$

is a complete solution of Navier's equation of motion. (This solution is known as the *Cauchy–Kovalevski–Somigliana solution*.) Hence, deduce the Galerkin solution of Exercise 32.

456   **9 EQUATIONS OF LINEAR ELASTICITY**

43. Let $\mathbf{p}$ be an arbitrary vector function obeying the equation $\square_2 \mathbf{p} = (-1/\mu)\mathbf{f}$ and p_0 is a scalar function satisfying the equation $\square_1 p_0 = 2 \operatorname{div} \mathbf{p} - \square_1(\mathbf{x} \cdot \mathbf{p})$. Show that

$$\mathbf{u} = \mathbf{p} - \frac{1}{4(1-\nu)} \nabla(p_0 + \mathbf{x} \cdot \mathbf{p})$$

is a complete solution of Navier's equation of motion. (This solution is known as the *Sternberg-Eubanks solution*.) Hence deduce the Papkovitch–Neuber solution of Exercise 30.

44. Let $\mathbf{k}$ be an arbitrary vector function obeying the equation $\square_1 \mathbf{k} = -(1/(\lambda + 2\mu))\mathbf{f}$ and $\mathbf{h}$ be a vector function obeying the equation $\square_2 \mathbf{h} = \operatorname{curl} \mathbf{k}$. Show that

$$\mathbf{u} = \mathbf{k} - \frac{1}{1-2\nu} \operatorname{curl} \mathbf{h}$$

is a complete solution of Navier's equation of motion. Hence deduce the solution indicated in Exercise 28.

45. Let ϕ be an arbitrary scalar function and $\boldsymbol{\psi}$ be an arbitrary vector function obeying the equations

$$\square_1 \phi = -\frac{1}{\lambda + 2\mu} \chi, \qquad \square_2 \boldsymbol{\psi} = -\frac{1}{\mu} \boldsymbol{\xi}$$

where χ and $\boldsymbol{\xi}$ are defined by $\mathbf{f} = (\nabla \chi + \operatorname{curl} \boldsymbol{\xi})$ with $\operatorname{div} \boldsymbol{\xi} = 0$. Show that $\mathbf{u} = \nabla \phi + \operatorname{curl} \boldsymbol{\psi}$ is a complete solution of Navier's equation of motion. (This solution is known as the *Green–Lamé solution*, and ϕ and $\boldsymbol{\psi}$ are called the *Lamé potentials*.) Show that the static counterpart of this solution is *not* complete.

46. If $u_3 = 0$ and u_1 and u_2 are functions of x_1, x_2 and t, show that

$$u_1 = \frac{\partial \phi}{\partial x_1} + \frac{\partial \psi}{\partial x_2}, \qquad u_2 = \frac{\partial \phi}{\partial x_2} - \frac{\partial \psi}{\partial x_1}$$

constitute a solution of Navier's equation of motion in the absence of body forces, provided $\square_1 \phi = 0$, $\square_2 \psi = 0$. Is this solution complete?

47. If ϕ and $\boldsymbol{\psi}$ are the Lamé potentials, show that

$$\operatorname{div} \mathbf{u} = \nabla^2 \phi, \qquad \operatorname{curl} \mathbf{u} = -\nabla^2 \boldsymbol{\psi}$$

Deduce that in the absence of body forces

$$\square_1(\operatorname{div} \mathbf{u}) = 0, \qquad \square_2(\operatorname{curl} \mathbf{u}) = \mathbf{0}$$

48. Show that every solution of Navier's equation of motion can be represented in the form $\mathbf{u} = \mathbf{u}^{(1)} + \mathbf{u}^{(2)}$, where

$$\Box_1 \mathbf{u}^{(1)} = -\frac{1}{\lambda + 2\mu} \nabla \chi, \quad \text{curl } \mathbf{u}^{(1)} = \mathbf{0}$$

$$\Box_2 \mathbf{u}^{(2)} = -\frac{1}{\mu} \text{curl } \boldsymbol{\xi}, \quad \text{div } \mathbf{u}^{(2)} = 0$$

with

$$\mathbf{f} = \nabla \chi + \text{curl } \boldsymbol{\xi}$$

49. Obtain expressions for the stress tensor associated with (i) the Cauchy-Kovalevski-Somigliana solution, (ii) the Sternberg-Eubanks solution, and the Green-Lamé solution, for the displacement vector.

50. Deduce the Cauchy-Kovalevski-Somigliana solution, the Sternberg-Eubanks solution and the Green-Lamé solution from each other.

51. Show that the stress system $\tau_{11} = \tau_{22} = \tau_{13} = \tau_{23} = \tau_{12} = 0$, $\tau_{33} = \rho g x_3$, where ρ and g are constants, satisfies the equations of equilibrium and the equations of compatibility. Find the corresponding displacements.

52. Show that the following stress system cannot be a solution of an elastostatic problem although it satisfies Cauchy's equations of equilibrium with zero body forces:

$$\tau_{11} = x_2^2 + v(x_1^2 - x_2^2), \quad \tau_{22} = x_1^2 + v(x_2^2 - x_1^2), \quad \tau_{33} = v(x_1^2 + x_2^2)$$
$$\tau_{12} = -2vx_1x_2, \quad \tau_{23} = \tau_{31} = 0$$

53. Determine whether or not the following stress components are a possible solution in elastostatics in the absence of body forces:

$$\tau_{11} = ax_2x_3, \quad \tau_{22} = bx_3x_1, \quad \tau_{33} = cx_1x_2, \quad \tau_{12} = dx_3^2,$$
$$\tau_{13} = ex_2^2, \quad \tau_{23} = fx_1^2$$

where a, b, c, d, e, f are all constants.

54. In an elastic body in equilibrium under the body force $\mathbf{f} = ax_1x_2\mathbf{e}_3$, where a is a constant, the stresses are of the form

$$\tau_{11} = ax_1x_2x_3, \quad \tau_{22} = bx_1x_2x_3, \quad \tau_{33} = cx_1x_2x_3$$
$$\tau_{12} = (ax_1^2 + bx_2^2)x_3, \quad \tau_{23} = (bx_2^2 + cx_3^2)x_1, \quad \tau_{13} = (cx_3^2 + ax_1^2)x_2$$

where a, b, c are constants. Find these constants.

55. In the absence of body forces, a general solution of Cauchy's equations of equilibrium is given by

$$\tau_{11} = \xi_{,22}, \quad \tau_{22} = \xi_{,11}, \quad \tau_{12} = -\xi_{,12}, \quad \tau_{12} = \tau_{13} = \tau_{33} = 0$$

Show that this is a possible solution in elasticity, provided $\nabla^2 \nabla^2 \xi = 0$.

9 EQUATIONS OF LINEAR ELASTICITY

56. If $\boldsymbol{\phi}$ is an arbitrary vector function satisfying the equation $\nabla^2 \boldsymbol{\phi} = \mathbf{f}$ and ω is a scalar function obeying the equation $\nabla^2 \omega = (1/(1-\nu)) \operatorname{div} \boldsymbol{\phi}$, show that

$$\mathbf{T} = \nabla\nabla\omega - \nu(\nabla^2 \omega)\mathbf{I} - (\nabla\boldsymbol{\phi} + \nabla\boldsymbol{\phi}^T)$$

is a complete solution of the Beltrami–Michell equation (9.11.7). (This is known as the *Schaefer's solution*.)

57. For the Beltrami–Michell equation in the static case, deduce the Schaefer solution by starting with (i) the Galerkin solution and (ii) the Papkovitch–Neuber solution of Navier's equation of equilibrium.

58. Specialize the Beltrami–Michell equations (9.11.7)' to the case where $f_i = \phi_{,i}$. Hence deduce that if $\nabla^2 \phi = 0$, then $\nabla^2 \tau_{kk} = 0$ and $\nabla^2 e_{kk} = 0$.

59. If h_i and ϕ are arbitrary functions satisfying the equations

$$\nabla^2 h_i = f_i, \qquad \nabla^2 \phi = \frac{1}{c_1^2(1-2\nu)} h_{k,k}$$

show that

$$\tau_{ij} = (h_{k,k} - 2c_2^2 \nabla^2 \phi)\delta_{ij} + 2c_2^2 \phi_{,ij} - (h_{i,j} + h_{j,i})$$

constitute a solution of the Beltrami–Michell equations (9.11.7)'.

60. If $\boldsymbol{\phi}$ is an arbitrary vector function obeying the equation $\Box_2 \boldsymbol{\phi} = \mathbf{f}$ and ω is a scalar function satisfying the equation $\Box_1 \omega = (1/(1-\nu)) \operatorname{div} \boldsymbol{\phi}$, show that

$$\mathbf{T} = \nabla\nabla\omega - \nu(\Box_2 \omega)\mathbf{I} - (\nabla\boldsymbol{\phi} + \nabla\boldsymbol{\phi}^T)$$

is a complete solution of the Beltrami–Michell equation (9.11.6). (This solution is known as the *Tedorescu's solution*.)

61. For the Beltrami–Michell equation in the dynamic case, deduce the Teodorescu solution by starting with (i) the Cauchy–Kovalevski–Somigliana solution, (ii) the Sternberg–Eubanks solution, and the (iii) Green–Lamé solution, of Navier's equation of motion.

62. Prove the following identities:

(i) $\displaystyle\int_V \mathbf{T}\, dV = \int_S \{\mu(\mathbf{u} \otimes \mathbf{n} + \mathbf{n} \otimes \mathbf{u}) + \lambda(\mathbf{u} \cdot \mathbf{n})\mathbf{I}\}\, dS$

(ii) $\displaystyle\int_S \mathbf{s}\, dS = \int_S \{(\lambda + 2\mu)(\operatorname{div} \mathbf{u})\mathbf{n} + \mu(\operatorname{curl} \mathbf{u}) \times \mathbf{n}\}\, dS$

63. For an elastic body in equilibrium, show that

$$\int_V \mathbf{E}\, dV = \frac{1}{2\mu}\left[\int_S (\mathbf{x} \otimes \mathbf{s})\, dS + \int_V (\mathbf{x} \otimes \mathbf{f})\, dV\right]$$
$$- \frac{\lambda}{2\mu(3\lambda + 2\mu)} \mathbf{I}\left[\int_S \mathbf{x} \cdot \mathbf{s}\, dS + \int_V \mathbf{x} \cdot \mathbf{f}\, dV\right]$$

64. For an elastic body in equilibrium, show that

$$\int_S \mathbf{u} \cdot \{(\lambda + 2\mu)(\text{div } \mathbf{u})\mathbf{n} + \mu(\text{curl } \mathbf{u}) \times \mathbf{n}\} \, dS + \int_V \mathbf{f} \cdot \mathbf{u} \, dV$$

$$= \int_V \{(\lambda + 2\mu)(\text{div } \mathbf{u})^2 + \mu|\text{curl } \mathbf{u}|^2\} \, dV$$

Hence deduce the uniqueness of solution of the displacement boundary value problem in elastostatics.

65. Consider a long circular cylinder rotating with a uniform angular speed ω about its axis. The cylinder is not free to deform longitudinally. The body force is ignored and the boundary is stress free. Find the displacements and stresses produced in the cylinder.

66. Consider a solid elastic sphere deforming statically due to mutual attraction between its parts. The force of gravity at any point inside the sphere is directly proportional to the distance of the point from the center and is directed toward the center. The surface of the sphere is stress free. Find the displacements and stresses produced in the sphere.

67. Find the displacements and stresses in a spherical shell whose inner boundary is fixed rigidly and outer boundary is subjected to a uniform pressure. Body forces are ignored.

68. Find the displacements and stresses in a long circular cylinder whose boundary is subjected to a uniform displacement in the radial direction. Body forces are ignored.

69. Find the displacement and stress fields developed in an elastic half-space with a plane boundary when the boundary is subjected to a uniform shear stress. Body forces are ignored.

70. Consider an unbounded elastic solid having a spherical cavity. The surface of the cavity is subjected to a time-dependent pressure $p(t)$ for time $t > 0^+$. Body forces are ignored. Find the displacement field generated.

71. Compute the displacements due to Rayleigh waves and analyze their nature on the boundary surface of the half-space.

72. Consider an elastic layer of uniform thickness and infinite lateral dimensions. The boundary surfaces of the layer are stress free. Investigate the existence of plane harmonic waves propagating parallel to the surfaces. Body forces are ignored.

CHAPTER 10
EQUATIONS OF FLUID MECHANICS

10.1
INTRODUCTION

In this chapter, we consider another important branch of continuum mechanics: the mechanics of nonviscous and Newtonian viscous fluids. Like the classical elasticity theory, this branch also uses linear constitutive equations. Many common fluids including water and air satisfy such constitutive equations. Nonviscous and Newtonian viscous fluids therefore serve as excellent models for studying the mechanical behavior of a wide variety of common liquids and gases.

The scope of fluid mechanics is vast. We restrict ourselves to the derivation of the governing equations for nonviscous and Newtonian viscous fluid flows and their immediate consequences. Some simple and standard applications are also discussed.

10.2
VISCOUS AND NONVISCOUS FLUIDS

The fundamental characteristic property of a fluid, which distinguishes it from a solid, is its inability to sustain shear stresses when it is at rest or in

462 10 EQUATIONS OF FLUID MECHANICS

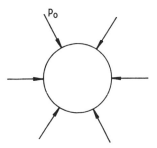

Figure 10.1. Hydrostatic pressure.

uniform motion. More specifically, whereas shear stresses can occur in a solid even when the solid is in static equilibrium or in uniform motion, shear stresses cannot occur in a fluid unless it undergoes a nonuniform motion. Experimental observations show that, in a fluid at rest or in uniform motion, the stress on a surface element consists only of a pressure (negative normal stress) and that this pressure has the same magnitude on all surface elements through a point. This pressure is usually called *hydrostatic pressure* or simply *static pressure*. Like density, static pressure is a physical property of the fluid and generally varies from one point to another.

Thus, on the basis of experimental observations, it is postulated that on a surface element in a fluid at rest or in uniform motion the stress vector **s** is given by

$$\mathbf{s} = -p_0 \mathbf{n} \qquad (10.2.1)$$

where p_0 is hydrostatic pressure; see Figure 10.1.

The Cauchy's law, (7.4.8), then yields (see Example 7.4.4)

$$\mathbf{T} = -p_0 \mathbf{I} \qquad (10.2.2)$$

Taking the trace of this equation, we find that

$$p_0 = \bar{p} \qquad (10.2.3)$$

where $\bar{p} = -(1/3)(\text{tr } \mathbf{T})$ is the mean pressure. Thus, the hydrostatic pressure is equal to the mean pressure.

Experimental observation also shows that fluids in nonuniform motion exert shear stresses in addition to normal stresses. This means that, when a fluid undergoes a nonuniform motion, the stress vector on a surface element in the fluid acts *obliquely* to the surface; see Figure 10.2. As such, for fluids in nonuniform motion relation (10.2.1) *does not* hold, in general.

However, in many practical problems, it is found that the effects of shear stresses are small and can even be neglected. But, important realistic situations occur where these effects are appreciable and cannot be ignored.

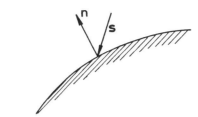

Figure 10.2. Stress vector acting obliquely to a surface.

In view of this observation, fluids are generally classified into two classes: (i) fluids that exert no or negligible shear stresses, and (ii) fluids that exert nonnegligible shear stresses. A fluid in which shear stresses are *not* negligible is called a *viscous fluid*. On the other hand, a fluid in which no or negligible shear stresses occur is called a *nonviscous* (or *inviscid* or *frictionless*) *fluid*. It should be pointed out that all fluids are viscous to a certain degree and that the concept of a nonviscous fluid is just an ideal one. Nevertheless, the study of nonviscous fluid flows is of great practical utility in engineering applications dealing with common liquids and gases (like water and air, for example) and an essential part of fluid mechanics.

10.3
STRESS TENSOR FOR A NONVISCOUS FLUID

From the definition of a nonviscous fluid, it follows that the stress vector on a surface element in such a fluid is of the form

$$\mathbf{s} = -p\mathbf{n} \qquad (10.3.1)$$

even when there is a nonuniform motion, where p is a scalar function that reduces to the hydrostatic pressure p_0 when there is no motion or when the motion is uniform. Relation (10.3.1) essentially implies that at a point of a nonviscous fluid every surface element is a principal plane of stress and every direction is a principal direction of stress, with $-p$ as the principal stress.

Consequently, Cauchy's law (7.4.8) gives the following equation for the stress tensor (as shown in Example 7.4.4):

$$\mathbf{T} = -p\mathbf{I} \qquad (10.3.2)$$

We note that $\mathbf{T}$, given by (10.3.2), is symmetric and isotropic. Thus, for a nonviscous fluid the symmetry of $\mathbf{T}$ is a physical property as in the case of an elastic solid governed by Hooke's law (9.2.3). Further, unlike in the case of an elastic solid, the isotropy of **T** is also a physical property of a nonviscous fluid. This means that there is no preferred direction in a nonviscous fluid. Relation (10.3.2), which describes the intrinsic material property of a nonviscous fluid, is taken as the material law for such a fluid.

It follows from (10.3.2) that

$$p = \bar{p} \tag{10.3.3}$$

where $\bar{p}$ is, as usual, the mean pressure. Thus, p present in the material law (10.3.2) represents the mean pressure in a nonviscous fluid in nonuniform motion. Like the static pressure p_0, the mean pressure p generally varies from one point to another and at any point it has the same value in all directions. When the fluid is brought to uniform motion or rest, p reduces to p_0, as already stated. Therefore p is called *dynamic pressure* or just *pressure*.

10.4
GOVERNING EQUATIONS FOR A NONVISCOUS FLUID FLOW

In Section 8.7, it was pointed out that the field equations (8.7.1) to (8.7.3) hold for all continua. Hence, these equations automatically hold for nonviscous fluids. Let us recall the spatial forms of these equations and record them here for ready reference.

Equation of continuity:

$$\frac{D\rho}{Dt} + \rho \operatorname{div} \mathbf{v} = 0 \tag{10.4.1}$$

Equations of motion:

$$\operatorname{div} \mathbf{T}^T + \rho \mathbf{b} = \rho \frac{D\mathbf{v}}{Dt} \tag{10.4.2a}$$

$$\mathbf{T} = \mathbf{T}^T \tag{10.4.2b}$$

Equation of energy:

$$\rho \frac{D\varepsilon}{Dt} = \mathbf{T} \cdot \nabla \mathbf{v} - \operatorname{div} \mathbf{q} + \rho h \tag{10.4.3}$$

To these field equations, let us append the material law (10.3.2), which is valid only for nonviscous fluids.

Material law:

$$\mathbf{T} = -p\mathbf{I} \tag{10.4.4}$$

As noted earlier, **T** is symmetric by virtue of this material law; therefore, equation (10.4.2b) is identically satisfied.

10.4 GOVERNING EQUATIONS FOR A NONVISCOUS FLUID FLOW

It is easily seen that 11 scalar equations are involved in the governing equations (10.4.1)–(10.4.4), whereas 15 unknown field functions are present in these equations, the unknowns being 3 velocity components v_i, 3 heat flux components q_i, 6 stress components $\tau_{ij} = (\tau_{ji})$ and 3 scalars ρ, p and ε. (The body force **b** and heat supply h are taken as known functions as usual.) Accordingly, equations (10.4.1) to (10.4.4) are inadequate to determine all the field functions. Therefore, we have to either reduce the number of unknowns or increase the number of governing equations in order to close the system of governing equations. These two possibilities lead us to consider the following two important particular cases.

10.4.1. INCOMPRESSIBLE FLUIDS

In Section 6.3, we noted that, for every motion of an incompressible continuum,

$$\operatorname{div} \mathbf{v} = 0 \tag{10.4.5}$$

In view of the equation of continuity (10.4.1), this condition is equivalent to the condition

$$\frac{D\rho}{Dt} \equiv 0 \tag{10.4.6}$$

which implies that ρ retains its initial value ρ_0 throughout the motion. A fluid for which this property of ρ holds is called an *incompressible fluid*, and (10.4.5) is employed as the equation of continuity for the motion of such a fluid.

In this case, equations (10.4.2) can be together rewritten as

$$\operatorname{div} \mathbf{T} + \rho_0 \mathbf{b} = \rho_0 \frac{D\mathbf{v}}{Dt} \tag{10.4.7}$$

Thus, for an incompressible fluid, equations (10.4.1) and (10.4.2) are replaced by equations (10.4.5) and (10.4.7).

From equations (10.4.4) and (10.4.5), we get

$$\mathbf{T} \cdot \nabla \mathbf{v} = -p(\mathbf{I} \cdot \nabla \mathbf{v}) = -p \operatorname{div} \mathbf{v} = 0 \tag{10.4.8}$$

Hence, for an incompressible nonviscous fluid, the energy equation (10.4.3) reduces to

$$\rho_0 \frac{D\varepsilon}{Dt} = -\operatorname{div} \mathbf{q} + \rho h \tag{10.4.9}$$

We note that 10 scalar equations are involved in (10.4.4), (10.4.5) and (10.4.7) and the number of unknown field functions is also 10, the functions being p, v_i and τ_{ij}. Equation (10.4.9) does not contain these field functions and is therefore not needed in their determination. Equations (10.4.4), (10.4.5) and (10.4.7) thus serve as a closed system of governing differential equations for the study of nonviscous, incompressible fluid flows. For such flows, energy considerations are generally not important and equation (10.4.9), which is purely an equation of balance of thermal energy, may be discarded.

10.4.2. COMPRESSIBLE FLUIDS

A fluid is said to be *compressible* if it is not incompressible; that is, if ρ is not constant during the motion. In this case, some constitutive relations from thermodynamics are appended to the system of equations (10.4.1) to (10.4.4) to close the system.

Since the density of a compressible fluid varies during the motion, the equation of continuity (10.4.1) shows that div $\mathbf{v} \neq 0$. Hence, fluid elements undergo a change in their volumes during the motion. A change in volume of a material is generally caused by a change in normal stress and a change in temperature. Thus, for a compressible fluid, it is postulated that p, ρ and T are related in a definite way, where T is absolute temperature; that is, there exists a functional relationship of the form

$$f(p, \rho, T) = 0 \qquad (10.4.10)$$

where f is a known function. An equation of this type is called a *kinetic equation of state*. The actual form of the equation depends on the physical properties of the fluid being dealt with.

The equation of energy (10.4.3) shows that ε is related to ρ, $\mathbf{T}$, $\mathbf{v}$ and $\mathbf{q}$ in a definite way. From the material law (10.4.4) and the equation of continuity (10.4.1), we find that

$$\mathbf{T} \cdot \nabla \mathbf{v} = -p\mathbf{I} \cdot (\nabla \mathbf{v}) = -p \operatorname{div} \mathbf{v} = \frac{p}{\rho} \frac{D\rho}{Dt} \qquad (10.4.11)$$

Thus, the stress-power $\mathbf{T} \cdot \nabla \mathbf{v}$ is directly related with p and ρ in a definite way. Consequently, it follows that ε is related to ρ, p and $\mathbf{q}$ in a definite way. If we employ *Fourier's law of heat conduction* (8.7.20), namely,

$$\mathbf{q} = -k \nabla T \qquad (10.4.12)$$

then $\mathbf{q}$ is determined by T. Also, the kinetic equation of state, (10.4.10), determines p as a function of ρ and T. Thus, ultimaltely ε can be related to ρ and T in a definite way. Hence, it is postulated that ε may be taken as a known function of ρ and T; that is,

$$\varepsilon = \varepsilon(\rho, T) \qquad (10.4.13)$$

10.4 GOVERNING EQUATIONS FOR A NONVISCOUS FLUID FLOW 467

An equation of this type is called a *caloric equation of state*. The actual form of the function $\varepsilon(\rho, T)$ depends on the physical properties of the fluid being dealt with.

If equations (10.4.10), (10.4.12) and (10.4.13) are appended to equations (10.4.1) to (10.4.4), then we have in all 16 scalar equations with 16 unknown field functions, the field functions being 3 velocity components v_i, 3 heat-flux components q_i, 6 stress components τ_{ij} and 4 scalars ρ, p, T and ε. Equations (10.4.1) to (10.4.4), (10.4.10), (10.4.12) and (10.4.13) therefore serve as a closed system of governing differential equations for the study of nonviscous compressible fluid flows. Note that whereas (10.4.1) to (10.4.3) are field equations, (10.4.4), (10.4.10), (10.4.12) and (10.4.13) are all constitutive equations. Unlike in the case of an incompressible fluid, where the material law (10.4.4) is the only constitutive equation, the constitutive equations for a compressible fluid consist of the equations of state (10.4.10) and (10.4.13) and a law of heat conduction (10.4.12), in addition to the material law (10.4.4). Also, in this case, the energy equation (10.4.3) is coupled with other governing equations.

Thus, for a nonviscous fluid flow, two sets of equations govern the flow—one set is applicable to incompressible fluids, and the other to compressible fluids. For an incompressible fluid, the density does not change with motion, and the pressure p, the velocity field $\mathbf{v}$ and the stress $\mathbf{T}$ serve as the field functions. On the other hand, for the case of a compressible fluid in which the density changes with motion, the field functions are the density ρ, the pressure p, the temperature T, the internal energy ε, the heat flux $\mathbf{q}$, the velocity $\mathbf{v}$ and the stress $\mathbf{T}$. Obviously, the study of compressible fluid flows is more difficult than that of incompressible fluid flows.

Experimental observations show that in the case of liquids and low-speed gases, the changes in density during the motion are very small and hence can be neglected. Liquids and low-speed gases are therefore generally considered as incompressible fluids. In the case of gases moving with high speed, the variations in density during the motion are considerably large and cannot be neglected. High-speed gases are therefore regarded as compressible fluids. In reality, no fluid is incompressible; every fluid is compressible to a certain degree. Like the concept of nonviscous fluid, the concept of an incompressible fluid is just an ideal one.

It is to be emphasized that fluids are classified as compressible or incompressible accordingly as the density changes or does not change *during the motion*. In a fluid, the density may change from one part of the fluid to the other, and such a change of density may occur in both compressible and incompressible fluids. A fluid in which the density is the same everywhere in the fluid (so that $\nabla \rho \equiv \mathbf{0}$) is called a *homogeneous fluid*. On the other hand, a fluid in which there is a change in density from one part of the fluid to the other (so that $\nabla \rho \neq \mathbf{0}$ in general) is called a *nonhomogeneous fluid*. Salt water with nonuniform concentrations with depth is a standard example of a nonhomogeneous fluid. The governing equations considered in the previous paragraphs hold for both homogeneous and nonhomogeneous nonviscous fluids.

A point that is to be clearly understood is the following. Recall that the initial density ρ_0 has been defined by $\rho_0 = \rho(\mathbf{x}^0, 0)$. Evidently, ρ_0 varies with $\mathbf{x}^0$ in general. It is only for homogeneous fluids that we have $\rho_0 = $ constant.

10.4.3. STANDARD FORMS OF EQUATIONS OF STATE

It was pointed out that the equations of state, namely, (10.4.10) and (10.4.13), serve as constitutive equations for compressible fluids and the exact forms of these equations depend on the physical properties of the fluid being dealt with. Here we give some standard forms of these equations that are needed in our further discussion.

The caloric equation of state (10.4.13) is intended to specify ε in terms of ρ and T. One of the simplest forms of this equation is

$$\varepsilon = c_v T \qquad (10.4.14)$$

where c_v is either a constant or a function of ρ. Then c_v is called the *specific heat at constant volume*.

The kinetic equation of state (10.4.10) is intended to specify a functional relationship between p, ρ and T. One of the simplest forms of this equation is the *Boyle's law* given by

$$p = \rho R T \qquad (10.4.15)$$

where R is a constant for a particular fluid under consideration.

A compressible fluid for which (10.4.14) with $c_v = $ constant and (10.4.15) hold is called a *perfect gas*, and R is then known as *gas constant*. Common gases like air, oxygen and nitrogen are regarded as perfect gases. For air, the value of R is approximately equal to 2.87×10^6.

An important particular case of equation (10.4.15) is

$$\frac{p}{\rho} = \text{constant} \qquad (10.4.16a)$$

$$T = \text{constant} \qquad (10.4.16b)$$

A compressible fluid for which (10.4.14) with $c_v = $ constant and (10.4.16) hold is called an *isothermal perfect gas*.

10.4 GOVERNING EQUATIONS FOR A NONVISCOUS FLUID FLOW

An important generalization of equations (10.4.16) is

$$\frac{p}{\rho^\gamma} = \text{constant} \tag{10.4.17a}$$

$$\frac{T}{\rho^{\gamma-1}} = \text{constant} \tag{10.4.17b}$$

where

$$\gamma = \left(1 + \frac{R}{c_v}\right) > 1 \tag{10.4.18}$$

A compressible fluid for which (10.4.14) with c_v = constant and (10.4.17) hold is called an *isentropic perfect gas*; the constant γ is called the *adiabatic constant*. If

$$c_p = c_v + R \tag{10.4.19}$$

then c_p is called the *specific heat at constant pressure*, and the following relations can be verified:

$$c_v = \frac{R}{\gamma - 1} \tag{10.4.20a}$$

$$c_p = \frac{\gamma R}{\gamma - 1} \tag{10.4.20b}$$

$$\frac{c_p}{c_v} = \gamma \tag{10.4.20c}$$

We often deal with what are called *barotropic fluids*. By a barotropic fluid, we mean a compressible fluid for which the kinetic equation of state, (10.4.10), is independent of T and can be expressed in *either* of the following forms:

$$p = p(\rho) \tag{10.4.21a}$$

$$\rho = \rho(p) \tag{10.4.21b}$$

A nonviscous barotropic fluid is often referred to as an *Eulerian fluid* or *elastic fluid*.

Equations of state that are more general than those just summarized are also found in the literature. For easy reference, the governing equations for incompressible and compressible nonviscous fluid flows are summarized in Tables 10.1 and 10.2 with the same equation numbers as in the text.

10 EQUATIONS OF FLUID MECHANICS

Table 10.1. Governing Equations for Nonviscous Incompressible Fluid Flows

1. Equation of continuity:
$$\text{div } \mathbf{v} = 0 \qquad (10.4.5)$$

2. Equation of motion:
$$\text{div } \mathbf{T} + \rho_0 \mathbf{b} = \rho_0 \frac{D\mathbf{v}}{Dt} \qquad (10.4.7)$$

3. Material law:
$$\mathbf{T} = -p\mathbf{I} \qquad (10.4.4)$$

Total number of scalar equations: 10
Total number of unknown field functions $(p, v_i \text{ and } \tau_{ij})$: 10

Table 10.2. Governing equations for Nonviscous Compressible Fluid Flows

1. Equation of continuity:
$$\frac{D\rho}{Dt} + \rho \text{ div } \mathbf{v} = 0 \qquad (10.4.1)$$

2. Equation of motion:
$$\text{div } \mathbf{T} + \rho \mathbf{b} = \rho \frac{D\mathbf{v}}{Dt} \qquad (10.4.2)$$

3. Equation of energy:
$$\rho \frac{D\varepsilon}{Dt} = \mathbf{T} \cdot \nabla \mathbf{v} - \text{div } \mathbf{q} + \rho h \qquad (10.4.3)$$

4. Material law:
$$\mathbf{T} = -p\mathbf{I} \qquad (10.4.4)$$

5. Fourier's law of heat conduction:
$$\mathbf{q} = -k \nabla T \qquad (10.4.12)$$

6. Equations of state
$$f(p, \rho, T) = 0 \qquad (10.4.10)$$
$$\varepsilon = \varepsilon(\rho, T) \qquad (10.4.13)$$

Total number of scalar equations: 16
Total number of unknown field functions $(\rho, p, T, \varepsilon, q_i, v_i \text{ and } \tau_{ij})$: 16

EXAMPLE 10.4.1 Assuming that the coefficient of thermal conductivity k is constant, show that the energy equation for a nonviscous compressible fluid can be expressed in the following alternative forms:

$$\frac{D\varepsilon}{Dt} = \frac{p}{\rho^2} \frac{D\rho}{Dt} + \frac{k}{\rho} \nabla^2 T + h \qquad (10.4.22)$$

$$\frac{D}{Dt}\left[\varepsilon + \frac{p}{\rho}\right] = \frac{k}{\rho} \nabla^2 T + \frac{1}{\rho} \frac{Dp}{Dt} + h \qquad (10.4.23)$$

10.4 GOVERNING EQUATIONS FOR A NONVISCOUS FLUID FLOW

Solution Substituting for $\mathbf{T} \cdot \nabla \mathbf{v}$ from (10.4.11) and for $\mathbf{q}$ from (10.4.12) in the energy equation (10.4.3) we immediately arrive at equation (10.4.22).
With the use of the identity

$$\frac{D}{Dt}\left(\frac{p}{\rho}\right) = \frac{1}{\rho}\frac{Dp}{Dt} - \frac{p}{\rho^2}\frac{D\rho}{Dt} \tag{10.4.24}$$

equation (10.4.22) reduces to equation (10.4.23). ∎

Note: The quantity $\varepsilon + (p/\rho)$ present on the lefthand side of (10.4.23) is called *specific enthalpy*.

EXAMPLE 10.4.2 A flow in which there is no heat flux is called an *adiabatic flow*. Assuming that the fluid is nonviscous and there is no heat supply, prove the following for an adiabatic flow:

(i) For a baratropic fluid

$$\varepsilon + \frac{p}{\rho} = \int \frac{dp}{\rho} + \text{constant} \tag{10.4.25}$$

(ii) For an isentropic perfect gas

$$\varepsilon + \frac{p}{\rho} = \frac{\gamma}{(\gamma - 1)}\frac{p}{\rho} + \text{constant} \tag{10.4.26}$$

Solution (i) For an adiabatic flow, we have $\mathbf{q} = 0$, so that $\nabla T = \mathbf{0}$ by the Fourier's law of heat conduction (10.4.12). Consequently, $\text{div}(\nabla T) = \nabla^2 T = 0$. Then in the absence of h, we obtain, from (10.4.23),

$$\frac{D}{Dt}\left[\varepsilon + \frac{p}{\rho}\right] = \frac{1}{\rho}\frac{Dp}{Dt} \tag{10.4.27}$$

For a barotropic fluid, we have $\rho = \rho(p)$ so that

$$\frac{d}{dp}\left\{\int \frac{1}{\rho}dp\right\} = \frac{1}{\rho} \tag{10.4.28}$$

Hence

$$\frac{1}{\rho}\frac{Dp}{Dt} = \frac{d}{dp}\left\{\int \frac{1}{\rho}dp\right\}\frac{Dp}{Dt} = \frac{D}{Dt}\left\{\int \frac{1}{\rho}dp\right\} \tag{10.4.29}$$

Using this in (10.4.27) and integrating the resulting equation, we obtain (10.4.25).

(ii) Recall that for an isentropic perfect gas, $p/\rho^\gamma = \text{constant} = \beta$, say. Then

$$\int \frac{dp}{\rho} = \beta\gamma \int \rho^{\gamma-2} d\rho = \frac{\beta\gamma}{(\gamma-1)}\rho^{\gamma-1} + \text{constant} = \frac{\gamma}{(\gamma-1)}\frac{p}{\rho} + \text{constant} \tag{10.4.30}$$

Using this in (10.4.25), we obtain (10.4.26). ∎

10 EQUATIONS OF FLUID MECHANICS

EXAMPLE 10.4.3 Show that for a nonviscous perfect gas flow, the energy equation can be expressed in the following alternative forms:

$$c_v \frac{DT}{Dt} = \frac{p}{\rho^2} \frac{D\rho}{Dt} + \frac{k}{\rho} \nabla^2 T + h \quad (10.4.31)$$

$$c_p \frac{DT}{Dt} = \frac{1}{\rho} \frac{Dp}{Dt} + \frac{k}{\rho} \nabla^2 T + h \quad (10.4.32)$$

Solution For a perfect gas, we get from (10.4.14), (10.4.15) and (10.4.20),

$$\varepsilon = c_v T = \frac{c_v}{R} \frac{p}{\rho} = \left(\frac{c_v}{R} - 1\right) \frac{p}{\rho} = c_p T - \frac{p}{\rho} \quad (10.4.33)$$

Substituting $\varepsilon = c_v T$ in (10.4.22), we immediately get (10.4.31). Substituting $\varepsilon = c_p T - (p/\rho)$ in (10.4.23), we obtain (10.4.32). ∎

EXAMPLE 10.4.4 (a) Show that for a nonviscous incompressible fluid flow, the equation of balance of mechanical energy is given by

$$\frac{DK}{Dt} = \int_V \rho(\mathbf{b} \cdot \mathbf{v}) \, dV - \int_S p\mathbf{v} \cdot \mathbf{n} \, dS \quad (10.4.34)$$

(b) If the body force is conservative so that $\mathbf{b} = -\nabla\chi$ for some scalar potential $\chi = \chi(\mathbf{x})$, show that (10.4.34) can be put in the form

$$\frac{D}{Dt}(K + \Omega) = -\int_S p\mathbf{v} \cdot \mathbf{n} \, dS \quad (10.4.35)$$

where

$$\Omega = -\int_V \rho\chi \, dV \quad (10.4.36)$$

is called the *potential energy* of the fluid contained in V.

Deduce that (i) if the fluid moves tangentially over the surface S, then $K + \Omega$ remains constant during the motion, and (ii) if $K + \Omega$ remains constant during the motion, then the stream lines lie on surfaces of constant pressure.

Solution (a) For a continuum, the equation of balance of mechanical energy is given by (8.6.27). For a nonviscous incompressible fluid, $\mathbf{T} \cdot \nabla \mathbf{v} = 0$, see (10.4.8). Recall also that $\mathbf{s} = -p\mathbf{n}$ on S and $\mathbf{T} \cdot \nabla \mathbf{v} = \mathbf{T} \cdot \mathbf{D}$. Using these results in (8.6.27), we immediately obtain equation (10.4.34).

Note: For an incompressible nonviscous fluid, the equation of balance of thermal energy is given by (10.4.9), which has no common term with the equation of balance of mechanical energy, (10.4.34). Hence the motion and the thermal state of such a fluid do not influence one another.

(b) For $\mathbf{b} = -\nabla\chi$, we have

$$\int_V \rho \mathbf{b} \cdot \mathbf{v}\, dV = -\int_V \rho(\nabla\chi \cdot \mathbf{v})\, dV = -\int_V \rho\left(\nabla\chi \cdot \frac{D\mathbf{x}}{Dt}\right) dV = -\int_V \rho \frac{D\chi}{Dt}\, dV \qquad (10.4.37)$$

which, by use of (8.2.10), becomes

$$\int_V \rho \mathbf{b} \cdot \mathbf{v}\, dV = -\frac{D}{Dt}\int_V \rho\chi\, dV \qquad (10.4.38)$$

Using (10.4.38) and (10.4.36) in (10.4.34) we immediately arrive at (10.4.35).

If the fluid moves tangentially over the surface S, we have $\mathbf{v} \cdot \mathbf{n} = 0$ on S. Consequently, (10.4.35) yields

$$\frac{D}{Dt}(K + \Omega) = 0 \qquad (10.4.39)$$

This implies that $K + \Omega$ is constant during the motion.

Conversely, if $K + \Omega$ is constant during the motion, then (10.4.39) holds and (10.4.35) yields

$$\int_S p(\mathbf{v} \cdot \mathbf{n})\, dS = 0 \qquad (10.4.40)$$

which, by use of the divergence theorem (3.6.1), becomes

$$\int_V \operatorname{div}(p\mathbf{v})\, dV = 0 \qquad (10.4.41)$$

Using the identity (3.4.14) and the equation of continuity (10.4.5), expression (10.4.41) reduces to

$$\int_V (\nabla p \cdot \mathbf{v})\, dV = 0 \qquad (10.4.42)$$

Since V is an arbitrary volume, it follows that $\nabla p \cdot \mathbf{v} = 0$ at all points of the fluid; that is, $\mathbf{v}$ is orthogonal to ∇p everywhere in the fluid. Since ∇p is orthogonal to surfaces of constant p (see Section 3.4), it follows that $\mathbf{v}$ is tangential to these surfaces. Thus, stream lines lie on surfaces of constant pressure at every point of the fluid. ∎

10.5
INITIAL AND BOUNDARY CONDITIONS

A boundary value problem in fluid mechanics consists in determining all the unknown field functions by solving the governing equations under

474 10 EQUATIONS OF FLUID MECHANICS

appropriate initial and boundary conditions. For incompressible nonviscous fluid flows, the unknown functions are p, $\mathbf{v}$ and $\mathbf{T}$ and the governing equations are (10.4.4), (10.4.5) and (10.4.7). In this case, the following initial and boundary conditions are prescribed:

$$\mathbf{v} = \mathbf{v}^{(0)} \quad \text{in} \quad V \quad \text{at} \quad t = 0 \tag{10.5.1}$$

$$\mathbf{v} = \mathbf{v}^* \quad \text{on} \quad S_v \quad \text{for} \quad t \geq 0 \tag{10.5.2a}$$

$$\mathbf{T}\mathbf{n} = \mathbf{s}^* \quad \text{on} \quad S_\tau \quad \text{for} \quad t \geq 0 \tag{10.5.2b}$$

where V is a *material volume* of the fluid bounded by a closed surface S, and S_v and S_τ are parts of S that are complementary to each other. As special cases S_v or S_τ can be equal to S. Also, $\mathbf{v}^{(0)}$, $\mathbf{v}^*$ and $\mathbf{s}^*$ are prescribed quantities in their respective domains.

The initial conditions (10.5.1) mean that the velocity of the fluid is known at every point of the initial volume. For a fluid motion starting from rest at time $t = 0$, $\mathbf{v}^{(0)} = \mathbf{0}$. The boundary conditions (10.5.2) mean that the velocity is prescribed on S_v and stress is specified on S_τ. The conditions (10.5.1) and (10.5.2ab) are similar to conditions (9.5.4) and (9.5.3), but there is one difference. In the case of linear elastic solids, S can be treated as the boundary surface in an initial configuration. In the case of fluids, S is the boundary surface in the current configuration.

For nonviscous compressible fluid flows, the unknown functions are ρ, p, T, ε, $\mathbf{v}$, $\mathbf{q}$ and $\mathbf{T}$; and the governing equations are (10.4.1)–(10.4.4), (10.4.10), (10.4.12) and (10.4.13). Since p, ε and $\mathbf{q}$ can be regarded as known functions of ρ and T through the equations of state (10.4.10) and (10.4.13) and the law of heat conduction (10.4.12), the initial conditions are usually specified in terms of $\mathbf{v}$, ρ and T and the boundary conditions in terms of $\mathbf{v}$, $\mathbf{T}$ and T. The conditions for $\mathbf{v}$ and $\mathbf{T}$ are usually taken as those given in (10.5.1) and (10.5.2ab), whereas the conditions for ρ and T are as follows:

$$\rho = \rho_0 \quad \text{in} \quad V \quad \text{at} \quad t = 0 \tag{10.5.3}$$

$$T = T_0 \quad \text{in} \quad V \quad \text{at} \quad t = 0 \tag{10.5.4}$$

$$\left. \begin{array}{l} T = T^* \quad \text{on} \quad S_T \\ \nabla T \cdot \mathbf{n} = q^* \quad \text{on} \quad S_q \end{array} \right\} \quad \text{for} \quad t \geq 0 \tag{10.5.5}$$

where S_T and S_q are parts of S that are complementary to each other (as special cases S_T and S_q can be equal to S). Also, ρ_0 is the initial density and T_0 is the initial temperature, and T^* and q^* are prescribed quantities in their respective domains in S.

10.5 INITIAL AND BOUNDARY CONDITIONS

The boundary conditions (10.5.5) mean that the temperature is prescribed on the part S_T of S and the normal derivative of temperature on the remaining part S_q of S. If $T^* = T_0$, then the points of S_T experience no change from the initial temperature, and S_T is then called an *isothermal surface*. On the other hand, if $q^* = 0$, then no heat flux occurs across S_q, and S_q is then called an *adiabatic surface*.

The boundary conditions (10.5.2) require special attention when the boundary surface is a rigid impermeable solid surface contacting a fluid and when it is a fluid surface exposed to atmosphere. A rigid solid surface contacting a fluid could be the wall of a container or the surface of a body immersed in the fluid. Since the fluid is nonviscous, shear stress on every surface element is 0; hence on a rigid surface S, no resistance to the motion of the fluid relative to S occurs and the fluid is free to slip over S. Therefore, no constraint can be imposed on the tangential component of $\mathbf{v}$ at a point of S. However, for the fluid and S to be in geometrical contact, the velocity of the fluid at the boundary must be such that its component normal to the boundary is equal to the normal component of velocity of the boundary. That is,

$$\mathbf{v} \cdot \mathbf{n} = \mathbf{v}_s \cdot \mathbf{n} \qquad (10.5.6)$$

on S, where $\mathbf{v}_s$ is the velocity of the rigid boundary and $\mathbf{n}$ is the unit normal to the boundary. In the particular case when the boundary is at rest, the condition (10.5.6) becomes

$$\mathbf{v} \cdot \mathbf{n} = 0 \qquad (10.5.7)$$

on S.

The condition (10.5.6) can be rewritten in an alternative form by using the geometrical equation of the boundary surface. Let $F(\mathbf{x}, t) = 0$ be the geometrical equation of the boundary surface. Since particles that initially lie on a boundary continue to lie on a boundary, $F(\mathbf{x}, t) = 0$ is a material surface, and the speed with which this surface advances perpendicular to itself is given by (6.2.62). That is,

$$\mathbf{v}_s \cdot \mathbf{n} = -\frac{(\partial F/\partial t)}{|\nabla F|} \qquad (10.5.8)$$

Consequently, the condition (10.5.6) can be rewritten as

$$\frac{\partial F}{\partial t} + (\mathbf{v} \cdot \mathbf{n})|\nabla F| = 0 \qquad (10.5.9)$$

on the boundary, $F(\mathbf{x}, t) = 0$.

When S is the boundary surface of a fluid that is exposed to atmosphere, the fluid pressure on S must balance the atmospheric pressure p_a.

Thus, on such a surface we impose the following boundary condition:

$$p = p_a \qquad (10.5.10)$$

on S. This condition is valid only if the *surface tension* effects are neglected.

EXAMPLE 10.5.1 For a certain flow of a nonviscous fluid in a region bounded by the fixed solid boundary $y = 0$, the velocity field is given by $\mathbf{v} = \nabla \phi$, where $\phi = x^3 - 3xy^2$. Verify that the boundary condition $\mathbf{v} \cdot \mathbf{n} = 0$ is satisfied and that at all points except the origin the fluid slips on the boundary.

Note that here and in the following discussion, (x, y) or (x, y, z) denote the Cartesian coordinates.

Solution From the given expression for $\mathbf{v}$, we find that

$$v_1 = 3(x^2 - y^2), \qquad v_2 = -6xy, \qquad v_3 = 0 \qquad (10.5.11)$$

From these components we note that on the boundary $y = 0$, we have $\mathbf{v} \cdot \mathbf{n} = v_2 = 0$. Thus, the boundary condition $\mathbf{v} \cdot \mathbf{n} = 0$ is verified.

Since $v_3 = 0$, v_1 is the only tangential component of velocity that occurs on the boundary $y = 0$. For $y = 0$, (10.5.11) yields $v_1 = 3x^2$, and thus v_1 vanishes only at the origin. Hence, except at the origin, everywhere else on the boundary $y = 0$ the fluid slips with speed $3x^2$ in the x direction. ∎

EXAMPLE 10.5.2 A long cylinder is fixed rigidly in the flow of a nonviscous fluid. Obtain the boundary condition for velocity on the lateral surface of the cylinder.

Solution Let us choose the coordinate axes such that the z axis is along the axis of the cylinder. Then at any point P on the lateral surface of the cylinder, the normal $\mathbf{n}$ is parallel to the xy plane so that $n_1 = \cos \theta$, $n_2 = \sin \theta$, $n_3 = 0$, where θ is the angle between $\mathbf{n}$ and the x direction;

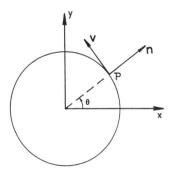

Figure 10.3. Example 10.5.2.

see Figure 10.3. Since the cylinder is stationary, the boundary condition (10.5.7) yields

$$\mathbf{v} \cdot \mathbf{n} \equiv v_1 \cos \theta + v_2 \sin \theta = 0 \qquad (10.5.12)$$

This is the condition to be satisfied by the velocity components of the fluid at a point P on the lateral surface of the cylinder. ∎

10.6
EULER'S EQUATION OF MOTION

In Section 10.4, it was shown that for a nonviscous fluid flow there are ten field functions (namely, p, v_i and τ_{ij}) for the incompressible case and sixteen field functions (namely, ρ, p, ε, T, v_i, q_i and τ_{ij}) for the compressible case, and as many governing equations. By combining the equations of motion (10.4.2) and the material law (10.4.4), it is possible to eliminate the stress components from the governing equations and thereby reduce the number of field functions and their governing equations. In this section, we derive the equation of motion through this method.

We start with the material law (10.4.4). Taking the divergence on both sides of this tensor equation and using the identity (3.5.35), we obtain

$$\text{div } \mathbf{T} = -\nabla p \qquad (10.6.1)$$

Substituting this into the equation of motion (10.4.2) we obtain the following equation of motion expressed in terms of ρ, p and $\mathbf{v}$:

$$\frac{D\mathbf{v}}{Dt} = -\frac{1}{\rho} \nabla p + \mathbf{b} \qquad (10.6.2)$$

This equation was first obtained by Euler in 1755 and is known as *Euler's equation of motion*. This equation holds for both incompressible and compressible fluids and is a basic governing differential equation in the theory of nonviscous fluid flows. In the case of incompressible fluid flows, $\rho = \rho_0$ and the equation contains $\mathbf{v}$ and p as the field functions. In the case of compressible fluid flows, the equation includes $\mathbf{v}$, ρ and p as the field functions. Thus the stresses are completely eliminated from the equation of motion.

For a fluid at rest (static equilibrium) or in uniform motion, Euler's equation (10.6.2) reduces to the *equation of equilibrium*:

$$-\frac{1}{\rho} \nabla p + \mathbf{b} = 0 \qquad (10.6.3)$$

where p denotes the *static pressure* and ρ denotes *time-independent density*.

478 **10 EQUATIONS OF FLUID MECHANICS**

By using the identities (6.2.10) and (6.2.11), that is,

$$\frac{D\mathbf{v}}{Dt} = \frac{\partial \mathbf{v}}{\partial t} + (\mathbf{v} \cdot \nabla)\mathbf{v} = \frac{\partial \mathbf{v}}{\partial t} + \mathbf{w} \times \mathbf{v} + \frac{1}{2}\nabla v^2 \qquad (10.6.4)$$

where $\mathbf{w} = \text{curl } \mathbf{v}$ is the vorticity vector, equation (10.6.2) can be rewritten in the following alternative forms:

$$-\frac{1}{\rho}\nabla p + \mathbf{b} = \frac{\partial \mathbf{v}}{\partial t} + (\mathbf{v} \cdot \nabla)\mathbf{v} \qquad (10.6.5a)$$

$$= \frac{\partial \mathbf{v}}{\partial t} + \frac{1}{2}\nabla v^2 + \mathbf{w} \times \mathbf{v} \qquad (10.6.5b)$$

Thus the Euler's equation of motion is a *nonlinear* partial differential equation of the first order, the nonlinearity being caused by the acceleration vector. This implies that in principle it is not possible to simply superpose solutions of Euler's equation.

In the case of an incompressible fluid flow, equation (10.6.2) may be solved under appropriate initial and boundary conditions described in Section 10.5 to find a solution for the velocity field $\mathbf{v}$ for a prescribed pressure p; generally p is specified through boundary data. Equation of continuity (10.4.5) serves as a constraint on $\mathbf{v}$.

For compressible fluid flows, p is usually specified by a kinetic equation of state, and equation (10.6.2) and (10.4.1) are to be solved *simultaneously* under appropriate initial and boundary conditions described in Section 10.5 to find $\mathbf{v}$ and ρ. Once $\mathbf{v}$ and ρ are so determined, the energy equation (10.4.3) may be employed under appropriate initial and boundary conditions to determine T. Note that ε is usually specified in terms of ρ and T by a caloric equation of state.

EXAMPLE 10.6.1 A rectangular tank containing a nonviscous liquid of constant density moves horizontally to the right with a constant acceleration. Gravitational force is the only external force. Find the pressure distribution in the liquid and the geometrical shape of the upper surface of the liquid.

Solution Choose the directions of the coordinate axes as indicated in Figure 10.4. Then, $D\mathbf{v}/Dt = a\mathbf{e}_1$ and $\mathbf{b} = -g\mathbf{e}_3$, where $a = |D\mathbf{v}/Dt|$ is a constant and g is the (constant) acceleration due to gravity. Euler's equation (10.6.2) now yields the following three equations for the three Cartesian (x, y, z) components of ∇p:

$$\frac{\partial p}{\partial x} = -a\rho \qquad (10.6.6a)$$

10.6 EULER'S EQUATION OF MOTION

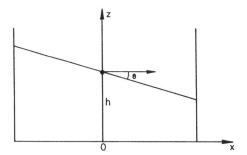

Figure 10.4. Example 10.6.1.

$$\frac{\partial p}{\partial y} = 0 \qquad (10.6.6b)$$

$$\frac{\partial p}{\partial z} = -g\rho \qquad (10.6.6c)$$

Equation (10.6.6b) shows that p is independent of y, and hence (10.6.6a) gives

$$p = -\rho ax + f(z) \qquad (10.6.7)$$

where $f(z)$ is an arbitrary function of z.

Equations (10.6.6c) and (10.6.7) give $f(z) = -\rho g z + C$, where C is a constant. Putting this $f(z)$ into (10.6.7), we get

$$p = -\rho(ax + gz) + C \qquad (10.6.8)$$

At the point where the z axis meets the upper surface of the liquid, we have $p = p_a$, where p_a is the atmospheric pressure. If this point is at a height h above the origin, (10.6.8) gives $C = p_a + \rho g h$. Thus the pressure distribution in the liquid is

$$p = p_a - \rho(ax + gz - gh) \qquad (10.6.9)$$

For $p = p_a$, equation (10.6.9) becomes

$$z = -\left(\frac{a}{g}\right)x + h$$

This is the equation of the upper surface of the liquid. Evidently, this surface is a plane making an acute angle $\theta = \tan^{-1}(a/g)$ with the horizontal. ∎

Note that in the limiting case when $a \to 0$, the liquid moves with constant velocity and the upper surface of the liquid becomes a horizontal plane.

480   10 EQUATIONS OF FLUID MECHANICS

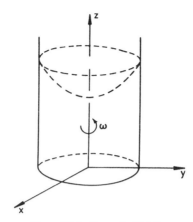

Figure 10.5. Example 10.6.2.

EXAMPLE 10.6.2 A column of a nonviscous liquid of constant density contained in a vertical circular vessel rotates like a rigid body about the axis of the vessel with a constant angular velocity ω. Gravitational force is the only external force. Find the pressure distribution in the liquid and the geometrical form of the upper surface of the liquid.

Solution For a fluid rotating like a rigid body, the acceleration is given by (6.2.20). Taking the axis of rotation as the z axis directed positive upward (see Figure 10.5) and noting $\boldsymbol{\omega} = \omega \mathbf{e}_3$ is a constant angular velocity, this relation becomes

$$\frac{D\mathbf{v}}{Dt} = \omega \mathbf{e}_3 \times (\omega \mathbf{e}_3 \times \mathbf{x}) = -\omega^2 \mathbf{r} \qquad (10.6.10)$$

where $\mathbf{r} = x\mathbf{e}_1 + y\mathbf{e}_2$.

Using (10.6.10) and the fact that $\mathbf{b} = -g\mathbf{e}_3$, Euler's equation of motion (10.6.2) reduces to

$$\nabla p = \rho(\omega^2 \mathbf{r} - g\mathbf{e}_3) \qquad (10.6.11)$$

Taking dot product with $\hat{\mathbf{r}} = \mathbf{r}/r$ (where $r = |\mathbf{r}|$) on both sides of this equation and noting that $\nabla p \cdot \hat{\mathbf{r}} = \partial p / \partial r$, we obtain

$$\frac{\partial p}{\partial r} = \rho \omega^2 r \qquad (10.6.12)$$

On the other hand, if we take dot product with $\mathbf{e}_3$ on both sides of (10.6.11), we find

$$\frac{\partial p}{\partial z} = -\rho g \qquad (10.6.13)$$

10.6 EULER'S EQUATION OF MOTION

Due to the axial symmetry of the problem, we assume that p depends only on r and z so that

$$dp = \frac{\partial p}{\partial r} dr + \frac{\partial p}{\partial z} dz \qquad (10.6.14)$$

Using (10.6.12) and (10.6.13) in (10.6.14) and integrating the resulting equation, we obtain

$$p = \tfrac{1}{2}\rho\omega^2 r^2 - \rho g z + C \qquad (10.6.15)$$

where C is a constant of integration. At the point where the z axis meets the upper surface of the liquid, we have $p = p_a$, where p_a is the atmospheric pressure. If this point is at a height h above the origin, (10.6.15) gives $C = p_a + \rho g h$. Putting this value of C into (10.6.15) we obtain the pressure distribution in the liquid as

$$p = p_a + \tfrac{1}{2}\omega^2 r^2 - g\rho(z - h) \qquad (10.6.16)$$

For $p = p_a$, equation (10.6.16) becomes

$$z = h - \frac{\omega^2}{2g} r^2 \qquad (10.6.17)$$

This is the equation of the upper surface of the rotating liquid; the surface is a paraboloid of revolution with the z axis as its axis and vertex downward. ∎

EXAMPLE 10.6.3 Treating the atmosphere around the earth as a nonviscous perfect gas at rest under constant gravitational field, show that the pressure at height z above the ground level is given by

$$p = p_1 \exp\left[-\frac{g}{R} \int_0^z \frac{dz}{T} \right] \qquad (10.6.18)$$

where p_1 is the pressure at the earth's surface. If $T = $ constant $= T_0$, deduce that

$$\frac{p}{p_1} = \frac{\rho}{\rho_1} = \exp\left(-\frac{gz}{RT_0} \right) \qquad (10.6.19)$$

where ρ_1 is the density at the earth's surface.

Solution With the positive z axis vertical upward, we have $\mathbf{b} = -g\mathbf{e}_3$, since the gravitational force is the only body force. The equation of equilibrium (10.6.3) now yields

$$\frac{\partial p}{\partial x} = 0, \quad \frac{\partial p}{\partial y} = 0, \quad \frac{\partial p}{\partial z} = -g\rho$$

482 10 EQUATIONS OF FLUID MECHANICS

The first two equations show that p is independent of x and y. Hence the last equation becomes

$$\frac{dp}{dz} = -\rho g \qquad (10.6.20)$$

Since the fluid under consideration is a perfect gas, we have $p = \rho RT$ by the equation state (10.4.15), and equation (10.6.20) can be rewritten as

$$\frac{dp}{dz} + \frac{g}{R}\frac{p}{T} = 0 \qquad (10.6.21)$$

Integration of this equation yields the pressure at a height z above the ground level ($z = 0$) as

$$p = A \exp\left(-\frac{g}{R}\int_0^z \frac{dz}{T}\right) \qquad (10.6.22)$$

where A is a constant of integration. Using the conditions $p = p_1$ at $z = 0$, we find $A = p_1$ and (10.6.18) follows immediately.

If $T = T_0$, then (10.6.18) becomes

$$p = p_1 \exp\left(-\frac{gz}{RT_0}\right) \qquad (10.6.23)$$

In view of the equation of state $p = \rho RT = \rho RT_0$, (10.6.23) gives

$$\rho = \frac{p_1}{RT_0} \exp\left(-\frac{gz}{RT_0}\right) \qquad (10.6.24)$$

The equation $p = \rho RT$ also gives $p_1 = \rho_1 RT_0$. Hence (10.6.24) becomes

$$\rho = \rho_1 \exp\left(-\frac{gz}{RT_0}\right) \qquad (10.6.25)$$

Expressions (10.6.23) and (10.6.25) yield (10.6.19). ∎

EXAMPLE 10.6.4 In a part of the atmosphere lying nearer the earth's surface, the temperature distribution is known to be of the form $T = T_1 - \alpha z$, where T_1 is the temperature at the ground level, z is the vertical distance from the ground level and α is a constant. Assuming that the atmosphere is a nonviscous perfect gas at rest under the gravitational field, prove the following:

(i) $$\frac{p}{p_1} = \left(\frac{T}{T_1}\right)^{(g/R\alpha)} \qquad (10.6.26)$$

(ii) $$\frac{\rho}{\rho_1} = \left(\frac{T}{T_1}\right)^{(g/R\alpha)-1} \tag{10.6.27}$$

(ii) $$\frac{\rho}{\rho_1} = \left(\frac{p}{p_1}\right)^{1-(R\alpha/g)} \tag{10.6.28}$$

where p_1 and ρ_1 are the pressure and the density at the ground level.

Solution Recall that (10.6.18) is the equation for the pressure p in a nonviscous perfect gas at rest under the gravitational field. Setting $T = T_1 - \alpha z$ in this equation and simplifying it with the condition, $T = T_1$ at $z = 0$, we get

$$p = p_1\left(1 - \frac{\alpha z}{T_1}\right)^{(g/R\alpha)} = p_1\left(\frac{T}{T_1}\right)^{(g/R\alpha)}$$

which is (10.6.26).

For $z = 0$, the equation of state $p = \rho RT$ yields $RT_1 = p_1/\rho_1$. Hence the equation $p = \rho RT$ can be rewritten as

$$p = \frac{\rho R T T_1}{T_1} = \frac{\rho p_1 T}{\rho_1 T_1}$$

so that

$$\frac{p}{p_1} = \frac{\rho}{\rho_1}\left(\frac{T}{T_1}\right) \tag{10.6.29}$$

Elimination of ρ/ρ_1 from (10.6.26) and (10.6.29) gives the result (10.6.27). Also, (10.6.28) follows by the elimination of T/T_1 from (10.6.26) and (10.6.29). ■

EXAMPLE 10.6.5 For an adiabatic flow of a nonviscous perfect gas with no heat supply, show that

$$\rho c_p \frac{DT}{Dt} = \frac{Dp}{Dt} = \frac{\gamma p}{\rho}\frac{D\rho}{Dt} \tag{10.6.30}$$

Deduce that

(i) in the absence of body force,

$$\frac{D}{Dt}\left(c_p T + \frac{1}{2}v^2\right) = \frac{1}{\rho}\frac{\partial p}{\partial t} \tag{10.6.31}$$

(ii) in the absence of body force and for steady flow,

$$\frac{\gamma}{\gamma - 1}\frac{p}{\rho} + \frac{1}{2}v^2 = \text{constant} \tag{10.6.32}$$

484   10 EQUATIONS OF FLUID MECHANICS

Solution   Recall that for a flow of a nonviscous perfect gas, the energy equation has been expressed in two alternative forms as given by equations (10.4.31) and (10.4.32). In the case of an adiabatic flow with no heat supply, these equations become

$$c_v \frac{DT}{Dt} = \frac{p}{\rho^2} \frac{D\rho}{Dt} \qquad (10.6.33)$$

$$c_p \frac{DT}{Dt} = \frac{1}{\rho} \frac{Dp}{Dt} \qquad (10.6.34)$$

Since $p = \rho RT$ for a perfect gas, we find

$$\frac{DT}{Dt} = \frac{1}{R} \frac{D}{Dt}\left(\frac{p}{\rho}\right) = \frac{1}{\rho R}\left[\frac{Dp}{Dt} - \frac{p}{\rho}\frac{D\rho}{Dt}\right]$$

Using this in (10.6.33) and taking note of (10.4.18) we get the equation

$$\frac{p\gamma}{\rho} \frac{D\rho}{Dt} = \frac{Dp}{Dt} \qquad (10.6.35)$$

Equations (10.6.34) and (10.6.35) together constitute equation (10.6.30).
In the absence of body force, Euler's equation of motion (10.6.2) gives

$$\mathbf{v} \cdot \frac{D\mathbf{v}}{Dt} = -\mathbf{v} \cdot \left(\frac{1}{\rho}\nabla p\right)$$

so that

$$\frac{D}{Dt}(v^2) = 2\mathbf{v} \cdot \frac{D\mathbf{v}}{Dt} = -2\mathbf{v} \cdot \left(\frac{1}{\rho}\nabla p\right) \qquad (10.6.36)$$

Also,

$$\frac{Dp}{Dt} = \frac{\partial p}{\partial t} + (\mathbf{v} \cdot \nabla)p \qquad (10.6.37)$$

Using (10.6.36) and (10.6.37) in (10.6.34), we obtain (10.6.31).
For steady flow, $\partial p/\partial t = 0$, and equation (10.6.31) yields

$$c_p T + \tfrac{1}{2}v^2 = \text{constant} \qquad (10.6.38)$$

Since $c_p = \gamma R/(\gamma - 1)$ by (10.4.20b) and $p = \rho RT$, we get

$$c_p T = \frac{\gamma}{\gamma - 1} \frac{p}{\rho} \qquad (10.6.39)$$

Results (10.6.38) and (10.6.39) together yield (10.6.32).   ∎

EXAMPLE 10.6.6 Show that the equation of motion of a nonviscous fluid can be expressed in terms of $\mathbf{x}^0$ and t as follows:

$$\nabla^0 p = \rho \mathbf{F}^T \left[\mathbf{b} - \frac{D^2 \mathbf{x}}{Dt^2} \right] \quad (10.6.40)$$

Solution Treating p as a function of x_i^0 and t, we have $p_{,i} = p_{,k} x_{k;i}$. That is,

$$\nabla^0 p = \mathbf{F}^T (\nabla p) \quad (10.6.41)$$

Substituting for ∇p from Euler's equation (10.6.2) in (10.6.41) and recalling that $D\mathbf{v}/Dt = D^2\mathbf{x}/Dt^2$, we get the equation (10.6.40). ∎

Note: Equation (10.6.40) is the equation of motion for a nonviscous fluid in the *material (Lagrangian) form*.

10.7
EQUATION OF MOTION OF AN ELASTIC FLUID

We now specialize Euler's equation of motion (10.6.2) to the case of an *elastic fluid* and analyze some consequences of it. Recall from Section 10.4 that an elastic fluid is a nonviscous compressible fluid for which the kinetic equation of state is given by (10.4.21a) or (10.4.21b).

Let us set

$$P = \int \frac{1}{\rho} dp \quad (10.7.1)$$

Using (10.4.21b), that is, $\rho = \rho(p)$, we find by virtue of the result (ii) of Example 3.4.1 that

$$\nabla P = \frac{1}{\rho} \nabla p \quad (10.7.2)$$

Consequently, Euler's equation of motion (10.6.2) becomes

$$\frac{D\mathbf{v}}{Dt} = -\nabla P + \mathbf{b} \quad (10.7.3)$$

This is the equation of motion for an elastic fluid. Evidently, this equation holds for a fluid of constant density also; in this case, $P = p/\rho$, where ρ is the constant density.

Equation (10.7.3) can be expressed in an alternative form. Since $p = p(\rho)$ for an elastic fluid, by (10.4.21a), we have

$$\nabla p = \left(\frac{dp}{d\rho} \right) \nabla \rho \quad (10.7.4)$$

Setting

$$c_s = \left(\frac{dp}{d\rho}\right)^{1/2} \tag{10.7.5}$$

we find from (10.7.2) and (10.7.4) that

$$\nabla P = \frac{c_s^2}{\rho}(\nabla \rho) \tag{10.7.6}$$

The physical meaning of c_s, which is evidently *not* a constant, will be given in Example 10.7.5.

Using (10.7.6) in (10.7.3), we get the following alternative version of (10.7.3):

$$\frac{D\mathbf{v}}{Dt} = -\frac{c_s^2}{\rho}(\nabla \rho) + \mathbf{b} \tag{10.7.7}$$

In many practical problems, the body force $\mathbf{b}$ is conservative; that is, $\mathbf{b} = -\nabla \chi$ for some scalar function χ (called the *potential of* $\mathbf{b}$). Equations (10.7.3) and (10.7.7) then become

$$\frac{D\mathbf{v}}{Dt} = -\nabla(P + \chi) \tag{10.7.8}$$

$$= -\frac{c_s^2}{\rho}(\nabla \rho) - \nabla \chi \tag{10.7.9}$$

10.7.1 CIRCULATION THEOREM

From equation (10.7.8) we note that the acceleration is the gradient of a scalar function. Hence, by Kelvin's circulation theorem proven in Section 6.6, it readily follows that the motion for which (10.7.8) is the governing differential equation is circulation preserving. Thus, *every motion of an elastic fluid under conservative body force is circulation preserving*. Consequently, all the properties of circulation-preserving motions, like the permanence of irrotational motion and transportation of vortex lines, as discussed in Section 6.6, hold for such a motion.

10.7.2 VORTICITY EQUATION

For a continuum, the vorticity vector $\mathbf{w}$ is governed by Beltrami's vorticity equation (8.2.16). For an elastic fluid, we find from (10.7.3) that

$$\operatorname{curl}\left(\frac{D\mathbf{v}}{Dt}\right) = \operatorname{curl} \mathbf{b}$$

10.7 EQUATION OF MOTION OF AN ELASTIC FLUID

Substituting this in (8.2.16), we obtain the equation

$$\frac{D}{Dt}\left(\frac{\mathbf{w}}{\rho}\right) = \left(\frac{\mathbf{w}}{\rho} \cdot \nabla\right)\mathbf{v} + \frac{1}{\rho}\operatorname{curl}\mathbf{b} \qquad (10.7.10)$$

This is the *vorticity equation for an elastic fluid*. This equation is by Helmholtz (1868) and Nanson (1874). If the external force is conservative, the second term on the righthand side of the equation vanishes identically.

10.7.3 EQUATIONS OF EQUILIBRIUM

For an elastic fluid at rest (static equilibrium) or in uniform motion, equations (10.7.3) and (10.7.7) become the *equations of equilibrium*:

$$\nabla P = \mathbf{b} \qquad (10.7.11)$$

$$\frac{c_s^2}{\rho}(\nabla \rho) = \mathbf{b} \qquad (10.7.12)$$

For $\mathbf{b} = -\nabla \chi$, equation (10.7.11) yields

$$P + \chi \equiv \int \frac{dp}{\rho} + \chi = \text{constant} \qquad (10.7.13)$$

EXAMPLE 10.7.1 For an elastic fluid in static equilibrium under the earth's gravitational field (taken as constant), show that

$$\int \frac{dp}{\rho} + gz = \text{constant} \qquad (10.7.14)$$

where z is the height above the surface of the earth. Hence deduce the following.

(i) For an isothermal perfect gas,

$$gz + \frac{p}{\rho}\log p = \text{constant} \qquad (10.7.15)$$

(ii) For an isentropic perfect gas,

$$gz + \frac{\gamma}{(\gamma - 1)}\frac{p}{\rho} = \text{constant} \qquad (10.7.16)$$

Solution Let us choose the positive z axis vertically upward from the earth's surface. Then, if gravity is the only body force that is taken as constant, we have $\mathbf{b} = -g\mathbf{e}_3 = -\nabla(gz)$, so that $\chi = gz$ is the potential of $\mathbf{b}$. Substituting $\chi = gz$ in the equation of equilibrium (10.7.13), we readily get (10.7.14).

10 EQUATIONS OF FLUID MECHANICS

For an isothermal perfect gas, we have by (10.4.16), $(p/\rho) = \text{constant} = A$, say, so that

$$\int \frac{dp}{\rho} = A \int \frac{dp}{p} = A \log p + \text{constant} = \frac{p}{\rho} \log p + \text{constant} \quad (10.7.17)$$

Using this in (10.7.14), we obtain (10.7.15).

For an isentropic perfect gas, we have, by (10.4.17a), $p = \beta \rho^\gamma$, where β and $\gamma (>1)$ are constants, so that

$$\int \frac{dp}{\rho} = \beta \int \frac{1}{\rho} \gamma \rho^{\gamma-1} \, d\rho = \beta \gamma \frac{\rho^{\gamma-1}}{(\gamma-1)} + \text{constant} = \frac{\gamma}{(\gamma-1)} \frac{p}{\rho} + \text{constant} \quad (10.7.18)$$

The use of this result in (10.7.14) gives (10.7.16). ∎

EXAMPLE 10.7.2 A vertical rectangular plate of width a and height b is exposed to an atmosphere that is an isothermal perfect gas in equilibrium under the earth's gravitational field (which is taken as constant). Show that the magnitude of the total force acting on a face of the plate is given by

$$f = \frac{RTa}{g} p(z_0) \left[1 - \exp\left(-\frac{g}{RT} b\right) \right] \quad (10.7.19)$$

where $p(z_0)$ is the pressure at the lower edge of the plate, and T is the constant temperature.

Solution Since the atmosphere to which the plate is exposed is an isothermal perfect gas in equilibrium under the earth's gravitational field, equation (10.7.15) holds. Making use of $p = \rho RT$ in (10.7.15), we find

$$gz + RT \log p = C \quad (10.7.20)$$

where C is a constant. If the lower edge of the plate is at a height z_0 above the earth's surface, see Figure 10.6, (in the particular case, z_0 can be 0), equation (10.7.20) gives

$$C = gz_0 + RT \log p(z_0)$$

Substituting this value of C into (10.7.20) and rewriting the resulting equation yields

$$p = p(z_0) \exp\left[-\frac{g}{RT}(z - z_0)\right] \quad (10.7.21)$$

The magnitude of the total force exerted on a face of the plate due to the pressure p is given by

$$f = \int_S p \, dS \quad (10.7.22)$$

10.7 EQUATION OF MOTION OF AN ELASTIC FLUID

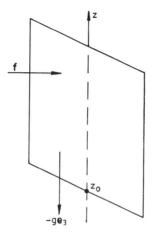

Figure 10.6. Example 10.7.2.

where S is the surface area of a face of the plate. Substituting for p from (10.7.21) into (10.7.22) and noting that p depends only on the vertical distance (as is evident from (10.7.21)), we obtain

$$f = \int_{z_0}^{z_0+b} ap(z)\, dz = \frac{aRT}{g} p(z_0)\left[1 - \exp\left(-\frac{gb}{RT}\right)\right]$$

which is the desired result (10.7.19). ■

EXAMPLE 10.7.3 In a steady flow of an elastic fluid, show that $\mathbf{v}$ satisfies the equation

$$c_s^2(\operatorname{div} \mathbf{v}) + \mathbf{b} \cdot \mathbf{v} = \mathbf{v} \cdot (\mathbf{v} \cdot \nabla)\mathbf{v} = \mathbf{v} \cdot \nabla(\tfrac{1}{2}v^2) \quad (10.7.23)$$

Deduce that, for potential flow, the velocity potential ϕ satisfies the equation

$$(c_s^2 \delta_{km} - \phi_{,k}\phi_{,m})\phi_{,km} + b_k v_k = 0 \quad (10.7.24)$$

Equations (10.7.23) and (10.7.24) are usually referred to as *gas dynamical equations*.

Solution Recalling that

$$\frac{D}{Dt} = \frac{\partial}{\partial t} + (\mathbf{v} \cdot \nabla)$$

the equation of motion (10.7.7) for an elastic fluid can be rewritten as

$$\frac{\partial \mathbf{v}}{\partial t} + (\mathbf{v} \cdot \nabla)\mathbf{v} = -\frac{c_s^2}{\rho}(\nabla \rho) + \mathbf{b} \quad (10.7.25)$$

490 **10 EQUATIONS OF FLUID MECHANICS**

Similarly, the equation of continuity (10.4.1) can be rewritten as (see equation (8.2.5a))

$$\frac{\partial \rho}{\partial t} + \mathbf{v} \cdot \nabla \rho + \rho \operatorname{div} \mathbf{v} = 0 \qquad (10.7.26)$$

For steady flow, $\partial \mathbf{v}/\partial t = \mathbf{0}$ and $\partial \rho/\partial t = 0$. In this case (10.7.25) and (10.7.26) give

$$\mathbf{v} \cdot (\mathbf{v} \cdot \nabla)\mathbf{v} = c_s^2 \operatorname{div} \mathbf{v} + \mathbf{b} \cdot \mathbf{v} \qquad (10.7.27)$$

By use of identity (3.4.29), we find that

$$\mathbf{v} \cdot (\mathbf{v} \cdot \nabla)\mathbf{v} = \tfrac{1}{2}\nabla(v^2) \cdot \mathbf{v} \qquad (10.7.28)$$

Identity (10.7.28) together with equation (10.7.27) yields the result (10.7.23).

In the suffix notation, equation (10.7.27) reads

$$v_k v_m v_{k,m} = c_s^2 v_{k,k} + b_k v_k$$

which can be rewritten as

$$(c_s^2 \delta_{km} - v_k v_m) v_{k,m} + b_k v_k = 0 \qquad (10.7.29)$$

For potential flow, we have $v_i = \phi_{,i}$, where ϕ is the velocity potential. Equation (10.7.29) then reduces to (10.7.24). ∎

EXAMPLE 10.7.4 Show that for an elastic fluid moving under conservative body force the equation of motion can be expressed in the form

$$\mathbf{F}^T \mathbf{v} = \mathbf{v}^0 - \nabla^0 \left\{ \int_0^t \left(\int \frac{dp}{\rho} + \chi - \tfrac{1}{2} v^2 \right) dt \right\} \qquad (10.7.30)$$

where $\mathbf{v}^0$ is the initial velocity. This equation is known as *Weber's equation*, after Weber who derived it in 1868.

Solution The equation of motion of an elastic fluid in the presence of conservative body force is given by (10.7.8). Operating by the tensor $\mathbf{F}^T$ on both sides of this equation, we obtain

$$\mathbf{F}^T \nabla(P + \chi) = -\mathbf{F}^T \frac{D\mathbf{v}}{Dt} = -\frac{D}{Dt}(\mathbf{F}^T \mathbf{v}) + \frac{D\mathbf{F}^T}{DT} \mathbf{v} \qquad (10.7.31)$$

But

$$\frac{D\mathbf{F}^T}{Dt} \mathbf{v} = \{(\nabla \mathbf{v})\mathbf{F}\}^T \mathbf{v}$$

by (6.2.44).

$$= \mathbf{F}^T (\nabla \mathbf{v})^T \mathbf{v} = \tfrac{1}{2} \mathbf{F}^T \nabla(\mathbf{v} \cdot \mathbf{v}) \qquad (10.7.32)$$

10.7 EQUATION OF MOTION OF AN ELASTIC FLUID

Substituting this into (10.7.31) and noting that $\mathbf{F}^T(\nabla\phi) = \nabla^0\phi$ for any function ϕ, we get the equation

$$\frac{D}{Dt}(\mathbf{F}^T\mathbf{v}) = -\nabla^0(P + \chi - \tfrac{1}{2}v^2) \tag{10.7.33}$$

Integrating this equation with respect to t from 0 to t, we find

$$[\mathbf{F}^T\mathbf{v}]_0^t = -\nabla^0\left\{\int_0^t (P + \chi - \tfrac{1}{2}v^2)\, dt\right\} \tag{10.7.34}$$

If we use the fact that $\mathbf{F} = \mathbf{I}$ and $\mathbf{v} = \mathbf{v}^0$ at $t = 0$, and note that $P = \int dp/\rho$, equation (10.7.34) reduces to equation (10.7.30). ∎

EXAMPLE 10.7.5 In the linearized case of an elastic fluid motion in the absence of body forces, show that ρ, p and div $\mathbf{v}$ all satisfy the same wave equation

$$c_s^2 \nabla^2 f = \frac{\partial^2 f}{\partial t^2} \tag{10.7.35}$$

Solution In the absence of body force, equations (10.7.2), (10.7.3) and (10.7.7) yield

$$\rho \frac{D\mathbf{v}}{Dt} = -\nabla p = -c_s^2 \nabla\rho \tag{10.7.36}$$

In the linearized case, we make the approximation

$$\rho \frac{D\mathbf{v}}{Dt} = \rho\left[\frac{\partial \mathbf{v}}{\partial t} + (\mathbf{v}\cdot\nabla)\mathbf{v}\right] \approx \rho_0 \frac{\partial \mathbf{v}}{\partial t} \tag{10.7.37}$$

Hence (10.7.36) becomes

$$\rho_0 \frac{\partial \mathbf{v}}{\partial t} = -\nabla p = -c_s^2 \nabla\rho \tag{10.7.38}$$

so that

$$\rho_0 \frac{\partial}{\partial t}(\text{div } \mathbf{v}) = -\nabla^2 p = -c_s^2 \nabla^2\rho \tag{10.7.39}$$

In obtaining the term on the far right, we have neglected $(\nabla c_s^2)\cdot\nabla\rho$.

In the linearized case, the equation of continuity (10.7.26) becomes

$$\frac{\partial \rho}{\partial t} + \rho_0 \text{ div } \mathbf{v} = 0 \tag{10.7.40}$$

Since $\rho = \rho(p)$ for an elastic fluid, (10.7.40) can be rewritten as

$$\frac{d\rho}{dp}\left(\frac{\partial p}{\partial t}\right) + \rho_0 \operatorname{div} \mathbf{v} = 0 \qquad (10.7.41)$$

or, on using (10.7.5),

$$\frac{1}{c_s^2}\frac{\partial p}{\partial t} + \rho_0 \operatorname{div} \mathbf{v} = 0. \qquad (10.7.42)$$

Substituting for div **v** from (10.7.40) and (10.7.42) in (10.7.39) we get

$$\frac{\partial^2 \rho}{\partial t^2} = \frac{1}{c_s^2}\frac{\partial^2 p}{\partial t^2} = \nabla^2 p = c_s^2 \nabla^2 \rho \qquad (10.7.43)$$

These show that both ρ and p satisfy the wave equation (10.7.35).

Taking the Laplacian of (10.7.40) and substituting for $\nabla^2 \rho$ from (10.7.39) in the resulting equation, we find

$$c_s^2 \nabla^2 (\operatorname{div} \mathbf{v}) = \frac{\partial^2}{\partial t^2}(\operatorname{div} \mathbf{v}) \qquad (10.7.44)$$

This shows that (div **v**) also satisfies the wave equation (10.7.35). ∎

Note: The wave equation (10.7.35) governing p, ρ and div **v** plays a fundamental role in the field of acoustics; it is called the *acoustic wave equation* for *small disturbances*. The speed $c_s = (dp/d\rho)^{1/2}$ associated with this equation represents the *local speed of sound*. It should be noted that c_s is *not* a constant, in general. For an isothermal nonviscous perfect gas (for which $(p/\rho) = $ constant), we find that $c_s = (p/\rho)^{1/2}$, which is a constant. For an isentropic, nonviscous perfect gas (for which $p = \beta\rho^\gamma$) we obtain $c_s = (\gamma p/\rho)^{1/2}$. Experiments on common gases show that $c_s = (\gamma p/\rho)^{1/2}$ is a good approximation for computing the local speed of sound. With $\gamma = 1.4$, the value of c_s for air at 0°C is known to be 331.3 m/s.

10.8

BERNOULLI'S EQUATIONS

It has been shown that for an elastic fluid moving under conservative body force the equation of motion is given by (10.7.8). Using relation (10.6.4), equation (10.7.8) can be rewritten as

$$\frac{\partial \mathbf{v}}{\partial t} + \mathbf{w} \times \mathbf{v} = -\nabla H \qquad (10.8.1)$$

where

$$H = P + \chi + \tfrac{1}{2}v^2 \qquad (10.8.2)$$

10.8 BERNOULLI'S EQUATIONS

It is possible to integrate the equation of motion (10.8.1) in three important special cases, considered next.

10.8.1 CASE I

Suppose the flow is of potential kind; that is, $\mathbf{v} = \nabla\phi$ for some scalar ϕ. Then $\mathbf{w} = \mathbf{0}$, and equation (10.8.1) becomes

$$\nabla\left(H + \frac{\partial\phi}{\partial t}\right) = 0 \tag{10.8.3}$$

This equation implies that

$$H + \frac{\partial\phi}{\partial t} = f(t) \tag{10.8.4}$$

where $f(t)$ is an arbitrary function of t. Substitution for H from (10.8.2) in (10.8.4) gives

$$P + \frac{1}{2}(\nabla\phi)^2 + \chi + \frac{\partial\phi}{\partial t} = f(t) \tag{10.8.5}$$

10.8.2 CASE II

Suppose the flow is steady, and the stream lines and vortex lines are noncoincident; that is, $\partial/\partial t \equiv 0$ and $\mathbf{v} \times \mathbf{w} \neq \mathbf{0}$. Then equation (10.8.1) becomes

$$\mathbf{w} \times \mathbf{v} = -\nabla H \tag{10.8.6}$$

We note that ∇H is a vector normal to the surface of constant H, see Section 3.4. Also, $\mathbf{v} \times \mathbf{w}$ is a vector perpendicular to both $\mathbf{v}$ and $\mathbf{w}$. Hence, from equation (10.8.6) it follows that $\mathbf{v}$ and $\mathbf{w}$ are tangential to a surface of constant H. Since $\mathbf{v}$ and $\mathbf{w}$ are tangential to the stream lines and vortex lines, respectively, it follows that these lines lie on a surface of constant H. In other words, H is constant along stream lines and vortex lines; that is, by (10.8.2),

$$P + \tfrac{1}{2}v^2 + \chi = \text{constant} \tag{10.8.7}$$

along stream lines and vortex lines.

A surface of constant H covered by a network of stream lines and vortex lines is known as *Lamb surface*, after H. Lamb (1878); this is illustrated in Figure 10.7.

10.8.3 CASE III

Suppose the flow is steady and is either an irrotational or a Beltrami flow. (A flow is called a *Beltrami flow* if the velocity vector is a Beltrami vector, as defined in Example 3.4.7.) Then $\partial/\partial t = 0$, and either $\mathbf{w} = \mathbf{0}$ or

494 10 EQUATIONS OF FLUID MECHANICS

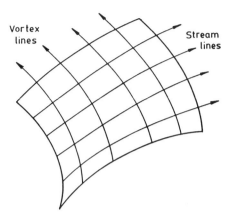

Figure 10.7. Lamb surface.

$\mathbf{v} \times \mathbf{w} = \mathbf{0}$. Then equation (10.8.1) gives $\nabla H = \mathbf{0}$. Since $\partial H/\partial t = 0$, it follows that

$$H \equiv P + \tfrac{1}{2}v^2 + \chi = \text{constant} \tag{10.8.8}$$

everywhere in the fluid.

Thus, under the assumed conditions, relations (10.8.5), (10.8.7) and (10.8.8) are integrals of the equation of motion of an elastic fluid. These relations are known as the *Bernoulli's equations*, after Daniel Bernoulli (1738). The function H, defined by (10.8.2), is known as the *Bernoulli function*.

Note: Since the Bernoulli's equations hold for an elastic fluid for which $\rho = \rho(p)$, they automatically hold in the special case of $\rho = $ constant as well.

EXAMPLE 10.8.1 For a certain flow of a nonviscous fluid of constant density under the earth's gravitational field, the velocity distribution is given by $\mathbf{v} = \nabla \phi$, where $\phi = x^3 - 3xy^2$. Find the pressure distribution.

Solution From the given $\mathbf{v}$, we find that curl $\mathbf{v} = \mathbf{0}$ and $\partial \mathbf{v}/\partial t = \mathbf{0}$. Further, since the body force is the gravitational force, it is conservative with $\chi = gz$, where z is measured vertically upward. Accordingly, the Bernoulli's equation (10.8.8) with $P = p/\rho$ holds; that is,

$$\frac{p}{\rho} + \frac{1}{2}v^2 + gz = C \tag{10.8.9}$$

where C is a constant.

From the given $\mathbf{v}$, we also find that

$$v_1 = \frac{\partial \phi}{\partial x} = 3(x^2 - y^2), \quad v_2 = \frac{\partial \phi}{\partial y} = -6xy, \quad v_3 = 0$$

10.8 BERNOULLI'S EQUATIONS

so that
$$v^2 = v_1^2 + v_2^2 = 9(x^2 + y^2)^2$$

Hence equation (10.8.9) becomes
$$\frac{p}{\rho} + \frac{9}{2}(x^2 + y^2)^2 + gz = C$$

From this result, it is evident that $C = p^0/\rho$, where p^0 is the pressure at the origin. Thus,
$$p = p^0 - \rho[\tfrac{9}{2}(x^2 + y^2)^2 + gz] \qquad (10.8.10)$$

is the required pressure distribution. ∎

EXAMPLE 10.8.2 A liquid flows out of a very large reservoir through a small opening (Figure 10.8). Assuming that the liquid is nonviscous and of constant density and that the flow is steady and irrotational, find the exit speed of the liquid jet. Assume that there is no external force apart from the gravitational force.

Solution For the given flow, the Bernoulli's equation (10.8.8) with $P = p/\rho$ and $\chi = gz$, where z is measured vertically upward, holds; that is,
$$\frac{p}{\rho} + \frac{1}{2}v^2 + gz = C \qquad (10.8.11)$$

where C is a constant.

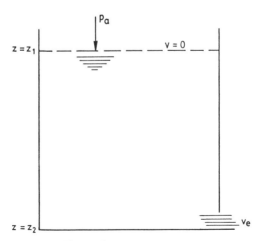

Figure 10.8. Example 10.8.2.

10 EQUATIONS OF FLUID MECHANICS

Since the reservoir is very large, at any given instant of time, the top surface of the liquid may be assumed to be at rest. Then at a point A on this surface (say, $z = z_1$) we have $v = 0$ and $p = p_a$, where p_a is the atmospheric pressure. Hence (10.8.11) gives

$$\frac{p_a}{\rho} + gz_1 = C \tag{10.8.12}$$

Also, at a point B on the surface of the jet (say, $z = z_2$) we have $v = v_e$ and $p = p_a$, where v_e is the exit speed of the jet. Hence (10.8.11) now gives

$$\frac{p_a}{\rho} + \frac{1}{2} v_e^2 + gz_2 = C \tag{10.8.13}$$

Relations (10.8.12) and (10.8.13) yield the following expression for the exit speed:

$$v_e = \sqrt{2g(z_1 - z_2)} \tag{10.8.14}$$

This expression is known as the *Torricelli's formula*, after Torricelli (1608-1647). We note that v_e is actually equal to the speed acquired by a body falling freely (under gravity) through a height $(z_1 - z_2)$. ∎

EXAMPLE 10.8.3 A liquid of constant density flows down under gravity in a long inclined pipe with a slowly tapering circular cross section. The length of the pipe is L, the angle of inclination is α and the radius of the upper end is N times the radius of the lower end, where $N > 1$. Also, the lower end is maintained at the atmospheric pressure p_a whereas the upper end is at pressure Mp_a, where M is a constant greater than 1. Assuming that the liquid is nonviscous and that the flow is steady and irrotational, find the exit speed of the liquid at the lower end.

Solution For the given flow, Bernoulli's equation (10.8.8) with $P = (p/\rho)$ and $\chi = gz$ holds, the positive z axis being taken upward as shown in Figure 10.9. Thus,

$$\frac{p}{\rho} + \frac{1}{2} v^2 + gz = C \tag{10.8.15}$$

where C is a constant.

Let v_1 be the speed of entry at the upper end ($z = L \sin \alpha$) and v_e be the speed of exit at the lower end ($z = 0$). Then, for $z = 0$ we have $v = v_e$ and $p = p_a$, and for $z = L \sin \alpha$ we have $v = v_1$ and $p = Mp_a$. Equation (10.8.15) therefore gives

$$\frac{1}{\rho} p_a + \frac{1}{2} v_e^2 = C \tag{10.8.16}$$

10.8 BERNOULLI'S EQUATIONS

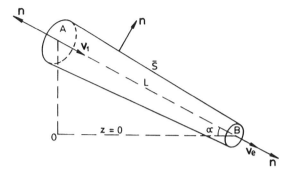

Figure 10.9. Example 10.8.3.

and

$$\frac{M}{\rho} p_a + \frac{1}{2} v_1^2 + gL \sin \alpha = C \tag{10.8.17}$$

so that

$$(M - 1)\frac{p_a}{\rho} + \frac{1}{2}(v_1^2 - v_e^2) + gL \sin \alpha = 0 \tag{10.8.18}$$

This does not give v_e unless v_1 is known. To find v_1, we use the equation of continuity, div $\mathbf{v} = 0$, which holds at every point of the liquid. If V is the volume of the liquid contained in the pipe (at any chosen instant of time), the equation of continuity yields

$$\int_V (\text{div } \mathbf{v}) \, dV = 0$$

By divergence theorem, this becomes

$$\int_S \mathbf{v} \cdot \mathbf{n} \, dS = 0 \tag{10.8.19}$$

where S is the surface enclosing V. We note that S is made up of the upper and lower end surfaces of the pipe, say, A and B, and the lateral surface of the pipe, say, $\bar{S}$. Since $\bar{S}$ is a solid surface that is at rest, we have $\mathbf{v} \cdot \mathbf{n} = 0$ on $\bar{S}$ by (10.5.7). Therefore equation (10.8.19) gives, on noting that $\mathbf{v} \cdot \mathbf{n} = -v_1$ on A and $\mathbf{v} \cdot \mathbf{n} = v_e$ on B,

$$-\int_A v_1 \, dS + \int_B v_e \, dS = 0 \tag{10.8.20}$$

We note that A and B are both circular areas such that if πa^2 is the area of B, then $\pi N^2 a^2$ is the area of A. If we take it that the speeds on A and B

498   **10 EQUATIONS OF FLUID MECHANICS**

are uniform, then (10.8.20) gives $(\pi N^2 a^2)v_1 = (\pi a^2)v_e$, so that

$$v_1 = \frac{1}{N^2} v_e \tag{10.8.21}$$

Substituting this back into (10.8.18), we get

$$v_e^2 = \frac{2N^4}{N^4 - 1}\left[(M-1)\frac{p_a}{\rho} + gL\sin\alpha\right] \tag{10.8.22}$$

which gives the exit speed at the lower end of the pipe. ∎

EXAMPLE 10.8.4 In a steady irrotational flow of a fluid of constant density under zero body force, show that the pressure is maximum at a stagnation point. (*A stagnation point* is a point at which the velocity is 0.)

Solution For a given flow, the Bernoulli's equation (10.8.8) with $P = p/\rho$ and $\chi = 0$, holds; that is,

$$\frac{p}{\rho} + \frac{1}{2}v^2 = C$$

where C is a constant. Consequently, since ρ is constant,

$$\frac{dp}{dv} = -\rho v, \qquad \frac{d^2p}{dv^2} = -\rho$$

Evidently, $dp/dv = 0$ for $v = 0$, and $d^2p/dv^2 < 0$ for all v. Hence p is maximum when $v = 0$. ∎

EXAMPLE 10.8.5 Consider the steady and irrotational flow of an elastic fluid for which the pressure-density relationship is $p = \beta\rho^\gamma$, where β and $\gamma(>1)$ are constants. Assuming that the body force is absent, show that

(i) $$\frac{1}{2}v^2 + \frac{\gamma}{(\gamma-1)}\frac{p}{\rho} = \text{constant} \tag{10.8.23}$$

(ii) $$v^2 = \frac{2\gamma}{(\gamma-1)}\frac{p}{\rho}\left[\left(\frac{p_0}{p}\right)^{(\gamma-1)/\gamma} - 1\right] \tag{10.8.24}$$

$$= \frac{2c_s^2}{\gamma-1}\left[\left(\frac{p_0}{p}\right)^{(\gamma-1)/\gamma} - 1\right] \tag{10.8.25}$$

(iii) $$p_0 = p\left[1 + \frac{1}{2}(\gamma-1)\frac{v^2}{c_s^2}\right]^{\gamma/(\gamma-1)} \tag{10.8.26}$$

where p_0 is the pressure at 0 speed.

10.8 BERNOULLI'S EQUATIONS

Solution (i) Under the given conditions, Bernoulli's equation (10.8.8) holds with $\chi = 0$; that is,

$$P + \tfrac{1}{2}v^2 = \text{constant} \qquad (10.8.27)$$

Also,

$$P \equiv \int \frac{dp}{\rho} = \beta\gamma \int \rho^{\gamma-2}\, d\rho$$

$$= \beta\gamma \frac{\rho^{\gamma-1}}{(\gamma-1)} + \text{constant}$$

$$= \frac{\gamma}{(\gamma-1)} \frac{p}{\rho} + \text{constant} \qquad (10.8.28)$$

Using this in (10.8.27), we readily get (10.8.23).

(ii) For $v = 0$, we have $p = p_0$. Also, the relation $p = \beta\rho^\gamma$ gives $\rho = ((1/\beta)p_0)^{(1/\gamma)}$ for $p = p_0$. Consequently, expression (10.8.23) yields

$$\frac{1}{2}v^2 + \frac{\gamma}{(\gamma-1)}\frac{p}{\rho} = \frac{\gamma}{(\gamma-1)} p_0 \left(\frac{\beta}{p_0}\right)^{1/\gamma} \qquad (10.8.29)$$

Substituting $\beta = p/\rho^\gamma$ on the righthand side of (10.8.29) and simplifying the resulting expression, we obtain (10.8.24).

Using the relation $p = \beta\rho^\gamma$, we find from (10.7.5) that

$$c_s^2 = \frac{d}{d\rho}(\beta\rho^\gamma) = \beta\gamma\rho^{\gamma-1} = \gamma\frac{p}{\rho} \qquad (10.8.30)$$

By use of (10.8.30), expression (10.8.24) immediately reduces to (10.8.25).

(iii) Expression (10.8.25) may be rewritten as

$$1 + \frac{(\gamma-1)}{2}\frac{v^2}{c_s^2} = \left(\frac{p_0}{p}\right)^{(\gamma-1)/\gamma}$$

so that

$$\frac{p_0}{p} = \left[1 + \frac{(\gamma-1)}{2}\frac{v^2}{c_s^2}\right]^{\gamma/(\gamma-1)}$$

which is (10.8.26). ∎

Note: In Example 10.6.5, expression (10.8.23) was obtained for a perfect gas with the use of the energy equation.

10.9
WATER WAVES

In Section 9.13 we considered Rayleigh waves that propagate near the surface of an elastic half-space. We now consider waves that propagate near the surface of a nonviscous, incompressible fluid, in particular, water. Water wave motions are of great importance; they range from waves generated by wind or solar heating in the oceans to flood waves in rivers, from waves caused by a moving ship in a channel to tsunami waves (tidal waves) generated by earthquakes, and from solitary waves on the surface of a canal caused by a disturbance to waves generated by underwater explosions, to mention only a few. We restrict ourselves to two simple cases, which are the starting points for the study of linear and nonlinear water waves.

We consider a body of water (nonviscous, incompressible fluid) occupying the region $-h < z < 0$ with the plane $z = -h$ as the *bottom boundary* and the plane $z = 0$ as the *upper boundary*, in the *undisturbed* (initial) state; see Figure 10.10. We suppose that the bottom boundary is a rigid solid surface and the upper boundary is the surface exposed to a constant atmospheric pressure p_a. We consider plane waves propagating in the x direction whose amplitude varies in the z direction, with the gravitational force as the only body force. Since the motion is supposed to start from rest, it is necessarily irrotational as a consequence of Kelvin's circulation theorem (see Section 10.7), and we take the velocity field to be the gradient of a potential $\phi = \phi(\mathbf{x}, t)$; that is, $\mathbf{v} = \nabla \phi$. The equation of continuity

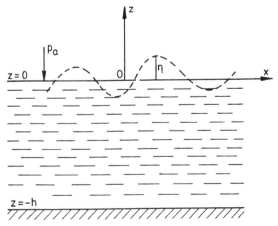

Figure 10.10. Water waves.

10.9 WATER WAVES

(10.4.5) then implies that ϕ is a harmonic function at every point of the fluid; that is,

$$\nabla^2 \phi \equiv \frac{\partial^2 \phi}{\partial x^2} + \frac{\partial^2 \phi}{\partial y^2} + \frac{\partial^2 \phi}{\partial z^2} = 0. \tag{10.9.1}$$

Since the gravitational force is the only body force, which acts in the negative z direction, we have $\mathbf{b} = -g\mathbf{e}_3 = -\nabla(gz)$, and Bernoulli's equation (10.8.5) yields the following equation for pressure p at every point of the fluid:

$$\frac{p}{\rho} + \frac{1}{2}(\nabla\phi)^2 + gz + \frac{\partial \phi}{\partial t} = f(t), \qquad t \geq 0 \tag{10.9.2}$$

Since the upper surface is exposed to the constant atmospheric pressure p_a, we have $p = p_a$ on this surface. In the undisturbed state, the equation of this surface is $z = 0$. After the motion is set up, let the equation of this surface, denoted S, be $z = \eta(x, t)$ where η is an *unknown* function of x and t that tends to 0 as $t \to 0$. The function $\eta(x, t)$ is referred to as the *surface elevation*. Thus, we have $p = p_a$ on S. Using (10.9.2), this condition reads

$$\frac{p_a}{\rho} + \frac{1}{2}(\nabla\phi)^2 + gz + \frac{\partial \phi}{\partial t} = f(t)$$

on S for $t \geq 0$. Absorbing (p_a/ρ) and $f(t)$ into $\partial\phi/\partial t$, this condition may be rewritten as

$$\frac{\partial \phi}{\partial t} + \frac{1}{2}(\nabla\phi)^2 + gz = 0 \tag{10.9.3}$$

on S for $t \geq 0$.

Since S is a boundary surface, it contains the same fluid particles for all time; that is, S is a material surface. Hence it follows from (10.9.3) that

$$\frac{D}{Dt}\left[\frac{\partial \phi}{\partial t} + \frac{1}{2}(\nabla\phi)^2 + gz\right] = 0$$

on S for $t \geq 0$, or equivalently

$$\left(\frac{\partial}{\partial t} + \nabla\phi \cdot \nabla\right)\left\{\frac{\partial \phi}{\partial t} + \frac{1}{2}(\nabla\phi)^2 + gz\right\}$$

$$\equiv \frac{\partial^2 \phi}{\partial t^2} + 2\nabla\phi \cdot \nabla\left(\frac{\partial \phi}{\partial t}\right) + \frac{1}{2}\nabla\phi \cdot \nabla(\nabla\phi)^2 + g\frac{\partial \phi}{\partial z} = 0 \tag{10.9.4}$$

on S for $t \geq 0$.

502   10 EQUATIONS OF FLUID MECHANICS

Since the lower boundary $z = -h$ is a rigid solid surface at rest, the condition to be satisfied at this boundary is, by (10.5.7),

$$\frac{\partial \phi}{\partial z} = 0 \qquad (10.9.5)$$

for $z = -h$, $t \geq 0$.

Thus, for the problem considered, Laplace's equation (10.9.1) serves as the partial differential equation satisfied by ϕ, and (10.9.4) and (10.9.5) serve as the upper and lower boundary conditions for ϕ. Once ϕ is determined, equation (10.9.3) gives the surface elevation, $z = \eta$.

Because of the presence of nonlinear terms in the boundary condition (10.9.4), the determination of ϕ in the general case is a difficult task. We restrict ourselves to two particular cases.

10.9.1   SMALL AMPLITUDE WAVES

We first consider the case where the motion is linear so that nonlinear terms in velocity components may be neglected. In this case no distinction is made between the initial and the current states of the upper boundary, and the boundary conditions (10.9.3) and (10.9.4) are taken in the linearized forms

$$\frac{\partial \phi}{\partial t} + g\eta = 0 \qquad (10.9.6)$$

for $z = 0$, $t > 0$;

$$\frac{\partial^2 \phi}{\partial t^2} + g\frac{\partial \phi}{\partial z} = 0 \qquad (10.9.7)$$

for $z = 0$, $t > 0$.

These conditions yield

$$\frac{\partial \phi}{\partial z} = \frac{\partial \eta}{\partial t} \qquad (10.9.8)$$

for $z = 0$, $t > 0$.

For a plane wave propagating in the x direction with frequency $\omega/2\pi$ and wavelength $2\pi/k$, we seek ϕ in the form

$$\phi = \text{Re } \Phi(z) \exp[i(\omega t - kx)] \qquad (10.9.9)$$

where $\Phi(z)$ is a function to be determined.

We find that ϕ given by (10.9.9) satisfies Laplace's equation (10.9.1) provided

$$\frac{d^2\Phi}{dz^2} - k^2\Phi = 0 \qquad (10.9.10)$$

The general solution of this ordinary linear differential equation is

$$\Phi(z) = Ae^{kz} + Be^{-kz} \qquad (10.9.11)$$

where A and B are arbitrary constants. Using the boundary condition (10.9.5) we get $Ae^{-kh} = Be^{kh}$. Consequently, solution (10.9.11) takes the form

$$\Phi = C\cosh k(z+h) \qquad (10.9.12)$$

where $C(=2Ae^{-kh} = 2Be^{kh})$ is an arbitrary constant so that the solution (10.9.9) becomes

$$\phi = \operatorname{Re} C\cosh k(z+h) \exp[i(\omega t - kx)] \qquad (10.9.13)$$

Using (10.9.6), we find

$$\eta = \operatorname{Re} a \exp[i(\omega t - kx)] \qquad (10.9.14)$$

where $a = (C\omega/ig)\cosh kh = \max|\eta|$ is the *amplitude*. Thus, the solution for ϕ has the final form

$$\phi = \operatorname{Re}\left(\frac{iag}{\omega}\right) \frac{\cosh k(z+h)}{\cosh kh} \exp[i(\omega t - kx)] \qquad (10.9.15)$$

Using (10.9.15) in (10.9.7), we obtain the following dispersion relation between the frequency and wave number:

$$\omega^2 = gk \tanh kh \qquad (10.9.16)$$

This *dispersion relation* can be rewritten in terms of the *phase velocity* $c = \omega/k$ as

$$c^2 = \frac{\omega^2}{k^2} = \frac{g}{k}\tanh(kh) \qquad (10.9.17)$$

This equation shows that the phase velocity c depends on the gravity g and depth h as well as the wavelength $2\pi/k$. Hence, the waves are dispersive in nature (like the Love waves considered in Section 9.13). This means that, as the time passes, the waves would disperse (spread out) into different groups such that each group would consist of waves having approximately the same wavelength. The quantity $d\omega/dk$ represents the velocity of such a group in the direction of propagation and is called *group velocity*, denoted $C(k)$. From (10.9.16) we find

$$C(k) = \frac{d\omega}{dk} = \left(\frac{g}{2\omega}\right)(\tanh kh + kh \operatorname{sech}^2 kh) \qquad (10.9.18)$$

which on using (10.9.17) becomes

$$C(k) = \frac{1}{2}c\left\{1 + \frac{2kh}{\sinh(2kh)}\right\} \qquad (10.9.19)$$

Evidently, the group velocity is different from the phase velocity.

In the case where the wavelength $2\pi/k$ is large compared with the depth, such waves are called *shallow water waves*, $kh \ll 1$ so that $\tanh kh \approx kh$ and $\sin 2kh \approx 2kh$. In such a situation, results (10.9.17) and (10.9.19) yield

$$C(k) \approx c \approx \sqrt{gh} \qquad (10.9.20)$$

Thus, shallow water waves are nondispersive, and their speed varies as the square root of the depth.

In the other limiting case where the wavelength is very small compared with the depth, such waves are called *deep water waves*, $kh \gg 1$. In the limit as $kh \to \infty$, $[\cosh k(z + h)]/(\cosh kh) \to e^{kz}$, and the corresponding solutions for ϕ and η become

$$\phi = \text{Re}\left(\frac{iag}{\omega}\right)\exp[kz + i(\omega t - kx)] = \left(\frac{ag}{\omega}\right)e^{kz}\sin(kx - \omega t) \qquad (10.9.21)$$

$$\eta = \text{Re}\, a \exp[i(\omega t - kx)] = a\cos(kx - \omega t) \qquad (10.9.22)$$

Results (10.9.16), (10.9.17) and (10.9.19) yield

$$\omega^2 = gk \qquad (10.9.23a)$$

$$c = (g/k)^{1/2} = (g\lambda/2\pi)^{1/2} \qquad (10.9.23b)$$

$$C(k) = \tfrac{1}{2}c \qquad (10.9.23c)$$

Thus, deep water waves continue to be dispersive and their phase velocity now is proportional to the square root of their wavelengths. Also, the group velocity is equal to one-half of the phase velocity.

10.9.2. FINITE AMPLITUDE WAVES (STOKES'S WAVES)

We now consider the case where the motion is nonlinear and the amplitude is not small. Let us recall (10.9.3) and (10.9.4) and write them for ready reference in the form

$$\eta = -\frac{1}{g}\left[\phi_t + \frac{1}{2}(\nabla\phi)^2\right]_{z=\eta} \qquad (10.9.24)$$

$$[\phi_{tt} + g\phi_z]_{z=\eta} + 2[\nabla\phi \cdot \nabla\phi_t]_{z=\eta} + \tfrac{1}{2}[\nabla\phi \cdot \nabla(\nabla\phi)^2]_{z=\eta} = 0 \qquad (10.9.25)$$

where, for simplicity, we have written ϕ_t for $\partial\phi/\partial t$, ϕ_z for $\partial\phi/\partial z$, ϕ_{tt} for $\partial^2\phi/\partial t^2$, etc.

A systematic procedure can be employed to rewrite these boundary conditions by using Taylor's series expansions of the potential ϕ and its derivatives in the typical form

$$\phi(x, y, z = \eta, t) = [\phi]_{z=0} + \eta[\phi_z]_{z=0} + \tfrac{1}{2}\eta^2[\phi_{zz}]_{z=0} + \cdots \quad (10.9.26)$$

$$\phi_z(x, y, z = \eta, t) = [\phi_z]_{z=0} + \eta[\phi_{zz}]_{z=0} + \tfrac{1}{2}\eta^2[\phi_{zzz}]_{z=0} + \cdots \quad (10.9.27)$$

Substituting these and similar Taylor's expansions into (10.9.24) gives

$$\eta = -\frac{1}{g}\left[\phi_t + \frac{1}{2}(\nabla\phi)^2\right]_{z=0} + \eta\left[-\frac{1}{g}\left\{\phi_t + \frac{1}{2}(\nabla\phi)^2\right\}_z\right]_{z=0} + \cdots$$

$$= -\frac{1}{g}\left[\phi_t + \frac{1}{2}(\nabla\phi)^2\right]_{z=0} + \frac{1}{g^2}\left[\left\{\phi_t + \frac{1}{2}(\nabla\phi)^2\right\}\left\{\phi_t + \frac{1}{2}(\nabla\phi)^2\right\}_z\right]_{z=0} + \cdots$$

$$= -\frac{1}{g}\left[\phi_t + \frac{1}{2}(\nabla\phi)^2 - \frac{1}{g}\phi_t\phi_{zt}\right]_{z=0} + O(\phi^3) \quad (10.9.28)$$

Similarly, condition (10.9.25) gives

$$[\phi_{tt} + g\phi_z]_{z=0} + \eta[(\phi_{tt} + g\phi_z)_z] + \tfrac{1}{2}\eta^2[(\phi_{tt} + g\phi_z)_{zz}]_{z=0} + \cdots$$
$$+ 2[\nabla\phi \cdot \nabla\phi_t]_{z=0} + 2\eta[\{\nabla\phi \cdot \nabla\phi_t\}_z]_{z=0} + \eta^2[\{\nabla\phi \cdot \nabla\phi_t\}_{zz}]_{z=0} + \cdots$$
$$+ \tfrac{1}{2}[\{\nabla\phi \cdot \nabla(\nabla\phi)^2\}]_{z=0} + \tfrac{1}{2}\eta[\{\nabla\phi \cdot \nabla(\nabla\phi)^2\}_z]_{z=0} + \tfrac{1}{4}\eta^2[\{\nabla\phi \cdot \nabla(\nabla\phi)^2\}_{zz}]_{z=0}$$
$$+ \cdots = 0 \quad (10.9.29)$$

We substitute (10.9.28) for η into (10.9.29) to obtain

$$[\phi_{tt} + g\phi_z]_{z=0} - \frac{1}{g}\left[\phi_t + \frac{1}{2}(\nabla\phi)^2 - \frac{1}{g}\phi_t\phi_{zt}\right]_{z=0}[(\phi_{tt} + g\phi_z)_z]_{z=0}$$
$$+ \frac{1}{2g^2}\left[\left\{\phi_t + \frac{1}{2}(\nabla\phi)^2 - \frac{1}{g}\phi_t\phi_{zt}\right\}^2\right]_{z=0}[(\phi_{tt} + g\phi_z)_{zz}]_{z=0}$$
$$+ 2[(\nabla\phi) \cdot \nabla\phi_t]_{z=0} - \frac{2}{g}\left[\phi_t + \frac{1}{2}(\nabla\phi)^2 - \frac{1}{g}\phi_t\phi_{zt}\right]_{z=0}[(\nabla\phi \cdot \nabla\phi_t)_z]_{z=0}$$
$$+ \frac{1}{2}[\nabla\phi \cdot \nabla(\nabla\phi)^2]_{z=0} - \frac{2}{g}\left[\phi_t + \frac{1}{2}(\nabla\phi)^2 - \frac{1}{g}\phi_t\phi_{zt}\right]_{z=0}[\{\nabla\phi \cdot \nabla(\nabla\phi)^2\}_z]_{z=0}$$
$$= 0 \quad (10.9.30)$$

The first-, second- and third-order boundary conditions on $z = 0$ are, respectively, given by

$$(\phi_{tt} + g\phi_z) = 0 + O(\phi^2) \tag{10.9.31}$$

$$(\phi_{tt} + g\phi_z) + 2[\nabla\phi \cdot \nabla\phi_t] - \frac{1}{g}\phi_t(\phi_{tt} + g\phi_z)_z = 0 + O(\phi^3) \tag{10.9.32}$$

$$(\phi_{tt} + g\phi_z) + 2[\nabla\phi \cdot \nabla\phi_t] + \frac{1}{2}[\nabla\phi \cdot \nabla(\nabla\phi)^2] - \frac{1}{g}\phi_t[\phi_{tt} + g\phi_z + 2(\nabla\phi \cdot \nabla\phi_t)]_z$$

$$- \frac{1}{g}\left[\frac{1}{2}(\nabla\phi)^2 - \frac{1}{g}\phi_t\phi_{zt}\right][\phi_{tt} + g\phi_z]_z + \frac{1}{2g^2}[\phi_t]^2[(\phi_{tt} + g\phi_z)_{zz}]$$

$$= 0 + O(\phi^4) \tag{10.9.33}$$

where $O(\ )$ indicates the order of magnitude of the neglected terms. These results can be used to determine the third-order expansion of plane progressive waves.

As indicated before, the first-order plane wave potential ϕ in deep water is given by (10.9.21). Direct substitution of the first-order velocity potential (10.9.21) in the second-order boundary condition (10.9.32) reveals that the second-order terms in (10.9.32) vanish. Thus the first-order potential is a solution of the second-order boundary-value problem, and we can state that

$$\phi = \left(\frac{ga}{\omega}\right)e^{kz}\sin(kx - \omega t) + O(a^3) \tag{10.9.34}$$

Substitution of this result into (10.9.28) leads to the second-order result for η in the form

$$\eta = a\cos(kx - \omega t) - \tfrac{1}{2}ka^2 + ka^2\cos^2(kx - \omega t) + \cdots$$

$$= a\cos(kx - \omega t) + \tfrac{1}{2}ka^2\cos 2(kz - \omega t) + \cdots \tag{10.9.35}$$

The second term in (10.9.35), which represents the second-order correction to the surface profile, is positive at the *crests* $kx - \omega t = 0, 2\pi, 4\pi, \ldots$, and negative at the *troughs* $kx - \omega t = \pi, 3\pi, 5\pi, \ldots$. But the crests are steeper, and the troughs flatter as a result of the nonlinear effect. The notable feature of solution (10.9.35) is that the wave profile is no longer sinusoidal. The actual shape of the wave profile is a curve known as a *trochoid* (see Figure 10.11), whose crests are steeper and the troughs are flatter than the sinusoidal wave.

Figure 10.11. The surface wave profile.

Substituting the wave potential (10.9.34) in the third-order boundary condition (10.9.33) reveals that all nonlinear terms vanish identically except but one term, $(1/2)\nabla\phi \cdot \nabla(\nabla\phi)^2$. Thus the boundary condition for the third-order plane-wave solution is given by

$$\phi_{tt} + g\phi_z + \tfrac{1}{2}\nabla\phi \cdot \nabla(\nabla\phi)^2 = 0 + O(\phi^4) \tag{10.9.36}$$

If the first-order solution (10.9.34) is substituted into the third-order boundary condition on $z = 0$, the *dispersion relation* with second-order effect is obtained in the form

$$\omega^2 = gk(1 + a^2k^2) + O(k^3a^3) \tag{10.9.37}$$

Note that this relation involves the amplitude in addition to frequency and wave number. This *nonlinear dispersion relation* can be expressed in terms of the phase velocity as

$$c = \frac{\omega}{k} = \left(\frac{g}{k}\right)^{1/2}(1 + k^2a^2)^{1/2} \approx \left(\frac{g}{k}\right)^{1/2}\left(1 + \frac{1}{2}a^2k^2\right) \tag{10.9.38}$$

Thus the phase velocity depends on the wave amplitude, and waves of large amplitude travel faster than smaller ones. The dependence of c on amplitude is known as the *amplitude dispersion* in contrast to the frequency dispersion as given by (10.9.23a, b). The nonlinear solutions for plane waves based on systematic power series in the wave amplitude are known as *Stokes's expansions*.

We conclude this section by discussing the phenomenon of breaking of water waves which is one of the most common observable phenomena in an ocean beach. A wave coming from deep ocean changes shape as it moves across a shallow beach. Its amplitude and wavelength also are modified. The wave train is very smooth some distance offshore, but as it moves inshore, the front of the wave steepens noticeably until, finally, it breaks. After breaking, waves continue to move inshore as a series of bores or hydraulic jumps, whose energy is gradually dissipated by means of the water turbulence. Of the phenomena common to waves on beaches, breaking is the physically most significant and mathematically least known. In fact, it is one of the most intriguing longstanding problems of water wave theory.

For waves of small amplitude in deep water, maximum particle velocity is $v = a\omega = ack$. But the basic assumption of small amplitude theory implies that $v/c = ak \ll 1$. Therefore, wave breaking can never be predicted by the small-amplitude wave theory, and the possibility arises only in the theory of finite-amplitude waves. It is to be noted that the Stokes's expansions are limited to relatively small amplitude and cannot predict the

508 10 EQUATIONS OF FLUID MECHANICS

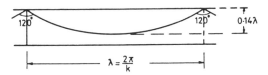

Figure 10.12. The steepest wave profile.

wavetrain of maximum height at which the crests are found to be very sharp. For a wave profile of a constant shape moving at a uniform velocity, it can be shown that the maximum total crest angle as the wave begins to break is 120°; see Figure 10.12.

10.10
STRESS TENSOR FOR A VISCOUS FLUID

In Sections 10.3 through 10.9 we dealt with equations governing the motion of nonviscous fluids and some of their consequences. Let us now turn our attention to viscous fluids. As mentioned in Section 10.2, a viscous fluid is characterized by the property that when the fluid undergoes nonuniform motion, a fluid surface experiences not only normal stress but shear stress as well. In view of this property, it is postulated that the stress tensor at a point in a viscous fluid is made up of two parts: one part due to normal stress and the other due to shear stress. Since the distinction between viscous and nonviscous fluids completely disappears when the fluid is at rest or in uniform motion, we make an additional postulate that the part due to normal stress is analogous to the stress tensor occurring in a nonviscous fluid as given by (10.3.2) and that the part due to shear stress is represented by a new tensor that vanishes when the motion is absent or uniform. We refer to this new tensor as the *viscous stress tensor*, denoted $\mathbf{T}^{(v)}$. Recall that in Section 6.3 we introduced the stretching tensor (strain-rate tensor) $\mathbf{D}$ and made the observation that $\mathbf{D} = \mathbf{0}$ if and only if the motion is absent or uniform. As such, the requirement that $\mathbf{T}^{(v)}$ should vanish when the motion is absent or uniform can be met by imposing the restriction: $\mathbf{T}^{(v)} = \mathbf{0}$ for $\mathbf{D} = \mathbf{0}$. Thus, the following generalization of the expression (10.3.2), proposed by George G. Stokes in 1845, is adopted in dealing with viscous fluids:

$$\mathbf{T} = -p\mathbf{I} + \mathbf{T}^{(v)} \qquad (10.10.1)$$

with $\mathbf{T}^{(v)} = \mathbf{0}$, for $\mathbf{D} = \mathbf{0}$, or in components,

$$\tau_{ij} = -p\delta_{ij} + \tau_{ij}^{(v)}, \qquad (10.10.1)'$$

10.10 STRESS TENSOR FOR A VISCOUS FLUID

with $\tau_{ij}^{(v)} = 0$ for $d_{ij} = 0$, where $\tau_{ij}^{(v)}$ are components of $\mathbf{T}^{(v)}$ and p is a scalar analogous to (but not necessarily identical with) the p of (10.3.2), which is also referred to as *dynamic pressure*.

The most direct way of accommodating the condition $\mathbf{T}^{(v)} = \mathbf{0}$ for $\mathbf{D} = \mathbf{0}$ is to postulate that $\mathbf{T}^{(v)}$ is a *homogeneous function of* $\mathbf{D}$. When this postulate is made, the fluid being dealt with is called a *Stokesian fluid*, after Stokes. In the particular case when $\mathbf{T}^{(v)}$ is taken as a *homogeneous linear function of* $\mathbf{D}$, the fluid is called a *linear viscous fluid*. Thus, for linear viscous fluids, it is postulated that

$$\tau_{ij}^{(v)} = a_{ijkm} d_{km} \tag{10.10.2}$$

where the coefficients a_{ijkm} are independent of d_{ij}. Note that the relations (10.10.2) are analogous to the relations (9.2.3) representing the generalized Hooke's law.

Since $\tau_{ij}^{(v)}$ and d_{ij} are components of tensors $\mathbf{T}^{(v)}$ and $\mathbf{D}$, respectively, it follows from (10.10.2) and the quotient law proven in Example 2.4.7 that a_{ijkm} are components of a fourth-order tensor. This tensor is analogous to the elasticity tensor and is called the *viscosity tensor*. Like the elasticity tensor, the viscosity tensor depends on the physical properties of the material, and its 81 components a_{ijkm} are called *coefficients of viscosity*. From (10.10.1)' and (10.10.2), we find that these coefficients have dimensions: (force × time)/area.

An isotropic linear elastic solid has been defined as the material for which the elasticity tensor is isotropic. Similarly, an isotropic linear viscous fluid is defined as the material for which the viscosity tensor is isotropic. This means that, like in an isotropic elastic solid, there is no preferred direction in an isotropic viscous fluid and the relations (10.10.2) represent a physical law that does not depend on the orientation of the axes. By virtue of the general representation (2.6.1) for a fourth-order isotropic tensor, it follows that, for a linear isotropic viscous fluid, a_{ijkm} are of the form

$$a_{ijkm} = \alpha \delta_{ij} \delta_{km} + \beta \delta_{ik} \delta_{jm} + \gamma \delta_{im} \delta_{jk} \tag{10.10.3}$$

where α, β and γ are scalars.

Substituting for a_{ijkm} from (10.10.3) in (10.10.2) and recalling that $d_{ij} = d_{ji}$, we obtain

$$\tau_{ij}^{(v)} = \lambda \delta_{ij} d_{kk} + 2\mu d_{ij} \tag{10.10.4}$$

where we have set $\lambda = \alpha$ and $(1/2)(\beta + \gamma) = \mu$.

The relations (10.10.4) are analogous to the stress-strain relations (9.2.6). It is to be noted that the symbols λ and μ appearing in (10.10.4) are *not* the same as those appearing in (9.2.6). While in (9.2.6), λ and μ denote the elastic moduli, in (10.10.4) these are the *coefficients of viscosity*. Thus, for a linear isotropic viscous fluid, there are only *two* viscosity coefficients.

Substituting for $\tau_{ij}^{(v)}$ from (10.10.4) into (10.10.1)', we obtain

$$\tau_{ij} = (-p + \lambda d_{kk})\delta_{ij} + 2\mu d_{ij} \qquad (10.10.5)$$

In the direct notation, these relations read

$$\mathbf{T} = \{-p + \lambda(\text{tr } \mathbf{D})\}\mathbf{I} + 2\mu\mathbf{D} \qquad (10.10.6)$$

This is the *stress-strain rate relation* valid for a linear, isotropic viscous fluid. This relation was obtained by Stokes in 1845 and is also known as *Stokes's law*. This law is adopted as the material law in the theory of linear, isotropic, viscous fluid flows. Note that, since **D** is symmetric, so is **T** by virtue of this law. Thus, as in the cases of a linear, isotropic elastic solid and a nonviscous fluid, the symmetry of **T** is a physical property of a linear, isotropic viscous fluid. A large class of liquids and gases possess this property. As such, the theory of viscous fluid flows obeying Stokes's law is very successful from practical point of view also. Henceforth we will refer to a linear, isotropic, viscous fluid simply as a *viscous fluid*.

It is sometimes convenient to have the Stokes's law (10.10.6) expressed in terms of **T** and **v**. Substitution for **D** from (6.3.3) in (10.10.6) yields

$$\mathbf{T} = (-p + \lambda \text{ div } \mathbf{v})\mathbf{I} + \mu(\nabla\mathbf{v} + \nabla\mathbf{v}^T) \qquad (10.10.7)$$

or, in components,

$$\tau_{ij} = (-p + \lambda v_{k,k})\delta_{ij} + \mu(v_{i,j} + v_{j,i}) \qquad (10.10.7)'$$

The *stress-velocity relation* (10.10.7) is analogous to the stress-displacement relation (9.8.4).

We note that in the limiting case when $d_{ij} \to 0$, that is, when the motion is absent or uniform, (10.10.6) reduces to the material law (10.3.2) valid for a nonviscous fluid; p then represents the mean pressure.

Comparison of Stokes's law (10.10.6) with Hooke's law (9.2.6) reveals an important difference between elastic solids and viscous fluids. Whereas in the case of an elastic solid the Hooke's law (9.2.6) shows that the stresses are all 0 when there is no deformation, in the case of a viscous fluid Stokes's law (10.10.6) shows that nonzero stresses specified by $-p\mathbf{I}$ do occur even when there is no deformation. The presence of the *residual stress* $-p\mathbf{I}$ makes a fundamental difference between elasticity and fluid mechanics.

EXAMPLE 10.10.1 By using Stokes's law, show that at every point of a viscous fluid, the principal directions of **T**, $\mathbf{T}^{(v)}$ and **D** are all coincident.

10.10 STRESS TENSOR FOR A VISCOUS FLUID

Solution At a point of a viscous fluid let **a** be an eigenvector of **D**. Then **a** is along a principal direction of **D** and $\mathbf{Da} = \Lambda \mathbf{a}$ for some scalar Λ. Stokes's law (10.10.6) then gives

$$\mathbf{Ta} = \bar{\Lambda}\mathbf{a} \qquad (10.10.8)$$

where $\bar{\Lambda} = -p + \lambda(\operatorname{tr} \mathbf{D}) + 2\mu\Lambda$, which is a scalar. Hence **a** is an eigenvector of **T** as well. Consequently, **a** is along a principal direction of **T**. Further, by use of (10.10.1) and (10.10.8), we get

$$\mathbf{T}^{(v)}\mathbf{a} = (\mathbf{T} + p\mathbf{I})\mathbf{a} = (\bar{\Lambda} + p)\mathbf{a} \qquad (10.10.9)$$

from which it is evident that **a** is along a principal direction of $\mathbf{T}^{(v)}$ also. ∎

Note: This result is analogous to that proven in Example 9.2.1.

EXAMPLE 10.10.2 Show that Stokes's law (10.10.6) is equivalent to the following relations taken together:

$$\mathbf{T}^{(d)} = 2\mu \mathbf{D}^{(d)} \qquad (10.10.10\text{a})$$

$$\operatorname{tr} \mathbf{T} = -3p + (3\lambda + 2\mu)(\operatorname{tr} \mathbf{D}) \qquad (10.10.10\text{b})$$

Solution Let us recall that $\mathbf{T}^{(d)}$ and $\mathbf{D}^{(d)}$ are the deviator parts of **T** and **D**, respectively, so that

$$\mathbf{T}^{(d)} = \mathbf{T} - \tfrac{1}{3}(\operatorname{tr} \mathbf{T})\mathbf{I} \qquad (10.10.11\text{a})$$

$$\mathbf{D}^{(d)} = \mathbf{D} - \tfrac{1}{3}(\operatorname{tr} \mathbf{D})\mathbf{I} \qquad (10.10.11\text{b})$$

Let us now consider Stokes's law (10.10.6). Taking the trace throughout in (10.10.6), we get (10.10.10b). Substituting for **T** and $(\operatorname{tr} \mathbf{T})$ from (10.10.6) and (10.10.10b) in (10.10.11a) and using (10.10.11b), we obtain (10.10.10a). Thus, (10.10.6) yields (10.10.10).

Conversely, with the use of (10.10.11), relations (10.10.10) yield

$$\mathbf{T} = \mathbf{T}^{(d)} + \tfrac{1}{3}(\operatorname{tr} \mathbf{T})\mathbf{I}$$

$$= 2\mu\{\mathbf{D} - \tfrac{1}{3}(\operatorname{tr}\mathbf{D})\mathbf{I}\} + \tfrac{1}{3}\{-3p + (3\lambda + 2\mu)(\operatorname{tr} \mathbf{D})\}\mathbf{I}$$

$$= -p\mathbf{I} + \lambda(\operatorname{tr} \mathbf{D})\mathbf{I} + 2\mu\mathbf{D}$$

which is (10.10.6). ∎

Note: This result is analogous to that proven in Example 9.2.2.

EXAMPLE 10.10.3 Let

$$\Phi = (\lambda + \tfrac{2}{3}\mu)(\operatorname{tr} \mathbf{D})^2 + 2\mu|\mathbf{D}^{(d)}|^2 \qquad (10.10.12)$$

512 10 EQUATIONS OF FLUID MECHANICS

Prove the following:

(i) $\qquad \dfrac{\partial \Phi}{\partial d_{ij}} = 2v_{ij}$ \hfill (10.10.13)

(ii) $\qquad \Phi = \mathbf{T}^{(v)} \cdot \mathbf{D}$ \hfill (10.10.14)

(iii) $\qquad \Phi = \mathbf{T} \cdot \mathbf{D} + (\operatorname{tr} \mathbf{D}) p$ \hfill (10.10.15)

(iv) $\qquad \Phi = \lambda (\operatorname{tr} \mathbf{D})^2 + 2\mu |\mathbf{D}|^2$ \hfill (10.10.16)

(v) $\qquad \Phi = \lambda (\operatorname{div} \mathbf{v})^2 + \mu (\nabla \mathbf{v} + \nabla \mathbf{v}^T) \cdot \nabla \mathbf{v}$ \hfill (10.10.17)

(vi) $\qquad \Phi = \lambda (\operatorname{div} \mathbf{v})^2 + 2\mu \left\{ \operatorname{div} \dfrac{D\mathbf{v}}{Dt} - \dfrac{D}{Dt}(\operatorname{div} \mathbf{v})^2 + \dfrac{1}{2}\mathbf{w}^2 \right\}$ \hfill (10.10.18)

(vii) $\displaystyle \int_V \Phi \, dV = \int_V \left\{ \lambda (\operatorname{div} \mathbf{v})^2 - 2\mu \dfrac{D}{Dt}(\operatorname{div} \mathbf{v})^2 + \mu \mathbf{w}^2 \right\} dV$

$\qquad \qquad + 2\mu \displaystyle \int_S \dfrac{D\mathbf{v}}{Dt} \cdot \mathbf{n} \, dS$ \hfill (10.10.19)

Expression (10.10.19) is known as the *Bobyleff–Forsythe formula*.

Solution (i) We first note that the given function Φ has the following expression in the suffix notation

$$\Phi = (\lambda + \tfrac{2}{3}\mu)(d_{kk})^2 + 2\mu d^{(d)}_{km} d^{(d)}_{km} \qquad (10.10.20)$$

Differentiating both sides of this expression w.r.t. d_{ij} and noting that

$$\dfrac{\partial}{\partial d_{ij}}(d_{kk}) = \delta_{ij}$$

and

$$\dfrac{\partial}{\partial d_{ij}}(d^{(d)}_{km}) = \dfrac{\partial}{\partial d_{ij}}\left(d_{km} - \tfrac{1}{3}\delta_{km} d_{rr}\right) = (\delta_{ik}\delta_{jm} - \tfrac{1}{3}\delta_{km}\delta_{ij})$$

we obtain

$$\dfrac{\partial \Phi}{\partial d_{ij}} = 2(\lambda + \tfrac{2}{3}\mu) d_{kk} \delta_{ij} + 4\mu(\delta_{ik}\delta_{jm} - \tfrac{1}{3}\delta_{km}\delta_{ij}) d_{km} = 2(\lambda \delta_{ij} d_{kk} + 2\mu d_{ij})$$
\hfill (10.10.21)

Expression (10.10.4) now yields (10.10.13).

(ii) Using (10.10.4) and (10.10.11b) we obtain

$$\mathbf{T}^{(v)} \cdot \mathbf{D} = \{\lambda (\operatorname{tr} \mathbf{D})\mathbf{I} + 2\mu \mathbf{D}\} \cdot \mathbf{D}$$
$$= \lambda (\operatorname{tr} \mathbf{D})^2 + 2\mu \mathbf{D}^{(d)} \cdot \mathbf{D}^{(d)} + \tfrac{2}{3}\mu(\operatorname{tr} \mathbf{D})^2$$
$$= (\lambda + \tfrac{2}{3}\mu)(\operatorname{tr} \mathbf{D})^2 + 2\mu |\mathbf{D}^{(d)}|^2$$

Expression (10.10.12) now yields (10.10.14).

10.10 STRESS TENSOR FOR A VISCOUS FLUID

(iii) Using (10.10.1) and (10.10.14) we get
$$\mathbf{T} \cdot \mathbf{D} = (-p\mathbf{I} + \mathbf{T}^{(v)}) \cdot \mathbf{D} = -(\operatorname{tr} \mathbf{D})p + \Phi$$
which is (10.10.15).

(iv) Expressions (10.10.15) and (10.10.6) yield
$$\Phi = \mathbf{T} \cdot \mathbf{D} + (\operatorname{tr} \mathbf{D})p = \{\lambda(\operatorname{tr} \mathbf{D})\mathbf{I} + 2\mu\mathbf{D}\} \cdot \mathbf{D}$$
$$= \lambda(\operatorname{tr} \mathbf{D})^2 + 2\mu|\mathbf{D}|^2$$
which is (10.10.16).

(v) Since $\mathbf{D} = \operatorname{sym} \nabla\mathbf{v} = (1/2)(\nabla\mathbf{v} + \nabla\mathbf{v}^T)$, we have $\operatorname{tr} \mathbf{D} = \operatorname{div} \mathbf{v}$ and
$$(\nabla\mathbf{v} + \nabla\mathbf{v}^T) \cdot \nabla\mathbf{v} = 2\mathbf{D} \cdot (\nabla\mathbf{v}) = 2\mathbf{D} \cdot (\operatorname{sym} \nabla\mathbf{v} + \operatorname{skw} \nabla\mathbf{v})$$
$$= 2\mathbf{D} \cdot \mathbf{D} = 2|\mathbf{D}|^2$$
Consequently, (10.10.16) yields (10.10.17).

(vi) By using result (i) of Example 6.3.1 in (10.10.16) and noting that $\operatorname{tr} \mathbf{D} = \operatorname{div} \mathbf{v}$, we immediately get (10.10.18).

(vii) Integrating both sides of (10.10.18) over a volume V and employing the divergence theorem (3.6.1), we obtain (10.10.19). ∎

Note: The function Φ is analogous to the strain-energy function of the elasticity theory and is called the *viscous dissipative function*. This function is nonnegative for $\lambda + (2/3)\mu \geq 0$ and $\mu \geq 0$, as is evident from (10.10.12).

EXAMPLE 10.10.4 In a certain flow obeying Stokes's law, the velocity field is given by
$$v_1 = 4x_1x_2x_3, \qquad v_2 = x_3^2, \qquad v_3 = -2x_2x_3^2 \qquad (10.10.22)$$
Find the strain-rate tensor and the stress tensor.

Solution Substituting for v_i from (10.10.22) in (6.3.4), we obtain
$$d_{11} = 4x_2x_3, \qquad d_{22} = 0, \qquad d_{33} = -4x_2x_3$$
$$d_{12} = 2x_1x_3, \qquad d_{13} = 2x_1x_2, \qquad d_{23} = 2x_3(1 - x_3) \qquad (10.10.23)$$
These are the components of the strain-rate tensor associated with the given velocity field.

Substituting for d_{ij} from (10.10.23) into the Stokes's law (10.10.5), we obtain
$$\tau_{11} = p + 8\mu x_2 x_3, \qquad \tau_{12} = 4\mu x_1 x_3$$
$$\tau_{22} = -p, \qquad \tau_{23} = 4\mu x_3(1 - x_3) \qquad (10.10.24)$$
$$\tau_{33} = -p - 8\mu x_2 x_3, \qquad \tau_{13} = 4\mu x_1 x_2$$

These are the components of the stress tensor associated with the given velocity field. ∎

Note that unless p is specified, normal stresses are not completely determined.

10.11
SHEAR VISCOSITY AND BULK VISCOSITY

In order to obtain the physical meanings of the coefficients of viscosity, let us consider the flow of a viscous fluid for which the velocity field is as follows:

$$v_1 = v_1(x_3), \qquad v_2 = 0, \qquad v_3 = 0 \qquad (10.11.1)$$

For this velocity field, Stokes's law (10.10.7) gives

$$\tau_{11} = \tau_{22} = \tau_{33} = -p, \qquad \tau_{12} = \tau_{23} = 0 \qquad (10.11.2)$$

$$\tau_{31} = \mu \frac{dv_1}{dx_3} \qquad (10.11.3)$$

A velocity field of the form (10.11.1) occurs, for example, when a viscous fluid bounded between two parallel horizontal plates is made to move due to a uniform movement of the upper plate in the x_1 direction, keeping the lower plate stationary (Figure 10.13). Relation (10.11.3) shows that there occurs a shear stress in the direction of the flow on planes parallel to the flow and that, this shear stress is directly proportional to the velocity gradient in the perpendicular direction, with μ serving as the proportionality factor. The coefficient μ thus represents the shear stress on a plane element parallel to the direction of flow due to a unit of velocity gradient in the perpendicular

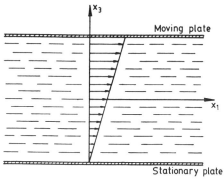

Figure 10.13. Shear viscosity.

direction. This coefficient is known as the *coefficient of shear viscosity*; it is analogous to the shear modulus in the elasticity theory. Experimental observations on common fluids show that μ is a nonnegative number.

Relation (10.11.3) that defines μ and is a particular case of Stokes's law (10.10.7) was proposed by Newton in 1687 on experimental grounds. This relation is known as *Newton's law of viscosity*. Stokes's law (10.10.6) was developed as a generalization of Newton's law. For this reason linear viscous fluids (obeying Stokes's law) are referred to as *Newtonian viscous fluids*.

Returning to the general case, we find, from relation (10.10.4), that

$$\tau_{kk}^{(v)} = 3\kappa d_{kk} \tag{10.11.4}$$

where

$$\kappa = \lambda + \tfrac{2}{3}\mu \tag{10.11.5}$$

Thus, κ is the proportionality factor relating the mean viscous stress $(1/3)\tau_{kk}^{(v)}$ to the rate of dilatation d_{kk}. (Recall the geometrical meaning of d_{kk} from Section 6.3.) The coefficient κ is thus analogous to the bulk modulus of the elasticity theory and is called the *coefficient of bulk viscosity*. Experimental observations on common fluids indicate that this coefficient may also be taken to be nonnegative.

From (10.11.5), it follows that λ represents the bulk viscosity diminished by two-thirds of the shear viscosity. It is called the *coefficient of dilatational viscosity* or *second coefficient of viscosity*. Thus, the coefficients λ and μ appearing in Stokes's law have definite physical meanings. Experiments show that these coefficients generally change with temperature and pressure. However, these changes are usually small and in most problems λ and μ are treated as constants.

10.11.1 STOKES'S CONDITION

Taking the trace of Stokes's law (10.10.7) we find that

$$p = \bar{p} + \kappa(\text{div } \mathbf{v}) \tag{10.11.6}$$

where $\bar{p} = -(1/3)(\text{tr } \mathbf{T})$ is as usual the mean pressure. Comparing (10.11.6) with (10.3.3), we find that the dynamic pressure p of the viscous case is not generally identical with the corresponding p of the nonviscous case. Whereas the p of the nonviscous case is equal to the mean pressure, the p of the viscous case represents the mean pressure plus the dilatation rate multiplied by the bulk viscosity. We find that $p = \bar{p}$ if and only if *one* of the following two conditions is satisfied:

(i) $\qquad\qquad\qquad \text{div } \mathbf{v} = 0 \qquad\qquad\qquad$ (10.11.7)

(ii) $\qquad\qquad\qquad \kappa = 0 \qquad\qquad\qquad$ (10.11.8)

516   10 EQUATIONS OF FLUID MECHANICS

Condition (10.11.7) holds whenever the volume of each element of the fluid remains constant during the motion. As mentioned in Section 6.3, such a motion is called an *isochoric motion* and materials for which every motion is isochoric are called *incompressible materials*. Thus, for an incompressible fluid, the geometrical condition (10.11.7) always holds. In fact (10.11.7) is the equation of continuity for an incompressible fluid as already noted in the nonviscous case.

In a compressible fluid, volumes of material elements generally change with time and condition (10.11.7) is an impossibility. However, in this case condition (10.11.8) can hold. Stokes studied compressible fluid flows by assuming condition (10.11.8), and this condition is known as *Stokes's condition*. Nearly all studies on compressible fluid flows make use of this condition, and a large number of theoretical results so obtained are verified even experimentally.

Thus, for incompressible viscous fluids and for compressible viscous fluids with zero bulk viscosity, the dynamic pressure p and the mean pressure $\bar{p}$ are identical as in the case of nonviscous fluids. For compressible viscous fluids not obeying Stokes's condition (10.11.8), the two pressures are not the same. Dense gases belong to such a class of compressible fluids.

It is to be pointed out that Stokes's law (10.10.6), upon which the theory of incompressible and compressible Newtonian viscous fluid flows is based, is a linear relationship between **T** and **D** and that the linearity of the relationship is merely a hypothesis. In recent years, relationships between **T** and **D** that are linear but more general than (10.10.6) and those that are even nonlinear in nature have also been postulated, and based upon such relationships some nonclassical theories of fluid flows have been developed. Fluids for which material laws other than (10.10.6) are postulated are referred to as *non-Newtonian fluids*, and the study of non-Newtonian fluid flows is called *rheology*. This topic is not treated in this text.

EXAMPLE 10.11.1   For a compressible fluid with zero bulk viscosity, show that Stokes's law (10.10.6) can be written in the following equivalent forms:

(i) $$\mathbf{T} = -(p + \tfrac{2}{3}\mu \operatorname{tr} \mathbf{D})\mathbf{I} + 2\mu \mathbf{D} \qquad (10.11.9)$$

(ii) $$\mathbf{T} = -p\mathbf{I} + 2\mu \mathbf{D}^{(d)} \qquad (10.11.10)$$

(iii) $$\mathbf{T}^{(d)} = 2\mu \mathbf{D}^{(d)} \quad \text{and} \quad \operatorname{tr} \mathbf{T} = -3p \qquad (10.11.11)$$

Solution   When $\kappa = 0$, we have $\lambda = -(2/3)\mu$ by (10.11.5) and Stokes's law (10.10.6) becomes (10.11.9).

10.11 SHEAR VISCOSITY AND BULK VISCOSITY

Relation (10.11.9) may be rewritten as

$$\mathbf{T} = -p\mathbf{I} + 2\mu\{\mathbf{D} - \tfrac{1}{3}(\operatorname{tr} \mathbf{D})\mathbf{I}\} = -p\mathbf{I} + 2\mu\mathbf{D}^{(d)}$$

which is (10.11.10). Thus (10.11.9) yields (10.11.10).

From relation (10.11.10), we get, on noting that $\operatorname{tr} \mathbf{D}^{(d)} = 0$,

$$\operatorname{tr} \mathbf{T} = -3p \qquad (10.11.12)$$

Hence

$$\mathbf{T}^{(d)} = \mathbf{T} - \tfrac{1}{3}(\operatorname{tr} \mathbf{T})\mathbf{I} = \{-p\mathbf{I} + 2\mu\mathbf{D}^{(d)}\} - \tfrac{1}{3}(-3p)\mathbf{I} = 2\mu\mathbf{D}^{(d)} \qquad (10.11.13)$$

Relations (10.11.12) and (10.11.13) constitute (10.11.11). Thus, (10.11.10) yields (10.11.11).

Relation (10.11.13) yields

$$\mathbf{T} - (\tfrac{1}{3}\operatorname{tr}\mathbf{T})\mathbf{I} = 2\mu\{\mathbf{D} - \tfrac{1}{3}(\operatorname{tr}\mathbf{D})\mathbf{I}\}$$

which on utilizing (10.11.12) becomes

$$\mathbf{T} = -(p + \tfrac{2}{3}\mu \operatorname{tr} \mathbf{D})\mathbf{I} + 2\mu\mathbf{D}$$

This is the relation (10.11.9). Thus, the relations (10.11.11) yield the relation (10.11.9).

This proves that the relations (10.11.9) to (10.11.11) are equivalent and hold for a compressible fluid obeying Stokes's condition $\kappa = 0$. ∎

EXAMPLE 10.11.2 For a compressible viscous fluid with zero bulk viscosity, show that the viscous stress tensor is identical with the stress deviator tensor.

Solution When $\kappa = 0$, we have $p = \bar{p} = -(1/3)\operatorname{tr} \mathbf{T}$. Consequently, we find from equations (10.10.1) and (10.10.11a) that

$$\mathbf{T}^{(v)} = p\mathbf{I} + \mathbf{T} = -\tfrac{1}{3}(\operatorname{tr} \mathbf{T})\mathbf{I} + \mathbf{T} = \mathbf{T}^{(d)} \qquad ∎ \qquad (10.11.14)$$

EXAMPLE 10.11.3 At a certain point of an incompressible viscous fluid, the stress matrix is

$$[\mathbf{T}] = \begin{bmatrix} -1 & 0 & 1 \\ 0 & -2 & 1 \\ 1 & 1 & 0 \end{bmatrix}$$

Find the pressure and the viscous stress tensor.

Solution For an incompressible fluid, we have $p = -(1/3)\tau_{kk}$. The given stress matrix yields $p = 1$.

Using the given τ_{ij} and the fact that $p = 1$ in (10.10.1), we get

$$[\mathbf{T}^{(v)}] = [\mathbf{T}] + p[\mathbf{I}] = \begin{bmatrix} 0 & 0 & 1 \\ 0 & -1 & 1 \\ 1 & 1 & 1 \end{bmatrix}$$

This is the matrix of the required viscous stress tensor. ∎

10.12
GOVERNING EQUATIONS FOR A VISCOUS FLUID FLOW

With the material law applicable to viscous fluids formulated and analyzed in Sections 10.10 and 10.11, let us summarize all the equations that serve as governing equations for flows of such fluids. Since the equation of continuity (10.4.1), the equations of motion (10.4.2a, b) and the equation of energy (10.4.3) hold for all continua, they hold not only for nonviscous fluids but also for viscous fluids. Let us record these equations for easy reference.

Equation of Continuity:

$$\frac{D\rho}{Dt} + \rho \operatorname{div} \mathbf{v} = 0 \qquad (10.12.1)$$

Equations of Motion:

$$\operatorname{div} \mathbf{T}^T + \rho \mathbf{b} = \rho \frac{D\mathbf{v}}{Dt} \qquad (10.12.2\text{a})$$

$$\mathbf{T} = \mathbf{T}^T \qquad (10.12.2\text{b})$$

Equation of Energy:

$$\rho \frac{D\varepsilon}{Dt} = \mathbf{T} \cdot \nabla \mathbf{v} - \operatorname{div} \mathbf{q} + \rho h \qquad (10.12.3)$$

Next, in place of the material law (10.4.4), which is valid only for nonviscous fluids, let us consider Stokes's law as given by (10.10.7) and record it.

Material Law:

$$\mathbf{T} = (-p + \lambda \operatorname{div} \mathbf{v})\mathbf{I} + \mu(\nabla \mathbf{v} + \nabla \mathbf{v}^T) \qquad (10.12.4)$$

As noted earlier, $\mathbf{T}$ is symmetric by virtue of this material law. Hence equation (10.12.2b) is identically satisfied (as in the case of nonviscous fluids).

10.12 GOVERNING EQUATIONS FOR A VISCOUS FLUID FLOW 519

It is easily seen that 11 scalar equations are involved in the governing equations (10.12.1)–(10.2.4) whereas 15 unknown field functions are present in these equations, the unknowns being 3 velocity components v_i, 3 heat flux components q_i, 6 stress components $\tau_{ij}(=\tau_{ji})$ and 3 scalars ρ, p and ε (as in the case of nonviscous fluids). The body force $\mathbf{b}$ and heat supply h are taken as known functions as usual. Accordingly, equations (10.12.1) to (10.12.4) are inadequate to determine all the field functions. Therefore, we have to either reduce the number of unknowns or increase the number of governing equations in order to close the system of governing equations. These two possibilities lead us to consider the cases of incompressible and compressible fluids as in the theory of nonviscous fluid flows.

The classification of fluids into the categories of compressible fluids and incompressible fluids holds for both nonviscous and viscous fluids. Note that the property of compressibility is associated directly with the change in density of fluid elements during the motion, and the property of viscosity is linked with the ability of the fluid to sustain shear stresses, again during the motion. Since the change in density is always accompanied by change in volume of elements, and shear stresses are always accompanied by change in shape of elements, the properties of compressibility and viscosity are not interlinked (at least theoretically). This means that a fluid can be (i) nonviscous and incompressible, (ii) nonviscous and compressible, (iii) viscous and incompressible and (iv) viscous and compressible. Thus, there exist four different classes of fluids. The first two classes of fluids, namely, incompressible and compressible nonviscous fluids, were analyzed in Section 10.4. Let us now consider the cases of incompressible and compressible viscous fluids to complete the task of obtaining a closed system of governing equations for flow of such fluids.

10.12.1 INCOMPRESSIBLE FLUIDS

In the case of incompressible fluids, for which ρ retains its initial value ρ_0 during the motion, the equations of continuity, motion and energy, namely, (10.12.1), (10.12.2) and (10.12.3), reduce to the following equations (respectively):

$$\text{div } \mathbf{v} = 0 \tag{10.12.5}$$

$$\text{div } \mathbf{T} + \rho_0 \mathbf{b} = \rho_0 \frac{D\mathbf{v}}{Dt} \tag{10.12.6}$$

$$\rho_0 \frac{D\varepsilon}{Dt} = -\text{div } \mathbf{q} + \rho h \tag{10.12.7}$$

Also, in this case, the material law (10.12.4) reduces to

$$\mathbf{T} = -p\mathbf{I} + \mu(\nabla \mathbf{v} + \nabla \mathbf{v}^T) \tag{10.12.8}$$

We note that ten scalar equations are involved in (10.12.5), (10.12.6) and (10.12.8) and the number of unknown field functions is also ten, the field functions being p, v_i and τ_{ij}. Equation (10.12.7), which is purely an equation of thermal energy does not contain these field functions and is therefore not needed in their determination. Equations (10.12.5), (10.12.6) and (10.12.8) thus serve as a closed system of governing equations for incompressible viscous fluid flows.

10.12.2 COMPRESSIBLE FLUIDS

As in the case of nonviscous compressible fluid flows, the equations of state and law of heat conduction are appended to the material law for obtaining a full set of constitutive equations for viscous compressible fluid flows. These additional constitutive equations are the same as those considered in Section 10.4. These equations are the *kinetic equation of state* (10.4.10), namely,

$$f(p, \rho, T) = 0 \qquad (10.12.9)$$

the *caloric equation of state* (10.4.13), namely,

$$\varepsilon = \varepsilon(\rho, T) \qquad (10.12.10)$$

and Fourier's law of heat conduction (10.4.12), namely,

$$\mathbf{q} = -k \nabla T \qquad (10.12.11)$$

If equations (10.12.9) to (10.12.11) are appended to equations (10.12.1) to (10.12.4), then we will have in all sixteen scalar equations for sixteen field functions, the field functions being three velocity components v_i, three heat-flux components q_i, six stress components τ_{ij} and four scalars ρ, p, T and ε. Equations (10.12.1) to (10.12.4) and (10.12.9) to (10.12.11) therefore serve as a closed system of governing differential equations for compressible viscous fluid flows. Note that, unlike the case of incompressible fluid flows, energy equation (10.12.3) is now fully coupled with other governing equations.

Thus, as in the case of nonviscous fluid flows, there are two sets of governing equations for viscous fluid flows. One of these sets is applicable to incompressible fluid flows and the other set to compressible fluid flows. Comparison of the governing equations of viscous fluid flows with those of nonviscous fluid flows reveals that the material law is the only governing equation that brings out the difference between the theory of viscous fluid flows and the theory of nonviscous fluid flows; (10.12.4) is the material law employed for viscous fluids and (10.4.4) is the material law for nonviscous fluids. The standard forms of equations of state, given by (10.4.14) to (10.4.21) are employed in the theory of viscous fluids also.

10.12 GOVERNING EQUATIONS FOR A VISCOUS FLUID FLOW

In Section 10.11, it was pointed out that compressible viscous fluids themselves fall into two classes depending on whether condition (10.11.8) holds or not. When condition (10.11.8) holds, we have $\lambda = -(2/3)\mu$ by (10.11.5) and the material law (10.12.4) reduces to

$$\mathbf{T} = -p\mathbf{I} + \mu\{(-\tfrac{2}{3})(\operatorname{div}\mathbf{v})\mathbf{I} + \nabla\mathbf{v} + (\nabla\mathbf{v})^T\} \qquad (10.12.12)$$

This relation together with (10.12.1) to (10.12.3) and (10.12.9) to (10.12.11) constitute the governing equations for compressible viscous fluid flows for which Stokes's condition (10.11.8) holds.

Thus, the theory of viscous fluid flows is generally studied in three different cases: (i) the incompressible case governed by equations (10.12.5), (10.12.6) and (10.12.8); (ii) compressible case with Stokes's condition, governed by equations (10.12.1) to (10.12.3), (10.12.12) and (10.12.9) to (10.12.11); and (iii) general compressible case governed by equations (10.2.1) to (10.12.4) and (10.12.9) to (10.12.11). The governing equations for viscous fluid flows valid in different cases are summarized in Tables 10.3 and 10.4 with the same equation numbers as in the text.

EXAMPLE 10.12.1 Show that the energy equation (10.12.3) can be rewritten in the form

$$\rho\frac{D\varepsilon}{Dt} = -p(\operatorname{div}\mathbf{v}) - \operatorname{div}\mathbf{q} + \rho h + \Phi \qquad (10.12.13)$$

where Φ is the viscous dissipative function considered in Example 10.10.3.

Solution By use of the material law (10.12.4), we find that

$$\mathbf{T} \cdot \nabla\mathbf{v} = -p(\operatorname{div}\mathbf{v}) + \lambda(\operatorname{div}\mathbf{v})^2 + \mu(\nabla\mathbf{v} + \nabla\mathbf{v}^T) \cdot \nabla\mathbf{v} \qquad (10.12.14)$$

Using (10.10.17), this becomes

$$\mathbf{T} \cdot \nabla\mathbf{v} = -p(\operatorname{div}\mathbf{v}) + \Phi \qquad (10.12.15)$$

Substituting this in (10.12.3), we obtain (10.12.13). ∎

Table 10.3. Governing Equations for Incompressible Viscous Fluid Flows

1. Equation of continuity:
$$\operatorname{div}\mathbf{v} = 0 \qquad (10.12.5)$$

2. Equation of motion:
$$\operatorname{div}\mathbf{T} + \rho_0\mathbf{b} = \rho_0\frac{D\mathbf{v}}{Dt} \qquad (10.12.6)$$

3. Material law:
$$\mathbf{T} = -p\mathbf{I} + \mu(\nabla\mathbf{v} + \nabla\mathbf{v}^T) \qquad (10.12.8)$$

Total number of scalar equations: 10
Total number of unknown field functions (namely, p, v_i and τ_{ij}): 10

10 EQUATIONS OF FLUID MECHANICS

Table 10.4. Governing Equations for Compressible Viscous Fluid Flows

1. Equation of continuity:
$$\frac{D\rho}{Dt} + \rho \operatorname{div} \mathbf{v} = 0 \tag{10.12.1}$$

2. Equation of motion:
$$\operatorname{div} \mathbf{T} + \rho \mathbf{b} = \rho \frac{D\mathbf{v}}{Dt} \tag{10.12.2}$$

3. Equation of energy:
$$\rho \frac{D\varepsilon}{Dt} = \mathbf{T} \cdot \nabla \mathbf{v} - \operatorname{div} \mathbf{q} + \rho h \tag{10.12.3}$$

4. Material law:
$$\mathbf{T} = \{-p + \lambda(\operatorname{div} \mathbf{v})\}\mathbf{I} + \mu(\nabla \mathbf{v} + \nabla \mathbf{v}^T), \quad \text{for } \kappa \neq 0 \tag{10.12.4}$$
$$\mathbf{T} = -p\mathbf{I} + \mu\{(-\tfrac{2}{3})(\operatorname{div} \mathbf{v})\mathbf{I} + (\nabla \mathbf{v} + \nabla \mathbf{v}^T)\}, \quad \text{for } \kappa = 0 \tag{10.12.12}$$

5. Equations of state:
$$f(p, \rho, T) = 0 \tag{10.12.9}$$
$$\varepsilon = \varepsilon(\rho, T) \tag{10.12.10}$$

6. Fourier's law of heat conduction:
$$\mathbf{q} = -k\nabla T \tag{10.12.11}$$

Total number of scalar equations: 16
Total number of unknown field functions (namely, $\rho, p, T, \varepsilon, q_i, v_i$ and τ_{ij}): 16

Note: By using Fourier's law (10.12.11), equation (10.12.13) can be expressed in terms of T as follows:

$$\rho \frac{D\varepsilon}{Dt} = -p(\operatorname{div} \mathbf{v}) + k\nabla^2 T + \rho h + \Phi \tag{10.12.16}$$

EXAMPLE 10.12.2 Show that, for a perfect gas, the energy equation (10.12.3) can be rewritten in the following forms:

$$\rho c_v \frac{DT}{Dt} = \frac{p}{\rho} \frac{D\rho}{Dt} + k\nabla^2 T + \rho h + \Phi \tag{10.12.17}$$

$$\rho c_p \frac{DT}{Dt} = \frac{Dp}{Dt} + k\nabla^2 T + \rho h + \Phi \tag{10.12.18}$$

Solution For a perfect gas, we have

$$\varepsilon = c_v T \tag{10.12.19a}$$

$$p = \rho R T \tag{10.12.19b}$$

Substituting for ε from (10.12.19a) and for div $\mathbf{v}$ from (10.12.1) in (10.12.16), we obtain (10.12.17).

Next, using (10.4.19), expression (10.12.19b) yields

$$c_v T = c_p T - \frac{p}{\rho}$$

so that

$$c_v \frac{DT}{Dt} = c_p \frac{DT}{Dt} - \frac{1}{\rho}\frac{Dp}{Dt} + \frac{p}{\rho^2}\frac{D\rho}{Dt} \qquad (10.12.20)$$

With the use of this expression in (10.12.17), we arrive at (10.12.18). ∎

Note: In the absence of viscous effects, we have $\Phi \equiv 0$, and equations (10.12.17) and (10.12.18) reduce to equations (10.4.31) and (10.4.32), respectively.

10.13
INITIAL AND BOUNDARY CONDITIONS

Since the field functions for viscous fluid flows are the same as those for nonviscous fluid flows, initial and boundary conditions (10.5.1) to (10.5.5) hold for the viscous case also.

However, because of the presence of shear stresses in a viscous fluid, the condition to be satisfied on a rigid boundary surface S contacting the fluid is different from the condition (10.5.6) employed in the nonviscous case. For a viscous fluid flow, the condition (10.5.6) is usually replaced by the stronger condition:

$$\mathbf{v} = \mathbf{v}_S \qquad (10.13.1)$$

on S, where $\mathbf{v}_S$ is the velocity with which the surface S moves. For a surface at rest, the condition becomes

$$\mathbf{v} = \mathbf{0} \qquad (10.13.1)'$$

on S.

Note that condition (10.5.6) is just a particular case of condition (10.13.1). Whereas condition (10.5.6) implies that the normal component of fluid velocity is the same as that of the solid boundary (at the point of contact), condition (10.3.1) implies such a restriction on the tangential component also. In other words, condition (10.13.1) implies that the fluid in contact with the solid surface must move with the surface. This amounts to saying that the fluid must adhere to the solid and therefore cannot slip over the surface of the solid. The condition (10.13.1) was first proposed by Stokes and is known as the *no-slip condition*. In order to satisfy this boundary condition, Prandtl (1905) made the hypothesis that within a thin layer of fluid adjacent to the boundary the relative fluid velocity increases rapidly from 0 at the solid boundary to the full value at its outer edge. This thin layer is called the *boundary layer* within which the viscosity effects are predominant. The condition (10.13.1) is employed as the standard boundary condition in common engineering problems. However, for high-altitude aerodynamical problems, this condition is known to be invalid, and one uses what are called *slip conditions*. We do not consider such conditions in this text.

EXAMPLE 10.13.1 Show that on a viscous fluid surface, the stress vector is given by (i) for incompressible fluid

$$\mathbf{s} = -p\mathbf{n} + 2\mu(\mathbf{n} \cdot \nabla)\mathbf{v} + \mu \mathbf{n} \times \mathbf{w} \qquad (10.13.2a)$$

and (ii) for compressible fluid

$$\mathbf{s} = (-p + \lambda \operatorname{div} \mathbf{v})\mathbf{n} + 2\mu(\mathbf{n} \cdot \nabla)\mathbf{v} + \mu \mathbf{n} \times \mathbf{w} \qquad (10.13.2b)$$

Solution For an incompressible viscous fluid, the material law is given by (10.12.8). Using this in Cauchy's law (7.4.8), we obtain

$$\mathbf{s} = \mathbf{T}\mathbf{n} = -p\mathbf{n} + \mu(\nabla\mathbf{v} + \nabla\mathbf{v}^T)\mathbf{n} = -p\mathbf{n} + 2\mu(\nabla\mathbf{v})\mathbf{n} + \mu(\nabla\mathbf{v}^T - \nabla\mathbf{v})\mathbf{n}$$
$$(10.13.3)$$

By use of identities (3.5.34) and (3.6.19), we find that

$$(\nabla\mathbf{v})\mathbf{n} = (\mathbf{n} \cdot \nabla)\mathbf{v}; \qquad (\nabla\mathbf{v}^T - \nabla\mathbf{v})\mathbf{n} = \mathbf{n} \times \operatorname{curl} \mathbf{v} \qquad (10.13.4)$$

Using these in (10.13.3), we obtain (10.13.2a).

For a compressible viscous fluid, the material law is given by (10.12.4). Using this in Cauchy's law (7.4.8) we obtain the following counterpart of (10.13.3):

$$\mathbf{s} = \mathbf{T}\mathbf{n} = -p\mathbf{n} + \lambda(\operatorname{div} \mathbf{v})\mathbf{n} + 2\mu(\nabla\mathbf{v})\mathbf{n} + \mu(\nabla\mathbf{v}^T - \nabla\mathbf{v})\mathbf{n} \quad (10.13.5)$$

Using (10.13.4) in (10.13.5), we obtain (10.12.2b). ■

Note: Expressions (10.13.2a) and (10.13.2b) can be utilized to rewrite the boundary condition (10.5.2b) entirely in terms of **v** and p as follows

$$-p\mathbf{n} + 2\mu(\mathbf{n} \cdot \nabla)\mathbf{v} + \mu(\mathbf{n} \times \operatorname{curl} \mathbf{v}) = \mathbf{s}^* \qquad (10.13.6a)$$

(incompressible case);

$$(-p + \lambda \operatorname{div} \mathbf{v})\mathbf{n} + 2\mu(\mathbf{n} \cdot \nabla)\mathbf{v} + \mu(\mathbf{n} \times \operatorname{curl} \mathbf{v}) = \mathbf{s}^* \qquad (10.13.6b)$$

(compressible case), on S_τ for $t \geq 0$.

10.13 INITIAL AND BOUNDARY CONDITIONS

EXAMPLE 10.13.2 Show that at a fixed rigid solid surface contacting a viscous fluid, the stress vector is given by

$$\mathbf{s} = -p\mathbf{n} + \mu \mathbf{w} \times \mathbf{n} \tag{10.13.7a}$$

(incompressible case);

$$\mathbf{s} = [-p + (\lambda + 2\mu)\, \text{div}\, \mathbf{v}]\mathbf{n} + \mu \mathbf{w} \times \mathbf{n} \tag{10.13.7b}$$

(compressible case). Hence, compute the normal and shear stresses exerted by the fluid on the solid surface.

Solution At a fixed rigid solid surface contacting a viscous fluid, we have $\mathbf{v} = \mathbf{0}$, by the no-slip boundary condition (10.13.1)'. Hence if C is an *arbitrary* simple closed curve chosen on this surface,

$$\int_C \mathbf{v} \times d\mathbf{x} = \mathbf{0} \tag{10.13.8}$$

By using a consequence of Stokes's theorem, given by (3.6.20), expression (10.13.8) becomes

$$\int_S [(\text{div}\, \mathbf{v})\mathbf{n} - (\mathbf{n} \cdot \nabla)\mathbf{v} - \mathbf{n} \times \mathbf{w}]\, dS = \mathbf{0} \tag{10.13.9}$$

where S is the area on the solid surface, bounded by C. Since C is arbitrary, it follows that the integrand of the surface integral in (10.13.9) should vanish identically; hence,

$$(\mathbf{n} \cdot \nabla)\mathbf{v} = (\text{div}\, \mathbf{v})\mathbf{n} + \mathbf{w} \times \mathbf{n} \tag{10.13.10}$$

on the solid surface considered.

Substituting for $(\mathbf{n} \cdot \nabla)\mathbf{v}$ from (10.13.10) in (10.13.2a) and noting that $\text{div}\, \mathbf{v} = 0$ for an incompressible fluid, we get the result (10.13.7a). Similarly, (10.13.10) and (10.13.2b) yield the result (10.13.7b).

The normal stress exerted by the fluid on the solid surface is given by

$$\sigma = -\mathbf{s} \cdot \mathbf{n} \tag{10.13.11}$$

where $\mathbf{n}$ is directed *into* the fluid. For an incompressible fluid, (10.13.7a) and (10.13.11) yield $\sigma = p$. Since p is equal to the mean pressure $\bar{p}$ for an incompressible fluid, it follows that

$$\sigma = \bar{p} \tag{10.13.12}$$

Thus, in the case of an incompressible viscous fluid, the fluid elements exert just the mean stress on the surface, along the normal.

For a compressible fluid, we find by using (10.13.7b) in (10.13.11) that

$$\sigma = p - (\lambda + 2\mu)(\text{div } \mathbf{v}) \qquad (10.13.13)$$

which on using (10.11.6) and (10.11.5) may be rewritten as

$$\sigma = \bar{p} - \tfrac{4}{3}\mu(\text{div } \mathbf{v}) \qquad (10.13.14)$$

Thus, in the case of a compressible viscous fluid the normal stress exerted by the fluid on the surface is caused by the mean pressure as well as the rate of change in volume of fluid elements.

If $\mathbf{t}$ is a unit vector tangential to the surface, the magnitude of the shear stress exerted by the fluid on the solid surface is

$$\tau = |\mathbf{s} \cdot \mathbf{t}| \qquad (10.13.15)$$

By using (10.13.7ab) in (10.13.15), we find that the following expression holds for both incompressible and compressible fluids:

$$\tau = \mu |[\mathbf{w}, \mathbf{n}, \mathbf{t}]| \qquad (10.13.16)$$

Thus, τ depends only on the vorticity vector $\mathbf{w}$ for both incompressible and compressible viscous fluids.

Form (10.13.16) it also follows that, in irrotational motion, no shear stress is exerted on a fixed solid boundary. ∎

EXAMPLE 10.13.3 Show that, at a fixed rigid solid surface contacting a viscous fluid, the vorticity vector lies tangential to the surface.

Solution Let C be an arbitrary simple closed curve on the surface and let S be the area of the surface enclosed by C. Then, by Stokes's theorem, we have

$$\int_C \mathbf{v} \cdot d\mathbf{x} = \int_S \mathbf{w} \cdot \mathbf{n} \, dS \qquad (10.13.17)$$

But by the no-slip boundary condition $\mathbf{v} = \mathbf{0}$ on S. Hence the lefthand side of (10.13.17) vanishes identically. Since C is arbitrary, it follows that $\mathbf{w} \cdot \mathbf{n} = 0$ or that $\mathbf{w}$ is tangential to the surface, at every point. ∎

10.14
NAVIER–STOKES EQUATION

We now deduce the equation of motion of a *viscous* fluid by eliminating the stresses from the governing equations.

Let us start with Stokes's law given by (10.12.4), namely,

$$\mathbf{T} = (-p + \lambda \, \text{div } \mathbf{v})\mathbf{I} + \mu(\nabla \mathbf{v} + \nabla \mathbf{v}^T) \qquad (10.14.1)$$

10.14 NAVIER-STOKES EQUATION

Taking the divergence on both sides of this tensor equation and using the identities (3.5.35), (3.5.13) and (3.5.14), we obtain

$$\text{div } \mathbf{T} = \nabla(-p + \lambda \text{ div } \mathbf{v}) + \mu(\nabla^2 \mathbf{v} + \nabla \text{ div } \mathbf{v}) \qquad (10.14.2)$$

Substituting this expression into Cauchy's equation of motion (10.12.2) we arrive at the following equation of motion expressed in terms of p, ρ and $\mathbf{v}$:

$$\mu \nabla^2 \mathbf{v} + (\lambda + \mu)\nabla(\text{div } \mathbf{v}) - \nabla p + \rho \mathbf{b} = \rho \frac{D\mathbf{v}}{Dt} \qquad (10.14.3)$$

This equation, essentially by Navier (1822) and Stokes (1845), is referred to as the *Navier–Stokes equation*. This equation holds for both compressible and incompressible viscous fluid flows; in the incompressible case, $\rho = \rho_0$ and div $\mathbf{v} = 0$.

In the absence of viscosity, that is, if λ and μ are negligibly small, equation (10.14.3) reduces to Euler's equation of motion (10.6.2).

Using the identity (10.6.4), namely,

$$\frac{D\mathbf{v}}{Dt} = \frac{\partial \mathbf{v}}{\partial t} + (\mathbf{v} \cdot \nabla)\mathbf{v} = \frac{\partial \mathbf{v}}{\partial t} + \mathbf{w} \times \mathbf{v} + \frac{1}{2}\nabla v^2 \qquad (10.14.4)$$

equation (10.14.3) can be rewritten in the following alternative forms:

$$\mu \nabla^2 \mathbf{v} + (\lambda + \mu)\nabla(\text{div } \mathbf{v}) - \nabla p + \rho \mathbf{b} = \rho \left\{ \frac{\partial \mathbf{v}}{\partial t} + (\mathbf{v} \cdot \nabla)\mathbf{v} \right\} \qquad (10.14.5a)$$

$$\mu \nabla^2 \mathbf{v} + (\lambda + \mu)\nabla(\text{div } \mathbf{v}) - \nabla p + \rho \mathbf{b} = \rho \left\{ \frac{\partial \mathbf{v}}{\partial t} + \mathbf{w} \times \mathbf{v} + \frac{1}{2}\nabla v^2 \right\} \qquad (10.14.5b)$$

For an incompressible viscous fluid, the Navier–Stokes equation (10.14.3) is often rewritten in the following form:

$$\nu \nabla^2 \mathbf{v} - \frac{1}{\rho}\nabla p + \mathbf{b} = \frac{D\mathbf{v}}{Dt} \qquad (10.14.6)$$

where

$$\nu = \mu/\rho. \qquad (10.14.7)$$

The coefficient ν has dimension area/time and is called *kinematic viscosity*.

By using (10.14.4), the equation (10.14.6) can be rewritten in two explicit forms as follows:

$$\nu \nabla^2 \mathbf{v} - \frac{1}{\rho}\nabla p + \mathbf{b} = \frac{\partial \mathbf{v}}{\partial t} + (\mathbf{v} \cdot \nabla)\mathbf{v} \qquad (10.14.8a)$$

$$\nu \nabla^2 \mathbf{v} - \frac{1}{\rho}\nabla p + \mathbf{b} = \frac{\partial \mathbf{v}}{\partial t} + \mathbf{w} \times \mathbf{v} + \frac{1}{2}\nabla v^2 \qquad (10.14.8b)$$

When the external force is conservative, that is, if $\mathbf{b} = -\nabla\chi$, and ρ is a constant we get the following useful form of equation (10.14.8b):

$$\nu\nabla^2\mathbf{v} - \nabla\left(\frac{p}{\rho} + \chi + \frac{1}{2}v^2\right) = \frac{\partial\mathbf{v}}{\partial t} + \mathbf{w}\times\mathbf{v} \qquad (10.14.9)$$

In the case of a compressible viscous fluid for which Stokes's condition (10.11.8) holds, we have $\lambda = -(2/3)\mu$; see (10.11.5). Then, equation (10.14.3) involves only the shear viscosity μ.

Like Euler's equation of motion for a nonviscous fluid, the Navier–Stokes equation is the fundamental equation of motion for a viscous fluid. Whereas the Euler's equation is a first-order partial differential equation in spatial derivatives of velocity, the Navier–Stokes equation is of second order in these derivatives.

It may be noted that the Navier–Stokes equation (10.14.3) is strikingly analogous to Navier's equation (9.8.6) of classical elasticity. While (9.8.6) is a *linear* partial differential equation containing the displacement $\mathbf{u}$ as the only unknown, (10.14.3) is a *nonlinear* partial differential equation that contains the velocity $\mathbf{v}$, the density ρ and the pressure p as unknowns, the nonlinearity of the equation stemming from the presence of $(\mathbf{v}\cdot\nabla)\mathbf{v}$ in the acceleration. Thus, whereas (9.8.6) provides a complete set of field equations for the determination of $\mathbf{u}$, (10.14.3) does not provide a complete set of field equations for the determination of $\mathbf{v}$; it has to be supplemented by the equation of continuity in the incompressible case and, further, equations of state and an equation of heat conduction in the compressible case.

10.14.1 REYNOLDS NUMBER

Due to nonlinearity of the Navier–Stokes equation, no general method of solution is available. Exact solutions are known only in special cases. Approximate solutions are often obtained either by neglecting the *convective term* $(\mathbf{v}\cdot\nabla)\mathbf{v}$ in comparison with the *viscous term* $\nu\nabla^2\mathbf{v}$ or by neglecting the viscous term in comparison with the convective term. Thus, the ratio of the order of the convective term to that of the viscous term plays a crucial role in solving the Navier–Stokes equation approximately.

Suppose that, in a given flow, V is a reference speed and L is a reference length. Then, the magnitude of the convective term $(\mathbf{v}\cdot\nabla)\mathbf{v}$ is of order $V(V/L) = V^2/L$ and the magnitude of the viscous term $\nu\nabla^2\mathbf{v}$ is of order $(\nu V/L^2)$. Thus,

$$\frac{|(\mathbf{v}\cdot\nabla)\mathbf{v}|}{|\nu(\nabla^2\mathbf{v})|} = O\left(\frac{VL}{\nu}\right)$$

10.14 NAVIER-STOKES EQUATION

Since v has dimension area/time, we note that (VL/v) is a dimensionless quantity. This quantity is called the *Reynolds number*, after O. Reynolds, and we denote it by Re. Thus, the convective term $(\mathbf{v} \cdot \nabla)\mathbf{v}$ is negligible in comparison with the viscous term $v(\nabla^2 \mathbf{v})$ when Re is very small, and the viscous term is negligible in comparison with the convective term when Re is very large. For small values of Re, the Navier–Stokes equation takes a linearized form; for example, equation (10.14.6) will reduce to equation (10.14.28) given later. For large values of Re, the effect of viscosity is significant only in boundary layers; at points outside a boundary layer the Navier–Stokes equation is approximated to Euler's equation valid for nonviscous fluids.

EXAMPLE 10.14.1 Find the pressure distribution such that the velocity field given by

$$v_1 = k(x_1^2 - x_2^2), \quad v_2 = 2kx_1x_2, \quad v_3 = 0, \quad (k = \text{constant}) \quad (10.14.10)$$

satisfies the Navier–Stokes equation for an incompressible fluid in the absence of body force.

Solution When written in the component form, the Navier–Stokes equation for an incompressible fluid given by (10.14.6) reads as follows, on using the identity (10.14.4) also taken in the component form:

$$\frac{\partial v_1}{\partial t} + \left(v_1 \frac{\partial}{\partial x_1} + v_2 \frac{\partial}{\partial x_2} + v_3 \frac{\partial}{\partial x_3} \right) v_1 = v\nabla^2 v_1 - \frac{1}{\rho}\frac{\partial p}{\partial x_1} + b_1$$

$$\frac{\partial v_2}{\partial t} + \left(v_1 \frac{\partial}{\partial x_1} + v_2 \frac{\partial}{\partial x_2} + v_3 \frac{\partial}{\partial x_3} \right) v_2 = v\nabla^2 v_2 - \frac{1}{\rho}\frac{\partial p}{\partial x_2} + b_2 \quad (10.14.11)$$

$$\frac{\partial v_3}{\partial t} + \left(v_1 \frac{\partial}{\partial x_1} + v_2 \frac{\partial}{\partial x_2} + v_3 \frac{\partial}{\partial x_3} \right) v_3 = v\nabla^2 v_3 - \frac{1}{\rho}\frac{\partial p}{\partial x_3} + b_3$$

Substituting the expressions for v_i from (10.14.10) in equations (10.14.11) and noting that $b_i = 0$, we obtain

$$\frac{\partial p}{\partial x_1} = -2k^2\rho x_1(x_1^2 + x_2^2) \quad (10.14.12)$$

$$\frac{\partial p}{\partial x_2} = -2k^2\rho x_2(x_1^2 + x_2^2) \quad (10.14.13)$$

$$\frac{\partial p}{\partial x_3} = 0 \quad (10.14.14)$$

Thus, the pressure-gradient in the x_3-direction should be 0. Accordingly, $p = p(x_1, x_2)$ so that

$$dp = \frac{\partial p}{\partial x_1} dx_1 + \frac{\partial p}{\partial x_2} dx_2$$

This yields, on using (10.14.12) and (10.14.13),

$$dp = -2k^2\rho(x_1^2 + x_2^2)(x_1\, dx_1 + x_2\, dx_2) = -k^2\rho d\{\tfrac{1}{2}(x_1^2 + x_2^2)^2\}$$

Hence

$$p = -\tfrac{1}{2}k^2\rho(x_1^2 + x_2^2)^2 + C \quad (10.14.15)$$

where C is an arbitrary constant. If p^0 is the pressure at the origin, we get

$$p = p^0 - \tfrac{1}{2}k^2\rho(x_1^2 + x_2^2)^2 \quad (10.14.16)$$

This is the required pressure distribution. ∎

EXAMPLE 10.14.2 For steady flow of an incompressible viscous fluid under a conservative body force, prove the following:

(i) $\quad \mathbf{v} \times \mathbf{w} = \nabla\left(\dfrac{p}{\rho} + \dfrac{1}{2}v^2 + \chi\right) + \nu\,\mathrm{curl}\,\mathbf{w} \quad (10.14.17)$

(ii) $\quad (\mathbf{v} \cdot \nabla)\mathbf{w} - (\mathbf{w} \cdot \nabla)\mathbf{v} = \nu\nabla^2\mathbf{w} \quad (10.14.18)$

(iii) If

$$\mathbf{v} = \psi_{,2}\mathbf{e}_1 - \psi_{,1}\mathbf{e}_2 \quad (10.14.19)$$

where $\psi = \psi(x_1, x_2)$, then

$$\begin{vmatrix} (\nabla^2\psi)_{,1} & (\nabla^2\psi)_{,2} \\ \psi_{,1} & \psi_{,2} \end{vmatrix} = \nu\nabla^4\psi \quad (10.14.20)$$

Solution (i) For an incompressible viscous fluid moving under a conservative force, the Navier–Stokes equation is given by (10.14.9).

For steady flow $\partial \mathbf{v}/\partial t = \mathbf{0}$. Also, for an incompressible fluid,

$$\mathrm{curl}\,\mathbf{w} = \mathrm{curl}\,\mathrm{curl}\,\mathbf{v} = \mathrm{grad}\,\mathrm{div}\,\mathbf{v} - \nabla^2\mathbf{v} = -\nabla^2\mathbf{v} \quad (10.14.21)$$

because $\mathrm{div}\,\mathbf{v} = 0$. Using these facts in equation (10.14.9), we obtain (10.14.17).

(ii) From (10.14.21), we get

$$\mathrm{curl}\,\mathrm{curl}\,\mathbf{w} = -\nabla^2\mathbf{w} \quad (10.14.22)$$

Also, since $\mathrm{div}\,\mathbf{v} = 0$ and $\mathrm{div}\,\mathbf{w} = 0$, we find from identity (3.4.25) that

$$\mathrm{curl}(\mathbf{v} \times \mathbf{w}) = (\mathbf{w} \cdot \nabla)\mathbf{v} - (\mathbf{v} \cdot \nabla)\mathbf{w} \quad (10.14.23)$$

10.14 NAVIER-STOKES EQUATION

Now, taking the curl on both sides of (10.14.17) and using (10.14.22), (10.14.23) and (3.4.16), we obtain (10.14.18).

(iii) When **v** is given by (10.14.19), we get

$$\mathbf{w} = \text{curl } \mathbf{v} = -\nabla^2 \psi \mathbf{e}_3 \qquad (10.14.24)$$

so that

$$(\mathbf{v} \cdot \nabla)\mathbf{w} = \left(\psi_{,2}\frac{\partial}{\partial x_1} - \psi_{,1}\frac{\partial}{\partial x_2}\right)(-\nabla^2 \psi)\mathbf{e}_3 \qquad (10.14.25)$$

and

$$(\mathbf{w} \cdot \nabla)\mathbf{v} = \mathbf{0} \qquad (10.14.26)$$

Using (10.14.24), (10.14.25) and (10.14.26) in (10.14.18), we obtain (10.14.20). ∎

Note: The function ψ defined by (10.14.19) is known as *two-dimensional stream function*.

EXAMPLE 10.14.3 A fluid motion for which the Reynolds number is small (so that nonlinear terms in velocity are negligible) is known as a *creeping flow* or *Stokes's flow*. For a steady creeping flow of an incompressible viscous fluid under zero body force, show that p is a harmonic function. Deduce that ψ defined by (10.14.19) is a biharmonic function in this case.

Solution For creeping flow,

$$\frac{D\mathbf{v}}{Dt} = \frac{\partial \mathbf{v}}{\partial t} + (\mathbf{v} \cdot \nabla)\mathbf{v} \approx \frac{\partial \mathbf{v}}{\partial t} \qquad (10.14.27)$$

Then the Navier-Stokes equation (10.14.6) becomes

$$\nu \nabla^2 \mathbf{v} - \frac{1}{\rho}\nabla p + \mathbf{b} = \frac{\partial \mathbf{v}}{\partial t} \qquad (10.14.28)$$

For steady flow with zero body force, this equation reduces to

$$\nu \nabla^2 \mathbf{v} = \frac{1}{\rho}\nabla p \qquad (10.14.29)$$

Taking the divergence of this equation and using the equation of continuity div **v** = 0, we get

$$\nabla^2 p = 0 \qquad (10.14.30)$$

Thus, p is a harmonic function.

For ψ defined through the relation (10.14.19), we have

$$v_1 = \psi_{,2}, \quad v_2 = -\psi_{,1}, \quad v_3 = 0$$

532   10 EQUATIONS OF FLUID MECHANICS

Using these in equation (10.14.29), we get

$$p_{,1} = \rho \nu \nabla^2(\psi_{,2}); \quad p_{,2} = -\rho \nu \nabla^2(\psi_{,1}) \tag{10.14.31}$$

From these, we obtain, respectively,

$$p_{,12} = \rho \nu \nabla^2(\psi_{,22}); \quad p_{,21} = -\rho \nu \nabla^2(\psi_{,11}) \tag{10.14.32}$$

so that

$$\nu \rho \nabla^2(\psi_{,11} + \psi_{,22}) = -(p_{,21} - p_{,12}) = 0 \quad \text{or} \quad \nabla^4 \psi = 0$$

Thus, ψ is a biharmonic function.   ∎

EXAMPLE 10.14.4 Consider a steady motion of an incompressible viscous fluid under a conservative body force. If

$$H_0 = \frac{1}{2} v^2 + \frac{p}{\rho} + \chi \tag{10.14.33}$$

prove the following.

(i) H_0 is constant along the field lines of the vector

$$\mathbf{f} = (\mathbf{v} \times \mathbf{w}) \times \operatorname{curl} \mathbf{w} \tag{10.14.34}$$

(ii) $\quad \mathbf{v} \cdot \nabla H_0 = \nu(\nabla^2 H_0 - \mathbf{w}^2) \tag{10.14.35}$

Solution For the motion considered, the relation (10.14.17) holds. Using (10.4.33), this relation can be rewritten as

$$\nabla H_0 = (\mathbf{v} \times \mathbf{w}) - \nu \operatorname{curl} \mathbf{w} \tag{10.14.36}$$

From (10.14.34) and (10.14.36) we readily see that $\mathbf{f} \cdot \nabla H_0 = 0$. Thus, ∇H_0 is orthogonal to $\mathbf{f}$ and hence to the field line of $\mathbf{f}$. But ∇H_0 is always orthogonal to the surfaces of constant H_0. Hence $\mathbf{f}$ must be tangential to a surface of constant H_0. That is, H_0 is constant along the field lines of $\mathbf{f}$.

From (10.14.36), we get

$$\nabla^2 H_0 = \operatorname{div}(\mathbf{v} \times \mathbf{w}) = \mathbf{w}^2 - \mathbf{v} \cdot \operatorname{curl} \mathbf{w} \tag{10.14.37}$$

and

$$\mathbf{v} \cdot \nabla H_0 = -\nu \mathbf{v} \cdot \operatorname{curl} \mathbf{w} \tag{10.14.38}$$

These relations together yield the relation (10.14.35).   ∎

EXAMPLE 10.14.5 For a nonsteady flow of an incompressible viscous fluid under conservative body force with $\operatorname{curl} \mathbf{w} = \nabla \xi$ for some scalar function ξ, show that the Navier-Stokes equation becomes

$$\frac{\partial \mathbf{v}}{\partial t} + \mathbf{w} \times \mathbf{v} = -\nabla H^* \tag{10.14.39}$$

where

$$H^* = \frac{p}{\rho} + \frac{1}{2}v^2 + \chi + \nu\xi \qquad (10.14.40)$$

Deduce the following.

(i) If the flow is of potential kind, then $H^* + (\partial\phi/\partial t) = f(t)$, where $f(t)$ is an arbitrary function of t.

(ii) If the flow is steady and $\mathbf{v} \times \mathbf{w} \neq \mathbf{0}$, then H^* is constant along stream lines and vortex lines.

(iii) If the flow is steady and $\mathbf{v} \times \mathbf{w} = \mathbf{0}$, then H^* is constant everywhere in the fluid.

Solution For an incompressible viscous fluid moving under a conservative body force, the Navier–Stokes equation is given by (10.14.9).

Since curl $\mathbf{w} = -\nabla^2\mathbf{v}$, the given condition, curl $\mathbf{w} = \nabla\xi$, yields $\nabla^2\mathbf{v} = -\nabla\xi$. Using this, equation (10.14.9) becomes (10.14.39), with H^* defined by (10.14.40).

We observe that equation (10.14.39) is strikingly similar to equation (10.8.1). Indeed, in the absence of viscous effects, function H^* defined by (10.14.40) reduces to Bernoulli's function H defined by (10.8.2). Following the steps that led to Bernoulli's equations (10.8.4), (10.8.7) and (10.8.8) from (10.8.1), we arrive at the desired results starting from the equation (10.14.39). ∎

EXAMPLE 10.14.6 For an incompressible viscous fluid moving under a conservative body force, prove the following:

(i) $$\nabla^2\left(\frac{p}{\rho} + \chi + \frac{1}{2}v^2\right) = \operatorname{div}(\mathbf{v} \times \mathbf{w}) \qquad (10.14.41)$$

(ii) $$\nabla^2\left(\frac{p}{\rho} + \chi\right) = \tfrac{1}{2}\mathbf{w}^2 - \mathbf{D}\cdot\mathbf{D} \qquad (10.14.42)$$

Further, if the motion is irrotational, deduce that

(iii) $$\nabla^2 v^2 = 2\mathbf{D}\cdot\mathbf{D} \geq 0 \qquad (10.14.43)$$

Solution (i) For an incompressible viscous fluid moving under a conservative body force, the Navier–Stokes equation is given by (10.14.9). If we take divergence of this equation and note that div $\mathbf{v} = 0$, we get the relation (10.14.41).

(ii) Substituting for div($\mathbf{v} \times \mathbf{w}$) from the identity (6.3.21) (with div $\mathbf{v} = 0$) in the righthand side of the relation (10.14.41), we get the relation (10.14.42).

(iii) For an irrotational motion, $\mathbf{w} = \mathbf{0}$. In this case, subtracting (10.14.42) from (10.14.41) we get $\nabla^2 v^2 = 2\mathbf{D}\cdot\mathbf{D}$. Since $\mathbf{D}\cdot\mathbf{D} = d_{ij}d_{ij} \geq 0$, we obtain the inequality $\nabla^2 v^2 \geq 0$. Thus, (10.14.43) is proven. ∎

534   10 EQUATIONS OF FLUID MECHANICS

EXAMPLE 10.14.7 For a flow of an incompressible viscous fluid under conservative body force, show that the vorticity equation is given by

$$\frac{D\mathbf{w}}{Dt} = (\mathbf{w} \cdot \nabla)\mathbf{v} + \nu\nabla^2\mathbf{w} \tag{10.14.44}$$

Deduce the following.

(i) If curl $\mathbf{w} = \nabla\xi$, equation (10.14.44) becomes

$$\frac{D\mathbf{w}}{Dt} = (\mathbf{w} \cdot \nabla)\mathbf{v} \tag{10.14.45}$$

(ii) If the motion is two-dimensional (where $v_3 \equiv 0$ and v_1 and v_2 are independent of x_3), equation (10.14.44) reduces to

$$\frac{Dw}{Dt} = \nu\nabla^2 w \tag{10.14.46}$$

where $w = w_3$.

(iii) If the motion is two-dimensional and in circles with centers on x_3 axis, equation (10.14.46) reduces to

$$\frac{\partial w}{\partial t} = \nu\nabla^2 w \tag{10.14.47}$$

Solution For the given flow, the Navier–Stokes equation is given by (10.14.6) with $\mathbf{b} = -\nabla\chi$. Taking curl on both sides of this equation, we get

$$\operatorname{curl}\left(\frac{D\mathbf{v}}{Dt}\right) = \nu\nabla^2\mathbf{w} \tag{10.14.48}$$

Using this expression in Beltrami's vorticity equation (8.2.16), we obtain the equation (10.14.44).

(i) If curl $\mathbf{w} = \nabla\xi$, then the identity (3.4.27) yields $\nabla^2\mathbf{w} = \mathbf{0}$. Consequently, equation (10.14.44) reduces to equation (10.14.45).

(ii) If $v_3 \equiv 0$ and v_1 and v_2 are independent of x_3, we readily find

$$w_1 = w_2 = 0, \qquad w_3 = v_{2,1} - v_{1,2} \tag{10.14.49}$$

Consequently,

$$(\mathbf{w} \cdot \nabla)\mathbf{v} = \mathbf{0} \tag{10.14.50}$$

and equation (10.14.44) reduces to equation (10.14.46).

(iii) If the motion is two-dimensional and in circles with centers on the x_3 axis, we have $\mathbf{v} = v\mathbf{e}_\theta$, with $v = v(R, t)$, where R is the radial distance parallel to the $x_1 x_2$ plane and $\mathbf{e}_\theta$ is the unit vector in the transverse direction (see Figure 10.14).

10.14 NAVIER–STOKES EQUATION

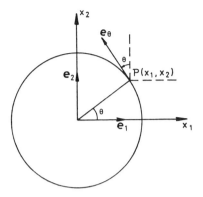

Figure 10.14. Example 10.14.7(iii).

From Figure 10.14, we find that

$$\mathbf{e}_\theta = (-\sin\theta)\mathbf{e}_1 + (\cos\theta)\mathbf{e}_2 = -\frac{x_2}{R}\mathbf{e}_1 + \frac{x_1}{R}\mathbf{e}_2$$

Thus,

$$\mathbf{v} = v\mathbf{e}_\theta = v[-x_2\mathbf{e}_1 + x_1\mathbf{e}_2]$$

so that

$$v_1 = -\frac{v}{R}x_2, \qquad v_2 = \frac{v}{R}x_1, \qquad v_3 = 0 \qquad (10.14.51)$$

These give, on noting that $v = v(R, t)$ and $R_{,i} = x_i/R$, $i = 1, 2$,

$$v_{1,2} = -\frac{v}{R} - \frac{1}{R}\frac{\partial v}{\partial R}R_{,2}x_2 + \frac{v}{R^2}R_{,2}x_2 = -\frac{v}{R} - \frac{1}{R^2}\frac{\partial v}{\partial R}x_2^2 + \frac{v}{R^3}x_2^2 \tag{10.14.52a}$$

$$v_{2,1} = \frac{v}{R} + \frac{1}{R}\frac{\partial v}{\partial R}R_{,1}x_1 - \frac{v}{R^2}R_{,1}x_1 = \frac{v}{R} + \frac{1}{R^2}\frac{\partial v}{\partial R}x_1^2 - \frac{v}{R^3}x_1^2 \tag{10.14.52b}$$

Using these in (10.14.49), we obtain

$$w = \frac{v}{R} + \frac{\partial v}{\partial R} \tag{10.14.53}$$

(where $w = w_3$), which yields

$$w_{,i} = \left[\frac{\partial^2 v}{\partial R^2} - \frac{1}{R}\frac{\partial v}{\partial R} - \frac{v}{R^2}\right]\frac{x_i}{R}, \qquad i = 1, 2 \tag{10.14.54}$$

From (10.14.51), (10.14.53) and (10.14.54) we find that

$$(\mathbf{v}\cdot\nabla)w = v_1 w_{,1} + v_2 w_{,2} = 0$$

536   10   EQUATIONS OF FLUID MECHANICS

Consequently,

$$\frac{Dw}{Dt} = \frac{\partial w}{\partial t} \tag{10.14.55}$$

Substituting this in (10.14.46), we obtain (10.14.47). ∎

Note: Using (10.14.53) it can be easily shown that

$$\nabla^2 w = w_{,kk} = \frac{\partial^2 w}{\partial R^2} + \frac{1}{R}\frac{\partial w}{\partial R} \tag{10.14.56}$$

Then equation (10.14.47) takes the polar form

$$\frac{\partial w}{\partial t} = v\left[\frac{\partial^2 w}{\partial R^2} + \frac{1}{R}\frac{\partial w}{\partial R}\right] \tag{10.14.47}'$$

It can be verified that a solution of this partial differential equation is

$$w = \frac{A}{8\pi vt}\exp\left[-\frac{R^2}{4vt}\right] \tag{10.14.57}$$

where A is a nonnegative constant. This solution shows that w decays rapidly with time, a phenomenon called *diffusion of vorticity*.

EXAMPLE 10.14.8   Show that the rate of decrease in kinetic energy due to viscosity in a finite volume V of an incompressible fluid is given by

$$W = \mu\int_V \mathbf{w}^2\, dV - \mu\int_S \mathbf{n}\cdot(\mathbf{v}\times\mathbf{w})\,dS \tag{10.14.58}$$

where S is the boundary of V.

If S is a rigid solid surface at rest, deduce that

$$W = \mu\int_V \mathbf{w}^2\, dV = \int_V \Phi\, dV \tag{10.14.59}$$

Solution   From (8.6.1), we recall that the kinetic energy of a continuum contained in a volume V is given by

$$K = \tfrac{1}{2}\int_V \rho(\mathbf{v}\cdot\mathbf{v})\,dV$$

Using (8.2.10), we obtain

$$\frac{DK}{Dt} = \frac{1}{2}\int_V \rho\frac{D}{Dt}(\mathbf{v}\cdot\mathbf{v})\,dV = \int_V \rho\mathbf{v}\cdot\frac{D\mathbf{v}}{Dt} \tag{10.14.60}$$

10.14 NAVIER-STOKES EQUATION

For an incompressible viscous fluid, expression (10.14.60) becomes, on using the Navier–Stokes equation (10.14.6),

$$\frac{DK}{Dt} = \mu \int_V \mathbf{v} \cdot \nabla^2 \mathbf{v}\, dV - \int_V \rho \mathbf{v} \cdot \nabla p\, dV + \int_V \rho \mathbf{v} \cdot \mathbf{b}\, dV \qquad (10.14.61)$$

The last two terms on the righthand side of this expression represents the rate of work done by the pressure and the body force, and the first term represents the contribution of viscosity to the rate of change of kinetic energy. Therefore, if W denotes the rate of *decrease* in kinetic energy due to viscosity, we have

$$W = -\mu \int_V \mathbf{v} \cdot \nabla^2 \mathbf{v}\, dV \qquad (10.14.62)$$

Since the fluid is incompressible, we note by using identities (3.4.27) and (3.4.24) that

$$\left.\begin{array}{l} \nabla^2 \mathbf{v} = -\text{curl curl } \mathbf{v} = -\text{curl } \mathbf{w} \\ \text{div}(\mathbf{v} \times \mathbf{w}) = \mathbf{w}^2 - \mathbf{v} \cdot \text{curl } \mathbf{w} \end{array}\right\} \qquad (10.14.63)$$

In view of these expressions, (10.14.62) becomes

$$W = \mu \int_V \mathbf{w}^2\, dV - \mu \int_V \text{div}(\mathbf{v} \times \mathbf{w})\, dV \qquad (10.14.64)$$

This expression yields (10.14.58), on using the divergence theorem (3.6.1).

If S is a rigid solid surface at rest, then by the no-slip boundary condition we have $\mathbf{v} = \mathbf{0}$ on S. Consequently, the surface integral on the righthand side of (10.14.58) becomes identically 0, and we obtain

$$W = \mu \int_V \mathbf{w}^2\, dV \qquad (10.14.65)$$

which is the first part of (10.14.59). To obtain the other part, we recall expression (10.10.19). For an incompressible fluid, this expression becomes

$$\int_V \Phi\, dV = \mu \int_V \mathbf{w}^2\, dV + 2\mu \int_S \frac{D\mathbf{v}}{Dt} \cdot \mathbf{n}\, dS \qquad (10.14.66)$$

If S is a rigid surface at rest, the surface integral in the expression (10.14.66) vanishes and we get

$$\int_V \Phi\, dV = \mu \int_V \mathbf{w}^2\, dV \qquad (10.14.67)$$

which is the second part of (10.14.59). ■

Expression (10.14.59) explicitly exhibits the remarkable relationship between the kinetic energy, the vorticity and the viscous dissipation of a fluid moving within a rigid enclosure.

538   **10 EQUATIONS OF FLUID MECHANICS**

EXAMPLE 10.14.9   For an incompressible viscous fluid moving under a conservative body force, show that the circulation I_c round a circuit c moving with the fluid is *not* constant in general. Deduce that I_c is constant if and only if curl $\mathbf{w} = \nabla \xi$, for some ξ.

Solution   For the given flow, the Navier–Stokes equation is given by (10.14.6) with $\mathbf{b} = -\nabla \chi$.

From (6.6.4), we recall that the rate of change of circulation round a material circuit is given by

$$\frac{DI_c}{Dt} = \oint_c \frac{D\mathbf{v}}{Dt} \cdot d\mathbf{x} \qquad (10.14.68)$$

Substituting for $D\mathbf{v}/Dt$ from (10.14.6) with $\mathbf{b} = -\nabla \chi$ in (10.14.68), we get

$$\frac{DI_c}{Dt} = \nu \oint_c \nabla^2 \mathbf{v} \cdot d\mathbf{x} \qquad (10.14.69)$$

Since the fluid is viscous, $\nu \neq 0$ and consequently $DI_c/Dt \neq 0$ when $\nabla^2 \mathbf{v} \neq \mathbf{0}$. Thus, I_c is *not* constant in general.

Since $\nabla^2 \mathbf{v} = -\text{curl } \mathbf{w}$ for an incompressible fluid, expression (10.14.69) can be rewritten as

$$\frac{DI_c}{Dt} = -\nu \oint_c \text{curl } \mathbf{w} \cdot d\mathbf{x} \qquad (10.14.70)$$

Evidently, I_c = constant if and only if curl $\mathbf{w} = \nabla \xi$ for some scalar ξ. (Thus, Kelvin's circulation theorem is not generally valid for viscous fluids.) ∎

10.15
SOME VISCOUS FLOW PROBLEMS

As remarked earlier, the presence of the nonlinear term in the Navier–Stokes equation makes the exact solution of the equation very difficult except for a very few cases. In this section we consider some simple examples of flows of an incompressible viscous fluid for which the Navier–Stokes equation admits exact solutions. The linearized Navier–Stokes equation is employed in the last example.

10.15.1 STEADY LAMINAR FLOW BETWEEN PARALLEL PLATES

First, we consider an incompressible viscous fluid bounded between two rigid infinite flat plates distance h apart. We suppose that the fluid flows steadily in a fixed direction parallel to the plates. Such a flow is called a

10.15 SOME VISCOUS FLOW PROBLEMS

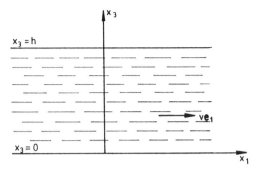

Figure 10.15. Steady laminar flow.

steady laminar flow. Assuming that there is no body force, the problem is to find the velocity field in the fluid.

We take the coordinate axes such that the x_1 axis is along the direction of flow and the x_3 axis is normal to the plates, with $x_3 = 0$ and $x_3 = h$ as the planes containing the plates; see Figure 10.15. It is convenient to refer to the plate in the plane $x_3 = 0$ as the lower plate and that in the plane $x_3 = h$ as the upper plate. Then the velocity is of the form $\mathbf{v} = v\mathbf{e}_1$, where $v = v(x_1, x_2, x_3)$, and the Navier-Stokes equation (10.14.6) gives the following equations for v and $p_{,i}$:

$$\nu \nabla^2 v - \frac{1}{\rho} p_{,1} = \frac{Dv}{Dt} \tag{10.15.1}$$

$$p_{,2} = 0, \tag{10.15.2a}$$

$$p_{,3} = 0 \tag{10.15.2b}$$

The equation of continuity (10.12.5) becomes

$$v_{,1} = 0 \tag{10.15.3}$$

If the fluid extends to infinity in the x_2 direction, then the variation of v in the x_2 direction may be neglected. Consequently, it follows that v varies only with x_3, and

$$\frac{Dv}{Dt} \equiv \frac{\partial v}{\partial t} + (\mathbf{v} \cdot \nabla)v = 0$$

Thus, under the assumptions made, inertia plays no role in the flow. From (10.15.2), it follows that $p = p(x_1)$, and equation (10.15.1) reduces to the ordinary linear differential equation

$$\mu \frac{d^2 v}{dx_3^2} = \frac{dp}{dx_1} \tag{10.15.4}$$

Thus the flow is induced entirely by the pressure gradient, $dp/dx_1 = p'(x_1)$ in the x_1 direction.

Integrating (10.15.4) successively twice, we obtain the following explicit expression for v:

$$\mu v = \tfrac{1}{2}x_3^2 p'(x_1) + x_2 f_1(x_1) + f_2(x_1) \tag{10.15.5}$$

where f_1 and f_2 are arbitrary functions of x_1. If we suppose that the lower plate ($x_3 = 0$) is stationary, then, by the no-slip boundary condition, we have $v = 0$ for $x_3 = 0$ and (10.15.5) gives $f_2(x_1) = 0$. If we compute $v_{,1}$ from (10.15.5) and use (10.15.3) we get

$$x_3 p''(x_1) = -2 f_1'(x_1) \tag{10.15.6}$$

Since this equation holds for all x_1 and x_3 in $0 \leq x_3 \leq h$, we must have $p''(x_1) = 0$ and $f_1'(x_1) = 0$ so that

$$f_1(x_1) = C_1 \tag{10.15.7a}$$

$$p(x_1) = C_2 - Gx_1 \tag{10.15.7b}$$

where C_1, C_2 and G are arbitrary constants. All these results allow us to rewrite (10.15.5) in the form

$$\mu v = C_1 x_3 - \tfrac{1}{2}Gx_3^2 \tag{10.15.8}$$

This shows that the flow occurs only when at least one of the constants C_1 and G is nonzero.

Since $v_1 = v(x_3)$, $v_2 = v_3 = 0$, we find from the material law (10.12.8) that

$$\tau_{31} = \mu \frac{dv}{dx_3} \tag{10.15.9}$$

which, by using (10.15.8), reduces to

$$\tau_{31} = C_1 - Gx_3 \tag{10.15.10}$$

When the constants C_1 and G are known, (10.15.8) gives the velocity distribution and (10.15.10) gives the shear stress on plane elements parallel to the plates. Evidently, (10.15.8) represents the parabolic profile of the flow between the plates. Expression (10.15.10) shows that the shear stress τ_{31} varies linearly with height (increasing x_3 direction) with

$$C_1 = [\tau_{31}]_{x_3 = 0} \tag{10.15.11}$$

as the shear stress at the lower plate. Thus the constant C_1 involved in the solution (10.15.8) has a definite physical interpretation. It can be seen from (10.15.7b) that the other constant G in (10.15.8) also has physical significance; in fact,

$$G = -p'(x_1) \tag{10.15.12}$$

Thus, G represents the negative of the pressure gradient in the direction of the flow. Since G is a constant, the pressure gradient must be a constant for the flow. Further, since the flow takes place in the x_1 direction, the pressure does not increase in that direction; that is, $G \geq 0$.

Two particular cases of this problem are of interest.

Case i: Plane Couette Flow Suppose the flow is generated by the movement of the upper plate with a constant speed v_0 in the x_1 direction. Such a flow is called *plane Couette flow*. Then by the no-slip boundary condition, we have $v = v_0$ for $x_3 = h$, and (10.15.8) gives

$$C_1 = \frac{1}{h}\mu v_0 + \frac{1}{2}Gh \tag{10.15.13}$$

Putting this expression back into (10.15.8), we obtain

$$\mu v = \frac{1}{2}Gx_3(h - x_3) + \frac{\mu v_0}{h}x_3 \tag{10.15.14}$$

This determines the velocity distribution completely, provided the constant pressure-gradient $G = -p'(x_1)$ is known beforehand. For example, if this pressure-gradient is 0, then

$$v = \frac{v_0}{h}x_3 \tag{10.15.15}$$

This represents a *linear* distribution of velocity depicted in Figure 10.16a. In the figure, a few velocity vectors are indicated to make the flow easier to visualize. Such a diagram is called a *velocity profile*. The velocity profile for $p'(x_1) < 0$ is given in Figure 10.16b.

Expression (10.15.14) can be used to find the *mass flow rate per unit of width*; that is the mass of fluid that passes the plane $x_1 = $ constant per unit of distance in the x_2 direction and per unit of time. This is given by

$$M = \int_0^h \rho v \, dx_3 = \frac{G}{2\nu}\int_0^h x_3(h - x_3)\, dx_3 + \frac{\rho v_0}{h}\int_0^h x_3 \, dx_3 = \frac{Gh^3}{12\nu} + \frac{\rho v_0 h}{2} \tag{10.15.16}$$

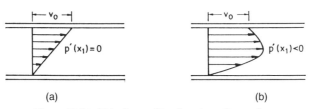

Figure 10.16. Velocity profiles for plane Coutte flow.

In the case of $p'(x_1) = 0$, we get

$$M = \frac{\rho h}{2} v_0 \tag{10.15.17}$$

Thus, in this case, M is independent of viscosity.
From (10.15.10) and (10.15.13), we get

$$\tau_{31} = \frac{\mu}{h} v_0 + G\left(\frac{h}{2} - x_3\right) \tag{10.15.18}$$

This expression gives the shear stress on a fluid element parallel to the plates (for Couette flow). The shear stress exerted on the plates is often called the *skin friction*. The skin friction on the lower plate is given by

$$[\tau_{31}]_{x_3=0} = \frac{\mu}{h} v_0 + \frac{Gh}{2} \tag{10.15.19}$$

and that on the upper plate is given by

$$[\tau_{31}]_{x_3=h} = \frac{\mu}{h} v_0 - \frac{Gh}{2} \tag{10.15.20}$$

When $p'(x_1) = 0$, (10.15.18) gives

$$\tau_{31} = \frac{\mu}{h} v_0 \tag{10.15.21}$$

Evidently, in this case, all fluid particles, including those in contact with the plates, experience the same shear stress. This stress is directly proportional to v_0 and inversely proportional to h and acts in the direction of the flow.

Case ii: Plane Poiseullie Flow Suppose the upper plate is also held stationary and the flow is generated only by the constant pressure gradient $p'(x_1) < 0$. Such a flow is called a *plane Poiseullie flow*. Then $v_0 = 0$ and $G > 0$, and (10.15.14) yields the following expression for velocity distribution:

$$\mu v = \tfrac{1}{2} G x_3 (h - x_3) \tag{10.15.22}$$

Evidently, the velocity distribution is now symmetrical about the midplane $x_3 = h/2$ and the velocity is maximum on this plane, with

$$v_{\max} = [v]_{x_3 = h/2} = \frac{G}{8\mu} h^2 \tag{10.15.23}$$

The velocity profile is depicted in Figure 10.17.

10.15 SOME VISCOUS FLOW PROBLEMS

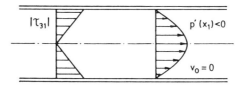

Figure 10.17. Plane Poiseuillie flow.

Substituting for v from (10.15.22) in (10.15.9) we get the following expression for the shear stress on a fluid element parallel to the plates:

$$\tau_{31} = G\left(\frac{h}{2} - x_3\right) \qquad (10.15.24)$$

We observe that $\tau_{31} = 0$ on the midplane $x_3 = h/2$ and that $\tau_{31} > 0$ for $0 \leq x_3 < h/2$ and $\tau_{31} < 0$ for $h/2 < x_3 \leq h$. That is, plane elements laying below the midplane experience a shear stress along the direction of the flow while those above experience a shear stress in the opposite direction, with the elements laying on the midplane remaining stress free.

From (10.15.24), we find that the skin friction on the lower plate is

$$[\tau_{31}]_{x_3 = 0} = \frac{Gh}{2} \qquad (10.15.25)$$

and that on the upper plate is

$$[\tau_{31}]_{x_3 = h} = -\frac{Gh}{2} \qquad (10.15.26)$$

The magnitudes of the skin frictions on both the plates are equal to $Gh/2$. But, while the lower plate experiences a friction along the direction of the flow, the upper plate experiences a (backward) *drag*.

Expression (10.15.22) can be employed to find the mass flow rate per unit of width. This is given by

$$M = \int_0^h \rho v \, dx_3 = \frac{G}{2\nu} \int_0^h x_3(h - x_3) \, dx_3 = \frac{Gh^3}{12\nu} \qquad (10.15.27)$$

Evidently, for given G and h, M varies inversely as ν. Expression (10.15.27) can be used to measure ν.

10.15.2. STEADY FLOW IN A STRAIGHT CONDUIT

We now consider the steady flow of an incompressible viscous fluid through a straight conduit of uniform cross section under a constant pressure gradient along the direction of flow. Such a flow is called the *Hagen-Poiseullie flow*. If we choose the axes such that the x_3 axis is along the direction of the conduit so that a right cross section of the conduit is bounded by a simple closed curve C parallel to the $x_1 x_2$ plane, then the velocity at any point of the fluid is of the form $\mathbf{v} = v\mathbf{e}_3$, where $v = v(x_1, x_2, x_3)$, and the equation of continuity (10.12.5) yields $v_{,3} = 0$. Consequently, $v = v(x_1, x_2)$ and $(\mathbf{v} \cdot \nabla)\mathbf{v} = 0$. Since the flow is steady, we get $(D\mathbf{v}/Dt) = 0$. Under these conditions the Navier–Stokes equation (10.14.6) yields the following three component equations in the absence of body force:

$$p_{,1} = 0 \tag{10.15.28a}$$

$$p_{,2} = 0 \tag{10.15.28b}$$

$$p_{,3} = \mu(v_{,11} + v_{,22}) \tag{10.15.28c}$$

The first two of these equations show that $p = p(x_3)$. It is assumed that the pressure-gradient is a constant; hence,

$$\frac{dp}{dx_3} = -G \tag{10.15.29}$$

where G is a constant. Note that, for the fluid to flow in the x_3 direction, G has to be positive.

Equations (10.15.28c) and (10.15.29) yield the following governing equation for v:

$$\frac{\partial^2 v}{\partial x_1^2} + \frac{\partial^2 v}{\partial x_2^2} = -\frac{G}{\mu} \tag{10.15.30}$$

If the conduit is stationary, the no-slip boundary condition yields

$$v = 0 \tag{10.15.31}$$

on C; see Figure 10.18.

We note that (10.15.30) is a two-dimensional Poisson equation. When solved under the boundary condition (10.15.31), this equation determines v uniquely. Once v is determined, the stresses developed in the fluid can be computed by use of the material law (10.12.8).

We now consider two particular cases for further analysis.

Case i: Circular Cross Section Suppose the conduit is of circular cross section. Then C is a circle whose equation can be taken as

$$x_1^2 + x_2^2 = a^2 \tag{10.15.32}$$

where a is the radius of the conduit.

Since $v = 0$ on C by (10.15.31), the equation (10.15.32) of C suggests that v is of the form

$$v = \alpha(x_1^2 + x_2^2 - a^2) \tag{10.15.33}$$

10.15 SOME VISCOUS FLOW PROBLEMS 545

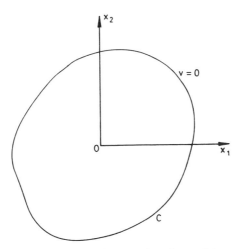

Figure 10.18. Cross section of a conduit.

where α is a constant. Substituting for v from (10.15.33) into equation (10.15.30), we find that $\alpha = -G/4\mu$. Thus,

$$v = \frac{G}{4\mu}(a^2 - x_1^2 - x_2^2) \tag{10.15.34}$$

is the expression for v that satisfies the governing equation (10.15.30) as well as the boundary condition (10.15.31). We note that the velocity distribution is in the form of a paraboloid of revolution; the velocity profile in a section parallel to the length of the pipe is depicted in Figure 10.19. We find that the velocity is maximum on the axis of the conduit, namely the x_3 axis, and

$$v_{\max} = \frac{G}{4\mu}a^2 \tag{10.15.35}$$

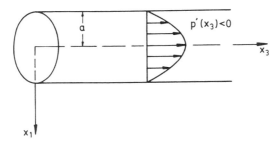

Figure 10.19. Velocity profile for case (i).

Expression (10.15.34) can be used to find the rate of mass of fluid that passes a cross section (x_3 = constant); this rate is given by

$$M = \int_A \rho v \, dA = 2\pi\rho \int_0^a vR \, dR \qquad (10.15.36)$$

where A is the area of cross section, and $R = (x_1^2 + x_2^2)^{1/2}$. Using (10.15.34) in (10.15.36), we get

$$M = \frac{\rho \pi G a^4}{8\mu} \qquad (10.15.37)$$

Consequently, the rate of volume of fluid that passes a cross section (x_3 = constant) is

$$\frac{M}{\rho} = \frac{\pi G a^4}{8\mu} \qquad (10.15.38)$$

This is known as *Poiseullie's formula*. This formula is generally used to measure the viscosity of a fluid.

If we consider a cylindrical fluid surface element dS coaxial with the conduit and having radius R, then the components of the exterior unit normal to this element are

$$n_1 = \frac{x_1}{R}, \qquad n_2 = \frac{x_2}{R}, \qquad n_3 = 0 \qquad (10.15.39)$$

The components of the stress vector on dS can be computed by use of Cauchy's law (7.4.9), the material law (10.12.8) and expressions (10.15.34) and (10.15.39). The stress components s_i thus obtained are

$$s_1 = -p\frac{x_1}{R} \qquad (10.15.40a)$$

$$s_2 = -p\frac{x_2}{R} \qquad (10.15.40b)$$

$$s_3 = -\frac{G}{2}R \qquad (10.15.40c)$$

Together with (10.15.39), expressions (10.15.40ab) yield the normal stress on dS as $\sigma = \mathbf{s} \cdot \mathbf{n} = -p$, as expected and expression (10.16.40c) gives the shear stress on dS along the conduit. Evidently, this shear stress acts opposite to the direction of flow. It is to be noted that all the three stress components s_i are independent of viscosity. If dS is taken on the boundary surface, we get $-(G/2)a$ as the skin friction along the direction of flow.

10.15 SOME VISCOUS FLOW PROBLEMS 547

Case ii: Elliptic Cross Section Suppose the conduit is of elliptic cross section. Then C is an ellipse whose equation can be taken as

$$\frac{x_1^2}{a^2} + \frac{x_2^2}{b^2} = 1 \tag{10.15.41}$$

In view of the boundary condition (10.15.31), v can be assumed in the form

$$v = \beta\left(\frac{x_1^2}{a^2} + \frac{x_2^2}{b^2} - 1\right) \tag{10.15.42}$$

where β is a constant. Substituting for v from (10.15.42) into equation (10.15.30), we find that

$$\beta = -\frac{a^2 b^2 G}{2\mu(a^2 + b^2)} \tag{10.15.43}$$

Thus,

$$v = \frac{a^2 b^2 G}{2\mu(a^2 + b^2)}\left[1 - \frac{x_1^2}{a^2} - \frac{x_2^2}{b^2}\right] \tag{10.15.44}$$

is the expression for v that satisfies governing equation (10.15.30) as well as boundary condition (10.15.31). We note that the velocity distribution is paraboloidal in nature. Also, the velocity is maximum on the axis of the conduit, and $v_{\max} = -\beta$.

Using (10.15.44), we find that the rate of mass of fluid that passes a cross section is given by

$$M = \int_A \rho v \, dA = \frac{\rho a^2 b^2 G}{2\mu(a^2 + b^2)}\int_A \left(1 - \frac{x^2}{a^2} - \frac{y^2}{b^2}\right) dA \tag{10.15.45}$$

where A is the area of cross section. Setting

$$x_1 = a\xi \cos\theta, \qquad x_2 = b\xi \sin\theta \tag{10.15.46}$$

and noting that ξ varies from 0 to 1 and θ varies from 0 to 2π as the point (x_1, x_2) varies over A, we find from (10.15.45) that

$$M = \frac{\rho a^2 b^2 G}{2\mu(a^2 + b^2)}\int_0^{2\pi}\int_0^1 (\xi^2 - 1)\xi \, d\xi \, d\theta = \frac{\rho\pi a^3 b^3 G}{4\mu(a^2 + b^2)} \tag{10.15.47}$$

This is a generalization of Poiseullie's formula (10.15.38). If we set $b = a$ in (10.15.47), we recover (10.15.38).

Let us consider a fluid element dS on an elliptic surface coaxial with and similar to the boundary surface of the conduit. On this surface element,

$$\phi(x_1, x_2) \equiv \frac{x_1^2}{a^2} + \frac{x_2^2}{b^2} = \text{constant} \tag{10.15.48}$$

The exterior unit normal to dS is therefore $\mathbf{n} = (\nabla\phi/|\nabla\phi|)$. From (10.15.48) and (10.15.46), we find that

$$\nabla\phi = \frac{2x_1}{a^2}\mathbf{e}_1 + \frac{2x_2}{b^2}\mathbf{e}_2 = \frac{2\xi}{ab}(b\cos\theta\,\mathbf{e}_1 + a\sin\theta\,\mathbf{e}_2)$$

$$|\nabla\phi| = \frac{2\xi}{ab}(b^2\cos^2\theta + a^2\sin^2\theta)^{1/2}$$

Hence the components of $\mathbf{n}$ are

$$n_1 = \frac{b}{N}\cos\theta, \qquad n_2 = \frac{a}{N}\sin\theta, \qquad n_3 = 0 \qquad (10.15.49)$$

where

$$N = (b^2\cos^2\theta + a^2\sin^2\theta)^{1/2} \qquad (10.15.50)$$

The components of the stress vector on dS can now be computed by using Cauchy's law (7.4.9), the material law (10.12.8) and expressions (10.15.44), (10.15.46) and (10.15.49). We find that

$$s_1 = \tau_{11}n_1 + \tau_{12}n_2 = -pn_1 = -p\frac{b}{N}\cos\theta \qquad (10.15.51\text{a})$$

$$s_2 = \tau_{21}n_1 + \tau_{22}n_2 = -pn_2 = -p\frac{a}{N}\sin\theta \qquad (10.15.51\text{b})$$

$$s_3 = \tau_{31}n_1 + \tau_{32}n_2 = \mu(v_{,1}n_1 + v_{,2}n_2) = -\frac{Gab\xi N}{a^2+b^2} \qquad (10.15.51\text{c})$$

Together with (10.15.49), expressions (10.15.51a, b) yield the normal stress on dS as $\sigma = \mathbf{s}\cdot\mathbf{n} = -p$, as expected. Expression (10.15.51c) gives the shear stress on dS along the conduit. Evidently, this stress acts opposite to the direction of flow. It may be observed that all the stress components s_i are independent of the viscosity.

If dS is chosen on the boundary surface, for which $\xi = 1$, we get $s_3 = \tau$, where

$$\tau = -\frac{Gab}{a^2+b^2}N \qquad (10.15.52)$$

This gives the skin friction at a point on the conduit and this varies from one point to another. By using (10.15.50), it can be shown that the skin friction is maximum at the endpoints of the minor axis and minimum at the endpoints of the major axis of the boundary curve (10.15.41). If $a > b$,

10.15 SOME VISCOUS FLOW PROBLEMS

the maximum and minimum values of $|\tau|$ are

$$\max|\tau| = \frac{Ga^2b}{a^2 + b^2}; \quad \min|\tau| = \frac{Gab^2}{a^2 + b^2} \quad (10.15.53)$$

When $b = a$, we have $N = a$ and $\xi = R/a$ and expressions (10.15.51) reduce to (10.15.40). Also, then, τ becomes equal to $-(G/2)a$ at all points of the boundary.

10.15.3 STEADY FLOW BETWEEN TWO COAXIAL ROTATING CYLINDERS

Here we consider the steady flow of an incompressible viscous fluid between two infinitely long coaxial circular cylinders due to the rotation of the cylinders with constant but different angular velocities about the common axis. Such a flow is called a *Couette flow*. Assuming that there is no body force, the problem is to find the velocity field.

We choose the axes such that the x_3 axis coincides with the axis of the cylinders and the x_1x_2 plane lies on a section of the tube formed by the cylindrical surfaces. Since the flow is caused by the rotation of the cylinders about the axis and the flow is steady, we assume that the fluid particles move in circular orbits centered on the axis and that the magnitude of velocity depends only on the distance of the particle from the axis. Accordingly, the velocity at any point of the fluid is taken in the form

$$\mathbf{v} = v(R)\mathbf{e}_\theta \quad (10.15.54)$$

where $R = (x_1^2 + x_2^2)^{1/2}$ is the distance of the point from the axis and $\mathbf{e}_\theta$ is the unit vector in the transverse direction (see Figure 10.20). If we denote the unit vector directed along the radial direction perpendicular to the axis by $\mathbf{e}_R$, then $\mathbf{e}_\theta = \mathbf{e}_3 \times \mathbf{e}_R$, so that

$$[\mathbf{e}_\theta]_i = \varepsilon_{ijk}[\mathbf{e}_3]_j[\mathbf{e}_R]_k = \varepsilon_{i3k}\frac{1}{R}x_k \quad (10.15.55a)$$

for $i, k = 1, 2$;

$$[\mathbf{e}_\theta]_3 = 0 \quad (10.15.55b)$$

Consequently, (10.15.54) gives

$$v_i = \varepsilon_{i3k} v(R) \frac{1}{R} x_k = \varepsilon_{i3k} \psi(R) x_k \quad (10.15.56a)$$

for $i, k = 1, 2$;

$$v_3 = 0 \quad (10.15.56b)$$

550 10 EQUATIONS OF FLUID MECHANICS

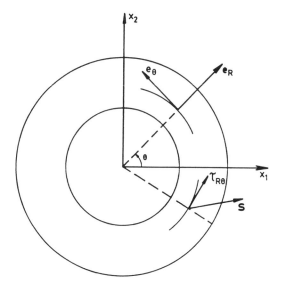

Figure 10.20. Cross section of a circular tube.

From (10.15.56a), we obtain the following relations valid for $i, j, k = 1, 2$:

$$v_{i,j} = \varepsilon_{i3j}\psi(R) + \varepsilon_{i3k}x_k x_j \frac{1}{R}\frac{d\psi}{dR} \qquad (10.15.57\text{a})$$

$$v_{i,jj} = \varepsilon_{i3k}\left\{\frac{3}{R}\frac{d\psi}{dR} + \frac{d^2\psi}{dR^2}\right\}x_k \qquad (10.15.57\text{b})$$

$$v_{i,j}v_j = \varepsilon_{i3j}\varepsilon_{j3k}x_k \psi^2 \qquad (10.15.57\text{c})$$

[The reader may note that explicit forms of (10.15.56a) and (10.15.57a) are contained in (10.14.51) and (10.14.52ab).]

From (10.15.57a) and (10.15.56b) we find that div $\mathbf{v} = 0$. Thus, the equation of continuity is satisfied. Also, (10.15.55ab), (10.15.56b) and (10.15.57bc) give

$$(\nabla^2 \mathbf{v}) \cdot \mathbf{e}_\theta = v_{i,jj}\left(\varepsilon_{i3k}\frac{1}{R}x_k\right) = \left[3\frac{d\psi}{dR} + R\frac{d^2\psi}{dR^2}\right]$$

$$= \frac{d}{dR}\left[\frac{1}{R}\frac{d}{dR}(R^2\psi)\right] = \frac{d}{dR}\left[\frac{1}{R}\frac{d}{dR}(Rv)\right] \qquad (10.15.58\text{a})$$

$$\{(\mathbf{v} \cdot \nabla)\mathbf{v}\} \cdot \mathbf{e}_\theta = v_{i,j}v_j\left(\varepsilon_{i3k}\frac{1}{R}x_k\right) = 0 \qquad (10.15.58\text{b})$$

10.15 SOME VISCOUS FLOW PROBLEMS

Because of the axisymmetric and steady nature of the flow, we suppose that the pressure is a function of R only so that

$$\nabla p = \frac{dp}{dR} \mathbf{e}_R$$

from which it follows that

$$(\nabla p) \cdot \mathbf{e}_\theta = 0 \tag{10.15.59}$$

Taking the scalar product with $\mathbf{e}_\theta$ on both sides of the Navier–Stokes equation (10.14.6) with body force equal to 0 and using expressions (10.15.58) and (10.15.59) in the resulting equation, we obtain the following ordinary linear differential equation for the function $v(R)$:

$$\frac{d}{dR}\left[\frac{1}{R}\frac{d}{dR}(Rv)\right] = 0 \tag{10.15.60}$$

Integration of this equation yields

$$v(R) = AR + \frac{B}{R} \tag{10.15.61}$$

where A and B are arbitrary constants.

We verify that the component of the Navier–Stokes equation (10.14.6) along $\mathbf{e}_R$ yields, on using (10.15.57bc),

$$\frac{dp}{dR} = \frac{\rho}{R} v^2 \tag{10.15.62}$$

and that the component along $\mathbf{e}_3$ is identically satisfied.

Thus, under the assumptions made, (10.15.61) gives the velocity field and (10.15.62) gives the pressure-gradient field when the constants A and B in (10.15.61) are determined by the boundary conditions.

Let a and b be the radii and ω_1 and ω_2 be the constant angular speeds of the inner and outer cylinders, respectively. Since the cylinders rotate rigidly about the x_3 axis, the velocities at points on these cylinders can be determined by (6.2.19). Thus, we find that $\omega_1 a \mathbf{e}_\theta$ and $\omega_2 b \mathbf{e}_\theta$ are the velocities at a point on the inner and outer cylinders, respectively. The no-slip boundary condition then yields $v(R) = \omega_1 a$ for $R = a$ and $v(R) = \omega_1 b$ for $R = b$. Using (10.15.61), these conditions yield the following two equations for the determination of the two constants A and B:

$$\begin{aligned} Aa + \frac{B}{a} &= \omega_1 a \\ Ab + \frac{B}{b} &= \omega_2 b \end{aligned} \tag{10.15.63}$$

552   10 EQUATIONS OF FLUID MECHANICS

Solving these equations we obtain

$$A = \frac{\omega_2 b^2 - \omega_1 a^2}{b^2 - a^2}, \qquad B = \frac{(\omega_1 - \omega_2)a^2 b^2}{b^2 - a^2} \qquad (10.15.64)$$

Substituting these into (10.15.61), we obtain

$$v(R) = \frac{1}{b^2 - a^2}\left[(\omega_2 b^2 - \omega_1 a^2)R + (\omega_1 - \omega_2)\frac{a^2 b^2}{R}\right] \qquad (10.15.65)$$

This determines the velocity field in the fluid. We immediately note that, if both the cylinders rotate with the same angular velocity $\omega \mathbf{e}_3$, then $v(R) = \omega R$. Hence, in this case, the fluid rotates like a rigid body along with boundary surfaces.

Substituting (10.15.65) in (10.15.62) we obtain the pressure-gradient at a point of the fluid.

Since the motion is essentially rotational, it is of interest to compute the vorticity at a point. By use of (10.15.56ab), we find that

$$w_i = [\operatorname{curl} \mathbf{v}]_i = \varepsilon_{ijk} v_{k,j} = \begin{cases} 0, & \text{for } i = 1, 2 \\ \dfrac{1}{R}\dfrac{d}{dR}(Rv), & \text{for } i = 3 \end{cases}$$

Consequently, with the use of (10.15.65), we obtain

$$\mathbf{w} = \operatorname{curl} \mathbf{v} = \frac{2(\omega_2 b^2 - \omega_1 a^2)}{b^2 - a^2}\mathbf{e}_3 \qquad (10.15.66)$$

Evidently, the vorticity is constant and is directed along the axis. Further, the flow is irrotational if and only if the angular speeds of the cylinders obey the relation

$$\frac{\omega_2}{\omega_1} = \frac{a^2}{b^2} \qquad (10.15.67)$$

The stresses produced at a point of the fluid can be computed by use of (10.15.56) in Stokes's law (10.12.8). Thus, we find

$$\tau_{ij} = -p\delta_{ij} + \mu(\varepsilon_{i3k}x_j + \varepsilon_{j3k}x_i)\frac{x_k}{R}\frac{d\psi}{dR}$$

for $i, j = 1, 2$; $\qquad (10.15.68)$

$$\tau_{31} = \tau_{32} = 0, \qquad \tau_{33} = -p$$

Cauchy's law (7.4.6) can be employed to find the stress vector $\mathbf{s}$ on a cylindrical surface $R = $ constant in the fluid. The tangential component of this stress vector is given by

$$\mathbf{s} \cdot \mathbf{e}_\theta = s_i [\mathbf{e}_\theta]_i = \left(\tau_{ij} \frac{x_j}{R} \right) \left(\varepsilon_{i3k} \frac{1}{R} x_k \right) \quad (10.15.69)$$

Substituting for τ_{ij} from (10.15.68) in (10.15.69), we arrive at the following simple formula for the shear stress $\tau_{R\theta} \equiv \mathbf{s} \cdot \mathbf{e}_\theta$:

$$\tau_{R\theta} = \mu R \frac{d\psi}{dR} = \mu R \frac{d}{dR}\left[\frac{v}{R}\right] \quad (10.15.70)$$

By use of (10.15.65) in (10.15.70), we get

$$\tau_{R\theta} = \frac{2\mu a^2 b^2}{b^2 - a^2} \frac{\omega_2 - \omega_1}{R^2} \quad (10.15.71)$$

Evidently, the shear stress $\tau_{R\theta}$ is tensile or compressive accordingly as the outer cylinder rotates faster or slower than the inner cylinder. Also, the magnitude of this stress is maximum on the inner cylinder and minimum on the outer cylinder. The stress vanishes when both the cylinders rotate with the same speed.

Expression (10.15.71) can also be employed to compute the couple exerted by the fluid on the cylinders due to viscous drag. For example, the magnitude M of the moment of the couple per unit length in the axial direction on the inner cylinder is given by

$$M = \int_0^{2\pi} [\tau_{R\theta} R^2 \, d\theta]_{R=a} = \frac{4\mu\pi a^2 b^2}{b^2 - a^2} |\omega_2 - \omega_1| \quad (10.15.72)$$

This result is used for measuring μ by a rotation viscometer.

Two interesting limiting cases are worth mentioning. For $a = 0$, the problem reduces to that of a flow inside a cylinder due to the rotation of the cylinder. In this case, we find from (10.15.65) that the motion is simply a rigid-body motion; the fluid rotates like a rigid body along with the cylinder. For $b \to \infty$ with a remaining finite and greater than 0, the problem reduces to that of a flow of an infinite fluid outside a cylinder due to the rotation of the cylinder. In this case, condition (10.15.67) is trivially satisfied and the flow is irrotational.

10.15.4 UNSTEADY FLOW NEAR A MOVING PLANE BOUNDARY

Here, we consider the unsteady motion of an incompressible viscous fluid occupying the half-space $x_3 > 0$ with a plane rigid plate at $x_3 = 0$ as its boundary. Suppose the fluid and the plate are initially at rest and that the flow is generated by the movement of the plate in a direction parallel to its own plane, say the x_1 direction; see Figure 10.21.

554   10 EQUATIONS OF FLUID MECHANICS

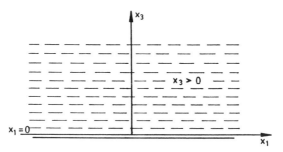

Figure 10.21. Unsteady flow near a plate.

When the plate moves in the x_1 direction it is natural to expect that the fluid also moves in the same direction so that the velocity of the fluid at any point is of the form $\mathbf{v} = v\mathbf{e}_1$. In the absence of body force, the Navier-Stokes equation (10.14.6) yields the following three component equations

$$\frac{Dv}{Dt} = -\frac{1}{\rho} p_{,1} + \nu \nabla^2 v \qquad (10.15.73)$$

$$p_{,2} = 0 \qquad p_{,3} = 0 \qquad (10.15.74)$$

The equation of continuity (10.12.5) becomes

$$\frac{\partial v}{\partial x_1} = 0 \qquad (10.15.75)$$

Since the fluid extends to infinity in the x_2 direction, the variation of v in the x_2 direction may be neglected. Then, in view of (10.15.75), v varies only with x_3 and t. Consequently,

$$\frac{Dv}{Dt} = \frac{\partial v}{\partial t} + (\mathbf{v} \cdot \nabla) v = \frac{\partial v}{\partial t} \qquad (10.15.76)$$

and equation (10.15.73) reduces to

$$\nu \frac{\partial^2 v}{\partial x_3^2} - \frac{\partial v}{\partial t} = \frac{1}{\rho} \frac{\partial p}{\partial x_1} \qquad (10.15.77)$$

Equations (10.15.74) imply that p is independent of x_2 and x_3. Since v does not depend on x_1 by (10.15.75), the lefthand side of (10.15.77) is a function of x_3 and t, while the righthand side is a function of x_1 and t. Therefore, each side of (10.15.77) must be a function of t alone; that is,

$$\nu \frac{\partial^2 v}{\partial x_3^2} - \frac{\partial v}{\partial t} = f(t) \qquad (10.15.78)$$

$$\frac{1}{\rho} \frac{\partial p}{\partial x_1} = f(t) \qquad (10.15.79)$$

where $f(t)$ is an arbitrary function of t. It follows from (10.15.79) that
$$p = \rho x_1 f(t) + f_1(t) \tag{10.15.80}$$
where $f_1(t)$ is again an arbitrary function of t. Since the fluid extends to infinity in the x_1 direction, $p \to \infty$ as $x_1 \to \infty$ unless $f(t) \equiv 0$. But the infinite pressure is physically unrealistic. Therefore, we assume $f(t) = 0$ so that (10.15.78) and (10.15.80) become
$$\frac{\partial v}{\partial t} = \nu \frac{\partial^2 v}{\partial x_3^2} \tag{10.15.81}$$
$$p = f_1(t) \tag{10.15.82}$$
Evidently, the velocity and pressure fields are now uncoupled; v can be determined independent of p. If desired, p can be obtained from a boundary condition prescribed at $x_3 = 0$.

Equation (10.15.81) is usually called the *diffusion equation*. To determine v we have to solve this equation under appropriate initial and boundary conditions. Since the fluid is assumed to be initially at rest, the initial condition is
$$v = 0 \tag{10.15.83}$$
in $x_3 > 0$ for $t = 0$.

As the plate is assumed to move in the x_1 direction, the no-slip boundary condition requires
$$v = V(t) \tag{10.15.84}$$
for $x_3 = 0$, $t > 0$, where $V(t)$ is the speed of the plate. This is a time-dependent boundary condition.

Since the flow is generated by the motion of the plate (at $x_3 = 0$), we may assume that the fluid particles far way from the plate remain unaffected by the movement of the plate. This leads to the condition
$$v \to 0 \tag{10.15.85}$$
as $x_3 \to \infty$ for $t > 0$. This is often called a *regularity condition* (or *boundary condition* at infinity).

We consider two particular cases of interest.

Case i: Oscillating Plate Suppose the plate oscillates with real constant amplitude V in its own plane with a given frequency ω so that
$$V(t) = V \cos \omega t \tag{10.15.86}$$
$t > 0$, and we are interested in the velocity field *strictly after* the flow is fully developed. Then the initial condition (10.15.83) is not required in our analysis. We now seek a solution of (10.15.81) in the form
$$v(x_3, t) = \text{Re}[f(x_3)e^{i\omega t}] \tag{10.15.87}$$

where Re stands for the real part. Substituting this solution into (10.15.81) yields the following ordinary differential equation for $f(x_3)$:

$$\frac{d^2 f}{dx_3^2} - \left(\frac{i\omega}{\nu}\right) f = 0 \qquad (10.15.88)$$

The general solution of this equation subject to the regularity condition (10.15.85) is

$$f(x_3) = A \exp[-(1 + i)\delta] x_3 \qquad (10.15.89)$$

where A is an arbitrary constant and $\delta = \sqrt{\omega/2\nu} > 0$. In view of (10.15.84), (10.15.86) and (10.15.87), we find that $A = V$ and hence the final form of the solution for $v(x_3, t)$ is

$$v(x_3, t) = \text{Re}\{V \exp[-(1 + i)\,\delta x_3 + i\omega t]\} = V e^{-\delta x_3} \cos(\omega t - \delta x_3) \qquad (10.15.90)$$

This shows that the fluid particles oscillate with the plate with the same frequency as that of the plate and that their amplitude, $V \exp(-\delta x_3)$, decreases exponentially with distance from the plate. In other words, in the fluid, shear waves that spread out from the oscillating plate propagate with exponentially decaying amplitude so that fluid oscillations are essentially confined to a layer adjacent to the plate. This layer is called the *Stokes's boundary layer* and the problem is called the *Stokes's problem*. The thickness of the layer is on the order $\delta^{-1} = (2\nu/\omega)^{1/2}$. Evidently, the thickness of the layer depends on the frequency of the plate: the layer becomes thicker as the frequency decreases and thinner as the frequency increases. In other words, a large amount of fluid oscillates with a slowly oscillating plate and only a small amount of fluid oscillates with a rapidly oscillating plate. The speed of oscillation is $(\omega/\delta) = \sqrt{2\nu\omega}$.

The use of the material law (10.12.8) and $\mathbf{v} = v\mathbf{e}_1$ gives the shear stress on a fluid element parallel to the plate as

$$\tau_{31} = \mu v_{,3} = \mu\,\delta V e^{-\delta x_3}\{\sin(\omega t - \delta x_3) - \cos(\omega t - \delta x_3)\}$$

$$= \sqrt{2}\,\mu\,\delta V e^{-\delta x_3} \sin\left(\omega t - \delta x_3 - \frac{\pi}{4}\right) \qquad (10.15.91)$$

This shear stress is maximum on the plate and decays exponentially with increasing distance from the plate, and

$$\max \tau_{31} = [\tau_{31}]_{x_3 = 0} = \sqrt{2}\,\alpha\mu V \sin\left(\omega t - \frac{\pi}{4}\right) \qquad (10.15.92)$$

This is the skin friction acting on the plate.

10.15 SOME VISCOUS FLOW PROBLEMS   557

Case ii: Impulsively Moved Plate   This case arises when the plate is started impulsively from rest with a constant velocity $V\mathbf{e}_1$ and this velocity is maintained for all time $t > 0$.

To determine v in this case, we transform governing equation (10.15.81) to an ordinary differential equation by using the substitution

$$\eta = \frac{x_3}{2\sqrt{vt}}, \qquad v = VF(\eta), \qquad t > 0 \qquad (10.15.93)$$

where η is called a *similarity variable*. Thus we find

$$\frac{\partial v}{\partial x_3} = \frac{\partial v}{\partial \eta} \cdot \frac{\partial \eta}{\partial x_3} = \frac{V}{2\sqrt{vt}} F'(\eta), \qquad \frac{\partial^2 v}{\partial x_3^2} = \frac{V}{4vt} F''(\eta)$$

$$\frac{\partial v}{\partial t} = \frac{\partial v}{\partial \eta} \cdot \frac{\partial \eta}{\partial t} = -\frac{Vvx_3}{4(vt)^{3/2}} F'(\eta)$$

so that (10.15.81) reduces to

$$F''(\eta) + 2\eta F'(\eta) = 0 \qquad (10.15.94)$$

The general solution of this first order ordinary linear differential equation for $F'(\eta)$ is

$$F'(\eta) = A\, e^{-\eta^2} \qquad (10.15.95)$$

This gives $F(\eta)$ as

$$F(\eta) = A \int_0^\eta e^{-\eta^2}\, d\eta + B \qquad (10.15.96)$$

where A and B are arbitrary constants.

From (10.15.84), (10.15.85) and (10.15.93), we note that $F(\eta)$ satisfies the following conditions:

$$F(\eta) = 1 \quad \text{for} \quad \eta = 0, \qquad F(\eta) \to 0 \text{ as } \eta \to \infty. \qquad (10.15.97\text{a, b})$$

Using these conditions in (10.15.96) and noting

$$\int_0^\infty \exp(-\eta^2)\, d\eta = \frac{\sqrt{\pi}}{2} \qquad (10.15.98)$$

we find $B = 1$ and $A = -2/\sqrt{\pi}$. Thus the final form of v is given by

$$v = V\left[1 - \frac{2}{\sqrt{\pi}} \int_0^\eta \exp(-\eta^2)\, d\eta\right] \qquad (10.15.99)$$

It can be verified that v decreases continuously from V to 0 as x_3 increases from 0 to infinity. As in the Stokes's problem, solution (10.15.99) exhibits a boundary layer nature; that is, the effect of the motion of the plate on the fluid flow decreases rapidly away from the plate. Here, the thickness of the boundary layer is on the order $\delta = \sqrt{vt}$, which grows with increasing time; the rate of growth is on the order $d\delta/dt \approx \sqrt{v/t}$, which decreases with increasing time. This problem is known as the *Rayleigh problem* and the associated boundary layer is called the *Rayleigh layer*.

Using the fact that $\mathbf{v} = v\mathbf{e}_1$, where v is given by (10.15.99), we find that the vorticity vector is directed along the x_2 axis with magnitude

$$|v_{,3}| = \frac{V}{\sqrt{\pi vt}} \exp\left(-\frac{x_3^2}{4vt}\right) \qquad (10.15.100)$$

Evidently, the vorticity is 0 at $t = 0$ except at $x_3 = 0$, and it decays exponentially to 0 with increasing x_3. This implies that the vorticity is generated at the plate and diffuses outward within the Rayleigh layer.

By using Stokes's law (10.12.8), we find the shear stress on a fluid element parallel to the plate as given by

$$\tau_{31} = \mu v_{,3} = -\frac{\mu V}{\sqrt{\pi vt}} \exp\left(-\frac{x_3^2}{4vt}\right) \qquad (10.15.101)$$

Evidently, the shear stress acts opposite to the direction of motion and its magnitude is maximum on the plate with

$$\max \tau_{31} = [\tau_{31}]_{x_3 = 0} = -\frac{\mu V}{\sqrt{\pi vt}} \qquad (10.15.102)$$

This is the skin friction on the plate. Its magnitude varies inversely as $\sqrt{t}$.

10.15.5 SLOW AND STEADY FLOW PAST A RIGID SPHERE

As already indicated, for sufficiently small Reynolds numbers, the viscous forces exceed the inertia forces and the nonlinear terms in the Navier-Stokes equation may be neglected. In this case the governing equations for an incompressible fluid flow are given by (10.12.5) and (10.14.28). As an example, we consider here the steady flow past a rigid, fixed sphere in the absence of body force.

We assume that the fluid particles far away from the sphere are undisturbed and move with a uniform velocity V_∞ in the x_1 direction. Then, the velocity field satisfies the following conditions:

$$v_1 = V_\infty, \qquad v_2 = v_3 = 0 \qquad \text{as } r \to \infty \qquad (10.15.103)$$

where r is the distance measured from the center of the sphere, which is taken as the origin (see Figure 10.22).

10.15 SOME VISCOUS FLOW PROBLEMS

Since the sphere is rigid and stationary, the condition to be satisfied on the surface of the sphere is the no-slip boundary condition

$$\mathbf{v} = \mathbf{0} \quad \text{or} \quad v_i = 0 \quad \text{for } r = a, \quad (10.15.104)$$

where a is the radius of the sphere.

Since the flow is steady and there is no body force, equations (10.14.29) and (10.14.30), which follow from equations (10.12.5) and (10.14.28), serve as the equations for determining p and v_i.

Noting that x_1/r^3 ($r \neq 0$) is a harmonic function, we take the pressure p, which is a harmonic function by (10.14.30), in the form

$$p = -\frac{A}{r^3} x_1 \quad (10.15.105)$$

where A is a constant. Substituting this value of p into (10.14.29) we obtain the following governing equations for v_i:

$$\mu \nabla^2 v_1 = \frac{A}{r^5}(3x_1^2 - r^2)$$

$$\mu \nabla^2 v_2 = \frac{3A}{r^5} x_1 x_2 \quad (10.15.106)$$

$$\mu \nabla^2 v_3 = \frac{3A}{r^5} x_1 x_3$$

It can be verified that the solutions for v_i are

$$v_1 = V_\infty - \frac{x_1^2}{2\mu r^5}(Ar^2 - C) + \frac{1}{6\mu r^3}(Br^2 - C) \quad (10.15.107a)$$

$$v_2 = -\frac{x_1 x_2}{2\mu r^5}(Ar^2 - D') \quad (10.15.107b)$$

$$v_3 = -\frac{x_1 x_3}{2\mu r^5}(Ar^2 - D') \quad (10.15.107c)$$

where A, B, C, D' are constants. In view of conditions (10.15.103) and (10.15.104) we obtain

$$A = \tfrac{3}{2}\mu a V_\infty, \quad B = -3A, \quad C = D' = Aa^2$$

Substituting these into (10.15.105) and (10.15.107), we obtain the following solution for the pressure and velocity fields:

$$p = -\frac{3\mu a V_\infty}{2r^3} x_1 \qquad (10.15.108)$$

$$v_1 = V_\infty - \frac{3aV_\infty}{4r^5}(r^2 - a^2)x_1^2 - \frac{aV_\infty}{4r^3}(3r^2 - a^2) \qquad (10.15.109a)$$

$$v_2 = -\frac{3aV_\infty}{4r^5}(r^2 - a^2)x_1 x_2 \qquad (10.15.109b)$$

$$v_3 = -\frac{3aV_\infty}{4r^5}(r^2 - a^2)x_1 x_3 \qquad (10.15.109c)$$

This solution was first obtained by Stokes in 1851. The velocity distribution shows that the stream lines of the flow are symmetric about the equatorial plane normal to the direction of the flow, as depicted in Figure 10.22.

Since the fluid essentially moves in the x_1 direction, the resultant force on the surface of the sphere acts in the x_1 direction. This force, called the *drag* on the sphere, is

$$D = \int_S s_1 \, dS \qquad (10.15.110)$$

where s_1 is, as usual, the x_1 component of the stress vector **s**, and S is the surface of the sphere.

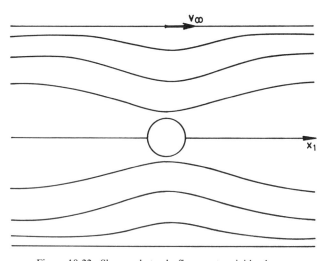

Figure 10.22. Slow and steady flow past a rigid sphere.

10.15 SOME VISCOUS FLOW PROBLEMS

By using Cauchy's law (7.4.9) and noting that $n_i = x_i/a$ on S, expression (10.15.110) can be rewritten as

$$D = \frac{1}{a}\int_S (\tau_{11}x_1 + \tau_{12}x_2 + \tau_{13}x_3)\,dS \qquad (10.15.111)$$

Computing τ_{1i} by using the material law (10.12.8) and results (10.15.108) and (10.15.109), substituting the resulting expressions onto the righthand side of (10.15.111) and subsequently evaluating the surface integral, we arrive at the following famous formula, known as *Stokes's formula*:

$$D = 6\mu\pi a V_\infty \qquad (10.15.112)$$

Computations leading to (10.15.112) from (10.15.111) reveal that one-third of the righthand side of (10.15.112) is due to pressure and two-thirds is due to skin friction.

A nondimensional *drag coefficient* for the sphere, C_D, is defined by the relation

$$D = \tfrac{1}{2}\rho V_\infty^2 C_D(\pi a^2) \qquad (10.15.113)$$

Using (10.15.112), we get

$$C_D = \frac{24}{Re} \qquad (10.15.114)$$

where $Re = (2aV_\infty/\nu)(<1)$ is the Reynolds number based on the diameter of the sphere.

It has to be pointed out that Stokes's solution (10.15.109) is not valid at long distances from the sphere. In order to check this point, we estimate the orders of magnitude of the inertia term $\rho(\mathbf{v}\cdot\nabla)\mathbf{v}$ and the viscous term $\mu\nabla^2\mathbf{v}$. According to Stokes's solution, the inertia term is on the order $\rho(aV_\infty^2/r^2)$, whereas the viscous term is on the order $\mu aV_\infty/r^3$. Hence Stokes's solution holds only when

$$\rho\frac{aV_\infty^2}{r^2} \ll \frac{\mu a V_\infty}{r^3}$$

that is, $r \ll \nu/V_\infty$.

The difficulty associated with Stokes's solution was first resolved by Oseen in 1910. Using the hypothesis that v_1 is approximately equal to V_∞ at long distances from the sphere, he assumed $(\mathbf{v}\cdot\nabla)\mathbf{v} \approx V_\infty(\partial\mathbf{v}/\partial x_1)$ in the Navier–Stokes equation and replaced equation (10.14.29) by the following equation to study the problem:

$$\rho V_\infty \frac{\partial \mathbf{v}}{\partial x_1} = -\nabla p + \mu\nabla^2\mathbf{v} \qquad (10.15.115)$$

We shall not attempt to find the solution of the problem on the basis of this equation, which is known as the *Oseen equation*. We just mention that the solution leads to the expression for the drag coefficient

$$C_D = \frac{24}{Re}\left(1 + \frac{3Re}{16}\right) \tag{10.15.116}$$

which agrees with experimental findings up to $Re = 2aV_\infty/\nu = 5$. Comparison of (10.15.114) and (10.15.116) shows that Stokes's formula (10.15.114) is a first approximation of Oseen's formula (10.15.116).

10.16

EXERCISES

1. If a fluid of uniform density is at rest under a constant gravitational field, show that the pressure varies linearly with depth.

2. For water, an approximate relationship between the pressure and density is $\rho = 1 + \alpha p$, where α is a constant. If ρ_a is the density at the surface of a lake, show that the density of water at depth h in the lake is given by $\rho = \rho_a e^{\alpha g h}$. Assume that p varies only with depth and gravity is the only body force.

3. At a height h above the earth's surface the density of air is ρ and pressure is p. At the surface, the corresponding values are ρ_0 and p_0. Assuming that the gravitational field is constant and $\rho = \rho_0 e^{-\alpha z}$, where α is a constant and z is the vertical distance above the earth's surface, show that

$$p = p_0 - \frac{\rho_0 g}{\alpha}(1 - e^{-\alpha z})$$

4. Assuming that the atmosphere around the earth is a perfect gas at rest in which the pressure-density relation at an altitude z is given by $(p/p_0) = (\rho/\rho_0)^\gamma$, where p_0 and ρ_0 are the pressure and density at a reference elevation z_0, determine expressions for T, p and ρ in terms of T_0, p_0, ρ_0, z and z_0. Here T and T_0 are temperatures at z and z_0, respectively.

5. A rigid sphere of radius a is fixed in an incompressible fluid at rest. The fluid extends to infinity in all directions and the pressure-density relation is $p = \alpha\rho$, where α is a constant. If every fluid particle is attracted towards the center O of the sphere by a force of magnitude β/r^2, where r is the distance from O, show that the pressure exerted on the surface of the sphere is proportional to $\exp(\beta/\alpha a)$.

6. Show that in an incompressible nonviscous fluid flow, the rate of work done by external forces is equal to the rate of change of kinetic energy and that the rate of change of internal energy is equal to the rate of heat input.

10.16 EXERCISES

7. If $e = \varepsilon + (p/\rho)$ is the specific enthalpy, show that $de/dT = c_p$ for a perfect gas.

8. Show that the energy equation for a compressible nonviscous fluid flow can be put in the following forms:

(i) $$\rho \frac{D\varepsilon}{Dt} = -p \operatorname{div} \mathbf{v} - \operatorname{div} \mathbf{q} + \rho h$$

(ii) $$\frac{D\varepsilon}{Dt} + p \frac{D}{Dt}\left(\frac{1}{\rho}\right) + \frac{1}{\rho} \operatorname{div} \mathbf{q} - h = 0$$

(iii) $$\operatorname{div}(p\mathbf{v} + \mathbf{q}) - \rho(\mathbf{v} \cdot \mathbf{b}) - \rho h = \rho \frac{De}{Dt} = \frac{\partial}{\partial t}(\rho e) + \operatorname{div}(\rho e)$$

(iv) $$\rho \frac{D\bar{e}}{Dt} = \frac{\partial p}{\partial t} - \operatorname{div} \mathbf{q} + \rho h + \rho(\mathbf{b} \cdot \mathbf{v})$$

where $e = \varepsilon + (p/\rho)$ and $\bar{e} = e + (v^2/2)$.

9. Write down Euler's equation of motion in the suffix notation.

10. Show that the velocity field given by

$$v_1 = \alpha \frac{x_1^2 - x_2^2}{R^4}, \quad v_2 = \frac{2\alpha x_1 x_2}{R^4}, \quad v_3 = 0$$

where α is a nonzero constant and $R^2 = x_1^2 + x_2^2 \neq 0$, obeys Euler's equation of motion for an incompressible fluid. Determine the pressure distribution associated with this velocity field.

11. The velocity field in an incompressible nonviscous fluid flow is given by

$$v_1 = v_0 \cos \frac{\pi x}{2a} \cos \frac{\pi z}{2a},$$

$$v_2 = v_0 \sin \frac{\pi x}{2a} \sin \frac{\pi z}{2a}, \quad v_3 = 0$$

where v_0 and a are nonzero constants. Show that the pressure is given by

$$p = \frac{1}{4} v_0^2 \left\{ \cos \frac{\pi z}{a} - \cos \frac{\pi x}{a} \right\} + \text{constant}$$

12. For a steady irrotational flow of a nonviscous fluid of constant density under gravity, show that

$$p = p_0 - \rho g z - \tfrac{1}{2} \rho v^2$$

where p_0 is a constant.

13. For an irrotational steady flow of a nonviscous fluid under zero body force, show that $dp = -\rho v \, dv$.

10 EQUATIONS OF FLUID MECHANICS

14. A liquid column of uniform density moves in a vertical direction (against gravity) with a constant acceleration **a**. Find the pressure at a point whose depth from the upper surface of the liquid is h.

15. For the liquid column considered in Example 10.6.2, show that the rise of the liquid above the vertex of the upper surface along the wall of the vessel is $\omega^2 a^2/2g$, where a is the radius of the vessel.

16. For a two-dimensional flow of an incompressible nonviscous fluid under zero body force, show that the vorticity is constant.

17. In the irrotational, two-dimensional flow of an incompressible nonviscous fluid, show that the velocity is constant in magnitude if and only if it is constant in direction as well.

18. The velocity field for a two-dimensional flow of an incompressible nonviscous fluid under zero body force is given by

$$v_1 = \psi_{,2}, \qquad v_2 = -\psi_{,1}, \qquad v_3 = 0$$

where $\psi = \psi(x_1, x_2, t)$. By using Euler's equation of motion show that ψ satisfies the equation

$$\frac{\partial}{\partial t}(\nabla^2 \psi) = \begin{vmatrix} \psi_{,1} & \psi_{,2} \\ \nabla^2 \psi_{,1} & \nabla^2 \psi_{,2} \end{vmatrix}$$

19. Show that in a compressible nonviscous fluid flow, the rate of change of circulation round a circuit c is given by

$$\frac{DI_c}{Dt} = -\int_S \left\{ \nabla\left(\frac{1}{\rho}\right) \times \nabla p - \operatorname{curl} \mathbf{b} \right\} \cdot \mathbf{n} \, dS$$

where S is a surface for which c is the rim.

20. Show that for an elastic fluid moving under conservative body force, Cauchy's vorticity equation (6.6.8) can be rewritten as

$$\frac{\mathbf{w}}{\rho} = \frac{\mathbf{w}_0}{\rho_0} \cdot \nabla^0 \mathbf{x}$$

21. For an elastic fluid moving under conservative force, show that

$$\frac{D}{Dt} \int_V \mathbf{w} \, dV = \int_S (\mathbf{w} \cdot \mathbf{n}) \mathbf{v} \, dS$$

where V is a material volume and S is its boundary surface.

22. Show that the equation (10.7.8) can be rewritten as

$$\frac{\partial \mathbf{v}}{\partial t} + \mathbf{w} \times \mathbf{v} = -\nabla(P + \chi + \tfrac{1}{2}v^2)$$

Deduce that $\nabla^2(P + \chi + \tfrac{1}{2}v^2) = w^2 + \mathbf{v} \cdot \nabla^2 \mathbf{v} - \mathbf{v} \cdot \nabla(\operatorname{div} \mathbf{v}) - (\partial/\partial t)(\operatorname{div} \mathbf{v})$.

23. Show that in the absence of body force the only possible steady flow of an elastic fluid for which $\mathbf{v} = v(x_1)\mathbf{e}_1$, $\rho = \rho(x_1)$, $P = P(x_1)$ is the one for which v, ρ and P are all constants, independent of x_1.

24. For a nonviscous perfect gas moving in the x_1 direction, show that

$$\frac{\partial^2 \rho}{\partial t^2} = \frac{\partial^2}{\partial x_1^2}\{\rho(v^2 + RT)\}$$

25. For a steady, irrotational flow of an adiabatic isentropic gas in the absence of body force, show that the speed of sound is given by

$$c_s^2 = c_0^2 - \frac{\gamma - 1}{2}v^2$$

where c_0 is the speed at a reference condition.

26. Show that for a perfect gas $c_s = \sqrt{\gamma RT}$.

27. Write equation (10.7.24) in the unabridged form.

28. Show that the Bernoulli's equation (10.8.7) reduces to

(i) $p \log(p/\rho) + (1/2)v^2 + \chi$ = constant, for a perfect gas.
(ii) $\gamma p/(\gamma - 1)\rho + (1/2)v^2 + \chi$ = constant, for an isentropic perfect gas.

29. A gas, in which p and ρ are related by an adiabatic relation $p = \beta \rho^n$, flows in a steady state in a horizontal conduit. Using Bernoulli's equation show that the maximum speed is given by

$$v_{max} = c_0 \left(\frac{2}{n - 1}\right)^{1/2}$$

where c_0 is the speed of sound at a reference condition.

30. A nonviscous incompressible fluid flows steadily around the outside of a fixed vertical cylinder of radius a so that the fluid particles move with speed a/R in horizontal circles concentric with the boundary of the cylinder, where R is the radial distance from the axis of the cylinder. Show that the flow is irrotational. If the surface of the fluid is open to the atmosphere, show, by using the Bernoulli's equation, that

$$2gz = 1 - (a^2/R^2)$$

for appropriate choice of the coordinate axes.

31. Find the stress matrix associated with the following velocity field occurring in a viscous fluid flow:

(i) $\quad v_1 = 0, \quad v_2 = 0, \quad v_3 = x_2$

(ii) $\quad v_1 = 0, \quad v_2 = x_2^2 - x_3^2, \quad v_3 = -2x_2 x_3$

(iii) $\quad v_1 = v_1(x_1, x_2), \quad v_2 = v_2(x_1, x_2), \quad v_3 \equiv 0$

566   10   EQUATIONS OF FLUID MECHANICS

32. For a certain motion of an incompressible viscous fluid, the velocity is of the form $\mathbf{v} = v(x_1, t)\mathbf{e}_1$. Verify that the motion is necessarily irrotational and nonisochoric. Find the stress components developed in the fluid. Deduce that the stress vector on planes parallel to and perpendicular to the direction of flow is a pure pressure.

33. For a viscous fluid, show that
$$\mathbf{T}^{(d)} = (\bar{p} - p)\mathbf{I} + \kappa(\text{tr } \mathbf{D})\mathbf{I} + 2\mu \mathbf{D}^{(d)}$$

34. Show that for a compressible viscous fluid,
$$\bar{p} = p + \frac{\kappa}{\rho}\frac{D\rho}{Dt}$$

35. Show that
$$\Phi = \kappa I_D^2 + 4\mu II_{\mathbf{D}^{(d)}} = \text{tr}(\mathbf{T}^{(v)}\mathbf{D})$$

36. The stress matrix at a given point of a viscous fluid with zero bulk viscosity is
$$[\mathbf{T}] = \begin{bmatrix} 0 & 2 & 0 \\ 2 & -4 & 1 \\ 0 & 1 & -2 \end{bmatrix}$$
Find the viscous stress matrix $[\mathbf{T}^{(v)}]$.

37. For a compressible viscous fluid with zero bulk viscosity contained in a volume V, show that the total surface force on the boundary S of V is given by
$$\int_S \mathbf{s}\, dS = \int_V (\text{div } \mathbf{T}^{(v)} - \nabla p)\, dV = \int_V [2\mu \text{ div } \mathbf{D}^{(d)} - \nabla p]\, dV = \int_V [\mu \nabla^2 \mathbf{v} - \nabla p]\, dV$$

38. Show that the energy equation (10.12.3) can be expressed in the form
$$\rho \frac{D\varepsilon}{Dt} = \frac{p}{\rho}\frac{D\rho}{Dt} + k\nabla^2 T + \rho h + \Phi$$

39. For an irrotational flow of a viscous fluid adjacent to a plane rigid wall, determine how the tangential velocity varies with the normal direction from the wall.

40. Show that the magnitude of normal stress on a stationary boundary with which a compressible viscous fluid is in contact is $\bar{p} + (4/3)\nu(D\rho/Dt)$.

41. For an irrotational flow of an incompressible viscous fluid in a region bounded by a rigid boundary, show that the velocity is uniquely determined by prescribing the normal component of velocity on the boundary.

42. Write down the Navier–Stokes equation (10.14.3) in the suffix notation.

43. For an irrotational flow of an incompressible fluid, show that the Navier–Stokes equation reduces to Euler's equation.

44. For an incompressible viscous fluid moving under conservative body force, show that

$$\frac{DI_c}{Dt} = \nu\nabla^2 I_c$$

45. For an incompressible viscous fluid moving under conservative body force, show that the vorticity obeys the equation

$$\frac{\partial \mathbf{w}}{\partial t} = \text{curl}(\mathbf{v} \times \mathbf{w}) + \nu\nabla^2\mathbf{w}$$

46. Generalize expression (10.14.20) to the nonsteady case, other conditions being the same.

47. Assuming that μ/ρ is a constant, show that, for an irrotational motion of a compressible fluid with zero bulk viscosity under a conservative body force, the Navier–Stokes equation has an integral in the form

$$\int \frac{dp}{\rho} + \frac{1}{2}v^2 + \chi - \frac{\partial \phi}{\partial t} + \frac{4}{3}\nu\nabla^2\phi = f(t)$$

48. In an incompressible fluid moving under zero body force, the velocity components are of the form

$$v_1 = -\tfrac{1}{2}\alpha x_1 - f(R)x_2$$
$$v_2 = -\tfrac{1}{2}\alpha x_2 + f(R)x_2$$
$$v_3 = \alpha x_3$$

where $R = (x_1^2 + x_2^2)^{1/2}$ and α is a positive constant. Show that the pressure distribution is of the form

$$p = p_0 - \tfrac{1}{8}\rho\alpha^2(R^2 + 4x_3^2) + \rho\int R\{f(R)\}^2 \, dR$$

where p_0 is a constant.

49. For the plane Poiseullie flow, show that the vorticity is a harmonic function.

50. An incompressible viscous fluid bounded between two inclined parallel, rigid plates flows down under gravity. If the upper plate is moving in the direction of flow with speed $v_0(t)$ and the lower plate is held stationary, obtain the velocity distribution in the fluid.

51. An incompressible viscous fluid flows steadily down an inclined plane under gravity. The fluid layer is of uniform thickness and the upper layer is exposed to atmospheric pressure. Obtain the velocity distribution in the fluid.

52. For the Hagen–Poiseullie flow in a circular conduit show that the vorticity is perpendicular to the conduit and that its magnitude varies directly with the radial distance from the axis.

568 10 EQUATIONS OF FLUID MECHANICS

53. For the steady flow of an incompressible viscous fluid through a straight triangular tube bounded by the planes $x = a$, $y = \pm(1/\sqrt{3})x$ under a constant pressure gradient and zero body force, show that the velocity distribution is given by

$$v = \frac{G}{4\mu a}(x - a)(3y^2 - x^2)$$

Deduce that the mass flow rate in a cross section is

$$M = \frac{Ga^4}{60\sqrt{3}\,\nu}$$

54. An incompressible viscous fluid moves steadily in a horizontal elliptic pipe of length L placed along the x axis due to a pressure difference P maintained across the ends. If $y^2 + 4z^2 = 4$ is the equation of the boundary of a cross section, show that the total volume of fluid delivered through the pipe per unit of time is $2\pi P/5\mu L$.

55. An incompressible viscous fluid flows steadily through a region bounded by a solid cylinder of radius a and a coaxial tube of radius b, $b > a$. The cylinder and the tube are stationary, there is no body force and the motion is due to a constant pressure gradient parallel to the axis. Show that the velocity distribution is given by

$$v = \frac{G}{4\mu}[(a^2 - R^2) + (b^2 - a^2)\{\log(R/a)/\log(b/a)\}]$$

where R is, as usual, the radial distance from the axis. Deduce that the shear stress and rate of mass flow are, respectively,

$$\tau = \frac{g}{4}\left[\frac{b^2 - a^2}{R\log(b/a)} - 2R\right], \quad M = \frac{\pi G}{8\nu}\left[b^4 - a^4 - \frac{(b^2 - a^2)^2}{\log(b/a)}\right]$$

56. An incompressible viscous fluid flows steadily through a region bounded by a solid cylinder of radius a and a coaxial tube of radius b, $b > a$. The motion is due to the movement of the cylinder with speed v_0 along its axis. The tube is stationary; there is no body force; and there is no pressure gradient. Show that the velocity distribution is given by $v = v_0 \log(b/R)/\log(b/a)$ and that the shear stress along the axis is $\tau = -\mu v_0/\{R \log(b/a)\}$, where R is, as usual, the radial distance from the axis.

57. Carry out the computations involved in obtaining (10.15.112) from (10.15.110).

ANSWERS AND HINTS TO SELECTED EXERCISES

Section 1.8

1. (i) meaningful; $a_{11} + a_{22} + a_{33}$
    (iii) meaningful; $a_{11}b_1 + a_{21}b_2 + a_{31}b_3$, $a_{12}b_1 + a_{22}b_2 + a_{32}b_3$, $a_{13}b_1 + a_{23}b_2 + a_{33}b_3$
    (iv) not meaningful
    (vi) not meaningful
    (vii) meaningful; $a_{11}b_{11} + a_{12}b_{21} + a_{13}b_{31} + a_{21}b_{12} + a_{22}b_{22} + a_{23}b_{32} + a_{31}b_{13} + a_{32}b_{23} + a_{33}b_{33}$
    (viii) not meaningful
2. $a_{ij}b_j$ and $a_{rs}b_s$ have the same meaning.
    $a_{ij}b_ib_j$, $a_{pq}b_pb_q$ and $a_{sr}b_sb_r$ have the same meaning.
3. (ii) not meaningful
    (iii) not meaningful
    (iv) meaningful
    (vi) meaningful
    (vii) not meaningful
    (viii) not meaningful
4. (ii), (iii), (iv) and (vi).
5. $a_{ij} = \frac{1}{2}(b_{ij} + b_{ji})$

6. Note that $a_{ij}x_ix_j = a_{ji}x_jx_i = -a_{ij}x_ix_j$.
8. Use the fact that $a_{ij}c_{ji} = a_{ji}c_{ij} = a_{ij}c_{ij}$.
10. Note that $\mathbf{a} \cdot \mathbf{e}_i = a_i$.
13. (i) Show that $(\mathbf{a} \times \mathbf{b}) \times \mathbf{c} - \mathbf{a} \times (\mathbf{b} \times \mathbf{c}) = \mathbf{b} \times (\mathbf{c} \times \mathbf{a})$.
(ii) $\mathbf{a}$ and $\mathbf{c}$ are collinear, or $\mathbf{b}$ is orthogonal to both $\mathbf{a}$ and $\mathbf{c}$.
14. $a_{ii} = 2$, $a_{ij}a_{ji} = 0$, $a_{ij}a_{ij} = 17$

15. $$[a_{ij}] = \begin{bmatrix} 2 & -3 & 6 \\ -3 & 8 & 5 \\ 6 & 5 & -8 \end{bmatrix} + \begin{bmatrix} 0 & 3 & -2 \\ -3 & 0 & -5 \\ 2 & 5 & 0 \end{bmatrix}$$

16. $$[a_{ij}][a_{ij}]^T = \begin{bmatrix} 20 & -12 & -16 \\ -12 & 100 & 32 \\ -16 & 32 & 228 \end{bmatrix}$$

$$[a_{ij}]^T[a_{ij}] = \begin{bmatrix} 104 & 32 & -56 \\ 32 & 164 & -80 \\ -56 & -80 & 80 \end{bmatrix}$$

$$[a_{ij}]^2 = \begin{bmatrix} 36 & 40 & -24 \\ -60 & 64 & -24 \\ -108 & 0 & 96 \end{bmatrix}$$

19. (i) $\begin{bmatrix} 1 & 0 & 0 \\ 0 & -\frac{3}{4} & \frac{1}{2} \\ 0 & \frac{1}{2} & 0 \end{bmatrix}$ (ii) $\begin{bmatrix} \frac{1}{4} & -\frac{1}{2} & 0 \\ \frac{1}{4} & \frac{1}{2} & 0 \\ 0 & 0 & -1 \end{bmatrix}$

22. (i) 0 (ii) $a_ib_jc_j$ (iii) $3 - a_{ij}a_{ij}$
23. (i) $[\alpha_{ij}][\alpha_{ij}]^T = [I]$
(ii) $[\alpha_{ij}]^T[\alpha_{ij}] = [I]$
(iii) $[a_{ij}] = \alpha b_{kk}[I] + \beta[b_{ij}]$
24. (i) $a_{ij} = -p\delta_{ij}$
(ii) $a_{ijj} + b_i = 0$
25. $a_1 = b_{23} - b_{32}$, $a_2 = b_{31} - b_{13}$, $a_3 = b_{12} - b_{21}$
26. Note that $\varepsilon_{ijk}b_j = -\varepsilon_{kji}b_j$.
27. (iii) Note that $\varepsilon_{pqr}\varepsilon_{rqs} = -\varepsilon_{pqr}\varepsilon_{sqr}$ and use (1.7.24)
28. $\mathbf{e}_i \cdot (\mathbf{e}_j \times \mathbf{e}_k) = (\mathbf{e}_i \times \mathbf{e}_j) \cdot \mathbf{e}_k = (\varepsilon_{ijm}\mathbf{e}_m) \cdot \mathbf{e}_k = \varepsilon_{ijm}\delta_{mk} = \varepsilon_{ijk}$
30. Use $\varepsilon_{ijk}a_ib_jc_k = [\mathbf{a}, \mathbf{b}, \mathbf{c}]$
31. $\varepsilon_{pqr}b_r = \frac{1}{2}\varepsilon_{pqr}\varepsilon_{jkr}a_{jk}$
$= \frac{1}{2}(\delta_{pj}\delta_{qk} - \delta_{pk}\delta_{qj})a_{jk}$
$= \frac{1}{2}(a_{pq} - a_{qp}) = a_{pq}$
32. Use $\varepsilon_{ijk}\varepsilon_{krs}n_jv_rn_s = (\delta_{ir}\delta_{js} - \delta_{is}\delta_{jr})n_jv_rn_s = (v_in_j - n_iv_j)n_j = v_i - n_iv_jn_j$

33. (i) $x_i = \alpha \varepsilon_{ijk} a_j b_k + \beta a_i$
(ii) $\varepsilon_{ijk}(a_i - b_i)c_j d_k = 0$
(iii) $x_i = b_i + \alpha \varepsilon_{ijk} \varepsilon_{krs}(a_r - b_r)c_j c_s$
(iv) $\varepsilon_{imn}\varepsilon_{ipq} a_m a_p b_n b_q = a_i a_i b_j b_j - a_i b_i a_j b_j$

Section 2.15

1. (a) $[\alpha_{ij}] = \begin{bmatrix} 1 & 0 & 0 \\ 0 & 0 & 1 \\ 0 & -1 & 0 \end{bmatrix}$ (b) $[\alpha_{ij}] = \begin{bmatrix} \frac{1}{2} & 0 & -\frac{\sqrt{3}}{2} \\ 0 & 1 & 0 \\ \frac{\sqrt{3}}{2} & 0 & \frac{1}{2} \end{bmatrix}$

4. $x_1' + (1 - \sqrt{3}/2)x_2' - (1/2 + \sqrt{3})x_3' = 1$

5. Use (1.7.21), (2.2.17) and orthogonal property of $[\alpha_{ij}]$.

7. The entity is a vector.

10. $a_1' = 0$, $a_2' = \dfrac{1 - \sqrt{3}}{2\sqrt{2}}$, $a_3' = -\dfrac{1 + \sqrt{3}}{2\sqrt{2}}$

$$[a_{ij}'] = \begin{bmatrix} 1 & -\frac{\sqrt{3}}{2} & -\frac{1}{2} \\ -\frac{\sqrt{3}}{2} & -\frac{1}{2} & -\frac{\sqrt{3}}{2} \\ -\frac{1}{2} & -\frac{\sqrt{3}}{2} & -\frac{1}{2} \end{bmatrix}$$

15. $\delta_{ij}' = \delta_{ij} = \alpha_{im}\alpha_{jm} = \alpha_{im}\alpha_{jn}\delta_{mn}$

21. Use (2.4.40).

24. If $\mathbf{a} \otimes \mathbf{b} = \mathbf{I}$, then for *any* $\mathbf{c}$, we get $\mathbf{c} = \mathbf{Ic} = (\mathbf{a} \otimes \mathbf{b})\mathbf{c} = (\mathbf{b} \cdot \mathbf{c})\mathbf{a}$, which is not necessarily true.

28. Show $(\mathbf{I} - 2\mathbf{a} \otimes \mathbf{a})(\mathbf{I} - 2\mathbf{a} \otimes \mathbf{a})^T = \mathbf{I}$; not a proper orthogonal tensor.

29. $[A] = \begin{bmatrix} 0 & 1 & 1 \\ 2 & 0 & 0 \\ 0 & 1 & 0 \end{bmatrix}$; $2\mathbf{e}_2 - \mathbf{e}_3$

31. Use a quotient law.

38. $[(\mathbf{b} \wedge \mathbf{I})\mathbf{c}]_i = \varepsilon_{irs} b_r \delta_{js} c_j = \varepsilon_{irs} b_r c_s$

40. Use $a_{ij} b_{ij} = \tfrac{1}{2} a_{ij}(b_{ij} - b_{ji})$.

43. (i) Use $\mathbf{a} \cdot (\text{skw } \mathbf{A})\mathbf{a} = (\text{skw } \mathbf{A})^T \mathbf{a} \cdot \mathbf{a} = -(\text{skw } \mathbf{A})\mathbf{a} \cdot \mathbf{a}$.
(ii) Use $\mathbf{a} \cdot (\mathbf{Aa}) = \mathbf{a} \cdot (\text{sym } \mathbf{A} + \text{skw } \mathbf{A})\mathbf{a} = \mathbf{a} \cdot (\text{sym } \mathbf{A})\mathbf{a}$.

44. Use (2.10.6) to get $\boldsymbol{\omega} = 3\mathbf{e}_1 + 5\mathbf{e}_2 + 2\mathbf{e}_3$.

45. Use (2.10.7).

49. (i) Use (2.10.4) to get
$$a_{ik}b_{kj} = \varepsilon_{ikm}u_m\varepsilon_{kjn}v_n = (\delta_{in}\delta_{mj} - \delta_{ij}\delta_{mn})u_m v_n = u_j v_i - \delta_{ij}u_m v_m$$
(ii) $a_{ik}b_{ki} = -2u_m v_m$

53. $(\mathbf{e}_k \otimes \mathbf{e}'_k)(\mathbf{e}_m \otimes \mathbf{e}'_m)^T = (\mathbf{e}_k \otimes \mathbf{e}'_k)(\mathbf{e}'_m \otimes \mathbf{e}_m) = (\mathbf{e}_k \cdot \mathbf{e}'_m)(\mathbf{e}_k \otimes \mathbf{e}_m)$
$$= \delta_{km}(\mathbf{e}_k \otimes \mathbf{e}_m) = \mathbf{e}_k \otimes \mathbf{e}_k = \mathbf{I}$$

55. (i) Use (2.13.25) and (2.11.23) to get
$$\mathbf{A}^3 - I_\mathbf{A}\mathbf{A}^2 + II_\mathbf{A}\mathbf{A} = (\det \mathbf{A})\mathbf{I} = (\mathbf{A}^*)^T\mathbf{A}$$
(ii) Use (2.11.24).

56. (i) Use (1.7.24) and (2.11.1).
(ii) Use (1.7.23) and (2.11.4).
(iii) Use (1.7.26) and (2.11.11).

57. For any Λ,
$$\det(\mathbf{BAB}^{-1} - \Lambda\mathbf{I}) = \det\{\mathbf{B}(\mathbf{A} - \Lambda\mathbf{I})\mathbf{B}^{-1}\}$$
$$= (\det \mathbf{B})\{\det(\mathbf{A} - \Lambda\mathbf{I})\}(\det \mathbf{B}^{-1}) = \det(\mathbf{A} - \Lambda\mathbf{I})$$

58. Use (2.11.23) and (2.11.25).

61. Use (2.11.1), (2.11.3) and (2.11.9).

63. Use (2.11.24) along with (2.11.22) and results of Exercise 56.

64. Use (2.12.2) and results of Exercise 56.

65. (ii) $I_\mathbf{A} = -2$, $II_\mathbf{A} = -1$, $III_\mathbf{A} = 0$; $\Lambda = 0$, $-1 \pm \sqrt{2}$

66. (ii) $1, 2, 4$; $\mathbf{e}_3$, $(1/\sqrt{2})(\mathbf{e}_1 + \mathbf{e}_2)$, $(1/\sqrt{2})(\mathbf{e}_1 - \mathbf{e}_2)$
(iv) $2, 1, -3$; $(1/\sqrt{5})(2\mathbf{e}_1 + \mathbf{e}_3)$, $\mathbf{e}_2$, $(1/\sqrt{5})(\mathbf{e}_1 - 2\mathbf{e}_3)$

68. Use $(\mathbf{a} \otimes \mathbf{b})\mathbf{a} = (\mathbf{a} \cdot \mathbf{b})\mathbf{a}$

72. (ii) $\begin{bmatrix} 1 & 0 & -1 \\ 0 & -2 & 0 \\ -1 & 0 & 2 \end{bmatrix}$

73. (i) $[\mathbf{Q}] = \begin{bmatrix} 0 & 1 & 0 \\ 1 & 0 & 0 \\ 0 & 0 & -1 \end{bmatrix}$, $[\mathbf{U}] = \begin{bmatrix} 2 & 1 & 0 \\ 1 & 0 & 0 \\ 0 & 0 & -1 \end{bmatrix}$, $[\mathbf{V}] = \begin{bmatrix} 0 & 1 & 0 \\ 1 & 2 & 0 \\ 0 & 0 & -1 \end{bmatrix}$

(iii) $[\mathbf{Q}] = \begin{bmatrix} \frac{1}{\sqrt{2}} & -\frac{1}{\sqrt{2}} & 0 \\ \frac{1}{\sqrt{2}} & \frac{1}{\sqrt{2}} & 0 \\ 0 & 0 & -1 \end{bmatrix}$, $[\mathbf{U}] = \begin{bmatrix} 2\sqrt{2} & 0 & 0 \\ 0 & \sqrt{2} & 0 \\ 0 & 0 & 1 \end{bmatrix}$,

$$[\mathbf{V}] = \begin{bmatrix} \dfrac{3}{\sqrt{2}} & \dfrac{1}{\sqrt{2}} & 0 \\ \dfrac{1}{\sqrt{2}} & \dfrac{3}{\sqrt{2}} & 0 \\ 0 & 0 & 1 \end{bmatrix}$$

Section 3.8

2. Start with $\mathbf{u} \cdot \mathbf{u} = u^2$ and differentiate both sides with respect to t.

3. Using $\mathbf{u} \cdot \mathbf{u} = 1$, show $\mathbf{u} \cdot (d\mathbf{u}/dt) = 0$, and use the definition of vector product to compute $|\mathbf{u} \times (d\mathbf{u}/dt)|$

8. $(u_{i,j} + u_{j,i}) = \begin{bmatrix} 4x_2 & 2x_1 & 3x_3 \\ 2x_1 & -2x_3 & -x_2 \\ 3x_3 & -x_2 & 6x_1 \end{bmatrix}$

11. $6/14$ and $-6/14$.

14. $f_{,i} = f'(r)r_{,i} = f'(r)\dfrac{x_i}{r}$.

17. (i) $\mathbf{e}_k \cdot \mathbf{u}_{,k} = \mathbf{e}_k \cdot (u_{1,k}\mathbf{e}_1 + u_{2,k}\mathbf{e}_2 + u_{3,k}\mathbf{e}_3) = \operatorname{div} \mathbf{u}$.
(ii) $\mathbf{e}_k \times \mathbf{u}_{,k} = \mathbf{e}_k \times (u_{1,k}\mathbf{e}_1 + u_{2,k}\mathbf{e}_2 + u_{3,k}\mathbf{e}_3)$;
$\mathbf{e}_k \times u_{1,k}\mathbf{e}_1 = -u_{1,2}\mathbf{e}_3 + u_{1,3}\mathbf{e}_2$, etc.

18. (i) $[\operatorname{curl}(\mathbf{u} \times \mathbf{v})]_i = \varepsilon_{ijk}\varepsilon_{kmn}(u_m v_n)_{,j}$
$= (\delta_{im}\delta_{jn} - \delta_{in}\delta_{jm})(u_{m,j}v_n + u_m v_{n,j})$
$= u_{i,j}v_j + u_i v_{j,j} - u_{j,j}v_i - u_j v_{i,j}$

20. $\nabla^2 |\mathbf{x}|^n = (r^n)_{,ii} = \left(nr^{n-1}\dfrac{x_i}{r}\right)_{,i} = n(n+1)r^{n-2}$

21. $\phi = A/r^3$, where A is a constant.

22. $\phi = A/r^2 + B/r$, where A and B are constants.

24. Note that $\nabla^2(\phi\psi) = \operatorname{div} \nabla(\phi\psi)$ and use (3.4.4) and (3.4.14).

26. Simplify $\operatorname{curl}\{(\operatorname{curl} \mathbf{u}) \times \mathbf{x}\}$ using (3.4.25).

27. $\operatorname{curl}(r^n \mathbf{x}) = r^n \operatorname{curl} \mathbf{x} + nr^{n-2}\mathbf{x} \times \mathbf{x} = \mathbf{0}$
$\operatorname{div}(r^n \mathbf{x}) = 3r^n + nr^{n-2}\mathbf{x} \cdot \mathbf{x} = (3 + n)r^n$

30. Use (3.4.25) and (3.4.26).

32. $\phi\mathbf{u} \cdot \operatorname{curl}(\phi\mathbf{u}) = \phi\mathbf{u} \cdot (\phi \operatorname{curl} \mathbf{u} + \nabla\phi \times \mathbf{u}) = 0$.

35. $[(\mathbf{v} \wedge \nabla\mathbf{u})\mathbf{w}]_i = \varepsilon_{irs} v_r u_{j,s} w_j$.

40. (ii) $[\operatorname{div} \nabla\mathbf{u}^T]_i = u_{k,ik} = (u_{k,k}\delta_{ij})_{,j}$.
(iv) $\operatorname{div}(\mathbf{A}\mathbf{u}) = (a_{ji}u_i)_{,j} = u_i a_{ji,j} + a_{ji}u_{i,j}$.
(ix) $\operatorname{div}[(\mathbf{u} \otimes \mathbf{v})\mathbf{w}] = (u_i v_j w_j)_{,i} = u_{i,i}v_j w_j + u_i(v_j w_j)_{,i}$.

41. If $f \neq 0$ at a point P of R, then for some infinitesimal region Δv containing P, we have
$$\int_{\Delta V} f \, dv = f(\Delta v) \neq 0$$
This is a contradiction.

42. Zero.

44. (i) Employ (3.6.1) to $\mathbf{u} = \phi(\nabla \psi)$.
(ii) Employ (3.6.1) to $\mathbf{u} = \phi(\nabla \psi) + \psi(\nabla \phi)$.

45. Use (3.6.3).

49. Take $\mathbf{A} = \mathbf{I}$ and note that $2\boldsymbol{\omega} = -\operatorname{curl} \mathbf{u}$ by Example 3.5.7.

53. (i) Employ divergence theorem to $\mathbf{u} \times \operatorname{curl} \mathbf{u}$ and use (3.4.24).
(iii) Employ divergence theorem to $(\mathbf{u} \cdot \mathbf{a})\nabla(\mathbf{u} \cdot \mathbf{a})$ and use (3.4.14).

55. Use (3.7.1) and (3.7.3) to get
$$\int_S [\{\mathbf{u} \otimes \mathbf{v} + \mathbf{v} \otimes \mathbf{u}\}\mathbf{n} - (\mathbf{u} \cdot \mathbf{v})\mathbf{n}] \, dS$$
$$= \int_V [(\operatorname{div} \mathbf{v})\mathbf{u} + (\operatorname{div} \mathbf{u})\mathbf{v} + (\mathbf{v} \cdot \nabla)\mathbf{u} + (\mathbf{u} \cdot \nabla)\mathbf{v} - \nabla(\mathbf{u} \cdot \mathbf{v})] \, dV$$
and then use (3.4.26).

Section 5.11

1. (i) $[\mathbf{F}] = [\mathbf{U}] = [\mathbf{V}] = \begin{bmatrix} \alpha & 0 & 0 \\ 0 & \beta & 0 \\ 0 & 0 & \gamma \end{bmatrix}$, $\mathbf{Q} = \mathbf{I}$

$[\mathbf{F}^{-1}] = \begin{bmatrix} 1/\alpha & 0 & 0 \\ 0 & 1/\beta & 0 \\ 0 & 0 & 1/\gamma \end{bmatrix}$; isochoric if $\alpha\beta\gamma = 1$

Principal stretches: α, β, γ.

(iii) $[\mathbf{F}] = \begin{bmatrix} \alpha & \beta & 0 \\ -\alpha & \beta & 0 \\ 0 & 0 & \gamma \end{bmatrix}$; $[\mathbf{F}^{-1}] = \begin{bmatrix} \dfrac{1}{2\alpha} & -\dfrac{1}{2\alpha} & 0 \\ \dfrac{1}{2\beta} & \dfrac{1}{2\beta} & 0 \\ 0 & 0 & 1/\gamma \end{bmatrix}$

$[\mathbf{U}] = \begin{bmatrix} \sqrt{2}\alpha & 0 & 0 \\ 0 & \sqrt{2}\beta & 0 \\ 0 & 0 & \gamma \end{bmatrix}$; $[\mathbf{V}] = \dfrac{1}{\sqrt{2}} \begin{bmatrix} \alpha+\beta & -\alpha+\beta & 0 \\ -\alpha+\beta & \alpha+\beta & 0 \\ 0 & 0 & \sqrt{2}\gamma \end{bmatrix}$

ANSWERS AND HINTS TO SELECTED EXERCISES 575

$$[Q] = \frac{1}{\sqrt{2}}\begin{bmatrix} 1 & 1 & 0 \\ -1 & 1 & 0 \\ 0 & 0 & \sqrt{2} \end{bmatrix}; \quad \text{isochoric if } \alpha\beta\gamma = 1/2$$

Principal stretches: $\sqrt{2}\alpha, \sqrt{2}\beta, \gamma$.

2. (i) The deformation is a uniform extension or contraction in the 1 direction according as $\alpha > 1$ or $\alpha < 1$. β measures the extension or contraction in the 2, 3 plane according as $\beta > 1$ or $\beta < 1$.
   (iv) The deformation displaces particles in the plane $x_3^0 = $ constant through a distance proportional to x_3^0 and to the distance $\{(x_1^0)^2 + (x_3^0)^2\}^{1/2}$ from the 3 axis; the direction of displacement is at right angles to the vector $x_1^0 \mathbf{e}_1 + x_2^0 \mathbf{e}_2$.

6. The hyperboloid of revolution: $x_1^2 + x_2^2 = a^2(1 + x_3^2)$.

9. (i) $[C] = \begin{bmatrix} 1+\alpha^2 & \alpha & 0 \\ \alpha & 1 & 0 \\ 0 & 0 & 1 \end{bmatrix}; \quad [B] = \begin{bmatrix} 1 & \alpha & 0 \\ \alpha & 1+\alpha^2 & 0 \\ 0 & 0 & 1 \end{bmatrix};$

$[G] = \frac{1}{2}\begin{bmatrix} \alpha^2 & \alpha & 0 \\ \alpha & 0 & 0 \\ 0 & 0 & 0 \end{bmatrix}$

(iv) $[C] = \begin{bmatrix} 1 & \tan\alpha & 0 \\ \tan\alpha & \sec^2\alpha & 0 \\ 0 & 0 & 1 \end{bmatrix}; \quad [B] = \begin{bmatrix} \sec^2\alpha & \sec\alpha & 0 \\ \tan\alpha & 1 & 0 \\ 0 & 0 & 1 \end{bmatrix};$

$[G] = \frac{1}{2}\begin{bmatrix} 0 & \tan\alpha & 0 \\ \tan\alpha & \tan^2\alpha & 0 \\ 0 & 0 & 0 \end{bmatrix}$

10. (ii) $[B^{-1}] = \begin{bmatrix} 1 & 0 & 0 \\ 0 & 1 & 0 \\ 0 & 0 & (1+\alpha\beta)^2 \end{bmatrix}; \quad [A] = \frac{1}{2}\begin{bmatrix} 0 & 0 & 0 \\ 0 & 0 & 0 \\ 0 & 0 & \alpha\beta(2+\alpha\beta) \end{bmatrix}$

14. $e = (1 + \alpha^2/2)^{1/2} - 1$

16. Arcs initially lying along the 3 axis.

21. (i) $\eta = \sqrt{2}$,   (ii) $\theta = \cos^{-1}(1/\sqrt{3})$

31. (ii) Material form: $v_1 = (x_1^0 + x_3^0)(e^t - 1)$, $u_2 = x_2^0(e^t - e^{-t})$, $u_3 = 0$.
    Spatial form: $u_1 = (x_1 + x_3)(1 - e^{-t})$, $u_2 = x_3(e^t - e^{-t})$, $u_3 = 0$.

576 CONTINUUM MECHANICS

39. (iii) $[E] = \begin{bmatrix} 2\alpha^2(x_1 - x_3) & 0 & \frac{\alpha}{2}\{2\alpha(x_3 - x_1) - x_2\} \\ 0 & 2\alpha^2(x_2 + x_3) & \frac{\alpha}{2}\{2\alpha(x_2 + x_3) - x_1\} \\ \frac{\alpha}{2}\{2\alpha(x_3 - x_1) - x_2\} & \frac{\alpha}{2}\{2\alpha(x_2 + x_3) - x_1\} & 0 \end{bmatrix}$

$[\Omega] = \begin{bmatrix} 0 & 0 & \frac{\alpha}{2}\{2\alpha(x_3 - x_1) + x_2\} \\ 0 & 0 & \frac{\alpha}{2}\{2\alpha(x_2 + x_3) + x_1\} \\ \frac{\alpha}{2}\{2\alpha(x_1 - x_3) - x_2\} & -\frac{\alpha}{2}\{2\alpha(x_2 + x_3) + x_1\} & 0 \end{bmatrix}$

$\boldsymbol{\omega} = -\frac{\alpha}{2}[\{x_1 + 2\alpha(x_2 + x_3)\}\mathbf{e}_1 - \{x_2 - 2\alpha(x_1 - x_3)\}\mathbf{e}_2]$

40. (i) $\frac{\alpha}{81}(58\alpha - 64)$, **(ii)** $\frac{2\alpha}{81}(119\alpha + 40)$

43. (ii) $u_1 = x_1^2 + x_2^2$, $u_2 = 2x_1x_2 + x_1^2$, $u_3 = x_3^2$

44. (i) $\alpha = 2\beta$

Section 6.7

1. (i) Material description:

$$v_1 = 2a^2tx_1^0, \quad v_2 = v_3 = 0; \quad \frac{Dv_1}{Dt} = 2a^2x_1^0, \quad \frac{Dv_2}{Dt} = \frac{Dv_3}{Dt} = 0$$

Spatial description:

$$v_1 = \frac{2a^2x_1t}{1 + a^2t^2}, \quad v_2 = v_3 = 0; \quad \frac{Dv_1}{Dt} = \frac{2a^2x_1}{1 + a^2t^2}, \quad \frac{Dv_2}{Dt} = \frac{Dv_3}{Dt} = 0$$

(iv) Material description:

$$v_1 = -\alpha(x_1^0 \sin \alpha t - x_2^0 \cos \alpha t), \quad v_2 = -\alpha(x_1^0 \cos \alpha t + x_2^0 \sin \alpha t), \quad v_3 = 0$$

$$\frac{Dv_1}{Dt} = -\alpha^2(x_1^0 \cos \alpha t + x_2^0 \sin \alpha t)$$

$$\frac{Dv_2}{Dt} = \alpha^2(x_1^0 \sin \alpha t + x_2^0 \cos \alpha t), \quad \frac{Dv_3}{Dt} = 0$$

Spatial description:

$$v_1 = \alpha x_2, \quad v_2 = -\alpha x_1, \quad v_3 = 0; \quad \frac{Dv_1}{Dt} = -\alpha^2 x_1, \quad \frac{Dv_2}{Dt} = -\alpha^2 x_2, \quad \frac{Dv_3}{Dt} = 0$$

3. $(-k\alpha e^{-\alpha t} + 3k^2 e^{-2\alpha t})\mathbf{e}_2$

ANSWERS AND HINTS TO SELECTED EXERCISES   577

6. $(D\mathbf{v}/Dt) = -\alpha^2(x_1\mathbf{e}_1 + x_2\mathbf{e}_2)$
10. Use the relation (6.3.21).
11. $J = e^t$.
14. $(x_1 + 2x_2 - 3x_3)e^{-t} - (x_1 - 2x_2 - 3x_3)e^{-2t}$
16. e^{-t}
17. For exercise 3:
  (i) $d_{11} = -k(3x_1^2 + x_2^2)e^{-\alpha t}$, $d_{22} = k(x_1^2 + 3x_2^2)e^{-\alpha t}$; other $d_{ij} = 0$.
  (ii) $w_{12} = -w_{21} = -2kx_1x_2 e^{-\alpha t}$; other $w_{ij} = 0$;
  $\mathbf{w} = 2kx_1x_2 e^{-\alpha t}\mathbf{e}_3$.
  (iii) $I_D = -2k(x_1^2 - x_2^2)e^{-\alpha t}$
  $II_D = -k^2(x_1^2 + 3x_2^2)(3x_1^2 + x_2^2)e^{-2\alpha t}$
  $III_D = 0$.
  (iv) $d_{11}^{(d)} = -\dfrac{k}{3}(7x_1^2 + 5x_2^2)e^{-\alpha t}$, $d_{22}^{(d)} = \dfrac{k}{3}(5x_1^2 + 7x_2^2)e^{-\alpha t}$; other $d_{ij}^{(d)} = 0$.

19. Principal stretchings: $(1/2)R\,df/dR$, $(-1/2)R(df/dR)$, 0;
Angular velocity: $(1/2)[2f + R\,df/dR]\mathbf{e}_3$;
Principal directions of stretching:
$$\cos\left(\theta + \dfrac{\pi}{4}\right)\mathbf{e}_1 + \sin\left(\theta + \dfrac{\pi}{4}\right)\mathbf{e}_2, \quad -\sin\left(\theta + \dfrac{\pi}{4}\right)\mathbf{e}_1 + \cos\left(\theta + \dfrac{\pi}{4}\right)\mathbf{e}_2, \quad \mathbf{e}_3,$$
where $\theta = \tan^{-1}(x_2/x_1)$.

20. Follow the method of Example 5.6.2.
24. Integrate the equation $\mathbf{D} = (1/2)(\nabla\mathbf{v} + \nabla\mathbf{v}^T)$ for the given [D].
25. Principal stretchings: $(1/2)\alpha$, $(-1/2)\alpha$, 0;
Principal directions of stretching: $\dfrac{1}{\sqrt{2}}(\mathbf{a} + \mathbf{b})$, $\dfrac{1}{\sqrt{2}}(\mathbf{a} - \mathbf{b})$, $\dfrac{\mathbf{a} \times \mathbf{b}}{|\mathbf{a} \times \mathbf{b}|}$.

43. (i) Both stream lines and path lines: $x_1 = a\cos\alpha t + b\sin\alpha t$, $x_2 = -a\sin\alpha t + b\cos\alpha t$, $x_3 = \beta t + c$, where a, b, c are constants
  (iii) Stream lines: $x_1 = ae^{s/(1+t)}$, $x_2 = be^s$, $x_3 = c$. Path lines: $x_1 = x_1^0(1 + t)$, $x_2 = x_2^0 e^t$, $x_3 = x_3^0$.
  (iv) Both stream lines and path lines: $x_i = (1 + t)x_i^0$.

Section 7.10

1. $\mathbf{s} = -2\mathbf{e}_2$.
2. $\mathbf{s} = (1/3)(5\mathbf{e}_1 + 6\mathbf{e}_2 + 5\mathbf{e}_3)$; $|\mathbf{s}| = \sqrt{86}/3$; $\theta = \cos^{-1}(9/\sqrt{86})$.
3. $\mathbf{s} = (2/\sqrt{6})(\mathbf{e}_1 + \mathbf{e}_3)$; $\sigma = 1$, $\tau = 1/\sqrt{3}$.
4. $a = b = c = -1/2$.
5. Plane perpendicular to the vector $\sqrt{2}\mathbf{e}_1 + \mathbf{e}_2$.
7. $a = 3/4$; plane perpendicular to the vector $\mathbf{e}_1 - 4\mathbf{e}_2 + \mathbf{e}_3$.
9. $\mathbf{s} = -(a/R)(2\cos^2\theta\,\mathbf{e}_1 + \sin 2\theta\,\mathbf{e}_2)$; $\sigma = -(2a/R)\cos\theta$; $\tau = 0$.

578 CONTINUUM MECHANICS

10. $\begin{bmatrix} 1 & 2 & 3 \\ 2 & 2 & 2 \\ 3 & 2 & -5 \end{bmatrix}$.

12. Symmetry of **T**.

13. Along the line of intersection of the elements.

17. $\mathbf{s} = T\{(\mathbf{b} \cdot \mathbf{n})\mathbf{a} + (\mathbf{a} \cdot \mathbf{n})\mathbf{b}\}$; $\mathbf{T} = T\{\mathbf{a} \otimes \mathbf{b} + \mathbf{b} \otimes \mathbf{a}\}$.

19. Principal stresses: 2, 1, −3;
Corresponding principal vectors: $(1/\sqrt{5})(2\mathbf{e}_1 + \mathbf{e}_3)$; $\mathbf{e}_2$, $(1/\sqrt{5})(\mathbf{e}_1 - 2\mathbf{e}_3)$.

21. $I_T = 0$, $II_T = -9$, $III_T = 8$.

22. Maximum shear stress of magnitude 3 occurs on a plane element perpendicular to the vector $2\mathbf{e}_1 + (1 - \sqrt{3})\mathbf{e}_2 + (1 + \sqrt{3})\mathbf{e}_3$.

24. $\mathbf{s} = (1/\sqrt{3})(4\mathbf{e}_1 + \mathbf{e}_2 - 2\mathbf{e}_3)$; $\sigma = 1$, $\tau = \sqrt{6}$.

27. Octahedral plane.

30. 0,   ±T.

34. $[\tau_{ij}^{(d)}] = \begin{bmatrix} -1 & -3 & \sqrt{2} \\ -3 & -1 & -\sqrt{2} \\ \sqrt{2} & -\sqrt{2} & 2 \end{bmatrix}$;   Principal deviator stresses: 0, ±4;

Stress deviator invariants: $II_{T^{(d)}} = -16$, $III_{T^{(d)}} = 0$.

35. Note that on the boundary surface, $n_1 = (1/a)x_1$, $n_2 = (1/a)x_2$, $n_3 = 0$.

36. Total force on each of the faces $x_1 = \pm a$, $x_2 = \pm a$ is $(8/3)ka^3 b$. Faces $x_3 = \pm b$ are stress-free.

38. $\tau_{11} \cos\theta + \tau_{12} \sin\theta = -p \cos\theta$; $\tau_{21} \cos\theta + \tau_{22} \sin\theta = -p \sin\theta$, where θ is the polar angle of the point.

Section 8.9

1. (i) and (ii): satisfied.

2. $\rho = \rho_0 \exp(-t^2)$.

3. $\rho = A/x_3$, where A is a constant.

4. $\mathbf{v} = v\mathbf{e}_3$, where v is independent of x_3.

6. $k = 1$.

9. Find J and use equation (8.2.8).

15. $\rho \mathbf{b} = (x_3^2 - 1)\mathbf{e}_1$.

18. Note that div $\mathbf{T} = (\mathbf{a} \cdot \nabla T)\mathbf{a}$.

19. $\phi(r) = (A/r^3) + B$, where A and B are constants.

20. Zero body force.

31. Note that $\mathbf{T} \cdot \mathbf{D} = -p \, \text{div} \, \mathbf{v}$ and use the equation of continuity.

32. Use equation (8.6.32).

33. Convert equations (8.6.27) and (8.6.28) to material form.

Section 9.14

4. Use relations (2.11.1), (2.11.4), (2.11.10) and (9.2.6).

6. $W = (1/4)\mu\alpha^2(x_1^2 + x_2^2)$.

8. (i) $\tau_{11} = \tau_{12} = \tau_{22} = \tau_{33} = 0$, $\quad \tau_{13} = k\theta_{,1}$, $\quad \tau_{23} = k\theta_{,2}$

(ii) $\tau_{11} = \tau_{22} = \tau_{33} = 0$, $\quad \tau_{12} = 2\mu k x_3$, $\quad \tau_{23} = 2\mu k x_1$, $\quad \tau_{31} = 2\mu k x_2$

(iii) $\tau_{11} = \tau_{22} = \tau_{33} = 0$, $\tau_{12} = 2\mu k x_3$, $\tau_{13} = \mu k(x_2 + 2x_1)$, $\tau_{23} = \mu k(x_1 - 2x_2)$

(iv) $\tau_{11} = \tau_{22} = \tau_{12} = \tau_{23} = \tau_{31} = 0$, $\quad \tau_{33} = -Ekx_1$

10.
$$e_{11} = e_{22} = e_{33} = \frac{3(1-v)}{E}(-p + \rho g x_3)$$

$$u_i = \frac{3(1-v)}{E}[-px_i + \rho g x_3 x_i], \quad i = 1, 2;$$

$$u_3 = \frac{3(1-v)}{E}\left[-px_3 + \frac{1}{2}\rho g x_3^2\right]$$

13. Use identities (3.5.34) and (3.6.19) to rewrite (9.8.26) as

$$\mathbf{s} = \lambda(\operatorname{div}\mathbf{u})\mathbf{n} + 2\mu(\mathbf{n}\cdot\nabla)\mathbf{u} - \mu(\nabla\mathbf{u} - \nabla\mathbf{u}^T)\mathbf{n}$$

Write this in the suffix notation and express λ and μ in terms of E and v.

15. Use relations (9.8.20) to (9.8.22).

21.
$$\tau_{11} = 2\mu A\left[\frac{x_2^2 + x_3^2}{r^3(r + x_3)} - \frac{x_1^2}{r^2(r + x_3)^2}\right]$$

$$\tau_{22} = 2\mu A\left[\frac{x_1^2 + x_3^2}{r^3(r + x_3)} - \frac{x_2^2}{r^2(r + x_3)^2}\right]$$

$$\tau_{12} = -2\mu A\frac{x_1 x_2(x_3 + 2r)}{r^3(r + x_3)^2}$$

$$\tau_{13} = -2\mu A\frac{x_1}{r^3}, \quad \tau_{23} = -2\mu A\frac{x_2}{r^3}, \quad \tau_{33} = -2\mu A\frac{x_3}{r^3}$$

On $r = $ constant, $\mathbf{s} = \{(-2\mu A)/r^2(r + x_3)\}[x_1\mathbf{e}_1 + x_2\mathbf{e}_2 + (r + x_3)\mathbf{e}_3]$.

33. $u_1 = -\frac{A}{r^5}x_2$, $\quad u_2 = \frac{A}{r^5}x_1$, $\quad u_3 = 0$

$\tau_{11} = \frac{10\mu A}{r^7}x_1 x_2$, $\quad \tau_{22} = -\frac{10\mu A}{r^7}x_1 x_2$, $\quad \tau_{12} = \frac{5\mu A}{r^7}(x_2^2 - x_1^2)$,

$\tau_{13} = \tau_{23} = \tau_{33} = 0$

41. $f(r) = A\cos(\omega/c_1)r + B\sin(\omega/c_1)r$, where A and B are constants.

48. Use the Green–Lamé solution.

55. Use the fact that τ_{kk} is a harmonic function.

62. (i) Use relations (9.8.4), (3.6.1) and (3.7.4).
(ii) Use relation (9.8.26) and simplify.

64. To obtain the required identity, start with the equilibrium counterpart of equation (9.8.8), take dot product with **u**, integrate over V, simplify using identities (3.4.14) and (3.4.24) and employ the divergence theorem (3.6.1).

Section 10.16

2. Solve equation (10.6.3) for p, with given ρ and $\mathbf{b} = g\mathbf{e}_3$. Express the solution in terms of p; note that $\rho = \rho_a$ for $z = 0$.

6. Consider equations (10.4.34) and (10.4.9).

12. Use the Bernoulli's equation (10.8.8).

14. $p = p_0 + \rho(g + |\mathbf{a}|)h$; p_0 is the pressure at the upper surface.

15. Use the fact that, at the point where the liquid contacts the wall, r is maximum and equal to a.

17. Use the fact that $v_3 = 0$, $v_{1,2} + v_{2,1} = 0$, $v_{2,1} - v_{1,2} = 0$.

19. $\dfrac{DI_c}{Dt} = \oint_c \dfrac{D\mathbf{v}}{Dt} \cdot d\mathbf{x} = \int_S \operatorname{curl} \dfrac{D\mathbf{v}}{Dt} \cdot \mathbf{n}\, dS = \int_S \operatorname{curl}\left\{-\dfrac{1}{\rho}\nabla p + \mathbf{b}\right\} \cdot \mathbf{n}\, dS.$

31. (ii) $[\mathbf{T}] = \begin{bmatrix} -p & 0 & 0 \\ 0 & -p + 4\mu x_2 & 0 \\ 0 & 0 & -p - 4\mu x_2 \end{bmatrix}$

34. Use equations (10.11.6) and (10.12.1).

36. $[\mathbf{T}^{(v)}] = \begin{bmatrix} 2 & 2 & 0 \\ 2 & -2 & 1 \\ 0 & 1 & 0 \end{bmatrix}$

39. Use relation (10.13.10).

40. Use relation (10.13.14) and the equation of continuity.

51. If the x_1 axis is along the direction of flow, the x_3 axis is perpendicular to the plane, θ is the inclination of the plane and h is the thickness of the layer, the velocity field is

$$\mathbf{v} = \dfrac{\rho g \sin\theta}{2\mu}(2h - x_3)x_3 \mathbf{e}_1$$

Bibliography

Achenbach, J. D. (1975). *Wave Propagation in Elastic Solids*. North-Holland, Amsterdam.
Aris, R. (1962). *Vectors, Tensors and Basic Equations of Fluid Mechanics*. Prentice-Hall, Englewood Cliffs, NJ.
Batchelor, G. K. (1967). *An Introduction to Fluid Mechanics*. Cambridge University Press, London.
Bowmen, R. M. (1989). *Introduction to Continuum Mechanics for Engineers*. Plenum Press, New York.
Bowmen, R. M., and Wang, C. C. (1976). *Introduction to Vectors and Tensors*, Vols. I and II. Plenum Press, New York.
Chadwick, P. (1976). *Continuum Mechanics*. George Allen and Unwin, London.
Debnath, L. (1994). *Nonlinear Water Waves*. Academic Press, Boston.
Erickson, J. L. (1960). "Tensor Fields," in *Handbuch der Physik*. Vol. III/3 (ed. S. Flügge), pp. 794–858. Springer-Verlag, Berlin.
Eringen, A. C. (1967). *Mechanics of Continua*. John Wiley and Sons, New York.
Eringen, A. C. (ed.) (1975). *Continuum Physics*, Vols. I and II. Academic Press, New York.
Eringen, A. C., and Suhubi, E. S. (1974–1975). *Elastodynamics*, Vols. I and II. Academic Press, New York.
Ewing, W. M., Jardetsky, W. S., and Press, F. (1957). *Elastic Waves in Layered Media*. McGraw-Hill Book Co., New York.
Frederick, D., and Chang, T. S. (1965). *Continuum Mechanics*. Allyn and Bacon, Boston.

Fung, Y. C. (1965). *Foundations of Solid Mechanics*. Prentice-Hall, Englewood Cliffs, NJ.
Goldstein, S. (1938). *Modern Developments in Fluid Dynamics*, Vols. I and II. Oxford University Press, London.
Gurtin, M. E. (1972). "The Linear Theory of Elasticity," in *Handbuch der Physik* (ed. S. Flügge), Vol. VI a/2, pp. 1–273. Springer-Verlag, Berlin.
Gurtin, M. E. (1981). *An Introduction to Continuum Mechanics*. Academic Press, New York.
Hunter, S. C. (1983). *Mechanics of Continuous Media*, 2nd ed. Ellis Horwood Ltd., Chichester, England.
Jaunzemis, W. (1967). *Continuum Mechanics*. Macmillan Publishing Co., New York.
Jefferys, H. (1952). *Cartesian Tensors*. Cambridge University Press, London.
Kolsky, H. (1953). *Stress Waves in Solids*. Clarendon Press, Oxford.
Lamb, H. (1932). *Hydrodynamics*, 6th ed. Cambridge University Press, London.
Landau, L. D., and Lifshitz, E. M. (1959). *Fluid Mechanics*. Pergamon Press, Oxford.
Landau, L. D., and Lifshitz, E. M. (1986). *Theory of Elasticity*, 3rd ed. Pergamon Press, Oxford.
Leigh, D. C. (1968). *Non-Linear Continuum Mechanics*. McGraw-Hill Book Co., New York.
Love, A. E. H. (1944). *The Mathematical Theory of Elasticity*. Dover Publications, New York.
Malvern, L. E. (1969). *Introduction to Mechanics of a Continuous Medium*. Prentice-Hall, Englewood Cliffs, NJ.
Milne-Thomson, L. M. (1960). *Theoretical Hydrodynamics*, 4th ed. Macmillan Publishing Company, New York.
Nowacki, W. (1963). *Dynamics of Elastic Systems*. Chapman and Hall, New York.
Prager, W. (1961). *Introduction to Mechanics of Continua*. Ginn and Co., Lexington, Mass.
Segal, L. A. (1977). *Mathematics Applied to Continuum Mechanics*. Macmillan, New York.
Serrin, J. (1959). "Mathematical Principles of Classical Fluid Mechanics," in *Handbuch der Physik* (ed. S. Flügge), Vol. III/1. Spring-Verlag, Berlin.
Sneddon, I. N., and Berry, D. S. (1958). "The Classical Theory of Elasticity," in *Handbuch der Physik* (ed. S. Flügge), Vol. VI, pp. 1–126. Springer-Verlag, Berlin.
Sokolnikoff, I. S. (1956). *Mathematical Theory of Elasticity*, 2nd ed. McGraw-Hill Book Co., New York.
Spencer, A. J. M. (1980). *Continuum Mechanics*. Longman, London.
Stoker, J. J. (1957). *Water Waves*. Interscience Publishers, New York.
Timoshenko, S. P., and Goodier, J. N. (1970). *Theory of Elasticity*, 2nd ed. McGraw-Hill Book Co., New York.
Truesdell, C. (1977). *A First Course in Rational Continuum Mechanics*, Vol. 1. Academic Press, New York.

Truesdell, C. (1966). *The Elements of Continuum Mechanics.* Springer-Verlag, New York.

Truesdell, C., and Noll, W. (1965). "The Non-Linear Field Theories of Mechanics," in *Handbuch der Physik* (ed. S. Flügge), Vol. III/3. Springer-Verlag, Berlin.

Truesdell, C., and Toupin, R. A. (1960). "The Classical Field Theories," in *Handbuch der Physik* (ed. S. Flügge), Vol III/1, pp. 226-790. Springer-Verlag, Berlin,

Wang, C. T. (1953). *Applied Elasticity.* McGraw-Hill Book Co., New York.

INDEX

A

Abnormality factor, 124
Absolute temperature, 351
Acceleration, 244
Acoustic wave equation, 492
Addition of
  matrices, 13
  tensors, 49
  vectors, 7
Adiabatic
  constant, 469
  flow, 471
  surface, 475
Adjoint of a
  matrix, 27
  tensor, 80
Adjugate of a
  matrix, 27
  tensor, 80
Airy's
  solution, 342–343
  stress function, 342
Almansi's strain tensor, 193
Alternating
  symbol, 20
  tensor, 57

Amplitude, 440, 503
Amplitude dispersion, 507
Angle between vectors, 9
Angular
  momentum, 336
  velocity, 260
Axial extension of a beam, 407
Axial vector, 74

B

Balance of
  angular momentum, 336
  energy, 344
  linear momentum, 300, 331
  mass, 376
  mechanical energy, 349, 472
  thermal energy, 349
Barotropic fluid, 469
Base vectors, 7
Beltrami's
  field, 124
  flow, 289, 493
  solution, 340
  vorticity equation, 330
Beltrami–Michell equation, 402–404

Bending of a beam, 413–416
Bernoulli's
 equations, 494
 function, 494
Bernoulli–Euler law, 416
Biharmonic function, 393
Bobyleff–Forsythe formula, 512
Body force, 294, 377
Body point, 158
Boundary conditions
 elasticity, 383
 non-viscous fluid flows, 474
 viscous fluid flows, 523
Boundary layer, 524
Boundary value problems
 in elastostatics, 384
 in elastodynamics, 387
Boussinesq's solution, 454
Boyle's law, 468
Bulk
 modulus, 371
 viscosity, 515

C

Caloric equation of state, 467, 520
Cayley–Hamilton theorem, 93
Cartesian
 axes, 7, 34
 coordinates, 33
 tensors, 39–41
Cauchy's
 deformation tensor, 193
 equation of equilibrium, 333
  general solutions of, 338
 equation of motion, 332
 law, 300
 reciprocal relation, 295
 strain tensor, 206, 364
 stress postulate, 295
 stress tensor, 300
Cauchy–Green deformation tensors, 193
Cauchy–Kovalevski–Somigliana solution, 455
Cauchy–Lagrange theorem, 281
Characteristic
 equation, 85
 polynomial, 85
Circulation
 material derivative of, 279

 preserving motion, 280
 theorem, 279–280, 486
Clapeyron's theorem, 381
Classical heat conduction inequality, 353
Clausius–Duhem inequality, 352
Clausius–Planck inequality, 353
Coefficient of
 shear viscosity, 515
 thermal conductivity, 355
 viscosity, 509–510
Cofactor of a
 matrix, 27
 tensor, 80
Collinear vectors, 7
Comma notation, 113
Compatibility conditions
 for strain, 219–220
 for stress, 403–404
Components of a
 tensor, 40, 41
 vector, 8
Compressible fluid, 466, 520
Compressive stress, 304
Configuration of a continuum, 157
 current, 162
 deformed (final), 168
 initial (reference), 161
Continuum hypothesis, 166
Conservation laws, 325
Conservation of mass, 325
Conservative
 force, 340
 vector, 139
Constitutive equations, 156, 326, 355
Continuity equation, 326
 material form, 327, 355
 spatial form, 326, 355
Continuum (continuous medium), 157
Continuum hypothesis, 156
Contraction of a
 material arc, 180
 tensor, 51
Convective
 part of acceleration, 246
 rate of change, 245, 246
 term, 528
Coordinate transformations, 33
Coordinates
 Cartesian, 7
 polar, 535

INDEX 587

Coplanar vectors, 11
Couette flow, 549
Couple stress, 296
Creeping flow, 531
Cross product of vectors, 10
Current position, 162
Curl of a
   tensor, 127
   vector, 120
Cylindrical tube under pressure, 424–428

D

D'Alembert's solution, 433, 436
Decomposition of deformation, 181
Deep water waves, 504
Deformable materials, 184
Deformation, 167, 241
   finite, 205
   homogeneous, 185
   infinitesimal, 205
   of a surface element, 171
      volume element, 174
Deformation-gradient
   material, 169
   spatial, 173
   tensor, 169
Deformation tensors
   Cauchy, 193
   Finger, 193
   Green, 188
Delta symbol, 16
Density, 159–160
Derivative following a particle, 243
Determinant of a
   matrix, 13
   tensor, 78
Deviator stresses, 313
Deviatoric tensor, 81–82
Diagonal matrix, 12
Differential operator, 113
Difference of
   tensors, 49
   vectors, 7
Diffusion equation, 555
Diffusion of vorticity, 536
Dilatation, 207
Dilatational wave, 439
Direction
   cosines, 10

   ratios, 10
Directions of stretching, 259
Directional derivative, 118
Dispersion relation, 503, 507
Displacement equation of motion, 401
Displacement formulation, 398
Displacement vector, 200
Displacement-gradient, 201
   material, 201
   spatial, 201
Divergence of a
   tensor field, 127
   vector field, 119
Divergence theorem for a
   tensor. 143
   vector, 136
Dot product of vectors, 9
Drag on a
   plate, 543
   sphere, 560–561
Drag coefficient, 561
Dual vector, 74
Dynamic pressure, 464, 509

E

Epsilon-delta identity, 25
Eigenvalue, 83
Eigenvector, 83
Elastic
   constants, 371–372
   fluid, 469, 485
   limit, 367
   moduli, 365
   potential, 378
   waves, 431–450
Elasticity
   tensor, 365
   nonlinear, 367
Energy
   equation, 346–349
   flux vector, 350
   internal, 345
   kinetic, 344
   mechanical, 349
   potential, 472
   thermal, 349
Entropy, 351
   flow, 352
   inequality, 351

source, 352
specific, 351
Equation of
  compatibility for
    strain, 219
    stress, 404
  continuity, 326, 518
  energy, 345-348, 518
  equilibrium, 333
    for an elastic body, 392
    for a fluid, 477, 487
  mechanical energy, 349, 472
  motion (Cauchy), 332
    for an elastic body, 391
    for an elastic fluid, 485
    for a perfect fluid, 477
    for a viscous fluid, 527
  thermal energy, 349
Euler's equation of motion, 477, 485
Euler's formulas, 253
Eulerian
  description, 163
  fluid, 469
  form, 326
  strain tensor, 195
Exterior product of vectors, 105
Extension of
  a beam, 410
  a material arc, 180

F

Field
  equations, 325
  line, 268
  scalar, 111
  tensor, 111
  tube, 271
  vector, 111
Finger's deformation tensor, 193
Finite deformation, 205
First law of thermodynamics, 345
Flow
  Couette, 541, 549
  creeping, 531
  potential, 261
  Poiseullie, 542
  Stokes', 531
Fluid
  barotropic, 469
  compressible, 466
  elastic, 469, 485
  homogeneous, 467
  incompressible, 465
  inviscid, 463
  nonviscous, 463
  viscous, 463
Force
  body, 294
  conservative, 340
  gravitational, 478, 494, 500
  surface, 294
Fourier's law of heat conduction, 355, 466
Fourier-Duhamel law, 354
Free
  energy, 354
  suffix, 3
  surface, 315
Fundamental invariants, 78
  strain, 228
  stress, 308
  tensor, 78

G

Galerkin solution, 454
Galerkin vector, 454
Gas
  constant, 468
  dynamical equations, 489
  perfect, 468
Generalized Hooke's law, 364
Generalized vorticity equation, 358
Governing equations for
  elasticity, 378
  compressible non-viscous fluid flows, 470
  compressible viscous fluid flows, 522
  incompressible non-viscous fluid flows, 470
  incompressible viscous fluid flows, 521
Gradient
  of a scalar field, 117
  of a tensor field, 126
  of a vector field, 125
  material deformation, 169
  spatial deformation, 173
Gravitational force, 478
Green's
  deformation tensor, 188
  strain tensor, 189
  theorem, 138

Green-Lamé solution, 456
Group velocity, 503

H

Hagen-Poiseullie flow, 543
Harmonic vector, 139, 340
Heat
  flux vector, 344
  supply, 345
Helmholtz representation, 139-140
Helmholtz free energy, 354
Helmholtz theorem, 281, 283
Homogeneous
  deformation, 185
  elastic solid, 365
  fluid, 467
  function, 509
  material, 365
Hooke's law, 364, 367
Hoop stress, 395
Hydrostatic
  pressure, 462
  stress, 303

I

Identity
  matrix, 15
  tensor, 54
Incompressible
  continuum, 175, 259
  elastic body, 371
  fluid, 465, 519
Index notation, 1
Infinitesimal
  deformation, 205
  normal strain, 206
  rotation tensor, 213
  rotation vector, 213
  shear strain, 206
  strain tensor, 205
  stretch, 211
Initial
  configuration, 161, 380
  coordinates, 164, 380
Instant coordinates, 164
Instantaneous
  configuration, 162
  position, 162

Internal dissipation, 361
Internal energy, 345
Internal surface force, 295
Invariants
  strain, 228
  stress, 308
  tensor, 77
Inverse
  function, 162
  matrix, 15
  tensor, 70
Invertible mapping, 162
Invertible tensor, 70
Inviscid fluid, 463
Irrotational
  motion, 261
    permanence of, 281
  vector, 139
Isentropic perfect gas, 487
Isochoric
  deformation, 175
  motion, 259, 516
Isothermal
  perfect gas, 468
  surface, 475
Isotropic
  elastic body, 365
  fluid, 509
  tensor, 58, 59
  thermal, 354

J

Jacobian, 165
  positiveness of, 165

K

Kelvin's circulation theorem, 279
Kelvin scale of temperature, 351
Kirchhoff's theorem, 208
Kinematic viscosity, 527
Kinetic energy, 344
Kinetic equation of state, 466, 520
Kronecker delta, 17

L

Lagrangian
  deformation tensor, 195

description, 163
strain tensor, 195
Lamb-Thomson formula, 348
Lamb's surfaces, 493
Lamé's
  moduli, 366
  potentials, 456
  pressure-vessel problems, 431
Laminar flow, 539
Laplace's equation, 501
Laplacian, 114, 129
Law of
  balance of angular momentum, 337
  balance of linear momentum, 331-332
  conservation of mass, 326, 329
  entropy, 351
Left
  Cauchy-Green deformation tensor, 193
  polar decomposition, 98
  stretch tensor, 184
Levi-Civita ε-symbol, 20
Linear
  combination of vectors, 7, 8
  elastic solid, 364
  momentum, 331
  motion, 246
  operators, 66
  rotation tensor, 213
  rotation vector, 213
  strain tensor, 205
  viscous fluid, 510
Local
  rate of change of velocity, 246
  speed of sound, 492
  time-derivative, 242
Localization theorem, 165
Longitudinal wave, 439
Love
  strain function, 455
  waves, 446-450

M

Macroscopic study, 157
Magnitude of a vector, 7
Mass, 159-160
Mass conservation law, 325-326, 329
Material
  arc, 159
  body, 159
  curve, 159

derivative, 242-243
  of a line integral, 274
  of a surface integral, 274
  of a volume integral, 274
  operator, 245
description, 163
Lagrangian form, 327, 485
law, 376, 518
point, 158
surface, 159
time derivative, 242
variables, 163
Matrix
  difference, 14
  nonsingular, 15
  orthogonal, 16
  of tensor, 43
  of transformation, 36
  product, 14
  singular, 15
  sum, 13
Maximum
  normal stress, 306
  shear stress, 306
Maxwell's solution, 341
Maxwell stress functions, 341
Mean pressure, 312
Microscopic study, 157
Mixed boundary value problem, 384
Mobile time derivative, 243
Modulus of
  compression, 371
  rigidity, 371
Moment of inertia, 414
Momentum
  angular, 336
  linear, 331
Morera's solution, 342
Morera's stress functions, 342
Motion
  descriptions of, 160
  Cauchy's equation of, 332
  Euler's equation of, 477
  Navier's equation of, 390-391
  Navier-Stokes equation of, 527

N

Nanson's formulae, 200
Navier's equation of

equilibrium, 392
motion, 391
Navier–Stokes equation, 527
Newton's law of viscosity, 515
Newtonian viscous fluid, 515
Nominal stress tensor, 317
Nominal stress vector, 317
Nonhomogeneous fluid, 468
Nonlinear dispersion relation, 507
Nonlinear elasticity, 367
NonNewtonian fluid, 516
Nonsingular matrix, 15
Non-viscous fluid, 463
Normal derivative, 118
Normal strain, 189
Normal stress, 304
No-slip condition, 523

O

Octahedral
  plane, 305
  normal stress, 312
  shear stress, 312
One-dimensional wave equation, 432
Orthogonal
  matrix, 16
  transformation, 36
  tensor, 70
  vectors, 9
Orthonormal relations, 36
Oseen equation, 562

P

Papkovitch–Neuber solution, 454
Parallelogram law, 7
Partial differential operator, 113
Particle, 157
Path line, 266
Perfect gas, 468
Peripheral stress, 395
Permutation
  identity, 25
  symbol, 20
  tensor, 57
Phase speed (velocity), 440, 503
Piola–Kirchhoff stress tensor
  first, 317
  second, 318

Plane
  Couette flow, 541
  flow, 265
  harmonic wave, 440
  Poiseullie flow, 542
  strain, 425
  stress problems, 343
  waves, 435–441
Poiseullie flow, 542–548
Poiseullie's formula, 546
Poisson's ratio, 369
Polar
  coordinates, 535
  decomposition, 94
    left, 98
    right, 98
Polar decomposition theorem, 97
Polar materials, 296
Position vector, 33
Positive definite tensor, 94
Potential, 486
  energy of a fluid, 472
  flow, 261
Power, 344
Prandtl
  boundary layer, 524
  solution, 343
  stress function, 343
Pressure, 464
  dynamic, 464
  mean, 312
  static, 462
Primary wave, 440
Principal
  axes of strain, 228
  axes of stress, 307
  axes of a tensor, 89
  deviator strains, 231
  directions of strain, 228
  directions of stretch, 182
  directions of stretching, 259
  directions of stress, 307
  direction of a tensor, 83
  invariants of a tensor, 77
  planes of strain, 228
  plane of stress, 307
  strains, 227
  stress, 307
  stretches, 182
  stretchings, 259

value of a tensor, 83
Product of two tensors, 51
Proper orthogonal tensor, 79
Proportional limit, 366
Pure shear stress, 304
P-wave, 440

Q

Quotient law, 54

R

Radial stress, 395
Range convention, 2
Rate of
  deformation tensor, 257
  rotation tensor, 262
  strain tensor, 262
  work, 344
Rayleigh
  layer, 558
  problem, 558
  waves, 441–445
Reciprocal theorem of
  Betti and Rayleigh, 383
  Cauchy, 295
Reference configuration, 161
Referential coordinates, 164
Relative
  displacement, 216
  velocity, 260
Residual stress, 510
Reynolds
  number, 529, 561
  transport formula, 276
Right
  Cauchy–Green deformation tensor, 193
  polar decomposition, 98
  stretch tensor, 184
Rigid-body transformation, 182
Rigid
  materials, 184
  rotation, 170
Rigid-translation, 169
Rotation-rate tensor, 262
Rotation tensor, 184

S

Saint-Venant's compatibility conditions, 220
Scalar
  field, 111
  invariant, 42
  multiple of a
    matrix, 13
    tensor, 48
    vector, 7
  potential, 139, 486
  product of
    tensors, 51
    vectors, 9
Schaefer's solution, 458
Second law of thermodynamics, 352
Secondary wave, 440
Semicolon notation, 165
Shallow water waves, 504
SH-wave, 440
Shear
  modulus, 371
  pure, 321
  strain, 191
  stress, 304–305
  viscosity, 515
  waves, 439
Shearing, 258
Similarity variable, 557
Singular matrix, 15
Skew-symmetric
  matrix, 13, 14
  part of a matrix, 14
Skew
  part of a tensor, 73
  tensor, 73
Skin friction, 542–543
Slip-condition, 524
Solenoidal vector, 137
Small deformations, 205
Spatial
  coordinates, 164
  deformation gradient, 173
  description, 163
  Eulerian form, 326
  variables, 163
Spectral representation of a tensor, 92
Specific enthalpy, 471
Specific entropy, 351

Specific heat at constant
  pressure, 469
  volume, 468
Speeds of elastic waves, 439
Spherical shell under pressure, 428–431
Spherical part of a tensor, 81
Square
  matrix, 12
  of a tensor, 52
Square-root of a tensor, 96
Stagnation point, 498
Static pressure, 462, 477
Steady laminar flow, 539
Steady motion, 246
Sternberg–Eubanks solution, 456
Stokes'
  boundary layer, 556
  condition, 516
  expansion, 507
  flow, 531
  formula, 561
  law, 510
  problem, 556
  theorem for a
    tensor, 145
    vector, 137
  waves, 504
Stokesian fluid, 509
Strain
  deviator, 230
  deviator invariants, 231
  energy function, 374, 378
  invariants, 228
  tensors, 188
Strain-displacement relations, 200, 376
Strain-energy functions, 374
Strain-rate tensor, 262
Stream function, 531
Stream
  function, 531
  lines, 269
  tube, 271
    strength of, 272
Stress
  components, 297
  deviator, 312
    invariants, 313
  equation of motion, 401
  invariants, 308
  matrix, 297

power, 346
tensor, 300, 337
  for an elastic solid, 364–365
  for a non-viscous fluid, 463
  for a viscous fluid, 508
vector, 294
Stress-displacement relations, 391
Stress formulation, 401
  uniqueness, 401
Stress-free surface, 315
Stress-strain relation, 364–365
Stress tensor for a non-viscous fluid, 463
Stress-velocity relation, 510
Stress waves, 432–433
Stretch
  of a material arc, 180
  tensors, 184
Stretching tensor, 257
Suffix
  dummy, 3
  free, 3
  notation, 3
Summation convention, 2
Surface
  elevation, 501
  force, 294–295
  traction, 295
  waves, 441–445
S-waves, 440
SV waves, 440
Symmetric
  matrix, 13, 14
  part of a matrix, 14
  part of a tensor, 73
  tensor, 72
Symmetry of stress tensor, 337

T

Table of direction cosines, 35
Tangential stress, 304
Temperature, absolute, 351
Tensile stress, 304
Tensor
  alternating, 57
  as linear operator, 66
  Cartesian, 39
  difference, 49
  equality, 47

594 **INDEX**

equation, 48
field, 111
invariants, 77
invertible, 70
isotropic, 58
multiplication, 50
orthogonal, 70
permutation, 57
product of tensors, 50
product of vectors, 40
skew, 72
symmetric, 72
sum, 49
unit, 54
vorticity, 260
Thermal conductivity, 354
  coefficient of, 355
  tensor, 354
Thermodynamics,
  first law, 345
  second law, 352
Teodorecu's solution, 458
Torricelli's formula, 496
Torsion of beams, 418
Torsional rigidity, 421
Total mass, 160
Trace of a tensor, 46
Traction, 294
Traction-free surface, 315
Transformation
  rule for a tensor, 41
  rule for a vector, 39
Translation, 169
Transport formulas, 273–278
Transpose of a
  matrix, 12
  tensor, 69
Transverse waves, 439
Triaxial stretch, 182
Triple product
  scalar, 11
  vector, 11
Trochoid, 506

U

Undeformed configuration, 168
Uniaxial stress, 302
Uniform motion, 246
Uniqueness theorem in

elastostatics, 385–386
elastodynamics, 387–388
Unit
  matrix, 15
  tensor, 54
  vector, 7

V

Vector
  axial, 74
  base, 7
  components, 8
  difference, 7
  field, 111
  gradient of, 125
  magnitude of, 7
  orthogonal, 9
  position, 33
  potential, 137
  product, 10
  stress, 294
  sum, 7
  triple product, 11
  unit, 7
  vorticity, 261
  zero, 7
Velocity, 243
  potential, 261
  profile, 541
Viscosity
  bulk, 515
  coefficients of, 509
  kinematic, 527
  shear, 515
  tensor, 509
Viscous
  dissipative function, 513
  fluid, 463, 510
  stress tensor, 508
  term, 528
Volume change, 207
Vortex
  lines, 269
    motion of, 281
  tube, 271
    strength of, 272
Vorticity
  equation (Beltrami), 330
  equation (Cauchy), 281

equation (Helmholtz), 487
equation for elastic fluid, 487
tensor, 260
vector, 261

W

Water waves, 500–508
Wave
 amplitude, 440
 dilatational, 439
 elastic, 431–450
 harmonic, 440
 length, 440
 longitudinal, 439
 Love, 446–450
 number, 440
 period, 440
 plane, 435
 primary (P), 440
 Rayleigh, 441–445
 resistance, 435
 secondary (S), 440
 secondary horizontal (SH), 440
 secondary vertical (SV), 440
 shear, 439
 surface waves, 441–445
Weber's equation, 490

Y

Young's modulus, 369

Z

Zero
 matrix, 14
 tensor, 47
 vector, 7
Zorawski's criterion, 278

ISBN 0-12-167880-6